FUNDAMENTALS OF SILICON CARBIDE TECHNOLOGY

FUNDAMENTALS OF SILICON CARBIDE TECHNOLOGY

GROWTH, CHARACTERIZATION, DEVICES, AND APPLICATIONS

Tsunenobu Kimoto

Kyoto University, Japan

James A. Cooper

Purdue University, USA

IEEE

WILEY

This edition first published 2014

Registered office
John Wiley & Sons Singapore Pte. Ltd., 1 Fusionopolis Walk, #07-01 Solaris South Tower, Singapore 138628.

For details of our global editorial offices, for customer services and for information about how to apply for permission to reuse the copyright material in this book please see our website at www.wiley.com.

Library of Congress Cataloging-in-Publication Data

Kimoto, Tsunenobu, 1963–
Fundamentals of silicon carbide technology : growth, characterization, devices and applications / Tsunenobu Kimoto, James A. Cooper.
pages cm
Includes bibliographical references and index.
ISBN 978-1-118-31352-7 (cloth)
1. Silicon carbide. 2. Semiconductors. 3. Integrated circuits. I. Cooper, James A., 1946– II. Title. III. Title: Fundamentals of SiC technology.
TK7871.15.S56K56 2014
621.3815′2 – dc23

2014016546

Typeset in 9/11pt TimesLTStd by Laserwords Private Limited, Chennai, India

2 2014

Contents

About the Authors

Tsunenobu Kimoto received the B.E. and M.E. degrees in Electrical Engineering from Kyoto University, Japan, in 1986 and 1988, respectively. He joined Sumitomo Electric Industries, LTD in 1988, where he conducted research and development of amorphous Si solar cells and semiconducting diamond material. In 1990, he started his academic career as a Research Associate at Kyoto University, and received the Ph.D. degree from Kyoto University in 1996, based on his work on SiC growth and device fabrication. In 1996–1997, he was a visiting scientist at Linköping University, Sweden. He is currently Professor at the Department of Electronic Science and Engineering, Kyoto University.

His main research activity includes SiC epitaxial growth, optical and electrical characterization, defect electronics, ion implantation, MOS physics, and high-voltage devices. He has also been involved in nanoscale Si, Ge devices, novel materials for nonvolatile memory, and GaN-based electron devices. He is a member of IEEE, MRS, JSAP, IEICE, and IEE.

James A. Cooper received his BSEE and MSEE degrees from Mississippi State University and Stanford University in 1968 and 1969, respectively. From 1968 to 1970 he was a staff member at Sandia National Laboratories. He attended Purdue University from 1970 to 1973, where he received his Ph.D. for generalizing the theory of the MOS conductance technique. He joined Bell Laboratories, Murray Hill, in 1973. At Bell Labs he designed CMOS integrated circuits, including AT&T's first microprocessor, and conducted research on high-field transport in silicon inversion layers. He joined the faculty of Purdue University in 1983 and is currently Professor of Electrical and Computer Engineering.

From 1983 to 1990 he explored dynamic memories in GaAs, and he began working in SiC in 1990. His group demonstrated the first SiC DMOSFETs and the first SiC digital integrated circuits, and contributed to the development of Schottky diodes, UMOSFETs, lateral DMOSFETs, BJTs, and IGBTs. They have also investigated a variety of other SiC devices, including thyristors, CCDs, MESFETs, SITs, and IMPATT diodes.

A Life Fellow of the IEEE, Professor Cooper has held the Charles William Harrison and Jai N. Gupta chairs in Electrical and Computer Engineering at Purdue, and was founding co-director of Purdue's Birck Nanotechnology Center.

Preface

Power semiconductor devices are attracting increasing attention as key components in a variety of power electronic systems. The major applications of power devices include power supplies, motor controls, renewable energy, transportation, telecommunications, heating, robotics, and electric utility transmission/distribution. The utilization of semiconductor power devices in these systems can enable significant energy savings, increased conservation of fossil fuels, and reduced environmental pollution.

Power electronics has gained renewed attention in the past decade due to the emergence of several new markets, including converters for photovoltaic and fuel cells, converters and inverters for electric vehicles (EVs) and hybrid-electric vehicles (HEVs), and controls for smart electric utility distribution grids. Currently, semiconductor power devices are one of the key enablers for global energy savings and electric power management in the future.

Silicon power devices have improved significantly over the past several decades, but these devices are now approaching performance limits imposed by the fundamental material properties of silicon, and further progress can only be made by migrating to more robust semiconductors. Silicon carbide (SiC) is a wide-bandgap semiconductor with superior physical and electrical properties that can serve as the basis for the high-voltage, low-loss power electronics of the future.

SiC is a IV–IV compound semiconductor with a bandgap of 2.3–3.3 eV (depending on the crystal structure, or polytype). It exhibits about 10 times higher breakdown electric field strength and 3 times higher thermal conductivity than silicon, making it especially attractive for high-power and high-temperature devices. For example, the on-state resistance of SiC power devices is orders-of-magnitude lower than that of silicon devices at a given blocking voltage, leading to much higher efficiency in electric power conversion. The wide bandgap and high thermal stability make it possible to operate certain types of SiC devices at junction temperatures of 300 °C or higher for indefinite periods without measurable degradation. Among wide-bandgap semiconductors, SiC is exceptional because it can be easily doped either p-type or n-type over a wide range, more than five orders-of-magnitude. In addition, SiC is the only compound semiconductor whose native oxide is SiO_2, the same insulator as silicon. This makes it possible to fabricate the entire family of MOS-based (metal-oxide-semiconductor) electronic devices in SiC.

Since the 1980s, sustained efforts have been directed toward developing SiC material and device technology. Based on a number of breakthroughs in the 1980s and 1990s, SiC Schottky barrier diodes (SBDs) were released as commercial products in 2001. The market for SiC SBDs has grown rapidly over the last several years. SBDs are employed in a variety of power systems, including switch-mode power supplies, photovoltaic converters, air conditioners, and motor controls for elevators and subways. Commercial production of SiC power switching devices, primarily JFETs (junction field-effect transistors) and MOSFETs (metal-oxide-semiconductor field-effect transistors), began in 2006–2010. These devices are well accepted by the markets and many industries are now taking advantage of the benefits of SiC power switches. As an example, the volume and weight of a power supply or inverter can be reduced by a factor of 4–10, depending on the extent to which SiC components are employed. In addition to the size and weight reduction, there is also a substantial reduction in power dissipation, leading to improved efficiency in electric power conversion systems due to the use of SiC components.

In recent years, the SiC professional community has grown rapidly in both academia and industry. More and more companies are developing SiC wafer and/or device manufacturing capabilities and the population of young scientists and engineers is increasing. Unfortunately, very few textbooks are yet available that cover the broad spectrum of SiC technology from materials to devices to applications. Thus, those scientists, engineers, and graduate students are potential readers of this text. The authors hope this book will be timely and beneficial for such readers, and will enable them to rapidly acquire the essential knowledge to practice in this field. Since this book covers both fundamentals and advanced concepts, a minimum knowledge of semiconductor physics and devices is assumed, but a graduate student majoring in material science or electrical engineering will have no difficulty in reading this book.

The main topics described in this book include SiC physical properties, bulk and epitaxial growth, characterization of electrical and optical properties, extended and point defects, device processing, design concepts of power rectifiers and switching devices, physics and features of unipolar/bipolar devices, breakdown phenomena, high-frequency and high-temperature devices, and system applications of SiC devices. Both fundamental concepts and state-of-art implementations are presented. In particular, we try to explain all the subjects with an in-depth treatment, including basic physics, present understanding, unaddressed issues, and future challenges.

Finally the authors acknowledge a number of colleagues and pioneers in this field, especially Prof. W. J. Choyke (University of Pittsburgh), Emeritus Prof. H. Matsunami (Kyoto University), the late Dr G. Pensl (University of Erlangen-Nürnberg), Prof. E. Janzén (Linköping University), and Dr J. W. Palmour (Cree) for their valuable contributions to the field and to our understanding. We also thank Mr. James Murphy and Ms. Clarissa Lim of Wiley for their guidance and patience. At last, we thank our family for their kind encouragement and support in writing this book. Without their support and understanding, this book would not have been published.

Kyoto and West Lafayette, September 2013
Tsunenobu Kimoto
James A. Cooper

1

Introduction

1.1 Progress in Electronics

Development of semiconductor materials and devices has been a strong driving force for a variety of revolutionary changes and innovations in modern society. Since the invention of germanium (Ge)-based bipolar transistors in 1947–1948 [1, 2] and the subsequent success of silicon (Si)-based metal-oxide-semiconductor field effect transistors (MOSFETs) [3], semiconductor devices have given rise to a new field, *solid state electronics*. The invention of integrated circuits (ICs) made by planar technology [4, 5] triggered rapid progress in *microelectronics*. Nowadays, Si-based large scale integrated circuits (LSIs) are the key components in almost all electrical and electronic systems. Despite predictions of physical limitations, remarkable progress continues to be made in Si-based LSIs, even today [6, 7]. Solar cells and various sensors are also mainly produced using silicon.

In the meantime, compound semiconductors have established unique positions in those applications where Si devices cannot exhibit good performance because of the inherent material properties. In particular, III–V semiconductors such as gallium arsenide (GaAs) and indium phosphide (InP) have been widely employed for high-frequency devices and light-emitting devices [8, 9]. In addition to the high electron mobility and direct band structure of most III–V semiconductors, bandgap engineering and formation of heterostructures can be utilized to enhance the performance of devices based on compound semiconductors. Success in making blue and green light-emitting devices using gallium nitride (GaN) and indium gallium nitride (InGaN) was also a great milestone in the history of semiconductors [10, 11]. Thus, *optoelectronics* is one of the most important fields of development, and relies on these III–V semiconductors.

As our society continues to advance technologically, various demands for new functionalities for semiconductor devices have arisen, such as high-temperature operation and flexibility. *High-temperature electronics* is a field where wide bandgap semiconductors possess much promise [12]. Conversely, organic semiconductors and oxide semiconductors have been developed for *flexible electronics* [13].

Improvement of energy efficiency (reduction of power consumption and dissipation) is one of the most basic problems we are facing. In 2010, the world average ratio of electrical energy consumption to total energy consumption is about 20% [14], and this ratio is expected to increase rapidly in the future. Independent of the means by which electrical power is generated, power conditioning and conversion are required for cost-effective and efficient delivery to the load. It is estimated that more than 50% of all electrical power flows through some form of power conversion.

Power electronics, the concept of which was introduced by Newell in 1973 [15], involves *conversion of electric power* using power semiconductor devices and circuits. Electric power is regulated and converted so that the power can be supplied to the loads in the best form. Electric power conversion includes

Fundamentals of Silicon Carbide Technology: Growth, Characterization, Devices, and Applications, First Edition.
Tsunenobu Kimoto and James A. Cooper.

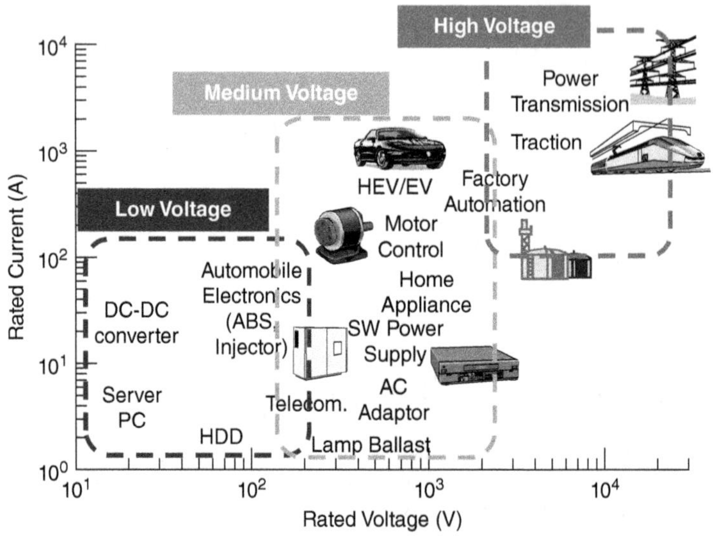

Figure 1.1 Major application areas of power devices plotted as a function of rated voltage.

AC–DC, DC–AC, DC–DC (voltage conversion), and AC–AC (voltage or frequency conversion) [16]. The efficiency of power conversion is typically 85–95% using currently available technology, which is not high enough, because approximately 10% of the electric power is lost as heat at every power conversion. In AC–DC and DC–AC conversions, which are very common, the efficiency becomes as low as about $(0.9)^2 \approx 0.8$.

In general, the efficiency of power electronics is limited by the performance of semiconductor devices, capacitors, inductors, and packaging. In particular, power semiconductor devices have attracted increasing attention as key components which limit the performance and size of power converters. As shown in Figure 1.1, the major applications of power devices include power supplies, motor control, telecommunications, heating, robotics, electric/hybrid vehicles, traction, lighting ballasts, and electric power transmission. Development of high-voltage and low-loss power devices is also essential for construction of future smart grids.

Realization of high-performance power devices will lead not only to enormous energy saving but also to conservation of fossil fuels and reduced environmental pollution. At present, Si is the most commonly used semiconductor for power devices. The performance of Si power switching devices has been significantly improved through development of power MOSFETs and IGBTs (insulated gate bipolar transistors) [17, 18]. Progress in Si LSI technology and in advanced simulation technology has had great impact on the development of Si power devices in recent decades. However, now that Si power device technology is relatively mature, it is not easy to achieve innovative breakthroughs using this technology. *Silicon carbide* (SiC) is an old but emerging semiconductor, which is promising for advanced power devices because it has superior physical properties. SiC devices are also promising for high-temperature and radiation-resistant operation. GaN is also attractive as a material for power devices, and the intrinsic potential of GaN is very similar to that of SiC (since they have almost the same bandgap and critical electric field strength). At present, however, growth and device-fabrication technologies for SiC are more advanced, and SiC power devices exhibit better performance and reliability. GaN-based lateral switching devices processed on heteroepitaxial GaN on Si show some promise for relatively low-voltage (100–300 V) applications. When the GaN technology becomes more mature, especially when large-diameter bulk growth is readily achieved, both SiC and GaN power devices will be widely employed, depending on the performance and cost. For high-voltage bipolar device applications, however, SiC should be inherently superior because SiC has an indirect band structure, leading to an inherently long carrier lifetime.

1.2 Features and Brief History of Silicon Carbide

Silicon carbide (SiC) is a IV–IV compound material with unique physical and chemical properties. The strong chemical bonding between Si and C atoms gives this material very high hardness, chemical inertness, and high thermal conductivity [19]. As a semiconductor, SiC exhibits a wide bandgap, high critical electric field strength, and high saturation drift velocity. Both n- and p-type control across a wide doping range is relatively easy in SiC; this makes SiC exceptional among wide bandgap semiconductors. The ability of SiC to form silicon dioxide (SiO_2) as a native oxide is an important advantage for device fabrication. Because of these properties, SiC is a promising semiconductor for high-power and high-temperature electronics [20–22]; subsequent chapters will describe in detail the fundamentals of SiC technologies, its properties, growth, characterization, device fabrication, and device characteristics.

The physical and chemical stability of SiC, however, has made crystal growth of SiC extremely difficult, and severely hampered development of SiC semiconductor devices and their electronic applications. The existence of various SiC structures with different stacking sequences (otherwise known as polytypism) [23] has also hampered growth of electronic-grade SiC crystals. SiC polytypes such as 3C-, 4H-, and 6H-SiC, are described in Section 2.1.

1.2.1 Early History

SiC itself is rare in nature, and synthesis of a compound material containing silicon–carbon bonds was first reported by Berzelius in 1824 [24]. Acheson invented a process for the synthesis of SiC from silica, carbon, and some additives (e.g., salt) in 1892 [25]. This process (*Acheson process*) provided volume production of SiC powders used for cutting, grinding, and polishing, which was the first industrial application of SiC. In the Acheson process, ingots which contain small single crystalline SiC platelets (mainly 6H-SiC) can be obtained as a by-product (Figure 1.2a). Although these SiC platelets are not pure, they were used for some basic studies on the physical and chemical properties of SiC. One of the highlights of this work was the first discovery of electroluminescence (emission of yellow light) from SiC by Round in 1907 [26]. In the meantime, Moissan discovered natural SiC and investigated this material as a mineral [27]. This is why SiC is named "Moissanite" in mineralogy or in the field of gem stones.

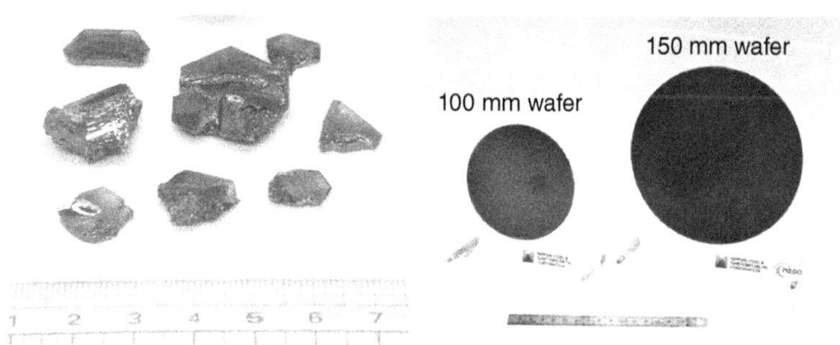

Figure 1.2 (a) SiC platelets (mainly 6H-SiC) obtained as a by-product in the Acheson process. (b) 4H-SiC wafers with 100 and 150 mm in diameter.

Lely successfully grew relatively pure SiC crystals by a sublimation technique (*Lely method*) in 1955 [28]. The crystals obtained are mostly 6H-SiC, but inclusions of foreign polytypes are often observed. Owing to the relatively high crystal quality of the Lely platelets, the first wave of research into SiC as a semiconductor emerged in the 1960s. During this period, the main target applications for semiconductor SiC were the development of high-temperature devices and blue light-emitting diodes [29, 30]. Shockley participated in an international conference on SiC, and emphasized the promise of SiC for high-temperature electronics [30]. Important academic studies on optical properties of SiC were extensively performed by Choyke [31]. However, because of the small size of Lely platelets and unsteady material supply, research and development of SiC semiconductors slowed down in the late 1970s, and the technology remained immature. Conversely, polycrystalline SiC technology was developed, and SiC-based ceramics, heating elements, passive components, and thermistors were commercialized.

1.2.2 Innovations in SiC Crystal Growth

In 1978–1981, Tairov and Tsvetkov invented a reproducible method for SiC boule growth [32, 33]. They introduced a 6H-SiC seed into a sublimation growth furnace, and designed an appropriate temperature gradient to control mass transport from the SiC source onto the seed crystal, based on thermodynamic and kinetic considerations. This growth process is called the *modified Lely method* or *seeded sublimation method*. Several groups followed and further developed the growth process to obtain SiC boules with a larger diameter and a reduced density of extended defects. Davis and Carter significantly refined this method [34]. The first commercialization of SiC (6H-SiC) wafers occurred in 1991 [35]. Through continuous efforts, reasonably high-quality SiC wafers, 100–150 mm in diameter, are commercially available from several vendors at present (Figure 1.2b). The availability of single crystalline wafers has driven rapid development of SiC-based electronic devices.

Concerning epitaxial growth of SiC, liquid phase epitaxy (LPE) of 6H-SiC on Lely platelets was investigated in the 1980s, in research targeting blue light-emitting diodes [36, 37]. Heteroepitaxial growth of 3C-SiC on a Si substrate by chemical vapor deposition (CVD) was developed [38, 39] in the early 1980s, but the performance of electronic devices (Schottky barrier diodes (SBDs), pn diodes, MOSFETs) was far below that expected. This result can be attributed to a high density of stacking faults and dislocations, which are generated because of large mismatches in the lattice constants and thermal expansion coefficients. Therefore, a few groups started CVD growth of 3C-SiC on 6H-SiC{0001} (Lely or Acheson platelets). Although the quality of 3C-SiC was much improved, it was still not satisfactory.

In 1987, Matsunami *et al.* discovered that high-quality 6H-SiC can be homoepitaxially grown by CVD at relatively low growth temperature, when a several degree off-angle is introduced into the 6H-SiC{0001} substrates ("*step-controlled epitaxy*") [40]. Davis *et al.* also reported homoepitaxial growth of 6H-SiC on off-axis substrates [41]. Homoepitaxial growth of 6H-SiC on off-axis 6H-SiC{0001} became a standard technique in the SiC community because it yielded high purity, good doping control, and uniformity. In 1993, a high mobility of over 700 $cm^2\ V^{-1}\ s^{-1}$ was first reported for 4H-SiC grown using this technique [42]. The combination of this result, the other superior physical properties of SiC, the commercial release of 4H-SiC wafers, and demonstration of excellent 4H-SiC devices made 4H-SiC the preferred choice for electronic device fabrication in the mid 1990s. In the meantime, the doping control was drastically improved by exploiting the "*site-competition*" concept proposed by Larkin *et al.* [43]. A *hot-wall CVD* reactor was proposed by Kordina *et al.* [44], and this reactor design is currently the standard, because it allows superior control of temperature distribution, has a much longer susceptor life, and better growth efficiency.

Since high-quality 4H- and 6H-SiC epitaxial layers (both n- and p-types) can be obtained, physical properties and defects of SiC have been extensively investigated in the University of Pittsburgh, the University of Erlangen-Nürnberg, Linköping University, Kyoto University, Ioffe Physical Technical Institute, Purdue University, the Naval Research Laboratory, the State University of New York at Stony Brook, Carnegie Mellon University, AIST, and so on.

1.2.3 Promise and Demonstration of SiC Power Devices

The outstanding potential of SiC-based power devices was suggested in 1989 by Baliga [45], and a systematic theoretical analysis of the performance was published in 1993 by the same group [46]. These papers have inspired and motivated scientists and engineers in this field.

As a result of the progress in homoepitaxial growth technology described above, lightly-doped hexagonal SiC epitaxial layers with reasonable quality became available in the early 1990s. Matus *et al.* reported a 1 kV 6H-SiC pn diode and its rectification operation up to 600 °C [47]. Urushidani *et al.* in 1993 demonstrated a 1 kV 6H-SiC SBD with a low specific on-resistance and 400 °C rectification [48]. In 1994, the on-resistance of high-voltage SiC SBDs was significantly reduced by using 4H-SiC [49]. After structure and process optimization, the first SiC SBD products were released in 2001 [50]. One of the typical applications of SiC SBDs was as fast diodes employed in a power-factor-correction circuit of switching-mode power supplies. Because of the negligibly small reverse recovery of SiC SBDs, the switching loss can be dramatically reduced and the switching frequency can be increased, leading to the downsizing of passive components. SiC SBDs are currently employed in a broad spectrum of applications, such as industrial motor control, photovoltaic converters, air conditioners, elevators, and traction (subway). In research and development, the maximum blocking voltage of SiC diodes exceeded 20 kV [51, 52].

In conjunction with development of high-voltage SiC diodes, fabrication of vertical SiC switching devices started in the early 1990s. In 1993, a vertical trench MOSFET of 6H-SiC was demonstrated by Palmour *et al.* [53]. Palmour and coworkers also extensively developed 4H-SiC trench MOSFETs, thyristors, and bipolar junction transistors (BJTs), as important steps toward high-power electronics [54]. In 1996 and 1997, the first planar double-implanted metal-oxide-semiconductor field effect transistor (DIMOSFET) of 4H-SiC with a blocking voltage of 760 V and low on-resistance was reported by Purdue University [55]. This group demonstrated a 1.4 kV–15 mΩ cm^2 4H-SiC trench MOSFET with a number of innovative design features in 1998 [56]. To avoid problems at the SiC MOS interface, vertical junction field effect transistors (JFETs) were also developed [57], leading to the commercialization of 4H-SiC power JFETs in the mid 2000s [50]. After steady improvement of MOS channel mobility and oxide reliability, 4H-SiC power DIMOSFETs have also been commercially available since 2010 [35, 58]. Figure 1.3 shows a picture of a 100 mm wafer after processing of SiC power MOSFETs. However, these

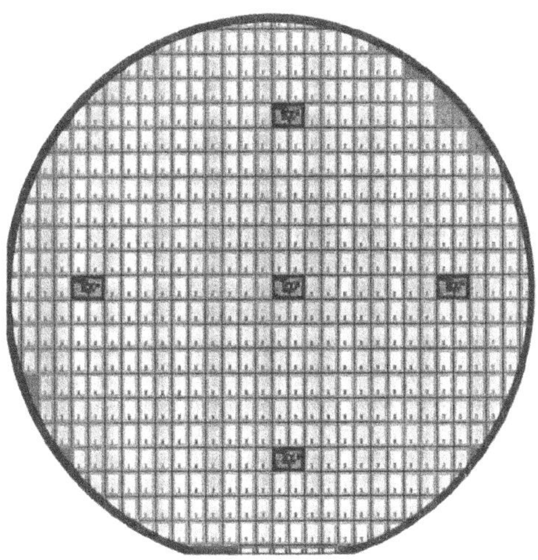

Figure 1.3 100 mm 4H-SiC wafer after processing of power MOSFETs. Reproduced by courtesy of T. Nakamura (Rohm).

SiC power switching devices require further improvement in performance and cost reduction. The market is slowly growing as these devices become more cost-effective. As far as ultrahigh-voltage switching devices are concerned, 12–21 kV thyristors, IGBTs, and BJTs have been demonstrated [59–62].

1.3 Outline of This Book

As a result of the rapid progress in SiC growth and device technologies in the last decade, some SiC power devices are now in commercial production. The major benefits of SiC devices include lower power dissipation, smaller size, and simplified cooling units of power converters. A number of academic studies on the materials science and device physics of SiC have been carried out, adding substantially to the scientific knowledge in this area. In this book, fundamental physics, present understanding, and unaddressed issues in SiC technology are summarized.

The outline of the chapters is as follows:

Chapter 2 describes the unique crystal structures and physical properties of SiC, and compares SiC with Si and other semiconductors.

Chapter 3 focuses on bulk growth of SiC for wafer production. The basic principles and technology development for sublimation growth are explained.

Chapter 4 gives the basics of homoepitaxial growth of hexagonal SiC by CVD. Doping control and defects in SiC epitaxial layers are presented.

Chapter 5 is devoted to techniques used to characterize the electrical and optical properties of SiC. Detection of various defects in SiC and the nature of these defects are also described.

Chapter 6 discusses device processing technologies, such as ion implantation, etching, MOS interface, and metallization. Both fundamental issues and practical considerations are given.

Chapter 7 describes the basic physics of power diodes, especially SBDs and pin diodes, and gives examples of SiC-based diodes and their performance.

Chapter 8 explains the structure, design, and performance of unipolar power switching devices, such as MOSFETs and JFETs. The oxide/SiC issues are also addressed.

Chapter 9 deals with bipolar power switching devices, such as BJTs, IGBTs, and thyristors.

Chapter 10 describes basic issues in the optimization of power devices, including design of blocking voltage, edge termination. A performance comparison of various Si, SiC, and GaN devices is also given.

Chapter 11 introduces applications of SiC devices in power systems. Basic circuits and operation of power conversion, motor drive, inverter, DC–DC converter, power supply are described.

Chapter 12 focuses on specialized SiC devices other than power devices. The devices include high-frequency devices, high-temperature devices, and sensors.

In a book this size it is difficult to completely cover the entire field of SiC materials and devices. The authors have tried to focus on the fundamental science and the state-of-the-art technology. For example, the description of solution growth of SiC boules, the heteroepitaxial growth of 3C-SiC, the theoretical study on defects in SiC, and latest device development is not very extensive. For additional detail, please see the related books [63–69], review papers, and conference proceedings.

References

[1] Bardeen, J. and Brattain, W.H. (1948) The transistor, a semi-conductor triode. *Phys. Rev.*, **74**, 230.

[2] Shockley, W. (1949) The theory of *p-n* junctions in semiconductors and *p-n* junction transistors. *Bell Syst. Tech. J.*, **28**, 435.

[3] Kahng, D. and Atalla, M.M. (1960) IRE-AIEEE Solid-State Device Research Conference Silicon-silicon dioxide field induced surface devices.
[4] Kilby, J. (1959) US Patent 3,138,743 Miniaturized electronic circuits.
[5] Noyce, R. (1959) US Patent 2,981,877 Semiconductor device-and-lead structure.
[6] Taur, Y. and Ning, T.H. (2009) *Fundamentals of Modern VLSI Devices*, 2nd edn, Cambridge University Press.
[7] International Technology Roadmap for Semiconductors http://www.itrs.net/reports.html (accessed 27 March 2014).
[8] Tiwari, S. (1991) *Compound Semiconductor Device Physics*, Academic Press.
[9] Schubert, E.F. (2003) *Light-Emitting Diodes*, Cambridge University Press.
[10] Edgar, J.H., Strite, S., Akasaki, I. *et al.* (eds) (1999) *Properties, Processing and Applications of Gallium Nitride and Related Semiconductors*, INSPEC.
[11] Nakamura, S. and Chichibu, S.F. (eds) (2000) *Introduction to Nitride Semiconductor Blue Lasers and Light Emitting Diodes*, Taylor & Francis.
[12] Willander, M. and Hartnagel, H.L. (2011) *High Temperature Electronics*, Chapman & Hall.
[13] Wong, W.S. and Salleo, A. (eds) (2010) *Flexible Electronics: Materials and Applications*, Springer.
[14] International Energy Agency http://www.iea.org (accessed 27 March 2014).
[15] Newell, W.E. (1973) Power Electronics Specialists Conference 1973, Pasadena, CA, Keynote Talk Power electronics-emerging from limbo.
[16] Kassakian, J.G., Schlecht, M.F. and Verghese, G.C. (1991) *Principles of Power Electronics*, Addison Wesley.
[17] Ghandhi, S.K. (1977) *Power Semiconductor Devices*, John Wiley & Sons, Inc., New York.
[18] Baliga, B.J. (2008) *Fundamentals of Power Semiconductor Devices*, Springer.
[19] Harris, G.L. (1995) *Properties of Silicon Carbide*, INSPEC.
[20] Davis, R.F., Kelner, G., Shur, M. *et al.* (1991) Thin film deposition and microelectronic and optoelectronic device fabrication and characterization in monocrystalline alpha and beta silicon carbide. *Proc. IEEE*, **79**, 677.
[21] Ivanov, P.A. and Chelnokov, V.E. (1992) Recent developments in SiC single-crystal electronics. *Semicond. Sci. Technol.*, **7**, 863.
[22] Morkoç, H., Strite, S., Gao, G.B. *et al.* (1994) Large-band-gap SiC, III-V nitride, and II-VI ZnSe-based semiconductor device technologies. *J. Appl. Phys.*, **76**, 1363.
[23] Verma, A.R. and Krishna, P. (eds) (1966) *Polymorphism and Polytypism in Crystals*, John Wiley & Sons, Inc., New York.
[24] Berzelius, J.J. (1824) *Ann. Phys. Chem. Lpz.*, **1**, 169.
[25] Acheson, E.G. (1892) English Patent 17911 Production of artificial crystalline carbonaceous materials, carborundum.
[26] Round, H.J. (1907) A note on carborundum. *Electr. World*, **19**, 309.
[27] Moissan, H. (1905) Étude du siliciure de carbone de la météorite de cañon diablo. *Compt. Rend.*, **140**, 405.
[28] Lely, J.A. (1955) Darstellung von einkristallen von siliziumcarbid und beherrschung von art und menge der eingebauten verunreinigungen. *Ber. Dtsch. Keram. Ges.*, **32**, 229.
[29] O'Connor, J.R. and Smiltens, J. (eds) (1960) *Silicon Carbide – A High Temperature Semiconductor*, Pergamon Press.
[30] Marshall, R.C., Faust, J.W. Jr., and Ryan, C.E. (1974) *Silicon Carbide 1973*, University of South Carolina Press.
[31] Choyke, W.J. (1969) Optical properties of polytypes of SiC: Interband absorption, and luminescence of nitrogen-exciton complexes. *Mater. Res. Bull.*, **4**, 141.
[32] Tairov, Y.M. and Tsvetkov, V.F. (1978) Investigation of growth processes of ingots of silicon carbide single crystalsInvestigation of growth processes of ingots of silicon carbide single crystals. *J. Cryst. Growth*, **43**, 209.
[33] Tairov, Y.M. and Tsvetkov, V.F. (1981) General principles of growing large-size single crystals of various silicon carbide polytypes. *J. Cryst. Growth*, **52**, 146.
[34] Davis, R.F., Carter, C.H., Jr.,, and Hunter, C.E. (1995) US Patent Re 34,861 Sublimation of silicon carbide to produce large, device quality single crystals of silicon carbide.
[35] Cree http://www.cree.com (accessed 27 March 2014).
[36] Ziegler, G., Lanig, P., Theis, D. and Weurich, C. (1980) Single crystal growth of SiC substrate material for blue light emitting diodes. *IEEE Trans. Electron. Devices*, **30**, 277.
[37] Ikeda, M., Hayakawa, T., Yamagiwa, S. *et al.* (1980) Fabrication of 6H-SiC light-emitting diodes by a rotation dipping technique: Electroluminescence mechanisms. *J. Appl. Phys.*, **50**, 8215.

[38] Matsunami, H., Nishino, S. and Ono, H. (1981) Heteroepitaxial growth of cubic silicon carbide on foreign substrates. *IEEE Trans. Electron. Devices*, **28**, 1235.

[39] Nishino, S., Powell, A. and Will, H.A. (1983) Production of large-area single-crystal wafers of cubic SiC for semiconductor devices. *Appl. Phys. Lett.*, **42**, 460.

[40] Kuroda, N., Shibahara, K., Yoo, W.S. *et al.* (1987) Extended Abstracts, 19th Conference on Solid State Devices and Materials, Tokyo, Japan, 1987, p. 227 Step controlled VPE growth of SiC single crystals at low temperatures.

[41] Kong, H.S., Kim, H.J., Edmond, J.A. *et al.* (1987) Growth, doping, device development and characterization of CVD beta-SiC epilayers on Si(100) and alpha-SiC(0001). *Mater. Res. Soc. Symp. Proc.*, **97**, 233.

[42] Itoh, A., Akita, H., Kimoto, T. and Matsunami, H. (1994) *Silicon Carbide and Related Materials 1993*, IOP, p. 59 Step-controlled epitaxy of 4H-SiC and its physical properties.

[43] Larkin, D.J., Neudeck, P.G., Powell, J.A. and Matus, L.G. (1994) Site-competition epitaxy for superior silicon carbide electronics. *Appl. Phys. Lett.*, **65**, 1659.

[44] Kordina, O., Hallin, C., Glass, R.C. *et al.* (1994) *Silicon Carbide and Related Materials 1993*, IOP, p. 41 A novel hot-wall CVD reactor for SiC epitaxy.

[45] Baliga, B.J. (1989) Power semiconductor device figure of merit for high-frequency applications. *IEEE Electron. Device Lett.*, **10**, 455.

[46] Bhatnagar, M. and Baliga, B.J. (1993) Comparison of 6H-SiC, 3C-SiC, and Si for power devices. *IEEE Trans. Electron. Devices*, **40**, 645.

[47] Matus, L.G., Powell, J.A. and Salupo, C.S. (1991) High-voltage 6H-SiC p-n junction diodes. *Appl. Phys. Lett.*, **59**, 1770.

[48] Urushidani, T., Kobayashi, S., Kimoto, T., and Matsunami, H. (1993) Extended Abstracts, 1993 International Conference on Solid State Devices and Materials, Chiba, Japan, 1993, p. 814 SiC Schottky barrier diodes with high blocking voltage of 1kV.

[49] Kimoto, T., Itoh, A., Akita, H. *et al.* (1995) Step-controlled epitaxial growth of α-SiC and application to high-voltage Schottky rectifiers, in *Compound Semiconductors – 1994*, IOP, Bristol, p. 437.

[50] Infineon http://www.infineon.com (accessed 27 March 2014).

[51] Agarwal, A., Das, M., Krishnaswami, S. *et al.* (2004) SiC power devices – An overview. *Mater. Res. Soc. Symp. Proc.*, **815**, 243.

[52] Niwa, H., Suda, J. and Kimoto, T. (2012) 21.7 kV 4H-SiC PiN diode with a space-modulated junction termination extension. *Appl. Phys. Exp.*, **5**, 064001.

[53] Palmour, J.W., Edmond, J.A., Kong, H.S. and Carter, C.H. Jr., (1994) Vertical power devices in silicon carbide, in *Silicon Carbide and Related Materials 1993*, IOP, p. 499.

[54] J.W. Palmour, V.F. Tsvetkov, L.A. Lipkin, and C.H. Carter, Jr.,, *Compound Semiconductors – 1994* (IOP, Bristol, 1995), p.377 Silicon carbide substrates and power devices.

[55] Shenoy, J.N., Cooper, J.A. and Melloch, M.R. (1997) High-voltage double-implanted power MOSFETs in 6H-SiC. *IEEE Electron Device Lett.*, **18**, 93.

[56] Tan, J., Cooper, J.A. Jr., and Melloch, M.R. (1998) High-voltage accumulation-layer UMOSFETs in 4H-SiC. *IEEE Electron Device Lett.*, **19**, 467.

[57] Friedrichs, P., Mitlehner, H., Kaltschmidt, R. *et al* (2000) Static and dynamic characteristics of 4H-SiC JFETs designed for different blocking categories. *Mater. Sci. Forum*, **338–342**, 1243.

[58] Nakamura, T., Miura, M., Kawamoto, N. *et al* (2009) Development of SiC diodes, power MOSFETs and intelligent power modules. *Phys. Status Solidi A*, **206**, 2403.

[59] Wang, X. and Cooper, J.A. (2010) High-voltage n-channel IGBTs on free-standing 4H-SiC epilayers. *IEEE Trans. Electron Devices*, **57**, 511.

[60] Zhang, Q.J., Agarwal, A., Capell, C. *et al.* (2012) 12 kV, 1 cm^2 SiC GTO thyristors with negative bevel termination. *Mater. Sci. Forum*, **717–720**, 1151.

[61] Ryu, S.-H., Cheng, L., Dhar, S. *et al.* (2012) Development of 15 kV 4H-SiC IGBTs. *Mater. Sci. Forum*, **717–720**, 1135.

[62] Miyake, H., Okuda, T., Niwa, H. *et al.* (2012) 21-kV SiC BJTs with space-modulated junction termination extension. *IEEE Electron Device Lett.*, **33**, 1598.

[63] Choyke, W.J., Matsunami, H. and Pensl, G. (eds) (1997) *Silicon Carbide, A Review of Fundamental Questions and Applications to Current Device Technology*, vol. 1 & 2, Akademie Verlag.

[64] Zetterling, C.M. (2002) *Process Technology for Silicon Carbide Devices*, INSPEC.

[65] Choyke, W.J., Matsunami, H. and Pensl, G. (eds) (2004) *Silicon Carbide – Recent Major Advances*, Springer.

[66] Feng, Z.C. and Zhao, J.H. (eds) (2004) *Silicon Carbide, Materials, Processing, and Devices*, Taylor & Francis Group.
[67] Baliga, B.J. (2006) *Silicon Carbide Power Devices*, World Scientific.
[68] Shur, M., Rumyantsev, S. and Levinshtein, M. (eds) (2006) *SiC Materials and Devices*, vol. 1 & 2, World Scientific.
[69] Friedrichs, P., Kimoto, T., Ley, L. and Pensl, G. (eds) (2010) *Silicon Carbide*, Vol. 1: Growth, Defects, and Novel Applications, Vol.2: Power Devices and Sensors, Wiley-VCH Verlag GmbH, Weinheim.

2

Physical Properties of Silicon Carbide

Silicon carbide (SiC) crystallizes in a wide variety of structures, each of which exhibits unique electrical, optical, thermal, and mechanical properties. The physical properties of SiC are very important subjects of academic study as well as critical parameters for accurate simulation of devices. This chapter briefly reviews the physical properties of SiC.

2.1 Crystal Structure

SiC is a compound semiconductor, which means that only a rigid stoichiometry, 50% silicon (Si) and 50% carbon (C), is allowed. The electronic structures of neutral Si and C atoms in their ground states are:

$$\text{Si}, 14\text{e}^- : 1\text{s}^2 2\text{s}^2 2\text{p}^6 3\text{s}^2 3\text{p}^2 \tag{2.1}$$

$$\text{C}, 6\text{e}^- : 1\text{s}^2 2\text{s}^2 2\text{p}^2 \tag{2.2}$$

Both Si and C atoms are tetravalent elements and have four valence electrons in their outermost shells. Si and C atoms are tetrahedrally bonded with covalent bonds by sharing electron pairs in sp^3-hybrid orbitals to form a SiC crystal. Each Si atom has exactly four C atom neighbors, and vice versa. The Si–C bond energy is very high (4.6 eV), which gives SiC a variety of outstanding properties, as described below.

From a crystallographic point of view, SiC is the best known example of *polytypism* [1–5]. Polytypism is the phenomenon where a material can adopt different crystal structures which vary in one dimension (that is, in stacking sequence) without changes in chemical composition. The variation in the occupied sites along the *c*-axis in a *hexagonal close-packed system* brings about different crystal structures, known as *polytypes*. Consider the occupied sites in the hexagonal close-packed system, shown schematically in Figure 2.1. There are three possible sites, denoted as A, B, and C. Two layers cannot successively occupy the same site; the next layer on top of an "A" layer must occupy either "B" or "C" sites (and, similarly, "A" or "C" is allowed over "B"). Though there exist, in principle, almost infinite variations of the stacking sequence when stacking a number of layers; for most materials, only one stacking structure (often either the zincblende or wurtzite structure) is usually stable. However, SiC crystallizes in surprisingly many (more than 200) polytypes.

In Ramsdell's notation, polytypes are represented by the number of Si-C bilayers in the unit cell and the crystal system (C for cubic, H for hexagonal, and R for rhombohedral). 3C-SiC is often called β-SiC, and

Fundamentals of Silicon Carbide Technology: Growth, Characterization, Devices, and Applications, First Edition. Tsunenobu Kimoto and James A. Cooper.

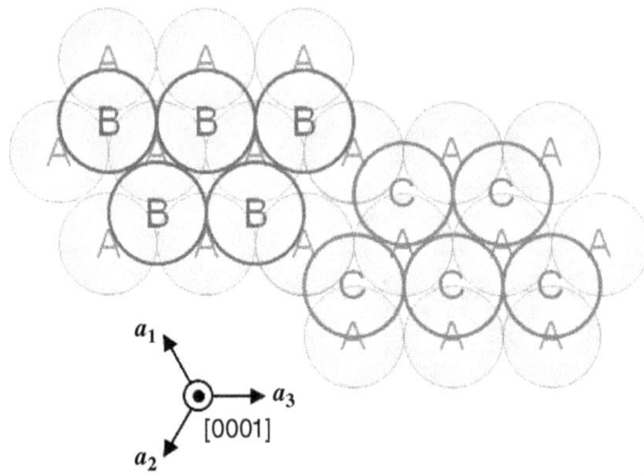

Figure 2.1 Occupation sites (A, B, and C) in the hexagonal close-packed system.

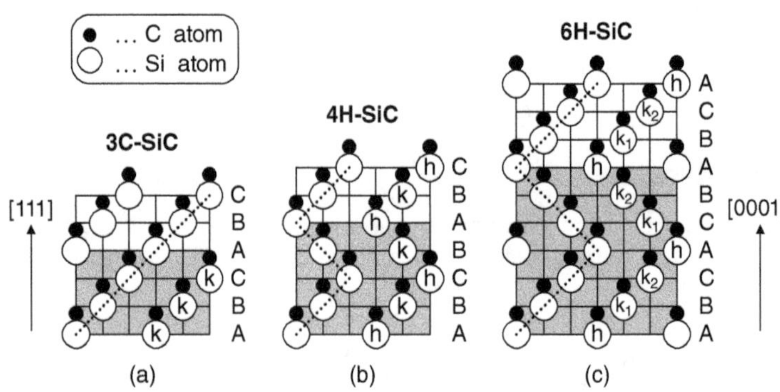

Figure 2.2 Schematic structures of popular SiC polytypes; (a) 3C-SiC, (b) 4H-SiC, and (c) 6H-SiC. Open and closed circles denote Si and C atoms, respectively.

other polytypes are referred to as α-SiC. The structures of popular SiC polytypes; 3C-SiC, 4H-SiC, and 6H-SiC, are shown schematically in Figure 2.2, where open and closed circles denote Si and C atoms, respectively. Here, A, B, and C are the potentially occupied sites in a hexagonal close-packed structure, as described above. These site names enable 3C-SiC to be described by the repeating sequence of ABCABC, or simply ABC. In a similar manner, 4H- and 6H-SiC can be described by ABCB (or ABAC) and ABCACB, respectively. The structures of these three SiC polytypes in a ball-stick model are shown in Figure 2.3. Because there are several popular notations to define stacking structures [1], the major SiC polytypes are described using Ramsdell's, Zhdanov's, and Jagodzinski's notations in Table 2.1. Figure 2.4 shows the primitive cells and fundamental translation vectors of (a) cubic (3C) SiC and (b) hexagonal SiC. The "3C" structure is equivalent to the zincblende structure, in which most III–V semiconductors like GaAs and InP crystallize. The wurtzite structure, also found in GaN and ZnS, can be denoted by "2H". However, it is still not fully understood why so many SiC polytypes exist. In general, crystals with strong covalent bonding crystallize in the zincblende structure, while the wurtzite structure is more stable for crystals with high ionicity. The intermediate ionicity of SiC (11% according to Pauling's definition) may be a possible reason for the occurrence of SiC polytypism [6, 7]. The space groups are T_d^2 for 3C-SiC, C_{6v}^4 for hexagonal polytypes, and C_{3v}^4 for rhombohedral polytypes [8]. Hexagonal and rhombohedral polytypes are uniaxial, and thus these polytypes exhibit unique polarized optical properties.

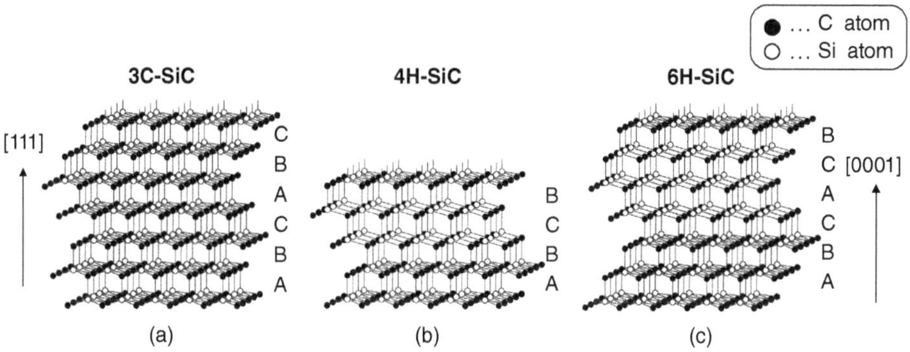

Figure 2.3 Structures of (a) 3C-SiC, (b) 4H-SiC, and (c) 6H-SiC in a ball-stick model.

Table 2.1 Ramsdell's, Zhdanov's, and Jagodzinski's notations of major SiC polytypes.

Ramsdell's notation	Zhdanov's notation	Jagodzinski's notation
2H	11	h
3C	∞	k
4H	22	hk
6H	33	hkk
15R	$(32)_3$	hkkhk

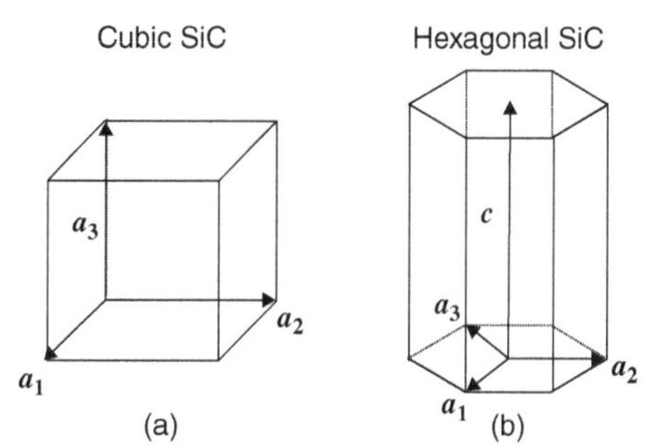

Figure 2.4 Primitive cells and fundamental translation vectors of (a) cubic (3C) SiC and (b) hexagonal SiC.

Because of the variety of ways to stack Si-C bilayers, there are several lattice sites in SiC, which differ in their structures of immediate neighbors. The lattice sites with hexagonal-structured surroundings are denoted "*hexagonal sites,*" and those with cubic-structured surroundings are denoted "*cubic sites.*" In Figure 2.2, hexagonal and cubic sites are indicated by "h" and "k," respectively. 4H-SiC has one hexagonal and one cubic site, and 6H-SiC one hexagonal and two inequivalent cubic sites, while 3C-SiC contains only cubic sites. Hexagonal and cubic sites differ in the location of the second-nearest neighbors, leading to different crystal fields. For example, the energy levels of dopants, impurities, and point defects (such as vacancies) depend on the lattice site (hexagonal/cubic). This is called the "*site effect*" [9–11].

Table 2.2 Lattice constants of major SiC polytypes at room temperature [26].

Polytype	*a* (Å)	*c* (Å)
3C	4.3596	–
4H	3.0798	10.0820
6H	3.0805	15.1151

The stability and nucleation probability of SiC polytypes depend strongly on temperature [12]. For example, 3C-SiC is not stable, and is transformed into hexagonal SiC polytypes such as 6H-SiC at very high temperatures, above 1900–2000 °C [13]. This instability of 3C-SiC makes it difficult to grow large 3C-SiC ingots at a reasonable growth rate. 2H-SiC is also unstable at high temperature, and large 2H-SiC crystals have not been obtained. Thus, 4H-SiC and 6H-SiC polytypes are very popular, and have been extensively investigated to date [14–20]. 3C-SiC is another popular polytype because 3C-SiC can be grown heteroepitaxially on Si substrates [21–23]. As well as these three main polytypes, 15R-SiC is occasionally obtained, and has been studied to some extent [24, 25].

Table 2.2 shows the lattice constants of major SiC polytypes at room temperature [26]. Though the lattice constants look very different for different SiC polytypes (because of their different crystal structures), all SiC polytypes possess almost the same Si-C bond length (1.89 Å). Thus the height of the Si-C bilayer along the *c*-axis (unit height) is 2.52 Å, although 3C-SiC and 2H-SiC have a slightly smaller height (2.50 Å). The lattice constants vary with temperature and doping density, as is also observed for other semiconductor materials. Figure 2.5 shows the *c*-axis lattice constant of 4H-SiC from room temperature

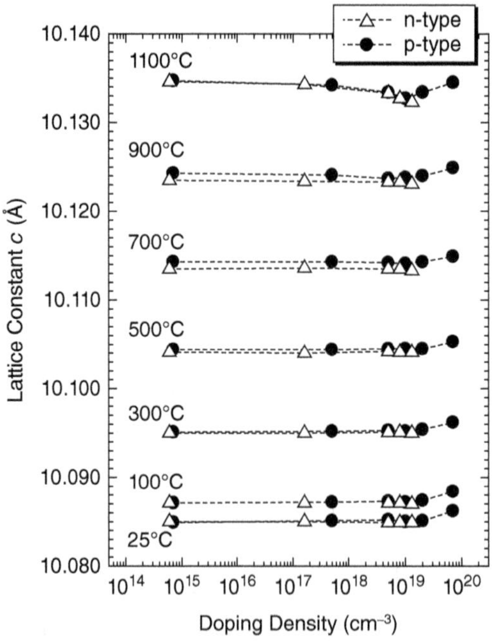

Figure 2.5 *c*-axis lattice constant of 4H-SiC from room temperature to 1100 °C as a function of doping density (nitrogen or aluminum).

to 1100 °C as a function of doping density (doping with nitrogen or aluminum) [27, 28]. In general, very high (>10^{19} cm^{-3}) nitrogen doping causes lattice contraction, and lattice expansion is induced by very high aluminum doping. This trend is more pronounced at temperatures above 1000 °C. Therefore, one should expect mismatch-induced stress at the $n^-/n^+, p^-/p^+, p^+/n^-, n^+/p^-$, and p^+/n^+ interfaces, which can lead to generation of extended defects such as basal plane dislocations. The axial thermal expansion coefficients of SiC perpendicular (α_{11}) and parallel (α_{33}) to the *c*-axis have been measured [29], and the temperature dependence for 4H-SiC is expressed by:

$$\alpha_{11} = 3.21 \times 10^{-6} + 3.56 \times 10^{-9}T - 1.62 \times 10^{-12}T^2(\mathrm{K}^{-1}) \quad (2.3)$$

$$\alpha_{33} = 3.09 \times 10^{-6} + 2.63 \times 10^{-9}T - 1.08 \times 10^{-12}T^2(\mathrm{K}^{-1}) \quad (2.4)$$

Here T is the absolute temperature. The thermal expansion coefficients of different SiC polytypes do not deviate very much from each other.

Because all SiC polytypes consist of similar Si-C bonds, mechanical properties such as hardness are very similar among different SiC polytypes [30]. However, different periodic potentials in different SiC polytypes result in very different electronic band structures, and thus significant variation in optical and electronic properties. This means that, for device applications, it is crucial to grow only the single desired SiC polytype; polytype control is a vital aspect of crystal growth of SiC.

Except for 3C-SiC, crystal planes and directions in SiC polytypes are usually expressed by using four Miller–Bravais indices [31]. A crystal plane $(h_1 h_2 h_3 l_h)$ is equivalent to a plane $(h\ k\ l)$, defined by three Miller indices in a monoclinic system, when the following relations are satisfied:

$$h_1 = h, \quad h_2 = k, \quad h_3 = -(h+k), \quad \text{and} \quad l_h = l. \quad (2.5)$$

In a similar manner, a crystalline direction $[u_1 u_2 u_3 w_h]$ is equivalent to a direction $[u\ v\ w]$, defined by three Miller indices in a monoclinic system, when the following relations are satisfied:

$$u_1 = (2u - v)/3, \quad u_2 = (2v - u)/3, \quad u_3 = -(u+v)/3, \quad \text{and} \quad w_h = w. \quad (2.6)$$

Because SiC is a compound semiconductor, the valence electrons are slightly localized near C atoms, which are more electronegative than silicon (C: 2.5, Si: 1.8). In this sense, Si atoms can be referred to as *cations* and C atoms as *anions*. This ionicity gives rise to *polarity* in SiC, which is of academic and technological importance. Schematic illustrations of bond configurations in a hexagonal SiC polytype are shown in Figure 2.6. In a hexagonal or rhombohedral structure, the (0001) face, where one bond from

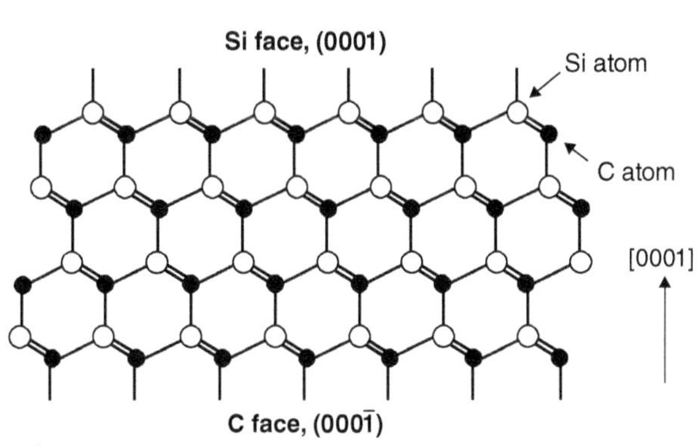

Figure 2.6 Schematic illustrations of bond configurations in a hexagonal SiC polytype. SiC{0001} is a polar face, either Si or C face.

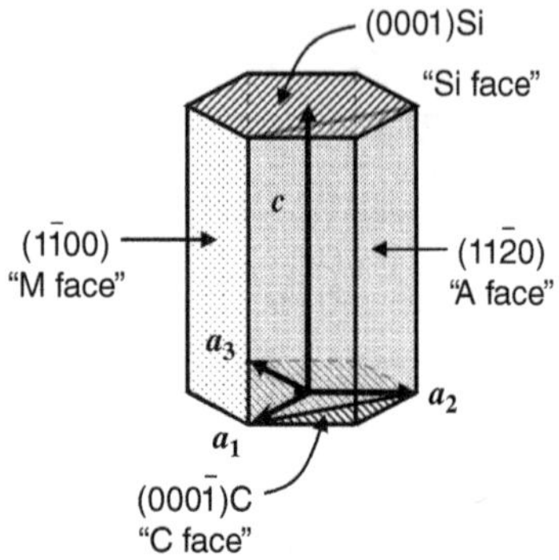

Figure 2.7 Definition of several major planes in a hexagonal SiC polytype.

a tetrahedrally-bonded Si atom is directed along the *c*-axis (<0001>), is called the "*Si face*," while the (000$\bar{1}$) face, where one bond from a tetrahedrally-bonded C atom is directed along the *c*-axis, is called the "*C face*." In 3C-SiC, the (111) and ($\bar{1}\,\bar{1}\,\bar{1}$) faces correspond to the Si face and C face, respectively; these faces are similar to the "A face" and "B face" in III–V semiconductors. The definition relies on the crystallographic orientation, and not on the terminating atoms on the surface. Figure 2.7 illustrates the definition of several major planes in a hexagonal SiC polytype. Other than the Si and C faces, the {11$\bar{2}$0} face is called the "*A face* (or *a*-face)," and the {1$\bar{1}$00} face the "*M face* (or *m*-face)." The surface energy, chemical reactivity, and electronic properties are significantly dependent on these crystal faces, details of which are described in the growth and device chapters. Standard wafers are SiC(0001) with several degrees off-axis toward <11$\bar{2}$0> [32]; these are described in detail in Section 3.5.

2.2 Electrical and Optical Properties

2.2.1 Band Structure

Figure 2.8 shows the first Brillouin zones of (a) 3C-SiC and (b) a hexagonal SiC polytype [26, 30]. Note that the height of the Brillouin zone shown in Figure 2.8b is different for different hexagonal polytypes because of their different values of the lattice parameter, *c*.

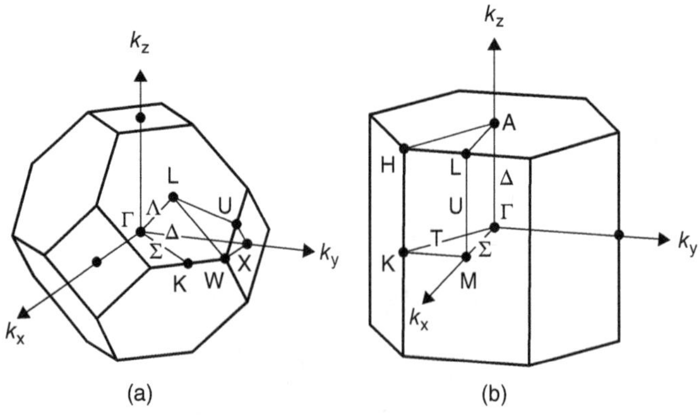

Figure 2.8 Brillouin zones of (a) 3C-SiC and (b) a hexagonal SiC polytype.

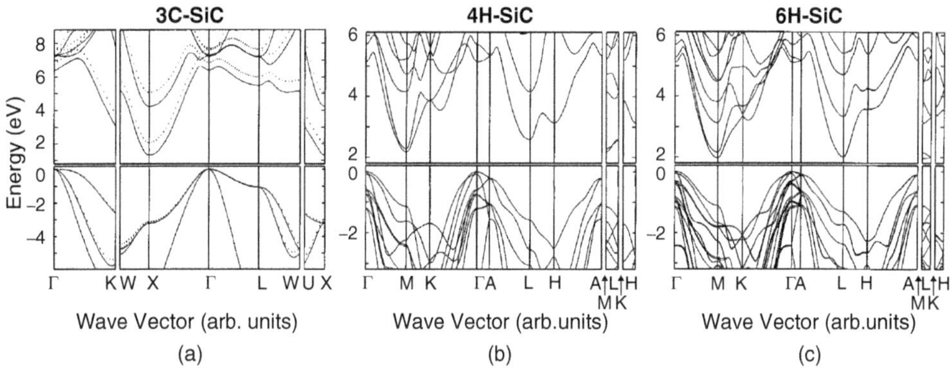

Figure 2.9 Electronic band structures of (a) 3C-SiC, (b) 4H-SiC, and (c) 6H-SiC [37]. Note that the absolute values of the bandgap are underestimated, due to a limitation of the theoretical calculation. ([37] Reproduced with permission from AIP Publishing LLC).

Figure 2.9 depicts the electronic band structures of (a) 3C-SiC, (b) 4H-SiC, and (c) 6H-SiC [33–37]. Note that the absolute values of the bandgap are underestimated in this figure, due to a limitation of the theoretical calculation (density functional theory). All the SiC polytypes have an *indirect band structure,* as is also the case for Si. The top of the valence band is located at the Γ point in the Brillouin zone, whereas the conduction band minima appear at the Brillouin zone boundary. The conduction band minima are located at the X point for 3C-SiC, M point for 4H-SiC, and U point (along the M–L line) for 6H-SiC. Thus, the number of conduction band minima in the first Brillouin zone (M_c) is 3 for 3C-SiC, 3 for 4H-SiC, and 6 for 6H-SiC. Because Si-C covalent bonds are common to all SiC polytypes, the valence band structure is similar amongst the different polytypes, except for the splitting. The top of the valence band is doubly degenerate in 3C-SiC, as a result of its cubic symmetry, and the next valence band is shifted 10 meV from the top by the spin–orbit interaction [38]. The crystal field, which exists in all hexagonal polytypes, splits the valence band degeneracy. The magnitudes of the spin–orbit splitting and crystal-field splitting for 4H-SiC are 6.8 and 60 meV, respectively [39].

Table 2.3 summarizes the effective masses of electrons and holes in 3C-, 4H-, and 6H-SiC [40–42]. The electron effective mass and its anisotropy depend strongly on the polytype, while the hole effective mass exhibits a weak polytype dependence. The former leads to large variation of electron mobility in different polytypes, and also to anisotropic electron transport, as explained in Section 2.2.4.

The exciton gaps of various SiC polytypes at 2 K are plotted as a function of "*hexagonality*" in Figure 2.10 [35, 43]. Here, hexagonality means the ratio of the number of hexagonal sites to the total number of Si-C bilayers (hexagonal and cubic sites) in a unit cell (the hexagonality is 1 for 2H-SiC, 0 for 3C-SiC, 1/2 for 4H-SiC, and 1/3 for 6H-SiC). It is interesting that the bandgap of SiC polytypes increases monotonically with increasing hexagonality. The bandgap at room temperature is 2.36 eV for 3C-SiC, 3.26 eV for 4H-SiC, and 3.02 eV for 6H-SiC. Figure 2.11 shows the temperature dependence of the bandgap for several SiC polytypes [44]. The bandgap (E_g) decreases with increasing temperature because of thermal expansion, and its temperature dependence can be semi-empirically expressed as [45]:

$$E_g(T) = E_{g0} - \frac{\alpha T^2}{T + \beta} \tag{2.7}$$

where E_{g0} is the bandgap at 0 K, T the absolute temperature, and α and β are fitting parameters ($\alpha = 8.2 \times 10^{-4}$ eV K^{-1}, $\beta = 1.8 \times 10^3$ K). Note that the bandgap also depends on the doping density; very high impurity doping, above 10^{19} cm^{-3}, causes the bandgap to shrink because of the formation of pronounced tail states near the band edges [46].

Table 2.3 Effective masses of electrons and holes in 3C-, 4H-, and 6H-SiC [40–42].

Polytype	Effective mass	Experiment (m_0)	Theory (m_0)
Electron effective mass			
3C-SiC	$m_{//}$	0.667	0.68
	$m_{\perp}$	0.247	0.23
4H-SiC	$m_{ML}(= m_{//})$	0.33	0.31
	$m_{M\Gamma}$	0.58	0.57
	m_{MK}	0.31	0.28
	$m_{\perp}\ (= (m_{M\Gamma} m_{MK})^{1/2})$	0.42	0.40
6H-SiC	$m_{ML}(= m_{//})$	2.0	1.83
	$m_{M\Gamma}$	–	0.75
	m_{MK}	–	0.24
	$m_{\perp}\ (= (m_{M\Gamma} m_{MK})^{1/2})$	0.48	0.42
Hole effective mass			
3C-SiC	$m_{\Gamma X}(= m_{[100]})$	–	0.59
	$m_{\Gamma K}(= m_{[110]})$	–	1.32
	$m_{\Gamma L}(= m_{[111]})$	–	1.64
4H-SiC	$m_{//}$	1.75	1.62
	$m_{\perp}$	0.66	0.61
6H-SiC	$m_{//}$	1.85	1.65
	$m_{\perp}$	0.66	0.60

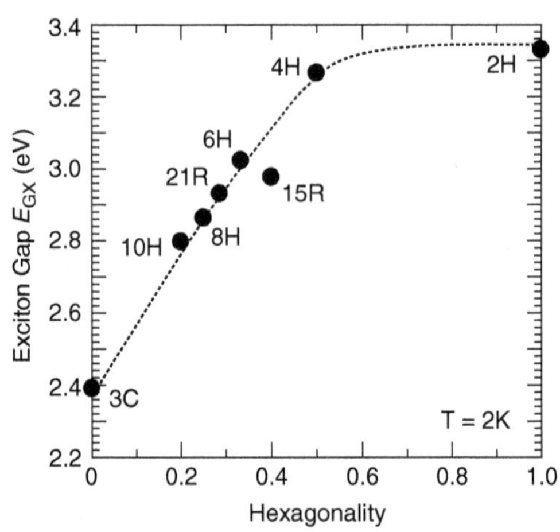

Figure 2.10 Exciton gaps of various SiC polytypes at 2 K versus hexagonality.

2.2.2 *Optical Absorption Coefficient and Refractive Index*

Figure 2.12 shows the optical absorption coefficients versus photon energy for the major SiC polytypes [47, 48]. Because of the indirect band structure of SiC, the absorption coefficient (α_{opt}) slowly increases,

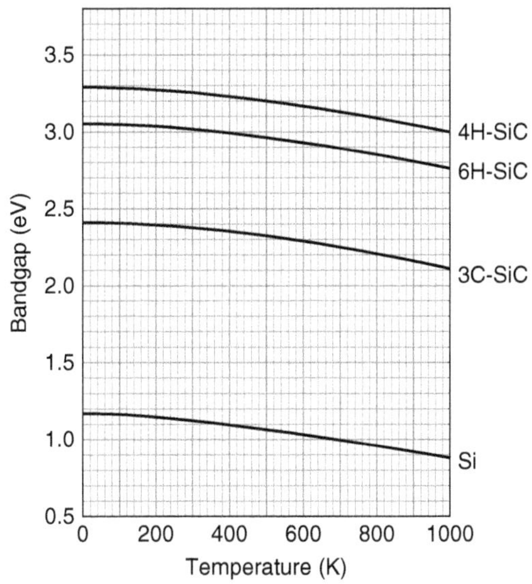

Figure 2.11 Temperature dependence of bandgap for several SiC polytypes.

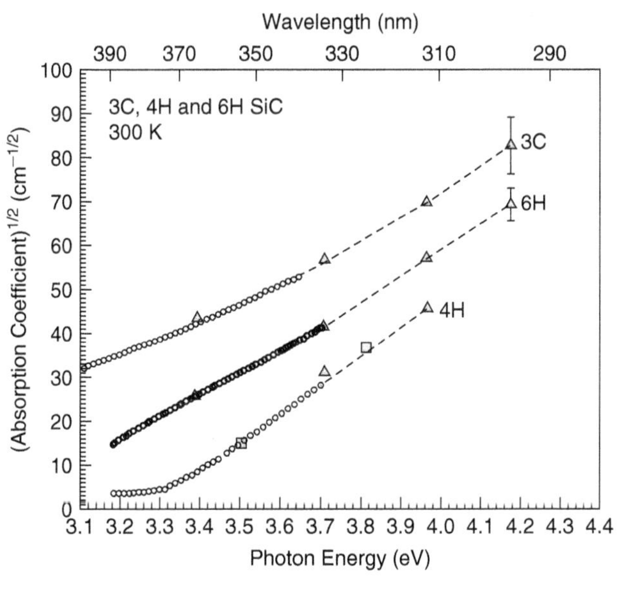

Figure 2.12 Optical absorption coefficients versus photon energy for major SiC polytypes. ([47, 48] Reproduced with permission from Springer-Verlag GmbH).

even when the photon energy exceeds the bandgap. Taking account of phonon absorption and emission, the absorption coefficient can be approximated as [49]:

$$\alpha_{\text{opt}} = \frac{A_{\text{ab}}}{h\nu}\left\{\frac{\left(h\nu - E_{\text{g}} + \hbar\omega\right)^2}{\exp(h\nu/kT) - 1} + \frac{(h\nu - E_{\text{g}} - \hbar\omega)^2}{1 - \exp(-h\nu/kT)}\right\} \tag{2.8}$$

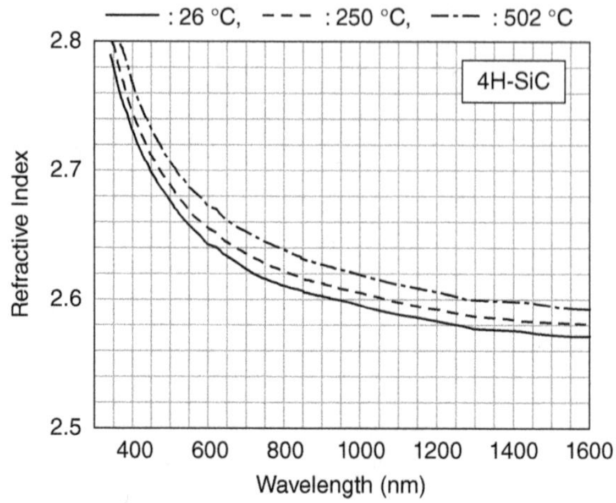

Figure 2.13 Refractive index of 4H-SiC versus wavelength across a wide range from ultraviolet to infrared at various temperatures.

Here $h\nu$ is the photon energy, $\hbar\omega$ the energy of a phonon involved, k the Boltzmann constant, and A_{ab} the parameter. When several different phonons are involved, the sum of those contributions must be calculated. The absorption coefficient of 4H-SiC at room temperature is 69 cm^{-1} at 365 nm (3.397 eV, Hg lamp), 210 cm^{-1} at 355 nm (3.493 eV, 3HG Nd-YAG laser), 1350 cm^{-1} at 325 nm (3.815 eV, He-Cd laser), and 14 200 cm^{-1} at 244 nm (5.082 eV, 2HG Ar ion laser). These values should be kept in mind when SiC materials are characterized by any optical technique, or when SiC-based photodetectors are fabricated. For example, the penetration depth, as defined by $1/\alpha_{opt}$, is 145 μm at 365 nm, 7.4 μm at 325 nm, and 0.7 μm at 244 nm for 4H-SiC at room temperature.

Figure 2.13 shows the refractive index of 4H-SiC versus wavelength across a wide range, from ultraviolet to infrared, at various temperatures [50]. This dispersion of the refractive index $n(\lambda)$ is described by a simple *Sellmeier equation* given by [51]:

$$n(\lambda) = A + \frac{B\lambda^2}{\lambda^2 - C^2} \tag{2.9}$$

where $A, B,$ and C are parameters. The refractive index at a wavelength of 600 nm is 2.64 for 4H-SiC. The thermo-optic coefficient, defined by dn/dT, is $(4.4\text{–}5.0) \times 10^{-4}$ K^{-1} in the visible–infrared region and increases to $(7\text{–}8) \times 10^{-4}$ K^{-1} near the ultraviolet region, due to the shrinkage of the bandgap at elevated temperature [50]. The relative dielectric constant has also been reported for several SiC polytypes [45, 52]. The relative dielectric constants in the high-frequency (100 kHz to 1 MHz) region for 4H-SiC (6H-SiC) at room temperature are 9.76 (9.66) perpendicular to the c-axis and 10.32 (10.03) parallel to the c-axis [52]. The dielectric constant of 3C-SiC is isotropic, 9.72.

2.2.3 *Impurity Doping and Carrier Density*

SiC is an exceptional wide bandgap semiconductor, in the sense that control of both n- and p-type doping over a wide range is relatively easy. Nitrogen or phosphorus are employed for n-type doping and aluminum for p-type doping. Although boron was also previously employed as an acceptor, it is currently not preferred because of its large ionization energy (~350 meV) [53], generation of a boron-related deep level (D center) [53, 54], and its abnormal diffusion [54, 55]. Gallium and arsenic work as acceptor and

Table 2.4 Nonpolar covalent radii of Si, C, and major dopants for SiC [56].

Atom	Si	C	N	P	B	Al
Radius (Å)	1.17	0.77	0.74	1.10	0.82	1.26

Table 2.5 Ionization energies and the solubility limits of nitrogen, phosphorus, aluminum, and boron in major SiC polytypes.

	Nitrogen	Phosphorus	Aluminum	Boron (shallow)
Ionization energy (meV)				
3C-SiC	55	–	250	350
4H-SiC (hexagonal/cubic)	61/126	60/120	198/201	280
6H-SiC (hexagonal/cubic)	85/140	80/130	240	350
Solubility limit (cm^{-3})	2×10^{20}	($\sim 1 \times 10^{21}$)	1×10^{21}	2×10^{19}

donor, respectively, in SiC. Their ionization energies are, however, relatively large, and their solubility limits are low. Nitrogen substitutes at the C sub-lattice site, while phosphorus, aluminum, and boron substitute at the Si sub-lattice site.

Table 2.4 shows the nonpolar covalent radii of Si, C, and major dopants for SiC [56]. The *ionization energies* and the *solubility limits* of nitrogen, phosphorus, and aluminum in major SiC polytypes are summarized in Table 2.5 [10, 11, 53, 57–62]. In SiC, the ionization energies of dopants depend on the lattice site, in particular, whether the site is hexagonal or cubic (site effect). In the case of nitrogen or phosphorus doping, the ionization energy of the donors is relatively small, and the ionization ratio of donors at room temperature is reasonably high, ranging from 50 to nearly 100%, depending on polytype and doping density. Conversely, the ionization energy of aluminum is large (200–250 meV), and *incomplete ionization* (5–30%) of acceptors is observed at room temperature. Note that the ionization energy decreases when the doping density is increased, as a result of bandgap shrinkage and formation of an impurity band. The dependence of dopant ionization energy, ΔE_{dopant}, on the dopant density is described by Efros *et al.* [63]:

$$\Delta E_{\text{dopant}} = \Delta E_{\text{dopant},0} - \alpha_{\text{d}} (N_{\text{dopant}})^{1/3} \tag{2.10}$$

Here $\Delta E_{\text{dopant},0}$ is the ionization energy in lightly-doped materials, N_{dopant} the dopant density, and α_{d} a parameter ($\alpha_{\text{d}} = (2\text{–}4) \times 10^{-8}$ eV cm). When the dopant density exceeds 10^{19} cm^{-3}, the ionization energy decreases sharply. As a result, near-perfect ionization is observed in heavily aluminum-doped SiC (>5×10^{20} cm^{-3}), in spite of the relatively large ionization energy of aluminum [64].

Because the band structure (bandgap, effective mass) is known, one can calculate the effective densities of states in the conduction band N_{C} and valence band N_{V} as well as the intrinsic carrier density n_{i} as follows [65]:

$$N_{\text{C}} = 2M_{\text{C}} \left(\frac{2\pi m_{\text{de}}^* kT}{h^2} \right)^{3/2} \tag{2.11}$$

$$N_{\text{V}} = 2 \left(\frac{2\pi m_{\text{dh}}^* kT}{h^2} \right)^{3/2} \tag{2.12}$$

$$n_{\text{i}} = \sqrt{N_{\text{C}} N_{\text{V}}} \exp\left(-\frac{E_{\text{g}}}{2kT} \right) \tag{2.13}$$

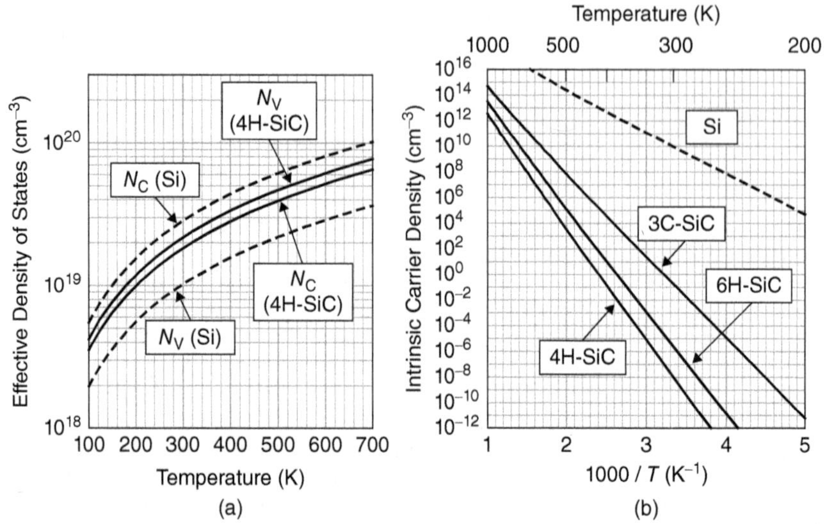

Figure 2.14 Temperature dependence of (a) the effective densities of states in the bands and (b) the intrinsic carrier density for major SiC polytypes, together with that of Si.

Here, M_C is the number of conduction band minima, m_{de}^* (m_{dh}^*) the density-of-state effective mass of electrons (holes), and h the Planck constant. By using the density-of-state effective mass of electrons (holes) and the number of conduction band minima, the N_C and N_V values for 4H-SiC at room temperature are calculated as 1.8×10^{19} and 2.1×10^{19} cm^{-3}, respectively. These values are important as they allow us to estimate whether the material will be degenerate when heavy impurity doping is performed. Figure 2.14 plots the temperature dependence of (a) the effective densities of states in the bands and (b) the intrinsic carrier density for major SiC polytypes, together with that of Si. Here, the temperature dependence of bandgaps is taken into account. The intrinsic carrier density at room temperature is extremely low in SiC, because of the wide bandgap, about 0.13 cm^{-3} for 3C-SiC, 5×10^{-9} cm^{-3} for 4H-SiC, and 1×10^{-6} cm^{-3} for 6H-SiC. This is the main reason why SiC electronic devices can operate at high temperatures with low leakage current.

Based on the Boltzmann approximation for a nondegenerate semiconductor, the neutrality equations in a semiconductor containing one type of donor or acceptor are given by [66]:

$$n + N_{\text{comp,A}} = \frac{N_D}{1 + \left(\frac{g_D n}{N_C}\right) \exp\left(\frac{\Delta E_D}{kT}\right)} \tag{2.14}$$

$$p + N_{\text{comp,D}} = \frac{N_A}{1 + \left(\frac{g_A p}{N_V}\right) \exp\left(\frac{\Delta E_A}{kT}\right)} \tag{2.15}$$

Here $n(p)$ is the free electron (hole) density, $N_{\text{comp,A}}$ ($N_{\text{comp,D}}$) the density of compensating acceptor (donor) levels, N_D (N_A) the donor (acceptor) density, ΔE_D (ΔE_A) the ionization energy of the donor (acceptor), and g_D (g_A) are the degeneracy factors for donors (acceptors), respectively. When multiple donor (or acceptor) levels exist, the sum for corresponding dopants must be considered in the right-hand term of the equation. This is the case for hexagonal SiC polytypes, because the donor (and acceptor) impurities at inequivalent lattice sites (e.g., i = k, h for 4H-SiC) exhibit different energy levels. The Arrhenius plots of the free carrier density in (a) nitrogen-doped and (b) aluminum-doped 4H-SiC are shown in Figure 2.15. Here, the temperature dependence of the bandgap and the doping-density dependence of the

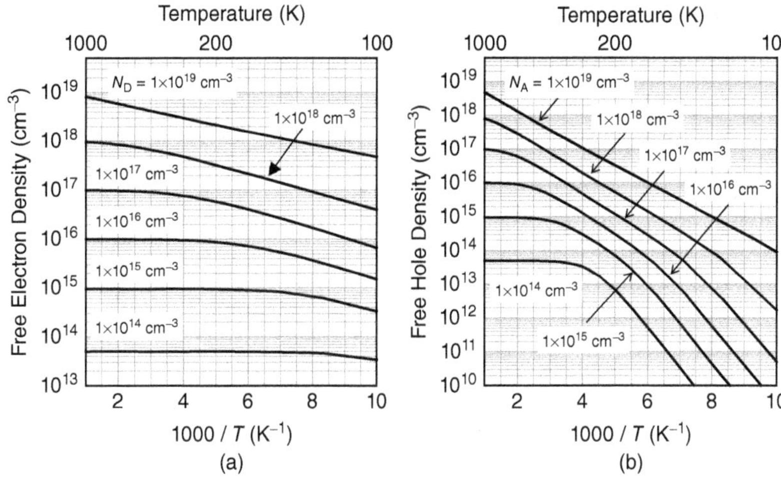

Figure 2.15 Arrhenius plots of the free carrier density in (a) nitrogen-doped and (b) aluminum-doped 4H-SiC. Here, the temperature dependence of the bandgap and the doping-density dependence of the ionization energies are taken into account. A compensating-level density of 5×10^{13} cm^{-3} is assumed.

ionization energies are taken into account. A compensating-level density of 5×10^{13} cm^{-3} is assumed. As shown in Figure 2.15, incomplete ionization is significant for p-type SiC. (See Appendix A.)

The position of the Fermi level E_F in nondegenerate semiconductors is calculated by [65]:

$$E_F = E_C - kT \ln\left(\frac{N_C}{n}\right) \tag{2.16}$$

$$E_F = E_V + kT \ln\left(\frac{N_V}{p}\right) \tag{2.17}$$

Here E_C (E_V) is the energy of the conduction (valence) band edge. Figure 2.16 shows the Fermi level for nitrogen- or aluminum-doped 4H-SiC as a function of temperature and impurity concentration, taking into account the temperature dependence of the bandgap and the incomplete ionization of dopants at low temperature. Because of the wide bandgap, the Fermi level does not approach the midgap (intrinsic level) even at a fairly high temperature of 700–800 K; this is as expected from the very low intrinsic carrier density shown in Figure 2.13.

2.2.4 *Mobility*

Figure 2.17 shows (a) the low-field electron mobility versus donor density and (b) the hole mobility versus acceptor density for 4H-SiC and 6H-SiC at room temperature. The electron mobility of 4H-SiC is almost double that of 6H-SiC at a given dopant density, and 4H-SiC exhibits a slightly higher hole mobility than 6H-SiC. The low-field electron and hole mobilities can be expressed by *Caughey–Thomas equations* as follows [64, 67–72]:

$$\mu_e(\text{4H-SiC}) = \frac{1020}{1 + \left(\frac{N_D + N_A}{1.8 \times 10^{17}}\right)^{0.6}} \ (\text{cm}^2\text{V}^{-1}\text{s}^{-1}) \tag{2.18}$$

$$\mu_e(\text{6H-SiC}) = \frac{450}{1 + \left(\frac{N_D + N_A}{2.5 \times 10^{17}}\right)^{0.6}} \ (\text{cm}^2\text{V}^{-1}\text{s}^{-1}) \tag{2.19}$$

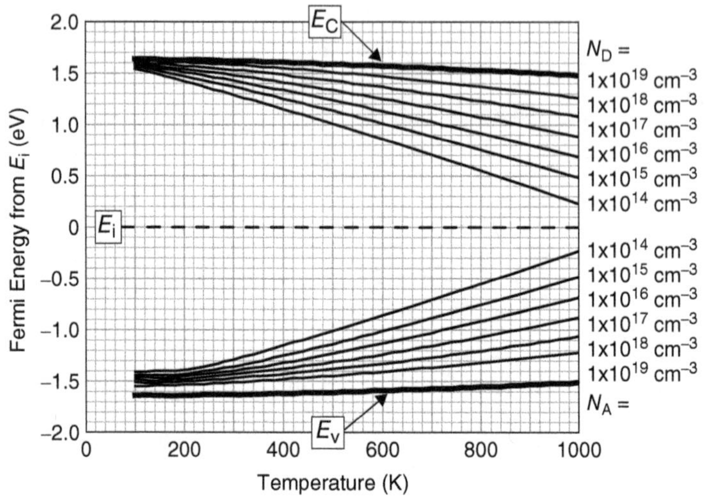

Figure 2.16 Fermi level for nitrogen- or aluminum-doped 4H-SiC as a function of temperature and impurity concentration, taking into account the temperature dependence of the bandgap and the incomplete ionization of dopants at low temperature.

$$\mu_h(\text{4H-SiC}) = \frac{118}{1 + \left(\frac{N_D + N_A}{2.2\times10^{18}}\right)^{0.7}} \ (\text{cm}^2\text{V}^{-1}\text{s}) \tag{2.20}$$

$$\mu_h(\text{6H-SiC}) = \frac{98}{1 + \left(\frac{N_D + N_A}{2.4\times10^{18}}\right)^{0.7}} \ (\text{cm}^2\text{V}^{-1}\text{s}^{-1}) \tag{2.21}$$

Here N_D and N_A are given in units of cm^{-3}. The slight differences in the doping-dependence parameters between 4H- and 6H-SiC originate from the differences in ionization energies of the dopants. It should be noted that hexagonal (and rhombohedral) SiC polytypes exhibit strong *anisotropy* in electron mobility [67, 73]. The data shown in Figure 2.17 are mobilities perpendicular to the *c*-axis. The anisotropy is particularly notable in 6H-SiC, where the electron mobility along the *c*-axis direction is only 20–25% of that perpendicular to the *c*-axis (the maximum electron mobility along the *c*-axis is about 100 cm^{2}V^{-1}s^{-1} in 6H-SiC at room temperature) [67]. The mobility anisotropy is relatively small in 4H-SiC, where the electron mobility along the *c*-axis direction is approximately 1200 cm^{2}V^{-1}s^{-1} at room temperature, which is 20% higher than that perpendicular to the *c*-axis. This is one of the major reasons why 4H-SiC is the most attractive polytype for vertical power devices fabricated on SiC{0001} wafers. The bulk mobility in 3C-SiC is isotropic. The electron mobility in lightly doped 3C-SiC is 750 cm^{2}V^{-1}s^{-1} in experiments [74] and is predicted to be 1000 cm^{2}V^{-1}s^{-1} in high-quality material [75]. In nondegenerate semiconductors, the diffusion coefficients of carriers (*D*) can be obtained by using the Einstein relation [65]:

$$D = \frac{kT}{q}\mu \tag{2.22}$$

Here q is the elementary charge. If the carrier lifetime τ is given, the diffusion length is given by $L = (D\tau)^{1/2}$.

Figure 2.18 shows (a) the low-field electron mobility versus donor density and (b) the hole mobility versus acceptor density for 4H-SiC at different temperatures [69–72]. At high temperature, the doping dependence of mobility becomes small, because the influence of impurity scattering decreases. In general, the temperature dependence of mobility is discussed by using a relationship of $\mu \sim T^{-n}$, where

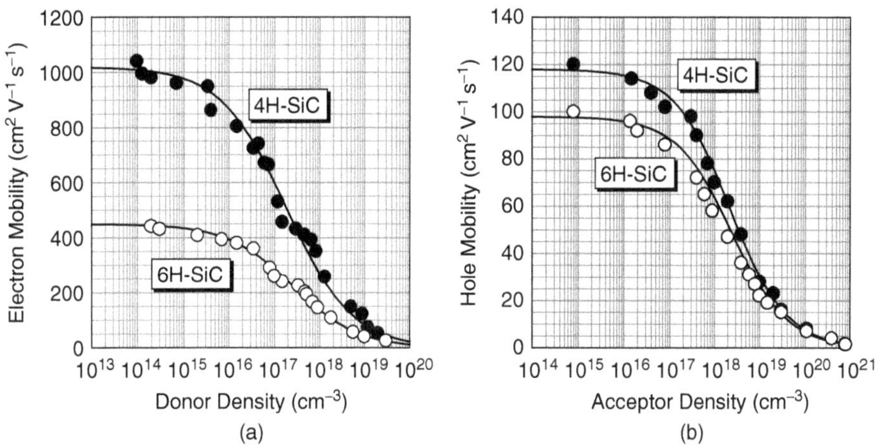

Figure 2.17 (a) Low-field electron mobility versus donor density and (b) hole mobility versus acceptor density for 4H-SiC and 6H-SiC at room temperature.

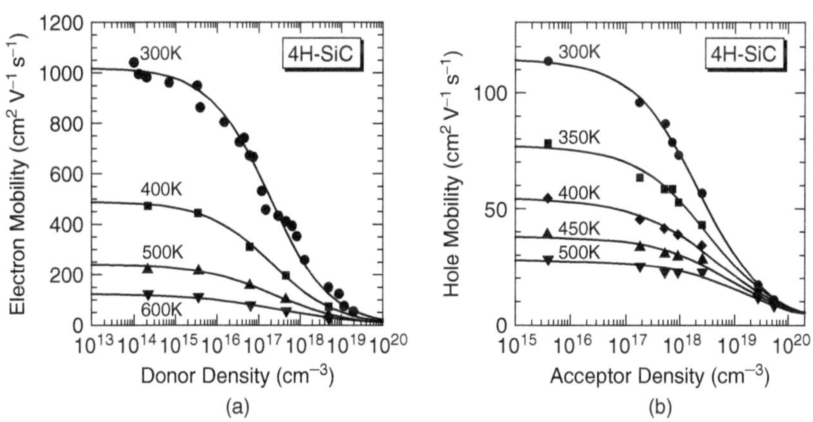

Figure 2.18 (a) Low-field electron mobility versus donor density and (b) hole mobility versus acceptor density for 4H-SiC at different temperatures ([69–72] reproduced with permission from AIP Publishing LLC).

μ is the mobility and T the absolute temperature. As seen from Figure 2.18, the value n depends strongly on the doping density, since the dominant scattering mechanism varies for SiC with different doping density. For example, the n value is 2.6 for lightly-doped and 1.5 for highly-doped n-type 4H-SiC [70].

Figure 2.19 shows the resistivity versus doping density at 300 K for nitrogen- or aluminum-doped 4H-SiC [64, 69–72]. In very heavily doped materials, the resistivity decreases to 0.003 Ω cm for n-type and 0.018 Ω cm for p-type. Note that the data shown in Figure 2.19 are obtained in high-quality epitaxial layers. In ion-implanted SiC, where a high density of point and extended defects is created, the resistivity is significantly higher than that shown in the figure for any given doping density. Substrates grown by sublimation (or other techniques) also show higher resistivities than those shown in Figure 2.19 because of a higher density of unwanted impurities and point defects.

The temperature dependence of electron mobility in nitrogen-doped 4H-SiC is shown in Figure 2.20, for donor densities of (a) 3.5×10^{15} cm^{-3} and (b) 7.5×10^{17} cm^{-3} [69]. Carrier scattering processes

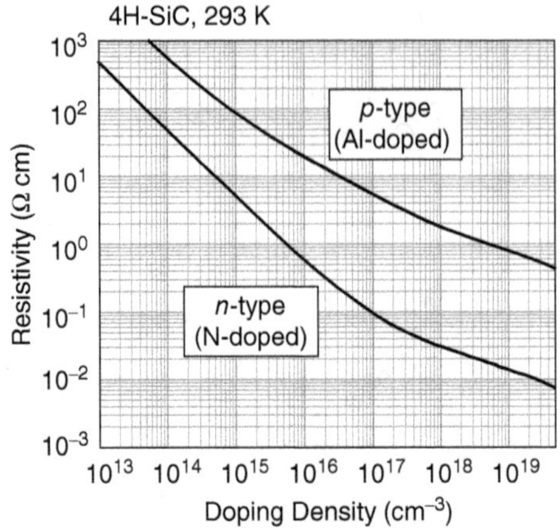

Figure 2.19 Resistivity versus doping density at 293 K for nitrogen- or aluminum-doped 4H-SiC.

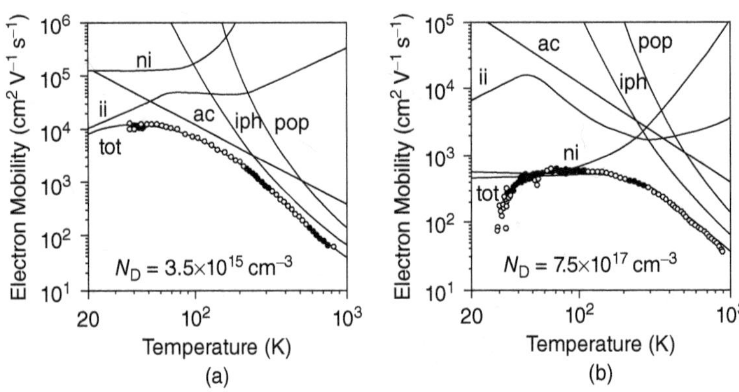

Figure 2.20 Temperature dependence of electron mobility in nitrogen-doped 4H-SiC for donor densities of (a) 3.5×10^{15} cm^{-3} and (b) 7.5×10^{17} cm^{-3} ([69] reproduced with permission from AIP Publishing LLC). Mobilities determined by several scattering processes are also plotted.

include acoustic-phonon scattering (ac), polar-optical-phonon scattering (pop), nonpolar-optical-phonon scattering (npo), intervalley scattering by phonons (iph), ionized-impurity scattering (ii), and neutral-impurity scattering (ni). In the figures, electron mobility determined by each scattering process is indicated, and the total mobility (μ) is approximately expressed according to *Matthiessen's rule* [76]:

$$\frac{1}{\mu} \cong \sum_{i} \frac{1}{\mu_{\mathrm{i}}} \tag{2.23}$$

In lightly-doped n-type SiC, the electron mobility is mainly determined by acoustic phonon scattering at low temperature (70–200 K) and by intervalley scattering at temperatures higher than 300 K, which is similar to the case of Si. In heavily-doped n-type SiC, the major scattering process is neutral impurity scattering at low temperature and intervalley scattering at high temperature.

Figure 2.21 shows the temperature dependence of hole mobility in aluminum-doped 4H-SiC with acceptor densities of (a) 1.8×10^{17} cm^{-3} and (b) 2.7×10^{19} cm^{-3} [72]. Mobilities determined by several scattering processes are also plotted. In moderately-doped p-type SiC, the hole mobility is mainly

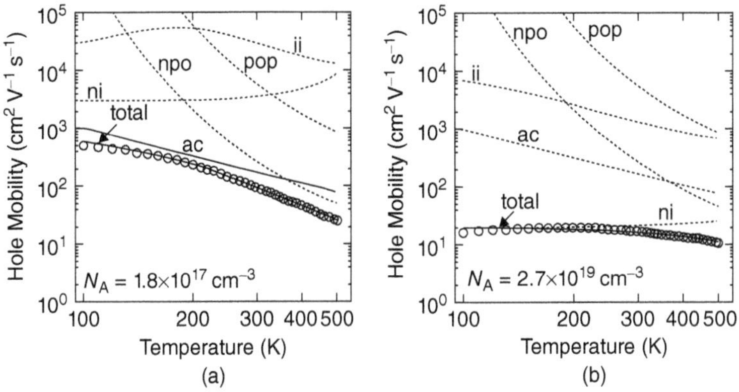

Figure 2.21 Temperature dependence of hole mobility in aluminum-doped 4H-SiC with acceptor densities of (a) 1.8×10^{17} cm^{-3} and (b) 2.7×10^{19} cm^{-3} ([72] reproduced with permission from AIP Publishing LLC. Mobilities determined by several scattering processes are also plotted.

determined by acoustic phonon scattering at or below room temperature, and by nonpolar optical phonon scattering at high temperature (>400 K). In heavily-doped p-type SiC, the major scattering process is neutral impurity scattering over a wide temperature range, since most Al acceptors remain neutral because of their large ionization energy.

2.2.5 *Drift Velocity*

At low electric fields, the drift velocity of carriers (v_d) is proportional to the electric field strength (E), $v_d = \mu E$. When the electric field is high, the accelerated carriers transfer more energy to the lattice by emitting more phonons, leading to nonlinear field dependence of drift velocity [76]. The electric field dependence of the drift velocity is expressed by [76]:

$$v_d = \frac{\mu E}{\left\{1 + \left(\frac{\mu E}{v_s}\right)^{\gamma}\right\}^{1/\gamma}} \tag{2.24}$$

where v_s is the sound velocity in a semiconductor and γ the parameter. At sufficiently high electric fields, carriers start to interact with optical phonons, and finally the drift velocity becomes saturated. The *saturated drift velocity* (v_{sat}) is approximately given by [65, 76]:

$$v_{sat} = \sqrt{\frac{8\hbar\omega}{3\pi m^*}} \tag{2.25}$$

where $\hbar\omega$ is the energy of the optical phonon (LO (longitudinal optical) phonon) emitted. Figure 2.22 shows the measured drift velocity of electrons versus applied electric field for n-type (a) 4H-SiC and (b) 6H-SiC [77]. The measurements were conducted in a structure carefully designed to minimize inaccuracy in potential distribution. For 4H-SiC, a low-field mobility of 450 cm^{2}V^{-1}s^{-1} was determined from the slope at low electric fields (<10^4 V cm^{-1}) at room temperature; this agrees with the data shown in Figure 2.17 for the donor density (2×10^{17} cm^{-3}) of this particular sample. The saturated drift velocity is determined as 2.2×10^7 cm s^{-1} at room temperature. This value is also in good agreement with that estimated from Equation 2.25. As indicated in Figure 2.22, the saturated drift velocity decreases with increasing temperature. Note that a so-called transferred-electron effect (Gunn effect) is not observed in SiC because of its indirect band structure. The saturated drift velocity of electrons in 6H-SiC is experimentally

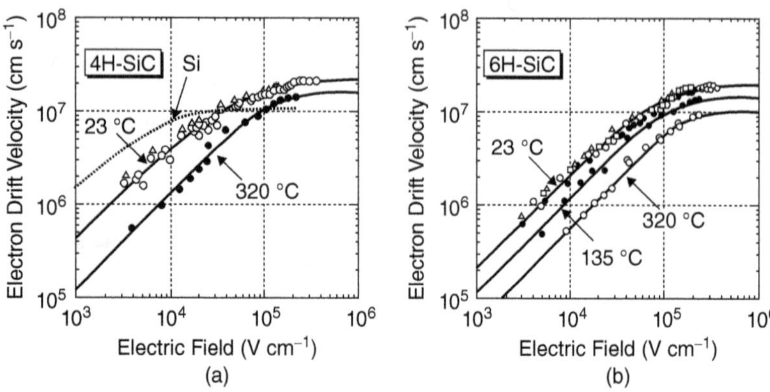

Figure 2.22 Drift velocity of electrons versus applied electric field for n-type (a) 4H-SiC and (b) 6H-SiC ([77] reproduced with permission from IEEE).

estimated as 1.9×10^7 cm s^{-1} [77, 78]. Although the saturated drift velocity of holes in SiC has not been experimentally studied, it can be estimated at 1.3×10^7 cm s^{-1} for 4H-SiC from Equation 2.25.

2.2.6 *Breakdown Electric Field Strength*

When a very high electric field is applied to a pn junction or Schottky barrier in the reverse-bias direction, the leakage current increases as a result of generation of electron–hole pairs, and the junction eventually breaks down. The breakdown mechanisms can be classified into (i) *avalanche breakdown* and (ii) *Zener (tunneling) breakdown* [65, 79]. For junctions with a lightly-doped region, avalanche breakdown is dominant; this is the case for most power devices. In avalanche breakdown, the carriers can gain enough energy under very high electric fields to excite electron–hole pairs by impact ionization. The generation of electron–hole pairs is multiplied inside the space-charge region of a junction, eventually leading to breakdown.

Avalanche breakdown is well described by using the *impact ionization coefficients* of electrons and holes [65, 79]. Breakdown can be defined as when the multiplication factor of the current approaches infinity, which has been shown to be equivalent to the following relationship [65, 79]:

$$\int_0^W \alpha_\mathrm{h} \exp\left\{-\int_0^x (\alpha_\mathrm{h} - \alpha_\mathrm{e})\mathrm{d}x'\right\} \mathrm{d}x = 1 \tag{2.26}$$

Here, α_e and α_h are the impact ionization coefficients for electrons and holes, respectively. Integration is performed in the space charge region extending from $x = 0$ to $x = W$. The integral term of the equation is called the *ionization integral*. Because the impact ionization coefficients depend strongly on the electric field strength, and the field strength is not uniform inside the space-charge region, numerical calculation is required to obtain the ionization integral given by Equation 2.26. Conversely, the impact ionization coefficients can be determined by measuring the multiplication factor as a function of electric field in properly designed pn junction diodes. In the measurements, light illumination is employed to increase the current at low reverse-bias voltages, and thereby to minimize the influence of nonideal leakage current. This is important for accurate determination of the multiplication factors. In general, the impact ionization coefficients are approximately expressed by the Chynoweth equation [80]:

$$\alpha_\mathrm{i} = a_\mathrm{i} \exp\left(-\frac{b_\mathrm{i}}{E}\right), \ (i : \mathrm{e\ or\ h}) \tag{2.27}$$

where a_i and b_i are the parameters and E the electric field strength.

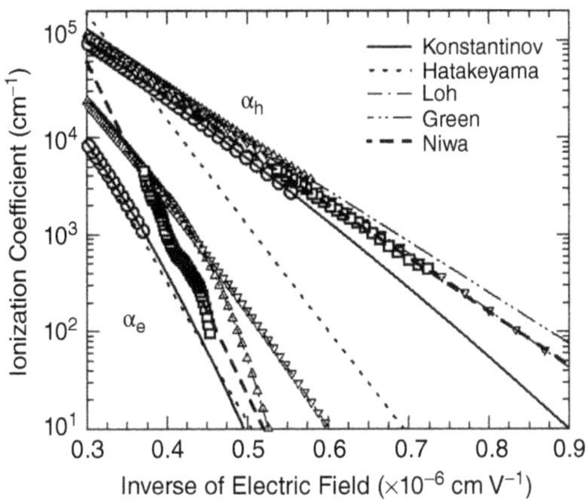

Figure 2.23 Impact ionization coefficients for electrons and holes in 4H-SiC versus the inverse of electric field strength [81–84].

Figure 2.23 shows the impact ionization coefficients for electrons and holes in 4H-SiC versus the inverse of electric field strength [81–84]. Different groups have reported similar but slightly different impact ionization coefficients. The ionization coefficients for 4H-SiC are considerably lower than those for Si owing to the wide bandgap of SiC. Another striking feature of Figure 2.23 is that the ionization coefficient for holes is much larger than that for electrons ($\alpha_h > \alpha_e$) in SiC, which is completely opposite to the case of Si ($\alpha_e > \alpha_h$). In 4H-SiC, the energy range of the conduction band is rather small because of the folding effect in the E–k relationship, and the highest energy of hot electrons is limited by the upper edge of the conduction band [85, 86]. This may be the reason why the ionization coefficient for electrons is unusually low in 4H-SiC (and in 6H-SiC). Note that the data shown in Figure 2.23 are extrapolated from several experimental data sets. In particular, the ionization coefficients at relatively low electric fields need more careful investigation. The temperature dependence of the coefficients has been recently reported [84]. It should also be noted that all data in Figure 2.23 are valid along the <0001> direction because they were obtained from 4H-SiC pn diodes on off-axis {0001} substrates. Since the carrier acceleration and scattering are strongly influenced by the energy band structure, the impact ionization coefficients depend on the crystallographic orientation. In particular, hexagonal SiC polytypes exhibit strong anisotropy in impact ionization and breakdown characteristics [82, 85, 86].

A semiconductor junction breaks down when the maximum electric field strength reaches a critical value which is inherent to the material. This critical value is called the *critical electric field strength* or *breakdown electric field strength*. The critical electric field strength E_B can be determined by calculation of the ionization integral using the impact ionization coefficients described above. Alternatively, it can be obtained experimentally from the breakdown characteristics of devices in which electric field crowding is perfectly suppressed. In n-type Schottky barrier diodes or a one-sided p$^+$n junction, the breakdown voltage V_B is given by [65, 79]:

$$V_B = \frac{\varepsilon_s {E_B}^2}{2qN_D} \tag{2.28}$$

Here a non-punchthrough structure is considered. ε_s is the dielectric constant of a semiconductor.

Figure 2.24 shows the critical electric field strength versus doping density for 4H-SiC <0001>, 6H-SiC <0001>, and 3C-SiC <111> [80, 81, 87, 88]. The data for Si are also shown for comparison. 4H- and 6H-SiC exhibit approximately eight times higher critical electric field strengths than Si at a given doping density, while the field strength of 3C-SiC is only three or four times higher because this polytype has a relatively small bandgap (similar to GaP). The high critical field strength of hexagonal SiC polytypes is the main reason why SiC is very attractive for power device applications [20, 89, 90]. One must be

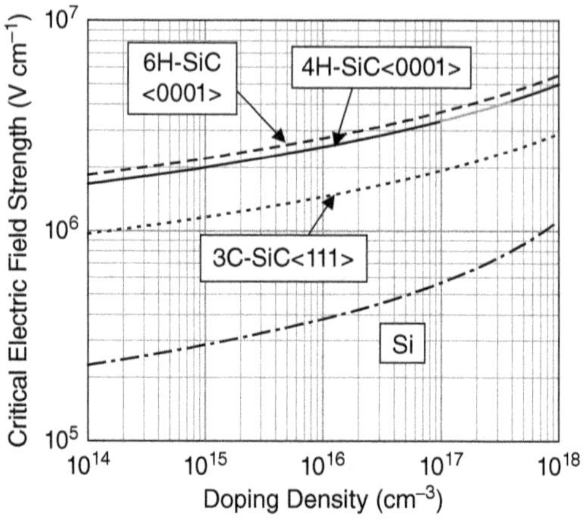

Figure 2.24 Critical electric field strength versus doping density for 4H-SiC <0001>, 6H-SiC <0001>, and 3C-SiC <111> [80, 81, 87, 88].

aware of the fact that the critical field strength is strongly dependent on the doping density, as shown in Figure 2.24. When the doping density is increased, the width of the space-charge region becomes small and the distance for carriers to be accelerated becomes short. Furthermore, the mobility is reduced in highly-doped materials because of enhanced impurity scattering. These are the reasons why the critical electric field strength apparently increases with increasing doping. As shown in Figure 2.24, the critical electric field of 6H-SiC <0001> is slightly higher than that of 4H-SiC <0001>, in spite of its smaller bandgap (E_g = 3.02 eV for 6H-SiC and 3.26 eV for 4H-SiC). As described in Section 2.2.4, 6H-SiC exhibits strong anisotropy in carrier transport, and the electron mobility along the <0001> direction is unusually low, about 100 $cm^2V^{-1}s^{-1}$ even in a high-purity material. The narrow width of the conduction band in 6H-SiC also helps to increase the critical electric field strength of 6H-SiC <0001>. Conversely, it is known that the critical field strength of 6H-SiC <$11\bar{2}0$> is only half that of 6H-SiC <0001> [85]. The anisotropy in critical field strength of 4H-SiC is smaller, and the field strength of 4H-SiC <$11\bar{2}0$> is only 20–25% lower than that of 4H-SiC <0001> [82, 86].

The critical field strength is a convenient physical property when the ideal breakdown voltage is estimated. However, it should be noted that the critical field strength is valid only for junctions with non-punchthrough structures. When punchthrough structures are considered, the critical field strength shown in Figure 2.24 does not give the correct breakdown voltage. In this case, simulation of leakage current or calculation of the ionization integral using a device simulator is required to determine the ideal breakdown voltage. Breakdown voltage is discussed in greater detail in Chapters 7 and 10.

2.3 Thermal and Mechanical Properties

2.3.1 Thermal Conductivity

Figure 2.25 shows the temperature dependence of thermal conductivity for SiC and Si [91, 92]. SiC, with its significant contribution from phonons, has a much higher thermal conductivity (4.9 W cm^{-1} K^{-1} for high-purity SiC at room temperature) than Si. It has been reported that the thermal conductivity is not sensitive to the SiC polytype, but depends on the doping density and the crystal direction [93]. In heavily-nitrogen-doped 4H-SiC substrates, which are usually employed as n^+-substrates for vertical power devices, the thermal conductivity along <0001> is 3.3 W cm^{-1} K^{-1} at room temperature [93].

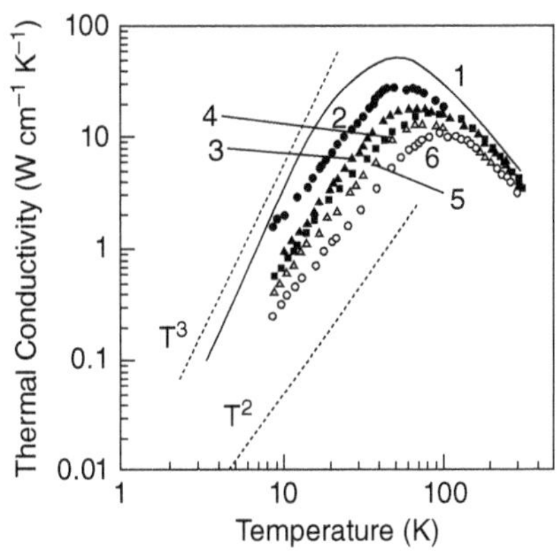

Figure 2.25 Temperature dependence of thermal conductivity for SiC and Si ([92] reproduced with permission from Institute of Physics (IOP)–Taylor & Francis).

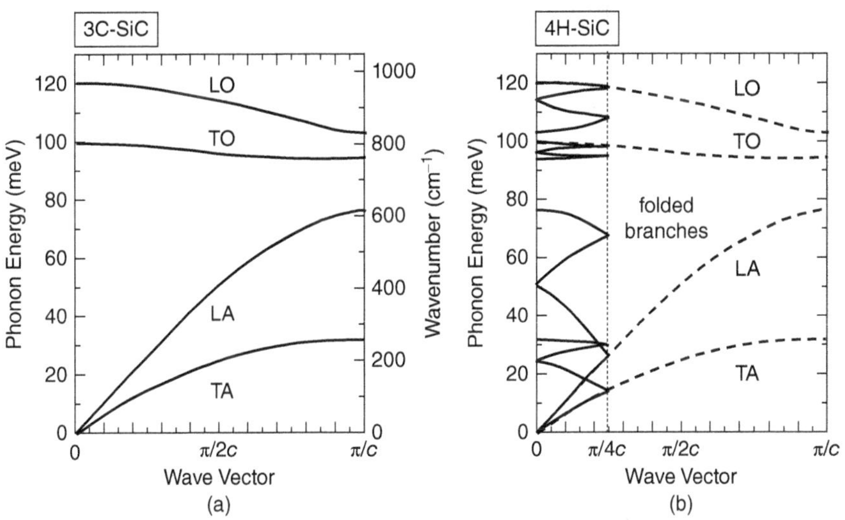

Figure 2.26 Phonon dispersion relationships for (a) 3C-SiC and (b) 4H-SiC.

2.3.2 *Phonons*

Figure 2.26 shows the phonon dispersion relationships for (a) 3C-SiC and (b) 4H-SiC [94, 95]. The basic branches consist of TA (transverse acoustic), LA (longitudinal acoustic), TO (transverse optical), and LO phonons, as in other semiconductors. Due to the large energy of Si-C bonds, the phonon frequencies in SiC are high. The unit cell length of the nH polytype ($n = 2, 4, 6 \ldots$) along the c-axis is n times larger than the unit length (Si-C bilayer). Thus, the Brillouin zone in the direction of Γ–L is reduced to $1/n$ of the basic Brillouin zone [31]. The dispersion curves of the phonons propagating along the <0001> direction in such polytypes can be approximated by folding the basic dispersion curve, as shown in Figure 2.24. This zone folding provides new phonon modes at the Γ point, which are called *"folded modes."* The number of atoms in the unit cell is 2 for 3C-SiC, 8 for 4H-SiC, and 12 for 6H-SiC. Therefore, the number of phonon branches is 6 for 3C-SiC, 24 for 4H-SiC, and 36 for 6H-SiC, neglecting the degeneracy.

Table 2.6 Major mechanical and thermal properties of SiC and Si at room temperature [30, 45].

Properties	4H- or 6H-SiC	Si
Density (g cm^{-3})	3.21	2.33
Young's modulus (GPa)	390–690	160
Fracture strength (GPa)	21	7
Poisson's ratio	0.21	0.22
Elastic constant (GPa)		
c_{11}	501	166
c_{12}	111	64
c_{13}	52	–
c_{33}	553	–
c_{44}	163	80
Specific heat (J g^{-1} K^{-1})	0.69	0.7
Thermal conductivity (W cm^{-1} K^{-1})	3.3–4.9	1.4–1.5

The major phonon energies (or wavenumber) can be directly observed by Raman scattering spectroscopy. Different phonon frequencies in different SiC polytypes enable identification of individual polytypes by Raman scattering measurements [96], detail of which is described in Section 5.1.2. It is known that the observed frequency of LO phonons increases with increasing carrier density because of a carrier–LO phonon coupling effect [97]. Phonon energies are also important in luminescence measurements. In particular, photoluminescence (PL) at low temperature is a powerful tool to characterize the purity and quality of SiC crystals [9, 10, 43, 98–102]. Because SiC has an indirect band structure, phonons are intensively involved in carrier recombination processes. As a result, strong multiple phonon replicas of a zero-phonon emission line are often observed in PL spectra of SiC. For example, the energies of major phonons which create phonon replicas in PL from 4H-SiC{0001} are 36 (TA), 46, 51, 77 (LA), 95, 96 (TO), 104, and 107 meV (LO). Real PL spectra are described in Section 5.1.1.

2.3.3 Hardness and Mechanical Properties

The mechanical properties of SiC are also unique; SiC is one of the hardest known materials. Table 2.6 shows the major mechanical properties of SiC and Si [30, 45], where the polytype dependence is small. The hardness and Young's modulus (380–700 GPa [103]) of SiC are much higher than those of Si, while the Poisson's ratio (0.21) of SiC is very similar to that of other semiconductors. SiC retains its high hardness and elasticity, even at very high temperatures. The yield (fracture) strength of SiC is as high as 21 GPa at room temperature and is estimated to be 0.3 GPa at 1000 °C, while the yield strength of Si falls to 0.05 GPa at 500 °C [104].

2.4 Summary

Table 2.7 summarizes the major physical properties of common SiC polytypes (also see Appendix C). The table includes the low-frequency Baliga's figure-of-merit (BFOM) $\varepsilon_s \mu E_B{}^3$ [105], normalized with respect to the value for Si. For 4H- and 6H-SiC, vertical devices on {0001} wafers are considered in the BFOM calculation. Owing to the high critical field strength and high electron mobility along the *c*-axis, 4H-SiC exhibits a significantly higher BFOM than other SiC polytypes. This is the main reason why 4H-SiC has been almost exclusively employed for power device applications [15, 17, 19, 20, 106–116]. Another advantage of 4H-SiC is that it has slightly smaller donor and acceptor ionization energies compared

Table 2.7 Major physical properties of common SiC polytypes at room temperature, including the low-frequency Baliga's figure-of-merit (BFOM) $\varepsilon_s \mu E_B{}^3$ normalized with respect to the value for Si.

Properties/polytype	3C-SiC	4H-SiC	6H-SiC
Bandgap (eV)	2.36	3.26	3.02
Electron mobility ($cm^2V^{-1}s^{-1}$)			
μ perpendicular to *c*-axis	1000	1020	450
μ parallel to *c*-axis	**1000**	**1200**	**100**
Hole mobility ($cm^2V^{-1}s^{-1}$)	100	120	100
Electron saturated drift velocity (cm s^{-1})	($\sim 2 \times 10^7$)	2.2×10^7	1.9×10^7
Hole saturated drift velocity (cm s^{-1})	($\sim 1.3 \times 10^7$)	($\sim 1.3 \times 10^7$)	($\sim 1.3 \times 10^7$)
Breakdown electric field (MV cm^{-1})			
E_B perpendicular to *c*-axis	1.4	2.2	1.7
E_B parallel to *c*-axis	**1.4**	**2.8**	**3.0**
Relative dielectric constant			
ε_s perpendicular to *c*-axis	9.72	9.76	9.66
ε_s parallel to *c*-axis	9.72	10.32	10.03
BFOM (**n-type**, parallel to *c*-axis) normalized by that of Si	61	**626**	63
BFOM (**p-type**, parallel to *c*-axis) normalized by that of Si, *taking account of incomplete ionization of acceptors*	2	25	19

with those of other SiC polytypes. Furthermore, the availability of single-crystalline 4H-SiC{0001} wafers with relatively large diameters and reasonable quality has driven fabrication of 4H-SiC-based electronic devices. In fact, the characteristics of commercial 4H-SiC power devices (Schottky barrier diodes and field effect transistors) have already outperformed the theoretical limits of 3C- and 6H-SiC unipolar devices. 3C-SiC is of academic interest to clarify the polytype dependence of physical properties. 3C-SiC may be attractive for relatively low-voltage (< 300 V) applications and high-temperature sensors. For more details of physical properties, refer to the following review papers and handbooks [30, 45, 117–125].

References

[1] Verma, A.R. and Krishna, P. (eds) (1966) *Polymorphism and Polytypism in Crystals*, John Wiley & Sons, Inc., New York.

[2] Schneer, C.S. (1955) Polymorphism in one dimension. *Acta Crystallogr.*, **8**, 279.

[3] Cheng, C., Needs, R.J. and Heine, V. (1988) Inter-layer interactions and the origin of SiC polytpes. *J. Phys. C*, **21**, 1049.

[4] Denteneer, P.J.H. and van Haeringen, W. (1986) Ground-state properties of polytypes of silicon carbide. *Phys. Rev. B*, **33**, 2831.

[5] Fisher, G.R. and Barnes, P. (1990) Towards a unified view of polytypism in silicon carbide. *Philos. Mag. B*, **61**, 217.

[6] Heine, V., Cheng, C. and Needs, R.J. (1991) The preference of silicon carbide for growth in the metastable cubic form. *J. Am. Ceram. Soc.*, **74**, 2630.

[7] Yoo, W.S. and Matsunami, H. (1992) Growth simulation of SiC polytypes and application to DPB-free 3C-SiC on alpha-SiC substrates, in *Amorphous and Crystalline Silicon Carbide IV*, Springer-Verlag, p. 66.

[8] Powell, R.C. (2010) *Symmetry, Group Theory, and the Physical Properties of Crystals*, Springer.

[9] Choyke, W.J. and Patrick, L. (1962) Exciton recombination radiation and phonon spectrum of 6H SiC. *Phys. Rev.*, **127**, 1868.

[10] Ikeda, M., Matsunami, H. and Tanaka, T. (1980) Site effect on the impurity levels in 4H, 6H, and 15R SiC. *Phys. Rev. B*, **22**, 2842.

[11] Suttrop, W., Pensl, G., Choyke, W.J. *et al.* (1992) Hall effect and infrared absorption measurements on nitrogen donors in 6H-silicon carbide. *J. Appl. Phys.*, **72**, 3708.

[12] Knippenberg, W.F. (1963) Growth phenomena in silicon carbide. *Philips Res. Rep.*, **18**, 161.

[13] Yoo, W.S. and Matsunami, H. (1991) Solid-state phase transformation in cubic silicon carbide. *Jpn. J. Appl. Phys.*, **30**, 545.

[14] Itoh, A., Akita, H., Kimoto, T. and Matsunami, H. (1994) High-quality 4H-SiC homoepitaxial layers grown by step-controlled epitaxy. *Appl. Phys. Lett.*, **65**, 1400.

[15] Kimoto, T., Itoh, A., Akita, H. *et al.* (1995) Step-controlled epitaxial growth of α-SiC and application to high-voltage Schottky rectifiers, in *Compound Semiconductors – 1994*, IOP, p. 437.

[16] Raghunathan, R., Alok, D. and Baliga, B.J. (1995) High voltage 4H-SiC Schottky barrier diodes. *IEEE Electron Device Lett.*, **16**, 226.

[17] Itoh, A., Kimoto, T. and Matsunami, H. (1995) High-performance of high-voltage 4H-SiC Schottky barrier diodes. *IEEE Electron Device Lett.*, **16**, 280.

[18] Kordina, O., Bergman, J.P., Hallin, C. and Janzen, E. (1996) The minority carrier lifetime of n-type 4H- and 6H-SiC epitaxial layers. *Appl. Phys. Lett.*, **69**, 679.

[19] Tan, J., Cooper, J.A. Jr., and Melloch, M.R. (1998) High-voltage accumulation-layer UMOSFETs in 4H-SiC. *IEEE Electron Device Lett.*, **19**, 467.

[20] Cooper, J.A. Jr.,, Melloch, M.R., Singh, R. *et al.* (2002) Status and prospects for SiC power MOSFETs. *IEEE Trans. Electron Devices*, **49**, 658.

[21] Nishino, S., Powell, A. and Will, H.A. (1983) Production of large-area single-crystal wafers of cubic SiC for semiconductor devices. *Appl. Phys. Lett.*, **42**, 460.

[22] Nishino, S., Suhara, H., Ono, H. and Matsunami, H. (1987) Epitaxial growth and electric characteristics of cubic SiC on silicon. *J. Appl. Phys.*, **61**, 4889.

[23] Yamanaka, M., Daimon, H., Sakuma, E. *et al.* (1987) Temperature dependence of electrical properties of n-and p-type 3C-SiC. *J. Appl. Phys.*, **61**, 599.

[24] Barrett, D.L. and Campbell, R.B. (1967) Electron mobility measurements in SiC polytypes. *J. Appl. Phys.*, **38**, 53.

[25] Schörner, R., Friedrichs, P., Peters, D. and Stephani, D. (1999) Significantly improved performance of MOSFETs on silicon carbide using the 15R-SiC polytype. *IEEE Electron Device Lett.*, **20**, 241.

[26] Levinshtein, M.E., Rumyantsev, S.L. and Shur, M.S. (2001) *Properties of Advanced Semiconductor Materials: GaN, AlN, InN, BN, SiC, SiGe*, John Wiley & Sons, Inc., New York.

[27] Sasaki, S., Suda, J. and Kimoto, T. (2012) Doping-induced lattice mismatch and misorientation in 4H-SiC crystals. *Mater. Sci. Forum*, **717–720**, 481.

[28] Stockmeier, M., Müller, R., Sakwe, S.A. *et al.* (2009) On the lattice parameters of silicon carbide. *J. Appl. Phys.*, **105**, 033511.

[29] Li, Z. and Bradt, R.C. (1986) Thermal expansion of the hexagonal (4H) polytype of SiC. *J. Appl. Phys.*, **60**, 612.

[30] Harris, G.L. (1995) *Properties of Silicon Carbide*, INSPEC.

[31] Kittel, C. (1995) *Introduction to Solid State Physics*, 7th edn, John Wiley & Sons, Inc, New York.

[32] SEMI http://ams.semi.org/ebusiness/ (accessed 28 March 2014).

[33] Lambrecht, W.R.L., Limpijumnong, S., Rashkeev, S.N. and Segall, B. (1997) Electronic band structure of SiC polytypes: A discussion of theory and experiment. *Phys. Status Solidi B*, **202**, 5.

[34] Käckel, P., Wenzien, R. and Bechstedt, F. (1994) Electronic properties of cubic and hexagonal SiC polytypes from ab initio calculations. *Phys. Rev. B*, **50**, 10761.

[35] van Haeringen, W., Bobbert, P.A. and Backes, W.H. (1997) On the band gap variation in SiC polytypes. *Phys. Status Solidi B*, **202**, 63.

[36] Wellenhofer, G. and Rössler, U. (1997) Global band structure and near-band-edge states. *Phys. Status Solidi B*, **202**, 107.

[37] Persson, C. and Lindefelt, U. (1997) Relativistic band structure calculation of cubic and hexagonal SiC polytypes. *J. Appl. Phys.*, **82**, 5496.

[38] Humphreys, R.G., Bimberg, D. and Choyke, W.J. (1981) Wavelength modulated absorption in SiC. *Solid State Commun*, **39**, 163.

[39] Sridhara, S.G., Bai, S., Shigiltchoff, O. *et al.* (2000) Differential absorption measurement of valence band splittings in 4H SiC. *Mater. Sci. Forum*, **338–342**, 567.

[40] Chen, W.M., Son, N.T., Janzen, E. *et al.* (1997) Effective masses in SiC determined by cyclotron resonance experiments. *Phys. Status Solidi A*, **162**, 79.
[41] Volm, D., Meyer, B.K., Hofmann, D.M. *et al.* (1996) Determination of the electron effective-mass tensor in 4H SiC. *Phys. Rev. B*, **53**, 15409.
[42] Son, N.T., Persson, C., Lindefelt, U. *et al.* (2004) Cyclotron resonance studies of effective masses and band structure in SiC, in *Silicon Carbide – Recent Major Advances* (eds W.J. Choyke, H. Matsunami and G. Pensl), Springer, p. 437.
[43] Choyke, W.J. and Devaty, R.P. (2004) Optical properties of SiC: 1997-2002, in *Silicon Carbide – Recent Major Advances* (eds W.J. Choyke, H. Matsunami and G. Pensl), Springer, p. 413.
[44] Choyke, W.J. (1969) Optical properties of polytypes of SiC: Interband absorption, and luminescence of nitrogen-exciton complexes. *Mater. Res. Bull.*, **4**, 141.
[45] Adachi, S. (2005) *Properties of Group-IV, III-V, and II-VI Semiconductors*, John Wiley & Sons, Ltd, Chichester.
[46] Persson, C., Lindefelt, U. and Sernelius, B.E. (1999) Band gap narrowing in n-type and p-type 3C-, 2H-, 4H-, 6H-SiC, and Si. *J. Appl. Phys.*, **86**, 4419.
[47] Choyke, W.J. and Patrick, L. (1960) Absorption of light in alpha silicon carbide near the band edge, in *Silicon Carbide* (eds J.P. O'Connor and J. Smiltens), Pergamon Press, p. 306.
[48] Sridhara, S.G., Devaty, R.P. and Choyke, W.J. (1998) Absorption coefficient of 4H silicon carbide from 3900 to 3250 Å. *J. Appl. Phys.*, **84**, 2963.
[49] Yu, P.Y. and Cardona, M. (2005) *Fundamentals of Semiconductors*, 3rd edn, Springer.
[50] Watanabe, N., Kimoto, T. and Suda, J. (2012) Thermo-optic coefficients of 4H-SiC, GaN, and AlN for ultraviolet to infrared regions up to 500 °C. *Jpn. J. Appl. Phys.*, **51**, 112101.
[51] Marple, D.T.F. (1964) Refractive index of ZnSe, ZnTe, and CdTe. *J. Appl. Phys.*, **35**, 539.
[52] Ninomiya, S. and Adachi, S. (1994) Optical constants of 6H–SiC single crystals. *Jpn. J. Appl. Phys.*, **33**, 2479.
[53] Troffer, T., Schadt, M., Frank, T. *et al.* (1997) Doping of SiC by implantation of boron and aluminum. *Phys. Status Solidi A*, **162**, 277.
[54] Kuznetsov, N.I. and Zubrilov, A.S. (1995) Deep centers and electroluminescence in 4H-SiC diodes with a p-type base region. *Mater. Sci. Eng. B*, **29**, 181.
[55] Negoro, Y., Kimoto, T., Matsunami, H. and Pensl, G. (2007) Abnormal out-diffusion of epitaxially doped boron in 4H-SiC caused by implantation and annealing. *Jpn. J. Appl. Phys.*, **46**, 5053.
[56] Purcell, K.F. and Kotz, J.C. (1977) *Inorganic Chemistry*, W.B. Saunders.
[57] Götz, W., Schöner, A., Pensl, G. *et al.* (1993) Nitrogen donors in 4H-silicon carbide. *J. Appl. Phys.*, **73**, 3332.
[58] Greulich-Weber, S. (1997) EPR and ENDOR investigations of shallow impurities in SiC polytypes. *Phys. Status Solidi A*, **162**, 95.
[59] Laube, M., Schmid, F., Semmelroth, K. *et al.* (2004) Phosphorus-related centers in SiC, in *Silicon Carbide – Recent Major Advances* (eds W.J. Choyke, H. Matsunami and G. Pensl), Springer, p. 493.
[60] Schöner, A., Nordell, N., Rottner, K. *et al.* (1996) Dependence of the aluminium ionization energy on doping concentration and compensation in 6H-SiC. *Inst. Phys. Conf. Ser.*, **142**, 493.
[61] Ivanov, I.G., Henry, A. and Janzén, E. (2005) Ionization energies of phosphorus and nitrogen donors and aluminum acceptors in 4H silicon carbide from the donor-acceptor pair emission. *Phys. Rev. B*, **71**, 241201.
[62] Vodakov, Y.A. and Mokhov, E.N. (1974) Diffusion and solubility of impurities in silicon carbide, in *Silicon Carbide* (eds R.C. Marshall, J.W. Faust Jr., and C.E. Ryan), University of South Carolina Press, p. 508.
[63] Efros, A.L., Lien, N.V. and Shklovskii, B.I. (1979) *J. Phys. C: Solid State Phys.*, **12**, 1869.
[64] Kimoto, T., Itoh, A. and Matsunami, H. (1997) Step-controlled epitaxial growth of high-quality SiC layers. *Phys. Status Solidi B*, **202**, 247.
[65] Sze, S.M. and Ng, K.K. (2007) *Physics of Semiconductor Devices*, 3rd edn, John Wiley & Sons, Inc, Hoboken, NJ.
[66] Pensl, G. and Choyke, W.J. (1993) Electrical and optical characterization of SiC. *Physica B*, **185**, 264.
[67] Schaffer, W.J., Negley, G.H., Irvine, K.G. and Palmour, J.W. (1994) Conductivity anisotropy in epitaxial 6H and 4H SiC. *Mater. Res. Soc. Symp. Proc.*, **339**, 595.
[68] Iwata, H., Itoh, K.M. and Pensl, G. (2000) Theory of the anisotropy of the electron Hall mobility in n-type 4H- and 6H-SiC. *J. Appl. Phys.*, **88**, 1956.
[69] Pernot, J., Zawadzki, W., Contreras, S. *et al.* (2001) Electrical transport in n-type 4H silicon carbide. *J. Appl. Phys.*, **90**, 1869.
[70] Kagamihara, S., Matsuura, H., Hatakeyama, T. *et al.* (2004) Parameters required to simulate electric characteristics of SiC devices for n-type 4H-SiC. *J. Appl. Phys.*, **96**, 5601.

[71] Matsuura, H., Komeda, M., Kagamihara, S. *et al.* (2004) Dependence of acceptor levels and hole mobility on acceptor density and temperature in Al-doped p-type 4H-SiC epilayers. *J. Appl. Phys.*, **96**, 2708.
[72] Koizumi, A., Suda, J. and Kimoto, T. (2009) Temperature and doping dependencies of electrical properties in Al-doped 4H-SiC epitaxial layers. *J. Appl. Phys.*, **106**, 013716.
[73] Schadt, M., Pensl, G., Devaty, R.P. *et al.* (1994) Anisotropy of the electron Hall mobility in 4H, 6H, and 15R silicon carbide. *Appl. Phys. Lett.*, **65**, 3120.
[74] Shinohara, M., Yamanaka, M., Daimon, H. *et al.* (1988) Growth of high-mobility 3C-SiC epilayers by chemical vapor deposition. *Jpn. J. Appl. Phys.*, **27**, L434.
[75] Nelson, W.E., Halden, F.A. and Rosengreen, A. (1966) Growth and properties of β-SiC single crystals. *J. Appl. Phys.*, **37**, 333.
[76] Hamaguchi, C. (2010) *Basic Semiconductor Physics*, 2nd edn, Springer.
[77] Khan, I.A. and Cooper, J.A. Jr., (2000) Measurement of high-field electron transport in silicon carbide. *IEEE Trans. Electron Devices*, **47**, 269.
[78] von Munch, W. and Pfaffeneder, I. (1977) Saturated electron drift velocity in 6H silicon carbide. *J. Appl. Phys.*, **48**, 4823.
[79] Baliga, B.J. (1996) *Physics of Semiconductor Power Devices*, JWS Publishing.
[80] Chynoweth, A.G. (1958) Ionization rates for electrons and holes in silicon. *Phys. Rev.*, **109**, 1537.
[81] Konstantinov, A.O., Wahab, Q., Nordell, N. and Lindefelt, U. (1997) Ionization rates and critical fields in 4H silicon carbide. *Appl. Phys. Lett.*, **71**, 90.
[82] Hatakeyama, T. (2009) Measurements of impact ionization coefficients of electrons and holes in 4H-SiC and their application to device simulation. *Phys. Status Solidi A*, **206**, 2284.
[83] Loh, W.S., Ng, B.K., Ng, J.S. *et al.* (2008) Impact ionization coefficients in 4H-SiC. *IEEE Trans. Electron Devices*, **55**, 1984.
[84] Niwa, H., Suda, J., and Kimoto, T. (2013) International Conference on Silicon Carbide and Related Materials 2013, Miyazaki, Th-2A-1 Measurement of impact ionization coefficient in 4H-SiC toward ultrahigh-voltage power devices.
[85] Dmitriev, A.P., Konstantinov, A.O., Litvin, D.P. and Sankin, V.I. (1983) Impact ionization and super-lattice in 6H-SiC. *Sov. Phys. Semicond.*, **17**, 686.
[86] Nakamura, S., Kumagai, H., Kimoto, T. and Matsunami, H. (2002) Anisotropy in breakdown field of 4H-SiC. *Appl. Phys. Lett.*, **80**, 3355.
[87] Palmour, J.W., Edmond, J.A., Kong, H.S. and Carter, C.H. Jr., (1993) 6H-silicon carbide devices and applications. *Physica B*, **185**, 461.
[88] Neudeck, P.G., Larkin, D.J., Starr, E. *et al.* (1994) Electrical properties of epitaxial 3C- and 6H-SiC p-n junction diodes produced side-by-side on 6H-SiC substrates. *IEEE Trans. Electron Devices*, **41**, 826.
[89] Bhatnagar, M. and Baliga, B.J. (1993) Comparison of 6H-SiC, 3C-SiC, and Si for power devices. *IEEE Trans. Electron Devices*, **40**, 645.
[90] Ruff, M., Mitlehner, H. and Helbig, R. (1994) SiC devices: physics and numerical simulation. *IEEE Trans. Electron Devices*, **41**, 1040.
[91] Slack, G.A. (1964) Thermal conductivity of pure and impure silicon, silicon carbide, and diamond. *J. Appl. Phys.*, **35**, 3460.
[92] Morelli, D., Heremans, J., Beetz, C. *et al.* (1994) Carrier concentration dependence of the thermal conductivity of silicon carbide. *Inst. Phys. Conf. Ser.*, **137**, 313.
[93] Hobgood, H.M., Brady, M., Brixius, W. *et al.* (2000) Status of large diameter SiC crystal growth for electronic and optical applications. *Mater. Sci. Forum*, **338–342**, 3.
[94] Feldman, D.W., Parker, J.H. Jr.,, Choyke, W.J. and Patrick, L. (1968) Phonon dispersion curves by Raman scattering in SiC, polytypes 3C, 4H, 6H, 15R, and 21. *Phys. Rev.*, **173**, 787.
[95] Bechstedt, F., Käckell, P., Zywietz, A. *et al.* (1997) Polytypism and properties of silicon carbide. *Phys. Status Solidi B*, **202**, 35.
[96] Nakashima, S. and Harima, H. (1997) Raman investigation of SiC polytypes. *Phys. Status Solidi A*, **162**, 39.
[97] Harima, H., Nakashima, S. and Uemura, T. (1995) Raman scattering from anisotropic LO-phonon–plasmon–coupled mode in n-type 4H–and 6H–SiC. *J. Appl. Phys.*, **78**, 1996.
[98] Dean, P.J., Choyke, W.J. and Patrick, L. (1977) The location and shape of the conduction band minima in cubic silicon carbide. *J. Lumin.*, **10**, 299.
[99] Devaty, R.P. and Choyke, W.J. (1997) Optical properties of silicon carbide polytypes. *Phys. Status Solidi A*, **162**, 5.

[100] Egilsson, T., Ivanov, I.G., Son, N.T. *et al.* (2004) Exciton and defect photoluminescence from SiC, in *Silicon Carbide, Materials, Processing, and Devices* (eds Z.C. Feng and J.H. Zhao), Taylor & Francis, p. 81.

[101] Janzen, E., Gali, A., Henry, A. *et al.* (2008) Defects in SiC, in *Defects in Microelectronic Materials and Devices*, Taylor & Francis Group, p. 615.

[102] Steeds, J., Sullivan, W., Furkert, S. *et al.* (2008) Creation and identification of the two spin states of dicarbon antisite defects in 4H-SiC. *Phys. Rev. B*, **77**, 195203.

[103] Zorman, C.A. and Parro, R.J. (2008) Micro- and nanomechanical structures for silicon carbide MEMS and NEMS. *Phys. Status Solidi B*, **245**, 1404.

[104] Suzuki, T., Yonenaga, I. and Kirchner, H.O.K. (1995) Yield strength of diamond. *Phys. Rev. Lett.*, **75**, 3470.

[105] Baliga, B.J. (1989) Power semiconductor device figure of merit for high-frequency applications. *IEEE Electron Device Lett.*, **10**, 455.

[106] Itoh, A. and Matsunami, H. (1997) Analysis of Schottky barrier heights of metal/SiC contacts and its possible application to high-voltage rectifying devices. *Phys. Status Solidi A*, **162**, 389.

[107] Powell, A.R. and Rowland, L.B. (2002) SiC materials-progress, status, and potential roadblocks. *Proc. IEEE*, **90**, 942.

[108] Cooper, J.A. Jr., and Agarwal, A. (2002) SiC power-switching devices-the second electronics revolution? *Proc. IEEE*, **90**, 956.

[109] Elassera, A. and Chow, T.P. (2002) Silicon carbide benefits and advantages for power electronics circuits and systems. *Proc. IEEE*, **90**, 969.

[110] Neudeck, P.G., Okojie, R.S. and Chen, L.Y. (2002) High-temperature electronics-a role for wide bandgap semiconductors? *Proc. IEEE*, **90**, 1065.

[111] Sugawara, Y. (2004) High voltage SiC devices, in *Silicon Carbide – Recent Major Advances* (eds W.J. Choyke, H. Matsunami and G. Pensl), Springer, p. 769.

[112] Agarwal, A., Ryu, S.H. and Palmour, J. (2004) Power MOSFETs in 4H-SiC: Device Design and Technology, in *Silicon Carbide – Recent Major Advances* (eds W.J. Choyke, H. Matsunami and G. Pensl), Springer, p. 785.

[113] Okumura, H. (2006) Present status and future prospect of widegap semiconductor high-power devices. *Jpn. J. Appl. Phys.*, **45**, 7565.

[114] Friedrichs, P. (2008) Silicon carbide power-device products–Status and upcoming challenges with a special attention to traditional, nonmilitary industrial applications. *Phys. Status Solidi B*, **245**, 1232.

[115] Nakamura, T., Miura, M., Kawamoto, N. *et al.* (2009) Development of SiC diodes, power MOSFETs and intelligent power modules. *Phys. Status Solidi A*, **206**, 2403.

[116] Kimoto, T. (2010) Technical Digest 2010 VLSI Technology Symposium, Honolulu, Hawaii, 2010, p. 9 SiC technologies for future energy electronics.

[117] Marshall, R.C., Faust, J.W. Jr., and Ryan, C.E. (eds) (1974) *Silicon Carbide 1973*, University of South Carolina Press.

[118] Davis, R.F., Kelner, G., Shur, M. *et al.* (1991) Thin film deposition and microelectronic and optoelectronic device fabrication and characterization in monocrystalline alpha and beta silicon carbide. *Proc. IEEE*, **79**, 677.

[119] Ivanov, P.A. and Chelnokov, V.E. (1992) Recent developments in SiC single-crystal electronics. *Semicond. Sci. Technol.*, **7**, 863.

[120] Morkoç, H., Strite, S., Gao, G.B. *et al.* (1994) Large-band-gap SiC, III-V nitride, and II-VI ZnSe-based semiconductor device technologies. *J. Appl. Phys.*, **76**, 1363.

[121] Choyke, W.J., Matsunami, H. and Pensl, G. (eds) (1997) *Silicon Carbide, A Review of Fundamental Questions and Applications to Current Device Technology*, vol. 1 & 2, Akademie Verlag.

[122] Zetterling, C.M. (2002) *Process Technology for Silicon Carbide Devices*, INSPEC.

[123] Choyke, W.J., Matsunami, H. and Pensl, G. (eds) (2004) *Silicon Carbide – Recent Major Advances*, Springer.

[124] Pensl, G., Ciobanu, F., Frank, T. *et al.* (2005) SiC material properties. *Int. J. High Speed Electron. Syst.*, **15**, 705.

[125] Friedrichs, P., Kimoto, T., Ley, L. and Pensl, G. (eds) (2010) *Silicon Carbide*, Vol. 1: Growth, Defects, and Novel Applications, Vol. 2: Power Devices and Sensors, Wiley-VCH Verlag GmbH, Weinheim.

3

Bulk Growth of Silicon Carbide

Bulk crystal growth is the essential technique for producing single-crystal wafers, the base material for device fabrication. Recent progress in SiC device development relies on the availability of relatively large SiC wafers with reasonable quality. At present, the standard technique for SiC bulk growth is the *seeded sublimation* (or *modified Lely*) method. However, a few alternative growth techniques have been intensively developed. This chapter describes fundamental aspects of SiC bulk growth and the associated technology development.

3.1 Sublimation Growth

3.1.1 Phase Diagram of Si-C

Figure 3.1 shows the phase diagram of the Si-C binary system [1, 2]. Because there exists no stoichiometric SiC liquid phase, it is impossible to employ congruent melt growth for SiC bulk growth at technically relevant system pressures. Instead, SiC sublimes at very high temperatures, above 1800–2000 °C. This is the key process of source supply in sublimation growth of SiC. The phase diagram indicates that up to 15% of carbon can be dissolved in a Si melt at about 2800 °C. Liquid phase (solution) growth, which exploits this phenomenon, is described in Section 3.6.

Sublimation growth of SiC consists of three steps: (i) *sublimation of the SiC source,* (ii) *mass transport of sublimed species,* and (iii) *surface reaction and crystallization.* Thus, this growth method is also called "*physical vapor transport (PVT)*" growth. Figure 3.2 shows the partial pressures of sublimed species in the (a) SiC + C and (b) SiC + Si systems at high temperature [3–5]. In the gas phase, the dominant species are not stoichiometric SiC molecules but Si_2C and SiC_2 molecules and atomic Si.

3.1.2 Basic Phenomena Occurring during the Sublimation (Physical Vapor Transport) Method

Growth of single crystalline SiC by sublimation was first achieved by Lely in 1955 [6]. Figure 3.3 shows a schematic illustration of a crucible used in the Lely method. The SiC source is placed along the inner walls of a cylindrical graphite crucible. The source material is usually SiC powder produced by the Acheson process [7]. By heating the crucible to the process temperature of about 2500 °C, the SiC source sublimes and is transported to the inner part of the crucible. Under this nearly isothermal condition, many SiC platelets nucleate randomly along the vapor transport paths in the growth cavity. The grown

Fundamentals of Silicon Carbide Technology: Growth, Characterization, Devices, and Applications, First Edition.
Tsunenobu Kimoto and James A. Cooper.

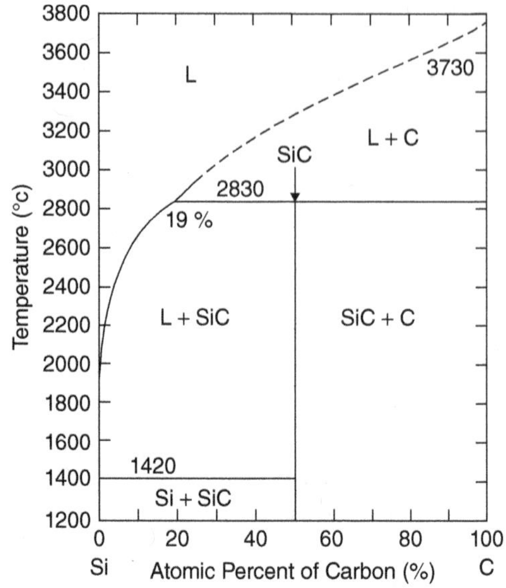

Figure 3.1 Phase diagram of the Si–C binary system [1, 2].

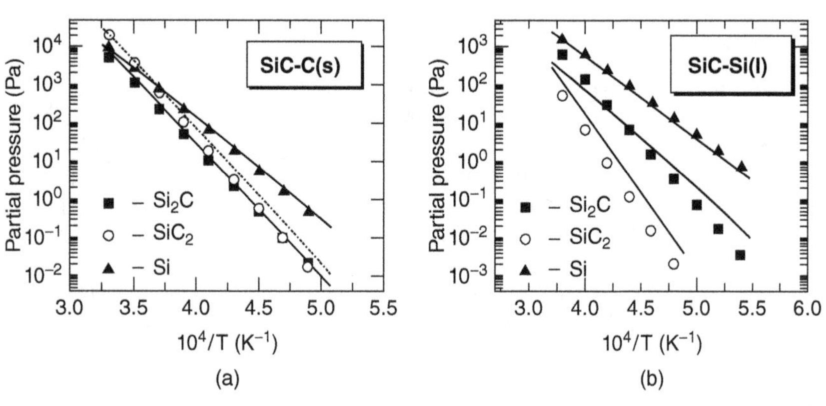

Figure 3.2 Partial pressures of sublimed species from (a) SiC + C and (b) SiC + Si systems at high temperature ([3–5] reproduced with permission from Wiley-VCH Verlag GmbH & Co. KGaA).

SiC platelets are of high quality, and the typical dislocation density of good platelets is only several 100 cm^{-2}. However, the platelets are very small and irregular in shape, with typical areas of 1–2 cm^2 and thickness of 0.3–0.5 mm. The polytype of the platelets is mainly 6H-SiC, but occasionally 4H- or 15R-SiC polytypes are mixed. Although these SiC platelets are not suitable for device development, in spite of their high quality, the platelets can be used as seed crystals for the early stage of the seeded sublimation growth, as described below.

Tairov and Tsvetkov developed the seeded sublimation (or modified Lely) method by placing a seed crystal at a slightly cooler place inside the crucible [8, 9]. Figure 3.4 shows a schematic illustration of a crucible used for seeded sublimation growth of SiC. The SiC source (SiC powder or sintered polycrystalline SiC) is placed at the bottom of a cylindrical dense graphite crucible, and a SiC seed crystal is placed near the lid of the crucible. The distance between the top of the SiC source and the seed crystal is

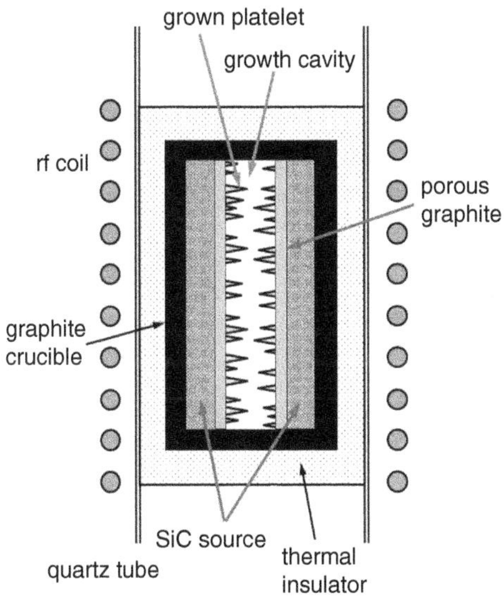

Figure 3.3 Schematic illustration of a crucible used in the Lely method.

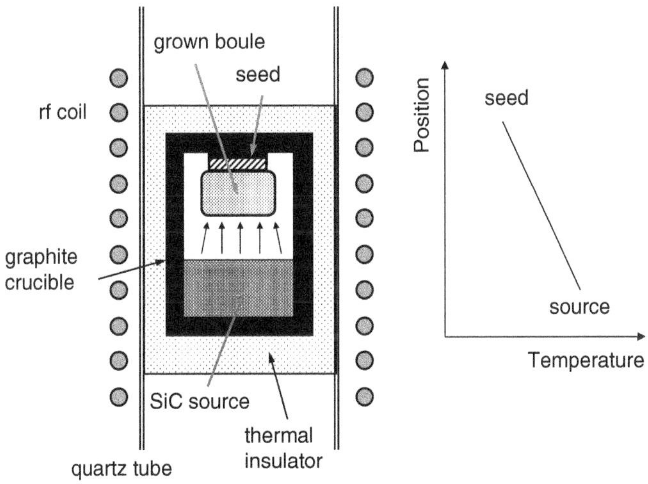

Figure 3.4 Schematic illustration of a crucible used for seeded sublimation growth of SiC.

typically 20–40 mm. The crucible is heated by radio frequency (rf) induction or resistive heating up to 2300–2400 °C. The crucible is thermally insulated by graphite felt or porous graphite; by choosing an appropriate frequency, direct heating of this insulation can be avoided. The seed temperature is fixed at about 100 °C lower than the source temperature, so that sublimed SiC species condense and crystallize on the seed. Growth is usually performed at low pressure to enhance the mass transport from the source to the seed. A high-purity Ar (or He) flow is employed for growth. Remarkable improvement in growth technology has been made in recent decades [10–24].

To grow high-quality SiC boule crystals, both *thermodynamic* and *kinetic* factors must be considered. The process control to maintain optimum thermal and chemical conditions is also very critical. Several aspects of each issue are described below.

3.1.2.1 Thermodynamic Considerations

As shown in Figure 3.2, the main species transported from the SiC source to the seed are Si, Si_2C, and SiC_2 at a growth temperature of 2300–2400 °C. Thus, the gas phase in sublimation growth is usually Si-rich because of preferential evaporation of Si from the SiC source. This leaves the source more and more C-rich, and graphitization of the source occurs during the growth. To avoid carbon inclusions in the growing crystal, silicon is added to the source to maintain a stoichiometric or Si-rich source surface. This is important because sufficient overpressure of Si is also required to avoid graphitization of the growing surface on the seed. Carbon in the graphite crucible can evaporate and participate in the growth (this can be reduced by using a TaC-coated crucible [19]). Taking into account these phenomena, the main reactions during sublimation growth are summarized as follows [5]:

$$\mathrm{Si_2C(g) + SiC_2(g) \leftrightarrow 3\ SiC(s)} \tag{3.1}$$

$$\mathrm{SiC_2(g) + 3\ Si(g) \leftrightarrow 2\ Si_2C(g)} \tag{3.2}$$

$$\mathrm{Si_2C(g) \leftrightarrow 2\ Si(g) + C(s)} \tag{3.3}$$

$$\mathrm{Si(g) \leftrightarrow Si(l)} \tag{3.4}$$

By using the thermodynamic properties of the relevant species, the equilibrium partial pressures of Si, Si_2C, and SiC_2 can be calculated as a function of temperature. The calculated results agree well with the experimental results shown in Figure 3.2 [5].

Figure 3.5 shows a diagram of the secondary phase formation during the sublimation growth of SiC with a fixed carbon flux of 7×10^{14} cm^{-2} s^{-1} (calculated by assuming thermodynamic equilibrium of the system) [5]. The area of stable growth appears in the region where a sufficient silicon flux is supplied. When the silicon flux is lower than a critical value, surface graphitization takes place; this critical value of silicon flux increases exponentially with temperature.

Glass *et al.* reported the temperature dependence of the partial pressure of the main components in the phase diagram (SiC-C and SiC-Si) using the latest thermodynamic data [17]. The obtained relationships between the partial pressure of Si and those of Si_2C and SiC_2 (in Pa) are:

$$P_{\mathrm{Si_2C}} = 2.85 \times 10^2 \exp(-1.79 \times 10^4/T) \times P_{\mathrm{Si}} \tag{3.5}$$

$$P_{\mathrm{SiC_2}} = 9.41 \times 10^{28} \exp(-14.35 \times 10^4/T)/P_{\mathrm{Si}} \tag{3.6}$$

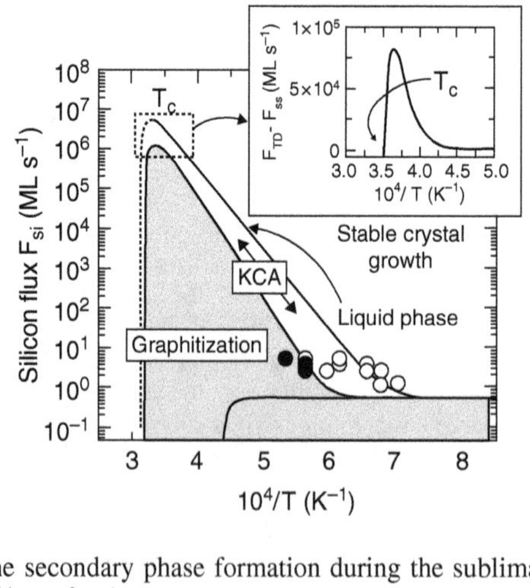

Figure 3.5 Diagram of the secondary phase formation during the sublimation growth of SiC with a fixed carbon flux of 7×10^{14} cm^{-2} s^{-1} (calculated by assuming thermodynamic equilibrium of the system) ([5] reproduced with permission from Wiley-VCH Verlag GmbH & Co. KGaA).

Based on these equations, one can estimate the temperature dependence of the ratio of Si to C atoms in the vapor phase. It should be noted that the condensation energy of SiC from the vapor phase is very large, about 580 kJ mol^{-1}, almost 10 times larger than that in Si melt growth. This factor must also be considered in growth simulation and process design.

3.1.2.2 Kinetic Considerations

The growth rate in sublimation growth is mainly determined by the flux of the source supply (sublimation rate) and the transport efficiency from the source to the seed. The sublimation rate is a function of the source temperature, while the transport efficiency depends strongly on the growth pressure, the temperature gradient, and the distance between the source and the seed. Figure 3.6 shows the pressure dependence of the growth rate during sublimation growth of SiC at various source temperatures [25]. Because the mass transport is *diffusion limited* in sublimation growth, the growth rate is almost inversely proportional to the growth pressure. In other words, as the pressure is reduced, the vapor diffusion rates increase and constituents move rapidly along the concentration gradient from the source toward the seed. Here, the concentration gradient is basically determined by the source and seed temperatures (*temperature gradient*). As a result, the growth rate is approximately proportional to the *supersaturation* on the seed surface σ_g, which is given by:

$$\sigma_g = \frac{P - P_e}{P_e} = \frac{P}{P_e} - 1 \tag{3.7}$$

where P and P_e are the vapor pressure on the seed and the equilibrium vapor pressure, respectively. In the case of SiC, of course, the vapor pressures for Si and Si-C compounds must be considered. For example, the seed temperature must be controlled to ensure that loss of Si from the surface is minimized by maintaining the overpressure of the Si vapor near the seed. The source temperature and pressure must be controlled to develop the appropriate temperature gradient and to ensure that the proper amounts of Si and Si-C compounds are transported from the source to the seed. Any fluctuations in the temperature profile and pressure can result in constitutional supercooling (e.g., Si droplet formation), surface graphitization, and C inclusions. All these phenomena lead to the formation of macro- and micro-defects in SiC boules. Defect formation in sublimation growth is described in Section 3.3.

In state-of-the-art growth technology, SiC boules are usually grown with a growth rate of 0.3–0.8 mm h^{-1} at several hundred pascals or even lower. Although the growth rate can be increased to 1–2 mm h^{-1}, this faster growth often results in significant generation of extended defects in the grown

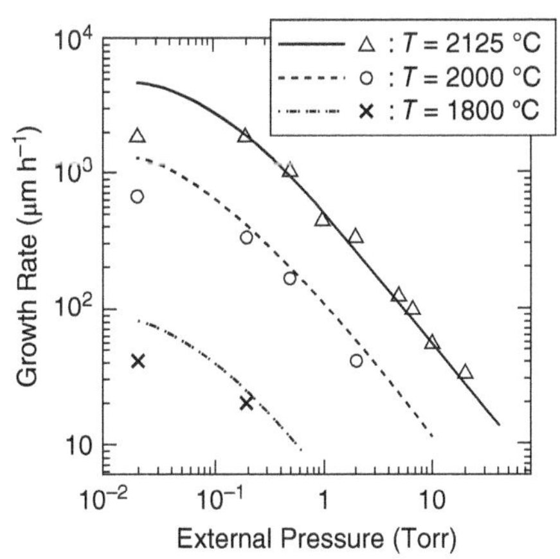

Figure 3.6 Pressure dependence of the growth rate during sublimation growth of SiC at various source temperatures ([25] reproduced with permission from Elsevier).

crystal. One obvious obstacle in sublimation growth is the limitation of growth time. The source material becomes C-rich as growth proceeds, for the reasons described above, and high-quality growth becomes impossible after several days. At present, this limits the length of SiC boule crystals to 30–50 mm. To overcome this problem, several modifications have been investigated. An additional gas pipe can be introduced into the crucible, and the C/Si ratio inside the crucible can be adjusted by supplying Si- or C-containing gas(es) [26]. Continuous feeding of polycrystalline SiC source during sublimation growth has also been investigated [27].

3.1.3 Modeling and Simulation

In seeded sublimation growth of SiC, the growth is carried out in a *quasi-closed* graphite crucible, and one can only control the process parameters such as the temperature and pressure from the outside, without monitoring the inside. Although careful *in-situ* X-ray imaging experiments have been conducted to visualize the phenomena inside the crucible [26], the insights obtained are still limited. Therefore, the experimental approach alone is not sufficient to develop a well-controlled sublimation growth process. To overcome this problem, modeling and simulation of SiC sublimation growth have been extensively investigated [21, 24, 28–33]. At present, calculation of heat and mass transport during sublimation growth of SiC is standard technology and provides us with very good insights into what is happening inside the crucible. These realistic simulations can be used to design crucible geometries and temperature profiles for the growth of high-quality SiC boules. The next step in the development of the simulations is to include chemistry inside the growth space. To treat chemical reactions in the gas phase and at the growing surface, a reliable database for a number of chemical reactions at very high temperatures is mandatory, but at present is a challenge. Nevertheless, a recent simulation package offers reasonably good agreement with experimental results such as the growth rate and shape of a grown boule, and has been employed as a powerful tool for crystal growers.

The temperature distribution is calculated, taking into account heat transfer by thermal conduction, gas phase convection, and radiation. In the heat transfer calculation, the large crystallization energy of SiC (that is, the latent heat of sublimation or condensation on the gas/solid interface) must be considered. Heat transfer by *radiation* is a dominant process at very high temperature in sublimation growth. SiC crystals are semitransparent to visible-infrared radiation, and the radiated light is absorbed by the growing SiC boule, to some extent. This absorption depends strongly on the free carrier density and thus the impurities present. For example, the temperature profile inside a growing boule heavily doped with nitrogen is very different from that inside a high-purity boule. The thermal conductivity of SiC at the growth temperature must be much smaller than that at room temperature [34]. Therefore, the absorption of radiation can form a significant thermoelastic field inside the growing boule. It is important to control this thermoelastic field to obtain a desirable boule shape and to reduce defect generation.

To predict the growth rate and boule shape, a mass transfer model must be coupled with the heat transfer calculation and the thermodynamic database. In the gas phase, fluid transport is based on the low pressure kinetic theory of gases. Diffusion coefficients, viscosity, conductivity, and specific heat of species are calculated as a local function of temperature, pressure, and composition. The Stefan flow caused by the phase change of SiC has to be considered. The thermodynamic calculation is performed by minimization of the total free energy of the Si–C–Ar system at high temperature. Nine possible gaseous molecules are usually considered (based on a literature survey): Si, Si_2, Si_3, C, C_2, C_3, SiC, Si_2C, SiC_2, in addition to Ar [21, 29]. Calculations indicate that three particular species, Si, Si_2C, and SiC_2, are indeed important to describe sublimation growth [21]. Modeling and simulation of chemistry during SiC sublimation growth has also been developed. Large sets of multi-step gas and surface reactions are handled, based on the most recent thermodynamic database. In the chemical simulation, "local thermodynamic equilibrium" is usually assumed. Deviation from thermodynamic equilibrium can also be included [21].

Figure 3.7 shows the effect of a radiation shield plate along the periphery of the growth front, demonstrating both experimental and simulation results [31]. By introducing the radiation shield plate ("flat

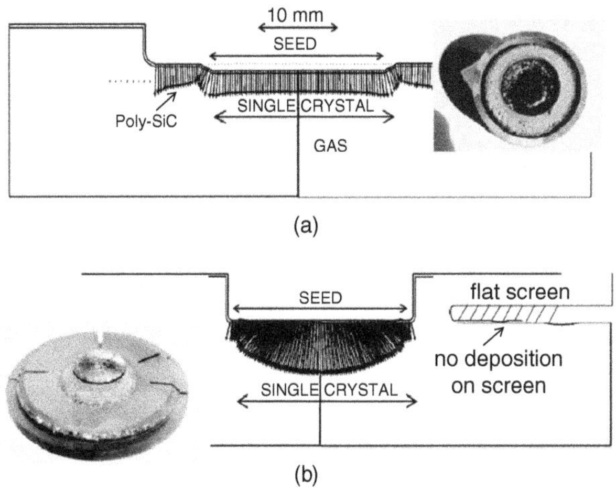

Figure 3.7 Effect of a radiation shield plate along the periphery of the growth front, demonstrating both experimental and simulation results, (a) without a shield and (b) with a shield ([31] reproduced with permission from Elsevier). By introducing the radiation shield plate ("flat screen" in the figure), the deposition of polycrystalline SiC around the main boule can be avoided.

screen" in the figure), the deposition of polycrystalline SiC around the main boule can be avoided because the shield plate is at a higher temperature. The polycrystalline SiC deposition, as well as the shape of the growth front, is reproduced well in the simulation. Figure 3.8 shows the improved geometry of the crucible used to grow a large boule with a nearly constant diameter [32]. By optimizing the shape and location of the "liner" inside the crucible, stable growth of a long boule with a constant diameter is established. The simulation gives good guidance in designing the crucible geometry and the temperature profile. Comparative studies of experiments and simulation have been done to investigate the optimization of thermal insulators, influence of ambient gases, and scaling up of the reactor. In particular, recent diameter enlargement of SiC wafers (boule crystals) is largely due to the development of the simulation technique.

Another important role of simulation in the sublimation growth of SiC is the simulation of stress in the boules. As will be described in Section 3.3, *thermal stress* plays a critical role in the generation of extended defects in SiC boules. The causes of thermal stress include different thermal expansion coefficients between SiC and the crucible (graphite), and radial or axial temperature inhomogeneity. When the resolved shear stress on a primary slip system exceeds a *critical resolved stress,* glide and multiplication of dislocations takes place, leading to a significantly increased dislocation density in the boule. If the thermal stress becomes very high, crystal cracking can also occur. Figure 3.9 shows the simulated thermal stress in SiC boules with (a) a convex growth front and (b) a flat growth front [35]. The convex front is formed when a radial temperature gradient exists (the peripheral temperature is higher than that at the center). The flat surface is obtained when the radial temperature gradient is very small. As shown in the figure, a very high peripheral component of the resolved shear stress ($\sigma_{\phi\phi}$) appears along the top-edge of the boule in the case of the convex growth. Indeed, this is the location where a crack is observed in a large-diameter boule. Conversely, the shear stress is more than one-order-of-magnitude smaller in the case of flat growth. Figure 3.10 shows (a) the distribution of calculated resolved shear stress (magnitude) inside a boule and (b) the experimentally obtained distribution of dislocation density [36]. In this particular case, the dislocation density is high near the center ($r/r_0 = 0$) and near the wafer edge ($r/r_0 = 1$). This distribution is consistent with the distribution of the shear stress. Thus, stress simulation is a powerful tool to predict trends in dislocation density.

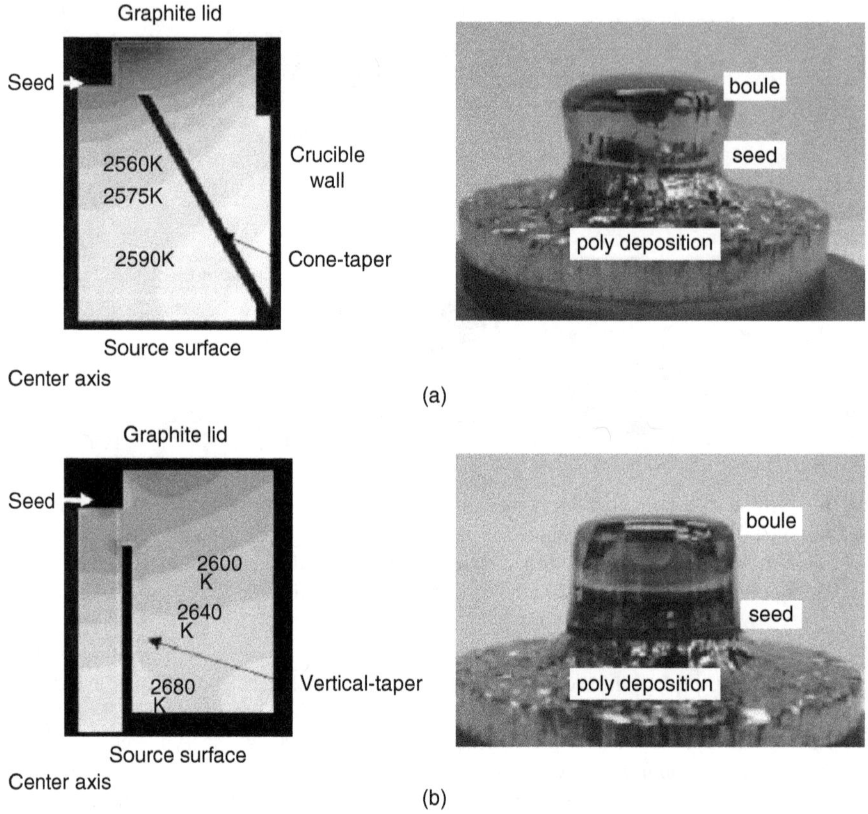

Figure 3.8 Improved geometry of the crucible used to grow a large boule with a nearly constant diameter, (a) with a cone-shaped liner and (b) with a vertical liner ([32] reproduced with permission from Elsevier). By optimizing the shape and location of the "liner" inside the crucible, stable growth of a long boule with a constant diameter is established.

3.2 Polytype Control in Sublimation Growth

For SiC wafers to be used in electronic applications, it is mandatory to grow a large SiC boule of a desired single polytype. Because of the low stacking fault energy of SiC, however, polytype mixing may happen during boule growth, when the growth conditions are not optimized. Knippenberg reported empirical observations of the relative stability (or occurrence) of individual polytypes in SiC bulk growth, as shown in Figure 3.11 [3]. According to this report, 3C-SiC is a metastable polytype, and 2H-SiC is believed to occur at relatively low temperatures, 1300–1600 °C. At high temperatures, above 2000 °C, at which sublimation growth is carried out, 6H-, 4H-, and 15R-SiC polytypes are often observed. However, from a materials science viewpoint, the kinetic and thermodynamic factors which determine the polytype actually grown are not well understood. Because SiC{0001} is usually employed as the seed crystal, polytype switching, or nucleation of foreign polytypes may occur during growth, unless intentionally controlled.

One obvious kinetic factor is the polytype replication through *spiral growth* around threading screw dislocations, after stable spiral growth has been established in the bulk. Figure 3.12 shows typical surface morphologies of a 6H-SiC boule taken with (a) an optical microscope and (b) an atomic force microscope.

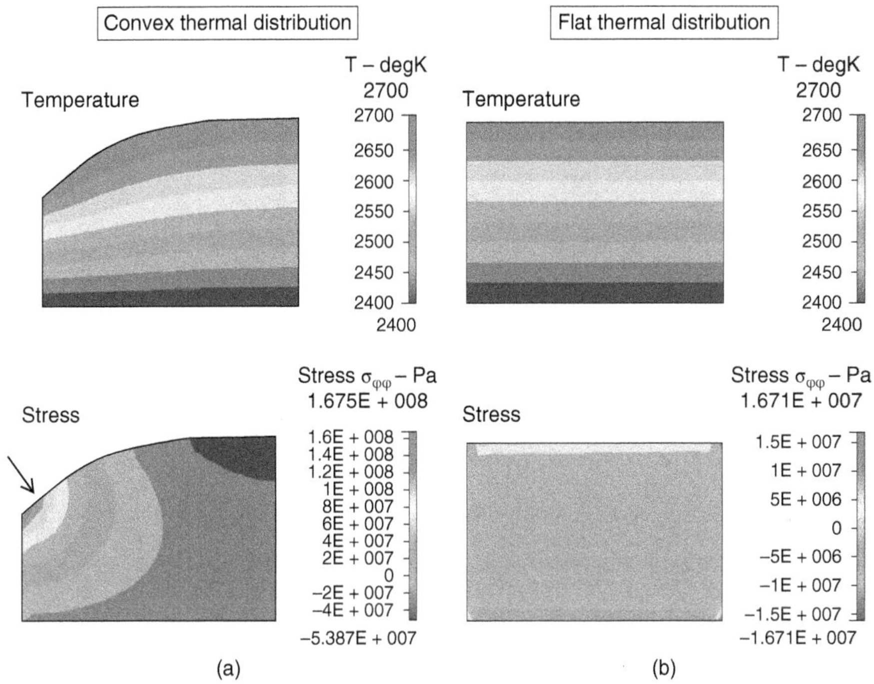

Figure 3.9 Simulated thermal stress in SiC boules with (a) a convex growth front and (b) a flat growth front ([35] reproduced with permission from Trans Tech Publications).

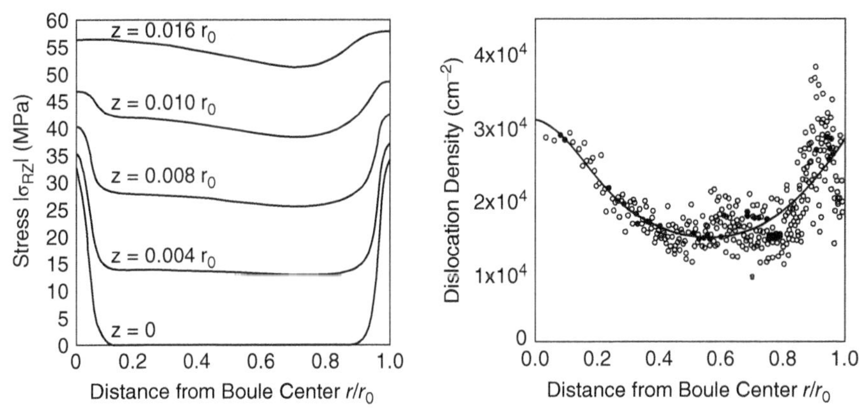

Figure 3.10 (a) Distribution of calculated resolved shear stress (magnitude) inside a boule and (b) experimentally obtained distribution of dislocation density ([36] reproduced with permission from Trans Tech Publications).

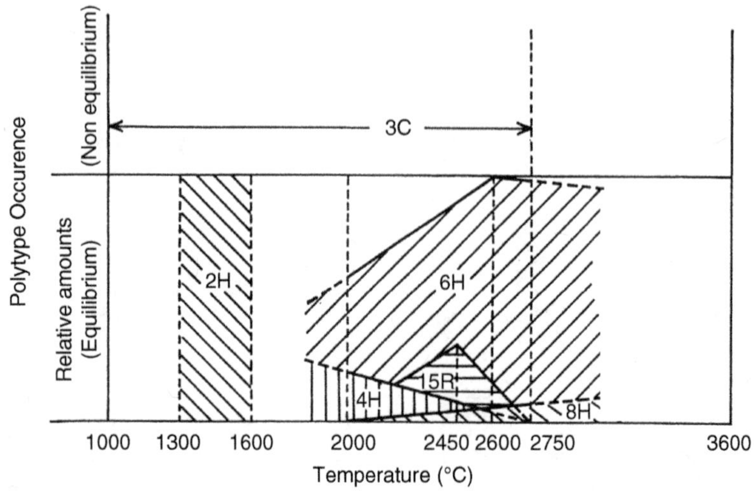

Figure 3.11 Empirical observations of relative stability (or occurrence) of individual polytypes in SiC bulk growth [3].

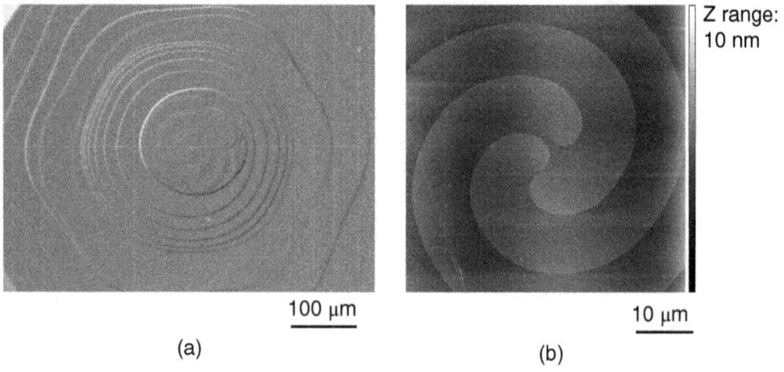

Figure 3.12 Typical surface morphologies of a 6H-SiC boule taken with (a) an optical microscope and (b) an atomic force microscope. (By courtesy of D. Nakamura and T. Mitsuoka, Toyota Central R&D Laboratories.).

These images indicate that spiral growth, via steps with a six-bilayer height, is dominant on the growing surface (in 6H-SiC growth). Along the step edges, the stacking information is provided, which ensures replication of that polytype in the growing crystal. Because the core of a threading screw dislocation acts as an infinite step source, this spiral growth is maintained throughout the growth, as long as the optimized growth conditions are maintained. In this sense, polytype replication via a spiral growth mechanism will become much more difficult in the future, when the threading screw dislocations in SiC boules are almost eliminated. Polytype replication by step-flow growth is described in greater detail in Chapter 4.

Although spiral growth is favorable for polytype replication, nucleation on the terraces (flat regions between steps) can naturally take place in the initial stages of growth as well as during growth. Therefore, it is essential to understand and to control the key factors which stabilize a desired polytype. It has

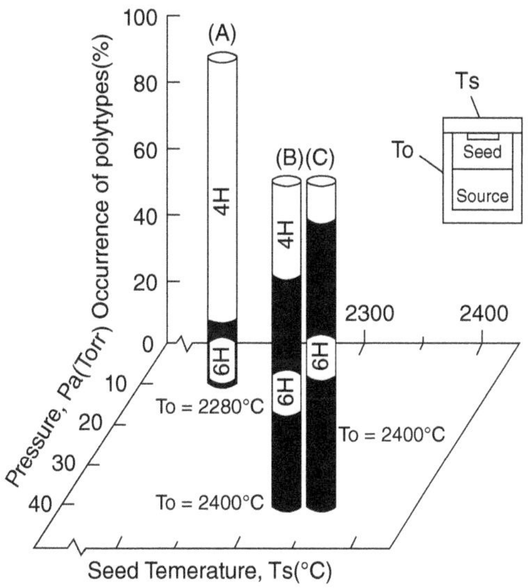

Figure 3.13 Influences of growth temperature and pressure on polytype stability in seeded sublimation growth of SiC ([12] reproduced with permission from AIP Publishing LLC). 4H-SiC is preferentially grown at relatively low temperature and low pressure.

been suggested that there exists a close relationship between the polytype stability and the *C/Si ratio (C enrichment)* in the growth ambient (or atmosphere) [37]. When the growth ambient is C-rich, a polytype with a higher hexagonality becomes stable. For example, 4H-SiC (hexagonality: 0.5) is more stable than 6H-SiC (hexagonality: 0.33) under C-rich growth conditions. In real experiments, the most striking parameter determining the polytype is the *polarity* of the seed crystal. Sublimation growth on a SiC(0001) (Si face) under adequate conditions gives a 6H-SiC boule, even if the seed is 4H-SiC(0001). In contrast, a 4H-SiC boule is grown on a SiC(000$\bar{1}$) (C face), irrespective of the seed polytype [13]. This result is explained by the difference in surface energy between the Si and C faces. The growth temperature and pressure also influence polytype stability, as shown in Figure 3.13 [12]. 4H-SiC is preferentially grown at relatively low temperature and low pressure, while relatively high temperature and high pressure result in 6H-SiC growth (the growth temperature and pressure, of course, influence the C/Si ratio on the growing surface). Another important factor is *impurity incorporation*. Nitrogen doping during the growth stabilizes 4H-SiC, while aluminum doping leads to preferential growth of 6H-SiC. Because nitrogen atoms occupy the carbon lattice sites, incorporation of nitrogen atoms will cause the growth environment to become slightly C-rich, which favors the growth of 4H-SiC. Furthermore, it is reported that impurity additives such as Sc and Ce stabilize 4H-SiC [37, 38]; this can also be explained by a shift in the growth ambient toward C-rich conditions.

In spite of poor understanding of the mechanism, 4H-SiC boules without polytype mixing can be reproducibly produced by sublimation growth on 4H-SiC(000$\bar{1}$) under optimized conditions with intentional nitrogen doping. Thus, it is easy to produce heavily-nitrogen-doped n-type 4H-SiC wafers, whereas production of low-resistivity *p*-type 4H-SiC wafers heavily doped with aluminum is a challenge in terms of polytype control.

When SiC(11$\bar{2}$0) or SiC(1$\bar{1}$00) is employed as the seed, instead of SiC{0001}, perfect polytype replication can be achieved across a wide range of growth conditions [18]. This is also explained by the mechanism that the stacking information appears on the (11$\bar{2}$0) and (1$\bar{1}$00) surfaces, and the grown crystal inherits this stacking sequence. Although generation of stacking faults was a critical issue in sublimation growth on these faces, this problem has been considerably suppressed by using a SiC$(11\bar{2}0)$ seed and optimizing growth conditions. Because of the limited availability of these seed crystals, boule growth on non-basal planes is not very common. This growth technique is, however, a key for reduction of extended defects, as described in Section 3.3.

Because 3C-SiC is only stable at relatively low temperature, it is not easy to grow 3C-SiC boules by sublimation. Sublimation growth on a 3C-SiC(001) or (111) seed at 1700–2100 °C has been investigated [39, 40]. The growth rate is low, 0.1–0.2 mm h^{-1}, as a result of the low growth temperature. When the temperature is increased above 1900–2000 °C, a polytype transformation into 6H-SiC takes place [41], making 3C-SiC bulk growth difficult [42].

3.3 Defect Evolution and Reduction in Sublimation Growth

SiC boule crystals and wafers contain a variety of crystal imperfections, both extended defects and point defects. This subsection describes the evolution and reduction of extended defects. The density of point defects in SiC boules is rather high, in the 10^{14}–10^{16} cm^{-3} range. The nature and properties of these point defects are described in Section 3.4 and Chapter 5.

Table 3.1 shows the major extended defects observed in SiC boules and wafers. The Burgers vector, the major direction, and the typical density of the extended defects in boules (wafers) prepared using state-of-art technology (for n-type 4H-SiC) are shown with additional comments. Note that, through recent efforts, three-dimensional defects such as large carbon inclusions and voids [43] are now eliminated.

3.3.1 Stacking Faults

Stacking faults are common defects because of the low stacking fault energy (14 mJ m^{-2} for 4H-SiC and 2.9 mJ m^{-2} for 6H-SiC) [44] and the occurrence of many polytypes in SiC. Typical stacking faults are 3C- or 6H-like laminar regions in 4H-SiC boules. Generation of double Shockley stacking faults observed in heavily-nitrogen-doped SiC is described in Section 3.4.2. Through the recent progress in polytype control, inclusions of foreign polytypes and stacking faults have been greatly reduced. The typical stacking fault density along the *c*-axis is well below 1 cm^{-1}. Generation of stacking faults *during SiC epitaxial growth* is one of the remaining issues.

Table 3.1 Major extended defects observed in SiC boules and wafers. The Burgers vector, major direction, and typical density of the extended defects in boules (wafers) prepared using state-of-art technology (for n-type 4H-SiC) are shown.

Dislocation	Burgers vector	Major direction	Typical density (cm^{-2})
Micropipe	$n\langle 0001\rangle$ ($n > 2$)	$\langle 0001\rangle$	0–0.1
Threading screw dislocation (TSD)	$n\langle 0001\rangle$ ($n = 1, 2$)	$\langle 0001\rangle$	300–600
Threading edge dislocation (TED)	$\langle 11\bar{2}0\rangle/3$	$\langle 0001\rangle$	2000–5000
(Perfect) Basal plane dislocation (BPD)	$\langle 11\bar{2}0\rangle/3$	in {0001} plane (preferably $\langle 11\bar{2}0\rangle$)	500–3000

3.3.2 Micropipe Defects

A micropipe defect is a hollow core associated with a *superscrew dislocation*. When the magnitude of the Burgers vector is very large, the strain field around the dislocation core becomes extremely high (proportional to $|\boldsymbol{b}|^2$, where $\boldsymbol{b}$: Burgers vector), and a microscopic pinhole is formed by breaking bonds [45, 46]. Micropipe defects are indeed located at the center of a large spiral on the surface of the SiC boule, and the diameters of the pinholes range from 0.5 μm to several micrometers. In SiC, the Burgers vector of an elementary threading screw dislocation (TSD) is already very large because the length of $1\boldsymbol{c}$ corresponds to 1.0 nm for 4H-SiC and 1.5 nm for 6H-SiC, which is much larger than that for Si (~0.24 nm). The magnitudes of the Burgers vector for micropipes have been investigated in great detail, and the minimum values were determined as $|3\boldsymbol{c}|$ for 4H-SiC and $|2\boldsymbol{c}|$ for 6H-SiC [47, 48], both of which correspond to 3 nm. In old wafers, a large micropipe with a Burgers vector of $8-12\boldsymbol{c}$ was also observed. Frank considered an energy balance between the elastic strain energy released by formation of a hollow core and the energy of the free surface created along the hollow core, and proposed that the radius of a hollow core (or in this case, micropipe) r_{MP}, assuming isotropic linear elasticity, is given by [45]:

$$r_{\mathrm{MP}} = \frac{\mu|\boldsymbol{b}|^2}{8\pi\gamma} \tag{3.8}$$

Here μ is the shear modulus and γ the surface energy of the hollow core. This equation is almost satisfied for micropipes in SiC [49]. Figure 3.14 shows an example of a micropipe in a 4H-SiC(0001) wafer, as observed by (a) optical microscopy and (b) atomic force microscopy. Near the center, a pinhole is discernible as a dark spot. When the wafer is observed in transmission mode, a dark line running along the c-axis can be traced.

Because a micropipe is a pinhole extending along the <0001> direction through the entire SiC wafer, it is not surprising that SiC devices which contain a micropipe exhibit severely degraded performance, such as excessive leakage current and premature breakdown [50, 51]. Micropipes also act as a source of impurity contamination in epitaxial growth and device processing. Thus, micropipes were identified as the most important killer defects, and growth technology has now been developed to eliminate micropipes.

Table 3.2 shows the possible causes of micropipe generation during the sublimation growth of SiC [17]. These possible causes can be classified into fundamental and technological issues. The fundamental issues include thermodynamic mechanisms such as thermoelastic stress arising from non-uniform

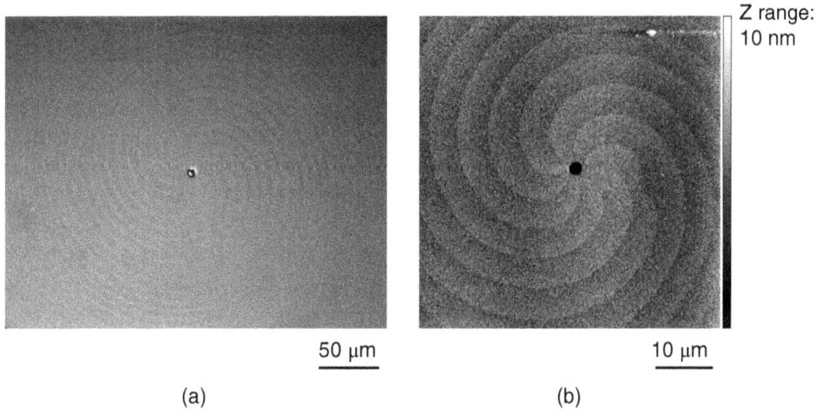

Figure 3.14 Micropipe in a 4H-SiC(0001) wafer, as observed by (a) optical microscopy and (b) atomic force microscopy. (By courtesy of D. Nakamura and T. Mitsuoka, Toyota Central R&D Laboratories.).

Table 3.2 Possible causes of micropipe generation during the sublimation growth of SiC ([17] reproduced with permission from Wiley-VCH Verlag KmbH).

Fundamental		
Thermodynamic	Kinetic	
Thermal field uniformity	Nucleation processes	
Dislocation formation	Inhomogeneous supersaturation	
Solid-state transformation	Constitutional supercooling	
Vapor phase composition	Growth face morphology	
Vacancy supersaturation	Capture of gas phase bubbles	
Technological		
Process instabilities	Seed preparation	Contamination

temperature distribution, and kinetic mechanisms such as an unwanted nucleation process. Technological issues such as process instabilities, imperfect surface preparation of a seed, and carbon inclusions also need to be considered. Inclusion of foreign polytypes causes severe mismatch in the stacking sequence when such an island meets the host polytype. This *stacking mismatch* and the associated large strains also trigger micropipe formation [52]. When elementary screw dislocations are introduced for some reason, the spiral steps emanating from them interact with each other. Because of the strong repulsive interaction between steps, the energetic bunching of spiral steps promotes the coalescence of adjacent screw dislocations, leading to micropipe formation [52]. Accumulation of screw dislocations around a surface depression [53], and the interaction between screw dislocations and twist-type misorientation [54] have also been suggested as mechanisms of micropipe formation.

Major approaches to micropipe elimination are summarized as follows:

1. *Micropipe-free seed*: Because micropipes in a seed crystal are basically replicated in the grown boule crystal, the use of a micropipe-free seed is a requirement to obtain a micropipe-free boule. The micropipe-free seed can be a selected SiC platelet produced by the Lely method. Another candidate is a SiC{0001} wafer prepared by slicing a boule grown on $(11\bar{2}0)$ or $(1\bar{1}00)$, which is inherently free of micropipes [18].
2. *Stable growth under optimized conditions*: As mentioned above, any fluctuations in the temperature profile or pressure in a crucible may cause unintentional supercooling and/or deviation of the C/Si ratio at the growing surface. Degradation of the source material (e.g., graphitization) also perturbs the control of the C/Si ratio. Because all these factors will result in micropipe generation, even on a micropipe-free seed, the growth conditions must be optimized and carefully maintained throughout the long sublimation growth.
3. *Micropipe closing (dissociation)*: In liquid phase epitaxy [55] and chemical vapor deposition (CVD) of SiC under Si-rich conditions [56], a micropipe with a Burgers vector of $n\boldsymbol{c}$ ($n = 3, 4, 5, \ldots$ in 4H-SiC) can dissociate into multiple elementary (closed-core) screw dislocations, each with a Burgers vector of $1\boldsymbol{c}$. As a result, the micropipe (pinhole) becomes closed during growth. Similar phenomena are observed during the sublimation growth of SiC [57], and in some cases, this micropipe closing is intentionally enhanced by adjusting the growth conditions. Because the elastic energy associated with a dislocation is proportional to $|\boldsymbol{b}|^2$ [58], a micropipe is not energetically favorable, based upon thermodynamic considerations. If we consider the Burgers vectors and elastic

energies of a micropipe ($\boldsymbol{b} = n\boldsymbol{c}$) and multiple elementary screw dislocations ($\boldsymbol{b} = 1\boldsymbol{c}$), the following equations are satisfied:

$$n\boldsymbol{c}(\text{micropipe}) = n \times 1\boldsymbol{c} \quad (n \times \text{elementary screw dislocations}) \tag{3.9}$$

$$|n\boldsymbol{c}|^2 > n \times |1\boldsymbol{c}|^2 \quad (n = 3, 4, 5, \ldots \text{ in 4H-SiC}) \tag{3.10}$$

This implies that a micropipe can be closed by dissociation into several elementary threading screw dislocations, if the potential barrier is overcome. The main driving force for micropipe dissociation is believed to be lateral growth, and subsequent interaction between macrosteps and the core of a micropipe.

Through these approaches, micropipe defects have been almost eliminated (micropipe density = 0 or <0.01 cm^{-2}) [35, 59, 60]. At present, micropipe-free wafers are commercially available from most vendors.

3.3.3 Threading Screw Dislocation

A threading screw dislocation (TSD) is located at the center of spiral growth during sublimation growth on a SiC{0001} surface. Figure 3.15 shows a schematic illustration of an elementary TSD in SiC. Although a threading dislocation in Si or GaAs usually creates a spiral with one-bilayer-height step, the step height of a spiral in 4H-SiC{0001} is four Si-C bilayers, corresponding to $|1\boldsymbol{c}|$ (the step often splits into two spiral steps with two-bilayer-height steps). The TSDs usually propagate almost along the <0001> direction, but occasionally they are bent toward the basal planes (and sometimes again bent back toward <0001>) [61]. When they are bent and lie in the basal plane, Frank-type stacking faults are formed because the Burgers vector ($1\boldsymbol{c}$ or $2\boldsymbol{c}$) must be conserved. Recent studies using synchrotron X-ray topography revealed that most TSDs possess a Burgers vector of $1\boldsymbol{c} + \boldsymbol{a}$ [62, 63]. This means that the majority of TSDs are not pure screw dislocations, but instead are *mixed dislocations*.

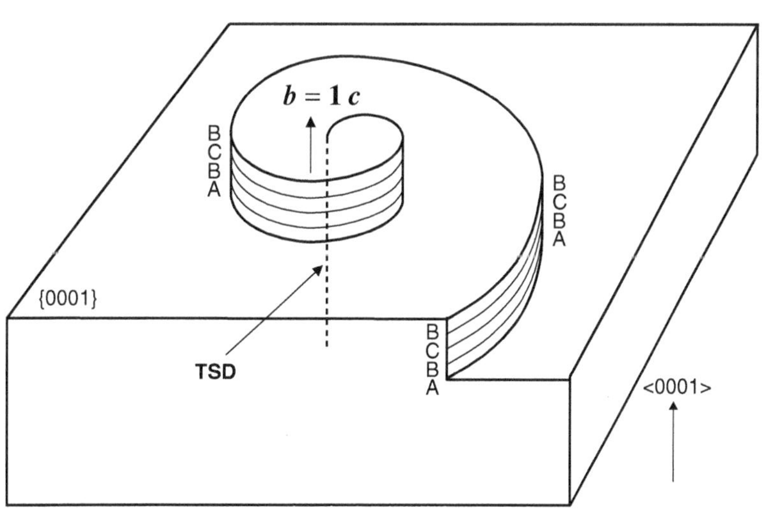

Figure 3.15 Schematic illustration of an elementary threading screw dislocation in SiC.

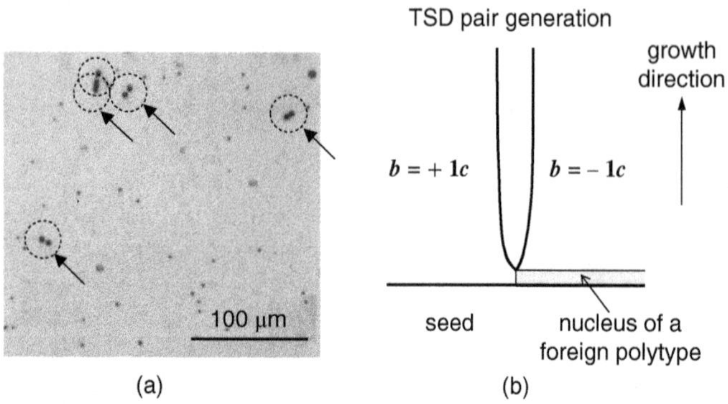

Figure 3.16 Generation of a half loop at the initial stage of bulk growth (a possible cause of threading-screw-dislocation nucleation in SiC sublimation growth). (a) Distribution of dislocation pits formed by KOH etching ([66] reproduced with permission from AIP Publishing LLC), (b) schematic illustration of a mechanism of threading-screw-dislocation nucleation.

Threading screw dislocations are basically replicated from a seed crystal, as also occurs for micropipes. A major cause of threading-screw-dislocation nucleation in SiC sublimation growth is the generation of a half loop at the initial stage of bulk growth, as shown in Figure 3.16. It is reported that the number of TSDs with a Burgers vector of $+1\boldsymbol{c}$ is almost the same as that with $-1\boldsymbol{c}$, and that $+1\boldsymbol{c}$ and $-1\boldsymbol{c}$ dislocations are often observed nearby as if they are a pair [64]. At the very initial stage of growth, stable spiral growth or layer-by-layer growth is not well established. When the growth conditions (e.g., effective C/Si ratio, temperature profile, and other factors) deviate from the optimum conditions, nucleation of foreign polytypes may occur at a microscopic scale. In this case, TSDs can be generated because of stacking mismatch (e.g., between a 4H-SiC host and a small 6H-SiC island), although the microscopic islands will eventually be overgrown [65]. In a similar manner, when a surface precipitate is overgrown, the growth fronts meet at the precipitate and coalesce with misalignment under the influence of stress. To accommodate this misalignment, a pair of screw dislocations of opposite signs is generated [65, 66]. The surface quality of the seed is also important. Polishing-induced damage and surface graphitization during temperature ramping should be completely eliminated. Furthermore, micropipe dissociation during growth is another source of TSDs, as described above. Under optimized conditions, the density of TSDs clearly decreases with increasing boule length [67]. There are two reasons for this: (i) pairs of $+1\boldsymbol{c}$ and $-1\boldsymbol{c}$ TSDs can merge and annihilate and (ii) TSDss can be bent to the basal planes and eventually reach the periphery of the boule.

3.3.4 Threading Edge Dislocation and Basal Plane Dislocation

A threading edge dislocation (TED) and basal plane dislocation (BPD) possess the same Burgers vector of $\boldsymbol{a}(\langle 11\bar{2}0\rangle/3)$. Figure 3.17 shows the two slip vectors, (a) $\langle 11\bar{2}0\rangle/3$ and (b) $\langle 1\bar{1}00\rangle/3$, in the basal plane of a close-packed system. A slip of $\langle 11\bar{2}0\rangle/3$ results in an *extra half plane* or a *missing half plane,* while keeping the stacking structure. Conversely, a slip of $\langle 1\bar{1}00\rangle/3$ causes a *fault in the stacking,* a change of the occupation site in a layer from "A" to "B," for example. This kind of defect is called a Shockley-type stacking fault (SSF) [58], and plays an important role in bipolar degradation phenomena, as described in Chapter 5. Figure 3.18a shows a schematic illustration of an extra (or missing) half plane introduced into a boule crystal. In this case, a dislocation with a Burgers vector of $[11\bar{2}0]/3$ exists along the edge of the extra half plane. As seen from the figure, the dislocation lying in the basal plane (line AB) is defined as a "BPD" (pure edge-type), and the dislocation lying along the $\langle 0001\rangle$ direction (line BC) is

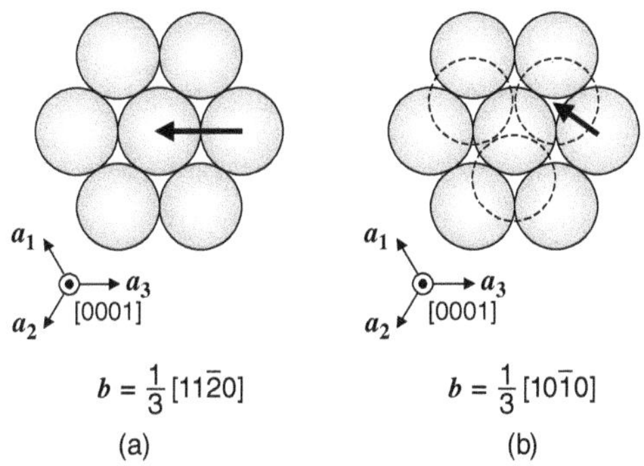

Figure 3.17 Two slip vectors in the basal plane of a close-packed system. (a) <11$\bar{2}$0>/3 and (b) <1$\bar{1}$00>/3.

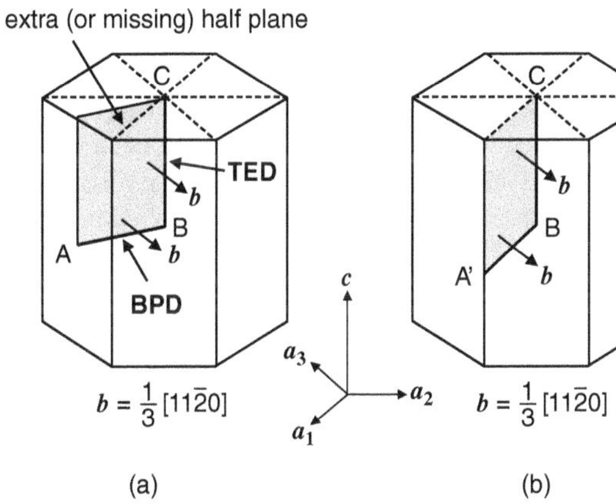

Figure 3.18 (a) Schematic illustration of an extra (or missing) half plane introduced into a SiC crystal. (b) Typical configuration of threading edge and basal plane dislocations, where the basal plane dislocation lies along a <11$\bar{2}$0> direction.

defined as a "TEDn". Therefore, "BPD" and "TED" have the same basic nature; the name simply differs depending on the dislocation direction. In fact, inside the boule crystals, conversion from BPD to TED and from TED to BPD is often observed [61, 68]. Note that pure edge-type BPDs are not very abundant, and quite often BPDs lie along the <11$\bar{2}$0> directions, as shown in Figure 3.18b, probably due to the *Peierls potential* [58]. In this particular case (Figure 3.18b), the BPD (line A′B) is a 60° dislocation. (The angle between the Burgers vector and the dislocation line is 60°.)

Both TEDs and BPDs in a seed are replicated in the SiC boule, though dislocation conversion may occur, as mentioned above. It is, however, more important to consider the nucleation of dislocations (*grown-in dislocations*) to reduce the dislocation density. As in the case of nucleation of screw dislocations, a pair of edge dislocations is generated at a faulted region or a surface precipitate, especially during the initial stage of sublimation growth [66]. When the partial dislocation encompassing the stacking fault is a Shockley type, the resulting threading dislocation will have an edge character. Conversely, a Frank-type partial dislocation will result in a TSD. In addition to the kinetics during the initial stage of growth, *thermal stress* is another important factor for dislocation nucleation. BPDs are relatively easily introduced into the growing boule when the resolved shear stress exceeds a certain (critical) value. A major source of the stress is thermal stress, which develops during sublimation growth when the temperature profile is not appropriate. Both radial and axial temperature gradients cause inhomogeneous thermal expansion inside the boule. Furthermore, the difference in the thermal expansion coefficients of the SiC and graphite parts causes severe thermal stress during cooling. Figure 3.19 shows a schematic illustration of the shear stress, the associated dislocations, and the bending of a basal plane in a growing SiC boule, taking into account typical radial and axial temperature gradients [69]. The temperature is higher along the periphery of the boule than in the center because of radiation from the crucible walls. The temperature of the growing surface is higher than the seed temperature because of the temperature gradient designed to promote mass transport from the source to the seed. Under these circumstances, thermal expansion is not uniform inside the boule; this causes significant thermal stress and bending of basal planes. Consider the components of stress inside a boule grown along <0001> (Figure 3.20). The resolved shear stress (σ_{RZ}) along <11$\bar{2}$0> inside a basal plane is expressed by [58]:

$$\sigma_{RZ} = (\sigma_{rr} + \sigma_{rz}) \cos\phi - \sigma_{\phi\phi} \sin\phi \tag{3.11}$$

where σ_{rr}, σ_{rz}, and $\sigma_{\phi\phi}$ are the components of shear stress, as shown in Figure 3.20 [20]. The resolved shear stress is a direct cause of dislocation nucleation, while too high a value of $\sigma_{\phi\phi}$ can also lead to cracking of SiC boules.

Figure 3.21 shows the critical stress in 6H-SiC as a function of temperature [70]. One has to consider two different critical stresses, (i) the *critical shear stress* resolved to a basal plane, which induces *basal plane slip* and (ii) the *critical normal stress,* which induces *prism plane slip.* In SiC, the critical shear stress along a basal plane is much smaller, and the value decreases greatly at high temperature. Therefore, at the temperature of sublimation growth (over 2200 °C), the critical shear stress becomes

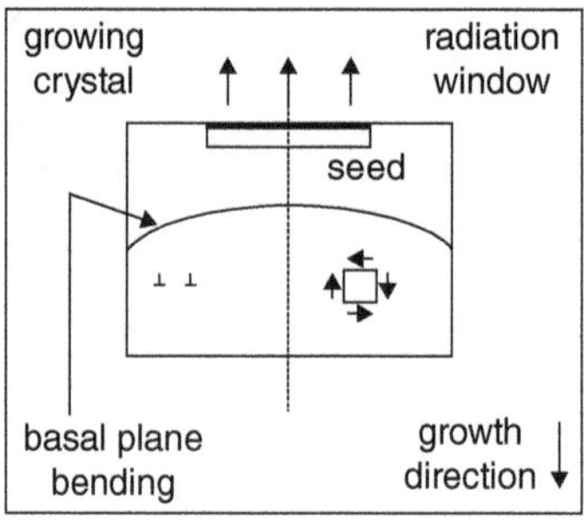

Figure 3.19 Schematic illustration of the shear stress, the associated dislocations, and the bending of a basal plane in a growing SiC boule, taking into account typical radial and axial temperature gradients ([69] reproduced with permission from Trans Tech Publications).

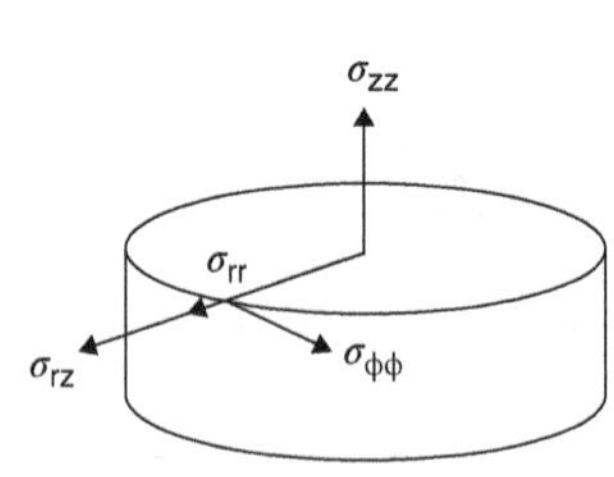

Figure 3.20 Components of stress inside a boule grown along <0001>.

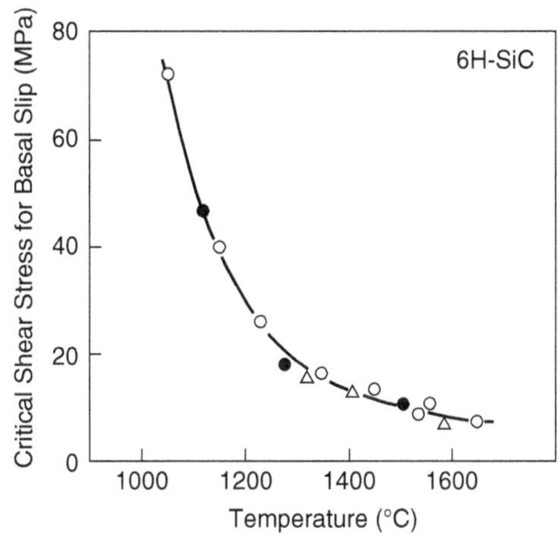

Figure 3.21 Critical stress in 6H-SiC as a function of temperature ([70] reproduced with permission from Taylor & Francis).

very low, and BPDs are easily introduced into SiC boules. In fact, arrays of BPDs are often observed in SiC boules; these can be ascribed to the basal slip bands [71]. Prismatic slip bands are also observed, but with smaller density [72]. However, Wellmann and coworkers discovered that the strain relaxation mechanism is different between n-type and p-type SiC crystals [73]. The BPD density is extremely low in p-type SiC, and thus thermoelastic strain relaxation in p-type SiC may take place, favoring the generation of TSDs instead of BPDs. This phenomenon is explained by doping-induced change of lattice hardness and electrostatic energy consideration [24]. Furthermore, multiplication of BPDs can take place via the Frank–Read mechanism [58], when the thermal stress is significant [74]. It is found that the densities of BPDs and TSDs exhibit a positive correlation; SiC crystals that contain more TSDs generally show a higher BPD density [74]. This correlation can be attributed to the multiplication process of BPDs via interaction between gliding BPDs and TSDs. For further reduction of the residual stress, ingot anneal or wafer anneal have been investigated. Most TEDs are formed by conversion from BPDs along the growth direction during growth.

Crystal mosaicity was commonly observed in SiC boules [75, 76]. Mosaicity comes from slightly misoriented domains bordered by regions with high dislocation density. When this is the case, edge dislocation walls aligned along the <$1\bar{1}00$> directions are present, especially near the periphery of SiC wafers [77]. Through improvement of crucible geometry and process conditions, mosaicity is now greatly reduced.

Figure 3.22 shows a schematic illustration of dislocation networks observed in SiC boules, as revealed by plane-view and cross-sectional synchrotron X-ray topography [68]. Threading dislocations are mostly propagating along the <0001> direction, and BPDs often connect neighboring TSDs. BPDs tend to align along the <$11\bar{2}0$> directions. TEDs and BPDs are transformed into each other, as described above.

3.3.5 *Defect Reduction*

One of the most striking techniques to reduce extended defects in SiC boules is the so-called "repeated *a*-face growth (RAF)" method [78]. Figure 3.23 schematically illustrates the RAF process. The main concept is the preparation of an almost dislocation-free seed and subsequent sublimation growth on the high-quality seed under stabilized conditions. The first step is normal sublimation growth on a SiC{0001} seed. By slicing the boule parallel to the growth direction, a SiC($11\bar{2}0$) (or ($1\bar{1}00$)) plate is obtained

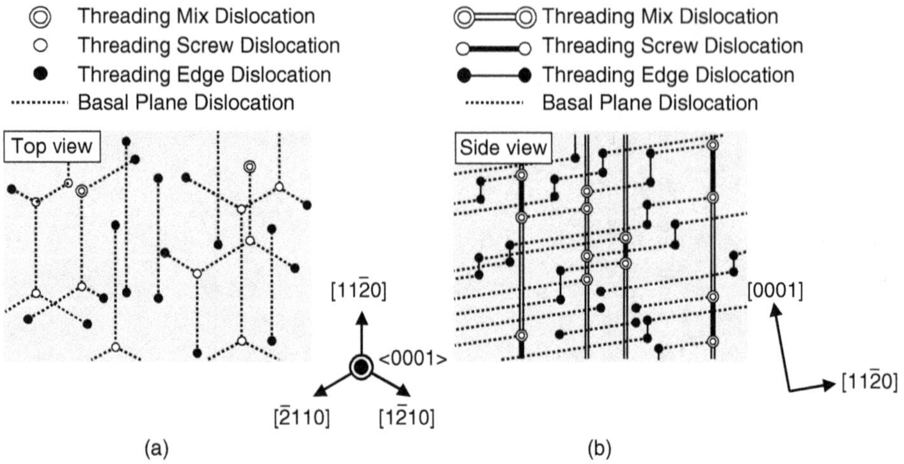

Figure 3.22 Schematic illustration of dislocation networks observed in SiC boules, as revealed by (a) plane-view and (b) cross-sectional synchrotron X-ray topography ([68] reproduced with permission from Elsevier).

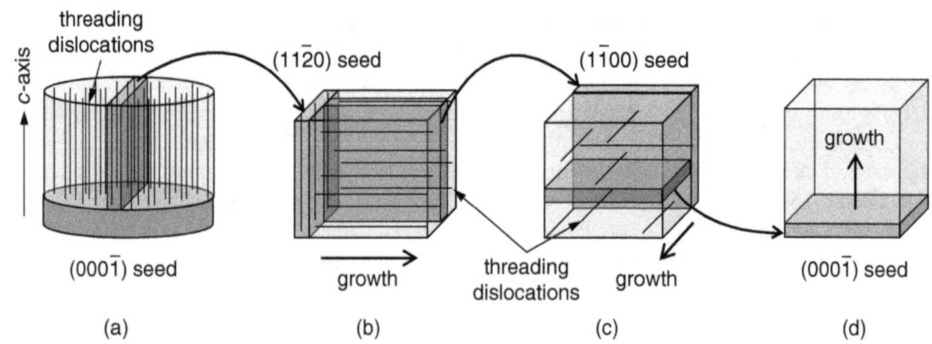

Figure 3.23 (a–d) Schematic illustration of the repeated *a*-face process in SiC boule growth.

(Figure 3.23a). Because most dislocations are propagating along the growth direction, only a limited number of dislocations (mainly BPDs) appear on the surface of this SiC(11$\bar{2}$0) (or (1$\bar{1}$00)) plate. Then, sublimation growth is performed using this plate as the seed, which gives a SiC(11$\bar{2}$0) (or (1$\bar{1}$00)) boule crystal. Inside this SiC boule, the density of elementary screw dislocations ($\boldsymbol{b} = 1\boldsymbol{c}$) and TEDs is low, and the majority of dislocations are BPDs propagating along the growth direction (<11$\bar{2}$0> or <1$\bar{1}$00>). Next, an SiC(1$\bar{1}$00) (or (11$\bar{2}$0)) plate is prepared by slicing the SiC(11$\bar{2}$0) (or (1$\bar{1}$00)) boule, as shown in Figure 3.23b. In this SiC plate, elementary screw dislocations ($\boldsymbol{b} = 1\boldsymbol{c}$) are almost eliminated, while some BPDs remain. And again, sublimation growth is carried out on this SiC plate, which results in a SiC boule grown with an extremely low dislocation density (though some BPDs and SFs remain). Then an off-axis SiC{0001} plate is obtained by slicing the SiC(1$\bar{1}$00) (or (11$\bar{2}$0)) boule (Figure 3.23c). This SiC plate still contains BPDs (and stacking faults) but contains almost no TSDs and TEDs (and no micropipes). As a final step, sublimation growth is performed on the off-axis SiC{0001} seed, under stable, and optimized conditions. As shown in Figure 3.23d, the BPDs in the off-axis SiC{0001} seed mostly propagate in the basal plane of the grown boule. Above this region, an almost dislocation-free crystal can be grown, as

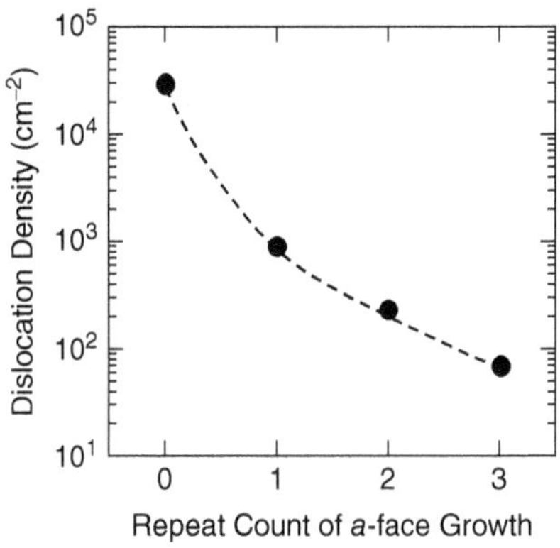

Figure 3.24 Total density of dislocations as a function of the number of *a*-face (or *m*-face) growth steps [78].

long as continuous supply of steps is assured. In particular, the density of threading screw dislocations can be extremely low. This is one reason why an off-axis {0001} seed is employed to ensure polytype replication in the boule crystal. Once this RAF process is successful, one does not have to repeat these complicated processes because high-quality SiC{0001} seed crystals can be directly obtained by slicing the high-quality boule. Figure 3.24 shows the total density of dislocations as a function of the number of *a*-face (or *m*-face) growth steps [78]. The dislocation density shows rapid decrease as the growth is repeated on $(11\bar{2}0)$ or $(1\bar{1}00)$ faces. An impressively low total dislocation density of 75 cm^{-2}, was achieved by employing this technique [78].

3.4 Doping Control in Sublimation Growth

3.4.1 Impurity Incorporation

For fabrication of vertical devices, low-resistivity wafers are required to minimize the series resistance, while high-resistivity wafers are desired for fabrication of lateral high-frequency devices, to reduce the parasitic impedance. In the sublimation growth of SiC, nitrogen and aluminum are the dopants of preference for growth of n- and p-type boules, respectively. Semi-insulating SiC boules can also be obtained. However, the purity or background impurity density, of SiC boules grown by the sublimation method should be carefully taken into account. The purity of SiC boules grown by sublimation is strongly dependent on the purity of the SiC source and graphite parts. The nature and amount of impurity atoms (i.e., those that are not the desired dopants), vary for different manufacturers; typical impurities include Ti, V, Cr, Fe, Co, Ni, and S [17]. The density of these metallic impurities is in the range of $10^{13}-10^{15}$ cm^{-3}. The density of these impurities in the grown boules is usually lower than that of the source by a factor of 2–100. As well as the metallic impurities, N, B, and Al are commonly incorporated in the boule crystals, at densities of $10^{14}-10^{16}$ cm^{-3}, even in nondoped growth. Therefore, nondoped SiC boules can be either n- or p-type, depending on the growth conditions and the environmental purity. The net doping density of nondoped boules ranges from a mid 10^{15} to a mid 10^{16} cm^{-3}.

In general, the dopant incorporation in SiC sublimation growth follows the trends observed for CVD of SiC: nitrogen incorporation is significantly higher for sublimation growth on $(000\bar{1})$ than on (0001), and the opposite tendency (higher on (0001)) occurs for aluminum incorporation. This polarity effect originates from the surface kinetics, and is independent of the gas-phase composition. Because nitrogen substitutes at the carbon lattice site, a nitrogen atom adsorbed onto a $(000\bar{1})$ surface is bound to three underlying silicon atoms, while it is only bound to one silicon atom on a (0001) surface. Thus, desorption of the nitrogen atoms absorbed on $(000\bar{1})$ must be much less than on (0001) (note that the nitrogen vapor pressure is very high at the growth temperature). This is the main reason why nitrogen incorporation is higher on $(000\bar{1})$ [79]. The higher aluminum incorporation on (0001) can be explained in a similar manner. Impurity incorporation is also influenced by the C/Si ratio, as in the case of SiC CVD (*site competition effect* [80]). It is, however, difficult to control the C/Si ratio during standard sublimation growth, independently of the other process parameters. For example, nitrogen incorporation usually decreases with increasing growth temperature while aluminum incorporation increases. This can be attributed to the shift in the effective C/Si ratio caused by changing the growth temperature.

Because of the wide bandgaps of SiC, some, or perhaps even all, visible light is not absorbed by SiC. The wavelength of absorbed light is determined by the bandgap, the major impurity levels, and the intra-band excitation levels. The carrier absorption in the conduction band is well known for most SiC polytypes [81], and occurs at about 460 nm (blue light) for n-type 4H-SiC and 620 nm (red light) for n-type 6H-SiC. High-purity 4H- and 6H-SiC, which possess very wide bandgaps, are colorless and transparent, like glass. However, n- or p-type doping causes carrier absorption in the visible region. Its individual light absorption and transmission characteristics give each SiC crystal a unique color. Table 3.3 summarizes the colors of major SiC polytypes, for different types of doping. Because the color of SiC crystals becomes darker when the dopant density is increased, the color and its darkness are good indications of the dopant type and density. The color of SiC crystals is also a good sign to identify the polytype for n-type materials.

The dopant density is usually higher in the center region of a SiC{0001} wafer, as observed as the darker color at the center. This is because of the enhanced impurity incorporation that occurs during *facet growth,* as shown in Figure 3.25. During sublimation growth of SiC{0001} boules, a {0001} facet appears near the center of the boule. On the {0001} facet, fast spiral growth takes place, but the growth rate along the <0001> direction is relatively slow. Therefore, impurity incorporation is enhanced on the {0001} facet region. Thus, the dopant density at the center (the facet region) of a wafer is usually 20–50%

Table 3.3 Color of major SiC polytypes.

Polytype	High-purity	n-type	*p*-type
3C-SiC	Yellow	Yellow	Gray-brown
4H-SiC	Colorless	Amber	Blue
6H-SiC	Colorless	Green	Blue
15R-SiC	Colorless	Yellow	Blue

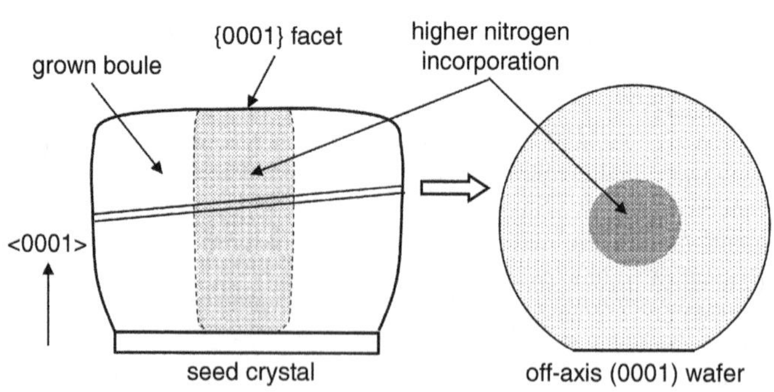

Figure 3.25 Enhanced impurity incorporation that occurs on the {0001} facet during SiC boule growth.

higher than that of the outer region of the wafer. This phenomenon means that radial doping uniformity of SiC{0001} wafers should be improved.

3.4.2 n-*Type Doping*

Nitrogen doping is performed simply by introducing some nitrogen gas into the growth ambient (or atmosphere), using, for example, a mixture of Ar and N_2. The nitrogen density in the grown boule is approximately proportional to the square root of the nitrogen partial pressure during growth, and is almost independent of growth rate. This result indicates that nitrogen incorporation is determined by the equilibrium between nitrogen in the gas phase and nitrogen adsorbed on the growing surface [82]. Figure 3.26 shows the resistivity versus nitrogen dopant density for 4H- and 6H-SiC boules [17, 83, 84]. The nitrogen dopant density can be increased to 10^{20} cm^{-3}, which results in a very low resistivity, 0.005 Ω cm. However, the typical resistivities of commercial n-type 4H-SiC wafers range from 0.015 to 0.025 Ω cm (nitrogen dopant density range: $6 \times 10^{18}-1.5 \times 10^{19}$ cm^{-3}). The electron mobility is rather low, 10–30 cm^2 V^{-1} s^{-1} for 4H-SiC because of the heavy doping. In low-resistivity *n*-type SiC wafers, there are several different deep levels or electron traps, at a relatively high density of $10^{14}-10^{15}$ cm^{-3} [85, 86].

It is known that stacking faults are formed when heavily-doped 4H-SiC wafers are oxidized or annealed in Ar at temperatures higher than 1000–1100 °C [84, 87–91]. Stacking fault formation becomes pronounced when the nitrogen density exceeds $(2-3) \times 10^{19}$ cm^{-3}, and the structure of the stacking fault has been identified as (6,2) in Zhdanov's notation [88, 89]. Another name for this defect is a "*double Shockley stacking fault*". It has been suggested that the electrostatic potential energy of the crystal can be lowered by formation of quantum-well-like stacking faults and subsequent electron trapping near these faults [89]. After several models were proposed and examined, the major trigger for generation of stacking faults is now believed to be stress in the crystal; in particular, polishing-induced damage can act as generation sites [91].

3.4.3 p-*Type Doping*

Aluminum doping is obtained by adding aluminum (or an aluminum-containing compound) to the SiC source. Aluminum doping is much more difficult than nitrogen doping during sublimation growth of SiC because severe depletion of the aluminum source occurs during growth. The aluminum incorporation is

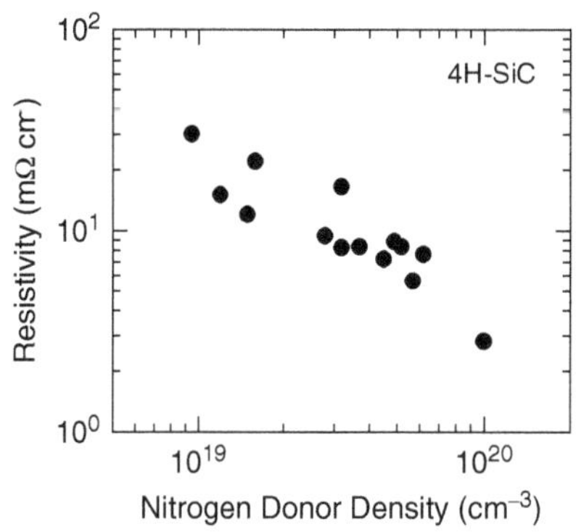

Figure 3.26 Resistivity versus nitrogen dopant density for 4H- and 6H-SiC boules [17, 83, 84].

almost in proportion to the aluminum vapor pressure inside the crucible. In 4H-SiC boule growth, heavy aluminum doping creates a condition favorable for 6H-SiC stabilization, as described in Section 3.2. Thus, sublimation growth of p^+-type 4H-SiC is a challenge. The dopant density and resistivity of typical p-type 4H-SiC wafers heavily doped with aluminum are $(0.5-2) \times 10^{18}$ cm^{-3} and 1–5 Ω cm, respectively, at present. Although deep levels in p-type SiC boules have not been studied in detail, persistent photoconductivity at room temperature is observed for p-type SiC wafers (but not for n-type at room temperature) [92, 93].

3.4.4 Semi-Insulating

High-resistivity wafers are required for fabrication of SiC- or GaN-based high-frequency devices to minimize the parasitic capacitances between the terminals, including the ground. The concept used to produce high-resistivity SiC wafers is basically the same as that employed for III–V compound semiconductors [94]. Because it is very difficult to reduce the background dopant density below 10^{10} cm^{-3} by purification processes, compensation of dopants (donors/acceptors) is used to decrease the density of free carriers in the bands. Consider an n-type material with donor density N_D. When a deep level (electron trap) energetically located at $E_C - E_T$ (E_C: the bottom of the conduction band) is introduced, the Fermi level E_F is changed, depending on the density of the deep level or electron trap (N_T). This is shown schematically in Figure 3.27. When the trap density is much smaller than the donor density ($N_D >> N_T$), as shown in Figure 3.27a, only a small portion of the electrons supplied from the donors is trapped, and the Fermi level is located slightly below the donor level. When the trap density is close to (but slightly lower than) the donor density, the Fermi level moves to an energy level between the donor level and the trap level. The situation changes dramatically when the trap density is sufficiently higher than the donor density ($N_D < N_T$): all the electrons supplied from the donors are captured by the traps, and the Fermi level is pinned near the trap level, as shown in Figure 3.27b. Although some electrons can be thermally excited from the deep levels to the conduction band, the free electron density is lower than the donor density by many orders of magnitude. When the trap level is deep enough, this compensated material shows *semi-insulating* properties.

In a nondegenerate semiconductor including a compensated material (such as that described above), the free electron density (n) can be estimated by using classical carrier statistics for a semiconductor [95]:

$$n = N_C \exp\left(-\frac{E_C - E_F}{kT}\right) \tag{3.12}$$

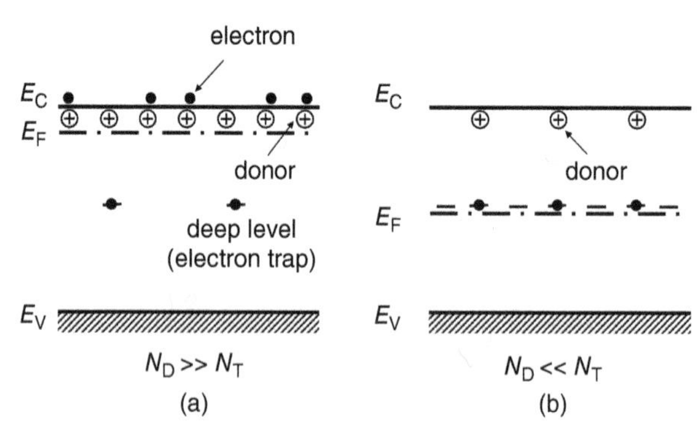

Figure 3.27 Schematic band diagram of an n-type semiconductor. (a) $N_D >> N_T$ and (b) $N_D < N_T$ (N_D: donor density, N_T: trap density).

where N_C is the effective density of states in the conduction band, k the Boltzmann constant, and T the absolute temperature. In this n-type semiconductor, the free hole density (p) is $p = n_i^2/n$, where n_i is the intrinsic carrier density. Thus, the resistivity of this semiconductor (ρ) is:

$$\rho = 1/(qn\mu_e + qp\mu_h) \tag{3.13}$$

here, μ_e and μ_h are the mobilities for electrons and holes, respectively. In the same manner, the free hole density (p) in a compensated p-type semiconductor is:

$$p = N_V \exp\left(-\frac{E_F - E_V}{kT}\right) \tag{3.14}$$

where N_V is the effective density of states in the valence band and E_V is the energy of the top of the valence band. Figure 3.28 shows the majority carrier density in the band and the estimated resistivity, as a function of the position of the Fermi level within the bandgap. 4H-SiC at a temperature of 300 K is considered in this figure; the mobilities for electrons and holes were taken as 500 and 50 $cm^2\ V^{-1}\ s^{-1}$, respectively (these values vary, depending on the impurity density, but this variation does not have a major impact on the estimated resistivity). One must bear in mind that the Fermi level is pinned at the deep level responsible for compensation, as described above. The estimated resistivity at room temperature is about 10^5 Ω cm when the Fermi level is located at $E_C - 0.5$ eV (or $E_V + 0.4$ eV), indicating semi-insulating properties. Owing to the wide bandgap of SiC, the resistivity can exceed 10^{15} Ω cm when the Fermi level is located at $E_C - 1.1$ eV (or $E_V + 1.0$ eV) or deeper. Therefore, the main approach to obtain semi-insulating boule crystals is (i) to reduce the background impurities as much as possible and (ii) to intentionally introduce deep levels, which can then act as efficient trap centers (energetically deeper levels are preferable).

In SiC, *vanadium* was the first element used as a compensating center to create semi-insulating wafers [96, 97]. Vanadium is an *amphoteric impurity* in SiC; it acts as an acceptor-like (−/0) trap in n-type SiC and a donor-like (0/+) trap in p-type SiC. The acceptor levels of vanadium are $E_C - 0.65/0.72$ eV in 6H-SiC and $E_C - (0.81\text{–}0.97)$ eV in 4H-SiC, while the donor levels of vanadium are estimated as $E_V + (1.3\text{–}1.5)$ eV for both 6H- and 4H-SiC [98–101]. Because the donor level is deeper for vanadium in SiC, vanadium doping was performed for slightly p-type SiC boules to achieve very high resistivity. The resistivity of the semi-insulating SiC wafers is about 10^{12}–10^{15} Ω cm at room temperature, and Arrhenius plots of the resistivity give activation energies of 1.2–1.4 eV [16, 96, 97].

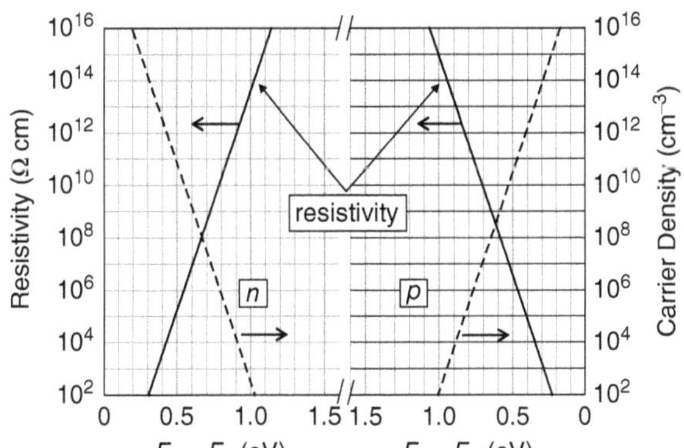

Figure 3.28 Majority carrier density in the band and the estimated resistivity, as a function of the position of the Fermi level within the bandgap of 4H-SiC.

The solubility limit of vanadium in SiC is not very high, in the mid 10^{17} cm^{-3}, while the residual dopant density in SiC boules can be 10^{16} cm^{-3}. Thus, precise control of vanadium doping is required during sublimation growth.

To overcome the difficulty in vanadium doping, use of intrinsic point defects, which create deep levels, has been investigated, and *high-purity semi-insulating* (HPSI) wafers have now been obtained [102, 103]. In this case, the density of residual dopants (especially nitrogen, boron, and aluminum) must be reduced by both source purification and the use of high-purity graphite. Intrinsic point defects can be introduced by adjusting the growth conditions or by high-energy particle irradiation after growth [104]. The obtained resistivity exceeds 10^{12} Ω cm at room temperature; the activation energy of the resistivity was found to vary significantly for different wafers, and ranges from 0.8 to 1.5 eV [102, 103]. Therefore, it is considered that there exist several different deep levels (point defects) responsible for the semi-insulating behavior. Recent studies based on electron paramagnetic resonance (EPR) measurements revealed that (i) silicon vacancies (V_{Si}) are the dominant traps in HPSI wafers with activation energies of 0.8–0.9 eV, (ii) carbon antisite–carbon vacancy pairs (C_{Si}-V_C) or carbon vacancies (V_C) are dominant in wafers with an activation energy of 1.1–1.3 eV, and (iii) V_C or divacancies (V_C-V_{Si}) are the main traps in wafers with an activation energy of 1.5 eV [105, 106]. After annealing HPSI wafers of types (i) and (ii) mentioned above at 1600 °C, the resistivity and the activation energy decrease to 10^5–10^6 Ω cm and about 0.6 eV, respectively [106]. However, the type (iii) HPSI wafers, which originally had an activation energy of 1.5 eV, are more thermally stable.

3.5 High-Temperature Chemical Vapor Deposition

To overcome the limitations of the sublimation method of SiC, high-temperature chemical vapor deposition (HTCVD) was proposed and has been developed [107–109]. Figure 3.29 schematically illustrates a reactor and an approximate temperature profile used for HTCVD of SiC. The SiC boule growth is performed in a vertical crucible made of graphite, where the precursor gases are fed upwards through

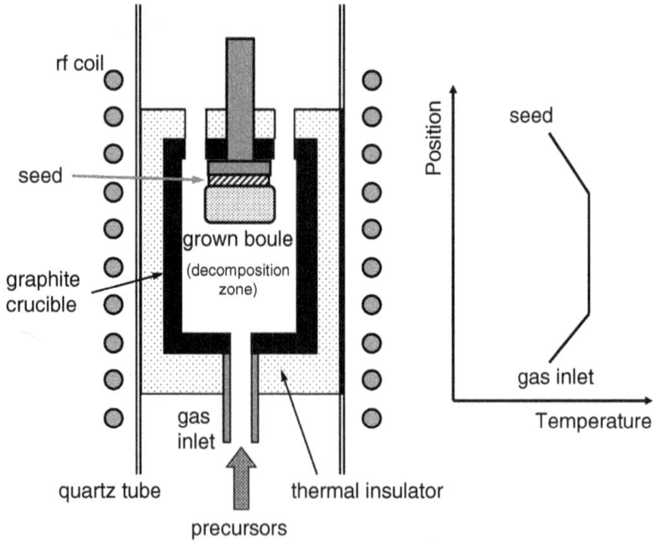

Figure 3.29 Schematic illustrations of a reactor and approximate temperature profile used for HTCVD of SiC.

a heating zone to the seed crystal holder placed at the top. The precursor gases are SiH_4 and a hydrocarbon, such as C_2H_4 and C_3H_8, diluted in a carrier gas. The geometry is similar to that of the vertical CVD reactors used for epitaxial growth, but the typical growth temperature is extremely high, 2100–2300 °C. Inside the hot zone, the precursors are completely decomposed and several reactions proceed. As a result, Si and SiC clusters are formed via homogeneous nucleation because of the high supersaturation in the gas phase. These clusters act as a virtual source for SiC boule growth on the seed. Therefore, it is important to establish an appropriate temperature gradient from the gas inlet to the seed. The temperature of the gas-cracking zone and the walls should be slightly higher than that of the seed crystal, to ensure mass transport and condensation on the seed. The choice of the carrier gas is also very important. The carrier gas should not attack nor react with the graphite walls, even at very high temperature. The carrier gas should be immediately heated so as not to form a cold jet onto the seed surface. Helium can fulfill these requirements, and is usually employed in HTCVD of SiC (hydrogen carrier gas is also employed by a few groups). The typical growth pressures and growth rates are 200–700 m bar and 0.3–0.7 mm h^{-1}, respectively [107–109]. The HTCVD process can benefit from the knowledge and insights already obtained for sublimation growth of SiC boules and epitaxial growth of SiC by normal CVD. Modeling and simulation of heat transfer, chemical reactions, and the growth process have been developed [110]. In some cases, chlorinated precursors are employed to reduce homogeneous nucleation in the gas phase; this growth process then becomes very similar to that found in pure CVD [111].

The major advantages of HTCVD compared with the sublimation method are summarized as follows:

1. *High purity*: The source purity is obviously much higher in HTCVD, and the residual impurity density can be reduced by 1–2 orders of magnitude. The density of residual impurities is in the mid 10^{14} cm^{-3} for nitrogen and boron, and is much lower for other impurities. Because of this purity, it is relatively easy to produce HPSI wafers by HTCVD [112].
2. *C/Si control*: In sublimation growth, the C/Si ratio cannot be changed as an independent process parameter. In HTCVD, however, the C/Si ratio, at least at the gas inlet, can be controlled independently across a relatively wide range. A low C/Si ratio is preferable for enhancing micropipe closing [56] or to increase nitrogen incorporation, while a high C/Si ratio is effective in decreasing nitrogen incorporation. Formation of intrinsic point defects can also be controlled by controlling the C/Si ratio during growth.
3. *Continuous supply of the source material*: In principle, the supply of source material in HTCVD is very stable in terms of both the absolute amount and the C/Si ratio, and it can be maintained for a very long period without source depletion. This should enable growth of very long SiC boules of uniformly high quality. In actual growth systems, however, a few technical problems remain to be solved, such as inlet and outlet closure during long growth. Stable supply of dopant impurities, which leads to uniform doping along the growth direction, is another advantage of HTCVD. In particular, very stable aluminum doping is achieved in HTCVD by using trimethylaluminum (TMA) as a dopant source. In n-type doping, nitrogen gas is mixed into the precursors.

Simulation of HTCVD has also been performed, including heat transfer, gas phase reactions, and surface reactions [110]. The main species in the gas phase obtained in the simulation are Si, Si_2C, SiC_2, and C_2H_2. Significant etching of SiC by hydrogen at very high temperature should be considered. Figure 3.30 shows the temperature dependence of the growth rate during HTCVD of SiC [110]. Because the simulated data show good agreement with experiments, simulation is a powerful tool for designing HTCVD reactors and processes.

Although reports on HTCVD of SiC are still limited compared with sublimation growth, this method possesses much potential for production of high-quality, long SiC boules. At present, at least one manufacturer is producing SiC wafers using this technique.

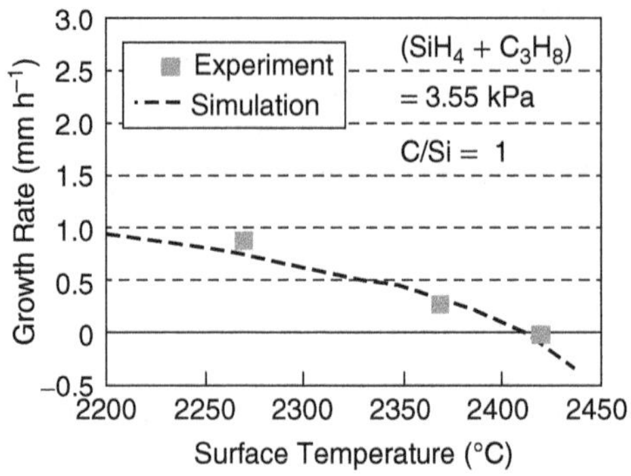

Figure 3.30 Temperature dependence of the growth rate during HTCVD of SiC ([110] reproduced with permission from Trans Tech Publications).

3.6 Solution Growth

Boule growth from the congruent melt (solution) is a standard technique for preparing ingots of other semiconductors [113]. One might expect solution growth of SiC to be a very promising way of obtaining high-quality boules because the supersaturation at the growing surface could be controlled well once the technology becomes mature. Figure 3.31 shows a schematic illustration of a furnace and temperature profile used for solution growth of SiC. A graphite crucible is filled with Si-based melt, and the seed crystal is placed in contact with the melt surface (*top seeded solution growth*, TSSG). The seed temperature is slightly lower than the melt, and this provides a driving force for growth. Growth is carried out in an inert atmosphere, such as under Ar. The seed and crucible are usually rotated in opposite directions. The growth temperature ranges from 1750 to 2100 °C.

In spite of the great promise, solution growth of SiC faces several difficulties. As described in Section 3.1.1, there exists no stoichiometric SiC liquid phase at atmospheric pressure, and the solubility

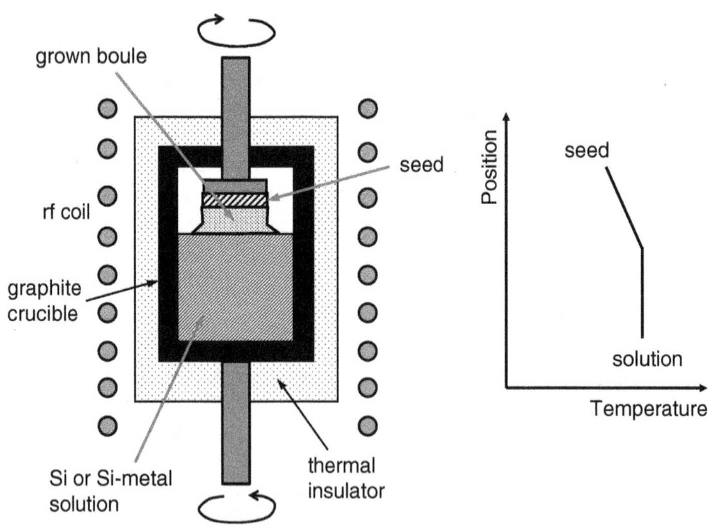

Figure 3.31 Schematic illustration of a furnace and temperature profile used for solution growth of SiC.

of carbon in the Si melt is only 15%, even at 2800 °C. At such high temperature, evaporation of Si is significant because of the high Si vapor pressure, making continuous growth almost impossible. Furthermore, the Si melt reacts significantly with the graphite crucible (which in turn acts as a carbon source for the growth), which presents another challenge for long growth. Solution growth of SiC boules is still in the early stages of development and has only been investigated intensively in recent years. Although large-diameter SiC wafers have not yet been produced using this technique, progress in this field is remarkable. A few approaches are briefly described below:

1. *High-pressure solution growth*: Because the Si evaporation can be suppressed at high pressure, solution growth under high Ar pressure (~100 bar) was investigated [114]. The growth rate is below 0.5 mm h^{-1} at 2200–2300 °C. The diameter and length of the SiC boule crystals grown are very limited at present.
2. *Solution growth with metal-added solvent*: It is known that the carbon solubility can be increased by adding rare-earth or transition metals such as Sc, Pr, Fe, Ti, or Cr. By using Si-Sc-C or Si-Ti-C solvent, for example, reasonable growth rates of about 0.3 mm h^{-1} were attained at relatively low temperatures of 1750–1900 °C [115–119]. These studies already demonstrate low dislocation density in the grown crystals. Micropipes are easily closed, and TSDs are mostly converted to Frank-type stacking faults in the basal planes [120]. Metallic contamination from the solvent could be a problem, and this subject is now being investigated.

By controlling the meniscus near the growing surface, remarkable diameter enlargement has been achieved without degradation of quality [121]. A trade-off between high growth rate and growth instability is currently being investigated.

Progress in solution growth of SiC bulk crystals is so rapid that the readers are encouraged to survey the latest papers in journals and conferences. For example, a "*dislocation free*" 4H-SiC boule was successfully demonstrated by solution growth on a $(000\bar{1})$ seed sliced from a $<1\bar{1}00>$-grown boule [122]. Though the crystal size is still small, this must be a milestone in the development of SiC crystal growth.

3.7 3C-SiC Wafers Grown by Chemical Vapor Deposition

Because 3C-SiC is unstable at very high temperature, growth of large 3C-SiC bulk crystals by the sublimation method is difficult. Instead, free-standing 3C-SiC films have been demonstrated by fast epitaxial growth of 3C-SiC on Si wafers, and subsequent etching of Si [123, 124]. In general, 3C-SiC layers grown on Si substrates contain a high density of stacking faults and microtwins, resulting from large mismatches of the crystal lattice (20%) and thermal expansion coefficient (8%) [125, 126]. Nagasawa and coworkers developed the technique to grow a very thick (~200 μm) 3C-SiC (001) layer with reduced density of stacking faults on a 6 in. Si wafer [124]. The growth of 3C-SiC is carried out by CVD at 1350 °C, using SiH_2Cl_2 and C_2H_2 as precursors, which yields a growth rate of 40 μm h^{-1}. The {111} stacking faults are considerably reduced by using an "undulant" Si substrate, where a submicron-scale ridge-valley structure is formed on the surface. The details of stacking fault reduction are described in the literature [123]. The crystal quality is much improved, compared with that of normal 3C-SiC films heteroepitaxially grown on Si, but further reduction in defects is required for production of high-performance electronic devices.

3.8 Wafering and Polishing

The process for production of SiC wafers from boule crystals is basically the same as that used for other semiconductors. SiC boules are usually grown on SiC {0001}, and the grown boules are cylindrical in shape, with a length of 20–50 mm. The crystallographic orientations of the boules, the exact $<0001>$, $<11\bar{2}0>$, and $<1\bar{1}00>$ directions, are accurately determined by X-ray diffraction. After this

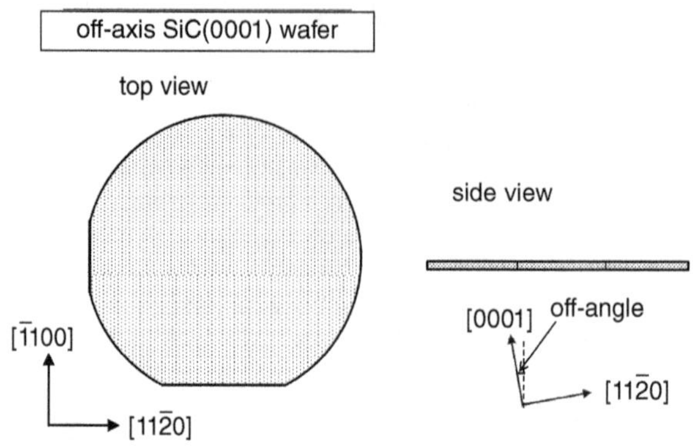

Figure 3.32 Schematic illustration of a standard SiC (0001) wafer defined by the SEMI standard.

process, the boules are sliced into a number of wafers which are carefully polished and cleaned before delivery.

Figure 3.32 shows a schematic illustration of a standard SiC (0001) wafer defined by the SEMI standard [127]. The prime and secondary orientation flats of the wafer are formed along [11$\bar{2}$0] and [1$\bar{1}$00], as shown in the figure. For subsequent SiC epitaxial growth and device fabrication, the off-axis (off-angle) is introduced to ensure high-quality homoepitaxial growth [128]. When the off-axis is introduced, the [0001] axis is tilted toward [$\bar{1}\bar{1}$20]. The off-axis is typically 4° toward [11$\bar{2}$0], and there is a trend that the off-angle becomes smaller with progress in epitaxial growth technology. Note that the SiC growth behavior and device performance are not perfectly isotropic inside the wafer because of the large anisotropy of the physical properties of SiC. It is important to keep it in mind that physical phenomena in epitaxial growth, device fabrication processes, and device operation depend on these orientations. Wafers to be used for other purposes, such as GaN growth, are usually cut on-axis (0001).

Because of the exceptional hardness and chemical inertness of SiC, both slicing and polishing are challenges. The SiC boules are usually sliced using a multi-wire saw embedded with diamond abrasives. Slicing using electric discharge is currently being investigated to improve the process speed and to reduce slicing-induced damage [129].

The quality of the wafer surface after the polishing process is critical for high-quality epitaxial growth with suppressed generation of extended defects. The surface quality includes the *flatness*, the *sub-surface dislocations*, and the *residual stress*. Many groups have reported and confirmed that improved surface flatness of SiC substrates naturally results in superior surface flatness of epitaxial layers grown on these substrates. To suppress defect generation during the initial stage of epitaxial growth, the surface must be stress-free, without sub-surface dislocations. If the residual damage near the surface is not sufficiently removed, epitaxial growth on the substrate leads to the generation of macroscopic defects, including 3C-SiC nucleation, generation of extended defects (dislocation half loops, etc.), and severe step bunching. It is difficult to completely remove the surface damage by *in situ* H_2 or HCl/H_2 etching prior to epitaxial growth because the defect generation and/or macrostep bunching have already begun during the heating process, and proceed further during high-temperature etching. Therefore, it is important to know the removal rate, the thickness of the removed layer, and the depth of any damaged layer for a particular polishing process. In general, the polishing process consists of several steps: it starts with a few mechanical polishing steps using abrasive powders of progressively smaller size, and finishes with *chemo-mechanical polishing* (CMP). The CMP process is particularly important for SiC because wet etching is almost impossible. CMP of SiC is carried out using a colloidal silica slurry at a slightly elevated temperature (~50 °C) and at high pH (>10) [130] or by adding H_2O_2 [131]. The recipe of the CMP process is modified in individual groups [132]. Note that SiC wafers are usually polished on both front and

Table 3.4 Latest results and technological aspects of major techniques for growth of SiC boules.

Aspects	Seeded sublimation	HTCVD	Solution growth
Diameter (mm)	+ + (150)	+(100)	−(<100)
Growth rate (mm h^{-1})	+(~0.5)	+(~0.4)	−(~0.1)
Boule length	+	+	−
Defect reduction	+	+	++
Purity (cm^{-3})	+(~10^{16})	+ + (~10^{14})	+(~10^{16})
n-type doping (cm^{-3})	+(~10^{19})	+(~10^{19})	+(~10^{19})
p-type doping (cm^{-3})	−(~10^{18})	+(~10^{18})	+ + (~10^{19})
Process control	+	++	+
Cost	++	−	+

back sides to minimize wafer warpage due to the *Twyman effect* [133]. In recent years, a catalyst-referred etching (CARE) method has been proposed [134]. In this technique, no abrasives are used; instead, SiC is removed by catalytic etching. Preparation of atomically flat surfaces without dislocation pits has been demonstrated [135].

3.9 Summary

Table 3.4 summarizes the latest results and technological aspects of major techniques for the growth of SiC boules [22–24, 136, 137]. At present, the seeded sublimation method is the most mature, and is the choice for SiC wafer production because of superior growth rate, process stability, and cost. Understanding of thermodynamic phenomena during sublimation growth has been rapidly updated [138]. However, progress in the other methods is also so rapid that one cannot predict which process will be adopted in the future. In particular, a reasonable growth rate and high quality have been achieved simultaneously using solution growth. If a stable and continuous supply of source materials can be established in solution growth, it will become a realistic method to produce long SiC boules with very low dislocation density. The main advantage of HTCVD is high purity; this method has much potential for production of HPSI wafers.

Growth of long SiC boules with low defect density is still one of the most important challenges in SiC technology. Because 150-mm-diameter SiC wafers are now commercially available, diameter enlargement will not be a major problem. Through recent efforts in boule growth, large SiC wafers with a very low dislocation density (TSD density <50 cm^{-2}, TED density <1000 cm^{-2}, BPD density <200 cm^{-2}) have been demonstrated. However, understanding of defect generation and reduction in SiC is still very limited. Generation, movement, and interaction of dislocations in SiC during high-temperature growth and cooling still need to be clarified. Further refinement of growth simulation is required to assist in controlling the thermal stress and surface stoichiometry to allow dislocation engineering in SiC.

References

[1] Scase, R.I. and Slack, G.A. (1960) The Si–C and Ge–C phase diagrams, in *Silicon Carbide – A High Temperature Semiconductor* (eds J.R. O'connor and J. Smiltens), Pergamon Press, p. 24.

[2] Olesinski, R.W. and Abbaschian, G.J. (1984) The C–Si (carbon–silicon) system. *Bull. Alloy Phase Diag.*, **5**, 486.

[3] Knippenberg, W.F. (1963) Growth phenomena in silicon carbide. *Philips Res. Rep.*, **18**, 161.

[4] Drowart, J. and de Maria, G. (1958) Thermodynamic study of SiC utilizing a mass spectrometer. *J. Phys. Chem.*, **29**, 1015.

[5] Kaprov, S.Y., Makarov, Y.N. and Ramm, M.S. (1997) Simulation of sublimation growth of SiC single crystals. *Phys. Status Solidi B*, **202**, 201.

[6] Lely, J.A. (1955) Darstellung von einkristallen von siliziumcarbid und beherrschung von art und menge der eingebauten verunreinigungen. *Ber. Dtsch. Keram. Ges.*, **32**, 229.

[7] Acheson, E.G. (1892) English Patent 17911 Production of artificial crystalline carbonaceous materials, carborundum.

[8] Tairov, Y.M. and Tsvetkov, V.F. (1978) Investigation of growth processes of ingots of silicon carbide single crystals. *J. Cryst. Growth*, **43**, 209.

[9] Tairov, Y.M. and Tsvetkov, V.F. (1981) General principles of growing large-size single crystals of various silicon carbide polytypes. *J. Cryst. Growth*, **52**, 146.

[10] Ziegler, G., Lanig, P., Theis, D. and Weurich, C. (1980) Single crystal growth of SiC substrate material for blue light emitting diodes. *IEEE Trans. Electron Devices*, **30**, 277.

[11] Koga, K., Nakata, T., and Niina, T. (1985) Extended Abstracts 17th Conference Solid State Devices and Materials, Tokyo, Japan, 1985, p. 249 Single crystal growth of 6H-SiC by a vacuum sublimation method.

[12] Kanaya, M., Takahashi, J., Fujiwara, Y. and Moritani, A. (1991) Controlled sublimation growth of single crystalline 4H-SiC and 6H-SiC and identification of polytypes by x-ray diffraction. *Appl. Phys. Lett.*, **58**, 56.

[13] Stein, R.A. and Lanig, P. (1992) Influence of surface energy on the growth of 6H-and 4H-SiC polytypes by sublimation. *Mater. Sci. Eng.,B,*, **11**, 69.

[14] Barrett, D.L., McHugh, J.P., Hobgood, H.M. *et al.* (1993) Growth of large SiC single crystals. *J. Cryst. Growth*, **128**, 358.

[15] Davis, R.F., Carter, C.H., Jr., and Hunter, C.E. (1995) US Patent 34861 Sublimation of silicon carbide to produce large, device quality single crystals of silicon carbide.

[16] Augustine, G., Hobgood, H.M., Balakrishna, V. *et al.* (1997) Physical vapor transport growth and properties of SiC monocrystals of 4H polytype. *Phys. Status Solidi B*, **202**, 137.

[17] Glass, R.C., Henshall, D., Tsvetkov, V.F. and Carter, C.H. Jr., (1997) SiC seeded crystal growth. *Phys. Status Solidi B*, **202**, 149.

[18] Takahshi, J. and Ohtani, N. (1997) Modified-Lely SiC crystals grown in $[1\bar{1}00]$ and $[11\bar{2}0]$ directions. *Phys. Status Solidi B*, **202**, 163.

[19] Vodakov, Y.A., Roenkov, A.D., Ramm, M.G. *et al.* (1997) Use of Ta-container for sublimation growth and doping of SiC bulk crystals and epitaxial layers. *Phys. Status Solidi B*, **202**, 177.

[20] Müller, S.G., Glass, R.C., Hobgood, H.M. *et al.* (2000) The status of SiC bulk growth from an industrial point of view. *J. Cryst. Growth*, **211**, 325.

[21] Pons, M., Madar, R. and Billon, T. (2004) Principles and limitations of numerical simulation of SiC boule growth by sublimation, in *Silicon Carbide – Recent Major Advances* (eds W.J. Choyke, H. Matsunami and G. Pensl), Springer, p. 121.

[22] Chaussende, D., Wellmann, P.J. and Pons, M. (2007) Status of SiC bulk growth processes. *J. Phys. D Appl. Phys.*, **40**, 6150.

[23] Ohtani, N. (2011) Toward the reduction of performance-limiting defects in SiC epitaxial substrates. *ECS Trans.*, **41**, 253.

[24] Sakwe, S.A., Stockmeier, M., Hens, P. *et al.* (2008) Bulk growth of SiC–review on advances of SiC vapor growth for improved doping and systematic study on dislocation evolution. *Phys. Status Solidi B*, **245**, 1239.

[25] Segal, A.S., Vorob'ev, A.N., Karpov, S.Y. *et al.* (1993) Transport phenomena in sublimation growth of SiC bulk crystalsTransport phenomena in sublimation growth of SiC bulk crystals. *Mater. Sci. Eng., B*, **61–62**, 40.

[26] Wellmann, P., Herro, Z., Winnacker, A. *et al.* (2005) In situ visualization of SiC physical vapor transport crystal growth. *J. Cryst. Growth*, **275**, e1807.

[27] Chaussende, D., Ucar, M., Auvray, L. *et al.* (2005) Control of the supersaturation in the CF-PVT process for the growth of silicon carbide crystals: Research and applications. *Cryst. Growth Des.*, **5**, 1539.

[28] Hofmann, D., Heinze, M., Winnacker, A. *et al.* (1995) On the sublimation growth of SiC bulk crystals: development of a numerical process model. *J. Cryst. Growth*, **146**, 214.

[29] Pons, M., Blanquet, E., Dedulle, J.M. *et al.* (1996) Thermodynamic heat transfer and mass transport modeling of the sublimation growth of silicon carbide crystals. *J. Electrochem. Soc.*, **143**, 3727.

[30] Herro, Z.G., Wellmann, P., Pusche, R. *et al.* (2003) Investigation of mass transport during PVT growth of SiC by ^{13}C labeling of source material. *J. Cryst. Growth*, **258**, 261.

[31] Pons, M., Anikin, M., Chourou, K. *et al.* (1999) State of the art in the modelling of SiC sublimation growth. *Mater. Sci. Eng., B*, **61–62**, 18.

[32] Nishizawa, S., Kato, T. and Arai, K. (2007) Effect of heat transfer on macroscopic and microscopic crystal quality in silicon carbide sublimation growth. *J. Cryst. Growth*, **303**, 342.
[33] Chaussende, D., Blanquet, E., Baillet, F. *et al.* (2006) Thermodynamic aspects of the growth of SiC single crystals using the CF–PVT process. *Chem. Vap. Deposition*, **12**, 541.
[34] Morelli, D., Heremans, J., Beetz, C. *et al.* (1993) Carrier concentration dependence of the thermal conductivity of silicon carbide. *Inst. Phys. Conf. Ser.*, **137**, 313.
[35] Nakabayashi, M., Fujimoto, T., Katsuno, M. *et al.* (2009) Carrier concentration dependence of the thermal conductivity of silicon carbide. *Mater. Sci. Forum*, **600–603**, 3.
[36] Hobgood, D., Brady, M., Brixius, W. *et al.* (2000) Status of large diameter SiC crystal growth for electronic and optical applications. *Mater. Sci. Forum*, **338–342**, 3.
[37] Vodakov, Y.A., Lomakina, G.A. and Mokhov, E.N. (1982) SiC nonstoichiometry and polytypism. *Sov. Phys. Solid State*, **24**, 1377.
[38] Itoh, A., Akita, H., Kimoto, T. and Matsunami, H. (1994) High-quality 4H-SiC homoepitaxial layers grown by step-controlled epitaxy. *Appl. Phys. Lett.*, **65**, 1400.
[39] Furukawa, K., Tajima, Y., Saito, H. *et al.* (1993) Bulk growth of single-crystal cubic silicon carbide by vacuum sublimation method. *Jpn. J. Appl. Phys.*, **32**, L645.
[40] Nishino, K., Kimoto, T. and Matsunami, H. (1995) Photoluminescence of homoepitaxial 3C-SiC on sublimation-grown 3C-SiC substrates. *Jpn. J. Appl. Phys.*, **34**, L1110.
[41] Yoo, W.S. and Matsunami, H. (1991) Solid-state phase transformation in cubic silicon carbide. *Jpn. J. Appl. Phys.*, **30**, 545.
[42] Furusho, T., Sasaki, M., Ohshima, S. and Nishino, S. (2003) Bulk crystal growth of cubic silicon carbide by sublimation epitaxy. *J. Cryst. Growth*, **249**, 216.
[43] Fujimoto, T., Tsuge, H., Katsuno, M. *et al.* (2013) A possible mechanism for hexagonal void movement observed during sublimation growth of SiC single crystals. *Mater. Sci. Forum*, **740–742**, 577.
[44] Hong, M.H., Samant, A.V. and Pirouz, P. (2000) Stacking fault energy of 6H-SiC and 4H-SiC single crystals. *Philos. Mag. A*, **80**, 919.
[45] Frank, F.C. (1951) Capillary equilibria of dislocated crystals. *Acta Crystallogr.*, **4**, 497.
[46] Sunagawa, I. and Bennema, P. (1981) Observations of the influence of stress fields on the shape of growth and dissolution spirals. *J. Cryst. Growth*, **53**, 490.
[47] Si, W., Dudley, M., Glass, R. *et al.* (1997) Hollow-core screw dislocations in 6H-SiC single crystals: A test of Frank's theory. *J. Electron. Mater.*, **26**, 128.
[48] Si, W., Dudley, M., Glass, R. *et al.* (1998) Experimental studies of hollow-core screw dislocations in 6H-SiC and 4H-SiC single crystals. *Mater. Sci. Forum*, **264–268**, 429.
[49] Heindl, J., Dorsch, W., Strunk, H.P. *et al.* (1998) Dislocation content of micropipes in SiC. *Phys. Rev. Lett.*, **80**, 740.
[50] Neudeck, P.G. and Powell, J.A. (1994) Performance limiting micropipe defects in silicon carbide wafers. *IEEE Electron Device Lett.*, **15**, 63.
[51] Koga, K., Fujikawa, Y., Ueda, Y. and Yamaguchi, T. (1992) Growth and characterization of 6H-SiC bulk crystals by the sublimation method, in *Amorphous and Crystalline Silicon Carbide IV*, Springer Proceedings of Physics, vol. 71 (eds C.Y. Yang, M.M. Rahman and G.L. Harris), Springer-Verlag, p. 96.
[52] Ohtani, N., Katsuno, M., Fujimoto, T. *et al.* (2001) Surface step model for micropipe formation in SiC. *J. Cryst. Growth*, **226**, 254.
[53] Giocondi, J., Rohrer, G.S., Skowronski, M. *et al.* (1997) An atomic force microscopy study of super-dislocation/micropipe complexes on the 6H-SiC(0001) growth surface. *J. Cryst. Growth*, **181**, 351.
[54] Pirouz, P. (1998) On micropipes and nanopipes in SiC and GaN. *Philos. Mag. A*, **78**, 727.
[55] Yakimova, R., Tuominen, M., Bakin, A.S. *et al.* (1996) Silicon carbide liquid phase epitaxy in the Si-Sc-C system. *Inst. Phys. Conf. Ser.*, **142**, 101.
[56] Kamata, I., Tsuchida, H., Jikimoto, T. and Izumi, K. (2000) Structural transformation of screw dislocations via thick 4H-SiC epitaxial growth. *Jpn. J. Appl. Phys.*, **39**, 6496.
[57] Balkas, C.M., Maltsev, A.A., Roth, M.D. *et al.* (2002) Reduction of macrodefects in bulk SiC single crystals. *Mater. Sci. Forum*, **389–393**, 59.
[58] Hull, D. and Bacon, D.J. (2001) *Introduction to Dislocations*, 4th edn, Butterworth-Heinemann.
[59] Leonard, R.T., Khlebnikov, Y., Powell, A.R. *et al.* (2009) 100 mm 4H N-SiC wafers with zero micropipe density. *Mater. Sci. Forum*, **600–603**, 7.
[60] Basceri, C., Khlebnikov, I., Khlebnikov, Y. *et al* (2006) Growth of micropipe-free single crystal silicon carbide (SiC) ingots via physical vapor transport (PVT). *Mater. Sci. Forum*, **527–529**, 39.

[61] Ohtani, N., Ohtani, N., Katsuno, M. *et al.* (2006) Propagation behavior of threading dislocations during physical vapor transport growth of silicon carbide (SiC) single crystals. *J. Cryst. Growth*, **286**, 55.

[62] Dudley, M. and Wu, F. (2011) Stacking faults created by the combined deflection of threading dislocations of Burgers vector c and $c + a$ during the physical vapor transport growth of 4H–SiC. *Appl. Phys. Lett.*, **98**, 232110.

[63] Wu, F., Dudley, M., Wang, H. *et al.* (2013) The nucleation and propagation of threading dislocations with c-component of Burgers vector in PVT-grown 4H-SiC. *Mater. Sci. Forum*, **740–742**, 217.

[64] Nakamura, D., Yamaguchi, S., Hirose, Y. *et al.* (2007) Direct determination of Burgers vector sense and magnitude of elementary dislocations by synchrotron white x-ray topography. *J. Appl. Phys.*, **103**, 013510.

[65] Dudley, M., Huang, X.R., Huang, W. *et al.* (1999) The mechanism of micropipe nucleation at inclusions in silicon carbide. *Appl. Phys. Lett.*, **75**, 784.

[66] Sanchez, E.K., Liu, J.Q., Graef, M.D. *et al.* (2002) Nucleation of threading dislocations in sublimation grown silicon carbide. *J. Appl. Phys.*, **91**, 1143.

[67] Powell, A.R., Leonard, R.T., Brady, M.F. *et al.* (2004) Large diameter 4H-SiC substrates for commercial power applications. *Mater. Sci. Forum.*, **457–460**, 41.

[68] Nakamura, D., Yamaguchi, S., Gunjishima, I. *et al.* (2007) Topographic study of dislocation structure in hexagonal SiC single crystals with low dislocation density. *J. Cryst. Growth*, **304**, 57.

[69] Ha, S., Rohrer, G.S., Skowronski, M. *et al.* (2000) Plastic deformation and residual stresses in SiC boules grown by PVT. *Mater. Sci. Forum*, **338–342**, 67.

[70] Fujita, S., Maeda, K. and Hyodo, S. (1987) Dislocation glide motion in 6H SiC single crystals subjected to high-temperature deformation. *Philos. Mag. A*, **55**, 203.

[71] Ha, S., Skowronski, M., Vetter, W.M. and Dudley, M. (2002) Basal plane slip and formation of mixed-tilt boundaries in sublimation-grown hexagonal polytype silicon carbide single crystals. *J. Appl. Phys.*, **92**, 778.

[72] Ha, S., Nuhfer, N.T., Rohrer, G.S. *et al.* (2000) Identification of prismatic slip bands in 4H SiC boules grown by physical vapor transport. *J. Electron. Mater.*, **29**, L5.

[73] Wellmann, P.J., Queren, D., Müller, R. *et al.* (2006) Basal plane dislocation dynamics in highly p-type doped versus highly n-type doped SiC. *Mater. Sci. Forum*, **527–529**, 79.

[74] Ohtani, N., Katsuno, M., Fujimoto, T. *et al.* (2009) Analysis of basal plane bending and basal plane dislocations in 4H-SiC single crystals. *Jpn. J. Appl. Phys.*, **48**, 065503.

[75] Glass, R.C., Kjellberg, L.O., Tsvetkov, V.F. *et al.* (1993) Structural macro-defects in 6H-SiC wafers. *J. Cryst. Growth*, **132**, 504.

[76] Tuominen, M., Yakimova, R., Glass, R.C. *et al.* (1994) Crystalline imperfections in 4H SiC grown with a seeded Lely method. *J. Cryst. Growth*, **144**, 267.

[77] Katsuno, M., Ohtani, N., Fujimoto, T. *et al.* (2000) Structural properties of subgrain boundaries in bulk SiC crystals. *J. Cryst. Growth*, **216**, 256.

[78] Nakamura, D., Gunjishima, I., Yamaguchi, S. *et al.* (2004) Ultrahigh-quality silicon carbide single crystals. *Nature*, **430**, 1009.

[79] Sugiyama, N., Okamoto, A. and Tani, T. (1996) Growth orientation dependence of dopant incorporation in bulk SiC single crystals. *Inst. Phys. Conf. Ser.*, **142**, 489.

[80] Larkin, D.J., Neudeck, P.G., Powell, J.A. and Matus, L.G. (1994) Site-competition epitaxy for superior silicon carbide electronics. *Appl. Phys. Lett.*, **65**, 1659.

[81] Dubrovskii, G.B., Lepneva, A.A. and Radovanova, E.I. (1973) Optical absorption associated with superlattice in silicon carbide crystals. *Phys. Status Solidi B*, **57**, 423.

[82] Ohtani, N., Katsuno, M., Takahashi, J. *et al.* (1998) Impurity incorporation kinetics during modified-Lely growth of SiC. *J. Appl. Phys.*, **83**, 4487.

[83] Onoue, K., Nishikawa, T., Katsuno, M. *et al.* (1996) Nitrogen incorporation kinetics during the sublimation growth of 6H and 4H SiC. *Jpn. J. Appl. Phys.*, **35**, 2240.

[84] Katsuno, M., Nakabayashi, M., Fujimoto, T. *et al.* (2009) Stacking fault formation in highly nitrogen-doped 4H-SiC substrates with different surface preparation conditions. *Mater. Sci. Forum*, **600–603**, 341.

[85] Jang, S., Kimoto, T. and Matsunami, H. (1994) Deep levels in 6H-SiC wafers and step-controlled epitaxial layers. *Appl. Phys. Lett.*, **65**, 581.

[86] Evwaraye, A.O., Smith, S.R. and Mitchel, W.C. (1996) Shallow and deep levels in n-type 4H-SiC. *J. Appl. Phys.*, **79**, 7726.

[87] Okojie, R.S., Xhang, M., Pirouz, P. *et al.* (2001) Observation of 4H-SiC to 3C-SiC polytypic transformation during oxidation. *Appl. Phys. Lett.*, **79**, 3056.

[88] Rost, H.-J., Doerschel, J., Irmscher, K. *et al.* (2003) Influence of nitrogen doping on the properties of 4H-SiC single crystals grown by physical vapor transport. *J. Cryst. Growth*, **257**, 75.
[89] Chung, H.J., Lin, J.Q., Henry, A. and Skowronski, M. (2003) Stacking fault formation in highly doped 4H-SiC epilayers during annealing. *Mater. Sci. Forum*, **433–436**, 253.
[90] Kuhr, T.A., Liu, J.Q., Chung, H.J. *et al.* (2002) Spontaneous formation of stacking faults in highly doped 4H-SiC during annealing. *J. Appl. Phys.*, **92**, 5863.
[91] Ohtani, N., Katsuno, M., Nakabayashi, M. *et al.* (2009) Investigation of heavily nitrogen-doped n^+ 4H-SiC crystals grown by physical vapor transport. *J. Cryst. Growth*, **311**, 1475.
[92] Kato, M., Kawai, M., Mori, T. *et al.* (2007) Excess carrier lifetime in a bulk p-type 4H-SiC wafer measured by the microwave photoconductivity decay method. *Jpn. J. Appl. Phys.*, **46**, 5057.
[93] Okuda, T., Miyake, H., Kimoto, T. and Suda, J. (2013) Long photoconductivity decay characteristics in p-type 4H-SiC bulk crystals. *Jpn. J. Appl. Phys.*, **52**, 010202.
[94] Willardson, R.K. and Beer, A.C. (eds) (1984) *Semiconductors and Semimetals*, vol. 20, Academic Press.
[95] Li, S.S. (2006) *Semiconductor Physical Electronics*, 2nd edn, Springer.
[96] Hobgood, H.M., Glass, R.C., Augustine, G. *et al.* (1995) Semi-insulating 6H-SiC grown by physical vapor transport. *Appl. Phys. Lett.*, **66**, 1364.
[97] Jenny, J.R., Skowronski, M., Mitchel, W.C. *et al.* (1996) Deep level transient spectroscopic and Hall effect investigation of the position of the vanadium acceptor level in 4H and 6H SiC. *Appl. Phys. Lett.*, **68**, 1963.
[98] Schneider, J., Muller, H.D., Maier, K. and Fuchs, F. (1990) Infrared spectra and electron spin resonance of vanadium deep level impurities in silicon carbide. *Appl. Phys. Lett.*, **56**, 1184.
[99] Kunzer, M., Kaufmann, U., Maier, K. and Schneider, J. (1995) Magnetic circular dichroism and electron spin resonance of the A^- acceptor state of vanadium, V^{3+}, in 6H-SiC. *Mater. Sci. Eng., B*, **29**, 118.
[100] Ilin, V.A. and Ballandovitch, V.S. (1993) EPR and DLTS of point defects in silicon carbide crystals. *Defect Diffus. Forum*, **103–105**, 633.
[101] Dalibor, T., Pensl, G., Matsunami, H. *et al.* (1997) Deep defect centers in silicon carbide monitored with deep level transient spectroscopy. *Phys. Status Solidi A*, **162**, 199.
[102] Müller, S.G., Brady, M.F., Brixius, W.H. *et al.* (2003) Sublimation-grown semi-insulating SiC for high frequency devices. *Mater. Sci. Forum*, **433–436**, 39.
[103] Jenny, J.R., Malta, D.P., Calus, M.R. *et al.* (2004) Development of large diameter high-purity semi-insulating 4H-SiC wafers for microwave devices. *Mater. Sci. Forum*, **457–460**, 35.
[104] Kaneko, H. and Kimoto, T. (2011) Formation of a semi-insulating layer in n-type 4H-SiC by electron irradiation. *Appl. Phys. Lett.*, **98**, 262106.
[105] Son, N.T., Carlsson, P., Magnusson, B. and Janzen, E. (2007) Intrinsic defects in semi-insulating SiC: Deep levels and their roles in carrier compensation. *Mater. Sci. Forum*, **556–557**, 465.
[106] Son, N.T., Carlsson, P., ul Hassan, J. *et al.* (2007) Defects and carrier compensation in semi-insulating 4H-SiC substrates. *Phys. Rev. B*, **75**, 155204.
[107] Kordina, O., Hallin, C., Ellison, A. *et al.* (1996) High temperature chemical vapor deposition of SiC. *Appl. Phys. Lett.*, **69**, 1456.
[108] Ellison, A., Zhang, J., Peterson, J. *et al.* (1999) High temperature CVD growth of SiC. *Mater. Sci. Eng., B*, **61–62**, 113.
[109] Ellison, A., Magnusson, B., Sundqvist, B. *et al.* (2004) SiC crystal growth by HTCVD. *Mater. Sci. Forum*, **457–460**, 9.
[110] Kitou, Y., Makino, E., Ikeda, K. *et al.* (2006) SiC HTCVD simulation modified by sublimation etching. *Mater. Sci. Forum*, **527–529**, 107.
[111] Fanton, M., Snyder, D., Weiland, B. *et al.* (2006) Growth of nitrogen-doped SiC boules by halide chemical vapor deposition. *J. Cryst. Growth*, **287**, 359.
[112] Ellison, A., Magnusson, B., Son, N.T. *et al.* (2003) HTCVD grown semi-insulating SiC substrates. *Mater. Sci. Forum*, **433–436**, 33.
[113] Elwell, D. and Scheel, H.J. (1975) *Crystal Growth from High-Temperature Solutions*, Academic Press, New York.
[114] Hofmann, D.H. and Muller, M.H. (1999) Prospects of the use of liquid phase techniques for the growth of bulk silicon carbide crystals. *Mater. Sci. Eng., B*, **61–62**, 29.
[115] Syväyärvi, M., Yakimova, R., Radamson, H.H. *et al.* (1999) Liquid phase epitaxial growth of SiC. *J. Cryst. Growth*, **197**, 147.
[116] Kusunoki, K., Munetoh, S., Kamei, K. *et al.* (2004) Solution growth of self-standing 6H-SiC single crystal using metal solvent. *Mater. Sci. Forum*, **457–460**, 123.

[117] Ujihara, T., Munetoh, S., Kusunoki, K. *et al.* (2004) Crystal quality evaluation of 6H-SiC layers grown by liquid phase epitaxy around micropipes using micro-Raman scattering spectroscopy. *Mater. Sci. Forum*, **457–460**, 633.
[118] Kamei, K., Kusunoki, K., Yashiro, N. *et al.* (2009) Solution growth of single crystalline 6H, 4H-SiC using Si–Ti–C melt. *J. Cryst. Growth*, **311**, 855.
[119] Danno, K., Saitoh, H., Seki, A. *et al.* (2010) High-speed growth of high-quality 4H-SiC bulk by solution growth using Si-Cr based melt. *Mater. Sci. Forum*, **645–648**, 13.
[120] Yamamoto, Y., Harada, S., Seki, K. *et al.* (2012) High-efficiency conversion of threading screw dislocations in 4H-SiC by solution growth. *Appl. Phys. Exp.*, **5**, 115501.
[121] Daikoku, H., Kado, M., Sakamoto, H. *et al.* (2012) Top-seeded solution growth of 4H-SiC bulk crystal using Si-Cr based melt. *Mater. Sci. Forum*, **717–720**, 61.
[122] Danno, K., Shirai, T., Seki, A. *et al.* (2013) Presented at 15th International Conference on Defects Recognition, Imaging and Physics in Semiconductors, Warsaw, Poland, 2013 Solution growth on 4H-SiC($1\bar{1}00$) for lowering density of threading dislocations.
[123] Nagasawa, H., Yagi, K., Kawahara, T. *et al.* (2004) Lowdefect 3C-SiC grown on undulant Si(001) substrates, in *Silicon Carbide – Recent Major Advances* (eds W.J. Choyke, H. Matsunami and G. Pensl), Springer, p. 207.
[124] Nagasawa, H., Abe, M., Yagi, K. *et al.* (2008) Fabrication of high performance 3C-SiC vertical MOSFETs by reducing planar defects. *Phys. Status Solidi B*, **245**, 1272.
[125] Nishino, S., Powell, A. and Will, H.A. (1983) Production of large-area single-crystal wafers of cubic SiC for semiconductor devices. *Appl. Phys. Lett.*, **42**, 460.
[126] Davis, R.F., Kelner, G., Shur, M. *et al.* (1991) Thin film deposition and microelectronic and optoelectronic device fabrication and characterization in monocrystalline alpha and beta silicon carbide. *Proc. IEEE*, **79**, 677.
[127] SEMI http://ams.semi.org/ebusiness/ (accessed 28 March 2014).
[128] Matsunami, H. and Kimoto, T. (1997) Step-controlled epitaxial growth of SiC: high quality homoepitaxy. *Mater. Sci. Eng., R*, **20**, 125.
[129] Kato, T., Noro, T., Takahashi, H. *et al.* (2009) Characterization of electric discharge machining for silicon carbide single crystal. *Mater. Sci. Forum*, **600–603**, 855.
[130] Zhou, L., Audurier, V., Pirouz, P. and Powell, A. (1997) Chemomechanical polishing of silicon carbide. *J. Electrochem. Soc.*, **144**, L161.
[131] Heydemann, V.D., Everson, W.J., Gamble, R.D. *et al.* (2004) Chemi-mechanical polishing of on-axis semi-insulating SiC substrates. *Mater. Sci. Forum*, **457–460**, 805.
[132] Hotta, K., Hirose, K., Tanaka, Y. *et al.* (2009) Improvements in electrical properties of SiC surface using mechano-chemical polishing. *Mater. Sci. Forum*, **600–603**, 823.
[133] Nikolova, E.G. (1985) On the Twyman effect and some of its applications. *J. Mater. Sci.*, **20**, 1.
[134] Hara, H., Sano, Y., Mimura, H. *et al.* (2006) Novel abrasive-free planarization of 4H-SiC(0001) using catalyst. *J. Electron. Mater.*, **35**, L11.
[135] Arima, K., Hara, H., Murata, J. *et al.* (2007) Atomic-scale flattening of SiC surfaces by electroless chemical etching in HF solution with Pt catalyst. *Appl. Phys. Lett.*, **90**, 202106.
[136] Powell, A., Jenny, J., Müller, S. *et al.* (2006) Growth of SiC substrates. *Int. J. High Speed Electron. Syst.*, **16**, 751.
[137] Burk, A.A., Tsvetkov, D., Barnhardt, D. *et al.* (2012) SiC epitaxial layer growth in a 6 × 150 mm warm-wall planetary reactor. *Mater. Sci. Forum*, **717–720**, 75.
[138] Fujimoto, T., Ohtani, N., Tsuge, H. *et al.* (2013) A thermodynamic mechanism for PVT growth phenomena of SiC single crystals. *ECS J. Solid State Sci. Technol.*, **2**, N3018.

4

Epitaxial Growth of Silicon Carbide

In SiC, epitaxial growth is essential to produce active layers with designed doping density and thickness. Homoepitaxial growth technology by chemical vapor deposition has shown remarkable progress, with polytype replication and wide-range doping control achieved by using step-flow growth and controlling the C/Si ratio, respectively. In this chapter, fundamental aspects and technological developments for hexagonal SiC homoepitaxial growth are described. Heteroepitaxial growth of 3C-SiC is also briefly introduced.

4.1 Fundamentals of SiC Homoepitaxy

Chemical vapor deposition (CVD) of a hexagonal SiC polytype on off-axis SiC{0001} of the identical polytype is the standard technology for SiC device development. Monosilane (SiH_4) and propane (C_3H_8) or ethylene (C_2H_4) are usually employed as the precursors. The carrier gas is hydrogen (H_2), and argon (Ar) is sometimes added. The typical growth temperature and growth rate are 1500–1650 °C and 3–15 μm h^{-1}, respectively. The CVD growth process for SiC usually consists of (i) *in situ* etching and (ii) main epitaxial growth. The *in situ* etching is performed with pure H_2, HCl/H_2, hydrocarbon/H_2, or SiH_4/H_2 at very high temperature, typically the same temperature as used for the main growth. The purpose of *in situ* etching is to remove the subsurface damage and to obtain regular step structures. Immediately after the etching, the main growth of n-type or p-type SiC (or their multilayers) is performed. In this section, fundamentals of SiC homoepitaxy (focusing on CVD) are described. Advanced growth technologies such as fast epitaxy are described in the subsequent sections.

4.1.1 Polytype Replication in SiC Epitaxy

As described in Chapter 2, 4H-SiC is the polytype chosen for major device applications. The polytypism of SiC means that perfect polytype control is an essential aspect of epitaxial growth of SiC. Early epitaxial growth of α-SiC (mainly 6H-SiC at that time) was performed by liquid phase epitaxy (LPE) [1, 2] or CVD methods [3–6]. Although advantages of CVD include precise control and uniformities of epilayer thickness and impurity doping, polytype mixing was a serious problem. For example, in the case of 6H-SiC CVD on {0001} basal planes of 6H-SiC Lely platelets, a growth temperature as high as 1800 °C was required for reproducible homoepitaxy, with twinned crystalline 3C-SiC grown at lower temperatures [3–6].

Fundamentals of Silicon Carbide Technology: Growth, Characterization, Devices, and Applications, First Edition.
Tsunenobu Kimoto and James A. Cooper.

The concept of perfect polytype replication in 6H-SiC epitaxial layers on 6H-SiC substrates was proposed by Matsunami *et al.* in 1987 [7, 8]. They investigated the polytype of SiC layers grown on SiC{0001} substrates with various off-angles and different off-directions, and proposed the optimum off-direction and off-angle to achieve high-quality homoepitaxy of 6H-SiC [7]. Davis *et al.* also reported homoepitaxial growth of 6H-SiC on an off-axis substrate in 1987 [9, 10]. Using step-flow growth on 2–6° off-axis 6H-SiC{0001}, homoepitaxial growth of 6H-SiC with a very smooth surface was achieved by CVD at 1450–1550 °C. In the same manner, homoepitaxial CVD growth of 4H-SiC was demonstrated on 5–6° off-axis 4H-SiC{0001} [11]. This growth technique is applicable to homoepitaxy of other polytypes such as 15R-SiC and 21R-SiC. The success of 4H-SiC homoepitaxy and demonstration of high-performance 4H-SiC Schottky barrier diodes on the homoepitaxial layers [12, 13] triggered the exclusive development of 4H-SiC for power device applications. This epitaxial growth technique was named "*step-controlled epitaxy*", since the polytype of the epilayers can be controlled by surface steps existing on the off-axis substrates. Homoepitaxial growth of high-quality SiC on off-axis SiC{0001} substrates has been reviewed in numerous papers [14–26].

Figure 4.1 shows the surface morphology (a, d), <$11\bar{2}0$>-azimuth reflection high-energy electron diffraction (RHEED) pattern (b, e), and surfaces after etching with molten KOH (c, f) for epilayers grown on on-axis and 6° off-axis 6H-SiC(0001) substrates at 1500 °C at a growth rate of 3 μm h^{-1} [15]. On an on-axis (0001) face, the epilayer exhibits a mosaic-like surface morphology, and relatively smooth domains are separated by step- or groove-like boundaries. From the RHEED analysis, the grown layer is identified as 3C-SiC(111) with twinning. Triangular etch pits indicating threefold symmetry are

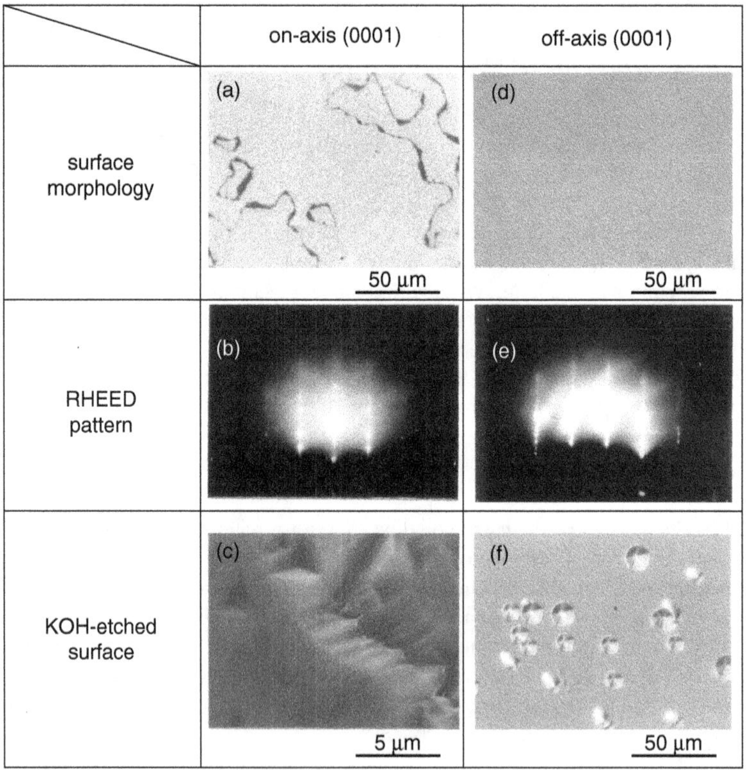

Figure 4.1 Surface morphology (a, d), <$11\bar{2}0$>-azimuth reflection high-energy electron diffraction (RHEED) pattern (b, e), and surfaces after etching with molten KOH (c, f) for SiC epitaxial layers grown on on-axis and 6° off-axis 6H-SiC(0001) substrates at 1500 °C at a growth rate of 3 μm h^{-1} ([15] reproduced with permission from Elsevier).

observed, which suggests the growth of cubic phase. Note that the etch pits are rotated by 180° relative to each other across the groove boundaries. This result means that the neighboring domains separated by the boundaries have a twin relationship, and form a so-called "*double positioning twin*" [27]. In contrast, the epilayers on off-axis substrates exhibit specular smooth surfaces, and the RHEED pattern shows the growth of single crystalline 6H-SiC(0001). The polytypes of the grown layers were verified by transmission electron microscope (TEM) observation, Raman scattering, and photoluminescence. Homoepitaxial layers with smooth surface morphology can also be obtained on an off-axis $(000\bar{1})$ face.

The preferred off-direction is $\langle 11\bar{2}0 \rangle$, because CVD growth on off-axis (0001) substrates inclined toward $\langle 1\bar{1}00 \rangle$ often exhibit a stripe-like morphology, which is caused by pronounced step bunching [8, 10]. The inclusion of 3C-SiC domains is occasionally observed after prolonged growth times on off-axis (0001) substrates inclined toward $\langle 1\bar{1}00 \rangle$. Although successful epitaxial growth on off-axis (0001) inclined toward $\langle 1\bar{1}00 \rangle$ was reported [28], several degree (typically 4°) off-axis toward $\langle 11\bar{2}0 \rangle$ is the standard used in the state-of-the-art technology.

Figure 4.2 shows a schematic illustration of growth modes and stacking sequences of the layers grown on (a) on-axis 6H-SiC(0001) and (b) off-axis 6H-SiC(0001). The bond configuration near an atomic step and on the (0001) terrace is also shown in Figure 4.2c. The precursors are heated and decomposed in the gas phase or near the substrate, and the source species diffuse toward the substrate surface. Adsorbed species migrate on the surface and are incorporated into a crystal at steps and/or kinks where the potential is low. However, there exists another competitive growth process of nucleation on terraces, which takes place when the supersaturation is high enough. The detailed descriptions of both growth processes follow:

1. On on-axis {0001}, the step density is very low, and large {0001} terraces exist. Then, crystal growth may initially occur on terraces through two-dimensional nucleation because of high supersaturation.

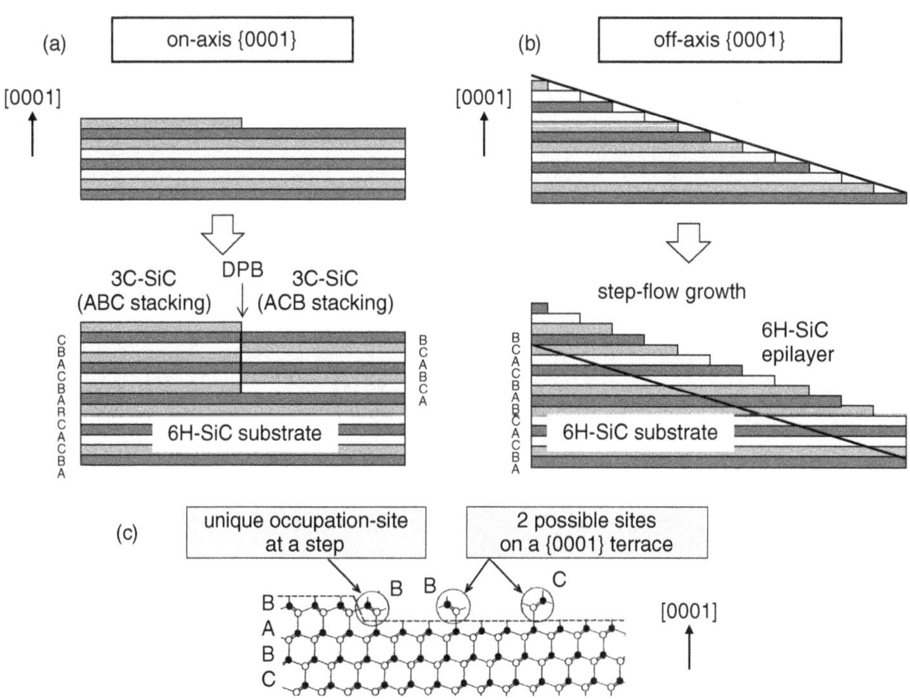

Figure 4.2 Schematic illustration of growth modes and stacking sequences of SiC layers grown on (a) on-axis 6H-SiC(0001) and (b) off-axis 6H-SiC(0001). (c) Bond configuration near an atomic step and on the (0001) terrace.

The polytype of the grown layers is determined by growth conditions, especially the growth temperature. This leads to the growth of 3C-SiC, which is stable at low temperature [29]. This phenomenon has been predicted by theoretical studies using a quantum-mechanical energy calculation [30] and an electrostatic model [31]. It is pointed out that defects such as polishing-induced damage on the surface can trigger 3C-SiC nucleation when the supersaturation is high [32]. Because the stacking order of 6H-SiC is ABCACB ..., the growing 3C-SiC can take two possible stacking orders of ABCABC ... and ACBACB ..., as shown in Figure 4.2a.

2. On off-axis {0001}, the step density is high, and the terrace width is narrow enough for adsorbed species to migrate and reach steps. At a step, the incorporation site (A, B, C) is uniquely determined by bonds from the step, as shown in Figure 4.2b. Hence, homoepitaxy can be achieved through the lateral growth from steps (step-flow growth), inheriting the stacking order of the substrate. Homoepitaxy with greatly improved crystal quality by using off-axis substrates has also been observed in sandwich-sublimation growth [33], LPE [34], and molecular beam epitaxy (MBE) [35] of hexagonal SiC. In general, step-flow growth on off-axis (vicinal) substrates has been extensively studied in many materials. In SiC epitaxy, the polytype of SiC epilayers can be controlled by the step density of substrates. The surface steps serve as a template, which forces the replication of the substrate polytype in the epilayer. This is the origin of the term "step-controlled epitaxy" [15].

Homoepitaxial growth of hexagonal SiC{0001} at lower temperature has been investigated. The growth temperature can be reduced to 1200 °C or even lower without 3C-SiC inclusions [36, 37], but the density of surface defects, such as triangular defects, increases significantly with decreasing growth temperature. At such a low growth temperature, the growth rate must be kept low, below a few $\mu m\ h^{-1}$, to minimize 3C-SiC inclusions. A larger off-angle is helpful to achieve SiC homoepitaxy at low temperature. These results are described in a quantitative manner in the next subsection. Another severe problem that occurs during low-temperature growth is the considerable increase in the background nitrogen doping density. For example, the donor density of undoped SiC(0001) epilayers grown at 1200 °C is approximately $(3-8) \times 10^{17}\ cm^{-3}$ [38], which is completely unacceptable for device fabrication.

4.1.2 *Theoretical Model of SiC Homoepitaxy*

A simple surface diffusion model based on the *BCF (Burton, Cabrera, and Frank) theory* [39] is considered, where steps with height h are separated by an equal distance λ_0 as shown in Figure 4.3. The adsorbed species diffuse on terraces toward steps. Some of the adsorbed species can reach steps and are incorporated into the crystal, and the others re-evaporate to the vapor. When nucleation on terraces does not occur, the continuity equation of adsorbed species is expressed by [39]:

$$-D_S \frac{d^2 n_S(y)}{dy^2} = J - \frac{n_S(y)}{\tau_S} \tag{4.1}$$

where $n_s(y)$ is the number of adsorbed species per unit area on the surface (hereafter, $n_s(y)$ is described as the adatom density), J the flux of reactants arriving at the surface, τ_s the mean residence time of adsorbed species, and D_s the surface diffusion coefficient. Here, the steps are assumed to be uniform and to act as perfect sinks for the incoming species, that is, the capture probability of adsorbed species at steps is unity, independent of the direction from which the adsorbed species approach the steps. Under the boundary condition that the supersaturation ratio $\alpha_s(= n_s/n_{s0})$ equals unity at steps: $n_s = n_{s0}$ at $y = \pm\lambda_0/2$, the adatom density on the terraces can be given as a solution of Equation 4.1:

$$n_S(y) = J\tau_S + (n_{S0} - J\tau_S)\frac{\cosh(y/\lambda_S)}{\cosh(\lambda_0/2\lambda_S)}, (-\lambda_0/2 \leq y \leq +\lambda_0/2) \tag{4.2}$$

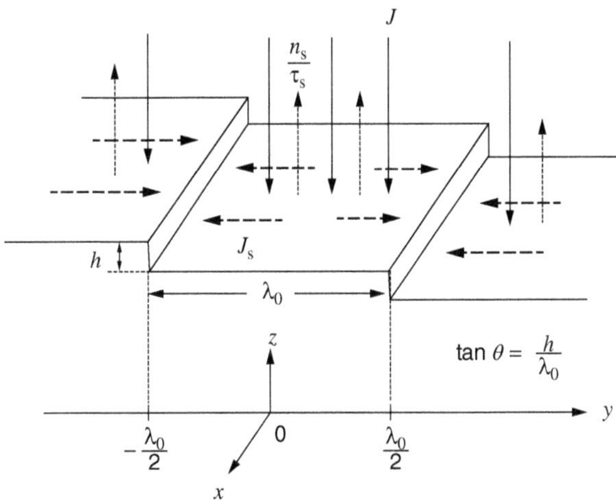

Figure 4.3 Surface diffusion model based on the BCF theory, where steps with height h are separated by an equal distance λ_0 ([40] reproduced with permission from AIP Publishing LLC).

where n_{s0} is the adatom density at equilibrium and λ_s is the surface diffusion length of adsorbed species, which is given by the following equation [39]:

$$\lambda_S = \sqrt{D_S\tau_S} = a\exp\left(\frac{E_{des} - E_{diff}}{2kT}\right) \tag{4.3}$$

Here, a, k, and T are the jump distance (interatomic distance), the Boltzmann constant, and the absolute temperature, respectively. E_{des} and E_{diff} are the activation energies for desorption and surface diffusion. This λ_s is an average length for adsorbed species to migrate on a "step-free" surface before desorption. In step-flow growth, the growth rate (R) is given by the product of the step velocity (v_{step}) and tan$\theta(= h/\lambda_0)$, where θ is the substrate's off-angle. Thus, the following equation is satisfied:

$$R = v_{step}\tan\theta = \frac{2h\lambda_S}{n_0\lambda_0}\left(J - \frac{n_{S0}}{\tau_S}\right)\tanh\left(\frac{\lambda_0}{2\lambda_S}\right) \tag{4.4}$$

Here, n_0 is the density of adsorption sites on the surface (1.21×10^{15} cm^{-2} for SiC{0001}).

Figure 4.4 shows the distribution of adatom density and the supersaturation ratio (α_s) on a surface. Since α_s takes a maximum value α_{s_max} at the center of a terrace, nucleation occurs most easily at this location. Based on Equations 4.2 and 4.4, α_{s_max} can be expressed by [39]:

$$\alpha_{s\_max} = 1 + \frac{\lambda_0 n_0 R\tau_S}{2\lambda_S n_{S0} h}\tanh\left(\frac{\lambda_0}{4\lambda_S}\right) \tag{4.5}$$

α_{s_max} depends on experimental conditions, such as the growth rate, growth temperature, and terrace width, and is an essential parameter that determines whether the growth mode is step-flow or two-dimensional nucleation. Because the two-dimensional nucleation rate J_{nuc} increases exponentially with the supersaturation ratio on a surface, nucleation becomes significant when α_{s_max} exceeds a critical value α_{s_crit}. Thus, the growth modes are determined according to the relationship between α_{s_max} and α_{s_crit} as follows:

$$\alpha_{s\_max} > \alpha_{s\_crit} : \textit{two-dimensional nucleation} \tag{4.6}$$

$$\alpha_{s\_max} < \alpha_{s\_crit} : \textit{step-flow} \tag{4.7}$$

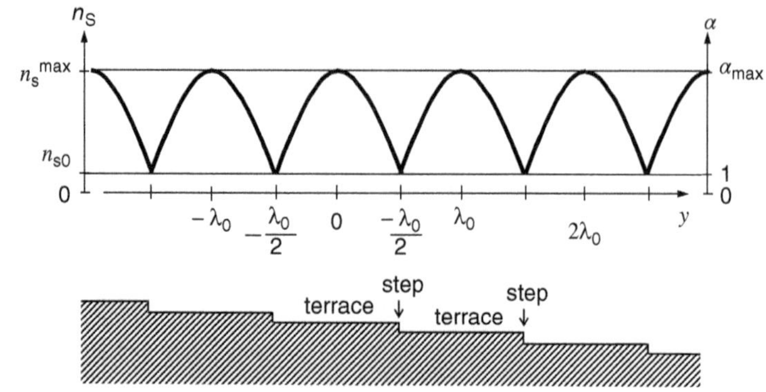

Figure 4.4 Distribution of adatom density and the supersaturation ratio (α_s) on a surface ([40] reproduced with permission from AIP Publishing LLC).

In the case of hexagonal SiC homoepitaxy on off-axis SiC{0001}, stable step-flow growth is required to ensure homoepitaxy, as described above. Two-dimensional nucleation on terraces or "defect sites" is likely to lead to inclusions of 3C-SiC or other foreign polytypes.

Under the *critical growth condition,* α_{s_max} should be equal to α_{s_crit}, which means that [40]

$$\frac{\lambda_0}{4\lambda_s}\tanh\left(\frac{\lambda_0}{4\lambda_s}\right) = \frac{(\alpha_{s\_crit}-1)h}{2n_0 R}\frac{n_{s0}}{\tau_s} \tag{4.8}$$

This is the basic equation that describes the growth mode on an off-axis substrate. In this equation, R and λ_0 are determined by growth conditions, h is also dependent on the growth condition, due to the occurrence of step bunching, while n_0 is an inherent parameter of a material. If the values of n_{s0}/τ_s and α_{s_crit} are known and several critical conditions are found experimentally, the surface diffusion length can be estimated from Equation 4.8. n_{s0}/τ_s can be calculated from the equilibrium vapor pressure P_0 using Knudsen's equation (from the kinetic theory of gases) [41]. Here, P_0 can be calculated from the chemical equilibrium constants of the reaction system [40]. Because the supply of Si species mainly controls the growth in the present case, P_0 in SiC CVD growth can be assumed to be an equilibrium vapor pressure of the Si species that undergo surface reactions [36, 42]. The equilibrium vapor pressure of Si (P_{Si}) as a function of temperature is obtained from the equilibrium equations for several reactions between Si and hydrocarbon species [40].

Nucleation on terraces becomes dominant when α_{s_max} exceeds a critical supersaturation ratio α_{s_crit}. When the critical nucleation rate J_{nuc} is assumed to be 10^{10} cm^{-2} s^{-1}, for example, which corresponds to one nucleation per unit time on a 100 nm × 100 nm area, α_{s_crit} for a disk-shaped nucleus is given by the following equation [43]:

$$\alpha_{s\_crit} = \exp\left\{\frac{\pi h_0 \sigma^2 \Omega}{(65 - \ln 10^{10})\, k^2 T^2}\right\} \tag{4.9}$$

where Ω and σ are the volume of a Si-C pair (2.07×10^{-23} cm^3) and the surface free energy. The surface free energy values of hexagonal SiC{0001} have been calculated as 2220 and 300 erg cm^{-2} for (0001) and (000$\bar{1}$), respectively [44]. The step height (h_0) can be determined by atomic force microscopy (AFM), and typically ranges from $c/2$ to several c. The temperature dependence of α_{s_crit} for (0001)Si and (000$\bar{1}$)C faces is shown in Figure 4.5. The α_{s_crit} for (000$\bar{1}$)C face takes very low values, indicating that nucleation occurs much more frequently on the (000$\bar{1}$)C face than on the (0001)Si face under the same supersaturation conditions.

Critical growth conditions can be found through CVD growth experiments under various growth conditions. The growth temperature and off-angle are varied in the range of 1100–1700 °C and 0.2–10°,

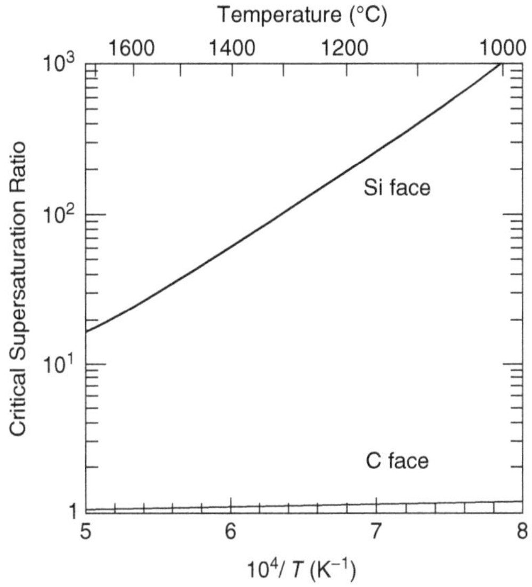

Figure 4.5 Temperature dependence of the critical supersaturation ratio (α_{s_crit}) for (0001)Si and (000$\bar{1}$)C faces ([40] reproduced with permission from AIP Publishing LLC).

Table 4.1 Some of the critical growth conditions at various growth temperatures for epitaxial growth of 4H-SiC(0001).

Growth temperature (°C)	Off-angle (°)	Growth rate (μm h^{-1})
1100	6	0.8
1200	4	2.4
1400	1	14
1500	0.2	6
1600	2	90

respectively. Table 4.1 summarizes some of the critical growth conditions at various growth temperatures for epitaxial growth of 4H-SiC(0001). These data are updated from an old reference [40]. Higher growth temperature, larger off-angle, and lower growth rate are preferable for homoepitaxy of 4H-SiC (step-flow growth). Furthermore, CVD growth with a low C/Si ratio (the ratio of carbon and silicon atoms in the supplied precursors) is beneficial to promote homoepitaxy [42].

Figure 4.6 shows the temperature dependence of surface diffusion length (λ_s) on 4H-SiC(0001), calculated from Equation 4.8 using several data points, as discussed above. In the figure, a dotted line denotes results of fitting to Equation 4.3. Since the surface diffusion lengths obtained in this study are the average lengths to migrate on a "step-free" surface before desorption, they decrease at high temperatures, where desorption is enhanced. The surface diffusion length on the C face has not been estimated due to the lack of various growth data. In an old study on 6H-SiC{0001}, the diffusion length on the C face is much longer than that on the Si face [40]. Although nucleation occurs much more easily on the C faces, the longer surface diffusion lengths on the C face may compensate for the frequent nucleation on the terraces.

Because the temperature dependences of n_{s0}/τ_s, α_{s_crit}, and λ_s have been obtained, critical growth conditions can be predicted using Equation 4.8. For example, if the growth temperature and off-angle of substrates (terrace width) are fixed, a critical growth rate (maximum growth rate to realize step-flow)

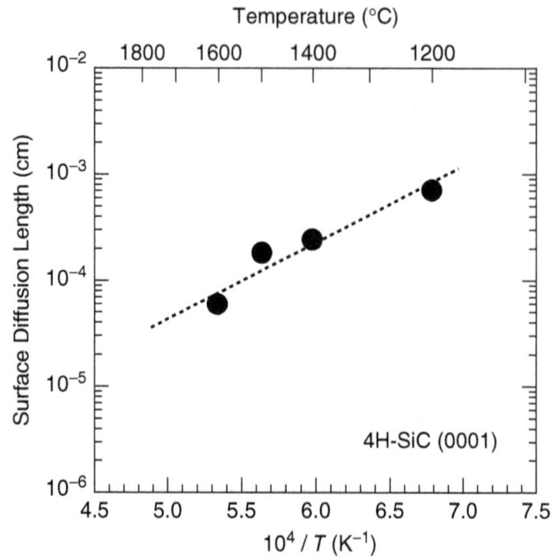

Figure 4.6 Temperature dependence of surface diffusion length (λ_s) on 4H-SiC(0001) calculated from Equation 4.8.

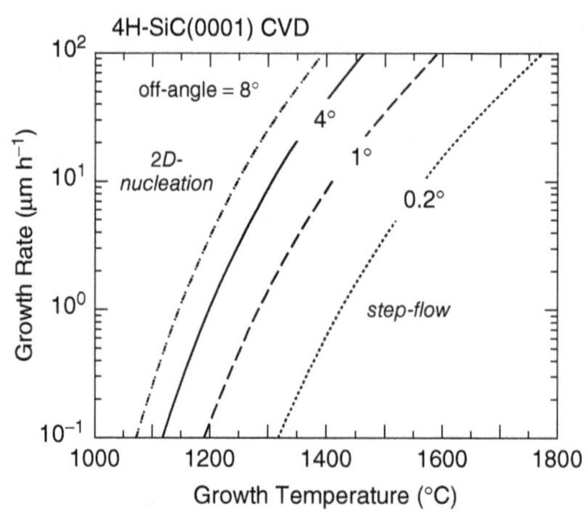

Figure 4.7 Critical growth conditions for CVD growth of 4H-SiC(0001), where the maximum growth rate is plotted as a function of growth temperature for a given off-angle of the substrate. The top-left and bottom-right regions separated by a curve correspond to the two-dimensional nucleation (severe 3C-SiC inclusions) and step-flow growth (homoepitaxy) conditions, respectively.

can be calculated. Curves for these critical growth conditions are shown in Figure 4.7 for substrates with off-angles of 0.2°, 1°, 4°, and 8°. Note that this chart has been significantly updated from one presented in reference [40], because recent experimental data were employed in the analyses. In the figure, the top-left and bottom-right regions separated by the curves correspond to the two-dimensional nucleation (severe 3C-SiC inclusions) and step-flow growth (homoepitaxy) conditions, respectively. Growth with a high growth rate and small off-angle can proceed via a step-flow mode at high growth temperatures.

At 1700 °C, a very small off-angle of 0.2°, which yields almost "on-axis" {0001}, is sufficient to achieve step-flow growth with a growth rate of about 50 μm h^{-1}. The role of surface defects in nucleation of 3C-SiC becomes important on SiC{0001} substrates with small off-angles [32]. Note that spiral growth around threading screw dislocations (TSDs) is not considered in the present model. Spiral growth naturally promotes homoepitaxial growth, and this effect is pronounced on SiC{0001} with very small off-angles. Conversely, large off-angles greater than 4° are needed to achieve homoepitaxy at a low temperature of 1200 °C with a reasonable growth rate (>1 μm h^{-1}).

4.1.3 Growth Rate and Modeling

Under typical conditions for CVD growth of SiC, differences in the growth rates on different SiC faces such as (0001), (000$\bar{1}$), and (11$\bar{2}$0) are very small, indicating that the SiC growth is *diffusion limited,* and the supply of source species onto the growing surface is the rate-determining step [36].

Figure 4.8a shows the C/Si ratio dependence of the growth rate for CVD of off-axis 4H-SiC(0001) [45]. In this particular experiment, the C/Si ratio was varied by changing the C_3H_8 flow rate while fixing the SiH_4 flow rate. When the C/Si ratio is lower than 1.0–1.3, the growth rate increases with increasing C/Si ratio (C_3H_8 flow rate). Above some C/Si ratio, the growth rate does not change if the C/Si ratio (C_3H_8 flow rate) is increased. In this saturation region, the growth rate is almost proportional to the SiH_4 flow rate. These are common trends in CVD growth of compound semiconductors. When the ambient is Si-rich, the growth rate is dominated by the carbon supply, whereas the growth rate is mainly determined by the silicon supply under C-rich conditions. Near the "kink point" (C/Si ratio = 1.0–1.3), a nearly stoichiometric condition must be established on the growing surface. In general, very good surface morphology is obtained near this stoichiometric condition. When the C/Si ratio is too low, the surface suffers from formation of severe macrosteps and Si droplets. In contrast, surface morphological defects, such as triangular defects, are easily generated when the C/Si ratio is too high. The change in surface stoichiometry and supersaturation by changing the ratio of C and Si source supply was modeled and shown in Figure 4.8b.

Figure 4.9 shows the growth rate at 1650 °C versus SiH_4 flow rate at a fixed C/Si ratio of 1.2 [46]. The growth rate increases almost in proportion to the SiH_4 flow rate, and reaches 50 μm h^{-1} at a SiH_4

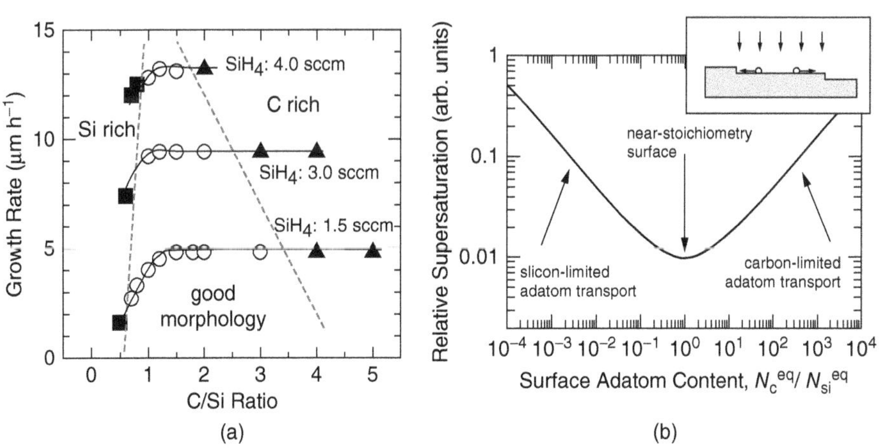

Figure 4.8 (a) C/Si ratio dependence of the growth rate for CVD of off-axis 4H-SiC(0001) ([45] reproduced with permission from The Japan Society of Applied Physics). (b) Change of surface stoichiometry and supersaturation by changing the ratio of C and Si source supply ([42] reproduced with permission from AIP Publishing LLC).

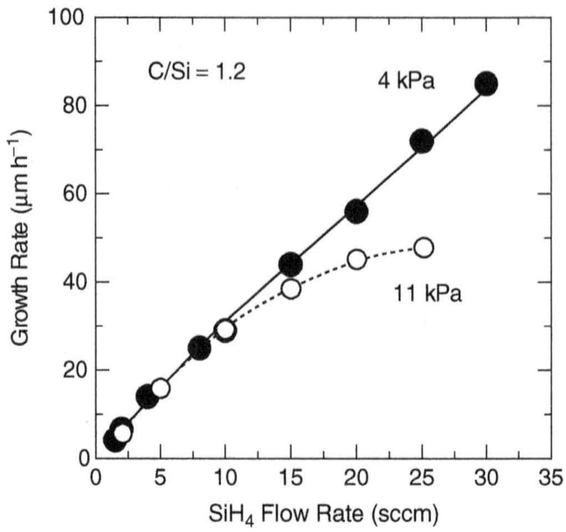

Figure 4.9 Growth rate of 4H-SiC at 1650 °C versus SiH_4 flow rate at a fixed C/Si ratio of 1.2.

flow rate of 20 sccm. However, at higher flow rates, the growth rate exhibits a sub-linear dependence on the SiH_4 flow rate and tends to saturate, when the growth was performed at 11 kPa. This is caused by Si polymerization in the gas phase, and is further discussed in the context of fast epitaxy (Section 4.4). This Si polymerization can be reduced by decreasing the growth pressure (4 kPa), and a high growth rate over 80 μm h^{-1} can be attained. The intercept of the vertical axis in Figure 4.9 gives a negative value, which implies that etching of SiC by H_2 is faster than growth when the SiH_4 flow rate is very small. The etching rate of SiC by H_2 at 1600–1650 °C is estimated at 0.2–0.4 μm h^{-1} when the pressure is 4 kPa.

Interaction between SiC and H_2 at high temperature is of academic and technological interest. At temperatures of 1500–1700 °C, silicon is removed by desorption because of its high equilibrium pressure, while carbon is removed as CH_4 (or CH_2) through reaction with H_2.

$$\mathrm{Si(s)} \rightarrow \mathrm{Si(g)} \tag{4.10}$$

$$\mathrm{C(s)} + 2\mathrm{H_2} \rightarrow \mathrm{CH_4(g)} \tag{4.11}$$

Here "g" and "s" mean the gas phase and solid phase, respectively. Therefore, the etching becomes faster as the process pressure is decreased (because silicon desorption is enhanced at lower pressures), as long as sufficient H_2 is supplied for carbon removal. Conversely, silicon desorption is hampered at relatively high pressures, such as atmospheric pressure, although carbon removal is enhanced further at higher H_2 pressures. Thus, preferential etching of carbon proceeds at high pressure (and high temperature), leading to formation of silicon droplets. These silicon droplets are, of course, harmful for subsequent epitaxial growth. Addition of a small amount of HCl [19, 20, 47, 48] or hydrocarbon [49, 50] during H_2 etching is an effective way to obtain a clean SiC surface without silicon droplets. However, carbon removal by H_2, as expressed by Equation 4.11, becomes extremely slow below 1200 °C. Therefore, when SiC is etched with H_2 at low pressure and at 900–1200 °C, preferential desorption of silicon takes place, leading to surface graphitization. Annealing SiC in vacuum at high temperature is one way to form graphene on the SiC surface [51–53]. By H_2 treatment of SiC at atmospheric pressure and about 1000 °C, a clean SiC surface passivated with hydrogen can be obtained [54].

Chemistries in SiC CVD have been studied intensively. Allendorf and Kee analyzed gas-phase and surface reactions at 1200–1600 °C in a SiH_4–C_3H_8–H_2 system [55]. Stinespring and Wohmhoudt also reported similar analysis of gas-phase kinetics [56]. Their analyses have shown that the dominant species that contribute to SiC growth are Si, SiH_2, and Si_2H_2 species from SiH_4, and CH_4, C_2H_2, and C_2H_4 molecules from C_3H_8. These simulation results suggest that Si (or SiH_2) may be preferentially adsorbed and then migrates on the surface. In fact, very little deposition of carbon films occurs in a standard CVD system if SiH_4 is not supplied.

Nishizawa and Pons [57–60] and Danielsson *et al.* [61, 62] established accurate simulation models for SiC CVD. In these models, the heat transfer and mass transfer equations are solved, as in the case of SiC boule growth (see Section 3.1.3). In simulation of CVD, however, a complete set of complicated chemical reactions must be taken into account. Although the CVD process is more easily controlled than the sublimation growth process, one has no way of knowing the real C/Si ratio on the growing surface, and how this varies when the C/Si ratio at the gas inlet is changed. Simulation of SiC CVD gives very important insights into the temperature distribution of wafers, surface kinetics (including actual C/Si ratio, doping efficiency), and guidelines for up-scaling a reactor. Tables 4.2 and 4.3 show the major chemical reactions in the gas phase (homogeneous) and on the surface (heterogeneous), respectively [57]. In the simulation, the temperature dependence of individual chemical reaction constants is, of course, considered. The effects of changes in the precursor supply, the C/Si ratio, the growth temperature, and the pressure on growth rate can be well predicted. Figure 4.10 shows the real C/Si ratio on the SiC surface simulated for SiC CVD growth at 25 kPa [57]. The figure shows the dependence of C/Si ratio on C_3H_8 flow rate for a few fixed SiH_4 flow rates. The real C/Si ratio is defined as the ratio of the C- and Si-molar fractions on the growing surface (not at the gas inlet) calculated by using the simulation model. Note that the increase in the C/Si ratio is nonlinear when the C_3H_8 flow rate is increased. Furthermore, the real C/Si ratio decreases significantly when the growth temperature is elevated (1783 K → 1823 K → 1873 K) at fixed SiH_4 and C_3H_8 flow rates. Figure 4.11 shows the real C/Si ratio as a function of the SiH_4 flow rate simulated for a few fixed C/Si ratios at the gas inlet [57]. The growth temperature and pressure are 1873 K

Table 4.2 Major chemical reactions in the gas phase (homogeneous) considered in the simulation of SiC CVD ([57] reproduced with permission from Wiley-VCH).

Major homogeneous reactions
$SiH_4 \leftrightarrow SiH_2 + H_2$
$Si_2H_6 \leftrightarrow SiH_2 + SiH_4$
$SiH_2 \leftrightarrow Si + H_2$
$2H + H_2 \leftrightarrow 2H_2$
$C_3H_8 \leftrightarrow CH_3 + C_2H_5$
$CH_4 + H \leftrightarrow CH_3 + H_2$
$C_2H_5 + H \leftrightarrow 2CH_3$
$2CH_3 \leftrightarrow C_2H_6$
$C_2H_4 + H \leftrightarrow C_2H_5$
$C_2H_4 \leftrightarrow C_2H_2 + H_2$
$H_3SiCH_3 \leftrightarrow SiH_2 + CH_4$
$H_3SiCH_3 \leftrightarrow HSiCH_3 + H_2$
$Si_2 \leftrightarrow 2Si$
$Si_2 + CH_4 \leftrightarrow Si_2C + 2H_2$
$SiH_2 + Si \leftrightarrow Si_2 + H_2$
$CH_3 + Si \leftrightarrow SiCH_2 + H$
$SiCH_2 + SiH_2 \leftrightarrow Si_2C + 2H_2$

Table 4.3 Major chemical reactions on the surface (heterogeneous) considered in the simulation of SiC(0001) CVD ([57] reproduced with permission from Wiley-VCH).

Major heterogeneous reactions
$C_{vol} + Si_{surf} + H_2 \leftrightarrow SiH_2 + C_{surf}$
$2Si_{vol} + 2C_{surf} + H_2 \leftrightarrow C_2H_2 + 2Si_{surf}$
$SiH_4 + C_{surf} \leftrightarrow SiH_{2surf} + H_2 + C_{vol}$
$SiH_{2surf} \leftrightarrow H_2 + Si_{surf}$
$SiH_2 + C_{surf} \leftrightarrow SiH_{2surf} + C_{vol}$
$Si + C_{surf} \leftrightarrow Si_{surf} + C_{vol}$
$C_2H_4 + Si_{surf} \leftrightarrow 2C_{surf} + 2H_2 + Si_{vol}$
$C_2H_4 + 2Si_{surf} \leftrightarrow 2C_{surf} + 2H_2 + 2Si_{vol}$
$CH_4 + Si_{surf} \leftrightarrow C_{surf} + 2H_2 + Si_{vol}$
$H_3SiCH_3 + C_{surf} \leftrightarrow Si_{surf} + H_2 + H + CH_3 + C_{vol}$
$CH_3 + Si_{surf} \leftrightarrow C_{surf} + 1.5H_2 + Si_{vol}$
$Si_2 + 2C_{surf} \leftrightarrow 2Si_{surf} + 2C_{vol}$
$Si_2C + Si_{surf} \leftrightarrow Si_2 + C_{surf} + Si_{vol}$
$SiCH_2 + C_{surf} \leftrightarrow Si_{surf} + CH_2 + C_{vol}$
$CH_2 + Si_{surf} \leftrightarrow C_{surf} + H_2 + Si_{vol}$

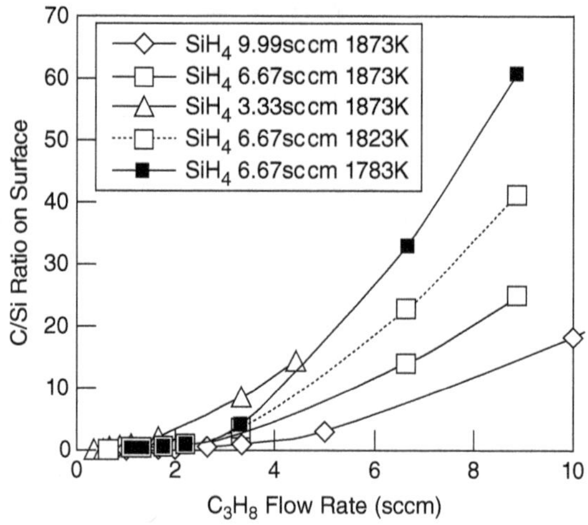

Figure 4.10 Dependence of real C/Si ratio at the growing surface on C_3H_8 flow rate for a few fixed SiH_4 flow rates simulated for SiC CVD growth at 25 kPa ([57] reproduced with permission from Wiley-VCH Verlag GmbH).

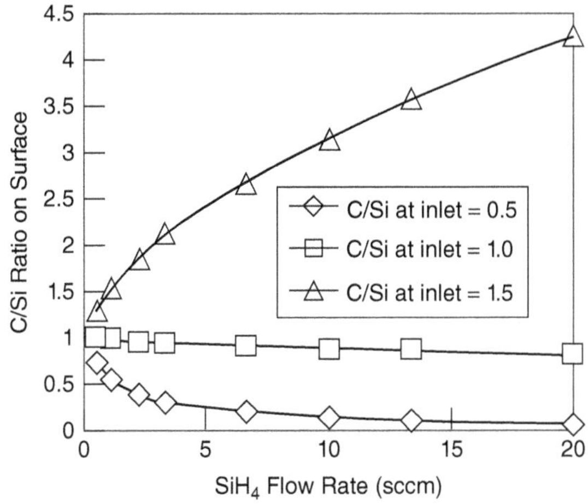

Figure 4.11 Real C/Si ratio as a function of the SiH_4 flow rate simulated for a few fixed C/Si ratios at the gas inlet ([57] reproduced with permission from Wiley-VCH Verlag GmbH). The growth temperature and pressure are 1873 K and 25 kPa, respectively.

and 25 kPa, respectively. Even if the C/Si ratio at the inlet is fixed at 1.0, the real C/Si ratio on the growing surface exhibits a continuous decrease as the precursor flows are increased to obtain high growth rates. When the C/Si ratio at the inlet is 0.5 (Si rich) or 1.5 (C rich), the real C/Si ratio is considerably shifted toward the Si-rich or C-rich condition, respectively, as the precursor flows increase. These results are important to understand the changes in surface morphology and impurity incorporation when the growth conditions are varied. The dependence of nitrogen and aluminum doping on the growth conditions has also been well described using simulation results [57, 60].

4.1.4 Surface Morphology and Step Dynamics

The surfaces of SiC homoepitaxial layers are smooth when the growth process is optimized [20, 22, 23, 48]. Figure 4.12 shows the surface morphology of a 4H-SiC(0001) epitaxial layer observed by (a) Nomarski microscopy and (b) AFM. The surface is mostly featureless, and the density of macroscopic

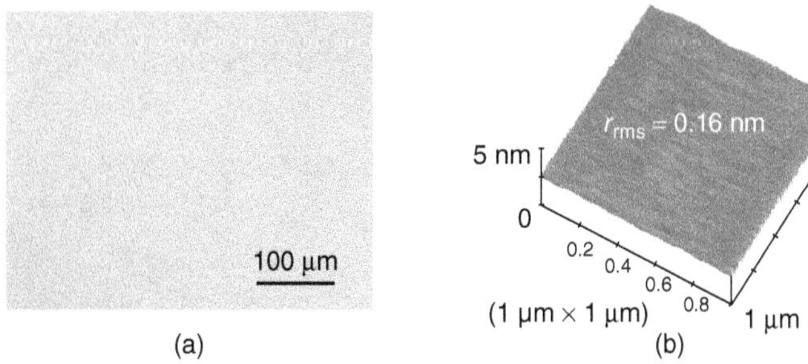

Figure 4.12 Surface morphology of a 4H-SiC(0001) epitaxial layer observed by (a) Nomarski microscopy and (b) atomic force microscopy (AFM).

surface defects is typically 0.1–0.5 cm^{-2}. The surface roughness defined by the root mean square (r_{rms}) is 0.14–0.22 nm for a scan area of 10 × 10 μm^2. Although the surface roughness tends to increase in thick (>50 μm) layers, chemical mechanical polishing of substrates, and optimized etching and growth conditions greatly improve the surface morphology, even for very thick (>100 μm) layers. When 4H-SiC(0001) with small off-angles (2–4°) is employed, formation of macrosteps is often observed [63–65]. Macrostep formation is not desirable because electric field crowding can take place, especially in gate oxides formed on such a surface. CVD growth under Si-rich conditions [64, 65] or the use of (000$\bar{1}$) substrates [65, 66] is effective at suppressing macrostep formation. Surface morphological defects are described in Section 4.3.1.

Step bunching in step-flow growth on off-axis substrates is an interesting and important aspect of both crystal growth and surface science. The step structure of 6H- and 4H-SiC{0001} homoepitaxial layers has been studied using AFM and TEM. In AFM observations, a distinctive difference in surface structures was observed on the (0001)Si and (000$\bar{1}$)C faces. SiC epitaxial growth on the off-axis (0001)Si face yields apparent macrosteps with a terrace width of 200–600 nm and a step height of 2–8 nm. In high-resolution observation, each macrostep is not a single multiple-height step but is instead composed of a number of microsteps as shown in Figure 4.13a [67]. This phenomenon is pronounced in 4H-SiC(0001) growth on substrates with a small off-angle (2–4°) or under C-rich conditions [64, 65]. On the off-axis (000$\bar{1}$)C face, however, the surface is rather flat, and macrosteps are rarely observed, as shown in Figure 4.13b. Although the mechanism of apparent macrostep formation is not very clear, the surface is similar to the so-called hill-and-valley (or faceted) structure often observed in step-flow growth of other materials [68, 69]. The off-axis SiC(0001) surface may spontaneously rearrange to minimize the total surface energy by increasing the area of a low-energy surface. When one looks at the heights of the steps, SiC exhibits unique step structures for each particular polytype. Figure 4.14 shows the histograms of step heights for the surfaces of (a) 4H-SiC and (b) 6H-SiC(0001) epitaxial layers grown on 3.5° or 8° off-axis substrates, respectively [48]. For 4H-SiC, the majority of the surface steps exhibit a two-bilayer height (half unit cell of 4H-SiC), though four-bilayer-height (full unit cell) steps are also observed. In the case of 6H-SiC, most steps possess a three-bilayer height (half unit cell of 6H-SiC). The distribution of step heights is, of course, dependent on the growth condition, but the appearance of half- or full-unit-cell height (c/2 or c) steps is common on SiC epitaxial layers grown by CVD.

Similar observations have been reported for 6H-SiC surfaces grown by the Lely method [70] and MBE [35]. 4H- and 6H-SiC surfaces around a TSD (or micropipe) also exhibit a similar step structure: the step

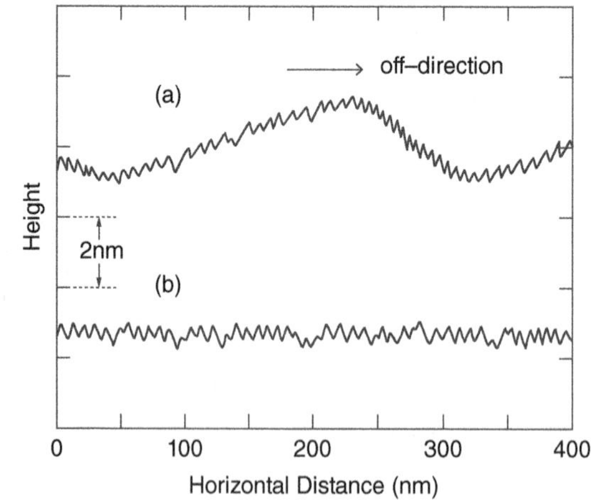

Figure 4.13 High-resolution section analyses of the surface of a 4H-SiC epitaxial layer grown on (a) (0001) and (b) (000$\bar{1}$) faces ([67] reproduced with permission from AIP Publishing LLC).

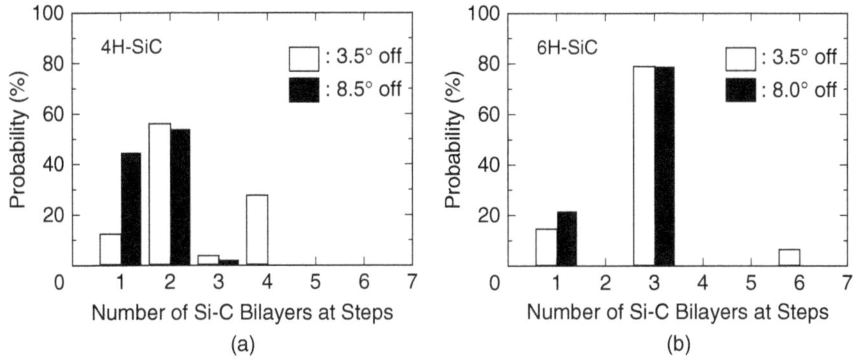

Figure 4.14 Histograms of step heights for the surfaces of (a) 4H-SiC and (b) 6H-SiC(0001) epitaxial layers grown on 3.5° or 8° off-axis substrates, respectively ([48] reproduced with permission from The Japan Society of Applied Physics).

height is *c* along the <1$\bar{1}$00> direction and *c*/2 along the <11$\bar{2}$0> direction [71, 72]. Thus, the formation of surface steps with a *c*/2 or *c* height seems to be an inherent aspect of 4H- and 6H-SiC growth. In 15R-SiC growth, the dominant step heights are two, three, and five bilayers, which corresponds to the zigzag stacking structure of 15R-SiC [48]. Heine *et al.* suggested that the surface energy is different for each SiC bilayer plane because of the peculiar stacking sequence [30]. Different surface energies may lead to different step velocities for each Si-C bilayer, thereby causing "structurally-induced macrostep formation" [73].

4.1.5 Reactor Design for SiC Epitaxy

Because very high temperature (1500–1700 °C) is required in SiC CVD, several unique reactor designs have been proposed. For example, buoyancy-driven convection is significant at these temperatures. A low growth pressure with a high carrier gas flow is an effective way to reduce the thermal convection. Note that radiation dominates as the main mechanism for temperature (heat) loss in this temperature range. Figure 4.15 shows schematic illustrations of several typical reactors employed for SiC CVD. Conventional horizontal [16, 19] and vertical [17] *cold-wall CVD* reactors are shown in Figure 4.15a,b, respectively. These conventional reactors are simple in configuration but possess some disadvantages when used for SiC CVD. Because of the high growth temperature, the temperature gradient normal to the wafer surface becomes very large (>100 K mm^{-1}), which causes severe warpage of SiC wafers [74]. It is difficult to establish uniform temperature distributions over a large scale at such high temperatures with this reactor configuration. The heating efficiency is very poor, since significant heat is lost by radiation.

These problems have been overcome by introduction of the *hot-wall CVD* concept by Kordina, Henry, Janzen, and coworkers [18, 23, 75]. In hot-wall CVD reactors, SiC wafers are placed inside a gas-flow channel formed in a susceptor. The susceptor is made of dense graphite coated with polycrystalline SiC or TaC. This susceptor is surrounded by a thermal insulator, such as porous graphite. By adjusting the frequency employed for rf-induction (rf: radio frequency) heating, the susceptor can be efficiently heated with minimal loss in the thermal insulator. The thermal insulation is so good that water cooling of the outer quartz tube is usually not necessary, in spite of the high growth temperature. In hot-wall CVD reactors, SiC wafers are heated from both sides, by radiation from the front side and by conduction (as well as radiation) from the back side. Therefore, the temperature gradient is considerably reduced (<10 K mm^{-1}), and good temperature uniformity can easily be established, which is critical for large-scale production of high-quality epitaxial wafers. The heating efficiency is also very high in hot-wall CVD (the required rf power is much smaller than that in cold-wall CVD). Since the horizontal

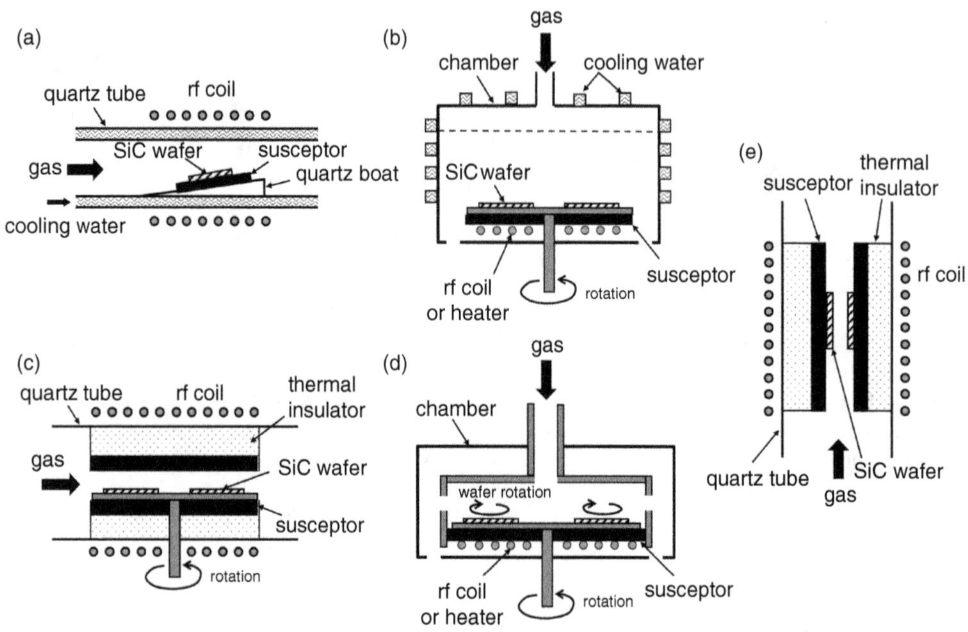

Figure 4.15 Schematic illustrations of several typical reactors employed for SiC CVD. (a) Horizontal cold-wall reactor, (b) vertical cold-wall reactor, (c) horizontal hot-wall reactor, (d) hot-wall (or warm-wall) planetary reactor, and (e) vertical chimney-type reactor.

hot-wall CVD reactor was proposed, several other hot-wall (or warm-wall) configurations have been proposed, as shown in Figure 4.15c [18, 21, 76, 77], (d) [78, 79], and (e) [80]. Among these reactor designs, a horizontal hot-wall CVD reactor with a rotating holder and a planetary warm-wall CVD reactor are commonly used for mass production of SiC epitaxial wafers.

4.2 Doping Control in SiC CVD

The *site-competition* effect discovered by Larkin *et al.* is a key concept for achieving wide-range doping control in SiC CVD [81, 82]. The doping efficiency of nitrogen is remarkably enhanced under Si-rich (low C/Si ratio) conditions and is reduced under C-rich (high C/Si ratio) conditions. This phenomenon can be explained by the competition between nitrogen and carbon atoms on the growing surface, because nitrogen atoms substitute the carbon lattice site in SiC. Low carbon-atom coverage on the growing surface promotes nitrogen incorporation into the lattice, while high carbon-atom coverage prevents nitrogen incorporation. Conversely, the doping of aluminum and boron, which substitute the silicon lattice site, shows the opposite trend: aluminum and boron incorporation are reduced under Si-rich conditions and enhanced under C-rich conditions.

4.2.1 Background Doping

Through process optimization and purification of source materials, the purity of nominally undoped (or unintentionally doped) SiC epitaxial layers is very high. The major source of unintentional dopants is nitrogen, for obvious reasons. Key ways to obtain high purity are to (i) increase the C/Si ratio [81, 82]

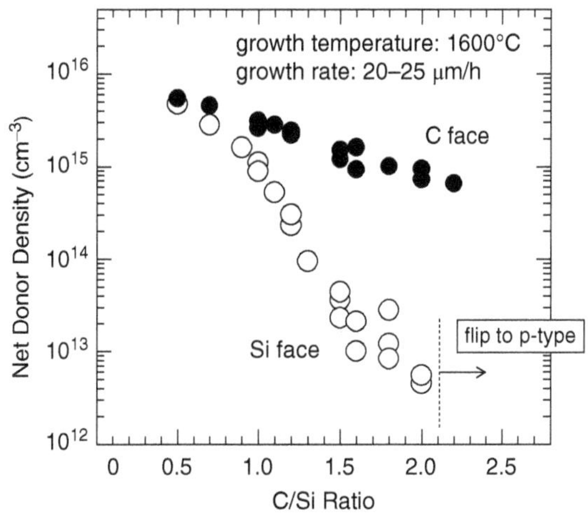

Figure 4.16 C/Si ratio dependence of the doping density of nominally undoped 4H-SiC{0001} epitaxial layers grown by hot-wall CVD.

and (ii) decrease the growth pressure [83, 84]. Figure 4.16 shows the C/Si ratio dependence of the doping density of nominally undoped 4H-SiC{0001} epitaxial layers grown by hot-wall CVD. In the case of a C/Si ratio of 0.5, the donor density is about 5×10^{15} cm^{-3}, irrespective of the substrate polarity. On the (0001)Si face, the donor density can be drastically reduced by increasing the C/Si ratio; for example, it reaches 5×10^{12} cm^{-3} for growth with a C/Si ratio of 2. Further increase in the C/Si ratio causes a switch of the conduction type from n-type to p-type in the nominally undoped epitaxial layers. Here, the p-type materials are obtained by reduction of nitrogen incorporation and enhancement of aluminum or boron incorporation, being consistent with the site competition concept. On the $(000\bar{1})$C face, however, the C/Si ratio dependence of the doping density is much smaller, and the lowest donor density is about 8×10^{14} cm^{-3} in this particular case. Higher nitrogen incorporation (and lower aluminum incorporation) on the C face is commonly observed in SiC CVD [81, 85–87] as well as in other growth techniques, including bulk growth. This result can be qualitatively explained by the different bond configurations on SiC(0001) and $(000\bar{1})$ faces, as described in Section 3.4. Figure 4.17 shows the growth pressure dependence of the doping density of nominally undoped and nitrogen-doped 4H-SiC{0001} epitaxial layers grown by hot-wall CVD [83]. Nitrogen incorporation is clearly suppressed when the growth pressure is decreased. This is partly attributed to an increase in the actual C/Si ratio on the growing surface at low pressure, mainly because of enhanced desorption of silicon atoms. Desorption of nitrogen atoms during surface migration may also be enhanced at low pressure.

4.2.2 n-*Type Doping*

In situ n-type doping is easily achieved by the introduction of N_2 during CVD growth. Figure 4.18 shows the donor density versus N_2 flow rate in hot-wall CVD of 4H-SiC(0001) at 1550 °C. The donor density determined from capacitance–voltage (C–V) characteristics is proportional to the N_2 flow rate over a wide range for CVD on both (0001)Si and $(000\bar{1})$C faces. When the growth temperature and pressure are fixed, the N_2 flow rate and the C/Si ratio are important parameters to achieve wide-range control of nitrogen doping (1×10^{14}–2×10^{19} cm^{-3}). The pressure dependence of nitrogen doping was investigated in detail [88]. The temperature dependence of nitrogen doping is more complicated. In cold-wall CVD of SiC, the nitrogen incorporation is suppressed on both the (0001) and $(000\bar{1})$ faces, when the growth temperature is elevated [89]. In hot-wall CVD, however, the nitrogen incorporation increases on (0001)

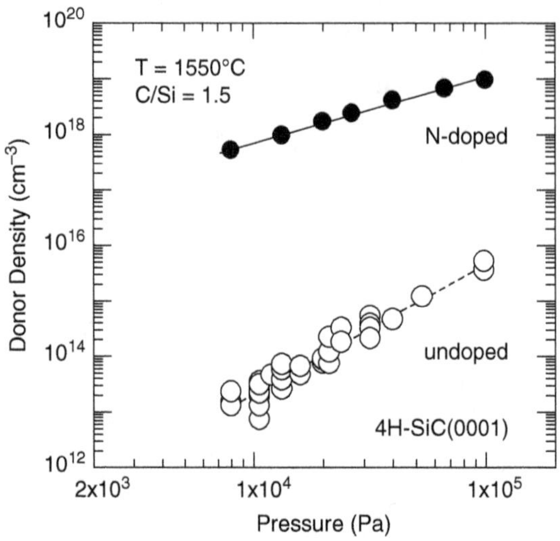

Figure 4.17 Growth pressure dependence of the doping density of nominally undoped and nitrogen-doped 4H-SiC{0001} epitaxial layers grown by hot-wall CVD ([83] reproduced with permission from AIP Publishing LLC).

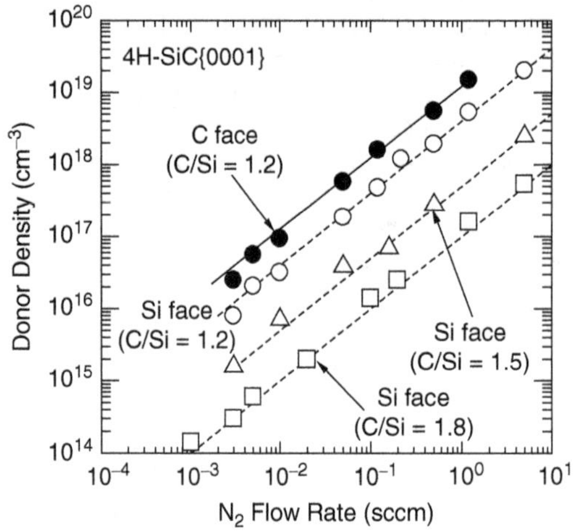

Figure 4.18 Donor density versus N_2 flow rate in hot-wall CVD of 4H-SiC(0001) at 1550 °C.

and decreases on (000$\bar{1}$) as the growth temperature is increased [87]. Growth simulation gives useful insights, but the detailed mechanism is still not fully understood. Although phosphorus doping has been investigated, it seems to be difficult to achieve high doping density [90].

4.2.3 p-*Type Doping*

The addition of a small amount of trimethylaluminum (TMA: $Al(CH_3)_3$) is effective for *in situ* p-type doping in SiC CVD [91]. Figure 4.19 shows the acceptor density versus the TMA flow rate in hot-wall CVD of 4H-SiC{0001} at 1550 °C. The acceptor density determined from $C-V$ measurements

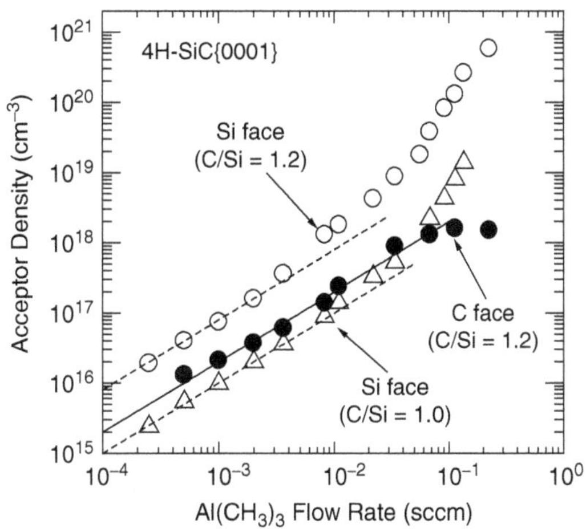

Figure 4.19 Acceptor density versus the TMA flow rate in hot-wall CVD of 4H-SiC{0001} at 1550 °C.

agrees well with the aluminum atom density determined by secondary ion mass spectrometry (SIMS) measurements. The doping efficiency is much higher (by a factor of 10–80) on the Si face than on the C face; again, this can be explained by the occupation site of aluminum atoms and the bond configuration. On the Si face, the acceptor density increases super-linearly with the TMA supply. This super-linearity may be attributed to the increased effective C/Si ratio under high TMA flow conditions, because the supply of TMA causes the growth conditions to become more C-rich as a result of release of CH_3 species from TMA molecules. The accessible range of aluminum doping is about 2×10^{14}–5×10^{20} cm^{-3} on SiC(0001). It is noted that heavily-doped p-type layers can be easily grown only on the Si face, while it is difficult to grow heavily-doped p-type layers on the C face. When the TMA supply is high, growth on the C face suffers from two- or three-dimensional nucleation, leading to a rough surface. Details of the dependence of aluminum doping on the growth conditions (temperature, pressure) are found in references [87, 92]. Other p-type dopants include boron (B) and gallium (Ga), which can be doped using B_2H_6 gas [93] or trimethylgallium (TMG), respectively [94]. However, boron- or gallium-doped SiC exhibits high resistivity, because of the large ionization energy of these dopants. Abnormal diffusion of boron atoms also causes problems in device processing [95].

Thus, the doping range can be greatly expanded by controlling the C/Si ratio during CVD. Furthermore, control of the C/Si ratio is an effective way to obtain a sharp transition between n-type and p-type epitaxial layers [96]. By synchronizing the change in the C/Si ratio to a dopant switch (from nitrogen to aluminum, for example), abrupt depth profiles of dopant atoms can easily be formed.

4.3 Defects in SiC Epitaxial Layers

4.3.1 Extended Defects

Various kinds of extended defects are present in SiC epitaxial layers. Some defects come from the substrates, while other defects are instead created during the epitaxial process. In this subsection, the classification and nature of extended defects observed in hexagonal SiC epitaxial layers are described.

4.3.1.1 Surface Morphological Defects

Except step bunching, SiC epitaxial layers grown on off-axis {0001} substrates exhibit several types of surface defects. Figure 4.20 shows the typical surface defects observed in 4H- and 6H-SiC{0001}

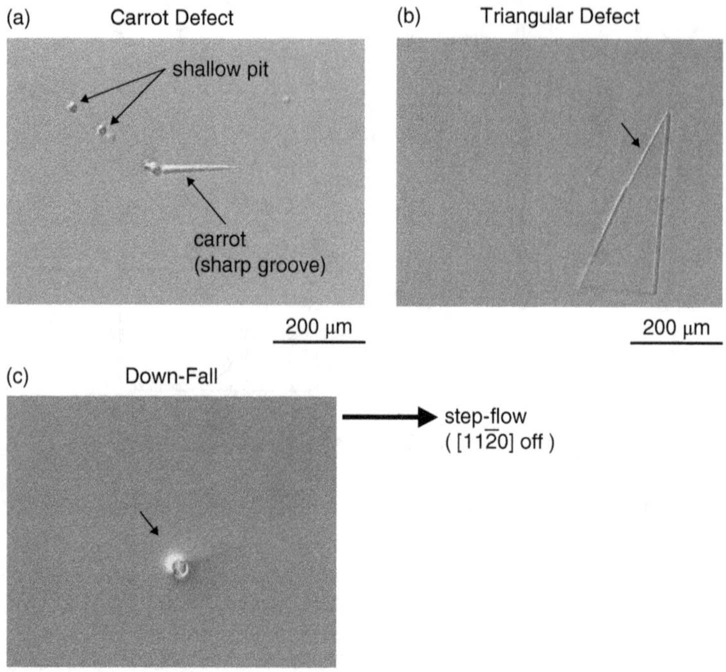

Figure 4.20 Typical surface defects observed in 4H- and 6H-SiC{0001} homoepitaxial layers: (a) "carrot" defect and shallow pit, (b) triangular defect, and (c) down-fall.

homoepitaxial layers, (a) "carrot" defect [23, 97–100] and shallow pit [20, 48], (b) triangular defect [99–101], and (c) down-fall. Although the exact formation mechanisms of these defects are not fully understood, they are usually created by technical issues such as incomplete removal of polishing damage or non-optimized growth processes. The down-fall is generated by a SiC particle initially formed on the susceptor wall falling down. The density of these defects is mostly influenced by the surface quality of the substrates and the conditions used for the growth process. The density of these defects depends only slightly on the substrate quality.

The carrot (in some case "comet" depending on the defect shape and structure), and triangular defects are usually elongated along the down-step direction of step-flow growth, which is a sign of disturbance of step-flow growth. As schematically shown in Figure 4.21, the defect length along the off-direction (L) is very close to the length of a basal plane in the epilayer when projected onto the surface, taking into account the substrate off-angle (θ):

$$L \approx d_{\text{epi}} / \tan\theta \tag{4.12}$$

where d_{epi} is the epilayer thickness. This observation has a very important implication – these defects are nucleated at the very initial stage of epitaxial growth. If these defects are observed (though they are not desirable), the epilayer thickness can be estimated from the defect length.

Careful TEM studies revealed that the carrot (and comet) defects contain both a basal plane fault and a prismatic plane fault [97–100]. It was also clarified that several types of carrot defects exist, where each has a slightly different arrangement of extended defects [98]. Figure 4.22 shows schematically the structure of a typical carrot defect [98], where a TSD ($\boldsymbol{b} = 1\boldsymbol{c}$) in a substrate is dissociated into two components of $\boldsymbol{c}/4$ and $3\boldsymbol{c}/4$. The $\boldsymbol{c}/4$ component is deflected into a basal plane and forms a Frank partial dislocation with the insertion of a single bilayer. This insertion naturally induces a basal slip, and

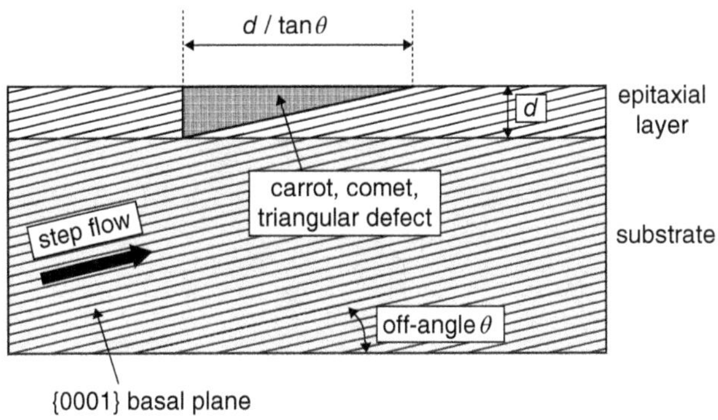

Figure 4.21 Schematic illustration of an epitaxially-induced defect in SiC. The defect length along the off-direction (L) is close to the length of a basal plane in the epitaxial layer when projected onto the surface, taking into account the substrate off-angle (θ).

the Burgers vector of this Frank partial dislocation is expected as $[0001]/4 + [1\bar{1}00]/3 = [4\bar{4}03]/12$. The other $3\boldsymbol{c}/4$ component penetrates into an epitaxial layer as a threading dislocation. Because of a stacking sequence mismatch, a prismatic fault is formed on $(1\bar{1}00)$ in the case of off-direction toward $[11\bar{2}0]$. Note that only a limited number (<1%) of TSDs in the substrate become nucleation sites for carrot (or comet) defects. Though the mechanism of the defect formation is not very clear, small pits created during *in situ* etching prior to CVD may play a key role in the carrot/comet defect formation [102]. It is also reported that a TSD in the substrate is not always necessary for nucleation of carrot (or comet) defects. Instead, generation of a TSD and a carrot (or comet) defect can occur simultaneously [98]. The triangular defects also exhibit a variety of structures. In some triangular defects, the triangular region is indeed 3C-SiC, while in other triangular defects only a 3C-like laminar region with a thickness of several Si-C bilayers is extended in the basal plane [100, 101]. In some cases, no 3C-SiC regions are observed, and a partial dislocation runs along the two sides of the triangular shape. So far, extended defects are not always observed beneath the shallow pits (typical depth: 20–100 nm), such as the one shown in Figure 4.20d. It is of interest that higher densities of carrot (or comet) defects and triangular defects are generated when the epitaxial layers are grown under Si-rich and C-rich conditions, respectively. Under Si-rich conditions, TSDs tend to be deflected into basal planes, forming Frank-type stacking faults. On the other hand, the migration length of adsorbed species decreases [103] and the supersaturation is increased under C-rich conditions [42]. These phenomena may influence the formation of carrot and triangular defects, respectively.

The typical density of these macroscopic defects in SiC homoepitaxial layers is approximately 0.02–2 cm^{-2}. The generation of shallow pits can be suppressed by minimizing the formation of Si droplets during the *in situ* etching process [20, 48]. When SiC devices include carrot (or comet) defects or triangular defects, the devices exhibit excessive leakage currents and significantly decreased breakdown voltages, while the impacts of shallow pits are negligibly small [104]. This is not surprising, because the stacking faults involved in carrot, comet, and triangular defects (in both basal and prismatic planes) should behave as leakage paths. Figure 4.23 shows the surface morphology of a 4H-SiC epitaxial layer observed by scanning electron microscopy (SEM) [105]. The image is taken with low acceleration voltage SEM to enhance the resolution. It is found that small depressions, 0.5–1 μm in size, are formed at the locations of TSDs and threading edge dislocations (TEDs) (it is hard to observe these depressions by normal optical microscopy). These depressions must be formed by disturbance of step flow at the dislocations. The depressions are shallow, about 3–20 nm in depth. Note that the shape and depth of these depressions depend greatly on the growth conditions including *in situ* etching and cooling processes. If deep depressions are formed, geometric effects, such as electric field crowding, significantly affect the device characteristics, as further described in Section 5.2.3.

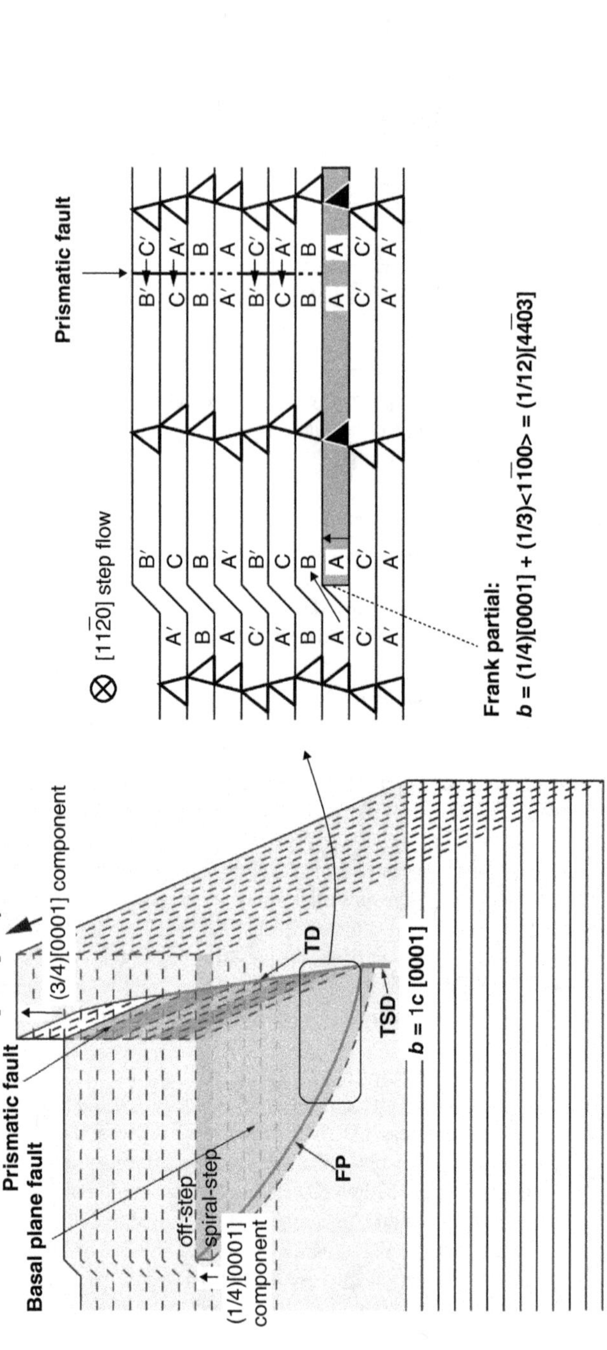

Figure 4.22 Schematic structure of a typical carrot defect observed in SiC epitaxial layers ([98] reproduced with permission from Wiley-VCH Verlag GmbH).

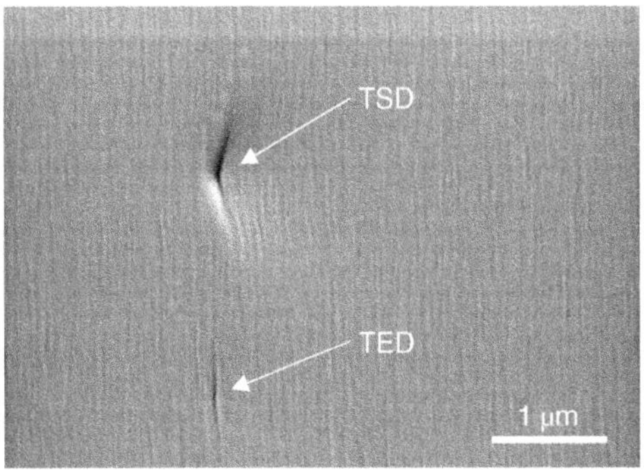

Figure 4.23 Surface morphology of a 4H-SiC epitaxial layer observed by scanning electron microscopy (SEM) ([105] reproduced with permission from The Electrochemical Society (ECS)). The image is taken with low acceleration voltage SEM to enhance the resolution.

4.3.1.2 Micropipes

In SiC CVD, most micropipes in a SiC substrate are replicated in the epitaxial layer grown on the substrate. Kamata *et al.* discovered that micropipes in an off-axis SiC(0001) substrate can be dissociated into several elementary closed-core screw dislocations during CVD growth, leading to "*micropipe closing*" [106, 107]. The mechanism is the same as that described for micropipe closing during bulk growth (Section 3.3.2). As shown in Figure 4.24, part of a superscrew dislocation, the core of a micropipe, is deflected into the basal plane through interaction with the lateral growth of steps. The deflected screw dislocation tends to change direction again and thread along the <0001> direction; the overall outcome is that one elementary screw dislocation has been separated from the superscrew dislocation. By repeating this conversion process several times, a micropipe is completely dissociated into individual elementary screw dislocations. The probability of micropipe closing depends strongl on the C/Si ratio during CVD; Si-rich conditions enhance micropipe closing [107]. By decreasing the C/Si ratio to 0.7, the micropipe closing probability reaches 99% or even higher. On the growing surface, competition exists between spiral growth around a micropipe and step-flow growth as a result of the off-axis substrate. It is believed that micropipe closing takes place when step-flow growth is dominant, even near the core of a micropipe, and the core is directly swept by steps proceeding in the lateral direction [107]. Under Si-rich growth conditions, step-flow growth is enhanced, while spiral growth is promoted under C-rich conditions [108]. Although micropipes are now almost eliminated in the substrates (wafers), micropipe closing during epitaxial growth is an interesting phenomenon from the viewpoint of defect engineering.

4.3.1.3 Dislocations

Now that the micropipe density in SiC wafers has been reduced down to well below 0.1 cm^{-2} (almost eliminated), normal dislocations and epi-induced defects such as carrot defects remain important quality issues in SiC epitaxial layers. Most dislocations in 4H-SiC homoepitaxial layers originate from dislocations in 4H-SiC substrates. Therefore, the dislocation density of a SiC homoepitaxial layer depends greatly on the quality of the substrate, assuming that the epitaxial growth process is sufficiently optimized. Major dislocations in SiC substrates include TSDs, TEDs, and basal plane dislocations (BPDs), as described in Section 3.3.

Figure 4.25 illustrates the dislocation replication and conversion typically observed in 4H-SiC epitaxial layers grown on off-axis {0001} by CVD [25, 77]. Almost all the TSDs in a substrate are replicated in an epilayer, but a small portion (typically <2%) of TSDs are converted to Frank-type partial dislocations [109]. A TSD in the substrate can act as a nucleation site for a carrot defect, as described above.

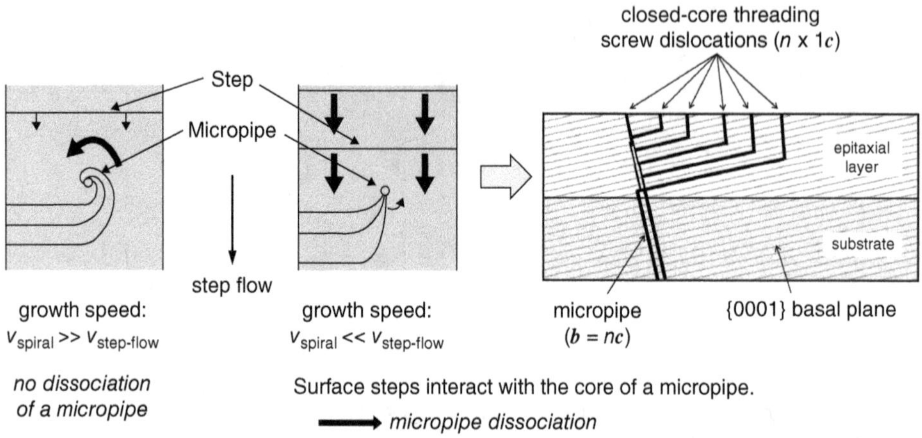

Figure 4.24 Schematic illustration of micropipe closing. A micropipe in an off-axis SiC(0001) substrate can be dissociated into several elementary closed-core screw dislocations during CVD growth.

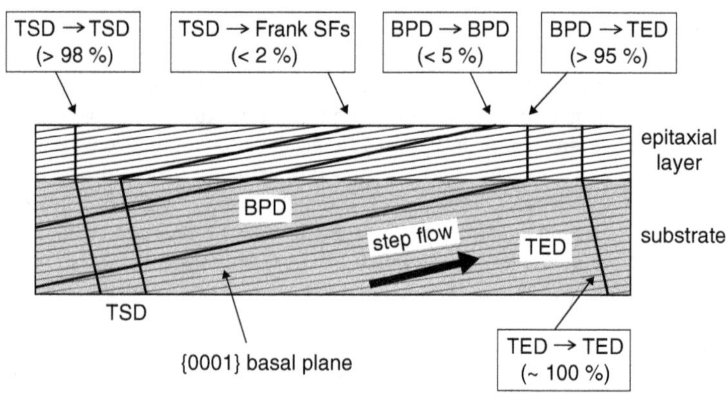

Figure 4.25 Schematic illustration of dislocation replication and conversion typically observed in 4H-SiC epitaxial layers grown on off-axis {0001} by CVD.

Behavior of BPDs during epitaxial growth is much more complicated. A BPD is a detrimental defect for SiC bipolar devices because it can be the source of a Shockley-type stacking fault upon carrier injection, and such a stacking fault causes local reduction of carrier lifetimes (increase of on-resistance) and increase in leakage current [110–112]. This is called "*bipolar degradation*", and is treated in a separate section (Section 5.2.2). Here, two important phenomena related to BPDs are described.

Conversion of BPD to TED

BPDs and TEDs possess the same Burgers vector $<11\bar{2}0>/3$); the naming is different depending on the dislocation direction (perpendicular or parallel to the *c*-axis). When a BPD is replicated in an epitaxial layer, it extends in a basal plane, which is inclined by several degrees from the crystal surface. Since the elastic energy of a dislocation is naturally proportional to the dislocation length [113], BPD replication in an epitaxial layer grown on off-axis {0001} results in a large increase in elastic energy of the dislocation. This energy is greatly decreased by conversion of a BPD to a TED, by which process the dislocation

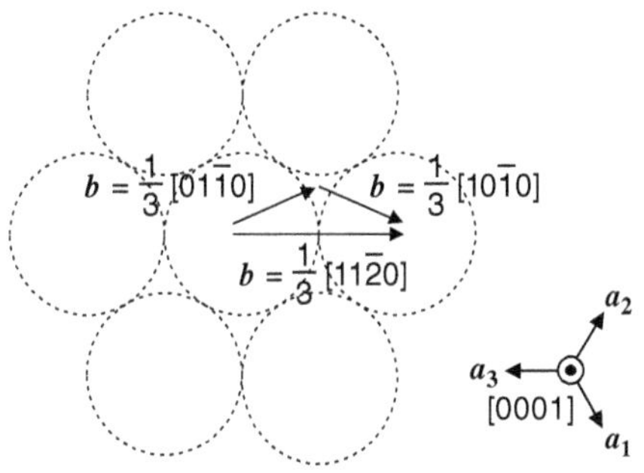

Figure 4.26 Split of a Burgers vector of a perfect BPD into two partial dislocations.

length is considerably shortened by a factor of cotθ (where θ is the off-angle). This dislocation conversion can be explained by a so-called "*image force*" applied to a BPD [113].

In reality, most (>90%) BPDs in the substrate are converted to TEDs within a few micrometers of an initial epitaxial layer without any special treatment [114–116]. Some BPDs are, however, replicated in a SiC epitaxial layer. It has been discovered that all these BPDs propagating in basal planes of an epitaxial layer are of a screw character [25, 117, 118]. It is known that a perfect BPD in SiC is dissociated into two partial dislocations, and a single Shockley-type stacking fault is created between the two partials [111, 119, 120]. This is expressed by using the Burgers vector as follows (also see Figure 4.26):

$$\frac{[11\bar{2}0]}{3}(\boldsymbol{b}_{\text{BPD}}) \rightarrow \frac{[10\bar{1}0]}{3}(\boldsymbol{b}_{\text{SSF1}}) + \frac{[01\bar{1}0]}{3}(\boldsymbol{b}_{\text{SSF2}}) \tag{4.13}$$

The schematic illustration of the BPD split is shown in Figure 4.27. The created stacking fault has a structure of (31) in Zhdanov's notation. This dissociation of a BPD takes place because of the energy gain. In a simple model, the energy balance per unit length is expressed by:

$$c_1|\boldsymbol{b}_{\text{BPD}}|^2 > c_2|\boldsymbol{b}_{\text{SSF1}}|^2 + c_2|\boldsymbol{b}_{\text{SSF2}}|^2 + \sigma_{\text{SF}}A_{\text{SF}} \tag{4.14}$$

where σ_{SF} is the stacking fault energy per unit area and A_{SF} the stacking fault area. c_1 and c_2 are the proportion constants associated with the shear modulus and defect geometry. Because the stacking fault energy in SiC is low, the relationship shown in Equation 4.14 is satisfied. The width of the Shockley-type stacking faults is typically 30–70 nm [119], and depends on the dislocation direction inside a basal plane as well as the doping density. The Shockley-type stacking fault width is usually wider for heavily-nitrogen-doped SiC; this arises from a gain in electrostatic energy caused by capture of free electrons at the localized level of the stacking faults. Because the two partial dislocations and the Shockley-type stacking fault between them cannot be directly deflected to the *c*-axis, it is supposed that the two partial dislocations must merge into a perfect BPD ($\boldsymbol{b} = \langle 11\bar{2}0\rangle/3$) before conversion to a TED. This has recently been confirmed by three-dimensional synchrotron X-ray topography analyses [121] and a careful TEM study [122].

Conversion from BPDs to TEDs is enhanced by several techniques, such as molten KOH etching [123–125] or H_2 etching [126] prior to epitaxial growth or interruption during growth [127]. These techniques form depressions at the location where BPDs intersect with the surface. Inside the depressions, the advance of surface steps is three-dimensional and two partial dislocations are forced to merge into a perfect BPD [125]. The use of a substrate with a lower off-angle is naturally effective to enhance the conversion, owing to the increased image force [128]. High-temperature (~1800 °C) annealing in Ar induces spontaneous conversion from a BPD to a TED near the surface (without growth) [129], which is consistent with the energy balance according to Equation 4.14. Furthermore, increasing the growth rate

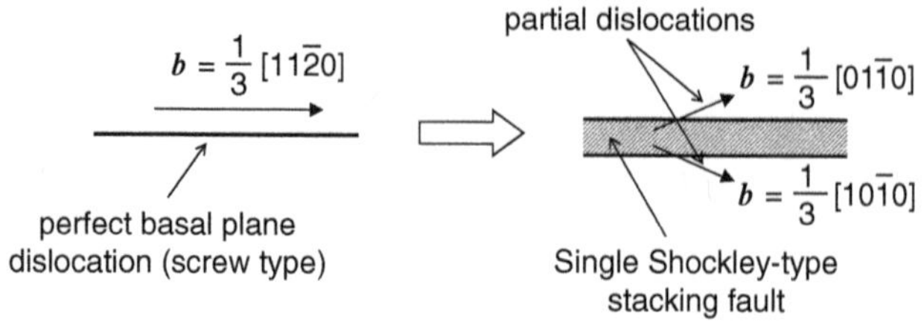

Figure 4.27 Schematic illustration of BPD split. A perfect BPD in SiC is dissociated into two partial dislocations, and a single Shockley-type stacking fault is created between the two partials.

also effectively enhances the BPD-TED conversion [46]. By combining these techniques, the conversion ratio has been increased to 99.8% or even higher. Thus, if the BPD density in a substrate is 1000 cm^{-2}, the density of BPDs replicated in an epitaxial layer is about $1-2$ cm^{-2}.

Formation of Interface Dislocation

BPDs easily glide under stress, because the critical resolved shear stress is relatively low in SiC, especially at high temperature, as described in Section 3.3. The glide motion of BPDs during epitaxial growth has in fact been observed by X-ray topography [130–133]. Figure 4.28a illustrates schematically the glide of a BPD during epitaxial growth of SiC [130]. A BPD replicated in an epitaxial layer is deflected to the direction normal to the step flow and the BPD often lies near the interface between a lightly-doped epitaxial layer and a heavily-doped substrate [130]. Thus, a BPD lying at the epitaxial layer/substrate interface is called an "*interface dislocation*". Note that this is not a pure *misfit dislocation* resulting from lattice relaxation caused by doping-induced misfit strain [134], because such interface dislocations are not observed, even for more than 100 μm-thick n^--type SiC epitaxial layers, when they are grown under appropriate conditions. A large temperature inhomogeneity can induce significant thermal stress, and when this stress is added to the misfit stress (i.e., acts in the same direction) pre-existing BPDs glide for lattice relaxation [133]. The direction of glide motion depends on the Burgers vector of the BPD and the stress direction. The situation is more complicated when a BPD is converted to a TED in the initial stage of epitaxial growth. The BPD component existing near the epitaxial layer/substrate interface glides, and a number of half-loop dislocations are created along the glide direction, as shown in Figure 4.28b [130, 135]. As a result, an array of half-loop dislocations is formed along $[1\bar{1}00]$ when the off-direction is $[11\bar{2}0]$. These dislocations can be identified as an array of TED pairs on the surface, as shown in Figure 4.28.

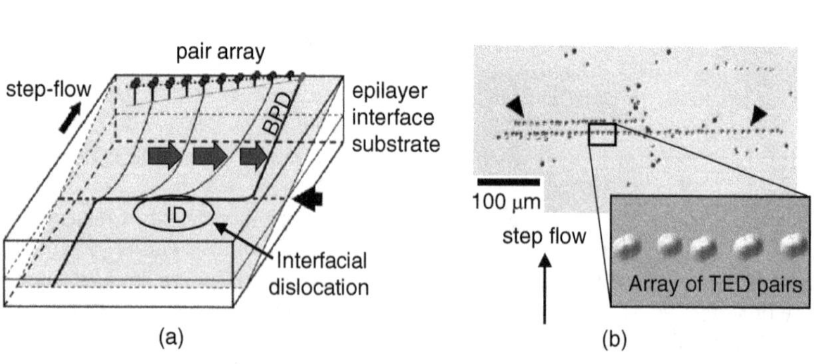

Figure 4.28 (a) Glide of a BPD during epitaxial growth of SiC and (b) formation of half-loop dislocations associated with the BPD glide ([130] reproduced with permission from AIP Publishing LLC).

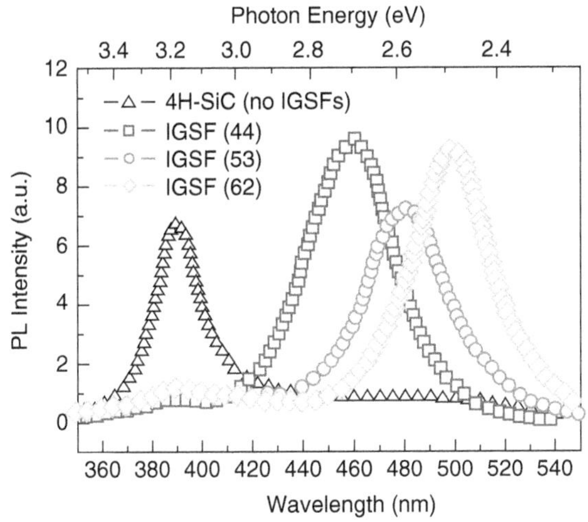

Figure 4.29 Micro-PL spectra acquired from several areas with and without in-grown stacking faults in a thick, high-purity 4H-SiC(0001) epitaxial layer at room temperature ([141] reproduced with permission from Elsevier).

4.3.1.4 In-Grown Stacking Faults

Nucleation of stacking faults (SFs) takes place during epitaxial growth, even if the substrate is free of stacking faults. In an early study, the existence of microscopic stacking faults in "high-quality" SiC epitaxial layers has been suggested, based on the observation of various abnormal photoluminescence peaks [136]. So far, several types of *in-grown SFs* have been identified by cross-sectional TEM. A majority of these SFs are caused by slips in basal planes (Shockley type). Note that most in-grown SFs are invisible in optical microscopy but photoluminescence (PL) mapping/imaging is a powerful method to detect these defects [137–140].

Figure 4.29 shows the micro-PL spectra acquired from several areas with and without in-grown SFs in a thick, high-purity 4H-SiC(0001) epitaxial layer at room temperature [141]. In the spectrum of the 4H-SiC matrix without SFs only one peak is observed, located at 390 nm and assigned to free excitons. In the spectra from the SF regions, however, distinct PL peaks at 455, 480, and 500 nm were observed, in addition to the weak band edge (free exciton) peak at 390 nm. Thus, PL-intensity mapping at a wavelength specific to each SF gives detailed information (location, shape, and density) about in-grown SFs. Figure 4.30 shows examples of PL-intensity maps taken at (a) 455, (b) 480, and (c) 500 nm from the same location [141]. The optical microscope image is also shown in Figure 4.30d. As shown in Figure 4.30, the shape of those SFs that show PL peaks at 455 nm is a right-angled triangle with its apex pointed toward the upstream side of the step flow. Conversely, the shape of the SF that shows 480 nm emission is an isosceles triangle, elongated along the off-direction. The lengths of all these SFs along the off-direction again agree with the projected length of a basal plane in the epitaxial layer, as was the case for carrot defects. This result implies that these SFs also nucleated in the initial stage of epitaxial growth. Although misalignment of atoms during step-flow growth has been suggested as the nucleation mechanism for SFs [142, 143], the detailed mechanism is not very clear at present. Figure 4.31 shows high-resolution TEM images taken from the major in-grown SFs that exhibit PL peaks at (a) 455, (b) 480, and (c) 500 nm [141]. The stacking sequences have been determined as (44), (53), and (62) types, respectively, in Zhdanov's notation. One-to-one correlation has been established between the PL peak position and the stacking sequence.

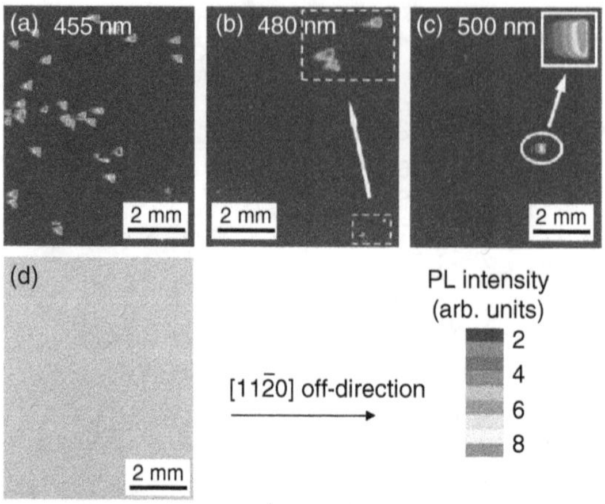

Figure 4.30 PL-intensity maps taken at (a) 455, (b) 480, and (c) 500 nm from the same location of a 4H-SiC (0001) epitaxial layer ([141] reproduced with permission from Elsevier). The surface morphology at the same location is shown in (d).

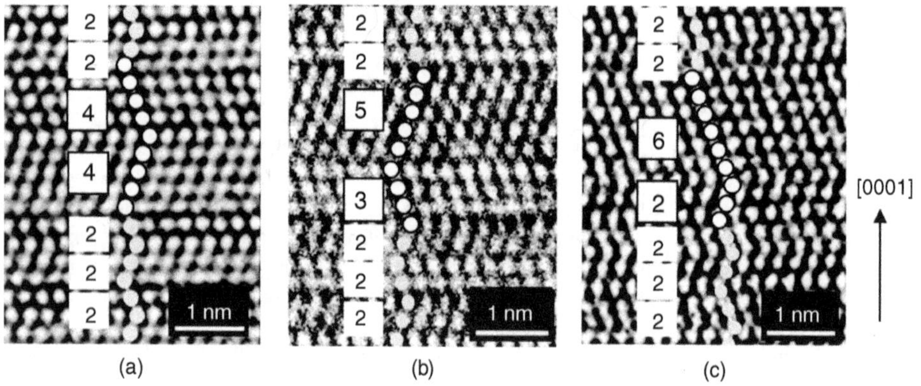

Figure 4.31 High-resolution TEM images taken from the major in-grown SFs that exhibit PL peaks at (a) 455, (b) 480, and (c) 500 nm, respectively ([141] reproduced with permission from Elsevier).

These in-grown SFs adversely affect device characteristics, for example, increase of leakage current [144]. Optimization of *in situ* H_2 etching and using lower growth rates at the start of epitaxial growth effectively reduce the density of these SFs. The total density of SFs is typically in the range 0.05–0.5 cm^{-2} for epilayers grown at a standard growth rate (5–15 μm h^{-1}); the SF density usually increases when the growth rate is increased [143]. Several kinds of Frank-type stacking faults (stacking sequences: (50), (42), and (41) in Zhdanov's notation) have also been identified [145]. Each Frank-type stacking fault also exhibits a luminescence peak at a particular wavelength. Elimination of Shockley- and Frank-type SFs remains an important issue in fast epitaxy of SiC.

4.3.2 Deep Levels

Another important type of defect in epitaxial layers is point defects, which create deep level(s) in a bandgap [146]. Deep levels are usually characterized by *deep level transient spectroscopy* (DLTS)

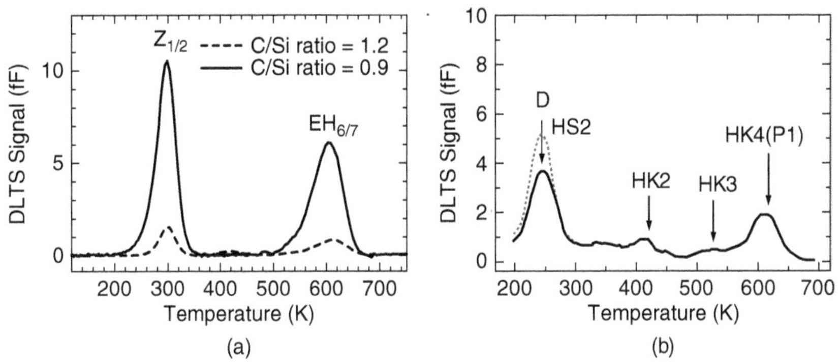

Figure 4.32 Typical DLTS spectra obtained from (a) n-type and (b) p-type 4H-SiC epitaxial layers with a doping density of about 2×10^{15} cm^{-3} ([168] reproduced with permission from Wiley-VCH Verlag GmbH).

measurements [147, 148] on SiC Schottky structures. The density of deep levels in lightly-doped as-grown 4H-SiC(0001) epitaxial layers is typically 5×10^{12}–2×10^{13} cm^{-3}, depending on the growth conditions. This value is fairly low, for a compound semiconductor and is acceptable for fabrication of unipolar devices. In fabrication of power metal-oxide-semiconductor field effect transistors (MOSFETs), for example, deep levels created by ion implantation are more important. For bipolar device applications, however, the above-mentioned density is not low enough, especially when a long carrier lifetime is required. Details are described in Section 5.3.

Figure 4.32 shows the typical DLTS spectra obtained from (a) n-type and (b) p-type 4H-SiC epitaxial layers with a doping density of about 2×10^{15} cm^{-3}. Figure 4.33 illustrates the energy levels of major deep levels observed in as-grown n-type and p-type 4H-SiC epitaxial layers [149–154]. Among these levels, the $Z_{1/2}$(E_c – 0.63 eV) [149] and EH6/7 (E_c – 1.55 eV) [150] centers are the dominant defects (commonly observed at densities of ((0.3–2) $\times 10^{13}$ cm^{-3}) in all as-grown epitaxial layers by CVD.

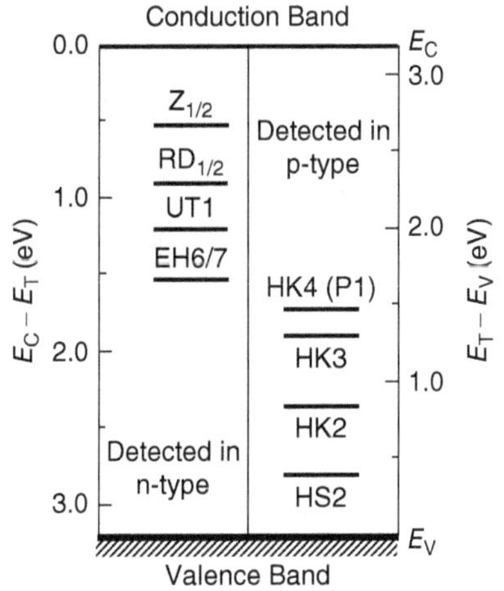

Figure 4.33 Energy levels of major deep levels observed in as-grown n-type and p-type 4H-SiC epitaxial layers.

Both centers are extremely stable against high-temperature (~1700 °C) annealing. In the lower half of the bandgap, the HK2 (E_v + 0.84 eV), HK3 (E_v + 1.24 eV), and HK4 (E_v + 1.44 eV) [153] centers are dominant deep levels. The densities of HK2, HK3, and HK4 centers are typically in the range (1–4) × 10^{12} cm^{-3}. Because the HK2, HK3, and HK4 centers almost disappear upon annealing at 1450–1550 °C [153], the $Z_{1/2}$ and EH6/7 centers are more important. Indeed, the $Z_{1/2}$ center has been identified as the dominant *lifetime killer*, at least for n-type 4H-SiC [155, 156]. Thus, it is very important to reduce and control the density of the $Z_{1/2}$ center to optimize the carrier lifetime in SiC bipolar devices. Note that the $Z_{1/2}$ and EH6/7 centers are also dominant in ion-implanted regions and dry-etched regions of 4H-SiC [149, 157–159]. As well as these levels, a few impurity-related levels are often observed in as-grown SiC epitaxial layers. For example, boron is a typical impurity that can be unintentionally doped from reactor parts (such as graphite susceptors). Boron (B) contamination creates the boron acceptor level (E_v + 0.35 eV) [160] and the boron-related "D center" (E_v + 0.55 eV) [160]. Another common impurity is titanium, which also comes from graphite parts as well as from pumping oil. Titanium (Ti) creates very shallow electron traps (E_c − 0.11/0.17 eV) in 4H-SiC [161]. The typical impurity density in CVD-grown 4H-SiC epitaxial layers is about (1–5) × 10^{13} cm^{-3} for boron and (0.5–5) × 10^{12} cm^{-3} for titanium.

The densities of $Z_{1/2}$ and EH6/7 centers are strongly dependent on the C/Si ratio and growth temperature; however, the growth rate only has a minor effect on defect generation, even if the growth rate is changed from 5 to 80 μm h^{-1} [83, 162–164]. Figure 4.34 shows the C/Si ratio dependence of the densities of $Z_{1/2}$ and EH6/7 centers in as-grown n-type 4H-SiC(0001) and (000$\bar{1}$) epitaxial layers [165]. Both the $Z_{1/2}$ and EH6/7 densities can be considerably reduced by increasing the C/Si ratio during CVD on (0001). In recent years, the origin of these centers has been identified as a carbon monovacancy with different charge states [166, 167]. Therefore, it is reasonable that the $Z_{1/2}$ density is high under Si-rich conditions and low under C-rich conditions. It is difficult to obtain a low $Z_{1/2}$ density in as-grown epilayers on the (000$\bar{1}$) face. The growth-temperature dependence of the $Z_{1/2}$ and EH6/7 densities is shown in Figure 4.35. The density of both defects exhibits significant increase with increasing growth temperature. This can be attributed to a higher equilibrium density of the carbon vacancy at higher temperatures, as described in detail in Section 5.3.1. In general, a higher growth temperature is preferable to obtain good surface morphology with a reduced density of extended defects (such as carrot-defects, in-grown SFs). However, high-temperature growth results in a high density of the $Z_{1/2}$ center, leading to a short

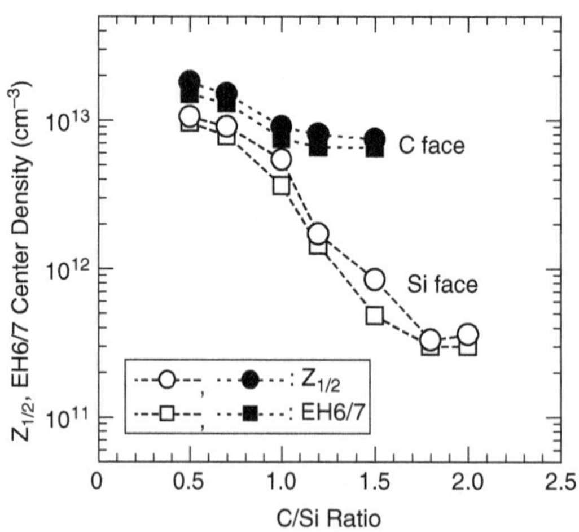

Figure 4.34 C/Si ratio dependence of the densities of $Z_{1/2}$ and EH6/7 centers in as-grown n-type 4H-SiC(0001) and (000$\bar{1}$) epitaxial layers ([165] reproduced with permission from The Japan Society of Applied Physics).

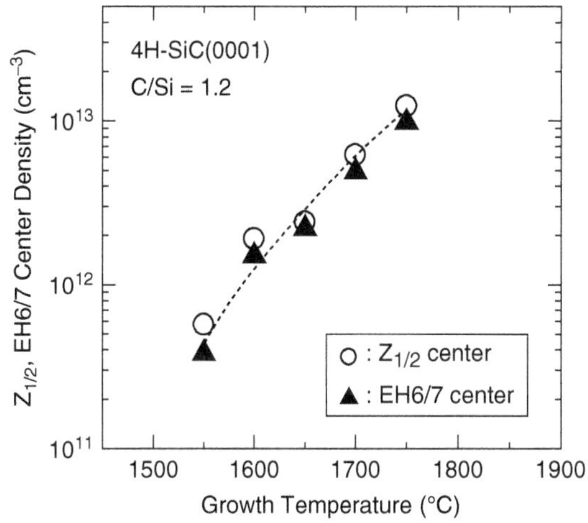

Figure 4.35 Growth-temperature dependence of the $Z_{1/2}$ and EH6/7 densities in CVD growth of 4H-SiC(0001).

carrier lifetime. Therefore, there is a trade-off between the low density of extended defects and a long carrier lifetime in as-grown epitaxial layers. After optimization of growth conditions, the typical density of the $Z_{1/2}$ center is about (3–6) × 10^{12} cm^{-3} in as-grown n-type 4H-SiC(0001) epitaxial layers, which gives a high-injection carrier lifetime of about 2–5 μs [168]. Enhancement of carrier lifetimes by defect reduction is described in Section 5.3.

4.4 Fast Homoepitaxy of SiC

Fast epitaxial growth of SiC is beneficial to increase the throughput, and thereby decrease the cost, of an epitaxial process. This is especially true for fabrication of very high voltage (>5 kV) devices, because the standard growth rate for mass production is 5–15 μm h^{-1} at present. As described in Chapter 7, lightly-doped voltage-blocking layers with thicknesses of about 40 and 80 μm are required to obtain blocking voltages of 5 and 10 kV, respectively. The CVD growth rate can, in principle, be increased by increasing the supply of precursors. In fast epitaxy of SiC, however, a few unique problems arise:

1. *Homogeneous nucleation of Si clusters in the gas phase.* In conventional CVD of SiC, SiH_4, and C_3H_8 (or C_2H_4) are employed as Si and C sources, respectively. Hydrocarbon molecules are rather stable and start to decompose at relatively high temperatures, above 1000–1200 °C. However, the decomposition temperature of SiH_4 molecules is much lower, as one can deposit amorphous or polycrystalline Si thin films by thermal CVD at 400 °C using SiH_4 as a precursor [169]. When the SiH_4 partial pressure becomes high (to increase the growth rate) in SiC CVD at 1550–1700 °C, the SiH_4 molecules introduced into a hot zone immediately decompose and start polymerizing via homogeneous nucleation (nSi → Si_n). This is called Si-cluster (in some cases, Si-droplet) formation. Because Si clusters can grow to a significant size (in excess of several tens of nm), such Si clusters have detrimental impacts on atomic-level epitaxial growth.
2. *Unstable step-flow growth on the growing surface.* Even if Si clusters are not formed in the gas phase, the supersaturation on the growing surface must be minimized to ensure stable step-flow growth in SiC. Very high supersaturation causes unwanted nucleation of a 3C-SiC phase on {0001} terraces or defect sites, which is also detrimental to high-quality homoepitaxy of 4H-SiC.

Several successful approaches that circumvent these problems have been reported. To overcome homogeneous nucleation in the gas phase: (i) decrease in the growth pressure [46, 170, 171], (ii) use of chlorine-based chemistry [24, 26, 172–174], and (iii) increase in the growth temperature [175–177] are effective. For stable step-flow growth, increase in growth temperature is also beneficial, but CVD growth at temperatures above 1750 °C can result in a high density of $Z_{1/2}$ centers, as described in Section 4.3.2. CVD growth on SiC{0001} substrates with a larger off-angle (8° rather than 4°) is another way to ensure stable step-flow growth, though this is not desirable in the recent trend.

In the first approach (decrease in growth pressure), the partial pressure of SiH_4 is reduced, which naturally hinders Si-cluster formation in the gas phase. As a result, more Si species are supplied onto the substrate surface, with minimal loss in the gas phase, resulting in a higher growth rate and improved morphology as the growth pressure is decreased. In fact, a remarkable improvement in surface morphology in fast epitaxy of 4H-SiC(0001) has been attained, with growth rates of 50–250 μm h^{-1} [46, 170, 171]. In this technique, all the knowledge obtained in conventional CVD processes can be transferred. Doping of nitrogen (n-type) and aluminum (p-type) across a wide range of densities (10^{14}–10^{19} cm^{-3}) is achieved, as is also the case for CVD at standard growth rates. A major drawback to this approach is the low growth efficiency. Because of the high gas flow rate and low pressure, the gas velocity is very high. As a consequence, most of the precursors and Si clusters formed in the gas phase are pushed away from the susceptor zone to the outlet. This may cause problems in the pumping and exhaust systems.

The second approach (using chlorine-based chemistry) can be more elegant, because the chemistry is more suitable for fast epitaxy. The larger bonding energy of Si-Cl compared with that of Si-H and Si-Si, means that appropriate precursors containing chlorine greatly suppress formation of Si clusters. Such precursors include $SiCl_4$ [178], $SiHCl_3$ [179], SiH_2Cl_2 [180], and SiH_3Cl [181]. These precursors do not decompose below 800 °C (cf. 400 °C for SiH_4) and start to form $SiCl_x$ (mainly $x = 2$) above about 1000 °C in a H_2 ambient [182]. CH_3Cl [183] and $SiCH_3Cl_3$ [184, 185] have also been used successfully for fast epitaxy of SiC. Another way is to simply add HCl into a conventional SiH_4-based chemistry [76, 172, 173, 186], though formation of Si clusters takes place to some extent. By using the chlorine-based chemistry, very high growth rates of 50–170 μm h^{-1} have been reported, while maintaining good morphology [24, 26]. The optimum Cl/Si ratio is dependent on the precursors and growth temperature, and is typically 1.5–3 (or 3–5 in the case of HCl addition). Because SiH_xCl_y is a standard precursor in epitaxial growth of Si at a mass production level [187], the purity and safety of these precursors should not be a concern. Incorporation of chlorine atoms or generation of chlorine-related deep levels has not been observed in the epitaxial layers. Nitrogen doping is easy (as is the case for more standard chemistry), but aluminum incorporation often exhibits a saturation in the range of 10^{18} cm^{-3} [188]. This may be ascribed to formation of stable $AlCl_3$ molecules in the gas phase when an aluminum source is introduced. Chlorine-based CVD of SiC has been reviewed in [26].

In the third approach, the growth temperature is increased from 1550–1650 °C to 1750–1900 °C [175–177], while using a conventional chemistry. At such high temperature, the formed Si clusters can decompose in a hot zone, leading to supply of more Si species. To avoid thermal convection at such a temperature, a chimney-type CVD reactor has been developed [175]. High growth rates of 30–70 μm h^{-1} and good morphology are obtained.

Figure 4.36 shows the growth rate versus the concentration of Si-precursors in the gas supply, for 4H-SiC growth using various gas chemistries reported in the literature. It is not easy to judge which chemistry is the best because different reactor designs with different sizes are employed. The process of fast epitaxy of SiC is still being developed, and further improvements will occur.

The extent of step bunching and other morphological instabilities in fast epitaxy of 4H-SiC have been considerably reduced by process optimization. When the growth rate is increased, the density of in-grown stacking faults tends to increase [142], most likely because the probability of misalignment of adatoms near surface steps increases. As an unexpected benefit of fast epitaxy, enhanced conversion from BPDs in a substrate to TEDs in an epitaxial layer is observed [46]. With regard to the deep levels, the growth-rate dependence of the $Z_{1/2}$ density is very small. However, a new deep level, UT1 (E_c – 1.39 eV), appears

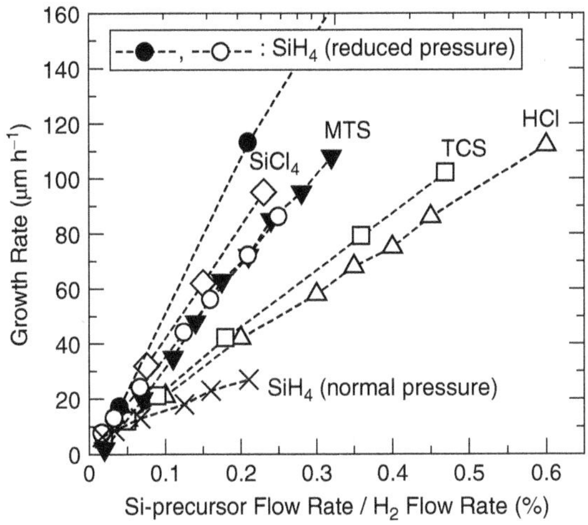

Figure 4.36 Growth rate versus concentration of Si-precursors in the gas supply, for 4H-SiC growth using various gas chemistries reported in literature.

and increases in number with increasing growth rate [46]. Thus, defect reduction is an important issue in fast epitaxy.

4.5 SiC Homoepitaxy on Non-standard Planes

4.5.1 SiC Homoepitaxy on Nearly On-Axis {0001}

A major driving force for developing SiC homoepitaxy on nearly on-axis {0001} substrates is elimination of BPDs in the epitaxial layers. The wafer cost may be reduced by using nearly on-axis {0001} wafers when the SiC boules are grown on on-axis {0001} seeds. It is, however, not very easy to ensure stable step-flow growth if the substrate's off-angle becomes smaller than 2°, as shown in Figure 4.7. This is especially true in fast epitaxy on nearly on-axis substrates. The disturbance or instability of step-flow growth can result in nucleation of 3C-SiC and/or stacking fault generation.

In an early study by Powell *et al.* [32], homoepitaxy of 6H-SiC on 0.2° off-axis 6H-SiC(0001) was demonstrated by removal of defective sites by appropriate *in situ* HCl/H_2 etching. This process was refined, and relatively large area homoepitaxy of 6H-SiC on nearly on-axis (0001) substrates was achieved under Si-rich conditions [189]. CVD growth under Si-rich (low C/Si ratio) conditions is a common approach in homoepitaxial growth of SiC on {0001} substrates with low off-angles. This is because Si-rich conditions increase the surface migration length of adsorbed species, and thus result in low supersaturation on the growing surface [103]. The major drawback of CVD under Si-rich conditions is greater nitrogen incorporation in SiC epitaxial layers. Neudeck *et al.* developed homoepitaxy of dislocation-free 4H- and 6H-SiC(0001) by using small (0.4 mm × 0.4 mm) mesa structures [190, 191].

More recently, Kojima *et al.* discovered that 4H-SiC without 3C-SiC inclusions can be homoepitaxially grown on nearly on-axis (about 0.3° off-axis) 4H-SiC(000$\bar{1}$) [66, 192]. Although the reasons for this success are unclear at present, the difference in the surface energy is suggested (this is also the reason that bulk 4H-SiC can be easily grown on a (000$\bar{1}$) seed by sublimation) [193]. Homoepitaxy on nearly on-axis 4H-SiC(0001) has been significantly improved by adopting *in-situ* etching with SiH_4 addition prior to CVD growth [194]. Figure 4.37 shows optical microscopy images of 4H-SiC epitaxial layers grown on (a) 4H-SiC(000$\bar{1}$) and (b) (0001) substrates with various off-angles from 0.24° to 0.79° [194].

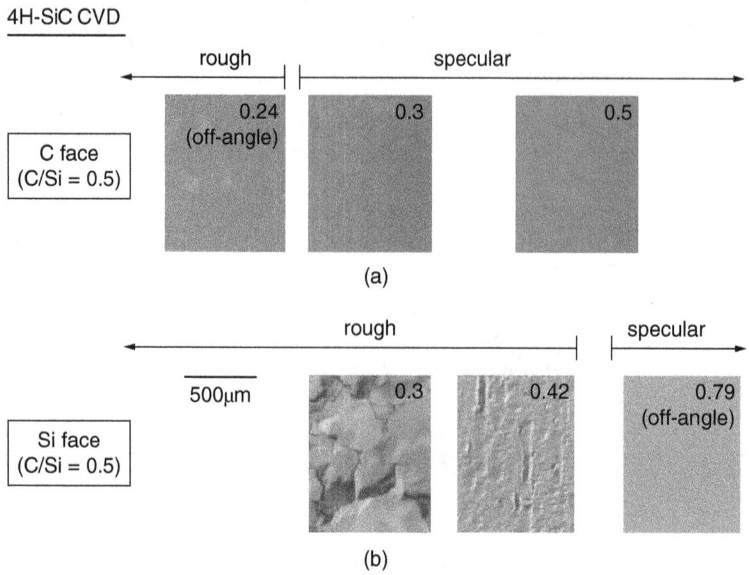

Figure 4.37 Optical microscopy images of 4H-SiC epitaxial layers grown on 4H-SiC (a) (000$\bar{1}$) and (b) (0001) substrates with various off-angles from 0.24° to 0.79° ([194] reproduced with permission from Trans Tech Publications).

Here, the growth temperature and growth rate are 1600 °C and about 5 μm h^{-1}, respectively. Homoepitaxy of 4H-SiC with good morphology is attained on 0.3° off-axis (000$\bar{1}$) and 0.79° off-axis (0001). The use of chlorine-based chemistry for 4H-SiC homoepitaxy on nearly on-axis (0001) substrates is also very effective, and homoepitaxy at high growth rates, over 100 μm h^{-1}, was realized [195–197]. Chlorine may preferentially etch defective islands and/or 3C-SiC inclusions during growth.

Achieving wide-range doping control capability and characterizing defects in more detail remain issues. Another concern is the role of TSDs. Spiral growth around TSDs naturally enables polytype replication in SiC epitaxial layers because the stacking sequence appears perfectly along the step edges created by the spiral. In the future, however, the density of TSDs will be greatly reduced by improvements in bulk growth technology. Therefore, homoepitaxy on nearly on-axis {0001} will become more difficult when TSDs in SiC wafers are almost eliminated.

4.5.2 *SiC Homoepitaxy on Non-basal Planes*

The typical non-basal planes used for epitaxy include 4H-SiC (11$\bar{2}$0), (1$\bar{1}$00), and {03$\bar{3}$8} (equivalent to {01$\bar{1}$4} in 6H-SiC) faces. Note that the (11$\bar{2}$0) face is crystallographically equivalent to the (110) face in a cubic crystal, and the {03$\bar{3}$8} face is semi-equivalent to the (001) face [198]. In an early study, low-temperature (1200–1350 °C) homoepitaxy by CVD was demonstrated on 6H-SiC(11$\bar{2}$0) [199] and (01$\bar{1}$4) [200]. The stacking sequence of 6H-SiC (ABCACB ...) appears directly on the surfaces of these faces, and no intentional off-angles are required to realize homoepitaxy of 6H- or 4H-SiC on these non-basal planes. As a consequence, the occupation site of adsorbed species is uniquely fixed by the bonds at any site on the surface, enabling homoepitaxy at low temperature. However, the range of optimum values for other conditions (such as the C/Si ratio) become narrow with decreasing growth temperature and macroscopic defects are easily generated. The residual nitrogen density also increases in

SiC epitaxial layers grown at low temperature [89]. Thus, low-temperature homoepitaxy is not attractive for power device applications.

The growth rates on the non-basal planes are almost the same as that on off-axis (0001) under the same conditions, indicating that the growth is controlled by mass transport [201]. The 4H-SiC$(11\bar{2}0)$, $(1\bar{1}00)$, and $\{03\bar{3}8\}$ epitaxial layers exhibit very good morphology and a small surface roughness of 0.12–0.22 nm in a $10 \times 10\ \mu m^2$ area, without any of the triangular defects and "carrot" defects that occasionally appear on off-axis SiC(0001) epitaxial layers [201–203]. No signs of "step-flow" growth (such as step bunching) have been observed for these epilayers, suggesting layer-by-layer growth.

The doping efficiencies of nitrogen and aluminum in CVD growth on these non-basal planes are in between those on off-axis (0001) and $(000\bar{1})$ faces. The typical background doping density is $2 \times 10^{13}\ cm^{-3}$ on off-axis (0001), $3 \times 10^{14}\ cm^{-3}$ on $(11\bar{2}0)$ and $(03\bar{3}8)$, and $8 \times 10^{14}\ cm^{-3}$ on off-axis $(000\bar{1})$. When nitrogen is introduced during CVD, the epitaxial layers grown on 4H-SiC $(11\bar{2}0)$, $(1\bar{1}00)$, and $(03\bar{3}8)$ always show higher donor densities than those of the layers grown on off-axis (0001). Figure 4.38 shows the C/Si ratio dependence of nitrogen density, determined by SIMS, for 4H-SiC epitaxial layers grown by hot-wall CVD on $(11\bar{2}0)$, $(03\bar{3}8)$, and off-axis (0001) and $(000\bar{1})$ faces. The nitrogen incorporation is significantly reduced by increasing the C/Si ratio on $(11\bar{2}0)$, $(03\bar{3}8)$, and off-axis (0001), as a result of the site competition effect. The C/Si ratio dependence is the largest on (0001), while $(11\bar{2}0)$ and $(03\bar{3}8)$ faces show smaller changes in nitrogen incorporation. On the $(000\bar{1})$ face, however, a higher C/Si ratio does not always lead to lower nitrogen density. In aluminum doping, the doping efficiency on these non-basal planes is lower than that on (0001) and higher than that on $(000\bar{1})$. Interestingly, the densities of the major deep levels, $Z_{1/2}$ and EH6/7 centers, in 4H-SiC epitaxial layers on $(11\bar{2}0)$, $(1\bar{1}00)$, and $(03\bar{3}8)$ are higher than those on (0001) and lower than those on $(000\bar{1})$ [165].

Major extended defects observed in 4H-SiC $(11\bar{2}0)$ and $(1\bar{1}00)$ epitaxial layers are BPDs replicated from the substrates. Stacking faults are also often observed in these epitaxial layers [204]. In CVD growth of 4H-SiC $\{03\bar{3}8\}$, the situation is more complicated. Threading screw and edge dislocations tend to deflect into basal planes under Si-rich conditions, whereas they propagate along nearly <0001> under C-rich conditions [205].

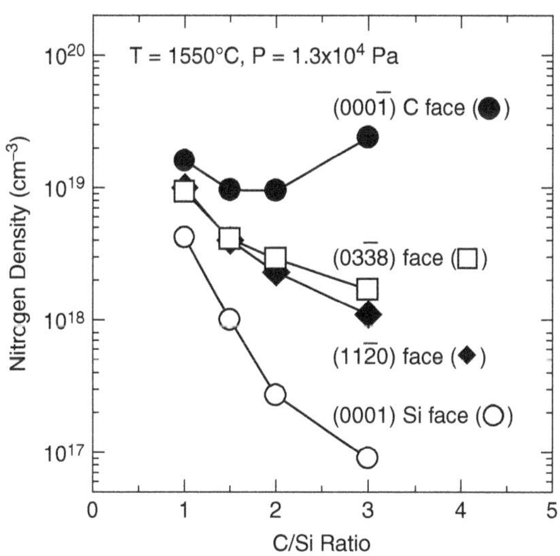

Figure 4.38 C/Si ratio dependence of nitrogen density, determined by SIMS, for nitrogen-doped 4H-SiC epitaxial layers grown by hot-wall CVD on $(11\bar{2}0)$, $(03\bar{3}8)$, and off-axis (0001) and $(000\bar{1})$ faces ([201] reproduced with permission from Elsevier.

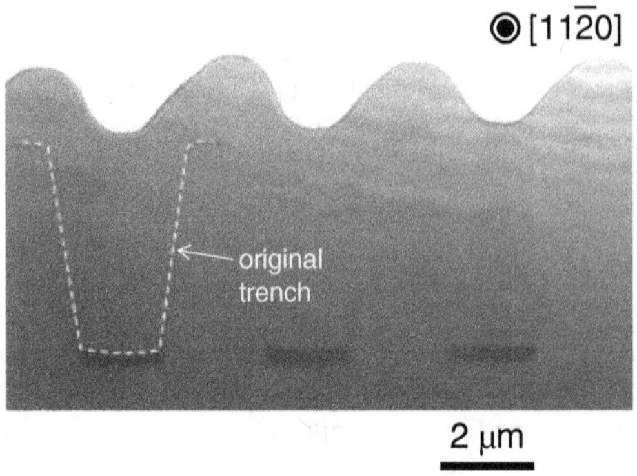

Figure 4.39 Cross-sectional TEM image of SiC trenches filled by embedded epitaxy ([211] reproduced with permission from Trans Tech Publications).

4.5.3 Embedded Homoepitaxy of SiC

Embedded homoepitaxial growth of SiC is useful for fabricating unique structures without needing ion implantation. Examples of such structures include the channel regions of vertical static induction transistors (or junction field effect transistors) [206, 207] and of lateral metal-semiconductor field effect transistors [208]. Reports of homoepitaxial growth of SiC on trench structures have been published [209–211]. When a trench is formed on off-axis SiC(0001), homoepitaxial growth takes place from the trench bottom, as well as from the trench sidewalls. In general, the growth rate and impurity incorporation for the layer grown from the bottom are different from those for the layers grown from the sidewalls. Furthermore, the crystal planes of trench sidewalls are important, because the crystalline quality and impurity incorporation also depend on these planes. When the off-axis is introduced along the [11$\bar{2}$0] direction, (1$\bar{1}$00) and ($\bar{1}$100) are typical planes of trench sidewalls (if the sidewall angle is nearly 90°). Another set of trench sidewalls can be (11$\bar{2}$0) and ($\bar{1}\bar{1}$20), but in this case, the actual sidewall planes are tilted from the original (11$\bar{2}$0) and ($\bar{1}\bar{1}$20) by the off-angle (typically 4°). It is reported that SiC pn junctions formed by embedded epitaxy exhibit lower leakage current when the sidewalls are {11$\bar{2}$0} rather than {1$\bar{1}$00} [212]. Selective embedded epitaxy by using carbon [213] or TaC [214] as masking materials has also been reported.

Figure 4.39 shows a cross-sectional TEM image of SiC trenches filled by embedded epitaxy [211]. When the CVD growth was carried out with a standard C/Si ratio (C/Si = 1.5) at a growth temperature of 1550 °C, the epitaxial layer near the top of a trench became thicker, leading to formation of a void inside the trench (not shown). In contrast, a low C/Si ratio (C/Si = 1.0) and a high growth temperature of 1650 °C were employed in Figure 4.39, where complete trench filling without voids is achieved. In CVD growth under Si-rich conditions at high temperature, surface migration of adsorbed atoms is enhanced. Thus, a trench is filled from the bottom region, because of the greater surface migration, leading to complete filling.

4.6 SiC Homoepitaxy by Other Techniques

Although CVD is the standard technique for epitaxial growth of SiC, other growth techniques have been also investigated. Such techniques include LPE, sublimation epitaxy, and MBE.

As described in Chapter 3, there is no stoichiometric SiC liquid phase, which means that it is impossible to employ congruent liquid phase epitaxial growth of SiC. LPE growth of SiC is performed by dipping SiC substrates into a Si-based melt contained in a graphite crucible [1, 2, 215, 216]. A small amount of the graphite is dissolved into the Si melt and transported to the SiC substrate, where it acts as a carbon source. Typical growth temperatures and growth rates are 1550–1700 °C and 5–30 μm h^{-1}, respectively. Micropipe closing during SiC epitaxial growth was first reported in LPE [216, 217], and later the micropipe-closing technique was refined in CVD. The capability of performing homoepitaxial growth of 4H-SiC on nearly on-axis {0001} substrates is a major advantage of LPE. However, control of the C/Si ratio during epitaxial growth is limited, and incorporation of impurities from the graphite crucible must be significantly reduced to obtain high-purity SiC epitaxial layers. Formation of macrosteps on the surface is another issue. Scale-up of the equipment and development of a multi-wafer growth system (such as one that can treat 6 × 150 mm. wafers), which is critical for volume production of SiC devices, will be a challenge in LPE.

A modified LPE technique, so-called *metastable solvent epitaxy* (MSE) has been reported [218]. In the MSE technique, a polycrystalline 3C-SiC plate is employed as the source material, and a Si melt is formed between the source and a SiC substrate. Because of the difference in chemical potential between 3C-SiC and 4H-SiC, 4H-SiC is homoepitaxially grown on the 4H-SiC substrate.

In sublimation epitaxy, the physical phenomena are similar to those for sublimation growth of SiC boule crystals, as described in Chapter 3. The epitaxial growth consists of three steps: (i) sublimation of the SiC source, (ii) mass transport of sublimed species, and (iii) surface reaction and crystallization [219]. In sublimation epitaxy, however, the distance between the SiC source and substrate is very small, typically 2–5 mm. Typical growth temperatures and pressure are 1600–1800 °C and 10^{-3} Pa, respectively. Fast epitaxial growth of high-quality SiC is possible on a nearly on-axis {0001} substrate with a growth rate of 30–80 μm h^{-1} [220, 221]. The purity of SiC epitaxial layers is mainly limited by the purity of the SiC source (usually polycrystalline SiC plate), and it is not easy to obtain a very low background doping density of 10^{13}–10^{14} cm^{-3}. Wide range doping control (both n- and p-types) has not yet been demonstrated using sublimation epitaxy. Formation of epitaxial pn junctions is also difficult with this technique.

The main benefits of MBE include atomic-level control of epitaxial growth and the ability to monitor the surface by *in situ* electron diffraction. In MBE growth of SiC, either solid sources (silicon, graphite) [222, 223] or gas sources (SiH_4, Si_2H_6, C_2H_4, C_3H_8) [224] are employed. Intentional polytype control during MBE growth of SiC has been investigated to realize hetero-polytypic junctions such as 4H/6H and 3C/4H [223]. However, growth at very high temperatures, above 1500 °C, as required for homoepitaxy of high-quality 4H-SiC, is technologically difficult in MBE, because of the limited ability in heating components. Furthermore, the surface of SiC is immediately graphitized in vacuum at temperatures above 1000 °C. Although high-quality homoepitaxy of 6H-SiC has been reported, the growth rate is very low, about 0.05–0.2 μm h^{-1} [224]. Nevertheless, MBE of SiC will be useful to obtain insights into the mechanisms of SiC epitaxy.

4.7 Heteroepitaxy of 3C-SiC

Because of the instability of 3C-SiC at very high temperature, growth of large 3C-SiC bulk crystals by the sublimation method is difficult, as described in Chapter 3. Instead, heteroepitaxial growth of SiC on a foreign substrate has been intensively investigated. The major substrates employed for 3C-SiC growth are silicon and hexagonal SiC. Heteroepitaxial growth is usually performed by a CVD technique.

4.7.1 Heteroepitaxial Growth of 3C-SiC on Si

Heteroepitaxial growth of 3C-SiC on Si must overcome large mismatches of the crystal lattice (20%) and thermal expansion coefficient (8%). The growth temperature is restricted to below 1350 °C because of the melting point of Si (1415 °C). Substantial slip lines can be introduced into the Si substrate during growth close to the melting point. Nevertheless, growth of monocrystalline 3C-SiC layers is achieved

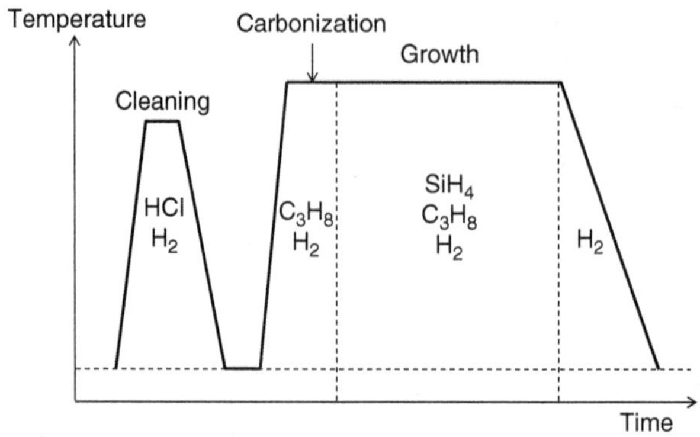

Figure 4.40 Typical growth program employed for CVD of 3C-SiC on Si. In the first step, the surface of Si is cleaned by *in situ* HCl/H_2 etching and a "carbonized buffer layer" is formed by introduction of hydrocarbon.

by a two-step growth program [225–233]. Figure 4.40 shows the typical growth program employed for CVD of 3C-SiC on Si. In the first step, the surface of Si is cleaned by *in situ* HCl/H_2 etching and a "*carbonized buffer layer*" is formed by introduction of hydrocarbon [225]. This step is immediately followed by the second step, which is the main growth of 3C-SiC. The carbonized buffer layer is actually very thin monocrystalline 3C-SiC [227, 230]. This layer must be free of pinholes and grain boundaries. If some pinholes and grain boundaries exist, Si atoms are supplied from the substrate underneath during subsequent CVD growth [234], leading to formation of voids below the 3C-SiC layer as well as to abnormal growth on the surface. To seal off the Si supply from the substrate, the nucleation density must be very high and the grown buffer layer should be of high crystal quality. To satisfy these requirements, hydrocarbon is introduced from relatively low temperatures, and the temperature ramp rate must be sufficiently high. By using this two-step growth, 3C-SiC can be heteroepitaxially grown on Si in a reproducible manner [225, 226]. When a Si(001) substrate is employed, a 3C-SiC(001) layer is obtained, and 3C-SiC(111) is grown on Si(111). The typical growth temperature and growth rates are 1350 °C and 1–5 $\mu m\ h^{-1}$, respectively.

Inside the buffer layer and the initial 3C-SiC layer grown on it, a high density of dislocations ($>10^9-10^{10}\ cm^{-2}$) and stacking faults exist. These extended defects meet each other and can be considerably reduced after the first several-100-nm-thick layer [230, 235, 236]. As a result, monocrystalline 3C-SiC with much improved quality is grown on Si. The density of dislocations in subsequent 3C-SiC layers is $10^7-10^8\ cm^{-2}$. Various types of stacking faults and microtwins are also observed in the 3C-SiC layers. Although these stacking faults can be reduced by employing "undulant Si substrates" [237, 238], as described in Chapter 3, the stacking fault density is still far from a satisfactory level. Another issue is severe wafer warpage because of the significant stress caused by the lattice and thermal expansion mismatch.

Another type of extended defect in 3C-SiC grown on Si is *anti-phase domains* (APDs). These are common defects that appear in many compound semiconductors grown on element semiconductors, such as GaAs on Si and GaAs on Ge [27]. Figure 4.41 shows a schematic illustration of the bond configuration near the 3C-SiC(001)/Si(001) interface, where the lattice mismatch has been neglected. On the Si(001) surface, atomic steps naturally exist after *in situ* cleaning prior to CVD growth, and the height of some steps can be that of a monoatomic layer. When SiC is grown on this face, the monolayers of carbon can misalign with each other because of the atomic steps. At the domain boundaries (called *anti-phase boundaries* (APBs)), a number of Si-Si or C-C bonds are formed. Remarkable reduction of APDs in 3C-SiC layers has been attained by the use of a Si(001) substrate with a few degrees off-axis toward the

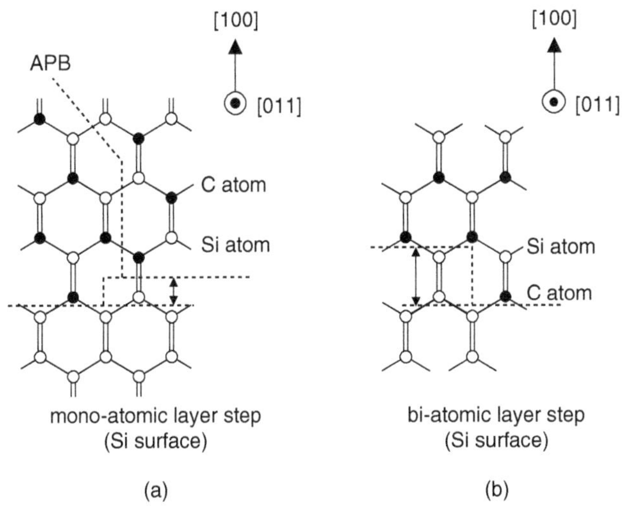

Figure 4.41 Schematic illustrations of the bond configuration near the 3C-SiC(001)/Si(001) interface. (a) 3C-SiC growth on on-axis Si (001) and (b) 3C-SiC growth on off-axis Si(001) ([228] reproduced with permission from Elsevier).

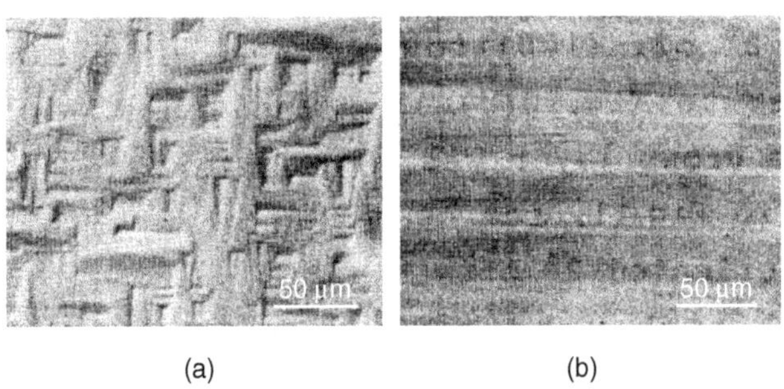

Figure 4.42 Nomarski microscope images of 3C-SiC layers grown on (a) on-axis Si(001) and 2° off-axis Si(001) ([239] reproduced with permission from AIP Publishing LLC).

[110] direction [239]. Figure 4.42 shows Nomarski microscope images of 3C-SiC layers grown on (a) on-axis Si(001) and (b) 2° off-axis Si(001). The cross-hatched morphology is changed to a stripe-like morphology by the introduction of an off-axis, and structural characterization by TEM and molten KOH etching revealed that APDs are mostly eliminated on the off-axis substrate. More recently, it has been reported that APDs can be almost eliminated by low-pressure CVD [240, 241].

In spite of these efforts, the density of extended defects in 3C-SiC grown on Si is still high. 3C-SiC pn junctions and Schottky barrier diodes fabricated on these heteroepitaxial layers exhibit large leakage currents and low breakdown voltages. The relatively high level of background nitrogen doping in 3C-SiC (typically in the mid 10^{15} cm^{-3}) because of the relatively low growth temperature is another issue that must be solved. Considerable efforts are being made, aiming at drastic improvement of the crystal quality. A thin 3C-SiC layer grown on Si is an attractive material for MEMS (microelectromechanical system) applications [242, 243].

4.7.2 *Heteroepitaxial Growth of 3C-SiC on Hexagonal SiC*

Hexagonal SiC{0001} is an attractive substrate for 3C-SiC growth, because it has a very small lattice mismatch. There is a definite difference in the lattice constants (or Si-C bond lengths) between 3C-SiC and hexagonal SiC (see Chapter 2), but the mismatch is much smaller than that for 3C-SiC/Si. From the crystallographic point of view, 3C-SiC(111) can be coherently grown on 4H- or 6H-SiC(0001). Another advantage of using SiC substrates is that high growth temperatures can be employed (without the restriction of the Si melting point), which makes it easier to obtain high-quality, high-purity 3C-SiC. Growth of 3C-SiC on 4H- or 6H-SiC(0001) has been performed using almost the same reactors and conditions as those employed for homoepitaxy of 4H- or 6H-SiC. Typical growth temperatures and growth rates are 1500–1600 °C and 2–5 μm h^{-1}, respectively. Of course, the 4H- and 6H-SiC(0001) substrates must be on-axis in this case, to suppress step-flow growth that would result in homoepitaxy of 4H- or 6H-SiC. Furthermore, hexagonal SiC(0001) is the substrate chosen, because three-dimensional growth often takes place on SiC(000$\bar{1}$), because of the much higher nucleation probability on this face [15, 40].

In addition to replication of threading dislocations from the hexagonal SiC substrates, new extended defects are generated during the heteroepitaxial growth. The most common type of defect is {111} stacking faults. In 3C-SiC, {111} is the main plane for stacking fault generation. Because of the low stacking fault energy of SiC, 3C-SiC always suffers from generation of a high density of {111} stacking faults, but the exact mechanism of stacking fault generation is not clear at present. These stacking faults are the main cause of large leakage currents in electronic devices fabricated on 3C-SiC(111) grown on hexagonal SiC(0001) (although leakage currents are much smaller than those in devices fabricated on 3C-SiC/Si).

Another type of extended defect observed in 3C-SiC grown on hexagonal SiC is *double positioning domains* (DPDs). As briefly explained in Figure 4.2a, 3C-SiC grown on on-axis 6H-SiC(0001) can take two possible stacking orders, ABCABC … and ACBACB … [244, 245], because the stacking order of 6H-SiC is ABCACB. The situation is similar when on-axis 4H-SiC(0001) is employed as a substrate. The domain with ABC stacking is rotated by 180° with respect to the domain with ACB stacking, and the domain boundary is called a *double positioning boundary* (DPB). When devices are fabricated on SiC that contains DPBs, unacceptably large leakage currents are observed. It has been revealed that the ABC and ACB stacking domains follow the stacking sequence of the top surface in the substrate; this phenomenon has been discussed, considering minimization of stacking energy in SiC polytypes [245]. The area of (0001) terraces with an ABC stacking must be identical to that with an ACB stacking for 6H-SiC (stacking: ABCACB), and the situation is similar for 4H-SiC. This area ratio can be 3 : 2 when 15R-SiC(0001) is employed as a substrate. In fact, remarkable reduction of DPBs has been reported in thick 3C-SiC growth on 15R-SiC(0001) [246]. However, so far, DPBs have not been completely eliminated. Ideally, DPBs can be eliminated if 3C-SiC is grown on a perfect SiC(0001) face, which is free of any atomic steps. Although this has been demonstrated on a small scale using small (0.4 × 0.4 mm) mesas [190, 191], it is not very realistic to prepare perfectly step-free surfaces across the entire area of 150-mm-diameter SiC(0001) wafers.

Thus, significant improvement in quality and doping control is required before 3C-SiC is suitable for electronic applications. In the future, an attractive approach may be to grow 3C-SiC on high-quality free-standing 3C-SiC substrates, produced by fast epitaxial growth of 3C-SiC on Si wafers and subsequent etching of Si.

4.8 Summary

Basic technologies of homoepitaxial growth of 4H- and 6H-SiC by CVD have been established, and recent developments in the epitaxial process have driven mass production of SiC power devices. Polytype replication in epitaxial layers is ensured by step-flow growth on off-axis SiC{0001} substrates. Growth of very pure SiC with a net doping density below 5×10^{13} cm^{-3} and wide-range control of both n-type (nitrogen doping: 10^{14}–10^{19} cm^{-3}) and p-type (aluminum doping: 10^{14}–10^{20} cm^{-3}) doping has been achieved using the site-competition effect and optimization of growth conditions. The density of deep levels is rather low as a compound semiconductor, in the range of $(3-10) \times 10^{12}$ cm^{-3}. The basic behavior of extended defects during SiC epitaxial growth has been intensively investigated. Though significant reduction of defect density has been attained, further reduction in defects and physical understanding of defect behavior are required. In particular, enhanced conversion from BPDs to TEDs during epitaxial growth and elimination of epitaxial-induced extended defects (such as triangular defects, carrot defects, and in-grown stacking faults) are important challenges, because relatively large chip areas (3×3 mm^2–10×10 mm^2) are needed for power devices. Industry is leading the drive to increase process throughput and improve uniformity of SiC epitaxial growth. Major advanced epitaxial technologies include very fast (>100 μm h^{-1}) epitaxial growth and homoepitaxial growth on nearly on-axis SiC{0001} substrates.

References

[1] Brander, R.W. and Sutton, R.P. (1965) Solution grown SiC p-n junctions. *Br. J. Appl. Phys.*, **2**, 24.

[2] Ikeda, M., Hayakawa, T., Yamagiwa, S. *et al.* (1980) Fabrication of 6H-SiC light-emitting diodes by a rotation dipping technique: Electroluminescence mechanisms. *J. Appl. Phys.*, **50**, 8215.

[3] Jennings, V.J., Sommer, A. and Chang, H.C. (1966) The epitaxial growth of silicon carbide. *J. Electrochem. Soc.*, **113**, 728.

[4] Campbell, R.B. and Chu, T.L. (1966) Epitaxial growth of silicon carbide by the thermal reduction technique. *J. Electrochem. Soc.*, **113**, 825.

[5] von Muench, W. and Phaffeneder, I. (1976) Epitaxial deposition of silicon carbide from silicon tetrachloride and hexane. *Thin Solid Films*, **31**, 39.

[6] Yoshida, S., Sakuma, E., Okumura, H. *et al.* (1987) Heteroepitaxial growth of SiC polytypes. *J. Appl. Phys.*, **62**, 303.

[7] Kuroda, N., Shibahara, K., Yoo, W.S. *et al.* (1987) Extended Abstract 19th Conference on Solid State Devices and Materials, Tokyo, Japan, 1987, p. 227 Step controlled VPE growth of SiC single crystals at low temperatures.

[8] Ueda, T., Nishino, H. and Matsunami, H. (1990) Crystal growth of SiC by step-controlled epitaxy. *J. Cryst. Growth*, **104**, 695.

[9] Kong, H.S., Kim, H.J., Edmond, J.A. *et al.* (1987) Growth, doping, device development and characterization of CVD beta-SiC epilayers on Si(100) and alpha-SiC(0001). *Mater. Res. Soc. Symp. Proc.*, **97**, 233.

[10] Kong, H.S., Glass, J.T. and Davis, R.F. (1988) Chemical vapor deposition and characterization of 6H-SiC thin films on off-axis 6H-SiC substrates. *J. Appl. Phys.*, **64**, 2672.

[11] Itoh, A., Akita, H., Kimoto, T. and Matsunami, H. (1994) High-quality 4H-SiC homoepitaxial layers grown by step-controlled epitaxy. *Appl. Phys. Lett.*, **65**, 1400.

[12] Kimoto, T., Itoh, A., Akita, H. *et al.* (1995) Step-controlled epitaxial growth of α-SiC and application to high-voltage Schottky rectifiers, in *Proceedings of the International Symposium on Compound Semiconductors 1994*, IOP, p. 437.

[13] Itoh, A., Kimoto, T. and Matsunami, H. (1995) High-performance of high-voltage 4H-SiC Schottky barrier diodes. *IEEE Electron Device Lett.*, **16**, 280.

[14] Davis, R.F., Kelner, G., Shur, M. *et al.* (1991) Thin film deposition and microelectronic and optoelectronic device fabrication and characterization in monocrystalline alpha and beta silicon carbide. *Proc. IEEE*, **79**, 677.

[15] Matsunami, H. and Kimoto, T. (1997) Step-controlled epitaxial growth of SiC: high quality homoepitaxy. *Mater. Sci. Eng., R*, **20**, 125.

[16] Burk, A. and Rowland, L.B. (1997) Homoepitaxial VPE growth of SiC active layers. *Phys. Status Solidi B*, **202**, 263.

[17] Rupp, R., Makarov, Y.N., Behner, H. and Wiedenhofer, A. (1997) Silicon carbide epitaxy in a vertical CVD reactor: experimental results and numerical process simulation. *Phys. Status Solidi B*, **202**, 281.

[18] Kordina, O., Hallin, C., Henry, A. *et al.* (1997) Growth of SiC by "Hot-Wall" CVD and HTCVD. *Phys. Status Solidi B*, **202**, 321.

[19] Kimoto, T., Itoh, A. and Matsunami, H. (1997) Step-controlled epitaxial growth of high-quality SiC layers. *Phys. Status Solidi B*, **202**, 247.

[20] Powell, J.A. and Larkin, D.J. (1997) Process-induced morphological defects in epitaxial CVD silicon carbide. *Phys. Status Solidi B*, **202**, 529.
[21] Schöner, A. (2004) New development in hot wall vapor phase epitaxial growth of silicon carbide, in *Silicon Carbide – Recent Major Advances* (eds W.J. Choyke, H. Matsunami and G. Pensl), Springer, p. 229.
[22] Burk, A. (2006) Development of multiwafer warm-wall planetary VPE reactors for SiC device production. *Chem. Vap. Deposition*, **12**, 465.
[23] Henry, A., ul Hassan, J., Bergman, J.P. *et al.* (2006) Thick silicon carbide homoepitaxial layers grown by CVD techniques. *Chem. Vap. Deposition*, **12**, 475.
[24] La Via, F., Galvagno, G., Foti, G. *et al.* (2006) 4H SiC epitaxial growth with chlorine addition. *Chem. Vap. Deposition*, **12**, 502.
[25] Tsuchida, H., Ito, M., Kamata, I. and Nagano, M. (2009) Formation of extended defects in 4H-SiC epitaxial growth and development of a fast growth technique. *Phys. Status Solidi B*, **246**, 1553.
[26] Pedersen, H., Leone, S., Kordina, O. *et al.* (2012) Chloride-based CVD growth of silicon carbide for electronic applications. *Chem. Rev.*, **112**, 2434.
[27] J.W. Matthews, ed. *Epitaxial Growth, Part B*, Chapter 5, Academic Press, New York, 1975.
[28] Landini, B.E. and Brandes, G.R. (1999) Characteristics of homoepitaxial 4H-SiC films grown on c-axis substrates offcut towards $\langle 1\bar{1}00 \rangle$ or $\langle 11\bar{2}0 \rangle$. *Appl. Phys. Lett.*, **74**, 2632.
[29] Knippenberg, W.F. (1963) Growth phenomena in silicon carbide. *Philips Res. Rep.*, **18**, 161.
[30] Heine, V., Cheng, C. and Needs, R.J. (1991) The preference of silicon carbide for growth in the metastable cubic form. *J. Am. Ceram. Soc.*, **74**, 2630.
[31] Yoo, W.S. and Matsunami, H. (1992) Growth simulation of SiC polytypes and application to DPB-free 3C-SiC on alpha-SiC substrates, in *Amorphous and Crystalline Silicon Carbide IV*, Springer-Verlag, Berlin, p. 66.
[32] Powell, J.A., Petit, J.B., Edgar, J.H. *et al.* (1991) Controlled growth of 3C-SiC and 6H-SiC films on low-tilt-angle vicinal (0001) 6H-SiC wafers. *Appl. Phys. Lett.*, **59**, 333.
[33] Tairov, Y.M., Tsvetkov, V.F., Lilov, S.K. and Safaraliev, G.K. (1976) Studies of growth kinetics and polytypism of silicon carbide epitaxial layers grown from the vapour phase. *J. Cryst. Growth*, **36**, 147.
[34] Matsushita, Y., Nakata, T., Uetani, T. *et al.* (1990) Fabrication of SiC blue LEDs using off-oriented substrates. *Jpn. J. Appl. Phys.*, **29**, L343.
[35] Tanaka, S., Kern, R.S. and Davis, R.F. (1994) Effects of gas flow ratio on silicon carbide thin film growth mode and polytype formation during gas-source molecular beam epitaxy. *Appl. Phys. Lett.*, **65**, 2851.
[36] Kimoto, T., Nishino, H., Yoo, W.S. and Matsunami, H. (1993) Growth mechanism of 6H-SiC in step-controlled epitaxy. *J. Appl. Phys.*, **73**, 726.
[37] Krishnan, B., Melnychuk, G. and Koshka, Y. (2010) Low-temperature homoepitaxial growth of 4H–SiC with CH_3Cl and $SiCl_4$ precursors. *J. Cryst. Growth*, **312**, 645.
[38] Kimoto, T., Nishino, H., Yamashita, A. *et al.* (1992) Low temperature homoepitaxial growth of 6H-SiC by VPE method, in *Amorphous and Crystalline Silicon Carbide IV*, Springer-Verlag, p. 31.
[39] Burton, W.K., Cabrera, N. and Frank, F.C. (1951) The growth of crystals and the equilibrium structure of their surfaces. *Philos. Trans. R. Soc. London, Ser. A*, **243**, 299.
[40] Kimoto, T. and Matsunami, H. (1994) Surface kinetics of adatoms in vapor phase epitaxial growth of SiC on 6H-SiC{0001} vicinal surfaces. *J. Appl. Phys.*, **75**, 850.
[41] Hirschfelder, J.O., Curties, F. and Bird, R.B. (1954) *Molecular Theory of Gases and Liquids*, John Wiley & Sons, Inc., New York.
[42] Konstantinov, A.O., Hallin, C., Kordina, O. and Janzen, E. (1996) Effect of vapor composition on polytype homogeneity of epitaxial silicon carbide. *J. Appl. Phys.*, **80**, 5704.
[43] Hirth, J.P. and Pound, G.M. (1963) *Condensation and Evaporation, Nucleation and Growth Kinetics*, Pergamon Press, Oxford.
[44] Pearson, E., Takai, T., Halicioglu, T. and Tiller, W.A. (1984) Computer modeling of Si and SiC surfaces and surface processes relevant to crystal growth from the vapor. *J. Cryst. Growth*, **70**, 33.
[45] Danno, K., Kimoto, T., Hashimoto, K. *et al.* (2004) Low-concentration deep traps in 4H-SiC grown with high growth rate by chemical vapor deposition. *Jpn. J. Appl. Phys.*, **43**, L969.
[46] Hori, T., Danno, K. and Kimoto, T. (2007) Fast homoepitaxial growth of 4H-SiC with low basal-plane dislocation density and low trap concentration by hot-wall chemical vapor deposition. *J. Cryst. Growth*, **306**, 297.
[47] Burk, A.A. Jr., and Rowland, L.B. (1996) The role of excess silicon and in situ etching on 4H-SiC and 6H-SiC epitaxial layer morphology. *J. Cryst. Growth*, **167**, 586.
[48] Kimoto, T., Chen, Z.Y., Tamura, S. *et al.* (2001) Surface morphological structures of 4H-, 6H-, and 15R-SiC(0001) epitaxial layers grown by chemical vapor deposition. *Jpn. J. Appl. Phys.*, **40**, 3315.

[49] Hallin, C., Konstantinov, A.O., Kordina, O. and Janzen, E. (1996) The mechanism of cubic SiC nucleation on off-axis substrates. *Inst. Phys. Conf. Ser.*, **142**, 85.

[50] Fujiwara, H., Danno, K., Kimoto, T. *et al.* (2005) Effects of C/Si ratio in fast epitaxial growth of 4H-SiC(0001) by vertical hot-wall chemical vapor deposition. *J. Cryst. Growth*, **281**, 370.

[51] Charrier, A., Coati, A., Argunova, T. *et al.* (2002) Solid-state decomposition of silicon carbide for growing ultra-thin heteroepitaxial graphite films. *J. Appl. Phys.*, **92**, 2479.

[52] Novoselov, K.S., Geim, A.K., Morozov, S.V. *et al.* (2004) Electric field effect in atomically thin carbon films. *Science*, **306**, 666.

[53] Seyller, T., Bostwick, A., Emtsev, K.V. *et al.* (2008) Epitaxial graphene: a new material. *Phys. Status Solidi B*, **245**, 1436.

[54] Tsuchida, H., Kamata, I. and Izumi, K. (1997) Infrared spectroscopy of hydrides on the 6H-SiC surface. *Appl. Phys. Lett.*, **70**, 3072.

[55] Allendorf, M.D. and Kee, R.J. (1991) A model of silicon carbide chemical vapor deposition. *J. Electrochem. Soc.*, **138**, 841.

[56] Stinespring, C.D. and Wormhoudt, J.C. (1988) Gas phase kinetics analysis and implications for silicon carbide chemical vapor deposition. *J. Cryst. Growth*, **87**, 481.

[57] Nishizawa, S. and Pons, M. (2006) Growth and doping modeling of SiC-CVD in a horizontal hot-wall reactor. *Chem. Vap. Deposition*, **12**, 516.

[58] Nishizawa, S. and Pons, M. (2006) Numerical modeling of SiC–CVD in a horizontal hot-wall reactor. *Microelectron. Eng.*, **83**, 100.

[59] Meziere, J., Ucar, M., Blanquet, E. *et al.* (2004) Modeling and simulation of SiC CVD in the horizontal hot-wall reactor concept. *J. Cryst. Growth*, **267**, 436.

[60] Nishizawa, S., Kojima, K., Kuroda, S. *et al.* (2005) Modeling of SiC-CVD on Si-face/C-face in a horizontal hot-wall reactor. *J. Cryst. Growth*, **275**, e515.

[61] Danielsson, Ö., Henry, A. and Janzen, E. (2002) Growth rate predictions of chemical vapor deposited silicon carbide epitaxial layers. *J. Cryst. Growth*, **243**, 170.

[62] Danielsson, Ö., Forsberg, U. and Janzen, E. (2003) Predicted nitrogen doping concentrations in silicon carbide epitaxial layers grown by hot-wall chemical vapor deposition. *J. Cryst. Growth*, **250**, 471.

[63] Tsvetkov, V.F., Allen, S.T., Kong, H.S. and Carter, C.H. Jr., (1996) Recent progress in SiC crystal growth. *Inst. Phys. Conf. Ser.*, **142**, 17.

[64] Chen, W. and Capano, M.A. (2005) Growth and characterization of 4H-SiC epilayers on substrates with different off-cut angles. *J. Appl. Phys.*, **98**, 114907.

[65] Wada, K., Kimoto, T., Nishikawa, K. and Matsunami, H. (2006) Epitaxial growth of 4H-SiC on 4 degrees off-axis (0001) and ($000\bar{1}$) substrates by hot-wall chemical vapor deposition. *J. Cryst. Growth*, **291**, 370.

[66] Kojima, K., Okumura, H., Kuroda, S. and Arai, K. (2004) Homoepitaxial growth of 4H-SiC on on-axis ($000\bar{1}$) C-face substrates by chemical vapor deposition. *J. Cryst. Growth*, **269**, 367.

[67] Kimoto, T., Itoh, A. and Matsunami, H. (1995) Step bunching in chemical vapor deposition of 6H- and 4H-SiC on vicinal SiC{0001} faces. *Appl. Phys. Lett.*, **66**, 3645.

[68] Herring, C. (1951) Some theorems on the free energies of crystal surfaces. *Phys. Rev.*, **82**, 87.

[69] W.A. Tiller, *The Science of Crystallization: Microscopic Interfacial Phenomena*, Chapter 2 (Cambridge University Press, Cambridge, 1991).

[70] Tyc, S. (1994) Structure of a 6H silicon carbide vicinal surface. *Inst. Phys. Conf. Ser.*, **137**, 333.

[71] Powell, J.A., Larkin, D.J., Abel, P.B. *et al.* (1996) Effect of tilt angle on the morphology of SiC epitaxial films grown on vicinal (0001)SiC substrates. *Inst. Phys. Conf. Ser.*, **142**, 77.

[72] Neudeck, P.G., Trunek, A.J. and Powell, J.A. (2004) Atomic force microscope observation of growth and defects on as-grown (111) 3C-SiC mesa surfaces. *Mater. Res. Soc. Symp. Proc.*, **815**, 59.

[73] Kimoto, T., Itoh, A., Matsunami, H. and Okano, T. (1997) Step bunching mechanism in chemical vapor deposition of α-SiC{0001}. *J. Appl. Phys.*, **81**, 3494.

[74] Thomas, B., Bartsch, W., Stein, R. *et al.* (2004) Properties and suitability of 4H-SiC epitaxial layers grown at different CVD systems for high voltage applications. *Mater. Sci. Forum*, **457–460**, 181.

[75] O. Kordina, C. Hallin, R.C. Glass, *et al.* (1994) *Inst. Phys. Conf. Ser.* **137**, 41 A novel hot-wall CVD reactor for SiC epitaxy.

[76] La Via, F., Izzo, G., Mauceri, M. *et al.* (2008) 4H-SiC epitaxial layer growth by trichlorosilane (TCS). *J. Cryst. Growth*, **311**, 107.

[77] Kimoto, T., Feng, G., Hiyoshi, T. *et al.* (2010) Defect control in growth and processing of 4H-SiC for power device applications. *Mater. Sci. Forum*, **645–648**, 645.

[78] Burk, A.A., O'Loughlin, M.J., Paisley, M.J. *et al.* (2005) Large area SiC epitaxial layer growth in a warm-wall planetary VPE reactor. *Mater. Sci. Forum*, **483–485**, 137.
[79] Thomas, B., Hecht, C., Stein, R. and Friedrichs, P. (2006) Challenges in large-area multi-wafer SiC epitaxy for production needs. *Mater. Sci. Forum*, **527–529**, 135.
[80] Ellison, A., Zhang, J., Henry, A. and Janzen, E. (2002) Epitaxial growth of SiC in a chimney CVD reactor. *J. Cryst. Growth*, **236**, 225.
[81] Larkin, D.J., Neudeck, P.G., Powell, J.A. and Matus, L.G. (1994) Site-competition epitaxy for superior silicon carbide electronics. *Appl. Phys. Lett.*, **65**, 1659.
[82] Larkin, D.J. (1997) SiC dopant incorporation control using site-competition CVD. *Phys. Status Solidi B*, **202**, 305.
[83] Kimoto, T., Nakazawa, S., Hahimoto, K. and Matsunami, H. (2001) Reduction of doping and trap concentrations in 4H-SiC epitaxial layers grown by chemical vapor deposition. *Appl. Phys. Lett.*, **79**, 2761.
[84] Tsuchida, H., Kamata, I., Jikimoto, T. and Izumi, K. (2002) Epitaxial growth of thick 4H–SiC layers in a vertical radiant-heating reactor. *J. Cryst. Growth*, **237–238**, 1206.
[85] Kimoto, T., Itoh, A. and Matsunami, H. (1995) Incorporation mechanism of N, Al, and B impurities in chemical vapor deposition of SiC. *Appl. Phys. Lett.*, **67**, 2385.
[86] Hallin, C., Ivanov, I.G., Egilsson, T. *et al.* (1998) The material quality of CVD-grown SiC using different carbon precursors. *J. Cryst. Growth*, **183**, 163.
[87] Kojima, K., Suzuki, T., Kuroda, S. *et al.* (2003) Epitaxial growth of high-quality 4H-SiC carbon-face by low-pressure hot-wall chemical vapor deposition. *Jpn. J. Appl. Phys.*, **42**, L637.
[88] Forsberg, U., Danielsson, Ö., Henry, A. *et al.* (2002) Nitrogen doping of epitaxial silicon carbide. *J. Cryst. Growth*, **236**, 101.
[89] Yamamoto, T., Kimoto, T. and Matsunami, H. (1998) Impurity incorporation mechanism in step-controlled epitaxy - Growth temperature and substrate off-angle dependence. *Mater. Sci. Forum*, **264–268**, 111.
[90] Wang, R., Bhat, I.B. and Chow, T.P. (2002) Epitaxial growth of n-type SiC using phosphine and nitrogen as the precursors. *J. Appl. Phys.*, **92**, 7587.
[91] Yoshida, S., Sakuma, E., Misawa, S. and Gonda, S. (1984) A new doping method using metalorganics in chemical vapor deposition of 6H–SiC. *J. Appl. Phys.*, **55**, 169.
[92] Forsberg, U., Danielsson, Ö., Henry, A. *et al.* (2003) Aluminum doping of epitaxial silicon carbide. *J. Cryst. Growth*, **253**, 340.
[93] Larkin, D.J., Sridhara, S.G., Devaty, R.P. and Choyke, W.J. (1995) Hydrogen incorporation in boron-doped 6H-SiC CVD epilayers produced using site-competition epitaxy. *J. Electron. Mater.*, **24**, 289.
[94] Kimoto, T., Yamashita, A., Itoh, A. and Matsunami, H. (1993) Step-controlled epitaxial growth of 4H-SiC and doping of Ga as a blue luminescent center. *Jpn. J. Appl. Phys.*, **32**, 1045.
[95] Negoro, Y., Kimoto, T., Matsunami, H. and Pensl, G. (2007) Abnormal out-diffusion of epitaxially doped boron in 4H-SiC caused by implantation and annealing. *Jpn. J. Appl. Phys.*, **46**, 5053.
[96] Nordell, N., Schöner, A. and Linnarsson, M.K. (1997) Control of Al and B doping transients in 6H and 4H SiC grown by vapor phase epitaxy. *J. Electron. Mater.*, **26**, 187.
[97] Benamara, M., Zhang, X., Skowronski, M. *et al.* (2005) Structure of the carrot defect in 4H-SiC epitaxial layers. *Appl. Phys. Lett.*, **86**, 021905.
[98] Tsuchida, H., Kamata, I. and Nagano, M. (2007) Investigation of defect formation in 4H-SiC epitaxial growth by X-ray topography and defect selective etching. *J. Cryst. Growth*, **306**, 254.
[99] Okada, T., Kimoto, T., Noda, H. *et al.* (2002) Correspondence between surface morphological faults and crystallographic defects in 4H-SiC homoepitaxial film. *Jpn. J. Appl. Phys.*, **41**, 6320.
[100] Okada, T., Kimoto, T., Yamai, K. *et al.* (2003) Crystallographic defects under device-killing surface faults in a homoepitaxially grown film of SiC. *Mater. Sci. Eng., A*, **361**, 67.
[101] Konstantinov, A.O., Hallin, C., Pecz, B. *et al.* (1997) The mechanism for cubic SiC formation on off-oriented substrates. *J. Cryst. Growth*, **178**, 495.
[102] Aigo, T., Ito, W., Tsuge, H. *et al.* (2013) Formation of epitaxial defects by threading screw dislocations with a morphological feature at the surface of 4° off-axis 4H-SiC substrates. *Mater. Sci. Forum*, **740–742**, 629.
[103] Kimoto, T. and Matsunami, H. (1995) Surface diffusion lengths of adatoms on 6H-SiC{0001} faces in chemical vapor deposition of SiC. *J. Appl. Phys.*, **78**, 3132.
[104] Kimoto, T., Miyamoto, N. and Matsunami, H. (1999) Performance limiting surface defects in SiC epitaxial p-n junction diodes. *IEEE Trans. Electron Devices*, **46**, 471.
[105] Ohtani, N. (2011) Toward the reduction of performance-limiting defects in SiC epitaxial substrates. *ECS Trans.*, **41**, 253.

[106] Kamata, I., Tsuchida, H., Jikimoto, T. and Izumi, K. (2000) Structural transformation of screw dislocations via thick 4H-SiC epitaxial growth. *Jpn. J. Appl. Phys.*, **39**, 6496.

[107] Kamata, I., Tsuchida, H., Jikimoto, T. and Izumi, K. (2002) Influence of 4H–SiC growth conditions on micropipe dissociation. *Jpn. J. Appl. Phys.*, **41**, L1137.

[108] Nakamura, S., Kimoto, T. and Matsunami, H. (2003) Effect of C/Si ratio on spiral growth on 6H-SiC(0001). *Jpn. J. Appl. Phys.*, **42**, L846.

[109] Tsuchida, H., Kamata, I. and Nagano, M. (2008) Formation of basal plane Frank-type faults in 4H-SiC epitaxial growth. *J. Cryst. Growth*, **310**, 757.

[110] Bergman, J.P., Lendenmann, H., Nilsson, P.A. *et al.* (2001) Crystal defects as source of anomalous forward voltage increase of 4H-SiC diodes. *Mater. Sci. Forum*, **353–356**, 299.

[111] Skowronski, M. and Ha, S. (2006) Degradation of hexagonal silicon-carbide-based bipolar devices. *J. Appl. Phys.*, **99**, 011101.

[112] Muzykov, P.G., Kennedy, R.M., Zhang, Q. *et al.* (2009) Physical phenomena affecting performance and reliability of 4H-SiC bipolar junction transistors. *Microelectron. Reliab.*, **49**, 32.

[113] Hull, D. and Bacon, D.J. (2001) *Introduction to Dislocations*, 4th edn, Butterworth-Heinemann.

[114] Ha, S., Mieszkowski, P., Skowronski, M. and Rowland, L.B. (2002) Dislocation conversion in 4H silicon carbide epitaxy. *J. Cryst. Growth*, **244**, 257.

[115] Ohno, T., Yamaguchi, H., Kuroda, S. *et al.* (2004) Influence of growth conditions on basal plane dislocation in 4H-SiC epitaxial layer. *J. Cryst. Growth*, **271**, 1.

[116] Jacobson, H., Bergman, J.P., Hallin, C. *et al.* (2004) Properties and origins of different stacking faults that cause degradation in SiC PiN diodes. *J. Appl. Phys.*, **95**, 1485.

[117] Tsuchida, H., Ito, M., Kamata, I. and Nagano, M. (2009) Fast epitaxial growth of 4H-SiC and analysis of defect transfer. *Mater. Sci. Forum*, **615–617**, 67.

[118] Jacobson, H., Birch, J., Yakimova, R. *et al.* (2002) Dislocation evolution in 4H-SiC epitaxial layers. *J. Appl. Phys.*, **91**, 6354.

[119] Hong, M.H., Samant, A.V. and Pirouz, P. (2000) Stacking fault energy of 6H-SiC and 4H-SiC single crystals. *Philos. Mag.*, **80**, 919.

[120] Pirouz, P., Demenet, J.L. and Hong, M.H. (2001) On transition temperatures in the plasticity and fracture of semiconductors. *Philos. Mag.*, **81**, 1207.

[121] Tanuma, R., Mori, D., Kamata, I. and Tsuchida, H. (2012) X-ray microbeam three-dimensional topography imaging and strain analysis of basal-plane dislocations and threading edge dislocations in 4H-SiC. *Appl. Phys. Express*, **5**, 061301.

[122] Chung, S., Wheeler, V., Myers-Ward, R. *et al.* (2011) Secondary electron dopant contrast imaging of compound semiconductor junctions. *J. Appl. Phys.*, **109**, 094906.

[123] Sumakeris, J.J., Bergman, J.P., Das, M.K. *et al.* (2006) Techniques for minimizing the basal plane dislocation density in SiC epilayers to reduce Vf drift in SiC bipolar power devices. *Mater. Sci. Forum*, **527–529**, 141.

[124] Sumakeris, J.J., Hull, B.A., O'Loughlin, M.J. *et al.* (2007) Developing an effective and robust process for manufacturing bipolar SiC power devices. *Mater. Sci. Forum*, **556–557**, 77.

[125] Zhang, Z. and Sudarshan, T.S. (2005) Basal plane dislocation-free epitaxy of silicon carbide. *Appl. Phys. Lett.*, **87**, 151913.

[126] Tsuchida, H., Kamata, I., Miyanagi, T. *et al.* (2006) Comparison of propagation and nucleation of basal plane dislocations in 4H-SiC (000$\bar{1}$) and (0001) epitaxy. *Mater. Sci. Forum*, **527–529**, 231.

[127] Starlbush, R.E., VanMil, B.L., Myers-Ward, R.L. *et al.* (2009) Basal plane dislocation reduction in 4H-SiC epitaxy by growth interruptions. *Appl. Phys. Lett.*, **94**, 041916.

[128] Tsuchida, H., Kamata, I., Miyanagi, T. *et al.* (2005) Growth of thick 4H-SiC (0001) epilayers and reduction of basal plane dislocations. *Jpn. J. Appl. Phys.*, **44**, L806.

[129] Zhang, X. and Tsuchida, H. (2012) Conversion of basal plane dislocations to threading edge dislocations in 4H-SiC epilayers by high temperature annealing. *J. Appl. Phys.*, **111**, 123512.

[130] Zhang, X., Skowronski, M., Liu, K.X. *et al.* (2007) Glide and multiplication of basal plane dislocations during 4H-SiC homoepitaxy. *J. Appl. Phys.*, **102**, 093520.

[131] Nagano, M., Tsuchida, H., Suzuki, T. *et al.* (2010) Annealing induced extended defects in as-grown and ion-implanted 4H-SiC epitaxial layers. *J. Appl. Phys.*, **108**, 013511.

[132] Zhang, N., Chen, Y., Zhang, Y. *et al.* (2009) Nucleation mechanism of dislocation half-loop arrays in 4H-silicon carbide homoepitaxial layers. *Appl. Phys. Lett.*, **94**, 122108.

[133] Zhang, X., Miyazawa, T. and Tsuchida, H. (2012) Critical conditions of misfit dislocation formation in 4H-SiC epilayers. *Mater. Sci. Forum*, **717–720**, 313.

[134] Matthews, J.W. and Blakeslee, A.E. (1974) Defects in epitaxial multilayers: I. Misfit dislocations. *J. Cryst. Growth*, **27**, 118.

[135] Ha, S., Chung, H.J., Nuhfer, N.T. and Skowronski, M. (2004) Dislocation nucleation in 4H silicon carbide epitaxy. *J. Cryst. Growth*, **262**, 130.

[136] Bai, S., Wagner, G., Choyke, W.J. *et al.* (2002) Spectra associated with stacking faults in 4H-SiC grown in a hot-wall CVD reactor. *Mater. Sci. Forum*, **389–393**, 589.

[137] Feng, G., Suda, J. and Kimoto, T. (2008) Characterization of stacking faults in 4H-SiC epilayers by room-temperature microphotoluminescence mapping. *Appl. Phys. Lett.*, **92**, 221906.

[138] Liu, K.X., Stahlbush, R.E., Lew, K.-K. *et al.* (2008) Examination of in-grown stacking faults in 8°- and 4°-offcut 4H-SiC epitaxy by photoluminescence imaging. *J. Electron. Mater.*, **37**, 730.

[139] Camassel, J. and Juillaguet, S. (2008) Optical properties of as-grown and process-induced stacking faults in 4H-SiC. *Phys. Status Solidi B*, **245**, 1337.

[140] Hassan, J., Henry, A., Ivanov, I.G. and Bergman, J.P. (2009) In-grown stacking faults in 4H-SiC epilayers grown on off-cut substrates. *J. Appl. Phys.*, **105**, 123513.

[141] Feng, G., Suda, J. and Kimoto, T. (2009) Characterization of major in-grown stacking faults in 4H-SiC epilayers. *Physica B*, **23–24**, 4745.

[142] Izumi, S., Tsuchida, H., Tawara, T. *et al.* (2005) Structure of in-grown stacking faults in the 4H-SiC epitaxial layers. *Mater. Sci. Forum*, **483–485**, 323.

[143] Izumi, S., Tsuchida, H., Kamata, I. and Tawara, T. (2005) Structural analysis and reduction of in-grown stacking faults in 4H-SiC epilayers. *Appl. Phys. Lett.*, **86**, 202108.

[144] Fujiwara, H., Kimoto, T. and Matsunami, H. (2005) Characterization of in-grown stacking faults in 4H-SiC (0001) epitaxial layers and its impacts on high-voltage Schottky barrier diodes. *Appl. Phys. Lett.*, **87**, 051912.

[145] Kamata, I., Zhang, X. and Tsuchida, H. (2010) Photoluminescence of Frank-type defects on the basal plane in 4H-SiC epilayers. *Appl. Phys. Lett.*, **97**, 172107.

[146] Milnes, A.G. (1973) *Deep Impurities in Semiconductors*, John Wiley & Sons, Inc., New York.

[147] Lang, D.V. (1974) Deep-level transient spectroscopy: A new method to characterize traps in semiconductors. *J. Appl. Phys.*, **45**, 3023.

[148] Weiss, S. and Kassing, R. (1988) Deep Level Transient Fourier Spectroscopy (DLTFS) – A technique for the analysis of deep level properties. *Solid State Electron.*, **31**, 1733.

[149] Dalibor, T., Pensl, G., Matsunami, H. *et al.* (1997) Deep defect centers in silicon carbide monitored with deep level transient spectroscopy. *Phys. Status Solidi A*, **162**, 199.

[150] Hemmingsson, C., Son, N.T., Kordina, O. *et al.* (1997) Deep level defects in electron-irradiated 4H SiC epitaxial layers. *J. Appl. Phys.*, **81**, 6155.

[151] Storasta, L., Bergman, J.P., Janzen, E. *et al.* (2004) Deep levels created by low energy electron irradiation in 4H-SiC. *J. Appl. Phys.*, **96**, 4909.

[152] Danno, K. and Kimoto, T. (2006) Investigation of deep levels in n-type 4H-SiC epilayers irradiated with low-energy electrons. *J. Appl. Phys.*, **100**, 113728.

[153] Danno, K. and Kimoto, T. (2007) Deep level transient spectroscopy on as-grown and electron-irradiated p-type 4H-SiC epilayers. *J. Appl. Phys.*, **101**, 103704.

[154] Storasta, L., Carlsson, F.H.C., Sridhara, S.G. *et al.* (2001) Pseudodonor nature of the defect in 4H-SiC. *Appl. Phys. Lett.*, **78**, 46.

[155] Klein, P.B., Shanabrook, B.V., Huh, S.W. *et al.* (2006) Lifetime-limiting defects in 4H-SiC epilayers. *Appl. Phys. Lett.*, **88**, 052110.

[156] Danno, K., Nakamura, D. and Kimoto, T. (2007) Investigation of carrier lifetime in 4H-SiC epilayers and lifetime control by electron irradiation. *Appl. Phys. Lett.*, **90**, 202109.

[157] Alfieri, G., Monakhov, E.V., Svensson, B.G. and Linnarsson, M.K. (2005) Annealing behavior between room temperature and 2000 °C of deep level defects in electron-irradiated n-type 4H silicon carbide. *J. Appl. Phys.*, **98**, 043518.

[158] Kawahara, K., Suda, J., Pensl, G. and Kimoto, T. (2010) Reduction of deep levels generated by ion implantation into n- and p-type 4H-SiC. *J. Appl. Phys.*, **108**, 033706.

[159] Kawahara, K., Krieger, M., Suda, J. and Kimoto, T. (2010) Deep levels induced by reactive ion etching in n- and p-type 4H-SiC. *J. Appl. Phys.*, **108**, 023706.

[160] Troffer, T., Schadt, M., Frank, T. *et al.* (1997) Doping of SiC by implantation of boron and aluminum. *Phys. Status Solidi A*, **162**, 277.

[161] Dalibor, T., Pensl, G., Nordell, N. and Schöner, A. (1997) Electrical properties of the titanium acceptor in silicon carbide. *Phys. Rev. B*, **55**, 13618.
[162] Zhang, J., Storasta, L., Bergman, J.P. *et al.* (2003) Electrically active defects in n-type 4H-silicon carbide grown in a vertical hot-wall reactor. *J. Appl. Phys.*, **93**, 4708.
[163] Danno, K., Hori, T. and Kimoto, T. (2007) Impacts of growth parameters on deep levels in n-type 4H-SiC. *J. Appl. Phys.*, **101**, 053709.
[164] Tsuchida, H., Ito, M., Kamata, I. *et al.* (2010) Low-pressure fast growth and characterization of 4H-SiC epilayers. *Mater. Sci. Forum*, **645–648**, 77.
[165] Kimoto, T., Hashimoto, K. and Matsunami, H. (2003) Effects of C/Si ratio in chemical vapor deposition of 4H-SiC(11$\bar{2}$0) and (0$\bar{3}$38). *Jpn. J. Appl. Phys.*, **42**, 7294.
[166] Son, N.T., Trinh, X.T., Løvlie, L.S. *et al.* (2012) Negative-U system of carbon vacancy in 4H-SiC. *Phys. Rev. Lett.*, **109**, 187603.
[167] Kawahara, K., Trinh, X.T., Son, N.T. *et al.* (2013) Investigation on origin of $Z_{1/2}$ center in SiC by deep level transient spectroscopy and electron paramagnetic resonance. *Appl. Phys. Lett.*, **102**, 112106.
[168] Kimoto, T., Danno, K. and Suda, J. (2008) Lifetime-killing defects in 4H-SiC epilayers and lifetime control by low-energy electron irradiation. *Phys. Status Solidi B*, **245**, 1327.
[169] Sze, S.M. (2002) *Semiconductor Devices, Physics and Technology*, 2nd edn, John Wiley & Sons, Inc., Hoboken, NJ.
[170] Ito, M., Storasta, L. and Tsuchida, H. (2008) Development of 4H–SiC epitaxial growth technique achieving high growth rate and large-area uniformity. *Appl. Phys. Express*, **1**, 015001.
[171] Ishida, Y., Takahashi, T., Okumura, H. *et al.* (2009) Development of a practical high-rate CVD system. *Mater. Sci. Forum*, **600–603**, 119.
[172] Crippa, D., Valente, G.L., Ruggerio, A. *et al.* (2005) New achievements on CVD based methods for SiC epitaxial growth. *Mater. Sci. Forum*, **483–485**, 67.
[173] Myers, R., Kordina, O., Shishkin, Z. *et al.* (2005) Increased growth rate in a SiC CVD reactor using HCl as a growth additive. *Mater. Sci. Forum*, **483–485**, 73.
[174] Pedersen, H., Leone, S., Henry, A. *et al.* (2007) Very high growth rate of 4H-SiC epilayers using the chlorinated precursor methyltrichlorosilane (MTS). *J. Cryst. Growth*, **307**, 334.
[175] Ellison, A., Zhang, J., Peterson, J. *et al.* (1999) High temperature CVD growth of SiC. *Mater. Sci. Eng., B*, **61**, 113.
[176] Zhang, J., Ellison, A., Danielsson, Ö. *et al.* (2002) Epitaxial growth of 4H SiC in a vertical hot-wall CVD reactor: Comparison between up- and down-flow orientations. *J. Cryst. Growth*, **241**, 421.
[177] Fujihira, K., Kimoto, T. and Matsunami, H. (2002) High-purity and high-quality 4H-SiC grown at high speed by chimney-type vertical hot-wall chemical vapor deposition. *Appl. Phys. Lett.*, **80**, 1586.
[178] Nishino, S., Matsunami, H. and Tanaka, T. (1978) Growth and morphology of 6H-SiC epitaxial layers by CVD. *J. Cryst. Growth*, **45**, 144.
[179] Leone, S., Mauceri, M., Pistone, G. *et al.* (2006) SiC-4H epitaxial layer growth using trichlorosilane (TCS) as silicon precursor. *Mater. Sci. Forum*, **527–529**, 179.
[180] Chowdhury, I., Chandrasekhar, M.V.S., Klein, P.B. *et al.* (2011) High growth rate 4H-SiC epitaxial growth using dichlorosilane in a hot-wall CVD reactor. *J. Cryst. Growth*, **316**, 60.
[181] MacMillan, M.F., Loboda, M.J., Chung, G. *et al.* (2006) Homoepitaxial growth of 4H-SiC using a chlorosilane silicon precursor. *Mater. Sci. Forum*, **527–529**, 175.
[182] Leone, S., Kordina, O., Henry, A. *et al.* (2012) Gas-phase modeling of chlorine-based chemical vapor deposition of silicon carbide. *Cryst. Growth Des.*, **12**, 1977.
[183] Koshka, Y., Lin, H.D., Melnychuk, G. *et al.* (2005) Homoepitaxial growth of 4H-SiC using CH_3Cl carbon precursor. *Mater. Sci. Forum*, **483-485**, 81.
[184] Lu, P., Edgar, J.H., Glembocki, O.J. *et al.* (2005) High-speed homoepitaxy of SiC from methyltrichlorosilane by chemical vapor deposition. *J. Cryst. Growth*, **285**, 506.
[185] Pedersen, H., Leone, S., Henry, A. *et al.* (2008) Very high crystalline quality of thick 4H-SiC epilayers grown from methyltrichlorosilane (MTS). *Phys. Status Solidi*, **2**, 188.
[186] Leone, S., Pedersen, H., Henry, A. *et al.* (2009) Improved morphology for epitaxial growth on 4° off-axis 4H-SiC substrates. *J. Cryst. Growth*, **311**, 3265.
[187] Pozzetti, V. (2001) in *Silicon Epitaxy*, Semiconductors and Semimetals, vol. **72** (eds D. Crippa, D.L. Rode and M. Masi), Academic Press.

[188] Pedersen, H., Beyer, F.C., Henry, A. and Janzen, E. (2009) Acceptor incorporation in SiC epilayers grown at high growth rate with chloride-based CVD. *J. Cryst. Growth*, **311**, 3364.

[189] Nakamura, S., Kimoto, T. and Matsunami, H. (2003) Homoepitaxy of 6H-SiC on nearly on-axis (0001) faces by chemical vapor deposition, Part I: Effect of C/Si ratio on wide-area homoepitaxy without 3C-SiC inclusions. *J. Cryst. Growth*, **256**, 341.

[190] Neudeck, P.G. and Powell, J.A. (2004) Homoepitaxial growth and heteroepitaxial growth on step-free SiC mesas, in *Silicon Carbide - Recent Major Advances* (eds W.J. Choyke, H. Matsunami and G. Pensl), Springer, p. 179.

[191] Neudeck, P.G., Trunek, A.J., Spry, D.J. *et al.* (2006) CVD growth of 3C-SiC on 4H/6H mesas. *Chem. Vap. Deposition*, **12**, 531.

[192] Kojima, K., Kuroda, S., Okumura, H. and Arai, K. (2006) Homoepitaxial growth on a 4H-SiC C-Face substrate. *Chem. Vap. Deposition*, **12**, 489.

[193] Stein, R.A. and Lanig, P. (1992) Influence of surface energy on the growth of 6H- and 4H-SiC polytypes by sublimation. *Mater. Sci. Eng., B*, **11**, 69.

[194] Kojima, K., Okumura, H. and Arai, K. (2009) Control of the surface morphology on low off angled 4H-SiC homoepitaxal growth. *Mater. Sci. Forum*, **615-617**, 113.

[195] Leone, S., Pedersen, H., Henry, A. *et al.* (2009) Thick homoepitaxial layers grown on on-axis Si-face 6H- and 4H-SiC substrates with HCl addition. *J. Cryst. Growth*, **312**, 24.

[196] Leone, S., Beyer, F.C., Pedersen, H. *et al.* (2010) High growth rate of 4H-SiC epilayers on on-axis substrates with different chlorinated precursors. *Cryst. Growth Des.*, **10**, 5334.

[197] Hassan, J., Bergman, J.P., Henry, A. and Janzen, E. (2008) On-axis homoepitaxial growth on Si-face 4H–SiC substrates. *J. Cryst. Growth*, **310**, 4424.

[198] Kimoto, T., Nakazawa, S., Fujihira, K. *et al.* (2002) Recent achievements and future challenges in SiC homoepitaxial growth. *Mater. Sci. Forum*, **389-393**, 165.

[199] Powell, J.A. and Will, H.A. (1973) Epitaxial growth of 6H SiC in the temperature range 1320-1390 °C. *J. Appl. Phys.*, **44**, 5177.

[200] Yamashita, A., Yoo, W.S., Kimoto, T. and Matsunami, H. (1992) Homoepitaxial chemical vapor deposition of 6H-SiC at low temperatures on $\{01\bar{1}4\}$ substrates. *Jpn. J. Appl. Phys.*, **31**, 3655.

[201] Kimoto, T., Hirao, T., Nakazawa, S. *et al.* (2003) Homoepitaxial growth of 4H-SiC($03\bar{3}8$) and nitrogen doping by chemical vapor deposition. *J. Cryst. Growth*, **249**, 208.

[202] Kimoto, T., Yamamoto, T., Chen, Z.Y. *et al.* (2001) Chemical vapor deposition and deep level analyses of 4H-SiC($11\bar{2}0$). *J. Appl. Phys.*, **89**, 6105.

[203] Hallin, C., Ellison, A., Ivanov, I.G. *et al.* (1998) CVD growth and characterisation of SiC epitaxial layers on faces perpendicular to the (0001) basal plane. *Mater. Sci. Forum*, **264-268**, 123.

[204] Kojima, K., Ohno, T., Senzaki, J. *et al.* (2002) Epitaxial growth of ($11\bar{2}0$) 4H-SiC using substrate grown in the [$11\bar{2}0$] direction. *Mater. Sci. Forum*, **389-393**, 195.

[205] Kimoto, T., Fujihira, K., Shiomi, H. and Matsunami, H. (2003) High-voltage 4H-SiC Schottky barrier diodes fabricated on ($03\bar{3}8$) with closed micropipes. *Jpn. J. Appl. Phys.*, **42**, L13.

[206] Tanaka, Y., Okamoto, M., Takatsuka, A. *et al.* (2006) 700-V 1.0-mΩcm^2 buried gate SiC-SIT (SiC-BGSIT). *IEEE Electron Device Lett.*, **27**, 908.

[207] Malhan, R.K., Bakowski, M., Takeuchi, Y. *et al.* (2009) Design, process, and performance of all-epitaxial normally-off SiC JFETs. *Phys. Status Solidi A*, **206**, 2308.

[208] Konstantinov, A.O., Harris, C.I. and Ray, I.C. (2005) High power lateral epitaxy MESFET technology in silicon carbide. *Mater. Sci. Forum*, **483-485**, 853.

[209] Nordell, N., Karlsson, S. and Konstantinov, A.O. (1998) Homoepitaxy of 6H and 4H SiC on nonplanar substrates. *Appl. Phys. Lett.*, **72**, 197.

[210] Chen, Y., Kimoto, T., Takeuchi, Y. *et al.* (2004) Homoepitaxy of 4H-SiC on trenched (0001) Si face substrates by chemical vapor deposition. *Jpn. J. Appl. Phys.*, **43**, 4105.

[211] Takeuchi, Y., Kataoka, M., Kimoto, T. *et al.* (2006) SiC migration enhanced embedded epitaxial (ME3) growth technology. *Mater. Sci. Forum*, **527-529**, 251.

[212] Negoro, Y., Kimoto, T., Kataoka, M. *et al.* (2006) Embedded epitaxial growth of 4H-SiC on trenched substrates and pn junction characteristics. *Microelectron. Eng.*, **83**, 27.

[213] Chen, Y., Kimoto, T., Takeuchi, Y. *et al.* (2005) Selective embedded growth of 4H-SiC trenches in 4H-SiC(0001) substrates using carbon mask. *Jpn. J. Appl. Phys.*, **44**, 4909.

[214] Li, C., Losee, P., Seiler, J. *et al.* (2005) Fabrication and characterization of 4H-SiC PN junction diodes by selective-epitaxial growth using TaC as the mask. *J. Electron. Mater.*, **34**, 450.

[215] Ziegler, G., Lanig, P., Theis, D. and Weurich, C. (1980) Single crystal growth of SiC substrate material for blue light emitting diodes. *IEEE Trans. Electron Devices*, **30**, 277.
[216] Koga, K., Fujikawa, Y., Ueda, Y. and Yamaguchi, T. (1992) Growth and characterization of 6H-SiC bulk crystals by the sublimation method, in *Amorphous and Crystalline Silicon Carbide IV*, Springer Proceedings of Physics, vol. **71** (eds C.Y. Yang, M.M. Rahman and G.L. Harris), Springer-Verlag, p. 96.
[217] Yakimova, R., Tuominen, M., Bakin, A.S. *et al.* (1996) Silicon carbide liquid phase epitaxy in the Si-Sc-C system. *Inst. Phys. Conf. Ser.*, **142**, 101.
[218] Nishitani, S.R. and Kaneko, T. (2008) Metastable solvent epitaxy of SiC. *J. Cryst. Growth*, **310**, 1815.
[219] Vodakov, Y.A., Roenkov, A.D., Ramm, M.G. *et al.* (1997) Use of Ta-container for sublimation growth and doping of SiC bulk crystals and epitaxial layers. *Phys. Status Solidi B*, **202**, 177.
[220] Syväjärvi, M., Yakimova, R., Tuominen, A, M. *et al.* (1999) Growth of 6H and 4H-SiC by sublimation epitaxy. *J. Cryst. Growth*, **197**, 155.
[221] Syväjärvi, M., Yakimova, R., Jacobsson, H. and Janzen, E. (2000) Structural improvement in sublimation epitaxy of 4H-SiC. *J. Appl. Phys.*, **88**, 1407.
[222] Motoyama, S., Morikawa, N. and Kaneda, S. (1990) Low-temperature growth and its growth mechanisms of 3C-SiC crystal by gas source molecular beam epitaxial method. *J. Cryst. Growth*, **100**, 615.
[223] Fissel, A., Schröter, B., Kaiser, U. and Richter, W. (2000) Advances in the molecular-beam epitaxial growth of artificially layered heteropolytypic structures of SiC. *Appl. Phys. Lett.*, **77**, 2418.
[224] Kern, R.S., Järrendahl, K., Tanaka, S. and Davis, R.F. (1997) Homoepitaxial SiC growth by molecular beam epitaxy. *Phys. Status Solidi B*, **202**, 379.
[225] Matsunami, H., Nishino, S. and Ono, H. (1981) Heteroepitaxial growth of cubic silicon carbide on foreign substrates. *IEEE Trans. Electron Devices*, **28**, 1235.
[226] Nishino, S., Powell, A. and Will, H.A. (1983) Production of large-area single-crystal wafers of cubic SiC for semiconductor devices. *Appl. Phys. Lett.*, **42**, 460.
[227] Nishino, S., Suhara, H., Ono, H. and Matsunami, H. (1987) Epitaxial growth and electric characteristics of cubic SiC on silicon. *J. Appl. Phys.*, **61**, 4889.
[228] Shibahara, K., Nishino, S. and Matsunami, H. (1986) Surface morphology of cubic SiC(100) grown on Si(100) by chemical vapor deposition. *J. Cryst. Growth*, **78**, 538.
[229] Yamanaka, M., Daimon, H., Sakuma, E. *et al.* (1987) Temperature dependence of electrical properties of n-and p-type 3C-SiC. *J. Appl. Phys.*, **61**, 599.
[230] Kong, H.S., Wang, Y.C., Glass, J.T. and Davis, R.F. (1988) The effect of off-axis Si (100) substrates on the defect structure and electrical properties of β-SiC thin films. *J. Mater. Res.*, **3**, 521.
[231] Davis, R.F. (1989) Epitaxial growth and doping of and device development in monocyrstalline β-SiC semiconductor thin films. *Thin Sold Films*, **181**, 1.
[232] Ishida, Y., Takahashi, T., Okumura, H. and Yoshida, S. (2003) Investigation of antiphase domain annihilation mechanism in 3C-SiC on Si substrates. *J. Appl. Phys.*, **94**, 4676.
[233] Kitabatake, M. (1997) Simulations and experiments of 3C-SiC/Si heteroepitaxial growth. *Phys. Status Solidi B*, **202**, 405.
[234] Mogab, C.J. and Leamy, H.J. (1974) Conversion of Si to epitaxial SiC by reaction with C_2H_2. *J. Appl. Phys.*, **45**, 1075.
[235] Pirouz, P., Chorey, C.M. and Powell, J.A. (1987) Antiphase boundaries in epitaxially grown β-SiC. *Appl. Phys. Lett.*, **50**, 221.
[236] Carter, C.H. Jr.,, Davis, R.F. and Nutt, S.R. (1987) Transmission electron microscopy of process-induced defects in beta-SiC thin films. *J. Mater. Res.*, **1**, 811.
[237] Nagasawa, H. and Yagi, K. (1997) 3C-SiC sngle-crystal films grown on 6-inch Si substrates. *Phys. Status Solidi B*, **202**, 335.
[238] Nagasawa, H., Yagi, K., Kawahara, T. and Hatta, N. (2006) Reducing planar defects in 3C-SiC. *Chem. Vap. Deposition*, **12**, 502.
[239] Shibahara, K., Nishino, S. and Matsunami, H. (1987) Antiphase-domain-free growth of cubic SiC on Si(100). *Appl. Phys. Lett.*, **50**, 1888.
[240] Ishida, Y., Takahashi, T., Okumura, H. and Yoshida, S. (2006) Effect of reduced pressure on 3C-SiC heteroepitaxial growth on Si by CVD. *Chem. Vap. Deposition*, **12**, 495.
[241] Ferro, G., Chassagne, T., Leycuras, A. *et al.* (2006) Strain tailoring in 3C-SiC heteroepitaxial layers grown on Si(100). *Chem. Vap. Deposition*, **12**, 483.
[242] Zorman, C. and Mehregany, M. (2004) Micromachining of SiC, in *Silicon Carbide - Recent Major Advances* (eds W.J. Choyke, H. Matsunami and G. Pensl), Springer, p. 671.

[243] Cheung, R. (2006) *Silicon Carbide Microelectromechanical Systems for Harsh Environments*, Imperial College Press.

[244] Kong, H.S., Glass, J.T. and Davis, R.F. (1989) Growth rate, surface morphology, and defect microstructures of β-SiC films chemically vapor deposited on 6H-SiC substrates. *J. Mater. Res.*, **4**, 204.

[245] Chien, F.R., Nutt, S.R., Yoo, W.S. *et al.* (1994) Terrace growth and polytype development in epitaxial β-SiC films on α-SiC (6H and 15R) substrates. *J. Mater. Res.*, **9**, 940.

[246] Nishino, K., Kimoto, T. and Matsunami, H. (1997) Reduction of double positioning twinning in 3C-SiC grown on α-SiC. *Jpn. J. Appl. Phys.*, **36**, 5202.

5

Characterization Techniques and Defects in Silicon Carbide

Characterization of the physical properties, doping densities, and defects in SiC epitaxial layers and substrates is a fundamental step in the development of SiC devices. Accurate determination of the physical properties and identification of defect structures as well as their behavior are important subjects of academic study. For example, many of the physical properties of SiC that should be used in device simulations are still unknown. It is essential to fully understand the nature of the various defects that are present in SiC, because any of these defects could affect device performance and reliability. In this chapter, the fundamentals of important material characterization techniques are described, and an overview of extended and point defects in SiC is given.

5.1 Characterization Techniques

Almost all the material characterization techniques commonly used are applicable to SiC although special care is often required. The major points that are unique to SiC characterization are summarized as follows:

1. *Sample structures (underlying substrate)*: Characterization of properties or defects in SiC epitaxial layers is often intended because they are likely to be directly linked to the device performance. However, SiC epitaxial layers are often relatively thin (5–10 μm), and the influence of the underlying substrates therefore cannot be neglected. Thus, it is important to be aware of the thickness that is actually monitored by each individual characterization technique. The effects of the substrate must often be taken into account even when the epitaxial layers are relatively thick (30–50 μm).
2. *Wide bandgap*: When optical excitation is required during characterization, the conventional systems that were designed for characterization of Si and GaAs must be modified substantially. For example, ultraviolet light with a wavelength that is shorter than 370 nm is required as an "above-bandgap" excitation source (e.g., lamp, laser) to characterize 4H-SiC at room temperature. Because SiC has an indirect band structure, the optical absorption coefficient is small, even when the photon energy is considerably larger than the bandgap, as described in Section 2.2.2. Also, when thermal excitation is used in the characterization process, very high temperatures (>450 °C) are required to monitor the midgap states because the bandgap of 4H-SiC is approximately three times larger than that of Si.

Fundamentals of Silicon Carbide Technology: Growth, Characterization, Devices, and Applications, First Edition.
Tsunenobu Kimoto and James A. Cooper.

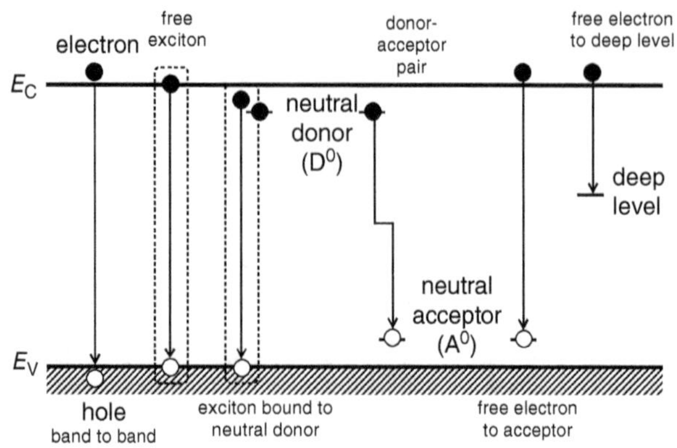

Figure 5.1 Schematic illustration of typical recombination paths of excess carriers in SiC.

In this section, the major techniques that can be used for characterization of the physical properties and detection of the defects in SiC are briefly introduced, together with typical example data. For a more detailed description of these characterization techniques, see the excellent book by Schroder [1].

5.1.1 Photoluminescence

Some of the excess carriers that are excited by appropriate light exhibit radiative recombination (*photoluminescence*, PL). The photon energy of the excitation source is usually larger than the bandgap (*above-gap excitation*). For example, He-Cd lasers (λ = 325 nm) or frequency-doubled Ar^+ lasers (λ = 244 nm) are often used. The absorption coefficients and penetration depths of the excitation light sources are described in Section 2.2.2. If the material under test is a perfect single crystal with extremely high purity, then only band-to-band or free exciton peaks are observed. However, because defects and/or impurities are present in real materials, multiple recombination paths compete with each other, leading to a variety of emission peaks in the PL spectra. PL is therefore a powerful technique for assessment of both the defects and the purity of semiconductor materials [1]. PL measurements are usually conducted at low temperatures (2, 4.2, or 77 K) to minimize peak broadening, and the temperature dependence of the PL spectra is helpful in the understanding of the recombination paths. Figure 5.1 shows schematically the typical recombination paths for excess carriers in SiC. The basic physics of the process is common to semiconductors, and has been well documented [2]. In SiC, however, an impurity or a point defect can substitute at certain inequivalent sites, leading to several different energy levels (*site effect*), as described in Section 2.1. This phenomenon makes the PL spectra of SiC more complex than those of other semiconductors. Several review papers on PL measurements of SiC have been published [3–7].

5.1.1.1 Free Excitons

Because SiC has an indirect band structure, direct band-to-band recombination without phonon involvement is prohibited. Additionally, because of the high binding energy of free excitons in SiC (E_x = 20–27 meV [3, 8]), the free excitons survive even at room temperature. Figure 5.2 shows typical PL spectra measured at 2 K for high-purity (a) 4H-SiC and (b) 6H-SiC epitaxial layers [7]. Here the PL peaks labeled by the I series originate from recombination of free excitons. In the peak assignment, the phonon energies (in meV) associated with the recombination processes are given as subscripts to the peak labels. The absence of a zero-phonon line of free exciton peaks reflects the indirect band structure, and

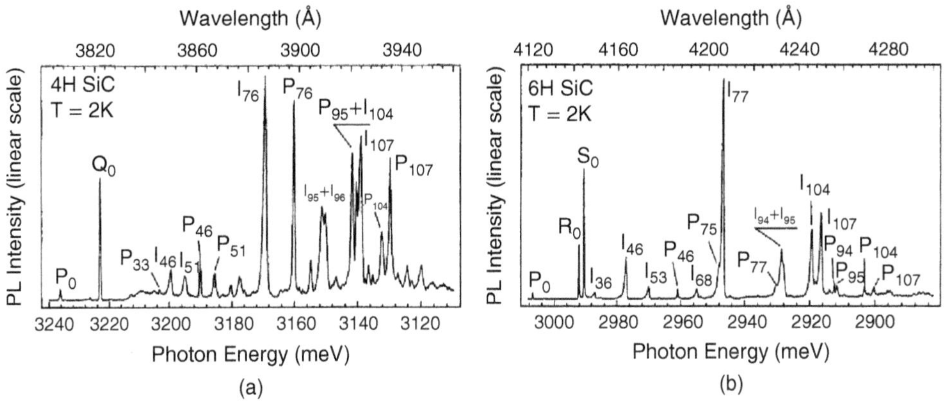

Figure 5.2 Typical photoluminescence (PL) spectra measured at 2 K for high-purity (a) 4H-SiC and (b) 6H-SiC epitaxial layers ([7] reproduced with permission from Taylor & Francis).

all free exciton peaks appear as phonon replicas. The energies of major phonons which create phonon replicas in PL from 4H-SiC{0001} are 36 (transverse acoustic (TA)), 46, 51, 77 (longitudinal acoustic (LA)), 95, 96 (transverse optical (TO)), 104, and 107 meV (longitudinal optical (LO)). The energy location of the zero-phonon line of free exciton peaks is called the *exciton gap*, and is often denoted by E_{gx}. The exciton gap is 2.390 eV for 3C-SiC, 3.265 eV for 4H-SiC, and 3.023 eV for 6H-SiC at 2–4 K [3, 6]. Note that the bandgap E_g is given by $E_g = E_{gx} + E_x$. The free exciton peaks are observed in lightly doped SiC (e.g., below 1×10^{17} cm^{-3}), and the relative intensity compared to that of an exciton peak bound to a dopant impurity can be used as a measure to estimate the purity of the material [9, 10]. The shapes of the free exciton peaks are highly asymmetrical, especially at temperatures higher than 10 K. When taking account of the kinetic energy and the density of states of the free excitons, the intensity of one free exciton peak (I_{FE}) as a function of the photon energy ($E = h\nu$) can be expressed approximately by the following equation [11]:

$$I_{FE} = C\sqrt{E}\exp\left(-\frac{E}{kT}\right) \tag{5.1}$$

Here, C is a proportional constant, k is the Boltzmann constant, and T is the absolute temperature. Figure 5.3 illustrates the change in the free-exciton-peak spectrum that was calculated with Equation 5.1 [12]. The peak energy increases with increasing temperature, and the peak broadening becomes more significant toward the high-energy side. These results originate from the increase in the kinetic energy of the free excitons.

5.1.1.2 Excitons Bound to Neutral Dopant Impurities

PL peaks that were attributed to excitons bound to neutral nitrogen donors have been investigated extensively in many SiC polytypes [13–15]. Typical PL spectra of excitons bound to neutral nitrogen donors in (a) 3C-SiC, (b) 4H-SiC, and (c) 6H-SiC and measured at 2 K are shown in Figure 5.4 [7]. The phonon energies (in meV) associated with the recombination processes are given as subscripts to the peak labels. Two series, denoted by *P* and *Q*, for 4H-SiC, and three series, denoted by *P*, *R*, and *S*, for 6H-SiC are observed because of the site effect (the *P* series corresponds to the hexagonal site). In the case of bound excitons, the wavefunction is spread out in the reciprocal space, and the direct recombination process is allowed, which leads to the appearance of strong zero-phonon line(s). The energy difference between the exciton gap and the zero-phonon line gives the *binding energy* of the excitons to the impurity/defect

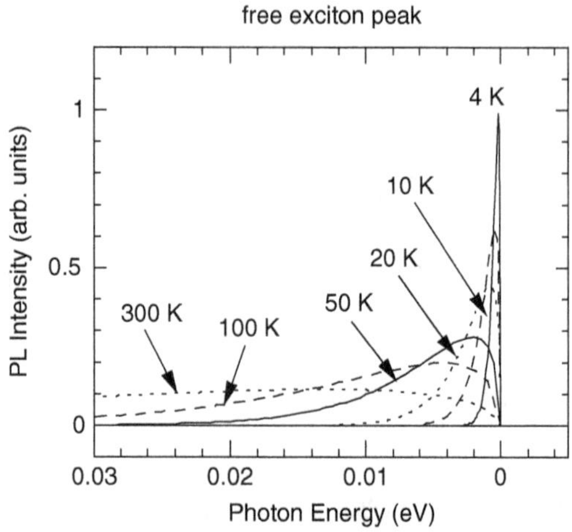

Figure 5.3 Change in the free-exciton-peak spectrum that was calculated with Equation 5.1 [12].

(a neutral nitrogen donor in the present case). As shown in Figure 5.4, the bound excitons with the smallest binding energy (the *P* series in both 4H- and 6H-SiC) are dominated by phonon replicas, while those with the largest binding energy (the *Q* series for 4H-SiC and the *S* series for 6H-SiC) are dominated by the zero-phonon lines. This can also be determined from the spreading of the wavefunction. Figure 5.5 shows the PL spectra measured for a high-purity 4H-SiC epitaxial layer at 4.2–300 K [16]. At a relatively low temperature, both the free exciton peaks and the nitrogen-bound exciton peaks are dominant. The nitrogen-bound exciton peaks are thermally quenched at approximately 10 K for the *P* series and 40 K for the *Q* series, forming more free excitons. In the temperature range from 50 to 60 K up to room temperature only the free exciton peaks dominate. Although PL peaks are also observed for the excitons that are bound to neutral phosphorus donors [17], the spectrum structure is less understood when compared with the nitrogen-bound exciton peaks.

Figure 5.6 shows the typical PL spectra of excitons that were bound to neutral aluminum or gallium acceptors in 4H-SiC, measured at 2 K [7, 18, 19]. The bound exciton spectrum consists of two zero-phonon lines that correspond to the hexagonal and cubic sites. The peak energies are slightly higher than that of Q_0 for aluminum, but slightly lower than that of Q_0 for gallium. Each zero-phonon line actually splits into three lines, as observed in high-resolution measurements, and this number is consistent with the result of a prediction based on group theory [7]. In high-purity aluminum-doped p-type SiC, the aluminum-acceptor bound exciton peaks (and the free exciton peaks) become dominant. In these materials, the PL peaks of excitons that are bound to *neutral* donors are hardly observed, because the donors act as compensating levels and are *ionized* even at low temperatures. The PL spectra of excitons that are bound to boron atoms require further investigation [20].

5.1.1.3 Donor–Acceptor Pair Recombination

When the wavefunction of an electron that is bound to a donor can interact with that of a hole that is bound to an acceptor, then *donor–acceptor pair* (DAP) recombination occurs. The photon energy of DAP luminescence is expressed using the following equation [2]:

$$h\upsilon = E_g - \Delta E_D - \Delta E_A + \frac{e^2}{4\pi\varepsilon_s r} \tag{5.2}$$

Here, ΔE_D and ΔE_A are the ionization energies of the donor and acceptor, respectively. The fourth term is the gain of the Coulomb potential, and is a function of the donor-acceptor distance (r). e and ε_s are the

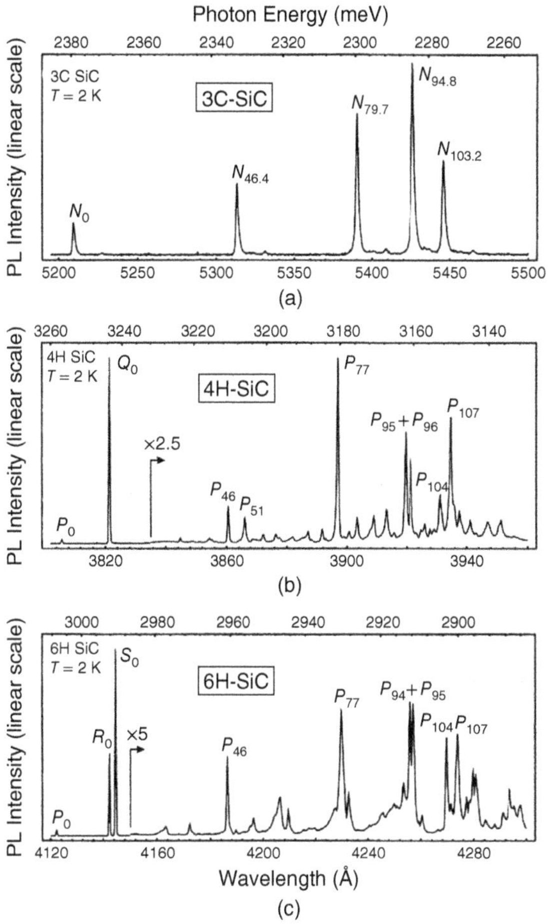

Figure 5.4 Typical PL spectra of excitons bound to neutral nitrogen donors in (a) 3C-SiC, (b) 4H-SiC, and (c) 6H-SiC, measured at 2 K ([7] reproduced with permission from Taylor & Francis).

elementary charge and the dielectric constant of the semiconductor, respectively. The recombination rate is high for DAPs separated by a short distance, and this rate decreases rapidly with increasing distance. As a result, the decay curves of the PL intensity of the DAP peaks generally show not a single exponential decay but a decay in the form of t^{-m}, where t and m are the time and the power factor of the decay [21], respectively.

DAP luminescence in various SiC polytypes was extensively studied in the 1970–1980s [22–24]. Figure 5.7 shows typical DAP luminescence spectra that were observed for 4H-SiC at (a) 4.2 and (b) 77 K [24]. In this figure, the nitrogen-aluminum, nitrogen-gallium, and nitrogen-boron DAP PL spectra are shown. Since the aluminum acceptor levels are relatively shallow and the boron levels very deep, the nitrogen-aluminum DAP PL peaks are located in the relatively high-energy region and the nitrogen-boron DAP appears in the low-energy region. As described in Section 2.2.3, boron incorporation into SiC induces two boron-related levels: shallow boron and deep boron. In the nitrogen-boron PL, the *deep boron* centers are mainly involved in radiative recombination [20, 25]. Because of the rather high ionization energy of the deep boron center, boron-related DAP, or free-to-acceptor (FA) PL survives at

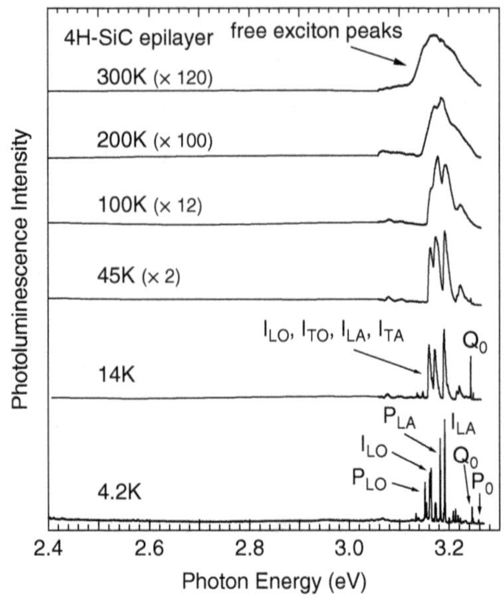

Figure 5.5 PL spectra measured for a high-purity 4H-SiC epitaxial layer at 4.2–300 K [16].

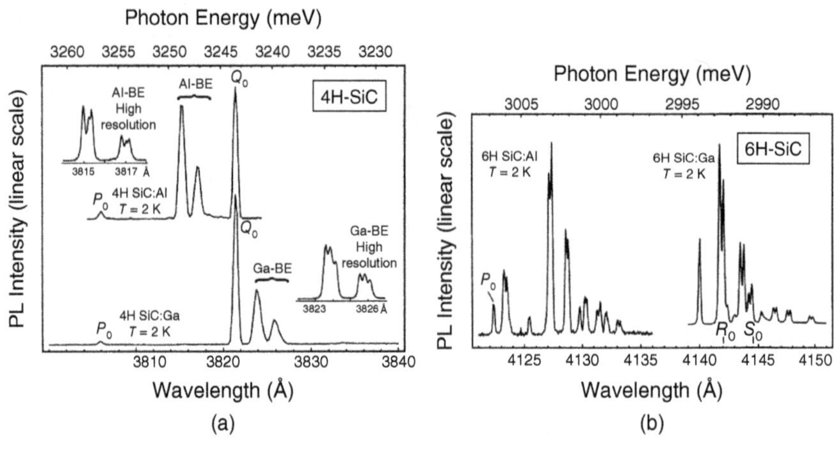

Figure 5.6 Typical PL spectra of excitons that are bound to neutral aluminum or gallium acceptors in (a) 4H-SiC and (b) 6H-SiC, measured at 2 K ([7] reproduced with permission from Taylor & Francis).

temperatures up to room temperature, yielding yellowish-green luminescence in 4H-SiC. Because of the large site effect of the nitrogen donors in hexagonal SiC polytypes, the DAP peaks of the nitrogen donors at the hexagonal site (*B* series) and those of nitrogen at the cubic site(s) (*C* series) appear at different photon energies. A fine nitrogen-aluminum DAP PL spectrum has been resolved and analyzed using Equation 5.2 [26]. Based on this analysis, the ionization energies of the nitrogen donors in 4H-SiC have been accurately determined to be 61 meV for the hexagonal site and 126 meV for the cubic site, and those of the aluminum acceptors to be 198 and 201 meV for the hexagonal and cubic sites, respectively.

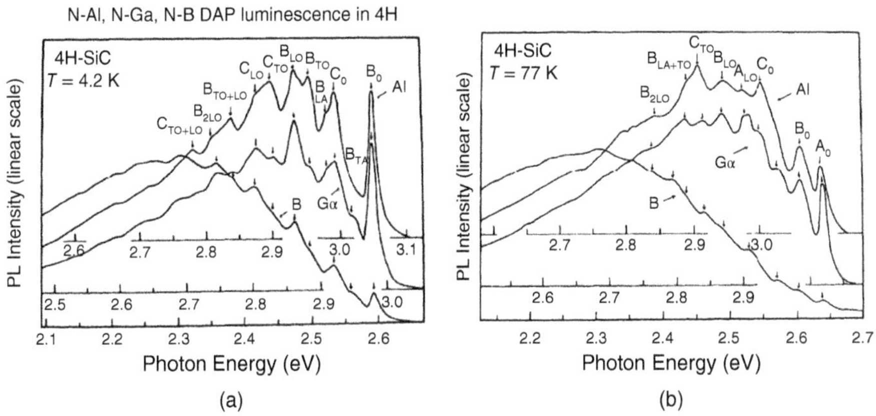

Figure 5.7 Typical donor–acceptor-pair (DAP) luminescence spectra observed for 4H-SiC at (a) 4.2 K and (b) 77 K ([24] reproduced with permission from American Physical Society.

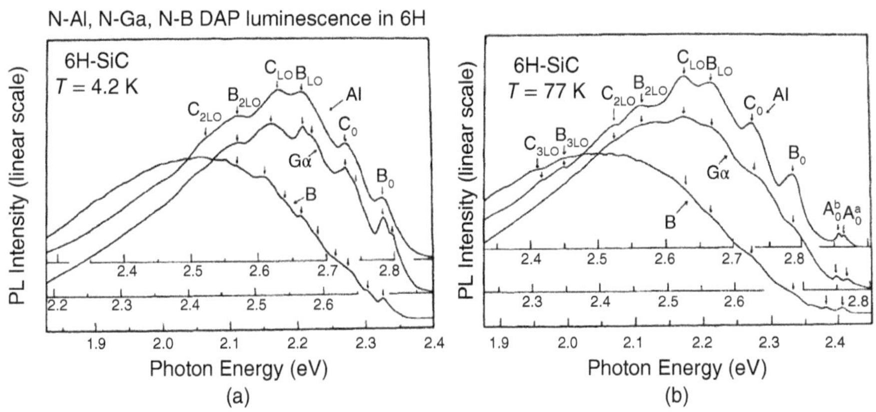

Figure 5.8 Typical donor–acceptor-pair (DAP) luminescence spectra observed for 6H-SiC at (a) 4.2 K and (b) 77 K ([24] reproduced with permission from American Physical Society).

When the measurement temperature is elevated, the ionization of the nitrogen donors proceeds while the acceptors remain unchanged, retaining holes, and thus recombination between the free electrons and holes that are bound to acceptors (FA) becomes dominant. Typical DAP luminescence spectra that were observed for 6H-SiC are shown in Figure 5.8. The trends are very similar to the case of 4H-SiC, but the PL peaks are shifted to the low-energy side, due to the smaller bandgap and the larger ionization energies of acceptors in 6H-SiC.

5.1.1.4 Other Impurities

Among the various impurities in SiC, titanium (Ti) incorporation, and its unique PL are well known [27–29]. Figure 5.9 shows typical PL spectra, showing the effects of Ti contamination of (a) 4H-SiC and (b) 6H-SiC, that were measured at 2 K [7]. Two zero-phonon lines (the A1 and B1 peaks) are observed

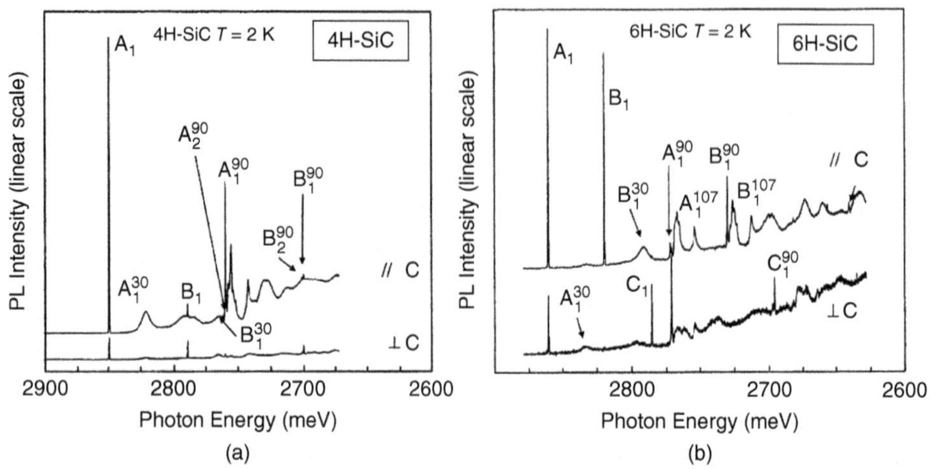

Figure 5.9 Example of PL spectra showing Ti contamination of (a) 4H-SiC and (b) 6H-SiC measured at 2 K ([7] reproduced with permission from Taylor & Francis).

Table 5.1 Energies of major impurity-related PL peaks observed in 4H- and 6H-SiC [7].

Defect	4H-SiC (eV)	6H-SiC (eV)
Nitrogen bound exciton	3.2580 (P_0), 3.2448 (Q_0)	3.006.9 (P_0), 2.9923 (R_0), 2.9905 (S_0)
Aluminum bound exciton	3.2465–3.2490	2.9989–3.0045
Titanium (Ti)	2.8492 (A_1), 2.7898 (B_1)	2.8608 (A_1), 2.8197 (B_1), 2.7854 (C_1)
Chromium (Cr)	1.1898, 1.1583	1.1886, 1.1797, 1.1557
Molybdenum (Mo)	1.1521	1.1057
Tungsten (W: formerly UD-1)	1.0595, 1.0586	1.0019, 1.0014, 0.9951
Vanadium (V)	0.9703, 0.9695, 0.9681, 0.9673, 0.9284	0.9482, 0.9475, 0.9460, 0.9453, 0.9170, 0.8931
Hydrogen-related	3.1611, 3.1446	3.0134, 3.0065, 2.9559

at 2.8492 and 2.7898 eV, and are assigned to Ti at the hexagonal and cubic Si-lattice sites, respectively. Unique phonon replicas with the involvement of a localized phonon (90 meV) are relatively intense in the Ti-related PL. A very broad band with a peak located at 2.58–2.48 eV (480–500 nm) can also be observed. These PL peaks originate from excitons that reside at an isoelectronic center of Ti in SiC, and exhibit a long decay time of about 100 μs [30].

Many PL peaks caused by other metallic impurities in SiC have been identified. These metallic impurities include vanadium (V) [31–33], chromium (Cr) [34], molybdenum (Mo) [35], and tungsten (W) [36], and they exhibit very sharp PL peaks in the near infrared region. Table 5.1 summarizes the energies of major PL peaks that have been observed in 4H- and 6H-SiC [7].

5.1.1.5 Intrinsic Point Defects

Extensive studies have been conducted to identify intrinsic point defects by PL, optical absorption, electron paramagnetic resonance (EPR), and optically detected magnetic resonance (ODMR) measurements,

Table 5.2 Energy positions of the optical transition for the major intrinsic point defects in 4H- and 6H-SiC at 2–4 K.

Defect	4H-SiC (eV)	6H-SiC (eV)
Silicon vacancy (V_{Si})	1.438, 1.352	1.438, 1.398, 1.366
Divacancy ($V_{Si}-V_C$)	1.0539, 1.0507, 1.0136, 0.9975	1.0746, 1.0487, 1.0300, 1.0119, 0.9887
D_I center	2.901 (L_1)	2.6245 (L_1), 2.589 (L_2), 2.569 (L_3)
D_{II} center	3.2045	2.9459
UD-2 (may contain impurity)	1.1496, 1.1194, 1.0968, 1.0953	1.1345, 1.1342, 1.1196, 1.1033, 1.0928, 1.0882
UD-3 (may contain impurity)	1.3557	1.3433
UD-4 (may contain impurity)	1.4646, 1.3952, 1.3944	1.426, 1.371

and through theoretical calculations [7]. The energy positions of the optical transition for the major intrinsic point defects in 4H- and 6H-SiC at 2–4 K are summarized in Table 5.2 [7, 37–40]. In the table, the peak positions of several unidentified defects (UDs) are also indicated. Note that the origin of these UD centers may contain an impurity.

Other than the PL peaks described above, three types of defect luminescence that are attributed to intrinsic point defects are often observed in SiC: these are the so-called *alphabet lines,* the D_I *center,* and the D_{II} *center* [4, 6, 7]. The alphabet lines consist of 12 zero-phonon lines in the 2.80–2.91 eV range and their phonon replicas in 4H-SiC [41–43]. These lines are not usually observed in as-grown materials, but emerge after particle bombardment, such as electron irradiation. Low-energy (<200 keV) electron irradiation, under which only carbon atoms are displaced in SiC, induces the appearance of these alphabet lines. The lines start to anneal out at approximately 650 °C, and almost completely disappear after annealing at 1000 °C [41]. Based on these observations, carbon Frenkel pairs or carbon interstitials have been suggested as the origins of the lines [41, 44].

After the alphabet lines are removed by annealing irradiated (or implanted) SiC at 900–1000 °C, strong luminescence caused by the D_I *center* is observed [45–47]. Figure 5.10a shows typical PL spectra of the D_I center for 4H- and 6H-SiC [7]. One strong zero-phonon line (L_1 peak: 2.901 eV) and its phonon replicas are observed in 4H-SiC, while three zero-phonon lines (L_1, L_2, and L_3 peaks) and their phonon replicas appear in 6H-SiC. Note that the D_I center is often observed, even in as-grown materials (both bulk crystals and epitaxial layers), and it has been shown that the intensity of the L_1 peak increases in SiC epitaxial layers that are grown at higher temperatures [48]. This defect center remains thermally stable at temperatures as high as 1800–2000 °C, and the L_1 peak intensity tends to be enhanced by high-temperature annealing [49]. This isoelectronic defect has a "*pseudo-donor*" property [50], and a level acting as a hole trap (HS1 center) located at 0.35 eV above the valence band edge has been proposed as being responsible for the luminescence, as shown in Figure 5.10b. Although the origin of the D_I center remains unclear, antisite pairs (Si_C-C_{Si}) [51] and an isolated silicon antisite (Si_C) [52] have been suggested as possible origins.

The D_{II} *center* yields a strong zero-phonon line at 3.2045 eV in 4H-SiC and 2.9459 eV in 6H-SiC [53, 54]. This center is also stable against high-temperature (>1600 °C) annealing. Because the D_{II} center appears only after ion implantation (and not after low-energy electron irradiation) and subsequent annealing, it has been suggested that the origin of the center must contain a large complex, such as a dicarbon antisite [54] or carbon antisite clusters [55].

Thus, PL is a very powerful technique for assessment of material quality. However, the origins of many PL peaks in SiC are still unassigned. To determine the absolute densities of the defects or impurities, other characterization techniques must be used in combination.

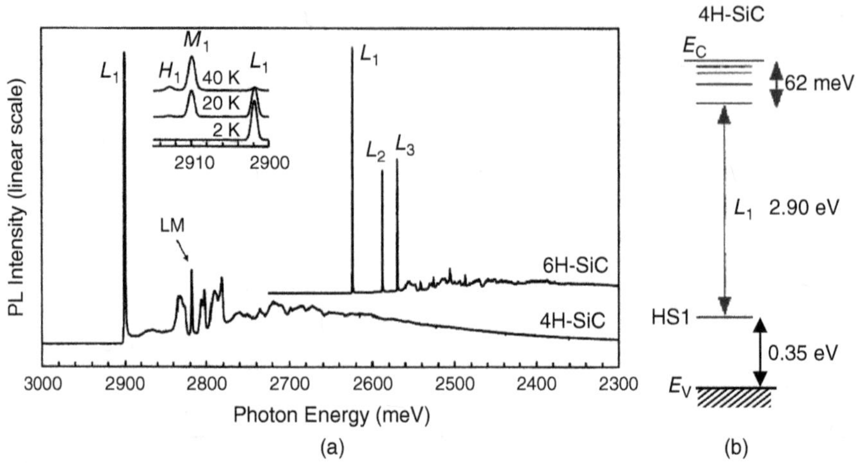

Figure 5.10 (a) Typical PL spectra of the D_I center for 4H- and 6H-SiC [7] and (b) pseudo-donor model of the luminescence taking account of a hole trap (HS1 center) [50] reproduced with permission from Taylor & Francis.

5.1.2 Raman Scattering

Raman scattering spectroscopy is a vibrational spectroscopic technique that enables a wide range of information about a material to be acquired. When light is scattered from a solid material, the scattered light mainly contains the wavelengths that were incident on the material, which is called *Rayleigh scattering*. However, the scattered light also contains different wavelengths at much lower intensities, caused by the interaction of the light with the material. A typical example is the interaction of the light with phonons, and this scattering is called *Raman scattering* [1]. The scattered light, which has an energy that is lower than that of the incident light by an amount equal to the energies of the emitted phonons (*Stokes-shifted Raman scattering*), is monitored in Raman scattering spectroscopy. An intense laser such as an argon ion laser is used as an excitation source, and a high-resolution double monochromator is used to exclude the Rayleigh scattered light.

Because Raman scattered light contains detailed information about the phonons, Raman scattering spectroscopy is a powerful tool for identification of the material polytype and characterization of the stress or the defects in SiC [56–59]. Because of the wide bandgaps of the SiC polytypes, 4H-SiC is completely transparent to conventional excitation light (e.g., 488 nm from Ar^+ laser). In this case, the Raman scattering signals have poor depth resolution. A confocal configuration and deep ultraviolet Raman scattering (using a frequency-doubled Ar^+ laser (244 nm)) have been proposed as methods to improve the depth resolution [60, 61]. In contrast, the lateral resolution is good, because the incident light can be focused to a small spot diameter (~1 μm), and mapping measurements are possible.

As briefly described in Section 2.3.2, group theory analysis predicts a number of phonon modes at the Γ point for the higher polytypes (*folded modes*). For the hexagonal (and rhombohedral) SiC polytypes, the phonon modes are divided into axial and planar modes, in which the atoms are displaced along the directions parallel and perpendicular to the *c*-axis, respectively. For *n*H-SiC, a folded mode at the Γ point corresponds to a phonon mode with a reduced wave number $x = q/q_B = 2\,m/n$ along the <111> direction in the basic Brillouin zone of 3C-SiC, where *m* is an integer that is less than or equal to $n/2$, and q_B is the wavenumber at the edge of the basic Brillouin zone [58]. The planar modes in *n*H-SiC consist of Raman active E_1 and E_2 modes, and the axial modes consist of Raman active A_1 and inactive B_1 modes. It should be noted that the Raman scattering spectra that are observed depend on the scattering

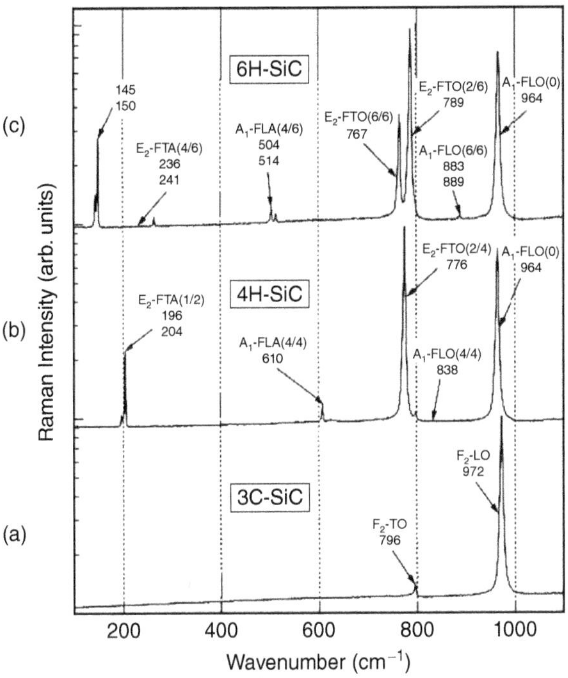

Figure 5.11 Raman scattering spectra of (a) 3C-, (b) 4H-, and (c) 6H-SiC polytypes observed with the back scattering geometries of {0001} crystals ([58] reproduced with permission from Wiley-VCH Verlag GmbH).

geometry. In the typical back scattering geometry for hexagonal SiC{0001}, E_2-type TO (folded transverse optical, FTO), and A_1-type LO (folded longitudinal optical, FLO) modes are allowed, while the E_1-type LO (FLO) mode is prohibited (the E_1-type LO peak is weakly observed when the sample has an off-angle).

Figure 5.11 shows the Raman scattering spectra of the 3C-, 4H-, and 6H-SiC polytypes, which are observed with the back scattering geometries of {0001} crystals. In the low-frequency region below 280 cm^{-1}, Raman peaks attributed to the TA (folded transverse acoustic, FTA) modes are observed, except in 3C-SiC. Intense Raman peaks caused by the TO and LO modes appear at wavenumbers of approximately 770–800 and 960–970 cm^{-1}, respectively, but the peak positions differ slightly for the different SiC polytypes. Table 5.3 summarizes the Raman wavenumbers of the TA, LA, TO, and LO modes observed for 3C-, 4H-, and 6H-SiC [58]. Raman scattering is thus a useful and nondestructive technique for unambiguous identification of the SiC polytypes. Raman modes for other SiC polytypes can be found in a review paper by Nakashima and Harima [58]. Because the Raman scattering spectra are sensitive to the stacking structures, the existence of stacking faults can cause the appearance of extra peaks and/or distortion of the Raman peaks.

In a polar semiconductor like SiC, free carriers interact with the LO phonons, forming the *longitudinal optical phonon-plasmon coupled (LOPC) mode*. Thus, the Raman peak due to the LO mode is affected by the carrier density in the monitored region. As the carrier density increases, the LO peak shifts toward the high wavenumber side with peak broadening. The carrier density can then be estimated by fitting this LO peak while taking the plasmon coupling into account [62]. The determination of the carrier density without use of contacts is a clear merit, although this method is only applicable for relatively highly doped n-type SiC.

Table 5.3 Major Raman wavenumbers of the TA, LA, TO, and LO modes observed for 3C-, 4H-, and 6H-SiC [58].

Polytype	TA (cm^{-1}) (planar acoustic)	LA (cm^{-1}) (axial acoustic)	TO (cm^{-1}) (planar optic)	LO (cm^{-1}) (axial optic)
3C-SiC	–	–	796	972
4H-SiC	196, 204	610	776, 796	964
6H-SiC	145, 150	504, 514	767, 789, 797	965

Because Raman scattering is sensitive to strain, local stresses can also be characterized. The stress can be quantitatively determined based on the shift of a Raman peak from its original position. Stress characterization by Raman scattering is receiving increasing attention in the Si and III–V semiconductor fields [63]. Similarly, Raman scattering is likely to be used extensively for stress characterization of SiC in the future.

5.1.3 *Hall Effect and Capacitance–Voltage Measurements*

Hall effect measurements give the *free carrier density* (electron density n or hole density p) and mobility of the material, whereas capacitance–voltage (C–V) measurements on Schottky structures give the *net doping density,* which is $N_D - N_A$ for n-type and $N_A - N_D$ for p-type, where N_D and N_A are the donor and acceptor densities, respectively. Because of the relatively large ionization energies of acceptors in SiC, Hall effect and C–V measurements on p-type SiC give very different values for the densities ($p < N_A - N_D$) at room temperature. Only the hole density changes when the measurement temperature is changed, while the net doping density remains almost constant, as expected.

For Hall effect measurements of SiC homoepitaxial layers, careful sample preparation is required. Figure 5.12 schematically shows sample structures that are suitable for Hall effect measurements.

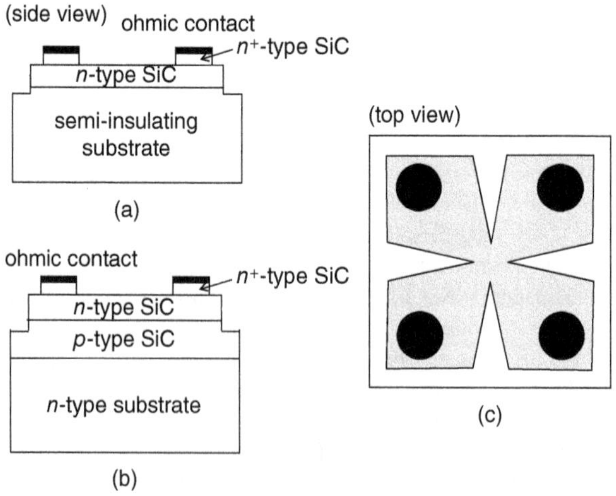

Figure 5.12 Schematic sample structures suitable for Hall effect measurements. Electrical isolation is achieved by using (a) semi-insulating substrate or (b) pn junction.

The SiC epitaxial layer to be measured must be electrically isolated from the substrate. One way to do this is to use semi-insulating substrates for electrical isolation, as shown in Figure 5.12a. Another way is to use a pn junction for the isolation, as shown in Figure 5.12b. In the former case, the resistance of the epitaxial layer to be measured must be carefully compared with that of the substrate at the measurement temperatures to ensure that the leakage current through the substrate is negligibly small. In the latter case, the underlying layer of the opposite conduction type must be heavily-doped and must be thick enough to avoid punch-through. The junction leakage over a relatively large area must be kept sufficiently low. It is also important to prepare a clover-leaf pattern for accurate measurements [1]. It is useful to form the heavily-doped regions only beneath the ohmic contacts to avoid the occurrence of nonohmic contacts in the low-temperature measurements.

The measured temperature dependence of the carrier density can be fitted using Equations 2.14 or 2.15. Here, at least two types of donor level (cubic/hexagonal sites) must be considered to fit the temperature dependence of the electron density in hexagonal SiC, and it is important to account for the compensating defects. The compensating defects are residual acceptors (donors) and deep levels in n-type (p-type) materials, and these defects have typical densities of 1×10^{13} to 5×10^{13} cm^{-3} in lightly-doped 4H-SiC epitaxial layers. Methods for analysis of the measured data have been proposed, such as DHES (differential Hall effect spectroscopy) [64] and FCCS (free carrier concentration spectroscopy) [65]. In these methods, the energy levels and the donor (or acceptor) density can be obtained analytically without ambiguous data fitting. A Hall scattering factor of unity is usually used to calculate the data in Hall effect measurements of SiC. Although a few basic studies of the Hall scattering factor have been reported [66], the scattering factor varies, depending on the doping density, the compensating defect density, and the temperature.

In C–V measurements, Schottky barriers are formed by deposition of metal electrodes on top of the SiC samples. The net doping density can be determined from the slope of a $1/C^2$–V plot [1]. If the doping density is not uniform throughout the depth, then the depth profile of the doping density can be obtained from the differential slope of the plot [1]. In this case, the monitored depth (d) is simply determined from $d = \varepsilon_s/C_d$. Here, ε_s and C_d are the dielectric constant and the depletion capacitance per unit area of the semiconductor, respectively. Because this sample structure consists of a Schottky barrier and a series resistance (R_S), selection of the probe frequency requires care. For accurate measurement of the capacitance of the space-charge region, C, it is important to satisfy the condition $\omega CR_S \ll 1$, where ω is the angular probe frequency in the C–V measurements. For example, the probe frequency must be lowered in C–V measurements of lightly-doped p-type SiC because of the relatively high series resistance. Basic equations for the C–V characteristics can be found in a book on semiconductor physics [67] and in Section 6.4.1 in this book.

5.1.4 Carrier Lifetime Measurements

The carrier lifetime is an important physical property that determines the performance of bipolar devices, and fundamental studies had already been performed on carrier lifetimes in SiC in the 1990s [68]. In this subsection, several lifetime measurement methods are summarized. Important issues in interpretation of the measured data are also described. For example, the decay curve of the measured signals ($I(t)$) can be expressed using the following equation:

$$I(t) = I_0 \exp\left(-\frac{t}{\tau_{\text{decay}}}\right) \tag{5.3}$$

Here, I_0 and τ_{decay} are the initial signal intensity and the decay time, respectively. However, the decay time (τ_{decay}) is not always equivalent to the carrier lifetime, as described below. Furthermore, the decay curve often does not follow a single exponential function (τ_{decay} is not constant).

Table 5.4 Relationship between the carrier lifetime (τ) and the decay time (τ_{decay}) obtained in several measurement techniques (The decay time obtained in individual measurements is indicated).

Injection level	Time-resolved PL	Photoconductance	Reverse recovery
High injection	$\tau/2$	τ	τ (often limited by surface recombination)
Low injection	τ	$\mu_n d(\Delta n)/dt + \mu_p d(\Delta p)/dt \approx \tau$ (when $\Delta n = \Delta p$)	τ (often limited by surface recombination)

5.1.4.1 Time-Resolved Photoluminescence (TRPL)

The PL intensity of the free exciton peaks is proportional to the number of excess free carriers, although the majority of the free carriers and excitons recombine via nonradiative recombination in SiC because of its indirect band structure. The carrier lifetime can therefore be determined from the decay curve of the PL intensity for the free exciton peaks (~390 nm at room temperature). No preparatory processing, such as contact formation, is required prior to the measurements, and it is easy to acquire maps of the carrier lifetimes of SiC wafers with high spatial resolution using a conventional PL mapping stage. The temperature dependence of the carrier lifetimes can also be measured easily. Assuming the density of the excess carriers as Δn, the intensity of the free exciton peaks is proportional to $(n + \Delta n)(p + \Delta n) \approx (\Delta n)^2$ when Δn is higher than the equilibrium carrier density (*high-injection conditions*), while it is proportional to Δn when Δn is lower than the equilibrium density (*low-injection conditions*). Therefore, the decay time obtained (τ_{decay}) gives $\tau/2$ under high-injection conditions and τ under low-injection conditions (where τ is the carrier lifetime) [69], as shown in Table 5.4.

5.1.4.2 Photoconductance Decay (PCD)

By measuring the transient electrical conductance of a semiconductor after a photo-excitation pulse, it is possible to determine the carrier lifetime. The conductance can be monitored by either direct measurement of the electrical properties after deposition of ohmic contacts or measurement of the intensity of an electromagnetic wave that has been reflected from the sample surface. In the latter case no processing is required, which enables mapping measurements. Microwaves are often used for the electromagnetic waves, and this method is called *microwave photoconductance decay* (μ-PCD) measurement. Figure 5.13 shows a schematic illustration of a differential μ-PCD measurement set-up. The traveling microwave (frequency of e.g., 26 GHz) is split into two by a magic-T, and a laser pulse is used to irradiate a sample surface that is facing one of the two waveguide openings. The signal-to-noise ratio can be greatly improved by using a differential signal of the reflected microwave intensity from a photo-irradiated region and a nonirradiated region. The photoconductance is a direct measure of the number of excess carriers, and this method has been used as a standard technique for lifetime measurements in Si [1]. The signal intensity, which corresponds to the photoconductance, is proportional to $\Delta n\mu_n + \Delta p\mu_p$, where Δn and Δp are the densities of the excess electrons and the excess holes, respectively. μ_n and μ_p are the electron and hole mobilities, respectively. When "above-bandgap" excitation is used and recombination of the electron–hole pairs is dominant (normal case), $\Delta n = \Delta p$ is basically satisfied. When *differential* μ-PCD measurements are performed, the obtained signal intensity must be in proportion to Δn (= Δp), and the obtained decay time (τ_{decay}) gives τ, irrespective of the injection conditions.

5.1.4.3 Reverse Recovery (RR)

During turn-off operation of a pn diode, the injected carriers do not disappear immediately and significant numbers of carriers are stored inside the diode for a certain period. During this period

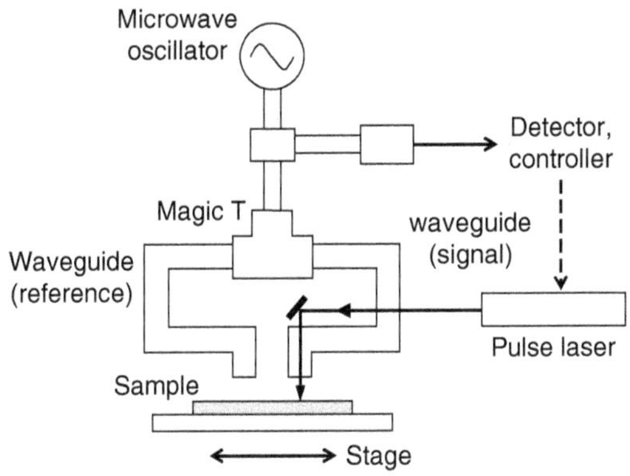

Figure 5.13 Schematic illustration of a differential μ-PCD measurement set-up.

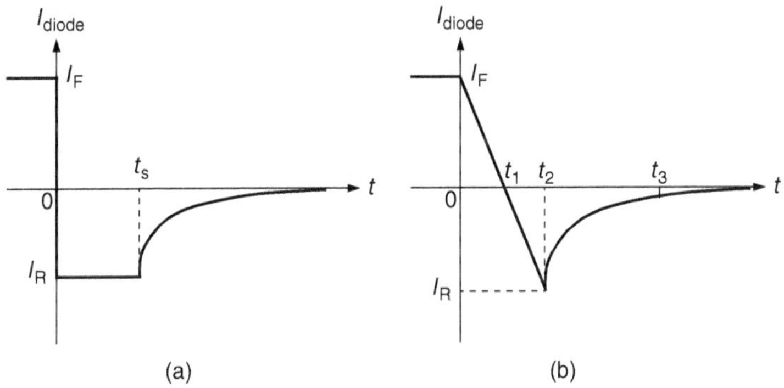

Figure 5.14 Turn-off characteristics (current waveform) of a pn diode. (a) Abruptly switching and (b) switching with a constant dI/dt.

(*storage time*), the diode remains conductive, and a reverse current passes through the diode until the stored carriers disappear by either recombination or diffusion (*reverse recovery characteristics*), as shown in Figure 5.14. The carrier lifetime can then be estimated from this current turn-off waveform [1]. When the current is abruptly switched, as shown in Figure 5.14a, the diode current transient $I(t)$ can be divided into the nearly constant current storage phase (where $0 \leq t \leq t_s$) and the reverse recovery phase (where $t > t_s$). The carrier lifetime (τ) is linked to the storage time (t_s) by the following equation [1]:

$$\mathrm{erf}\sqrt{\frac{t_s}{\tau}} = \frac{1}{1 + I_R/I_F} \tag{5.4}$$

Here, "erf," I_F, and I_R are the error function, the forward current, and the reverse current during the storage time, respectively, as shown in Figure 5.14a. A diode is often turned off with a constant current-damping rate (dI/dt), as shown in Figure 5.14b. In this case, the lifetime can be approximately given by [1]:

$$\tau \cong \sqrt{(t_2 - t_1)(t_3 - t_1)} \tag{5.5}$$

Here, t_1 and t_2 are as indicated in Figure 5.14b. t_3 is defined as the time where $I(t) = 0.1I_R$.

Although full processing of the pn diodes is required, this method can measure the carrier lifetime inside the completed device structures. The carrier lifetimes obtained are, however, often underestimated because of the influence of surface recombination. Also, it must be remembered that the obtained τ value contains information on the lifetime in the heavily-doped emitter region. Thus, the deduced τ value is always much shorter than the bulk lifetime in the lightly-doped region. As a modification of this method, the *open-circuit voltage decay* (OCVD) method has been proposed [1]. In this method, the open-circuit voltage of a pn diode, rather than the reverse recovery current, is monitored. The merits and demerits of this method are almost the same as those for measurements of reverse-recovery current.

5.1.4.4 Effects of Carrier Recombination at the Surface and in the Substrate

As described above, the decay times obtained in carrier lifetime measurements are not always the true carrier lifetimes for several reasons. In carrier lifetime measurements of SiC homoepitaxial layers, the effects of carrier recombination at the surface and in the substrate must be taken adequately into account [70].

Numerical simulations based on an excess carrier diffusion equation are useful for interpretation of the obtained decay curves [71]. In the simulation, a two-layer model of a SiC epitaxial layer (thickness : W_{epi}) on a substrate (thickness : W_{sub}) is considered, and the distribution (i.e., the depth profile) of the excess carrier density is calculated as a function of time, while taking account of both bulk recombination and surface recombination. The diffusion equation and the boundary conditions are given by the following equations [70, 71]:

$$\frac{\partial n_e}{\partial t} = D\frac{\partial^2 n_e}{\partial x^2} - \frac{n_e}{\tau} \tag{5.6}$$

Boundary conditions:

$$D\frac{dn_e}{dx}|_{x=0} = S_R n_e \tag{5.7}$$

Here, n_e is the excess carrier density in the semiconductor ($= \Delta n$), D is the diffusion constant, and τ is the bulk lifetime. The origin of the x coordinates is taken to be at the surface of the epitaxial layer. S_R is the *surface recombination velocity* (SRV) at the surface of the epitaxial layer. The recombination velocity on the back of the substrate is not important because of the large thickness and short carrier lifetimes of the substrates. Note that the bulk lifetime in the SiC substrate (τ_{sub}) is very short, with typical values of 0.01–0.04 μs [70]. Under high-injection conditions, carrier diffusion in the epitaxial layer can be assumed to proceed in an ambipolar mode [1]. An ambipolar diffusion constant of 4.2 cm^2 s^{-1} [72] is used for the epitaxial layer in the simulation, while a standard hole diffusion constant of 0.3 cm^2 s^{-1} is assumed for the substrate, because of the high doping density.

Figure 5.15 shows depth profiles of the excess carrier density in a 4H-SiC epitaxial layer on a substrate as a function of time (from $t = 0$ to $t = 5$ μs, at time intervals of 0.5 μs) [70]. Here, a YLF-third harmonic generation laser ($\lambda = 349$ nm) is assumed to be the excitation source. The photon density during excitation is 1×10^{14} cm^{-2}. In this case, the bulk lifetime in the epitaxial layer and the SRV (S_R) are assumed to be 5.0 μs and 1×10^3 cm s^{-1}, respectively. The epitaxial layer thicknesses are (a) 50 and (b) 200 μm. Because of the short carrier lifetime in the substrate (0.01–0.04 μs), the excess carriers generated in the substrate disappear almost completely within the first 0.1–0.2 μs, and the carriers in the epitaxial layer near the substrate diffuse toward the substrate, because of a large carrier density gradient. For the 50-μm-thick epitaxial layer, a significant number (about 30–35%) of the generated carriers recombine within the substrate in the short time of 0.5 μs after the excitation pulse. Because of the relatively large SRV, approximately 15% of the generated carriers are killed at the surface in the first 0.5 μs. The carrier profile continues to be dominated heavily by recombination at the surface and in the substrate as time passes, as shown in Figure 5.15a. After 5 μs, which corresponds to the bulk lifetime of the epitaxial layer, about 98% of the generated carriers have already been lost. However, in the 200-μm-thick epitaxial layer, the effect of substrate recombination on the carrier profile is much smaller. Although fast recombination

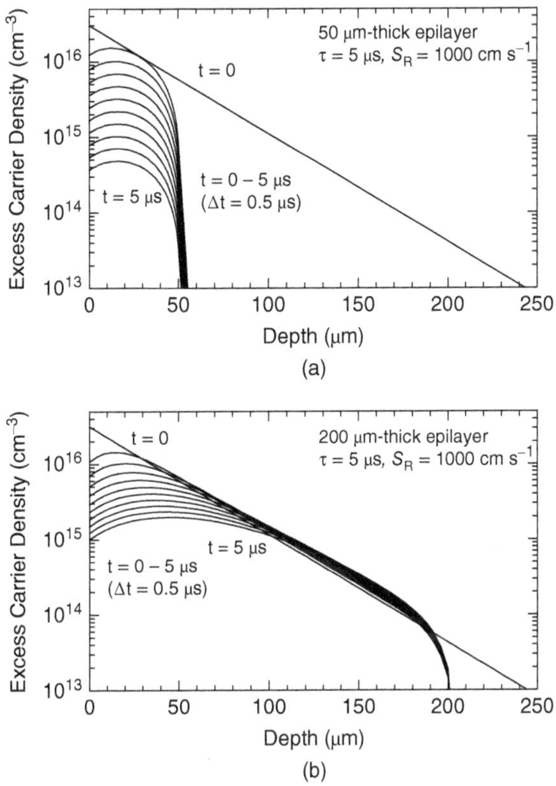

Figure 5.15 Depth profiles of the excess carrier density in a 4H-SiC epitaxial layer on a substrate as a function of time (from $t = 0$ to $t = 5$ μs, time intervals: 0.5 μs) [70]. In this case, the bulk lifetime in the epitaxial layer and the surface recombination velocity (S_R) are assumed to be 5.0 μs and 1×10^3 cm s^{-1}, respectively. The epitaxial layer thicknesses are (a) 50 and (b) 200 μm ([70] reproduced with permission from AIP Publishing LLC).

in the substrate also occurs, the number of carriers that recombine in the substrate is only 1–2% of the total number of generated carriers in the first 0.5 μs, as shown in Figure 5.15b. The surface recombination has a comparatively large influence on the carrier profile when the epitaxial layer is very thick, as expected.

Based on the time-dependent depth profiles of the excess carriers, the decay curves can be calculated by integration of the excess carrier density in 4H-SiC as a function of time. Figure 5.16 shows the decay curves of the excess carrier density in 4H-SiC for different surface recombination velocities [70]. In this particular case, a carrier lifetime of 5.0 μs was used as the bulk lifetime in the (a) 50-μm-thick and (b) 200-μm-thick epitaxial layers. All the decay curves are normalized to the integrated carrier density at $t = 0$. In these figures, a single exponential decay with a lifetime of 5.0 μs is plotted using a broken line as a reference. When the epitaxial layer is 50 μm thick (Figure 5.16a), the decay curves exhibit fast decay with much steeper slopes than that of the broken line (5.0 μs exponential curve) at the beginning, even if the SRV is 0 cm s^{-1}. Note that the difference between the decay curve for the zero-SRV and the 5.0 μs exponential curve is attributed to carrier recombination in the substrate. Thus, the effect of recombination in the substrate can be clearly separated from that of surface and bulk recombination. In addition to substrate recombination, surface recombination severely affects the decay curves when the

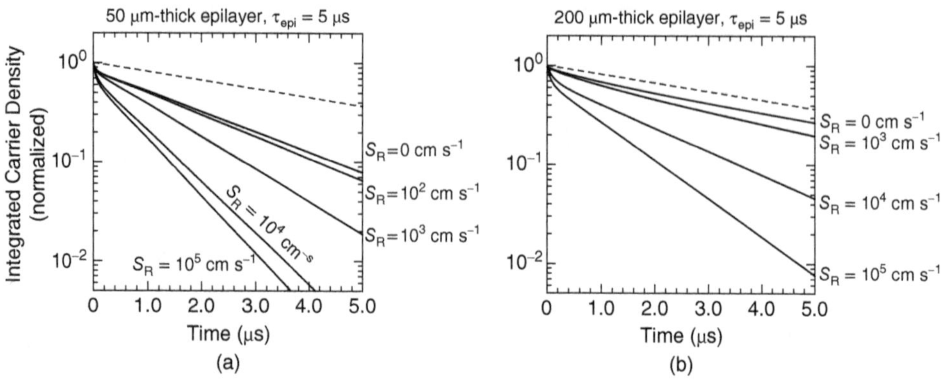

Figure 5.16 Decay curves of the excess carrier density in 4H-SiC for different surface recombination velocities ([70] reproduced with permission from AIP Publishing LLC). In this particular case, a carrier lifetime of 5.0 μs was used as the bulk lifetime in the (a) 50-μm-thick and (b) 200-μm-thick epitaxial layers.

SRV is higher than 1000 cm s^{-1}. The "effective carrier lifetime," which is defined from the slope of the simulated main decay curve (after the fast initial decay), is estimated to be 2.1, 1.3, and 0.77 μs for S_R values of 0, 10^3, and 10^5 cm s^{-1}, respectively. The situation is quite different in the case where there are very thick epitaxial layers. As shown in Figure 5.16b, the simulated decay curve is close to the 5.0 μs exponential curve (broken line) when the SRV is 0 cm s^{-1}, indicating that the substrate recombination has little influence. However, the surface recombination governs the effective lifetime, especially when the SRV is higher than 10^3 cm s^{-1}.

The effective lifetimes obtained from the simulated decay curves are plotted as a function of the bulk lifetime of the epitaxial layers in Figure 5.17 [70]. In this simulation, the epitaxial layer thickness is also varied as a parameter, while the SRV is fixed at 1000 cm s^{-1}. When the epitaxial layer is 50 μm thick and the bulk lifetime of the epitaxial layer is shorter than 0.5 μs, the effective lifetime is nearly equal to the bulk lifetime of the epitaxial layer. However, the effective lifetime saturates at a value of 1.8 μs for the 50-μm-thick epitaxial layers when the bulk lifetime exceeds 30 μs. At a 10 μs bulk lifetime, for example, the effective lifetime is only 1.5 μs, which indicates almost detrimental underestimation. When the bulk lifetime is long, for example, 10 μs, then the effective lifetime increases with increasing epitaxial layer thickness, and reaches 8.5 μs for a thickness of 300 μm. If a very long bulk lifetime is achieved, then a very thick epitaxial layer with low SRV is required for accurate evaluation of the carrier lifetimes. In other words, it is very difficult to obtain accurate bulk lifetimes for 10–30-μm-thick epitaxial layers. These are the major problems in carrier lifetime measurements of SiC. From Figure 5.17, one can estimate the minimum epitaxial layer thickness required for accurate lifetime measurements. Because the measured lifetimes are always underestimated, it is assumed that 20% underestimation is acceptable. The minimum epilayer thickness required to obtain 80% of the bulk lifetime (0.8 τ_{epi}) can be estimated as being approximately four times larger than the ambipolar diffusion length ($W_{epi} > 4\,L_a$). Note that the required epilayer thickness decreases slightly if the SRV decreases, and vice versa.

5.1.5 *Detection of Extended Defects*

5.1.5.1 Chemical Etching

Chemical etching techniques are commonly used to create dislocation-induced etch pits [73]. SiC is an extremely inert material, but can be etched by molten KOH, NaOH, or Na_2O_2 at 450–600 °C. In these melts, the SiC surface is oxidized and the oxide that forms is subsequently removed by the melt [74].

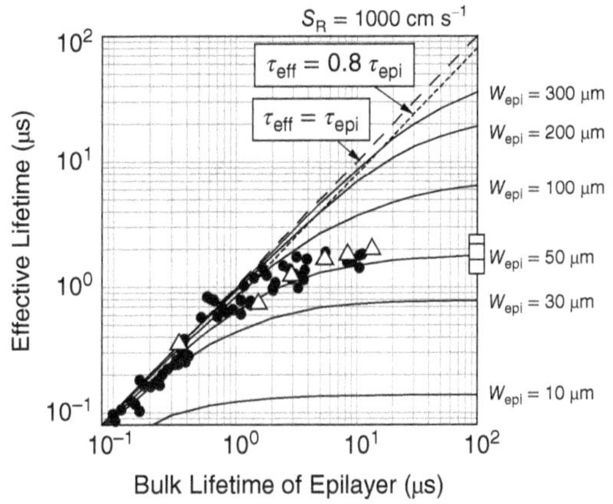

Figure 5.17 Effective lifetimes obtained from the simulated decay curves are plotted as a function of the bulk lifetime of the epitaxial layers ([70] reproduced with permission from AIP Publishing LLC).

At the intersection points of the dislocations or the stacking faults with the surface, the etching rate is different (usually faster), because of the high strain, leading to the formation of the etch pits.

Figure 5.18 shows an off-axis 4H-SiC(0001) surface after etching in molten KOH at 500 °C for 10 min. A large hexagonal hole is created at the location of a micropipe defect [75], but micropipe defects are rarely observed in state-of-the-art wafers. In Figure 5.18, at least three types of etch pits are observed, that is, (i) a large hexagonal pit, (ii) a small hexagonal pit, and (iii) an oval (or shell-shaped) pit. It has been found that these etch pits correspond to the threading screw dislocation (TSD), the threading edge dislocation (TED), and the basal plane dislocation (BPD) [75, 76], respectively. Figure 5.19 shows cross-sections near the etch pits which were formed on off-axis SiC(0001), where the lines for the TSD, TED, and BPD are indicated by broken lines. The shapes of the etch pits are determined by the angle of the dislocation with respect to the SiC surface and also by the lattice symmetry. The TSD and TED, which propagate almost along the *c*-axis, intersect with the surface at an angle that is close to 90°, although the dislocation line is slightly tilted because of the off-angle (typically 4°). Thus, the hexagonal etch pits are formed at the locations of the TSDs and TEDs, and the pit size is much larger for the TSD, because the magnitude of the Burgers vector, and thereby the strain field, is much larger for TSDs. In contrast, the BPD intersects the surface at the off-angle of the substrate. Therefore, etching creates an oval pit at the BPD location. The direction along which the oval pit exhibits the long depression corresponds

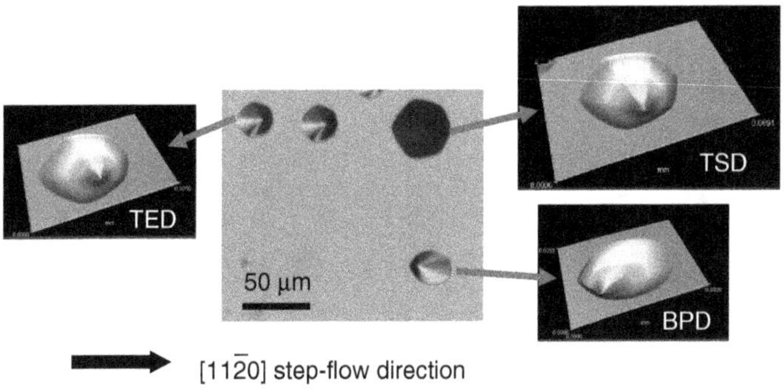

Figure 5.18 Off-axis 4H-SiC(0001) surface after etching in molten KOH at 500 °C for 10 min.

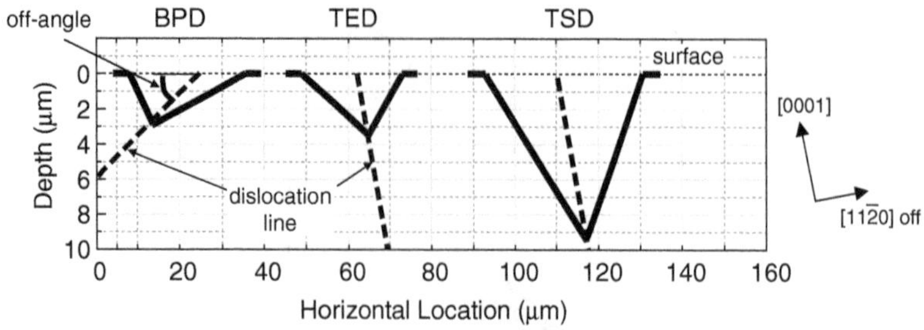

Figure 5.19 Cross-sections near the etch pits which were formed on off-axis SiC(0001), where the lines for the TSD, TED, and BPD are indicated by broken lines.

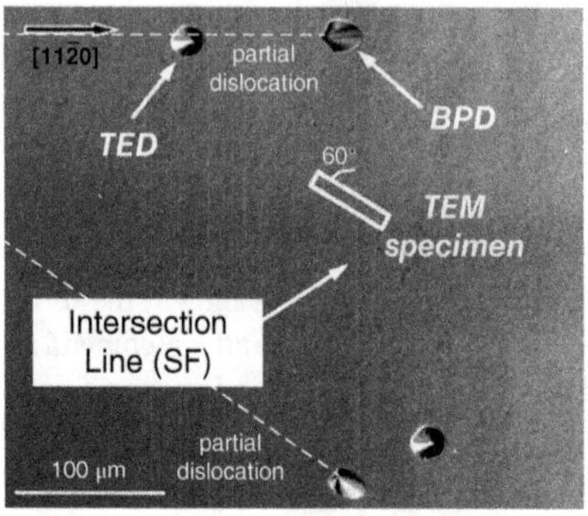

Figure 5.20 Surface of a 4H-SiC epitaxial layer on off-axis (0001), after etching in molten KOH. When a stacking fault (SF) intersects with the surface, a groove is formed by the etching process, and at the end of the groove, an oval etch pit that corresponds to a partial dislocation is created.

to the direction of the BPD near the surface. However, it is difficult to identify the Burgers vector of a dislocation (e.g., pure TSD or mixed dislocation) through simple observation of the etch pits.

When the donor density is very high, as in the case of the n^+-type substrates, the discrimination of TSDs from TEDs then becomes difficult, because both dislocations yield etch pits of similar size and shape. In this case, etching using a mixture of KOH and Na_2O_2 is useful for discrimination of the dislocations [77]. It is known that a long wet oxidation process creates surface pits at the locations of dislocations, because of the enhanced oxidation near the dislocation cores [78, 79].

When a stacking fault (SF) in either the basal plane or the prismatic plane intersects with the surface, a groove is formed by the etching process, and at the end of the groove, an oval etch pit that corresponds to a partial dislocation is created, as shown in Figure 5.20. It is obvious that the BPDs and basal-plane SFs cannot be detected by etching of *on-axis* SiC{0001}.

On the SiC(000$\bar{1}$) face, hillocks are formed at the dislocation locations after etching with molten KOH at 450–500 °C [80]. When vaporized KOH is used at 950–1000 °C, three types of etch pit are formed that correspond to the TSDs, TED, and BPDs [81].

5.1.5.2 X-Ray Topography

X-ray topography is a nondestructive technique that is useful for the characterization of extended defects in crystals [82]. An X-ray beam with a linear cross-section irradiates the crystal, and the X-ray that is diffracted from the crystal is recorded on a high-resolution film. By moving the crystal and the film in

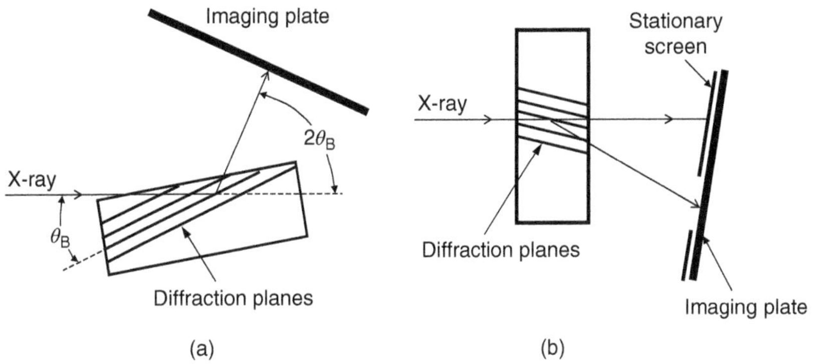

Figure 5.21 Schematic configurations for (a) grazing-incidence reflection X-ray topography and (b) transmission X-ray topography.

the appropriate manner, the entire picture of the diffracted X-ray image for the crystal (i.e., a topography image) can be acquired. Either *reflection topography* (e.g., the Berg–Barrett method) or *transmission topography* (e.g., the Lang method) techniques are used, depending on the purpose of the characterization process [1, 82]. The X-ray penetration depth (typically 10–50 μm) must also be taken into account when analyzing the topography images obtained. Figure 5.21 shows schematics of the configurations for (a) grazing-incidence reflection X-ray topography and (b) transmission X-ray topography. In grazing-incidence X-ray topography of hexagonal SiC{0001}, a diffraction vector $\boldsymbol{g} = 11\bar{2}8$ has often been used, which takes the structures of the major extended defects and the imaging sensitivity into account [83]. By using synchrotron X-rays, the spatial resolution and the exposure time can be improved significantly [83–85].

As in the case of transmission electron microscopy (TEM), the Burgers vector of a dislocation can be determined by using the following *invisibility criteria* [82]:

$$\boldsymbol{g} \bullet \boldsymbol{b} = 0 \tag{5.8}$$

$$(\boldsymbol{t} \times \boldsymbol{g}) \bullet \boldsymbol{b} = 0 \tag{5.9}$$

Here, $\boldsymbol{b}$ and $\boldsymbol{t}$ are the Burgers vector and the unit vector along the dislocation, respectively. Figure 5.22 shows transmission X-ray topography images that were taken with different diffraction vectors of (a) $\boldsymbol{g} = 11\bar{2}0$, (b) $\boldsymbol{g} = 1\bar{2}10$, (c) $\boldsymbol{g} = \bar{2}110$, (d) $\boldsymbol{g} = 1\bar{1}00$, (e) $\boldsymbol{g} = \bar{1}010$, and (f) $\boldsymbol{g} = 01\bar{1}0$ at the same location on a 4H-SiC(0001) free-standing epitaxial layer [86]. The dark spots and dark curves in the images correspond to threading dislocations and BPDs, respectively. Note that all the BPDs are invisible for $\boldsymbol{g} = 1\bar{1}00$ (Figure 5.22d), while they are in contrast for the other conditions. This result means that all these BPDs possess Burgers vectors that are parallel to the $[11\bar{2}0]$ step-flow direction. Because the BPDs in the epitaxial layer propagate roughly toward the $[11\bar{2}0]$ step-flow direction, these BPDs have a screw-type character [86, 87].

Figure 5.23 shows another synchrotron grazing-incidence X-ray topography image that was taken with $\boldsymbol{g} = 11\bar{2}8$ for a 4H-SiC(0001) epitaxial layer [88]. In the image, a relatively large, circular contrast corresponds to a TSD. Smaller dot-like contrasts labeled (a–f) correspond to the TEDs. The six different contrasts observed for the TEDs originate from the difference in the Burgers vector of the TEDs. A topographic image of the dislocations can be simulated by *ray-tracing simulation*. The simulated topography images ($\boldsymbol{g} = 11\bar{2}8$) for TEDs in 4H-SiC(0001) are shown in Figure 5.24 [88], where the direction of the Burgers vector ($\boldsymbol{b}$) is rotated every 60° with respect to the $\boldsymbol{g}$ vector in the basal plane. The vectors $\boldsymbol{b}$ and $\boldsymbol{g}$ and the angle between them are indicated in the figures. As shown in these images,

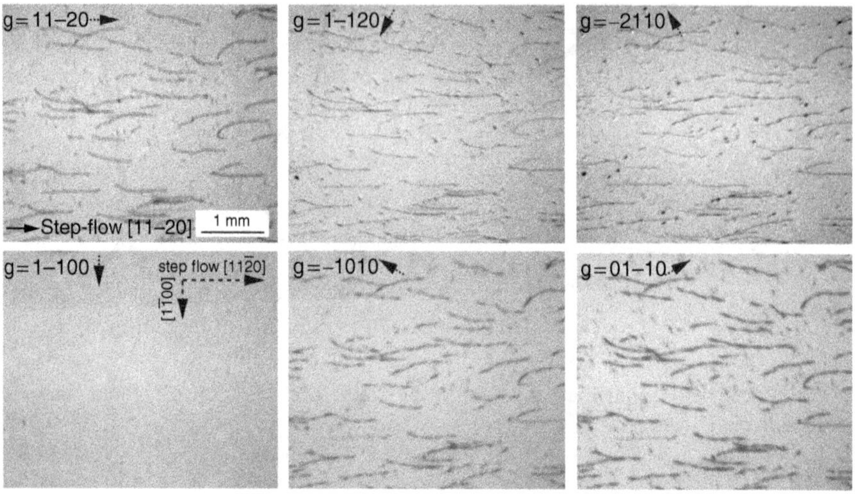

Figure 5.22 Transmission X-ray topography images that were taken with different diffraction vectors: $\boldsymbol{g} = 11\bar{2}0, 1\bar{2}10, \bar{2}110, 1\bar{1}00, \bar{1}010$, and $01\bar{1}0$ at the same location on a 4H-SiC(0001) free-standing epitaxial layer ([86] reproduced with permission from Trans Tech Publications).

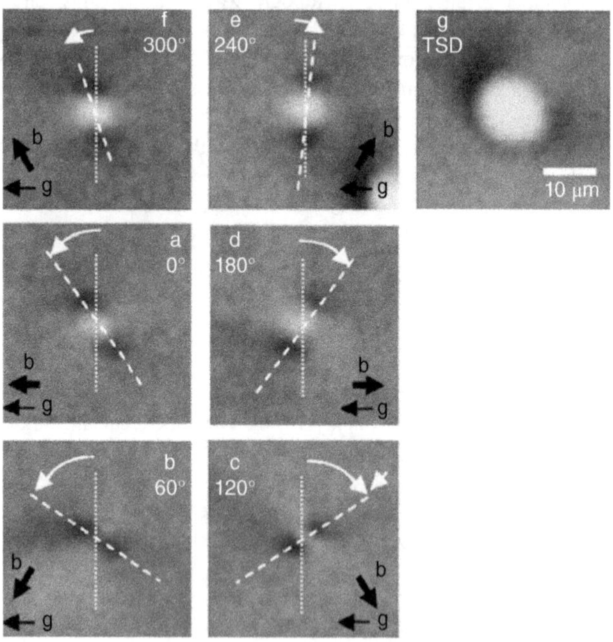

Figure 5.23 Synchrotron grazing-incidence X-ray topography image that was taken with $\boldsymbol{g} = 11\bar{2}8$ for a 4H-SiC(0001) epitaxial layer ([88] reproduced with permission from Elsevier). In the image, a relatively large, circular contrast corresponds to a TSD (g). Smaller dot-like contrasts labeled (a–f) correspond to the TEDs.

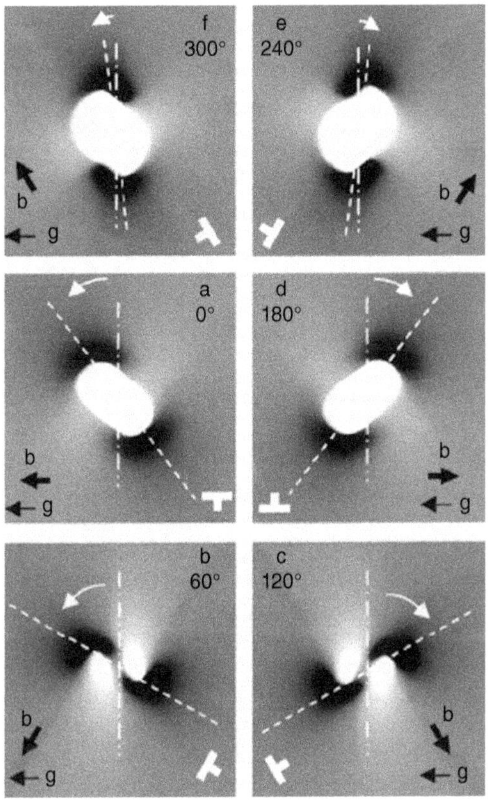

Figure 5.24 (a–f) Simulated topography images ($g = 11\bar{2}8$) for TEDs in 4H-SiC(0001) ([88] reproduced with permission from Elsevier), which agree with the experimental observations shown in Figure 5.23.

the Burgers vector of individual TEDs can be determined simply from the contrast shape shown in the topography images without the need to apply the invisibility criteria described above.

The Burgers vector magnitude and sense ($+1\boldsymbol{c}$ or $-1\boldsymbol{c}$) of the TSDs in SiC can be determined unambiguously by synchrotron X-ray section topography [89]. Three-dimensional synchrotron X-ray topography analysis has also been proposed for visualization of the propagation behavior of dislocations [90].

5.1.5.3 Photoluminescence Mapping/Imaging

Because any kind of active defect can affect carrier recombination, PL measurement with high spatial resolution provides a powerful method to detect extended defects in semiconductors [1]. In this technique, full wafer characterization is possible without any chemical etching or contact formation. The measurement set-up is rather simple, and this method can be used as a routine wafer characterization process. Two different approaches have been demonstrated, called *PL mapping* and *PL imaging*; schematics of the measurement set-ups of these approaches are shown in Figure 5.25a and b, respectively.

PL Mapping

In this technique, the excitation laser beam is focused to a spot size of a few micrometers, and the sample is selectively excited with the laser beam. The detection set-up is similar to that used for normal PL

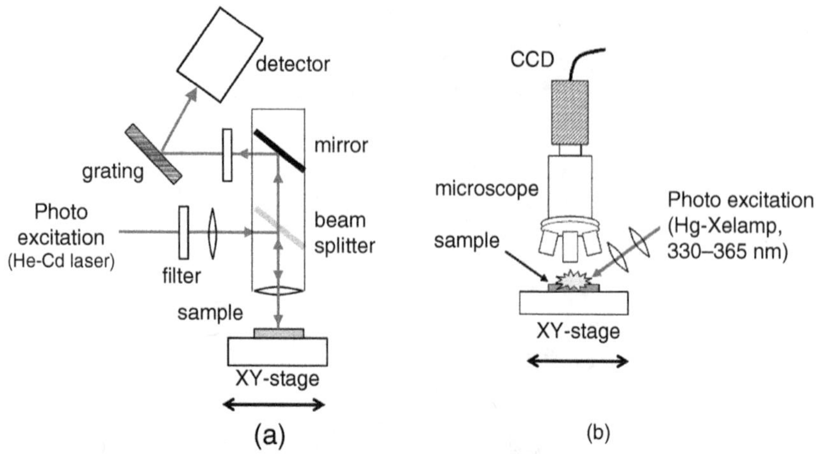

Figure 5.25 Schematics of the measurement set-ups for (a) PL mapping and (b) PL imaging techniques.

measurements. PL from the excited area is typically dispersed with a monochromator and detected using a photomultiplier. The optical alignment is usually fixed, and a high-resolution $X-Y$ stage is moved by digital motor drives to enable PL mapping. The typical step (pitch) of the stage movement is 1–20 μm. When the PL intensity at a specific wavelength is acquired, this technique is called *PL-intensity mapping* [91–94]. A PL spectrum can also be acquired at each point, and this is called *PL-spectrum mapping*. Obtaining the PL decay curves at individual points means that *carrier-lifetime mapping* is also possible [68, 95]. Although PL mapping is quite time-consuming when the measurement pitch is reduced to 1–2 μm to improve the spatial resolution, the PL information that is acquired from each point perfectly reflects the radiative recombination taking place at that location.

Figure 5.26a shows PL-intensity mapping images taken from a 72-μm-thick n-type 4H-SiC(0001) with a doping density of 1×10^{15} cm^{-3} [96]. The PL intensity at 390 nm (for the free exciton peaks) at room temperature was mapped with a pitch of 1 μm. In the PL mapping image, three circular areas are observed with reduced PL intensity, in contrast to the matrix, indicating the locations of significant nonradiative recombination centers. Among the three circular areas, two areas are of a larger size than the other area. Figure 5.26b presents an optical microscopy image of the sample surface (at the same location) after

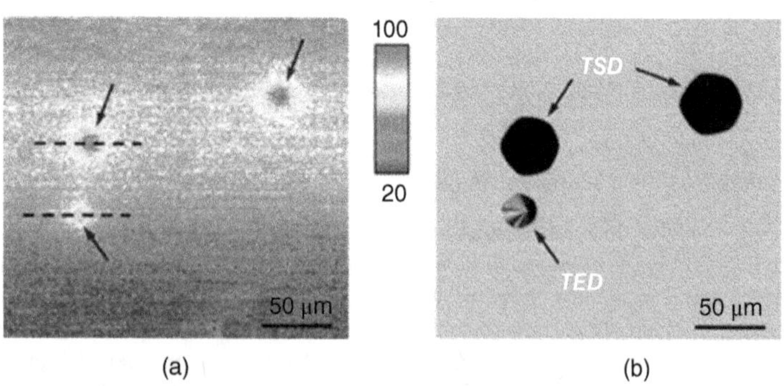

Figure 5.26 (a) PL-intensity mapping images at 390 nm taken from a 72-μm-thick n-type 4H-SiC(0001) with a doping density of 1×10^{15} cm^{-3}. (b) Optical microscopy image of the sample surface (at the same location) after molten KOH etching at 480 °C for 10 min ([96] reproduced with permission from APL Publishing LLC).

molten KOH etching at 480 °C for 10 min. There is a one-to-one correlation between the circular areas with reduced PL intensity and the threading dislocations. A TSD is indicated in PL-intensity mapping by a larger and darker circular area than that of a TED, suggesting that TSDs have a more pronounced impact on nonradiative recombination activity than TEDs. BPD detection is easy, because long dark lines along the off-direction can be observed during PL mapping; in many cases, a BPD is dissociated into two partial dislocations and a Shockley-type stacking fault (SSF) forms between them during PL measurements. It should be noted that these contrasts are more pronounced in epitaxial layers with longer carrier lifetimes [96].

PL Imaging

In PL imaging, a relatively large area (~ 1 to 5 mm^2) of a sample is illuminated with ultraviolet (UV) light, and "PL pictures" with a typical size of 1 mm × 1 mm (this size depends on the magnification of the objective lens) are acquired with a highly-sensitive charge-coupled device (CCD) camera, as in the case of optical microscopy [97–100]. The spatial resolution is about 1 μm (1 mm/1024 pixels). To acquire PL images at a specific wavelength, a band-pass filter is attached to the objective lens of the microscope. A sample (or wafer) is placed on an $X-Y$ stage, and the movement pitch is adjusted to match the picture size (e.g., 1 mm × 1 mm), which is much larger than that used in PL mapping. A short exposure time of 0.5 s is sufficient to obtain a clear PL image, and thus this technique provides very fast characterization. The PL images obtained contain "continuous" PL data from the targeted area, while the data is discrete (at every $X-Y$ pitch-size) in PL mapping. However, one must be aware that the PL-image data at any (x, y) point in the picture also contains some information about the PL that is emitted in the vicinity of the (x, y) point.

Figure 5.27 shows (a) a typical PL image taken from a 180-μm-thick n-type 4H-SiC(0001) epitaxial layer at 880 nm and (b) the optical photograph of the same location after molten KOH etching. In PL images taken at the band-edge emission (390 nm), threading dislocations, BPDs, and stacking faults appear as dark spots, dark lines, and dark areas (not shown), respectively, which is consistent with the results of PL-intensity mapping at 390 nm. In contrast, the infrared PL image shows bright spots and bright lines, and these features have been identified as the locations of threading dislocations and BPDs, as shown in Figure 5.27b. TSDs produce much more intense PL in the infrared region (750–900 nm) than TEDs, which makes it possible to discriminate between TSDs and TEDs. While the mechanism of luminescence from dislocations is not clear at present, localized states that are generated near the

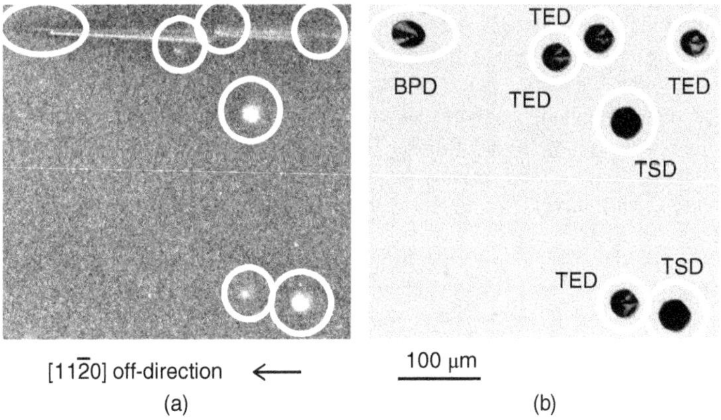

Figure 5.27 (a) Typical PL image taken from a 180-μm-thick n-type 4H-SiC(0001) epitaxial layer at 880 nm. (b) Optical photograph of the same location after molten KOH etching.

dislocation cores [101] may be responsible for the infrared luminescence. PL spectra observed near the dislocation cores in SiC have been investigated [102, 103]. PL imaging is also powerful for the detection of stacking faults, because each stacking fault has a unique PL band that depends on the stacking structure, as described in Section 4.3.1.

5.1.5.4 High-Resolution Mapping of Surface Morphology

Extended defects in SiC epitaxial layers usually cause some surface irregularities, because the step-flow growth is disturbed by extended defects. Because it is difficult to detect sub-micron-sized surface irregularities by conventional optical microscopy, an advanced laser scattering technique has been developed for this purpose. The very small surface depressions caused by TSDs and TEDs can also be detected with this technique in a quick and nondestructive manner [104].

5.1.6 Detection of Point Defects

Another important defect type that is present in semiconductors is a point defect, which often creates a deep level in the bandgap [105]. Electrically active levels can be monitored by *deep level transient spectroscopy* (DLTS). If a point defect has a spin (unpaired electron), then it is detectable by *electron paramagnetic resonance* (EPR) or *electron paramagnetic resonance* (EPR).

5.1.6.1 Deep Level Transient Spectroscopy

In DLTS measurements, the capacitance transients of a junction are monitored, and the transient signals are analyzed numerically to obtain the density, the energy position, and the capture cross-section of a deep level [106]. Figure 5.28 shows the typical bias voltage sequence that is applied during DLTS measurements of an n-type semiconductor (Schottky structure), along with the energy band diagrams, which correspond to (a) the steady state under a reverse bias voltage, (b) during application of a pulse voltage, and (c) after application of the pulse voltage, respectively. In the steady state condition, the capacitance is determined by the width of the space charge region (x_R), which, in turn, depends on the doping density (N_D) and the bias voltage (V). In this condition, the deep levels inside the space charge region are empty except in the region near the edge of the space charge region, as shown in Figure 5.28a. The deep levels are then filled with electrons by the applied pulse voltage, from the deep region to that near the surface, as shown in Figure 5.28b. Immediately after the pulse voltage is removed, the space charge region expands to a depth that is larger than the steady state value (x_R), because the trapped electrons cannot be emitted immediately, and the net doping density in the space charge region is reduced. The trapped electrons are gradually emitted to the conduction band with time, as shown in Figure 5.28c. Therefore, the width of the space charge region gradually decreases and the capacitance gradually increases with time, and both reach steady-state values given sufficient time. Thus, the capacitance transient after the pulse voltage contains information about the *thermal emission* of the carriers from the deep levels to the band.

The *emission time constants* (τ_e) for electrons and holes are given by the following equations [1, 105]:

For electrons:

$$\tau_e = \frac{1}{N_C v_{thn} \sigma_e} \exp\left(\frac{E_C - E_T}{kT}\right) \tag{5.10}$$

For holes:

$$\tau_e = \frac{1}{N_V v_{thp} \sigma_h} \exp\left(\frac{E_T - E_V}{kT}\right) \tag{5.11}$$

Here, N_C and N_V are the effective density of states in the conduction and valence bands, respectively. v_{thn}/v_{thp} and σ_e/σ_h are the thermal velocities of electrons/holes and the capture cross-sections of deep

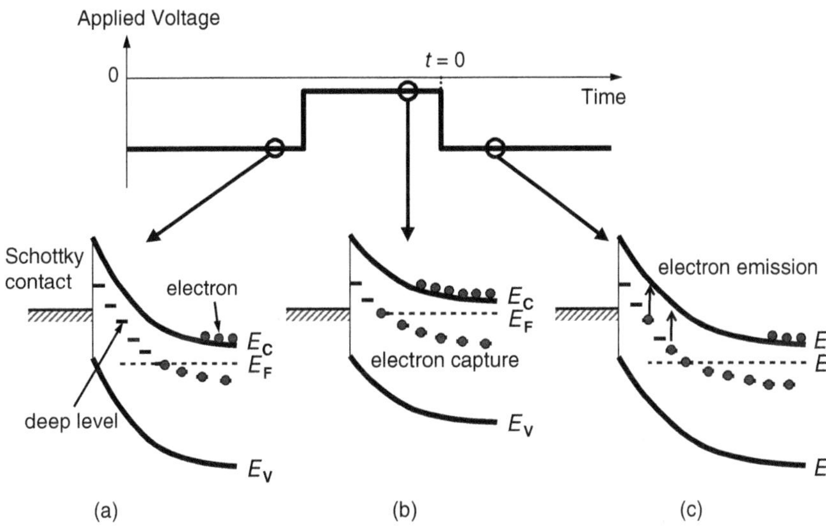

Figure 5.28 Typical bias voltage sequence that is applied during DLTS measurements of an n-type semiconductor (Schottky structure), along with the energy band diagrams, which correspond to (a) the steady state under a reverse bias voltage, (b) during application of a pulse voltage, and (c) after application of the pulse voltage, respectively.

levels for electrons/holes, respectively. E_T is the energy position of the deep level, k is the Boltzmann constant, and T is the absolute temperature. The capture cross-section depends strongly on the charge state and the trapping process of the deep level, and usually has a value of 10^{-16} to 10^{-14} cm^2 for neutral deep levels and 10^{-13} to 10^{-12} cm^2 for deep levels where Coulomb attraction works between the carriers and the deep levels.

When an n-type Schottky structure is considered, the capacitance transient after the pulse voltage is given by the following equation [1, 106]:

$$C(t) = \sqrt{\frac{\varepsilon_S q\{N_D - n_T(t)\}}{2(V_d - V_R)}} = C_{st}\sqrt{1 - \frac{n_T(t)}{N_D}} = C_{st}\sqrt{1 - \frac{N_T \exp(-t/\tau_e)}{N_D}} \tag{5.12}$$

where C_{st} is the steady state capacitance given by:

$$C_{st} = \sqrt{\frac{\varepsilon_S q N_D}{2(V_d - V_R)}} \tag{5.13}$$

Here, N_D and $n_T(t)$ are the donor density and the density of the electrons trapped at the deep levels, respectively. V_d and ε_s are the built-in potential and the dielectric constant of the semiconductor, respectively. V_R is the bias voltage (usually in reverse bias) under the steady state. In Equation 5.12, it is assumed that only one type of deep level contributes to the electron emission under the condition studied. When the density of the deep levels is much smaller than the donor density ($N_T \ll N_D$), Equation 5.12 can then be expressed approximately by:

$$C(t) \cong C_{st}\left\{1 - \frac{N_T \exp\left(-t/\tau_e\right)}{2N_D}\right\} \tag{5.14}$$

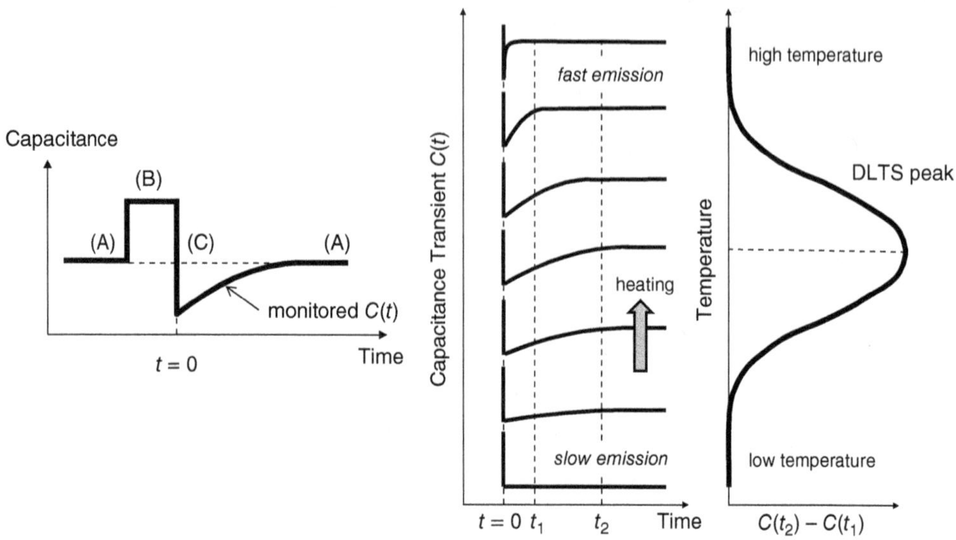

Figure 5.29 Schematic illustrations of the capacitance transient of an n-type Schottky structure and the DLTS signal $\Delta C = C(t_2) - C(t_1)$, where $t_2 > t_1$.

Figure 5.29 illustrates schematically the capacitance transient of an n-type Schottky structure and the DLTS signal $\Delta C = C(t_2) - C(t_1)$, where $t_2 > t_1$. The DLTS signal ΔC is close to zero at sufficiently low temperatures, because almost no electron emission occurs. The DLTS signal is also nearly zero at sufficiently high temperatures because almost all the trapped electrons have already been emitted at time t_1. As a result, the DLTS signal has a maximum value when the following condition is satisfied:

$$\tau_e = \frac{t_2 - t_1}{\ln(t_2/t_1)} \tag{5.15}$$

After DLTS measurements with a number of (t_1, t_2) values, the sets of the emission time constant (τ_e) and the temperature of the DLTS peak (T_p) that were obtained are analyzed using an Arrhenius plot; an example of this is shown in Figure 5.30. Because N_C (or N_V) is proportional to $T^{3/2}$ and v_{thn} (or v_{thp}) is proportional to $T^{1/2}$, $\ln(\tau_e T^2)$ is usually plotted against the inverse of the absolute temperature. Based on Equations 5.10 or 5.11, the slope and ordinate intercept of this plot yield the activation energy (the energy level from the band edge) and the capture cross-section of the deep level, respectively, by assuming a temperature-independent capture cross-section. Note that the capture cross-section is not actually constant but is a function of temperature. Additionally, the capture cross-section that is estimated from the intercept is highly inaccurate. To obtain accurate capture cross-sections, the trap filling by changing the filling-pulse width must be analyzed [1]. From this basic concept, a variety of analysis methods, such as Fourier transform DLTS [107], Laplace DLTS [108], and ICTS (isothermal capacitance transient spectroscopy) [109] have been proposed.

In the DLTS spectra, the signal intensity of each peak is proportional to the N_T/N_D ratio (and not to N_T itself), and the peak temperature roughly corresponds to the energy position (i.e., the energetic depth) of the deep levels. Thus, the detection limit of defect density is determined by the N_T/N_D ratio. In standard DLTS, this ratio is approximately 10^{-4}, which means that the detection limit is approximately 1×10^{11} cm^{-3} when the doping density of the sample is 1×10^{15} cm^{-3}. The deep level density obtained in this way is highly accurate, and all kinds of electrically active defects can be detected as long as the appropriate measurement conditions are used (in PL, only the defects that exhibit radiative recombination can be detected). The monitored depth from the surface is determined by the space charge region, the

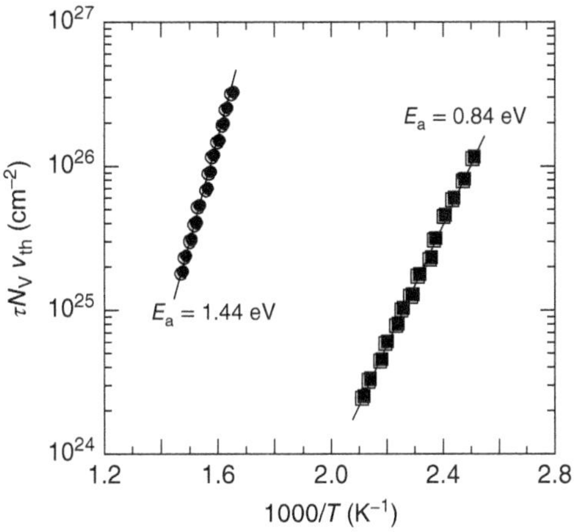

Figure 5.30 Example of Arrhenius plot of emission time constant (τ_e) in DLTS measurements. Because N_C (or N_V) is proportional to $T^{3/2}$ and υ_{thn} (or υ_{thp}) is proportional to $T^{1/2}$, $\ln(\tau_e T^2)$ is usually plotted against the inverse of the absolute temperature.

width of which, in turn, depends on both the doping density and the bias voltage. Double-correlated deep level transient spectroscopy (DDLTS) [110] has proved useful in obtaining accurate depth profiles of deep levels.

When DLTS measurements are conducted on an n-type Schottky structure, the DLTS signals reflect the capture of electrons at the deep levels and the emission of electrons from the deep levels to the conduction band. Thus, the DLTS peaks correspond to "*electron traps*". In contrast, DLTS peaks acquired from a p-type Schottky structure originate from hole interactions between the valence band and the deep levels, and thus correspond to "*hole traps*". In the case of DLTS measurements on pn junctions, the nature of the DLTS signals depends on the bias voltage of the filling pulse. When a reverse bias is maintained, then only the majority-carrier traps can be detected. If a sufficiently large forward voltage is applied by the filling pulse, then the minority carriers are injected during the DLTS measurements. In this case, the minority-carrier traps (hole traps in n-type materials and electron traps in p-type materials) are detectable in addition to the majority-carrier traps. The DLTS signals of these minority-carrier traps exhibit the opposite sign (+/−) to those of the majority-carrier traps, and can thus be identified immediately [1, 106].

In DLTS measurements of SiC, the transient curves must be acquired over a wide temperature range, typically from 100 to 750 K, to monitor both shallow and deep levels, because of the wide bandgap of SiC. To obtain accurate DLTS signals, the leakage current of the diodes (Schottky diodes or pn diodes) must be minimized. Schottky metals that are suitable for n-type and p-type SiC Schottky diodes are, for example, Ni and Ti, respectively, which give large Schottky barrier heights and thermal stability up to at least 800 K. Examples of typical DLTS spectra are shown in Section 4.3.2. However, it is necessary to be aware of another limitation of the energy range that is monitored in DLTS. When the Schottky barrier used for DLTS has a barrier height ϕ_B, any deep levels that lie deeper than ϕ_B cannot be detected. Because these deep levels are energetically located below the quasi-Fermi level (in the n-type case), electron emission from these levels does not occur, leading to the absence of the corresponding DLTS peaks [111]. The situation is similar for deep hole traps in p-type.

5.1.6.2 Electron Paramagnetic Resonance

When an unpaired electron exists at a point defect, the electron spin can be detected by EPR measurements. The application of an external magnetic field provides a magnetic potential energy which

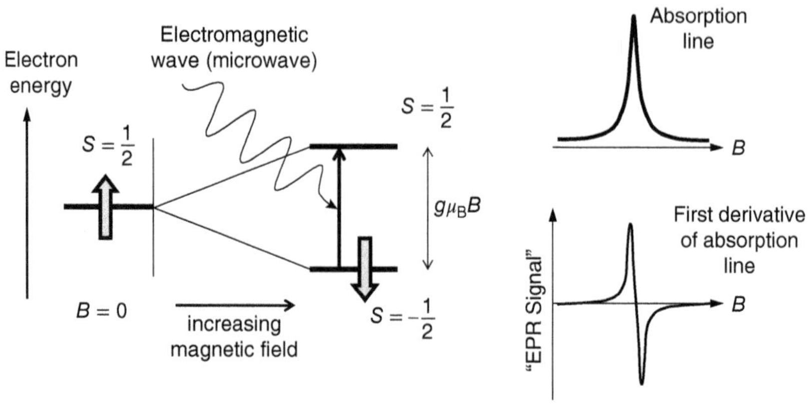

Figure 5.31 Principle of electron paramagnetic resonance measurement. The application of an external magnetic field provides a magnetic potential energy which splits the spin states into two, depending on the spin value (*Zeeman effect*). The magnetic potential energy (U) of an electron spin is proportional to the strength of the magnetic field (B).

splits the spin states into two, depending on the spin value (*Zeeman effect*). The magnetic potential energy (U) of an electron spin is proportional to the strength of the magnetic field (B), as shown in Figure 5.31, and is expressed by [112]:

$$U = \pm\frac{1}{2}g\mu_B B \tag{5.16}$$

Here, g is the *electron spin g-factor* (or *gyromagnetic ratio*) and μ_B is the *Bohr magneton*. Thus, the radiation of an electromagnetic wave with a suitable frequency (usually in the microwave range) causes a transition from one spin state to another, which is observed as the absorption of the incident electromagnetic wave. The g-factor obtained by EPR provides important information that helps to identify the origin of the defect, because this value can be directly compared with that determined by theoretical calculations for each point defect. The EPR signals from isotope atoms such as ^{29}Si and ^{13}C are also useful for defect identification. Additionally, by measuring the angular dependence of the EPR signals, the symmetry of a measured defect can be determined, which provides another important insight for identification of the microscopic structure. Detailed discussion on the hyperfine splitting and fine splitting structures can be found in the literature [112].

Although the absolute spin density (defect density) can be determined by EPR measurements, this technique has almost no spatial resolution. The spin number obtained is the total number of spins present in the entire volume of the sample under test. Thus, either bulk materials or thick free-standing epilayers are required for EPR measurements. If the sample under test consists of a SiC homoepitaxial layer grown on a SiC substrate, all the spin information from both the epitaxial layer and the substrate are mixed, and the resulting data are less useful. Another deficiency of EPR measurements is the relatively poor detection limit for spins (point defects). The detection limit of point defect density is roughly in the $10^{15}-10^{16}$ cm^{-3} range, depending on the sample size. This is often close to the doping density of the samples used. Point defects that do not have an unpaired electron are, of course, EPR-inactive and are thus not detectable.

Major point defects in 4H-SiC have been identified using EPR measurements. These defects include the carbon vacancy (V_C) [113], the silicon vacancy (V_{Si}) [114], the carbon antisite-carbon vacancy pair (C_{Si}-V_C), the divacancy (V_{Si}-V_C) [115], the dicarbon antisite ($(C_2)_{Si}$), the carbon split interstitial ($(C_2)_C$) [116], and several metallic impurities. Figure 5.32 shows the typical EPR spectra obtained from three different high-purity semi-insulating (HPSI) 4H-SiC samples [40]. The measurements were conducted under illumination ($h\nu$ = 2.0–2.8 eV) and at 77 K. It is known that HPSI 4H-SiC can be categorized into at least three material types, because of the three different activation energies of the resistivity. The

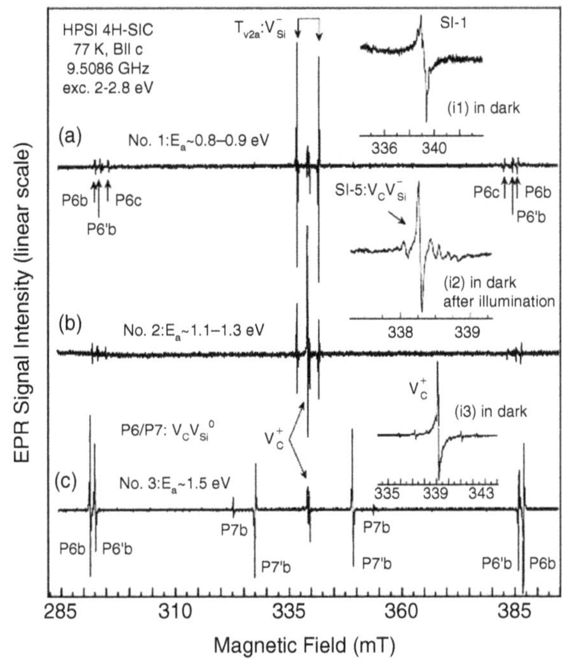

Figure 5.32 (a–c) Typical EPR spectra obtained from three different high-purity semi-insulating (HPSI) 4H-SiC samples ([40] reproduced with permission from American Physical Society). The measurements were conducted under illumination ($h\nu$ = 2.0–2.8 eV) and at 77 K.

activation energy (E_a) values observed were (a) 0.8–0.9, (b) 1.1–1.3, and (c) ~ 1.5 eV, which correspond to the EPR spectra in Figure 5.32a–c, respectively. For the HPSI sample with E_a = 0.8–0.9 eV, the negatively charged V_{Si} exhibits the highest density (low 10^{15} cm^{-3}). In the HPSI sample with E_a = 1.1–1.3 eV, however, the positively charged V_C or the negatively charged C_{Si}-V_C are dominant. In the HPSI sample with E_a = 1.5 eV, the positively charged V_C and V_C-V_{Si} are dominant (mid 10^{15} cm^{-3}). These EPR data suggest the main defects which are responsible for the compensation. The resistivity of HPSI sample (a) (E_a = 0.8–0.9 eV) significantly decreases after annealing at 1600 °C, while the resistivity of HPSI sample (c) (E_a = 1.5 eV) is stable, even after high-temperature annealing. Note that these activation energies are well correlated with the energy levels of the individual point defects. From the defect engineering viewpoint, an increase in V_C density and reduction of the V_{Si} density are important to enable thermally stable and high-resistivity HPSI 4H-SiC to be obtained [40]. For further reference, a recent comprehensive review describes EPR analysis of defects in SiC in detail [117].

As described in Section 5.1.1.5, PL is also useful in the detection of point defects, as far as the defects exhibit radiative carrier recombination. Thus, a comparison study based on EPR, DLTS, and PL is desirable for fundamental defect characterization. It should be noted that EPR can also be useful in the identification of defects at the oxide/SiC interface [118].

5.2 Extended Defects in SiC

5.2.1 Major Extended Defects in SiC

Classification of the major extended defects (micropipes, TSDs, TEDs, and BPDs) in SiC wafers and their generation mechanism were summarized in Section 3.3. Basic information on the major extended

Table 5.5 Basic information on the major extended defects in 4H-SiC epitaxial layers.

Dislocation	Burgers vector	Major direction	Typical density (cm^{-2})
Micropipe	$n\langle 0001\rangle$ (n>2)	$\langle 0001\rangle$	0–0.02
Threading screw dislocation (TSD)	$n\langle 0001\rangle$ ($n = 1, 2$)	$\langle 0001\rangle$	300–600
Threading edge dislocation (TED)	$\langle 11\bar{2}0\rangle/3$	$\langle 0001\rangle$	3000–6000
(Perfect) Basal plane dislocation (BPD)	$\langle 11\bar{2}0\rangle/3$	in {0001} plane (predominantly $\langle 11\bar{2}0\rangle$)	1–10
Stacking faults (SF)	Shockley: $\langle 1\bar{1}00\rangle/3$ Frank: $\langle 0001\rangle/n$	in {0001} plane	0.1–1

defects in 4H-SiC epitaxial layers is summarized in Table 5.5. The dislocation behavior during epitaxial growth and the macroscopic defects generated during epitaxial growth (including triangular defects, carrot defects, in-grown stacking faults, and particles) were explained in Section 4.3.1. In the following subsections, the bipolar degradation caused by BPDs and the more general effects of extended defects on SiC device performance are described.

5.2.2 *Bipolar Degradation*

After carrier injection (or excitation) followed by recombination, nucleation and expansion of a *single* SSF (Shockley Stacking Fault) takes place at the location of a BPD or at a basal plane segment of other dislocations [119–121]. The expanded SSF causes a significant reduction in the carrier lifetimes (and probably in the formation of potential barriers for carrier transport) [119], leading to an increase in the forward voltage drop in SiC bipolar devices such as pin diodes, bipolar junction transistors, and thyristors. The expanded SSF also acts as a severe leakage current path in the reverse-biased junction [122]. These phenomena are called "*bipolar degradation*", which is detrimental to the reliability of SiC-based bipolar devices. Note that pure SiC unipolar devices, such as Schottky barrier diodes, do not exhibit this degradation because of the lack of carrier injection. A well-documented overview of bipolar degradation in SiC is available [123].

Figure 5.33 shows examples of bipolar degradation that were observed in 4H-SiC pin diodes, where (a) the forward current–voltage characteristics and (b) the change in the forward voltage drop at 100 A cm^{-2} as a function of the stress time. In this experiment, the diodes were stressed under constant current conditions (100 A cm^{-2}). The forward voltage drop exhibits irregular increases at several different times, and it can reach 6 V or higher in some diodes. Even a low current density of 1–10 A cm^{-2} is sufficient to cause this degradation. Figure 5.34 shows a panchromatic cathodoluminescence image of a degraded pin diode measured at low temperature [120]. The dark triangular regions correspond to the expanded stacking fault planes, clearly indicating the reduced carrier lifetimes in the stacking faults.

The exact stacking structure of the expanded stacking faults has been identified by cross-sectional TEM [124]. Figure 5.35 shows (a) a high-resolution TEM image of the fault region of a 4H-SiC(0001) pin diode, (b) the schematic stacking sequence of the observed structure, and (c) that of a perfect 4H-SiC. The stacking fault can be denoted by the (31) structure using Zhdanov's notation, while the perfect 4H-SiC has the (22) structure. In the case of 6H-SiC (the (33) structure), the fault region exhibits the (42) structure. Thus, the observed degradation can be directly linked to the expansion of the SSFs.

The nucleation sites for SSFs were carefully investigated, and it was found that a BPD or a basal plane segment of a dislocation is always present at the origin of the SSF expansion. As described in

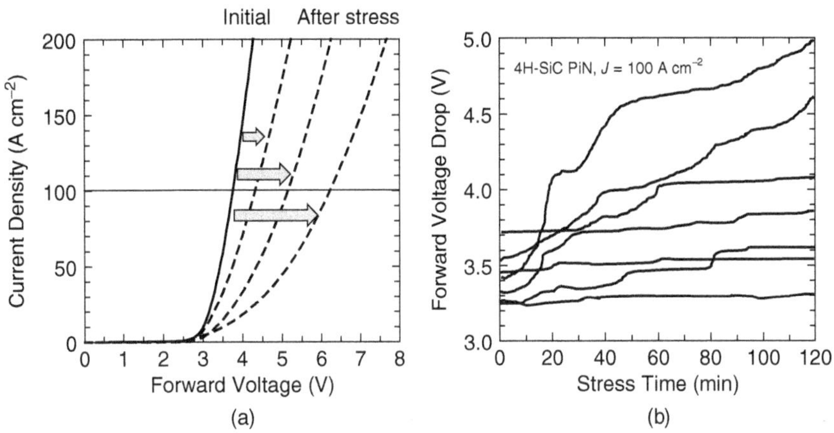

Figure 5.33 Examples of bipolar degradation that were observed in 4H-SiC pin diodes. (a) Forward current-voltage characteristics and (b) change in the forward voltage drop at 100 A cm^{-2} as a function of the stress time.

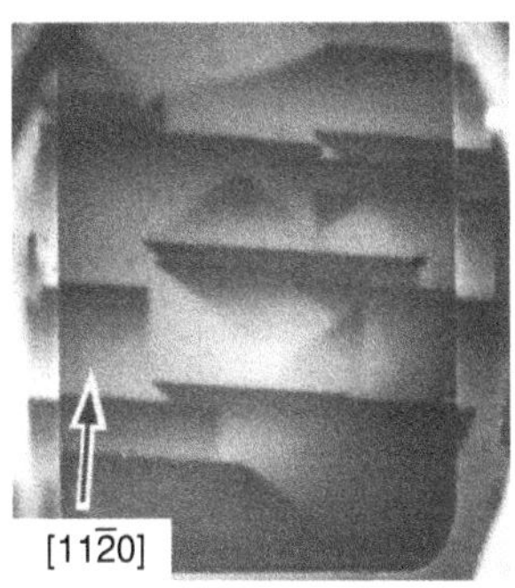

Figure 5.34 Panchromatic cathodoluminescence image of a degraded pin diode measured at low temperature ([120] reproduced with permission from Trans Tech Publications).

Section 4.3.1, a BPD is already dissociated into two partial dislocations with a SSF between them in SiC [125]. Expansion of the SSF can be interpreted as the *recombination-enhanced glide* of partial dislocation(s). When the initial stage of SSF expansion is observed, a common morphology of the SSFs is a rhombus shape with sides along the <11$\bar{2}$0> directions and angles of 60° and 120°, as shown in Figure 5.36a [126]. The fault region is bounded by a 30° partial dislocation loop. Figure 5.36b shows schematically the evolution of the SSF into the rhombus shape from a point source. It should be noted that among the four bounding partial dislocations of the rhombus-shaped fault, only two of the partials move and exhibit bright red-infrared luminescence (peak wavelength ~ 700 nm). The other two partials are immobile and only emit very weak luminescence. Extensive TEM studies revealed that the core of the mobile (and bright) partials is composed of Si-Si bonds, while that of the immobile (and dark) partials is composed of C-C bonds [127]. Thus, the former are called *Si-core partial dislocations* and the latter *C-core partial dislocations*. In general, a partial dislocation loop in hexagonal SiC polytypes can be illustrated as shown in Figure 5.37a, where each segment lies along the <11$\bar{2}$0> directions because of the Peierls potential [73]. Because the mobility of 90° partials is usually higher than that of 30° partials, the 90° partials disappear, leaving only the 30° partials, as shown in Figure 5.37b. This is the reason why

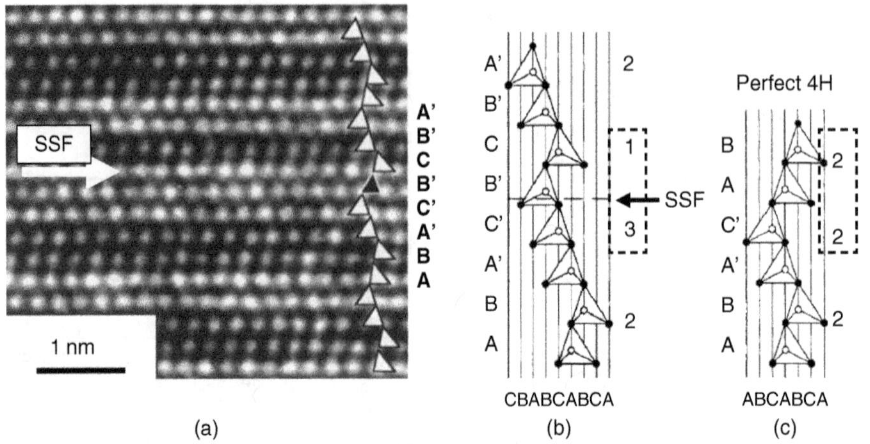

Figure 5.35 (a) High-resolution TEM image of the fault region of a 4H-SiC(0001) pin diode, (b) schematic stacking sequence of the observed structure, and (c) stacking sequence of a perfect 4H-SiC [124] reproduced with permission from AIP Publishing LLC). The stacking fault can be denoted by the (31) structure using Zhdanov's notation.

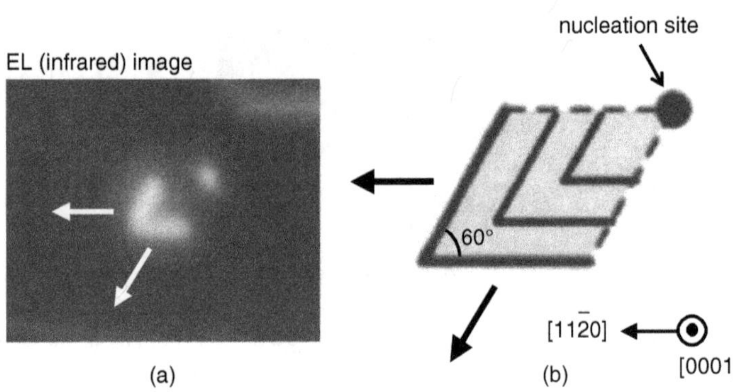

Figure 5.36 Expansion of a single Shockley-type stacking fault (SSF) from a point source. (a) Infrared electroluminescence image and (b) schematic illustration of SSF evolution into a rhombus shape ([123] reproduced with permission from AIP Publishing LLC).

the glide motion of the partials results in the evolution of the rhombus-shaped SSFs [123]. In Figure 5.37c, the bond configurations for Si-core and C-core partial dislocations are shown. At the Si-core partial, Si-Si bonds are formed along the core, while C-C bonds are aligned in the C-core partial. The bonding energy of Si-Si is much smaller than that of C-C, and this is a main reason why the Si-core partials can glide more easily.

The expansion of SSFs occurs because the activation energy for the glide motion of the partials is significantly reduced by the energy transfer of the electron–hole recombination process. This phenomenon is thus not only observed during the on-state operation of SiC bipolar devices but also during PL or cathodoluminescence measurements. In real SiC bipolar devices, however, the morphology of the expanded SSFs is influenced by the device structure. In SiC pin ($p^+/i/n^+$ structure) diodes, for example, carrier recombination mainly occurs within the lightly-doped i-region, where a high

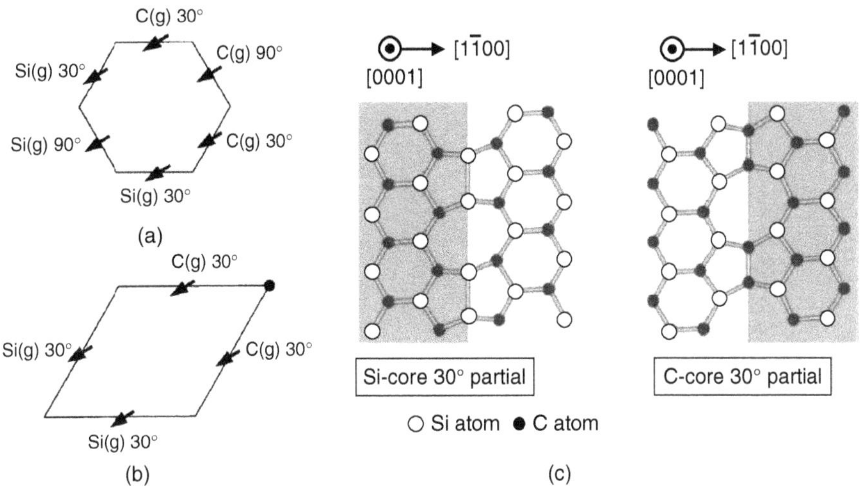

Figure 5.37 (a) Partial dislocation loop in a hexagonal SiC polytype, (b) glide motion of the partials results in the evolution of a rhombus-shaped SSF, and (c) bond configurations for a Si-core and C-core partial dislocations ([123] reproduced with permission from AIP Publishing LLC).

density electron–hole plasma is created by forward biasing (carrier recombination also occurs inside the p^+-anode and n^+-cathode, roughly within a distance equivalent to the diffusion length in individual highly-doped regions). Therefore, the expansion of the SSFs almost ends when the fronts of the gliding partials reach the p^+/i or i/n^+ interfaces. Schematic illustrations of SSF expansion are shown in Figure 5.38. In this figure, SSF expansion from a screw-type BPD ($\boldsymbol{b}$ = [11$\bar{2}$0]/3, dislocation line //[11$\bar{2}$0]) is shown in (a), while SSF expansion inside a SiC pin diode is shown in (b). In a pin diode,

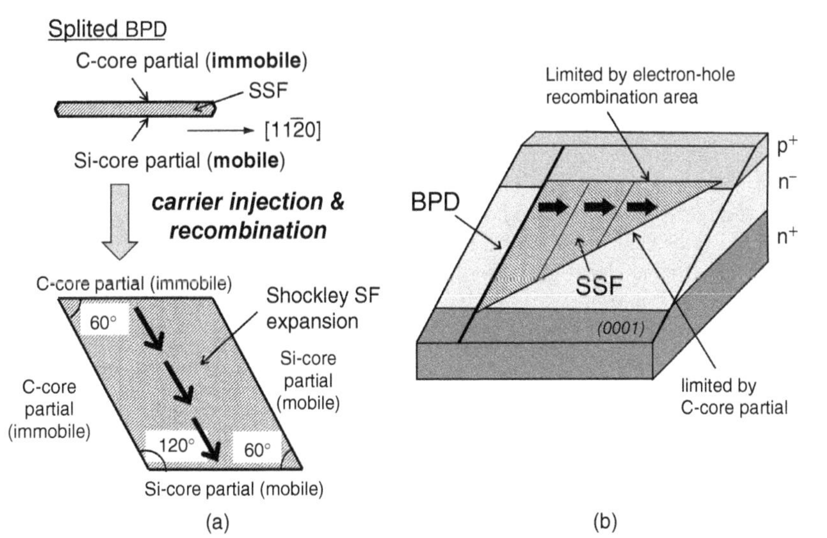

Figure 5.38 (a) SSF expansion from a screw-type BPD ($\boldsymbol{b}$ = [11$\bar{2}$0]/3, dislocation line //[11$\bar{2}$0]) and (b) schematic illustration of SSF expansion in a SiC pin diode.

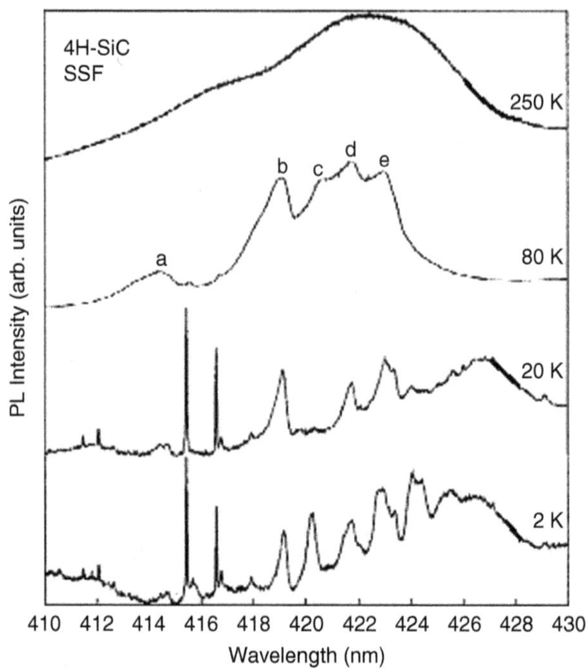

Figure 5.39 Photoluminescence spectra measured at a SSF location at various temperatures ([128] reproduced with permission from AIP Publishing LLC).

the SSF expansion is restricted by not only the C-core partial but also the electron–hole recombination region. For the reasons described above, the majority of the expanded SSFs in pin diodes (or other devices) is triangular in shape.

The Si-core partials exhibit strong red-infrared luminescence with a peak wavelength of about 700 nm (1.8 eV) at room temperature. The SSF area forms a quasi-quantum well, and bright violet luminescence is observed from the faulted region. Figure 5.39 shows PL spectra that were measured at a SSF location at various temperatures [128]. The PL peak is located at 422–424 nm at room temperature. The energy position of the electronic states formed by a SSF in 4H-SiC is calculated to be $E_C - 0.22$ eV [129], which is consistent with the experimentally observed value. Recombination of excess carriers via this state may be responsible for the reduced lifetimes inside the fault region. Trapping of electrons at these states naturally causes the formation of a space-charge region in the vicinity of the SSF, which may create a potential barrier to electron transport. It has been reported that the expanded SSFs can shrink by annealing at moderate temperatures above 250–300 °C [130, 131]. This interpretation takes account of the electrostatic energy of the electrons [131]. However, the shrunken SSFs expand again upon carrier injection at room temperature.

The driving force behind the SSF expansion has been studied, and seems to be intrinsic to the material and not caused by stress [123]. Therefore, complete elimination of the nucleation sites is essential for the development of SiC bipolar devices. Possible nucleation sites include BPDs, dislocation half-loop arrays [132], and the basal plane segments of other types of dislocations.

Bipolar degradation is the most serious problem in the development of SiC bipolar devices, and can also be harmful to SiC field effect transistors when the body diodes (p^+ body/n^- drift layer) are forward-biased. Reduction of BPDs in bulk growth, enhanced conversion from BPDs to TEDs during epitaxial growth, and elimination of BPD nucleation during device processing are the most important remaining issues in SiC technology.

5.2.3 Effects of Extended Defects on SiC Device Performance

Extended defects can affect the performance and reliability of semiconductor devices [133]. In SiC, however, the knowledge and understanding of the effects of extended defects are still limited, despite intensive studies over the last decade. In this subsection, the current understanding of the effects of extended defects on SiC device performance and reliability is summarized (Table 5.6).

All kinds of macroscopic defects that are generated during epitaxial growth cause considerable increase in leakage current and reduction of blocking voltage, and are detrimental to SiC devices. These defects include triangular defects, carrot defects [134], and macro-defects induced by particles ("down-falls"). These defects usually contain stacking faults in a basal plane and/or a prismatic plane, or contain a 3C-lamella, as described in Section 4.3.1. The density of these defects is typically 0.1 – 1 cm^{-2}, and tends to increase in thick epitaxial layers.

In-grown stacking faults also cause increased leakage current and reduce the blocking voltage of SiC devices. It has been predicted theoretically that the bandgap is locally reduced at an in-grown stacking fault [135]. When SiC Schottky barrier diodes contain an in-grown stacking fault, the barrier height is locally reduced, leading to excessive leakage currents [136]. The density of the in-grown stacking faults is 0.1 – 3 cm^{-2}, and increases with increasing growth rate, especially at rates above 50 μm h^{-1}.

A micropipe defect associated with a superscrew dislocation was previously a typical device-killing defect in SiC [137], but this defect has almost been eliminated and is no longer a concern for device development.

Table 5.6 Current understanding of impacts of extended defects on SiC device performance and reliability.

Defects/Devices	SBD	MOSFET, JFET	pin, BJT, Thyristor, IGBT
TSD (without pit)	No	No	No, but causes local reduction of carrier lifetime
TED (without pit)	No	No	No, but causes local reduction of carrier lifetime
BPD (including interface dislocation, half-loop array)	No, but can cause degradation of MPS diode	No, but can cause degradation of body diode	Bipolar degradation (increase of on-resistance and leakage current)
In-grown SF	V_B reduction (by 20–50%)	V_B reduction (by 20–50%)	V_B reduction (by 20–50%)
Carrot, triangular defects	V_B reduction (by 30–70%)	V_B reduction (by 30–70%)	V_B reduction (by 30–70%)
Down-fall	V_B reduction (by 50–90%)	V_B reduction (by 50–90%)	V_B reduction (by 50–90%)

SBD = Schottky barrier diode; MOSFET = metal-oxide-semiconductor field effect transistor; JFET = junction field effect transistor; BJT = bipolar junction transistor; IGBT = insulated gate bipolar transistor; MPS = merged pin-Schottky.

Any kind of BPDs (including interface dislocations and a basal plane segment of a dislocation half-loop) act as nucleation sites for SSFs upon minority-carrier injection and recombination, as described in the previous subsection. The expanded stacking fault causes a significant reduction in the carrier lifetime and a severe increase in the leakage current, as described in Section 5.2.2.

The effects of TSDs and TEDs on device characteristics have been investigated, but some conflicting results have been reported. Neudeck *et al.* clarified the effects of a TSD on SiC(0001) pn diodes by direct comparison between the diode characteristics and the locations of dislocations that were obtained by synchrotron X-ray topography [138]. Figure 5.40a shows the reverse characteristics of 4H-SiC pn diodes with or without a TSD. When a diode contains one TSD, the leakage current is abruptly increased at a bias voltage that is slightly lower than the breakdown voltage. Along with the increase in leakage current, *microplasma* is simultaneously observed at the location of the TSD. However, the breakdown voltage itself is hardly affected by the existence of a single TSD. The local reduction of the breakdown electric field strength (by enhanced impact ionization) at the TSD may cause the onset of avalanche, leading to the formation of the microplasma. This microplasma may cause the diode to be locally heated. However, this local heating does not result in the physical destruction of the diode. This may be attributed to efficient heat conduction because of the high thermal conductivity of SiC and/or a positive temperature dependence of the breakdown voltage (the breakdown voltage increases at elevated temperature). Extensive comparisons between the characteristics of 4H-SiC Schottky barrier diodes and the locations of their dislocations have also been conducted by several groups [139, 140]. However, no direct evidence for the negative effects of TSDs and TEDs on the diode characteristics was obtained. An example of such results is shown in Figure 5.40b. The leakage current

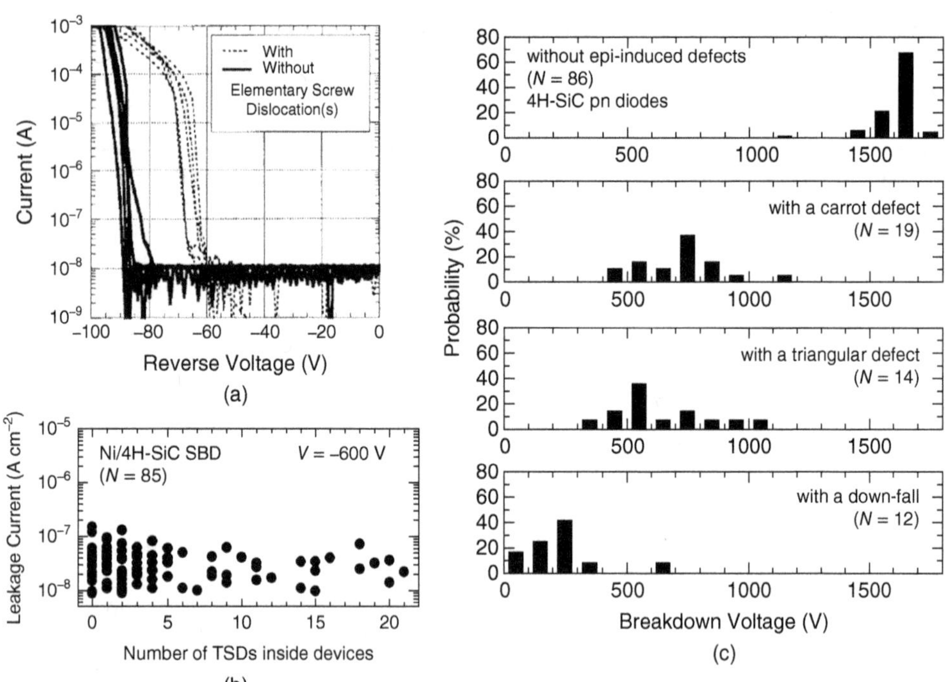

Figure 5.40 (a) Reverse characteristics of 4H-SiC pn diodes with or without a TSD ([138] reproduced with permission from IEEE), (b) leakage current density of Ni/SiC Schottky barrier diodes versus number of TSDs inside the diodes, and (c) histograms of breakdown voltage of 1500 V class 4H-SiC pn diodes which contain epitaxially-induced defects.

of Ni/4H-SiC Schottky barrier diodes which contain 15–20 TSDs is almost identical to that of TSD-free diodes. Similar results are also obtained for SiC pn diodes, where the quantum tunneling effect is very small.

In early studies, a few groups reported that the breakdown voltage and leakage currents of SiC diodes are adversely affected by TSDs (but not by TEDs) [141, 142]. However, one should be cautious before making a conclusive judgment for two reasons:

1. *Influence of other defects*: In early papers on the subject, variation of the number of dislocations inside SiC devices was often linked to the variation of the active area (different sizes). When we consider devices that contain large numbers of dislocations, the area of the corresponding devices is actually large. In this case, other types of defects, such as triangular defects and particles, can be incorporated. This type of investigation can therefore provide misleading results.
2. *Influence of surface pits*: In recent years, the SiC community has learned that surface pits can be formed at the locations of the TSDs and TEDs. The depth of these pits is typically 3–20 nm, and the pits are deeper for TSDs than for TEDs [143]. When the pits are created, local electric field crowding occurs, simply because of a geometric effect [144]. Fujiwara *et al.* clarified that the negative effects of TSDs on the diode characteristics can almost be eliminated when the formation of dislocation-induced pits is suppressed [145]. The size and depth of these dislocation-induced pits depend strongly on the growth conditions and processing conditions. This may be the main reason why slightly different results for the effects of dislocations have been observed by different groups. In summary, the negative effects of dislocations can be minimized to a satisfactory level by suppressing pit formation or by preparing pit-free surfaces via polishing.

When the pure effects of dislocations (without surface pits) on SiC devices are considered, a change in the electronic states near the dislocation cores must be taken into account. Chung *et al.* conducted electron holography measurements for TSDs in n-type 4H-SiC and reported that a deep state ($E_C - 0.89$ eV) is formed near a TSD core [101]. Because the chemical bonds are severely distorted (or broken) near a dislocation core, which is a major disturbance in the periodic potential inherent to SiC, localized states and/or local change of the bandgap are expected to occur near the dislocation core [133]. The states associated with these dislocations must be, of course, active under high electric fields, making them potential candidates for excessive leakage currents. The total leakage current of a SiC junction ($I_{leakage}$) can be expressed approximately using the following equation:

$$I_{leakage} = I_0 + I_{gen} + \sum_i I_i(\text{dislocation}) \tag{5.17}$$

Here, I_0 is the ideal leakage current that was determined for a defect-free SiC junction. I_{gen} is the generation current, which is dominated by bulk carrier generation via the deep levels and by surface carrier generation via the surface states. I_i(dislocation) is the leakage (generation) current caused by the *i*th dislocation inside the junction. The ideal leakage current (I_0) of a pn diode is proportional to n_i^2 (n_i: the intrinsic carrier density), while the generation current via the point defects, surface states, and dislocations is basically in proportion to n_i [67, 133]. Because of the very low intrinsic carrier density of 4H-SiC ($\sim 10^{-9}$ cm^{-3}at 297 K), the ideal leakage current (I_0) of 4H-SiC pn diodes is estimated to be in the 10^{-40} A cm^{-2} range or lower. Therefore, the observed leakage current of SiC pn diodes must definitely be governed by the generation current (via the point defects, surface states, and dislocations). The dominant leakage path depends on the densities of the defects (point defects, surface states, and dislocations). In state-of-the-art SiC epitaxial wafers, the contribution of dislocations to the leakage current is not always a major factor because of the relatively small numbers of dislocations in these wafers, unless a very high electric field that is close to the breakdown field is applied. In other words, each dislocation (TSD or TED) adds a small component of generation current to the leakage current, but these dislocations do not have a detrimental impact on SiC devices because of their relatively low density. The third term of Equation 5.17 can be unacceptably large when the dislocation density is much higher ($\gg 10^6$ cm^{-2}), as in the case of heteroepitaxial materials. For SiC Schottky barrier diodes, the ideal leakage current is much higher than that of pn diodes, because of the quantum tunneling effect, as explained in Section 6.4.1.

Thus, the generation current caused by the dislocations tends to be hidden by the relatively large ideal leakage current as far as pit formation at the dislocations is suppressed.

It is quite often found that the leakage current density of SiC devices increases significantly while the breakdown voltage decreases rapidly when the device area is scaled up. Although a larger device contains more dislocations, the relative contribution of the generation current (or the leakage current density (A cm^{-2})) caused by the dislocations does not change much because the dislocation density (cm^{-2}) remains almost the same. In such large devices, the probability that the device contains macroscopic defects which were generated by epitaxial growth (triangular defects, carrot defects, in-grown stacking faults, and particles) increases. In fact, many groups have observed that those epitaxially-induced defects actually have detrimental effects on SiC devices. Figure 5.40c shows the histograms of breakdown voltage of 1500 V class 4H-SiC pn diodes which contain these epitaxially-induced defects. One can immediately see how detrimental these defects are. This is the main reason why increasing numbers of devices exhibit high leakage current densities and low breakdown voltages as the device area is scaled up. Therefore, the dominant device-killing defects are not TSDs or TEDs but those epitaxially-induced defects in the state-of-the-art SiC epitaxial wafers.

The correlation between gate-oxide reliability and the dislocations in SiC has been investigated by several groups [146–149]. In the early studies, it was found that TSDs and BPDs severely degrade the oxide reliability (reduction of the mean failure time under high electric fields or reduction of the charge-to-breakdown) [146–149]. In more recent studies, it was found that these negative effects can be minimized by removing the surface pits that were created at the dislocation sites [150]. The epitaxially-induced defects are far more detrimental to the dielectric properties of the gate oxides.

Thus, TSDs and TEDs are not considered to be detrimental to most SiC device types, as long as the maximum junction temperature is not very high (roughly <250–300 °C). However, there is a lack of basic knowledge about the effects of dislocations if device operation is considered at much higher temperatures (>400–500 °C). BPDs, however, are clearly detrimental to any SiC bipolar devices, because BPDs cause the bipolar degradation phenomenon following carrier injection.

Assuming uniform distribution of the device-killing defects, the *yield*(Y), which is defined as the number of good devices divided by the number of fabricated devices, can be expressed using the following equation [151]:

$$Y = \exp(-DA) \tag{5.18}$$

Here, D and A are the device-killing defect density and the device area, respectively. Figure 5.41 shows the device yield versus the device area that was calculated from Equation 5.18 by varying the device-killing

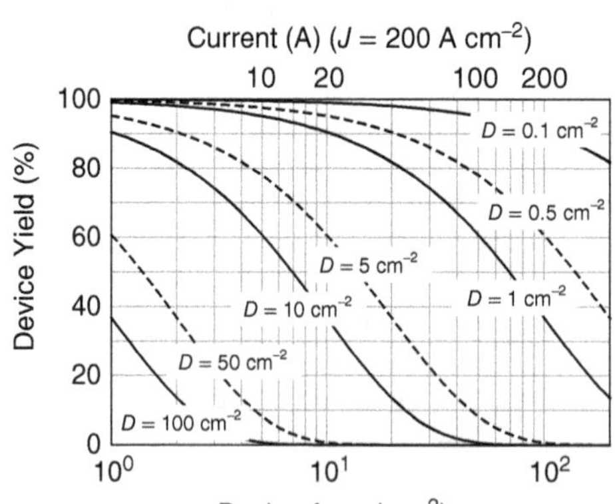

Figure 5.41 Device yield versus device area calculated from Equation 5.18 by varying the device-killing defect density (D) as a parameter.

defect density (D) as a parameter. In the figure, the rough value of the rated current is plotted on the upper horizontal axis, assuming a current density of 200 A cm^{-2}. If the device-killing defect density is 10 cm^{-2}, then to attain a high yield of 80% the maximum device area must be less than 2 mm^2, which corresponds to a maximum rated current of only 4 A. To make 100 A devices (with device area ~ 50 mm^2) with an 80% yield, the device-killing defect density must be reduced to 0.4 cm^{-2}. Based on the discussion above, the main device-killing defects in state-of-the-art SiC epitaxial wafers can thus be assumed to be the epitaxially-induced defects (including triangular defects, carrot defects, in-grown stacking faults, and particles) for unipolar devices, and the sum of the epitaxially-induced-defects and the BPDs for bipolar devices.

5.3 Point Defects in SiC

5.3.1 Major Deep Levels in SiC

5.3.1.1 Intrinsic Defects

The major deep levels that are observed in as-grown n-type and p-type 4H-SiC epitaxial layers are mostly intrinsic defects, and their energy positions in the bandgap are as shown in Figure 4.33. Among them, the $Z_{1/2}$ (E_C – 0.63 eV) [152] and EH6/7 (E_C – 1.55 eV) [153] centers are the dominant thermally stable defects that are most commonly observed with the highest concentration (2×10^{12} to 2×10^{13} cm^{-3}) in all as-grown epitaxial layers. These levels are also dominant in ion-implanted, plasma-etched, or particle-irradiated SiC, as described in Chapter 6. Although a few hole traps, which may originate from the intrinsic defects, are observed, such as HK4 (or P1) (E_V + 1.44 eV) [154], they can be annealed out at 1450–1550 °C. Thus, the $Z_{1/2}$ and EH6/7 centers are the most common and important deep levels in SiC. Note that the signals from the EH6 and EH7 centers usually overlap severely in normal DLTS spectra, and these two defect levels are thus often treated as the single EH6/7 center, though the EH7 component is dominant in the "EH6/7 peak". Laplace-DLTS measurements succeeded in resolving this peak, demonstrating that, in terms of generation upon particle irradiation and thermal stability, there are slight differences between the EH6 and EH7 centers [155]. Here, the features of the $Z_{1/2}$ and EH6/7 centers are summarized.

1. The $Z_{1/2}$ center exhibits the so-called "*negative-U*" property, as indicated in Figure 5.42 [156]. One DLTS peak usually reflects the carrier emission from a single deep level unless multiple peaks are overlapping. In the case of the $Z_{1/2}$ center, however, two electrons are emitted simultaneously from the defect level. In the carrier capture process, two electrons are captured almost simultaneously against the Coulomb repulsion between the two electrons, because of the large *Jahn–Teller effect* [157]. Thus, the intensity of the DLTS signal from the $Z_{1/2}$ center is exactly double that of the real density of the $Z_{1/2}$ center.
2. The $Z_{1/2}$ center and the EH6/7 center (or at least the EH7 center) are observed with almost identical density in almost all the 4H-SiC samples. Figure 5.43 plots the densities of the $Z_{1/2}$ center versus those of the EH6/7 center for as-grown, electron-irradiated, and annealed 4H-SiC [158, 159]. The densities of both deep levels are varied by changing the epitaxial growth conditions, the electron energy and fluence during irradiation, and the annealing conditions. The $Z_{1/2}$ center density is almost identical to the EH6/7 density for each sample over a wide range from 10^{11} to 10^{14} cm^{-3}. This result suggests that the $Z_{1/2}$ center and the EH6/7 (at least the EH7) centers originate from the same defect center but with different charge states.
3. The origin of the $Z_{1/2}$ center and the EH7 center was unambiguously identified as a carbon vacancy with different charge states, based on the results of theoretical calculations [160] and a comparison study using DLTS and EPR [161, 162]. Figure 5.44a shows the area density of a single-negatively-charged carbon vacancy ($V_C(-)$) that was determined by EPR versus the area density of the $Z_{1/2}$ center taken from the depth profile that was determined by DLTS measurements

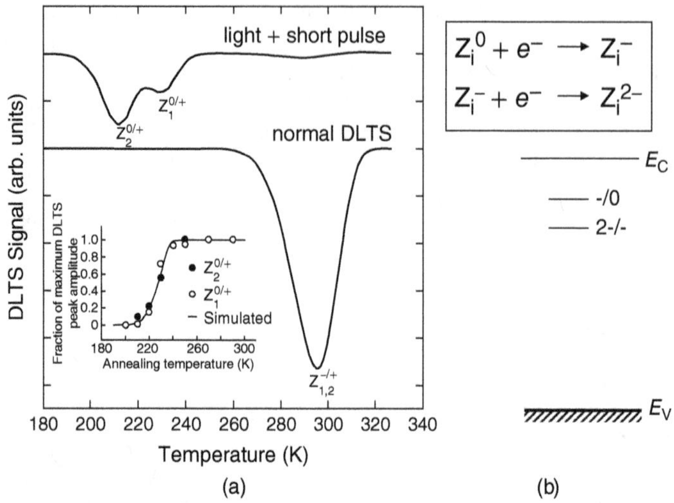

Figure 5.42 (a) DLTS peaks originating from the $Z_{1/2}$ center in n-type 4H-SiC. The "negative-U" property is revealed by using light illumination and short filling pulse ([156] reproduced with permission from American Physical Society). (b) Energy levels of the $Z_{1/2}$ center in n-type 4H-SiC.

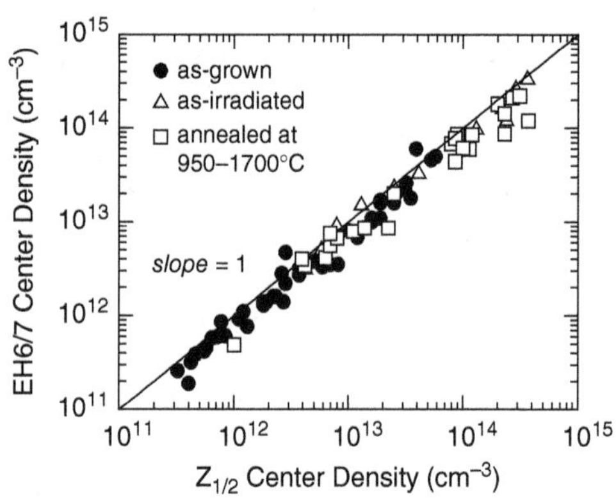

Figure 5.43 Densities of the $Z_{1/2}$ center versus those of the EH6/7 center for as-grown, electron-irradiated, and annealed 4H-SiC [158, 159]. The densities of both deep levels are varied by changing the epitaxial growth conditions, the electron energy and fluence during irradiation, and the annealing conditions.

[163]. Here, photo-excitation is necessary to observe $V_C(-)$, because the carbon vacancy also has a negative-U nature and stabilizes as $V_C(2-)$ by capturing two electrons. Because $V_C(2-)$ is not EPR-active (because it contains no unpaired electrons), the V_C density is estimated by producing $V_C(-)$ through photo-excitation of n-type SiC. As shown in Figure 5.44a, the V_C density is very close to the $Z_{1/2}$ center density for a series of samples. Considering this correspondence and the negative-U nature described above, the origin of the $Z_{1/2}$ center was identified as the acceptor levels of a carbon vacancy. In Figure 5.44b, the acceptor and donor levels of the carbon vacancy in 4H-SiC that were determined by photo-EPR and those from theoretical calculations are also shown. As

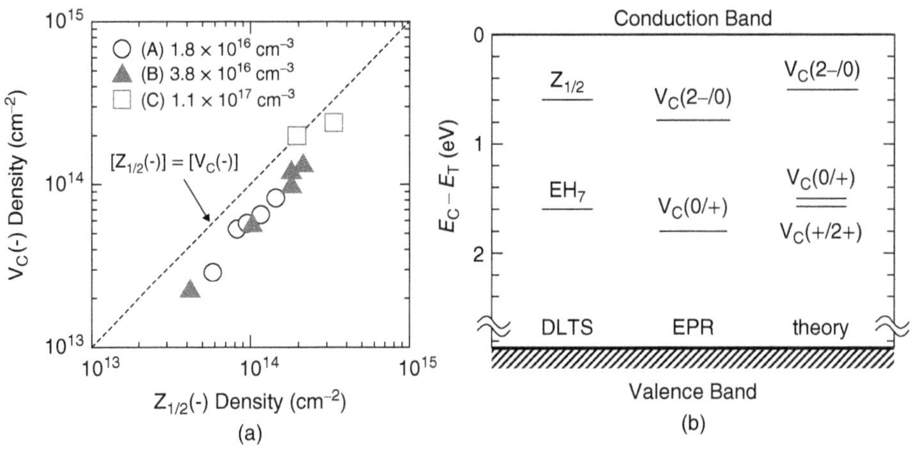

Figure 5.44 (a) Area density of a single-negatively-charged carbon vacancy ($V_C(-)$) determined by EPR versus that of the $Z_{1/2}$ center taken from the depth profile determined by DLTS measurements [163]. (b) Comparison of energy levels of the $Z_{1/2}$ and EH6/7 centers obtained by DLTS and that those of carbon vacancy in 4H-SiC (photo-EPR and theory).

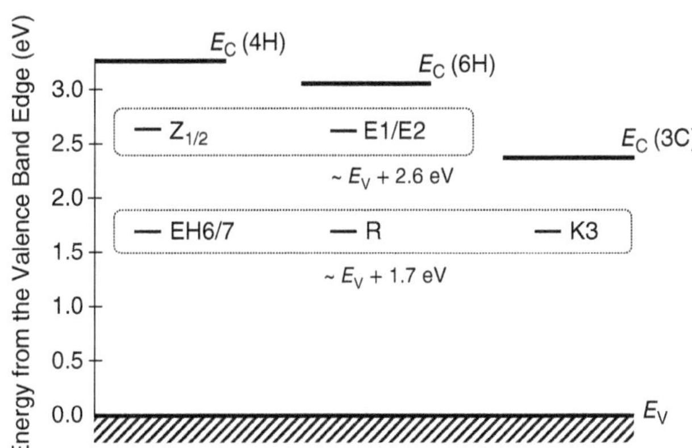

Figure 5.45 Energy positions of deep levels originating from a carbon vacancy in 4H-, 6H-, and 3C-SiC.

shown in Figure 5.44b, the EH7 center has been identified as the donor level of a carbon vacancy, based on these experimental results and theoretical study [160].

4. Systematic DLTS studies have shown that the E1/E2 (E_C – 0.45 eV) and R centers (E_C – 1.25 eV) that are observed in 6H-SiC are equivalent to the $Z_{1/2}$ and EH7 centers in 4H-SiC, respectively [159]. Additionally, the K3 center (E_C – 0.73 eV) that is observed in 3C-SiC should have the same origin as that of the EH7 center in 4H-SiC [164], that is, a carbon vacancy. Figure 5.45 shows the energy positions of these deep levels in three SiC polytypes. The energy positions for the different charge states of carbon vacancies in the different SiC polytypes are aligned with respect to the valence band edge, and are thus scaled with respect to the conduction band edge according to the bandgap. This is expected, because the vacancy causes a localized disturbance of the valence electrons.
5. The $Z_{1/2}$ and EH6/7 centers in 4H-SiC are generated through low-energy electron irradiation, by which only the carbon atoms are displaced, and no thermal treatment is required after the irradiation

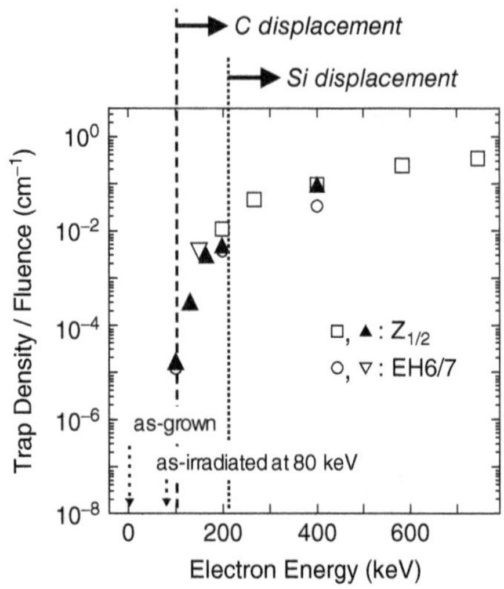

Figure 5.46 Generation rate of the $Z_{1/2}$ and EH6/7 centers versus irradiation energy of electrons ([166] reproduced with permission fromAIP Publishing LLC).

process to form the defect centers [158, 165]. The threshold electron energy for the generation of these levels is around 95–100 keV, which agrees with the threshold energy that was theoretically estimated for carbon-atom displacement (102 keV), as shown in Figure 5.46 [166]. Although several carbon-interstitial-related deep levels are also created by the electron irradiation, these levels are annealed out by thermal treatment at 1000 °C, probably because of the out-diffusion of the carbon interstitials. The densities of both the carbon-vacancy-related and carbon-interstitial-related defects are almost in proportion to the electron fluence, and the defect density can exceed that of any impurities in the SiC epitaxial layers, indicating the exclusion of the involvement of impurities.

6. In the as-grown epilayers, the densities of the $Z_{1/2}$ and EH6/7 centers increase significantly when the epilayer is grown under Si-rich conditions, but decrease when grown under C-rich conditions [167, 168], which is consistent with the identified defect origin (carbon vacancy).
7. The $Z_{1/2}$ and EH6/7 centers in 4H-SiC are thermally stable [152, 169], but the densities of the $Z_{1/2}$ and EH6/7 centers in the epitaxial layers are found to increase rapidly through thermal annealing in Ar at temperatures above 1750 °C [170]. Figure 5.47a shows the density of the $Z_{1/2}$ center in several 4H-SiC epitaxial layers annealed at various temperatures. The Arrhenius plot of the density of the $Z_{1/2}$ center as a function of the inverse of the annealing temperature is also shown in Figure 5.47b. Here, the annealing was performed in pure Ar for 30 min. When an epitaxial layer with a very low initial $Z_{1/2}$ center density ($\sim 10^{11}$ cm^{-3}) was used, the $Z_{1/2}$ density had already started to increase when the annealing temperature was 1600 °C. When the initial $Z_{1/2}$ density was rather high ($\sim 10^{14}$ cm^{-3}), the $Z_{1/2}$ density gradually decreased until the temperature reached 1700 °C, but then increased remarkably after reaching a minimum value. After annealing at temperatures of more than 1750 °C, the $Z_{1/2}$ densities for all samples fall into the same line, as shown in Figure 5.47. The activation energy obtained from the slope is approximately 5.8 eV.
8. As described in Section 4.3.2, the densities of the $Z_{1/2}$ and EH6/7 centers in the epitaxial layers depend strongly on the growth temperature. Figure 5.48 shows the $Z_{1/2}$ density for as-grown 4H-SiC(0001) epitaxial layers that were grown at various temperatures. The densities of the $Z_{1/2}$ (and EH6/7) centers are reduced with increasing the high C/Si ratio, and the minimum $Z_{1/2}$ density obtained at each growth temperature actually coincides with the data obtained from the annealing experiment (Figure 5.47): the dotted line in Figure 5.48 is the same as that plotted in Figure 5.47a. Therefore, this line, with an activation energy of about 5.8 eV, is universal in SiC, and may correspond to the equilibrium density of the carbon vacancy in SiC.

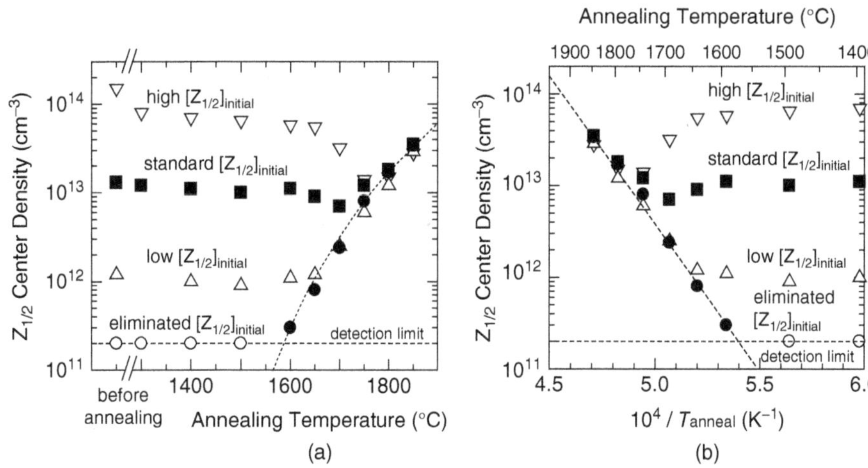

Figure 5.47 (a) Density of the $Z_{1/2}$ center in several 4H-SiC epitaxial layers annealed at various temperatures and (b) Arrhenius plot of the density of the $Z_{1/2}$ center as a function of the inverse of the annealing temperature ([170] reproduced with permission from AIP Publishing LLC).

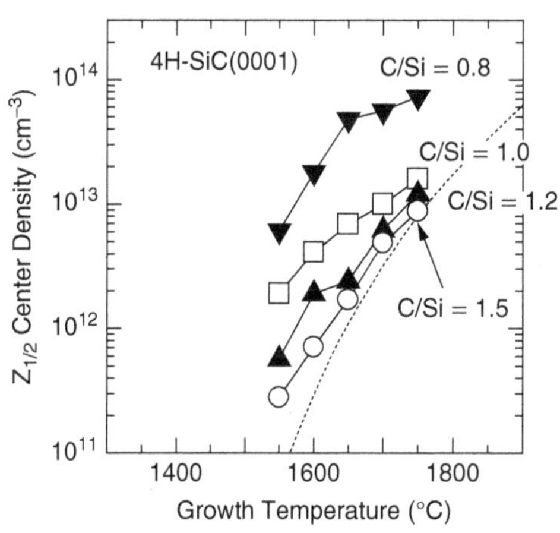

Figure 5.48 $Z_{1/2}$ density for as-grown 4H-SiC(0001) epitaxial layers grown at various temperatures.

The $Z_{1/2}$ center is the most important defect, because of its abundance and because of its property of being a major carrier-lifetime killer [171, 172]. To obtain a long carrier lifetime, which is beneficial for reduction of the on-resistance in bipolar devices, the density of the $Z_{1/2}$ center must be decreased to the order of 3×10^{12} cm^{-3} or lower. Thus far, two successful techniques have been proposed to eliminate the $Z_{1/2}$ center.

In the first technique, excess carbon atoms are introduced from the outside and the diffusion of these carbon atoms into the bulk region is promoted by high-temperature annealing [173, 174]. This can actually be achieved by carbon ion-implantation and subsequent annealing in Ar at 1650–1700 °C. Figure 5.49 shows the DLTS spectra obtained from n-type 4H-SiC epitaxial layers before and after carbon ion-implantation followed by high-temperature Ar annealing. Both the $Z_{1/2}$ and EH6/7 centers are eliminated by this process down to the detection limit (approximately 1×10^{11} cm^{-3} in this case). Because

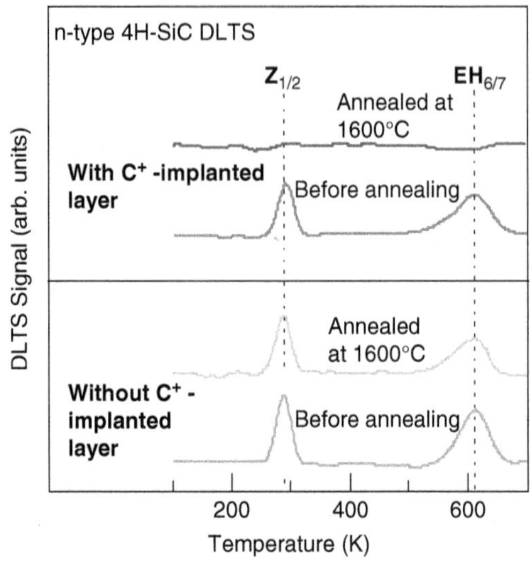

Figure 5.49 DLTS spectra obtained from n-type 4H-SiC epitaxial layers before and after carbon ion-implantation followed by high-temperature Ar annealing [173].

the created carbon interstitials have large diffusion coefficients above 1400–1500 °C, the excess carbon interstitials diffuse and fill the carbon vacancies, leading to the elimination of the $Z_{1/2}$ and EH6/7 centers from the surface region to the bulk. After annealing, the defective implanted region near the surface is removed by plasma etching.

In the second technique, thermal oxidation of SiC under appropriate conditions results in the elimination of the $Z_{1/2}$ and EH6/7 centers. During thermal oxidation, the carbon atoms are mostly removed by the creation of carbon monoxide (CO). However, some portion of the carbon atoms is emitted to the bulk region and diffuses into a deep region of SiC. Because silicon atoms are emitted during thermal oxidation of Si [175], it is expected that carbon (and also silicon) atoms will be emitted during the thermal oxidation of SiC. As in the case of the carbon ion implantation, the diffusion coefficient of the carbon interstitials is so large that the depth to which the $Z_{1/2}$ and EH6/7 centers are eliminated can exceed 100 μm, by either the combination of oxidation and subsequent high-temperature Ar annealing or by high-temperature oxidation (1300–1400 °C) [176–178]. Note that the thermal oxide is removed before Ar annealing. It was also reported that the migration energy of the carbon interstitials is much smaller than that of the silicon interstitials [179], and the self-diffusion coefficient of carbon is much larger than that of silicon [180]. Figure 5.50 shows the depth profile of the $Z_{1/2}$ density for the 4H-SiC epitaxial layers after various thermal treatments (e.g., oxidation, Ar annealing) [178]. By extending the oxidation time or raising the oxidation temperature, the "$Z_{1/2}$–free" region extending from the surface becomes thicker (100 μm or even thicker).

In both the deep-level reduction techniques (carbon ion implantation and thermal oxidation), the depth of the "$Z_{1/2}$ – free" region depends not only on the process conditions but also on the initial $Z_{1/2}$ density. When the initial $Z_{1/2}$ density is relatively low ($<5 \times 10^{12}$ cm^{-3}), it is easy to eliminate these deep levels in 100–200 μm-thick regions. However, a thick "$Z_{1/2}$-free" region can hardly be obtained if the initial $Z_{1/2}$ density is high ($>1 \times 10^{14}$ cm^{-3}). The depth profiles of the $Z_{1/2}$ density after these defect-reduction processes can be predicted through numerical simulation of the carbon diffusion process [178]. It was noted that carbon diffusion into the SiC bulk region results in the generation of a few new deep levels, such as the ON1 (E_C – 0.84 eV), ON2 (E_C – 1.1 eV), and HK0 (E_V + 0.78 eV) centers [181, 182], which are commonly observed near the surface in both carbon-implanted SiC and oxidized SiC. The HK0

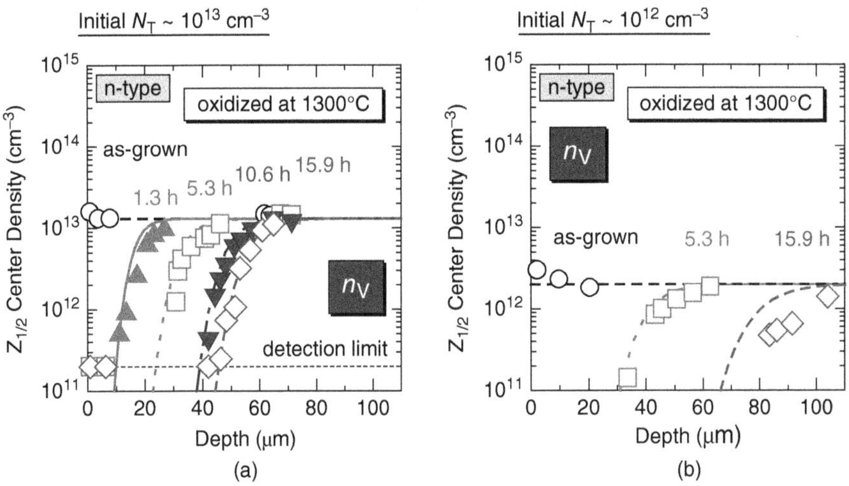

Figure 5.50 (a,b) Depth profile of the $Z_{1/2}$ density for the 4H-SiC epitaxial layers after various thermal treatments (e.g., oxidation, Ar annealing) ([178] reproduced with permission from AIP Publishing LLC).

center is annealed out by thermal treatment at temperatures above 1350–1400 °C, while the ON1/ON2 centers (the same defect, but with different charge states) survive even after annealing at 1800 °C. A carbon-interstitial-related defect such as the carbon di-interstitial $(C_i)_2$ was suggested as the origin of the HK0 center [182].

For details of other intrinsic deep levels in SiC, such as $RD_{1/2}$ (E_C − 0.93 eV) [152], UT1 (E_C − 1.45 eV) [183], HS2 (E_V + 0.47 eV) [165], and HK4 (E_V + 1.44 eV) [154], please see the individual references provided or review papers [152, 184]. The exact correlation between the observed deep levels and a specific defect structure (e.g., silicon vacancy, antisite) has not yet been established. Figure 5.51 shows

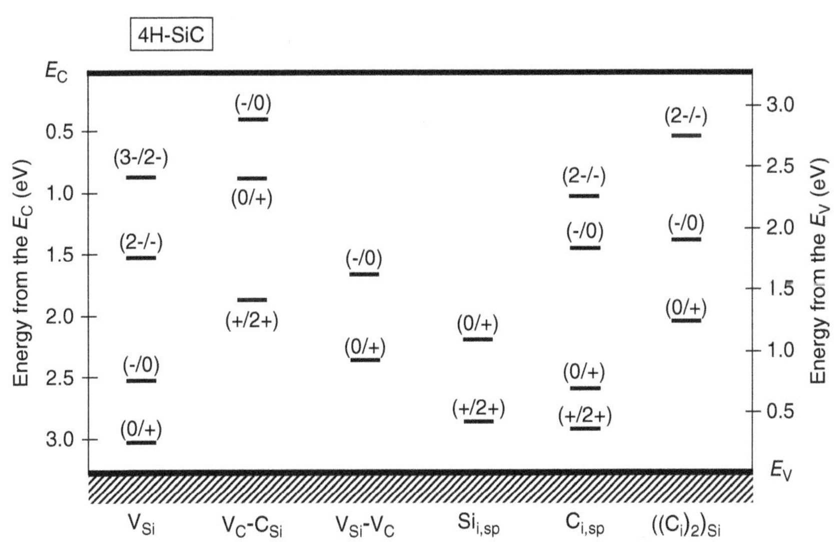

Figure 5.51 Energy positions of the various intrinsic defects that are expected in 4H-SiC, which were obtained through theoretical calculations [179, 185, 186].

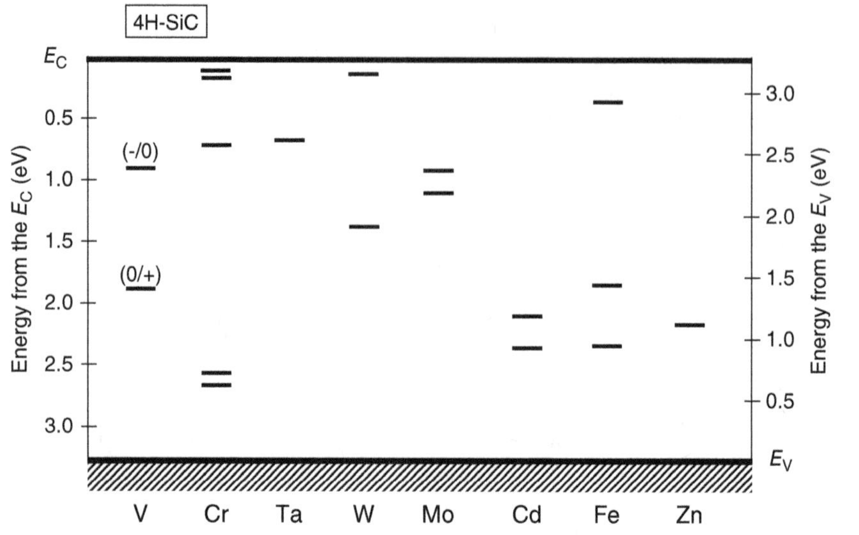

Figure 5.52 Energy positions of the major metallic-impurity-related deep levels experimentally observed in 4H-SiC.

the energy positions of the various intrinsic defects that are expected in 4H-SiC, which were obtained through theoretical calculations [179, 185, 186]. The energy levels of the silicon vacancy (V_{Si}), the carbon antisite–carbon vacancy pair ($C_{Si}-V_C$), the carbon di-interstitial (C_i-C_i), and other defects have been reported. The diffusion of these intrinsic defects has also been investigated through calculations [179]. However, the experimental assignment of actual deep levels to these defects has not yet been achieved.

5.3.1.2 Impurities

The purity of the SiC epitaxial layers grown by chemical vapor deposition (CVD) is rather high, although the substrates do contain various impurities, including heavy metals. Other than the intentional dopants, the major impurities that are present in SiC epitaxial layers are boron and titanium, as described in Section 4.3.2. The typical densities of these impurities are $(1-5) \times 10^{13}$ cm^{-3} for boron and $(0.5-5) \times 10^{12}$ cm^{-3} for titanium. Several other impurities are also occasionally observed. Figure 5.52 shows the energy positions of the major metallic impurity-related deep levels that are observed in 4H-SiC.

Boron (B) atoms substituting the silicon lattice sites create boron acceptors ($E_V + 0.35$ eV) [187]. However, some boron atoms form pairs with carbon vacancies ($B_{Si}-V_C$) [188], leading to the formation of a "deep boron" level; this is the D center in 4H-SiC ($E_V + 0.45$ eV) [187]. The energy position of the D center is deeper in 6H-SiC ($E_V + 0.55$ eV). The D center is also acceptor-like [189], and contributes to the DAP luminescence at relatively long wavelengths. The generation of the D center in SiC epitaxial layers is enhanced under Si-rich growth conditions and is reduced under C-rich conditions [20]. The D center is thermally stable, but can be reduced by thermal oxidation or by carbon ion implantation, which is reasonable because interaction of the D center ($B_{Si}-V_C$) with a carbon interstitial will naturally annihilate the defect.

Oxygen (O) is a common impurity in almost all semiconductor materials, including Si and GaAs. Because nitrogen is a dominant residual impurity in high-purity SiC, oxygen contamination is naturally expected. However, it was found that oxygen is hardly incorporated in SiC during epitaxial growth, even when an oxygen-containing gas is intentionally introduced into the CVD reactor [190, 191]. Etching

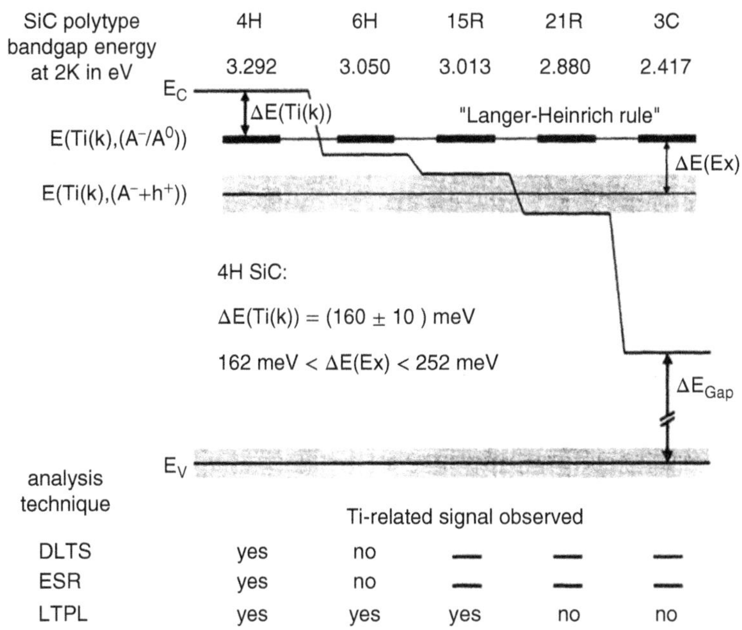

Figure 5.53 Energy positions of the Ti levels in the different SiC polytypes ([29] reproduced with permission from American Physical Society).

of Si-O or Si-C-O by hydrogen that is supplied as a carrier gas, or decomposition of SiO_2 at the high growth temperatures may suppress oxygen incorporation. No deep levels that could be attributed to pure oxygen-related defects have been identified. The oxygen density in SiC seems to be very low ($\sim 10^{12}$ cm^{-3} or lower).

Titanium (Ti) is another common impurity in SiC, and it creates very shallow electron traps near the conduction band edge (E_C − 0.11/0.17 eV) in 4H-SiC [29]. Figure 5.53 shows schematically the energy positions of the Ti levels in the different SiC polytypes. The Ti levels with respect to E_V are almost identical when plotted in the bandgaps of the different polytypes, indicating that the *Langer–Heinrich rule* [192] is valid for Ti in SiC. Ti is electrically active in 4H-SiC, but its acceptor levels are resonant in the conduction band for 6H-, 15R-, and 3C-SiC. As described in Section 5.1.1, Ti exhibits a unique greenish luminescence in 4H-, 6H-, and 15R-SiC, but not in 3C-SiC.

Vanadium (V) is an amphoteric impurity, and it forms an acceptor level in the upper half of the bandgap and a donor level in the lower half of the bandgap. In 4H-SiC, the acceptor level and the donor level are located at E_C − (0.81–0.97) eV and E_V + 1.4 eV, respectively [31]. Thus, vanadium creates a very deep level, especially in p-type SiC, and is thermally stable. This is why vanadium doping can be used to form semi-insulating SiC, as described in Section 3.4.4. In SiC epitaxial layers grown by CVD, however, the vanadium density is usually less than 1×10^{12} cm^{-3}.

The deep levels of chromium (Cr), molybdenum (Mo), tungsten (W), and iron (Fe) in 4H-SiC have been identified [7, 193–195]. In infrared PL spectra from 4H-SiC, the sharp PL peaks at 1170 and 1171 nm (1.0586 and 1.0595 eV) are well known and have been assigned to a UD center (UD1). In recent years, systematic investigations of this defect center by PL, EPR, DLTS, and theoretical calculations have revealed that the origin of UD1 is actually tungsten [36]. In high-purity SiC epitaxial layers, however, the densities of Cr, Mo, W, and Fe are lower than 1×10^{12} cm^{-3}. It was demonstrated that some transition metals such as Ni and Cr exhibit significant diffusion in SiC at temperatures higher than 1500 °C [196].

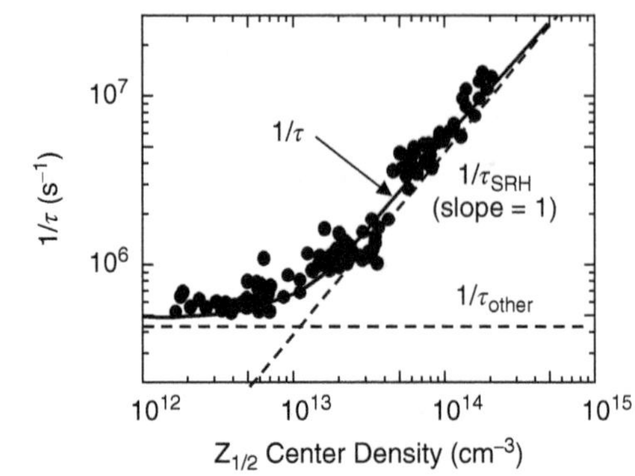

Figure 5.54 Inverse of the carrier lifetime versus the measured density of the $Z_{1/2}$ center for 50-μm-thick n-type 4H-SiC epitaxial layers [172, 198].

5.3.2 Carrier Lifetime Killer

Correlation between the carrier lifetimes and the density of the $Z_{1/2}$ center was suggested at an early stage of the studies of SiC [167, 197] and, in recent years, the $Z_{1/2}$ center has been unambiguously identified as a major carrier lifetime killer, at least, in n-type 4H-SiC [171, 172]. Figure 5.54 shows the inverse of the carrier lifetime versus the measured density of the $Z_{1/2}$ center for 50-μm-thick n-type 4H-SiC epitaxial layers [172, 198]. When the $Z_{1/2}$ density is higher than $(1-2) \times 10^{13}$ cm^{-3}, the inverse of the carrier lifetime is proportional to the $Z_{1/2}$ density, indicating that the lifetime is governed by Shockley–Read–Hall (SRH) recombination via the $Z_{1/2}$ center. However, the correlation between the lifetime and the $Z_{1/2}$ density is unclear when the $Z_{1/2}$ density is in the $10^{11}-10^{12}$ cm^{-3} range. As described in Section 5.1.4, multiple recombination processes exist, including SRH recombination and several other recombination processes. Thus, the carrier lifetime (τ) can be expressed by the following equation:

$$\frac{1}{\tau} = \frac{1}{\tau_{SRH}} + \frac{1}{\tau_{other}} \tag{5.19}$$

where τ_{SRH} is the SRH lifetime governed by the recombination centers, and τ_{other} is the carrier lifetime that is governed by the other recombination processes, such as surface recombination, recombination in the substrate, Auger recombination, and recombination at extended defects. Here, the inverse of τ_{SRH} is proportional to the density of the recombination centers ($1/\tau_{SRH} = a\, N_{Z1/2}$, where a is a constant, and $N_{Z1/2}$ is the density of the $Z_{1/2}$ center), while τ_{other} can be assumed to be independent of the $Z_{1/2}$ density. By using the model expressed by Equation 5.19, the experimental data can be fitted. The fitted result is shown in Figure 5.54 as a solid line for $1/\tau_{SRH} + 1/\tau_{other}$, and as two broken lines for $1/\tau_{SRH}$ and $1/\tau_{other}$. Here the $1/\tau_{other}$ line is determined by recombination at the surface and in the substrate, due to the limited thickness of epitaxial layers (50 μm in this plot), and this component decreases by using thicker epitaxial layers. The detail was explained in Section 5.1.4.4.

Based on the results described above, a simple relationship between the high-injection bulk lifetime (τ_{SRH}) and the $Z_{1/2}$ density ($N_{Z1/2}$) can be established as follows:

$$\tau_{SRH}[\mu s] = \frac{1.6 \times 10^{13}}{N_{Z1/2}[cm^{-3}]} \tag{5.20}$$

Note that this equation is valid when Auger recombination and the recombination at extended defects are less important, and it is independent of both the epilayer thickness and the SRV (surface recombination velocity). Instead, the factor of 1.6×10^{13} is a function of the excitation intensity (i.e., the injection level) and the temperature.

In semiconductor textbooks, it is stated that a midgap level is an effective recombination center and can thus be a lifetime killer [157]. In 4H-SiC, however, the midgap level of the EH6/7 (E_C − 1.55 eV) center does not work as a carrier lifetime killer [199]. This can be deduced by accounting for the properties of the $Z_{1/2}$ and EH6/7 (EH7) centers. As described in Section 5.3.1, the origin of both the $Z_{1/2}$ and EH7 centers is a carbon vacancy (V_C) with different charge states:

$Z_{1/2}$: $V_C(2-/0)$, negative U (trap and emit two electrons), $E_C-0.63$ eV
EH7 : $V_C(0/+)$, $E_C-1.55$ eV.

In n-type 4H-SiC, the Fermi level is well above the $Z_{1/2}$ center, and the V_C defect is charged to 2−. When excess electron–hole pairs are generated, a hole is first captured at the V_C defect, because of the large hole capture cross-section (the attractive Coulomb force between V_C (2−) and the hole). Then, the V_C defect changes its charge state as follows: $V_C(2-)$ + hole → $V_C(-)$. If the hole density is extremely high, then $V_C(-)$ + hole → $V_C(0)$ may proceed. Under these circumstances, the capture cross-sections of the electrons for $V_C(-)$ and $V_C(0)$ are very large, because of its negative-U property. Thus, the V_C defect immediately tends to revert to the original state; $V_C(-)$ + electron → $V_C(2-)$ or $V_C(0)$ + electron → $V_C(-)$, and subsequently $V_C(-)$ + electron → $V_C(2-)$. These capture and recombination processes are repeated upon electron–hole excitation. Therefore, only the $Z_{1/2}$ center (2−/0) is involved in the recombination process. The EH7 center ($V_C(0/+)$ defect) does not appear during this process. This is why the EH7 center is not important in carrier recombination in n-type 4H-SiC. The EH7 center may, however, be important in p-type 4H-SiC, where the Fermi level is close to the valence band and the V_C defect is charged to +.

Figure 5.55 shows the μ-PCD decay curves at room temperature that were obtained from a 220-μm-thick n-type 4H-SiC epitaxial layer [100]. The decay curves for the as-grown material and that after $Z_{1/2}$-center reduction through thermal oxidation at 1400 °C for 48 h are shown. For the as-grown epitaxial layer, the measured lifetime is 1.1 μs, while the lifetime has improved remarkably to 26.1 μs after the defect reduction process. The depth of the $Z_{1/2}$-eliminated region of this particular sample is estimated to be about 230 μm, indicating that the defect is eliminated throughout the entire thickness of the epitaxial layer. After lifetime measurement of the defect-eliminated sample, the surface was passivated with a 20-nm-thick deposited oxide and annealed in nitric oxide (NO) at 1250 °C for 30 min. After this surface passivation, the lifetime increased to 33.2 μs, and the lifetime increased further to 47 μs

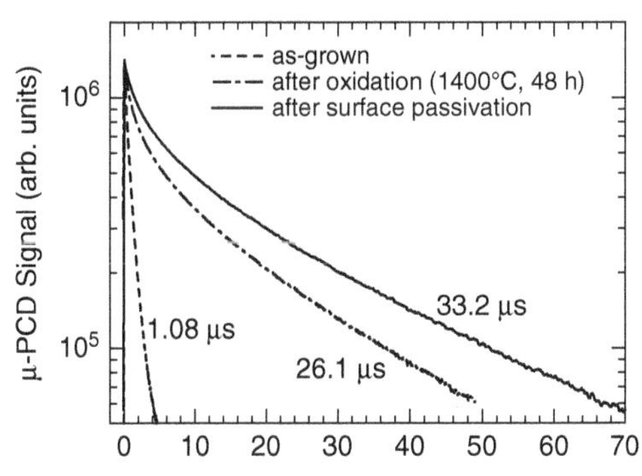

Figure 5.55 μ-PCD decay curves at room temperature obtained from a 220-μm-thick n-type 4H-SiC epitaxial layer ([100] reproduced with permission from The Japan Society of Applied Physics). The decay curves for the as-grown material and that after $Z_{1/2}$-center reduction through thermal oxidation at 1400 °C for 48 h are shown.

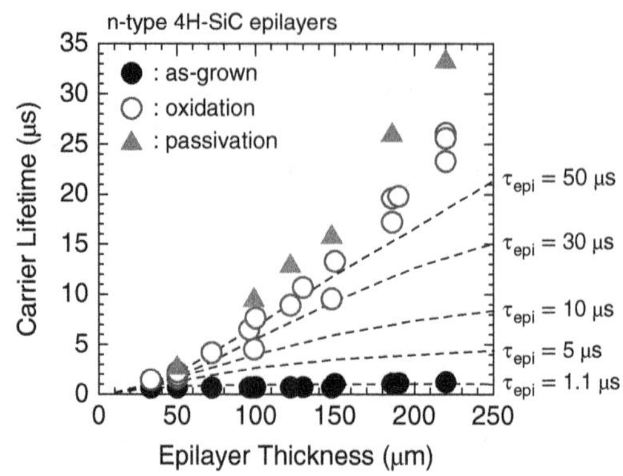

Figure 5.56 Measured lifetimes versus epitaxial layer thickness. The results for samples taken before oxidation (as-grown), after oxidation, and with the subsequent surface passivation are indicated by closed circles, open circles, and closed triangles, respectively.

at a measurement temperature of 200 °C. Because a relatively low interface state density is obtained over the whole energy range of the bandgap by this passivation process [200], surface recombination may be suppressed, leading to a longer carrier lifetime. Similar carrier lifetimes can be achieved by carbon ion implantation and subsequent Ar annealing [201].

In general, the experimentally determined carrier lifetimes are underestimated, because of parasitic recombination paths, as described in Section 5.1.4. In Figure 5.56, the measured lifetimes are plotted as a function of the epitaxial layer thickness. The results for samples taken before oxidation (as-grown), after oxidation, and with the subsequent surface passivation are indicated by closed circles, open circles, and closed triangles, respectively. In the oxidized samples, sufficiently long-term oxidation was performed at 1300–1400 °C to ensure that the $Z_{1/2}$ center is eliminated throughout the entire thickness of the epitaxial layers. The lifetimes of the oxidized samples increase rapidly with increasing epitaxial layer thickness, while the thickness dependence is very small for the as-grown samples. This result indicates that the measured lifetime is severely affected by the carrier recombination in the substrate (or near the epitaxial layer/substrate interface) when the bulk lifetime becomes very long because of $Z_{1/2}$ elimination, which is consistent with Figure 5.17. In Figure 5.56, the simulated dependence of the lifetime on the epilayer thickness is also indicated by dashed lines for the various bulk lifetimes of the epitaxial layers. The SRV was assumed to be 1000 cm s^{-1}. As determined from Figure 5.56, the real bulk lifetime (τ) will be longer than 50 μs. The ambipolar diffusion length L_a is given by the following equation [1, 151]:

$$L_a = \sqrt{D_a \tau} \tag{5.21}$$

$$D_a = \frac{2D_n D_p}{D_n + D_p} \tag{5.22}$$

Here, D_n and D_p are the diffusion coefficients of the electrons and the holes, respectively. Ambipolar diffusion is described in greater detail in Section 7.3.1. By accounting for the diffusion coefficients and assuming a long carrier lifetime of 50 μs, the diffusion length of the carriers is estimated to be longer than 150 μm in the $Z_{1/2}$-eliminated region.

In Si, heavy metallic impurities, such as Au, Pt, and Fe, create midgap levels that act as efficient recombination centers [151]. Although the deep levels of several metallic impurities (V, Cr, W) are known in SiC, the densities of these impurity levels determined by DLTS measurements are usually close to the detection limit (about 5×10^{11} cm^{-3}). Therefore, it can be concluded that the SRH recombination center (i.e., the lifetime killer) is indeed the $Z_{1/2}$ center in n-type 4H-SiC.

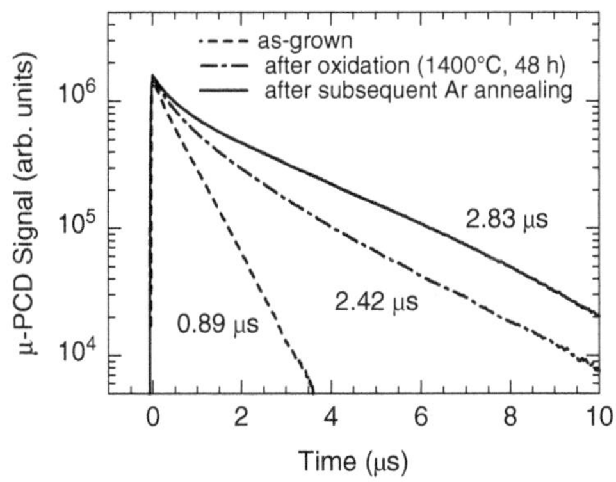

Figure 5.57 μ-PCD decay curves at room temperature obtained from a 147-μm-thick, lightly-doped p-type 4H-SiC epitaxial layer [202, 203]. The decay curves for the as-grown material and for the material after carbon-vacancy reduction via thermal oxidation at 1400 °C for 48 h are shown.

The carrier lifetime in p-type SiC is more complicated. Because the Fermi level is close to the valence band in p-type SiC, the carbon vacancy defect must be positively charged ($V_C(+)$: EH7 center) in equilibrium. Thus, it is expected that any excess electrons are quickly trapped by the defect, and the charge state will change to neutral. However, accurate capture cross-sections for the electrons and holes are not yet known for this neutral defect. Figure 5.57 shows the μ-PCD decay curves at room temperature that were obtained from a 147-μm-thick, lightly-doped p-type 4H-SiC epitaxial layer [202, 203]. The decay curves for the as-grown material and for the material after carbon-vacancy reduction via thermal oxidation at 1400 °C for 48 h are shown. The measured lifetime of the as-grown epilayer is 0.9 μs, and the lifetime increased to 2.4 μs after the defect reduction process. The lifetime improved slightly to 2.6 μs following surface passivation with a nitrided oxide [202, 203]. Carbon ion implantation followed by Ar annealing at 1650 °C, instead of thermal oxidation, provides a very similar result. Thus, reduction of the carbon vacancies (i.e., the $Z_{1/2}$ and EH6/7 centers) is effective for improving the carrier lifetimes in p-type SiC, but the level of improvement is much smaller than that in n-type SiC. Further investigations are required to identify the lifetime killer and enhance the carrier lifetime in p-type SiC.

The effects of extended defects on carrier lifetimes have also been investigated [68, 204, 205]. All the extended defects cause *local reduction of the carrier lifetimes* near the defects. This is already indicated in the PL mapping data shown in Figure 5.26, where the PL intensities of the free exciton peaks are greatly reduced near the dislocations. The magnitude of the reduction in PL intensity is much larger near TSDs than TEDs. Figure 5.58 shows the PL decay curves that were measured from a 180-μm-thick n-type 4H-SiC(0001) epitaxial layer at room temperature. The measurements were conducted at a defect-free region, near a TSD core, a TED core, and inside a single SSF expanded from a BPD. The carrier lifetime obtained from the PL decay was 4.6 μs from the defect-free region, 1.8 μs near the TED, 1.2 μs near the TSD, and 0.3 μs for the SSF. Thus, the stacking faults (nucleated from BPDs) and the TSDs are the most detrimental defects for the carrier lifetimes in SiC. The effects of these extended defects depend strongly on the crystal quality, and are relatively larger in high-quality SiC with long carrier lifetimes.

5.3.2.1 Lifetime Control

Because the density of the $Z_{1/2}$ (and the EH6/7) center can be intentionally increased by low-energy electron irradiation [158, 165], lifetime control is easily achieved in n-type SiC. Although several deep levels, which may be ascribed to the interstitial-related defects, are also generated by electron irradiation,

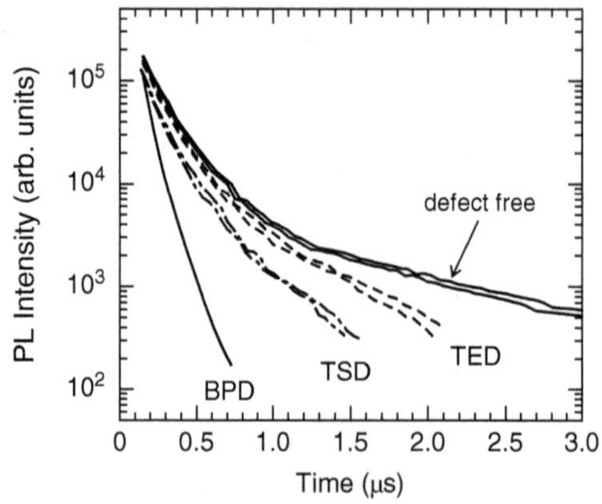

Figure 5.58 PL decay curves that were measured from a 180-μm-thick n-type 4H-SiC(0001) epitaxial layer at room temperature. The measurements were conducted at a defect-free region, near a TSD core, a TED core, and inside a single SSF expanded from a BPD.

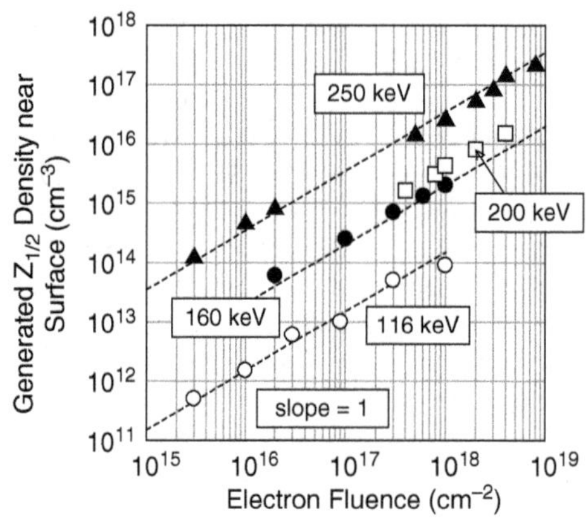

Figure 5.59 Density of the $Z_{1/2}$ center generated by electron irradiation (after 950 °C annealing) versus the electron fluence [158, 163]. The electron irradiation was carried out at energies of 116, 160, 200, or 250 keV.

any deep levels other than the $Z_{1/2}$ and EH6/7 centers are eliminated by annealing at 900–1000 °C [159, 165]. Figure 5.59 shows the density of the $Z_{1/2}$ center generated by electron irradiation (after 950 °C annealing) versus the electron fluence [158, 163]. The electron irradiation was carried out at energies of 116, 160, 200, or 250 keV. In either case, the $Z_{1/2}$ center density increases almost in proportion to the electron fluence. By increasing the fluence from 3×10^{15} to 1×10^{19} cm^{-2} along with the irradiation energy, the $Z_{1/2}$ center density can be controlled over a very wide range from approximately 1×10^{12} to 3×10^{17} cm^{-3}.

Figure 5.60 shows a map of the carrier lifetimes for n-type 4H-SiC epitaxial layers after electron irradiation and subsequent annealing in Ar at 950 °C for 30 min. The starting material was a 70-μm-thick 4H-SiC epitaxial layer, grown on an n-type substrate and doped to 1×10^{15} cm^{-3}. Electron irradiation was performed at 160 keV with the fluence range from 1×10^{15} to 1.5×10^{17} cm^{-2} on the sample

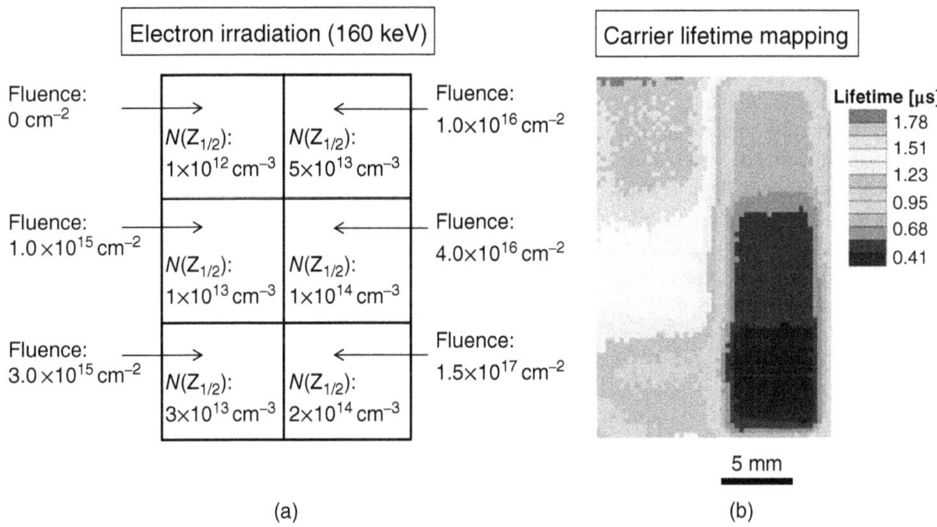

Figure 5.60 (a) Electron fluence and resulting $Z_{1/2}$ center density in a 70-μm-thick n-type 4H-SiC epitaxial layer doped to 1×10^{15} cm^{-3} and (b) map of the carrier lifetimes for the selectively irradiated 4H-SiC epitaxial layers after electron irradiation and subsequent annealing in Ar at 950 °C for 30 min.

being controlled by using a copper mask during irradiation. Because selective electron irradiation was performed, different $Z_{1/2}$-center densities ranging from 1×10^{12} to 2×10^{14} cm^{-3} over six different areas were realized in the sample. The carrier lifetime is shorter in the area where the electron fluence is higher (i.e., the $Z_{1/2}$-center density is higher). More importantly, a highly uniform carrier lifetime can be realized in each area. These results indicate that the carrier lifetime can be controlled reasonably well by the electron irradiation process. This technique is useful for obtaining very uniform carrier lifetimes or for the reduction of the switching losses in SiC bipolar devices. Control of the carrier lifetimes by similar low-energy electron irradiation is also possible for p-type 4H-SiC [203], suggesting that a carbon vacancy (e.g., the $Z_{1/2}$ or EH6/7 centers) can be a lifetime killer when its density is sufficiently high ($>3 \times 10^{13}$ cm^{-3}).

5.4 Summary

Basic characterization techniques have been described in this chapter with emphasis placed on the special care that is required for the characterization of SiC. PL is a powerful technique for characterization of optical properties, and can reflect the incorporation of specific impurities (including dopants), point defects, and any other localized levels. Raman scattering is useful for the identification of the SiC polytypes and for characterization of strains. C–V measurements give the net doping density ($N_D - N_A$ or $N_A - N_D$), while the carrier density (n or p) can be measured directly only by Hall effect measurements. The carrier lifetimes can be determined by several techniques, but the decay time obtained does not always mean the carrier lifetime. Special attention must also be paid to the effects of surface recombination and recombination in the underlying substrate.

Extended defects have been detected via destructive methods such as molten KOH etching. High-resolution X-ray topography and PL mapping/imaging serve as attractive techniques for non-destructive identification of the locations and types of extended defects. Deep levels, which originate from point defects or impurities, can be characterized by DLTS, and high-temperature (up to 800 K) measurements are required to monitor the midgap levels in SiC. To identify the origin of a deep level, a detailed comparison study of the results of DLTS, EPR measurements, and theoretical calculations is required.

Because micropipe defects have been eliminated, the macroscopic defects that are generated during epitaxial growth, such as triangular defects, carrot defects, in-grown stacking faults, and particles, are the most detrimental defects for any SiC device. A BPD is split into two partial dislocations with a single SSF in between them. This stacking fault is then expanded upon carrier injection and recombination, leading to a considerable reduction of the carrier lifetimes and increased leakage current. Therefore, BPDs cause the degradation of bipolar-type SiC devices (but not unipolar devices). The effects of TSDs on the breakdown voltage and leakage current are very small when no surface pits are formed at the TSD locations. The effects of TEDs are mostly negligibly small. The major carrier lifetime killer has been identified as the $Z_{1/2}$ center ($E_C - 0.63$ eV), which is the acceptor level of a carbon vacancy, in n-type 4H-SiC. In p-type 4H-SiC, however, the carrier-lifetime killers have not yet been conclusively identified. TSDs and TEDs cause local reduction of the carrier lifetimes, while BPDs induce expansion of the SSFs upon carrier injection, leading to significant reduction of the carrier lifetimes in the fault region. However, further investigations are required to fully clarify and understand the behavior of the extended and point defects in SiC.

The present chapter did not describe the characterization of SiC by other techniques, such as standard X-ray diffraction, TEM, secondary ion mass spectrometry (SIMS), X-ray photoelectron spectroscopy (XPS), and Auger electron spectroscopy (AES). These techniques are frequently used for material characterization but special care and unique analyses are not always required for SiC. Please see individual textbooks for details of these techniques.

References

[1] Schroder, D.K. (2006) *Semiconductor Material and Device Characterization*, 3rd edn, Wiley-IEEE.

[2] Perkowitz, S. (1993) *Optical Characterization of Semiconductors*, Academic Press.

[3] Choyke, W.J. (1969) Optical properties of polytypes of SiC: Interband absorption, and luminescence of nitrogen-exciton complexes. *Mater. Res. Bull.*, **4**, 141.

[4] Devaty, R.P. and Choyke, W.J. (1997) Optical characterization of silicon carbide polytypes. *Phys. Status Solidi A*, **162**, 5.

[5] Choyke, W.J. and Devaty, R.P. (2004) Optical properties of SiC: 1997–2002, in *Silicon Carbide – Recent Major Advances* (eds W.J. Choyke, H. Matsunami and G. Pensl), Springer, p. 413.

[6] Egilsson, T., Ivanov, I.G., Son, N.T. *et al.* (2004) Exciton and defect photoluminescence from SiC, in *Silicon Carbide, Materials, Processing, and Devices* (eds Z.C. Feng and J.H. Zhao), Taylor & Francis Group, p. 81.

[7] Janzen, E., Gali, A., Henry, A. *et al.* (2008) Defects in SiC, in *Defects in Microelectronic Materials and Devices*, Taylor & Francis Group, p. 615.

[8] Ivanov, I.G., Zhang, J., Storasta, L. and Janzen, E. (2002) Photoconductivity of lightly-doped and semi-insulating 4H-SiC and the free exciton binding energy. *Mater. Sci. Forum*, **389–393**, 613.

[9] Ivanov, I.G., Hallin, C., Henry, A. *et al.* (1996) Nitrogen doping concentration as determined by photoluminescence in 4H- and 6H-SiC. *J. Appl. Phys.*, **80**, 3504.

[10] Henry, A., Forsberg, U., Linnarsson, M.K. and Janzen, E. (2005) Determination of nitrogen doping concentration in doped 4H-SiC epilayers by low temperature photoluminescence. *Phys. Scr.*, **72**, 254.

[11] Bebb, H.B. and Williams, E.W. (1972) Photoluminescence, in *Semiconductors and Semimetals*, vol. **8** (eds R.K. Willardson and A.C. Beer), Academic Press.

[12] Itoh, A., Kimoto, T. and Matsunami, H. (1996) Exciton-related photoluminescence in 4H-SiC grown by step-controlled epitaxy. *Jpn. J. Appl. Phys.*, **35**, 4373.

[13] Choyke, W.J., Hamilton, D.R. and Patrick, L. (1964) Optical properties of cubic SiC: luminescence of nitrogen-exciton complexes, and interband absorption. *Phys. Rev.*, **133**, A1163.

[14] Patrick, L., Choyke, W.J. and Hamilton, D.R. (1965) Luminescence of 4H SiC, and location of conduction-band minima in SiC polytypes. *Phys. Rev.*, **137**, A1515.

[15] Choyke, W.J. and Patrick, L. (1962) Exciton recombination radiation and phonon spectrum of 6H SiC. *Phys. Rev.*, **127**, 1868.

[16] Kimoto, T., Nakazawa, S., Hashimoto, K. and Matsunami, H. (2001) Reduction of doping and trap concentrations in 4H-SiC epitaxial layers grown by chemical vapor deposition. *Appl. Phys. Lett.*, **79**, 2761.
[17] Henry, A. and Janzen, E. (2006) Photoluminescence of phosphorous doped SiC. *Mater. Sci. Forum*, **527–529**, 589.
[18] Clemen, L.L., Devaty, R.P., MacMillan, M.F. *et al.* (1993) Aluminum acceptor four particle bound exciton complex in 4H, 6H, and 3C SiC. *Appl. Phys. Lett.*, **62**, 2953.
[19] Pedersen, H., Henry, A., Hassan, J. *et al.* (2007) Growth and photoluminescence study of aluminium doped SiC epitaxial layers. *Mater. Sci. Forum*, **556–557**, 97.
[20] Sridhara, S.G., Clemen, L.L., Devaty, R.P. *et al.* (1998) Photoluminescence and transport studies of boron in 4H SiC. *J. Appl. Phys.*, **83**, 7909.
[21] Suzuki, A., Matsunami, H. and Tanaka, T. (1977) Photoluminescence spectra of Ga-doped and Al-doped 4H-SiC. *J. Phys. Chem. Solids*, **38**, 693.
[22] Choyke, W.J. and Patrick, L. (1970) Luminescence of donor-acceptor pairs in cubic SiC. *Phys. Rev. B*, **2**, 4959.
[23] Choyke, W.J. (1990) Optical and electronic properties of SiC, in *The Physics and Chemistry of Carbides, Nitrides, and Borides* (ed R. Freer), Kluwer Academic Publishers, p. 563.
[24] Ikeda, M., Matsunami, H. and Tanaka, T. (1980) Site effect on the impurity levels in 4H, 6H, and 15R SiC. *Phys. Rev. B*, **22**, 2842.
[25] Yamada, S. and Kuwabara, S. (1974) Photoluminescence of β-SiC doped with boron and nitrogen, in *Silicon Carbide 1973* (eds R.C. Marshall, J.W. Faust Jr., and C.E. Ryan), University of South Carolina Press, p. 305.
[26] Ivanov, I.G., Henry, A. and Janzen, E. (2005) Ionization energies of phosphorus and nitrogen donors and aluminum acceptors in 4H silicon carbide from the donor-acceptor pair emission. *Phys. Rev. B*, **71**, 241201(R).
[27] van Kemenade, A.W.C. and Hagen, S.H. (1974) Proof of the involvement of Ti in the low-temperature abc luminescence spectrum of 6H SiC. *Solid State Commun.*, **14**, 1331.
[28] Lee, K.M., Dang, L.S., Watkins, G.D. and Choyke, W.J. (1985) Optically detected magnetic resonance study of SiC: Ti. *Phys. Rev. B*, **32**, 2273.
[29] Dalibor, T., Pensl, G., Nordell, N. and Schöner, A. (1997) Electrical properties of the titanium acceptor in silicon carbide. *Phys. Rev. B*, **55**, 13618.
[30] Henry, A. and Janzen, E. (2006) Titanium related luminescence in SiC. *Superlattices Microstruct.*, **40**, 328.
[31] Schneider, J., Muller, H.D., Maier, K. and Fuchs, F. (1990) Infrared spectra and electron spin resonance of vanadium deep level impurities in silicon carbide. *Appl. Phys. Lett.*, **56**, 1184.
[32] Lauer, V., Bremond, G., Souifi, A. *et al.* (1999) Electrical and optical characterisation of vanadium in 4H and 6H-SiC. *Mater. Sci. Eng., B*, **61–62**, 248.
[33] Prezzi, D., Eberlein, T.A.G., Filhol, J.-S. *et al.* (2004) Optical and electrical properties of vanadium and erbium in 4H-SiC. *Phys. Rev. B*, **69**, 193202.
[34] Kunzer, M., Dombrowski, K.F., Fuchs, F. *et al.* (1996) Identification of optically and electrically active molybdenum trace impurities in 6H-SiC substrates. *Inst. Phys. Conf. Ser.*, **142**, 385.
[35] Son, N.T. (1999) Photoluminescence and Zeeman effect in chromium-doped 4H and 6H SiC. *J. Appl. Phys.*, **86**, 4348.
[36] Gällström, A., Magnusson, B., Beyer, F. *et al.* (2012) Optical identification and electronic configuration of tungsten in 4H- and 6H-SiC. *Physica B*, **407**, 1462.
[37] Magnusson, B. and Janzen, E. (2005) Optical characterization of deep level defects in SiC. *Mater. Sci. Forum*, **483–485**, 341.
[38] Wagner, M., Magnusson, B., Chen, W.M. *et al.* (2000) Electronic structure of the neutral silicon vacancy in 4H and 6H SiC. *Phys. Rev. B*, **62**, 16555.
[39] Janzen, E., Ivanov, I.G., Son, N.T. *et al.* (2003) Defects in SiC. *Physica B*, **340–342**, 15.
[40] Son, N.T., Carlsson, P., ul Hassan, J. *et al.* (2007) Defects and carrier compensation in semi-insulating 4H-SiC substrates. *Phys. Rev. B*, **75**, 155204.
[41] Egilsson, T., Henry, A., Ivanov, I.G. *et al.* (1999) Photoluminescence of electron-irradiated 4H-SiC. *Phys. Rev. B*, **59**, 8008.
[42] Steeds, J.W., Evans, G.A., Danks, L.R. *et al.* (2002) Transmission electron microscope radiation damage of 4H and 6H SiC studied by photoluminescence spectroscopy. *Diamond Relat. Mater.*, **11**, 1923.
[43] Steeds, J.W. and Sullivan, W. (2008) Identification of antisite carbon split-interstitial defects in 4H-SiC. *Phys. Rev. B*, **77**, 195204.
[44] Mattausch, A., Bockstedte, M., Pankratov, O. *et al.* (2006) Thermally stable carbon-related centers in 6H-SiC: Photoluminescence spectra and microscopic models. *Phys. Rev. B*, **73**, 161201(R).

[45] Patrick, L. and Choyke, W.J. (1972) Photoluminescence of radiation defects in ion-implanted 6H SiC. *Phys. Rev. B*, **5**, 3253.
[46] Haberstroh, C., Helbig, R. and Stein, R.A. (1994) Some new features of the photoluminescence of SiC(6H), SiC(4H), and SiC(15R). *J. Appl. Phys.*, **76**, 509.
[47] Egilsson, T., Bergman, J.P., Ivanov, I.G. *et al.* (1999) Properties of the D_1 bound exciton in 4H-SiC. *Phys. Rev. B*, **59**, 1956.
[48] Hori, T., Danno, K. and Kimoto, T. (2007) Fast homoepitaxial growth of 4H-SiC with low basal-plane dislocation density and low trap concentration by hot-wall chemical vapor deposition. *J. Cryst. Growth*, **306**, 297.
[49] T. Dalibor, C. Peppermuller, G. Pensl, *et al.*, *Inst. Phys. Conf. Ser.* **142** 1996), 517 Defect centers in ion-implanted 4H silicon carbide.
[50] Storasta, L., Carlsson, F., Sridhara, S.G. *et al.* (2001) Pseudodonor nature of the D_I defect in 4H-SiC. *Appl. Phys. Lett.*, **78**, 46.
[51] Gali, A., Deak, P., Rauls, E. *et al.* (2003) Correlation between the antisite pair and the D_I center in SiC. *Phys. Rev. B*, **67**, 155203.
[52] Eberlein, T.A.G., Jones, R., Öberg, S. and Briddon, P.R. (2006) Density functional theory calculation of the D_I optical center in SiC. *Phys. Rev. B*, **74**, 144106.
[53] Patrick, L. and Choyke, W.J. (1973) Localized vibrational modes of a persistent defect in ion-implanted SiC. *J. Phys. Chem. Solids*, **34**, 565.
[54] Sridhara, S.G., Nizhner, D.G., Devaty, R.P. *et al.* (1998) D_{II} revisited in a modern guise – 6H and 4H SiC. *Mater. Sci. Forum*, **264–268**, 493.
[55] Mattausch, A., Bockstedte, M. and Pankratov, O. (2004) Carbon antisite clusters in SiC: A possible pathway to the D_{II} center. *Phys. Rev. B*, **69**, 045322.
[56] Feldman, D.W., Parker, J.H. Jr.,, Choyke, W.J. and Patrick, L. (1968) Phonon dispersion curves by Raman scattering in SiC, polytypes 3C, 4H, 6H, 15R, and 21R. *Phys. Rev.*, **170**, 698.
[57] Dubrovskii, G.B. and Lepneva, A.A. (1983) Energy band structure and optical spectra of silicon carbide crystals. *Sov. Phys. Solid State*, **25**, 1330.
[58] Nakashima, S. and Harima, H. (1997) Raman investigation of SiC polytypes. *Phys. Status Solidi A*, **162**, 39.
[59] Nakashima, S. and Harima, H. (2004) Characterization of defects in SiC crystals by Raman scattering, in *Silicon Carbide – Recent Major Advances* (eds W.J. Choyke, H. Matsunami and G. Pensl), Springer, p. 585.
[60] Nakashima, S. (2004) Raman imaging of semiconductor materials: characterization of static and dynamic properties. *J. Phys. Condens. Matter*, **16**, S25.
[61] Nakashima, S., Okumura, H., Yamamoto, T. and Shimidzu, R. (2004) Deep-ultraviolet Raman microspectroscopy: characterization of wide-gap semiconductors. *Appl. Spectrosc.*, **58**, 224.
[62] Harima, H., Nakashima, S. and Uemura, T. (1995) Raman scattering from anisotropic LO-phonon-plasmon–coupled mode in n-type 4H- and 6H-SiC. *J. Appl. Phys.*, **78**, 1996.
[63] Lockwood, D.J. and Baribeau, J.-M. (1992) Strain-shift coefficients for phonons in $Si_{1-x}Ge_x$ epilayers on silicon. *Phys. Rev. B*, **45**, 8565.
[64] Hoffmann, H.J. (1979) Defect-level analysis of semiconductors by a new differential evaluation of n (1/T)-characteristics. *Appl. Phys.*, **19**, 307.
[65] Matsuura, H., Komeda, M., Kagamihara, S. *et al.* (2004) Dependence of acceptor levels and hole mobility on acceptor density and temperature in Al-doped p-type 4H-SiC epilayers. *J. Appl. Phys.*, **96**, 2708.
[66] Schmidt, F., Krieger, M., Laube, M. *et al.* (2004) Hall scattering factors for electrons and holes in SiC, in *Silicon Carbide – Recent Major Advances* (eds W.J. Choyke, H. Matsunami and G. Pensl), Springer, p. 517.
[67] Sze, S.M. and Ng, K.K. (2007) *Physics of Semiconductor Devices*, 3rd edn, John Wiley & Sons, Inc..
[68] Bergman, J.P., Kordina, O. and Janzen, E. (1997) Time resolved spectroscopy of defects in SiC. *Phys. Status Solidi A*, **162**, 65.
[69] Klein, P.B. (2008) Carrier lifetime measurement in n^- 4H-SiC epilayer. *J. Appl. Phys.*, **103**, 033702.
[70] Kimoto, T., Hiyoshi, T., Hayashi, T. and Suda, J. (2010) Impacts of recombination at the surface and in the substrate on carrier lifetimes of n-type 4H-SiC epilayers. *J. Appl. Phys.*, **108**, 083721.
[71] Galeckas, A., Linnros, J., Frischholz, M. and Grivickas, V. (2001) Optical characterization of excess carrier lifetime and surface recombination in 4H/6H–SiC. *Appl. Phys. Lett.*, **79**, 365.
[72] Grivickas, P., Linnros, J. and Grivickas, V. (2001) Carrier diffusion characterization in epitaxial 4H–SiC. *J. Mater. Res.*, **16**, 524.
[73] Hull, D. and Bacon, D.J. (2001) *Introduction to Dislocations*, 4th edn, Butterworth-Heinemann.
[74] Katsuno, M., Ohtani, N., Takahashi, J. *et al.* (1999) Mechanism of molten KOH etching of SiC single crystals: comparative study with thermal oxidation. *Jpn. J. Appl. Phys.*, **38**, 4661.

[75] Koga, K., Fujiwara, Y., Ueda, Y. and Yamaguchi, T. (1992) Growth and characterization of 6H-SiC bulk crystals by the sublimation method, in *Amorphous and Crystalline Silicon Carbide IV* (eds C.Y. Yang, M.M. Rahman and G.L. Harris), Springer-Verlag, p. 96.
[76] Takahashi, J., Kanaya, M. and Fujiwara, Y. (1994) Sublimation growth of SiC single crystalline ingots on faces perpendicular to the (0001) basal plane. *J. Cryst. Growth*, **135**, 61.
[77] Yao, Y.Z., Ishikawa, Y., Sugawara, Y. *et al.* (2011) Molten KOH etching with Na_2O_2 additive for dislocation revelation in 4H-SiC epilayers and substrates. *Jpn. J. Appl. Phys.*, **50**, 075502.
[78] Powell, J.A., Petit, J.B., Edgar, J.H. *et al.* (1991) Application of oxidation to the structural characterization of SiC epitaxial films. *Appl. Phys. Lett.*, **59**, 183.
[79] Nakano, Y., Nakamura, T., Kamisawa, A. and Takasu, H. (2009) Investigation of pits formed at oxidation on 4H-SiC. *Mater. Sci. Forum*, **600–603**, 377.
[80] Suda, J., Shoji, H. and Kimoto, T. (2011) Origin of etch hillocks formed on on-aixs SiC $(000\bar{1})$ surfaces by molten KOH etching. *Jpn. J. Appl. Phys.*, **50**, 038002.
[81] Yao, Y.Z., Ishikawa, Y., Sato, K. *et al.* (2012) Dislocation revelation from $(000\bar{1})$ carbon-face of 4H-SiC by using vaporized KOH at high temperature. *Appl. Phys. Express*, **5**, 075601.
[82] Bowen, D.K. and Tanner, B.K. (1998) *High-Resolution X-Ray Diffractometry and Topography*, Taylor & Francis Group.
[83] Ohno, T., Yamaguchi, H., Kuroda, S. *et al.* (2004) Direct observation of dislocations propagated from 4H–SiC substrate to epitaxial layer by X-ray topography. *J. Cryst. Growth*, **260**, 209.
[84] Huang, X.R., Dudley, M., Vetter, W.M. *et al.* (1999) Direct evidence of micropipe-related pure superscrew dislocations in SiC. *Appl. Phys. Lett.*, **74**, 353.
[85] Tsuchida, H., Kamata, I. and Nagano, M. (2007) Investigation of defect formation in 4H-SiC epitaxial growth by X-ray topography and defect selective etching. *J. Cryst. Growth*, **306**, 254.
[86] Tsuchida, H., Ito, M., Kamata, I. and Nagano, M. (2009) Fast epitaxial growth of 4H-SiC and analysis of defect transfer. *Mater. Sci. Forum*, **615–617**, 67.
[87] Jacobson, H., Birch, J., Yakimova, R. *et al.* (2002) Dislocation evolution in 4H-SiC epitaxial layers. *J. Appl. Phys.*, **91**, 6354.
[88] Kamata, I., Nagano, M., Tsuchida, H. *et al.* (2009) Investigation of character and spatial distribution of threading edge dislocations in 4H-SiC epilayers by high-resolution topography. *J. Cryst. Growth*, **311**, 1416.
[89] Nakamura, D., Yamaguchi, S., Hirose, Y. *et al.* (2008) Direct determination of Burgers vector sense and magnitude of elementary dislocations by synchrotron white x-ray topography. *J. Appl. Phys.*, **103**, 013510.
[90] Tanuma, R., Mori, D., Kamata, I. and Tsuchida, H. (2012) X-ray microbeam three-dimensional topography imaging and strain analysis of basal-plane dislocations and threading edge dislocations in 4H-SiC. *Appl. Phys. Express*, **5**, 061301.
[91] Tajima, M., Tanaka, M. and Hoshino, N. (2002) Characterization of SiC epitaxial wafers by photoluminescence under deep UV excitation. *Mater. Sci. Forum*, **389–393**, 597.
[92] Kamata, I., Tsuchida, H., Miyanagi, T. and Nakamura, T. (2005) Development of non-destructive in-house observation techniques for dislocations and stacking faults in SiC epilayers. *Mater. Sci. Forum*, **527–529**, 415.
[93] Camassel, J. and Juillaguet, S. (2008) Optical properties of as-grown and process-induced stacking faults in 4H-SiC. *Phys. Status Solidi B*, **245**, 1337.
[94] Feng, G., Suda, J. and Kimoto, T. (2008) Characterization of stacking faults in 4H-SiC epilayers by room-temperature microphotoluminescence mapping. *Appl. Phys. Lett.*, **92**, 221906.
[95] Ito, M., Storasta, L. and Tsuchida, H. (2009) Development of a high rate 4H-SiC epitaxial growth technique achieving large-area uniformity. *Mater. Sci. Forum*, **600–603**, 111.
[96] Feng, G., Suda, J. and Kimoto, T. (2011) Nonradiative recombination at threading dislocations in 4H-SiC epilayers studied by micro-photoluminescence mapping. *J. Appl. Phys.*, **110**, 033525.
[97] Isono, H., Tajima, M., Hoshino, N. and Sugimoto, H. (2009) Rapid characterization of SiC crystals by full-wafer photoluminescence imaging under below-gap excitation. *Mater. Sci. Forum*, **600–603**, 545.
[98] Stahlbush, R.E., Liu, K.X., Zhang, Q. and Sumakeris, J.J. (2007) Whole-wafer mapping of dislocations in 4H-SiC epitaxy. *Mater. Sci. Forum*, **556–557**, 295.
[99] Nagano, M., Kamata, I. and Tsuchida, T. (2013) Photoluminescence imaging and discrimination of threading dislocations in 4H-SiC epilayers. *Mater. Sci. Forum*, **740–742**, 653.
[100] Ichikawa, S., Kawahara, K., Suda, J. and Kimoto, T. (2012) Carrier recombination in n-type 4H-SiC epilayers with long carrier lifetimes. *Appl. Phys. Express*, **5**, 101301.
[101] Chung, S., Berechman, R.A., McCartney, M.R. and Skowronski, M. (2011) Electronic structure analysis of threading screw dislocations in 4H–SiC using electron holography. *J. Appl. Phys.*, **109**, 034906.

[102] Liu, K.X., Zhang, X., Stahlbush, R.E. *et al.* (2009) Differences in emission spectra of dislocations in 4H-SiC epitaxial layers. *Mater. Sci. Forum*, **600–603**, 345.

[103] Caldwell, J.D., Giles, A., Lepage, D. *et al.* (2013) Experimental evidence for mobile luminescence center mobility on partial dislocations in 4H-SiC using hyperspectral electroluminescence imaging. *Appl. Phys. Lett.*, **102**, 242109.

[104] Kitabatake, M. (2013) The International Conference on Silicon Carbide and Related Materials 2013, Miyazaki, Japan, We-1A-1. Electrical characteristics/reliability affected by defects analysed by the integrated evaluation platform for SiC epitaxial films

[105] Milnes, A.G. (1973) *Deep Impurities in Semiconductors*, John Wiley & Sons, Inc.

[106] Lang, D.V. (1974) Deep-level transient spectroscopy: A new method to characterize traps in semiconductors. *J. Appl. Phys.*, **45**, 3023.

[107] Weiss, S. and Kassing, R. (1988) Deep Level Transient Fourier Spectroscopy (DLTFS) - A technique for the analysis of deep level properties. *Solid State Electron.*, **31**, 1733.

[108] Dobaczewski, L., Kaczor, P., Hawkins, I.D. and Peaker, A.R. (1994) Laplace transform deep-level transient spectroscopic studies of defects in semiconductors. *J. Appl. Phys.*, **76**, 194.

[109] Okushi, H. and Tokumaru, Y. (1981) Isothermal capacitance transient spectroscopy. *Jpn. J. Appl. Phys.*, **20** (Suppl. 20-1), 261.

[110] Lefevre, H. and Shultz, M. (1977) Double correlation technique (DDLTS) for the analysis of deep level profiles in semiconductors. *Appl. Phys.*, **12**, 45.

[111] Reshanov, S.A., Pensl, G., Danno, K. *et al.* (2007) Effect of the Schottky barrier height on the detection of midgap levels in 4H-SiC by deep level transient spectroscopy. *J. Appl. Phys.*, **102**, 113702.

[112] Weil, J.A. and Bolton, J.R. (2007) *Electron Paramagnetic Resonance: Elementary Theory and Practical Applications*, Wiley-Interscience.

[113] Umeda, T., Isoya, J., Morishita, N. *et al.* (2004) EPR and theoretical studies of positively charged carbon vacancy in 4H-SiC. *Phys. Rev. B*, **70**, 235212.

[114] Itoh, H., Hayakawa, N., Nashiyama, I. and Sakuma, E. (1989) Electron spin resonance in electron-irradiated 3C-SiC. *J. Appl. Phys.*, **66**, 4529.

[115] Son, N.T., Carlsson, P., Hassan, J. *et al.* (2006) Divacancy in 4H-SiC. *Phys. Rev. Lett.*, **96**, 055501.

[116] Umeda, T., Isoya, J., Morishita, N. *et al.* (2009) Dicarbon antisite defect in n-type 4H-SiC. *Phys. Rev. B*, **79**, 115211.

[117] Isoya, J., Umeda, T., Mizuochi, N. *et al.* (2008) EPR identification of intrinsic defects in SiC. *Phys. Status Solidi B*, **245**, 1298.

[118] Umeda, T., Kosugi, R., Sakuma, Y. *et al.* (2012) MOS interface states: Similarity and dissimilarity from silicon. *ECS Trans.*, **50**, 305.

[119] Lendenmann, H., Dahlquist, F., Johansson, N. *et al.* (2001) Long term operation of 4.5 kV PiN and 2.5 kV JBS diodes. *Mater. Sci. Forum*, **353–356**, 727.

[120] Bergman, J.P., Lendenmann, H., Nilsson, P.A. *et al.* (2001) Crystal defects as source of anomalous forward voltage increase of 4H-SiC diodes. *Mater. Sci. Forum*, **353–356**, 299.

[121] Lendenmann, H., Bergman, J.P., Dahlquist, F. and Hallin, C. (2003) Degradation in SiC bipolar devices: sources and consequences of electrically active dislocations in SiC. *Mater. Sci. Forum*, **433–436**, 901.

[122] Muzykov, P.G., Kennedy, R.M., Zhang, Q. *et al.* (2009) Physical phenomena affecting performance and reliability of 4H-SiC bipolar junction transistors. *Microelectron. Reliab.*, **49**, 32.

[123] Skowronski, M. and Ha, S. (2006) Degradation of hexagonal silicon-carbide-based bipolar devices. *J. Appl. Phys.*, **99**, 011101.

[124] Liu, J.Q., Skowronski, M., Hallin, C. *et al.* (2002) Structure of recombination-induced stacking faults in high-voltage SiC pn junctions. *Appl. Phys. Lett.*, **80**, 749.

[125] Pirouz, P., Demenet, J.L. and Hong, M.H. (2001) On transition temperatures in the plasticity and fracture of semiconductors. *Philos. Mag.*, **81**, 1207.

[126] Skowronski, M., Liu, J.Q., Vetter, W.M. *et al.* (2002) Recombination-enhanced defect motion in forward-biased 4H–SiC p-n diodes. *J. Appl. Phys.*, **92**, 4699.

[127] Ha, S., Benamara, M., Skowronski, M. and Lendenmann, H. (2003) Core structure and properties of partial dislocations in silicon carbide pin diodes. *Appl. Phys. Lett.*, **83**, 4957.

[128] Sridhara, S.G., Carlsson, F.H.C., Bergman, J.P. and Janzen, E. (2001) Luminescence from stacking faults in 4H SiC. *Appl. Phys. Lett.*, **79**, 3944.

[129] Iwata, H., Lindefelt, U., Oberg, S. and Briddon, P.R. (2001) Localized electronic states around stacking faults in silicon carbide. *Phys. Rev. B*, **65**, 033203.

[130] Miyanagi, T., Tsuchida, H., Kamata, I. *et al.* (2006) Annealing effects on single Shockley faults in 4H-SiC. *Appl. Phys. Lett.*, **89**, 062104.
[131] Caldwell, J.D., Stahlbush, R.E., Hobart, K.D. *et al.* (2007) Reversal of forward voltage drift in 4H-SiC pin diodes via low temperature annealing. *Appl. Phys. Lett.*, **90**, 143519.
[132] Ha, S., Skowronski, M. and Lendenmann, H. (2004) Nucleation sites of recombination-enhanced stacking fault formation in silicon carbide p-i-n diodes. *J. Appl. Phys.*, **96**, 393.
[133] Matare, H.F. (1971) *Defect Electronics in Semiconductors*, John Wiley & Sons, Inc.
[134] Kimoto, T., Miyamoto, N. and Matsunami, H. (1999) Performance limiting surface defects in SiC epitaxial p-n junction diodes. *IEEE Trans. Electron Devices*, **46**, 471.
[135] Lindefelt, U. and Iwata, H. (2004) Electronic properties of stacking faults and thin cubic inclusions in SiC polytypes, in *Silicon Carbide – Recent Major Advances* (eds W.J. Choyke, H. Matsunami and G. Pensl), Springer, p. 89.
[136] Fujiwara, H., Kimoto, T., Tojo, T. and Matsunami, H. (2005) Characterization of in-grown stacking faults in 4H–SiC (0001) epitaxial layers and its impacts on high-voltage Schottky barrier diodes. *Appl. Phys. Lett.*, **87**, 051912.
[137] Neudeck, P.G. and Powell, J.A. (1994) Performance limiting micropipe defects in silicon carbide wafers. *IEEE Electron Device Lett.*, **15**, 63.
[138] Neudeck, P.G., Huang, W. and Dudley, M. (1999) Study of bulk and elementary screw dislocation assisted reverse breakdown in low-voltage (< 250 V) 4H-SiC p^+-n junction diodes. Part I: DC properties. *IEEE Trans. Electron Devices*, **46**, 478.
[139] Morisette, D.T. and Cooper, J.A. Jr., (2002) Impact of material defects on SiC Schottky barrier diodes. *Mater. Sci. Forum*, **389–393**, 1133.
[140] Saitoh, H., Kimoto, T. and Matsunami, H. (2004) Origin of leakage current in SiC Schottky barrier diodes at high temperature. *Mater. Sci. Forum*, **457–460**, 997.
[141] Wahab, Q., Ellison, A., Henry, A. *et al.* (2000) Influence of epitaxial growth and substrate-induced defects on the breakdown of 4H–SiC Schottky diodes. *Appl. Phys. Lett.*, **76**, 2725.
[142] Tsuji, T., Izumi, S., Ueda, A. *et al.* (2002) Analysis of high leakage currents in 4H-SiC Schottky barrier diodes using optical beam-induced current measurements. *Mater. Sci. Forum*, **389–393**, 1141.
[143] Ohtani, N. (2011) Toward the reduction of performance-limiting defects in SiC epitaxial substrates. *ECS Trans.*, **41**, 253.
[144] Fujiwara, H., Katsuno, T., Ishikawa, T. *et al.* (2012) Relationship between threading dislocation and leakage current in 4H-SiC diodes. *Appl. Phys. Lett.*, **100**, 242102.
[145] Fujiwara, H., Naruoka, H., Konishi, M. *et al.* (2012) Impact of surface morphology above threading dislocations on leakage current in 4H-SiC diodes. *Appl. Phys. Lett.*, **101**, 042104.
[146] Maranowski, M.M. and Cooper, J.A. Jr., (1999) Time-dependent-dielectric-breakdown measurements of thermal oxides on n-type 6H-SiC. *IEEE Trans. Electron Devices*, **46**, 520.
[147] Senzaki, J., Kojima, K. and Fukuda, K. (2004) Effects of n-type 4H-SiC epitaxial wafer quality on reliability of thermal oxides. *Appl. Phys. Lett.*, **85**, 6182.
[148] Senzaki, J., Kojima, K., Kato, T. *et al.* (2006) Correlation between reliability of thermal oxides and dislocations in n-type 4H-SiC epitaxial wafers. *Appl. Phys. Lett.*, **89**, 022909.
[149] Tanimoto, S. (2006) Impact of dislocations on gate oxide in SiC MOS devices and high reliability ONO dielectrics. *Mater. Sci. Forum*, **527–529**, 955.
[150] Senzaki, J., Shimozato, A., Kojima, K. *et al.* (2012) Challenges of high-performance and high-reliablity in SiC MOS structures. *Mater. Sci. Forum*, **717–720**, 703.
[151] Sze, S.M. (2002) *Semiconductor Devices, Physics and Technology*, 2nd edn, John Wiley & Sons, Inc.
[152] Dalibor, T., Pensl, G., Matsunami, H. *et al.* (1997) Deep defect centers in silicon carbide monitored with deep level transient spectroscopy. *Phys. Status Solidi A*, **162**, 199.
[153] Hemmingsson, C., Son, N.T., Kordina, O. *et al.* (1997) Deep level defects in electron-irradiated 4H SiC epitaxial layers. *J. Appl. Phys.*, **81**, 6155.
[154] Danno, K. and Kimoto, T. (2007) Deep level transient spectroscopy on as-grown and electron-irradiated p-type 4H-SiC epilayers. *J. Appl. Phys.*, **101**, 103704.
[155] Alfieri, G. and Kimoto, T. (2013) Resolving the $EH_{6/7}$ level in 4H-SiC by Laplace-transform deep level transient spectroscopy. *Appl. Phys. Lett.*, **102**, 152108.
[156] Hemmingsson, C.G., Son, N.T., Ellison, A. *et al.* (1998) Negative-U centers in 4H silicon carbide. *Phys. Rev. B*, **58**, R10119.
[157] Grundmann, M. (2006) *The Physics of Semiconductors* Chapter 7, Springer.

[158] Danno, K. and Kimoto, T. (2006) Investigation of deep levels in n-type 4H-SiC epilayers irradiated with low-energy electrons. *J. Appl. Phys.*, **100**, 113728.
[159] Sasaki, S., Kawahara, K., Feng, G. *et al.* (2011) Major deep levels with the same microstructures observed in n-type 4H-SiC and 6H-SiC. *J. Appl. Phys.*, **109**, 013705.
[160] Hornos, T., Gali, A. and Svensson, B.G. (2011) Negative-U system of carbon vacancy in 4H-SiC. *Mater. Sci. Forum*, **679–680**, 261.
[161] Son, N.T., Trinh, X.T., Løvlie, L.S. *et al.* (2012) Large-scale electronic structure calculations of vacancies in 4H-SiC using the Heyd-Scuseria-Ernzerhof screened hybrid density functional. *Phys. Rev. Lett.*, **109**, 187603.
[162] Kawahara, K., Trinh, X.T., Son, N.T. *et al.* (2013) Investigation on origin of $Z_{1/2}$ center in SiC by deep level transient spectroscopy and electron paramagnetic resonance. *Appl. Phys. Lett.*, **102**, 112106.
[163] Kawahara, K., Trinh, X.T., Son, N.T. *et al.* (2014) Quantitative comparison between $Z_{1/2}$ center and carbon vacancy in 4H-SiC. *J. Appl. Phys*, **115**, 143705.
[164] Alfieri, G. and Kimoto, T. (2009) Capacitance spectroscopy study of midgap levels in n-type SiC polytypes. *Mater. Sci. Forum*, **615–617**, 389.
[165] Storasta, L., Bergman, J.P., Janzen, E. *et al.* (2004) Deep levels created by low energy electron irradiation in 4H-SiC. *J. Appl. Phys.*, **96**, 4909.
[166] Kaneko, H. and Kimoto, T. (2011) Formation of a semi-insulating layer in n-type 4H-SiC by electron irradiation. *Appl. Phys. Lett.*, **98**, 262106.
[167] Zhang, J., Storasta, L., Bergman, J.P. *et al.* (2003) Electrically active defects in n-type 4H-silicon carbide grown in a vertical hot-wall reactor. *J. Appl. Phys.*, **93**, 4708.
[168] Kimoto, T., Hashimoto, K. and Matsunami, H. (2003) Effects of C/Si ratio in chemical vapor deposition of 4H-SiC($11\bar{2}0$) and ($03\bar{3}8$). *Jpn. J. Appl. Phys.*, **42**, 7294.
[169] Alfieri, G., Monakhov, E.V., Svensson, B.G. and Linnarsson, M.K. (2005) Annealing behavior between room temperature and 2000 °C of deep level defects in electron-irradiated n-type 4H silicon carbide. *J. Appl. Phys.*, **98**, 043518.
[170] Zippelius, B., Suda, J. and Kimoto, T. (2012) High temperature annealing of n-type 4H-SiC: Impact on intrinsic defects and carrier lifetime. *J. Appl. Phys.*, **111**, 033515.
[171] Klein, P.B., Shanabrook, B.V., Huh, S.W. *et al.* (2006) Lifetime-limiting defects in 4H-SiC epilayers. *Appl. Phys. Lett.*, **88**, 052110.
[172] Danno, K., Nakamura, D. and Kimoto, T. (2007) Investigation of carrier lifetime in 4H-SiC epilayers and lifetime control by electron irradiation. *Appl. Phys. Lett.*, **90**, 202109.
[173] Storasta, L. and Tsuchida, H. (2007) Reduction of traps and improvement of carrier lifetime in 4H-SiC epilayers by ion implantation. *Appl. Phys. Lett.*, **90**, 062116.
[174] Storasta, L., Tsuchida, H., Miyazawa, T. and Ohshima, T. (2008) Enhanced annealing of the $Z_{1/2}$ defect in 4H–SiC epilayers. *J. Appl. Phys.*, **103**, 013705.
[175] Dunham, S.T. and Plummer, J.D. (1986) Point-defect generation during oxidation of silicon in dry oxygen. I. Theory. *J. Appl. Phys.*, **59**, 2541.
[176] Hiyoshi, T. and Kimoto, T. (2009) Reduction of deep levels and improvement of carrier lifetime in n-type 4H-SiC by thermal oxidation. *Appl. Phys. Express*, **2**, 041101.
[177] Hiyoshi, T. and Kimoto, T. (2009) Elimination of the major deep levels in n- and p-type 4H-SiC by two-step thermal treatment. *Appl. Phys. Express*, **2**, 091101.
[178] Kawahara, K., Suda, J. and Kimoto, T. (2012) Analytical model for reduction of deep levels in SiC by thermal oxidation. *J. Appl. Phys.*, **111**, 053710.
[179] Bockstedte, M., Mattausch, A. and Pankratov, O. (2004) Ab initio study of the annealing of vacancies and interstitials in cubic SiC: Vacancy-interstitial recombination and aggregation of carbon interstitials. *Phys. Rev. B*, **69**, 235202.
[180] Hong, J.D., Davis, R.F. and Newbury, D.E. (1981) Self-diffusion of silicon-30 in α-SiC single crystals. *J. Mater. Sci.*, **16**, 2485.
[181] Kawahara, K., Suda, J. and Kimoto, T. (2013) Deep levels generated by thermal oxidation in p-type 4H-SiC. *J. Appl. Phys.*, **113**, 033705.
[182] Kawahara, K., Suda, J. and Kimoto, T. (2013) Deep levels generated by thermal oxidation in n-type 4H-SiC. *Appl. Phys. Express*, **6**, 051301.
[183] Danno, K., Hori, T. and Kimoto, T. (2007) Impacts of growth parameters on deep levels in n-type 4H-SiC. *J. Appl. Phys.*, **101**, 053709.
[184] Lebedev, A.A. (1999) Deep level centers in silicon carbide: A review. *Semiconductors*, **33**, 107.

[185] Zywietz, A., Furthmüller, J. and Bechstedt, F. (1999) Vacancies in SiC: Influence of Jahn-Teller distortions, spin effects, and crystal structure. *Phys. Rev. B*, **59**, 15166.

[186] Torpo, L., Marlo, M., Staab, T.E.M. and Nieminen, R.M. (2001) Comprehensive ab initio study of properties of monovacancies and antisites in 4H-SiC. *J. Phys. Condens. Matter*, **13**, 6203.

[187] Troffer, T., Schadt, M., Frank, T. *et al.* (1997) Doping of SiC by implantation of boron and aluminum. *Phys. Status Solidi A*, **162**, 277.

[188] Duijn-Arnold, A., Ikoma, T., Poluektov, O.G. *et al.* (1998) Doping of SiC by implantation of boron and aluminum. *Phys. Rev. B*, **57**, 1607.

[189] Negoro, Y., Kimoto, T. and Matsunami, H. (2005) Carrier compensation near tail region in aluminum- or boron-implanted 4H-SiC(0001). *J. Appl. Phys.*, **98**, 043709.

[190] Dalibor, T., Trageser, H., Pensl, G. *et al.* (1999) Oxygen in silicon carbide: Shallow donors and deep acceptors. *Mater. Sci. Eng., B*, **61–62**, 454.

[191] Klettke, O., Pensl, G., Kimoto, T. and Matsunami, H. (2001) Oxygen-related defect centers observed in 4H/6H-SiC epitaxial layers grown under CO_2 ambient. *Mater. Sci. Forum*, **353–356**, 459.

[192] Langer, J.M. and Heinrich, H. (1985) Deep-level impurities: a possible guide to prediction of band-edge discontinuities in semiconductor heterojunctions. *Phys. Rev. Lett.*, **55**, 1414.

[193] Kunzer, M., Kaufmann, U., Maier, K. and Schneider, J. (1995) Magnetic circular dichroism and electron spin resonance of the A^- acceptor state of vanadium, V^{3+}, in 6H-SiC. *Mater. Sci. Eng., B*, **29**, 118.

[194] Achtziger, N. and Witthuhn, W. (2004) Radiotracer deep level transient spectroscopy, in *Silicon Carbide – Recent Major Advances* (eds W.J. Choyke, H. Matsunami and G. Pensl), Springer, p. 537.

[195] Beyer, F., Hemmingsson, C., Leone, S. *et al.* (2011) Deep levels in iron doped n- and p-type 4H-SiC. *J. Appl. Phys.*, **110**, 123701.

[196] Danno, K., Saitoh, H., Seki, A. *et al.* (2012) Diffusion of transition metals in 4H-SiC and trials of impurity gettering. *Appl. Phys. Express*, **5**, 031301.

[197] Tawara, T., Tsuchida, H. and Izumi, S. (2004) Evaluation of free carrier lifetime and deep levels of the thick 4H-SiC epilayers. *Mater. Sci. Forum*, **457–460**, 565.

[198] Kimoto, T., Danno, K. and Suda, J. (2008) Lifetime-killing defects in 4H-SiC epilayers and lifetime control by low-energy electron irradiation. *Phys. Status Solidi B*, **245**, 1327.

[199] Reshanov, S.A., Bartsch, W., Zippelius, B. and Pensl, G. (2009) Lifetime investigations of 4H-SiC pin power diodes. *Mater. Sci. Forum*, **615–617**, 699.

[200] Noborio, M., Suda, J., Beljakowa, S. *et al.* (2009) 4H-SiC MISFETs with nitrogen-containing insulators. *Phys. Status Solidi A*, **206**, 2374.

[201] Miyazawa, T., Ito, M. and Tsuchida, H. (2010) Evaluation of long carrier lifetimes in thick 4H silicon carbide epitaxial layers. *Appl. Phys. Lett.*, **97**, 202106.

[202] Hayashi, T., Asano, T., Suda, J. and Kimoto, T. (2011) Temperature and injection level dependencies and impact of thermal oxidation on carrier lifetimes in p-type and n-type 4H-SiC epilayers. *J. Appl. Phys.*, **109**, 114502.

[203] Hayashi, T., Asano, K., Suda, J. and Kimoto, T. (2012) Enhancement and control of carrier lifetimes in p-type 4H-SiC epilayers. *J. Appl. Phys.*, **112**, 064503.

[204] Maximenko, S.I., Freitas, J.A. Jr.,, Klein, P.B. *et al.* (2009) Cathodoluminescence study of the properties of stacking faults in 4H-SiC homoepitaxial layers. *Appl. Phys. Lett.*, **94**, 092101.

[205] Hassan, J. and Bergman, J.P. (2009) Influence of structural defects on carrier lifetime in 4H-SiC epitaxial layers: Optical lifetime mapping. *J. Appl. Phys.*, **105**, 123518.

6

Device Processing of Silicon Carbide

After epitaxial growth of SiC several processing steps are performed to fabricate electronic devices. Such processing steps include doping by ion implantation, etching, oxidation, and metallization. Figure 6.1 shows a schematic of a trench-type vertical metal-oxide-semiconductor field-effect transistor (MOSFET). The contribution of each individual process to the complete structure can be seen. The process flow in SiC device fabrication is similar to that in Si technology, but several unique technologies, with particular requirements, are also needed because of the unique physical and chemical properties of SiC.

6.1 Ion Implantation

Ion implantation is a key process in fabrication of almost all kinds of SiC devices. Wide-range doping control of both n- and p-type conductivity can be achieved by ion implantation. The major differences between ion implantation technologies for SiC and for Si are summarized as follows:

1. Because of the extremely low diffusion constants of dopants within SiC, selective doping by a diffusion process is not realistic. Diffusion of most implanted impurities during post-implantation annealing is negligibly small.
2. If the as-implanted lattice damage is of near-amorphous level, lattice recovery (that is, repairing this damage) is very difficult. Therefore, implantation at elevated temperature is often employed, especially when the implant dose is very high. On the other hand, implantation at room temperature is sufficient when the implant dose is not very high, as long as appropriate annealing is carried out.
3. Irrespective of the implant dose and the implantation temperature, post-implantation annealing at very high temperature (>1500–1600 °C) is required to attain lattice recovery and high electrical activation ratios. Such high-temperature annealing may cause incongruent evaporation of silicon and surface roughening.

One can find review papers and books on ion implantation into SiC [1–6]. In this section, common features observed in ion implantation of SiC are described.

Fundamentals of Silicon Carbide Technology: Growth, Characterization, Devices, and Applications, First Edition.
Tsunenobu Kimoto and James A. Cooper.

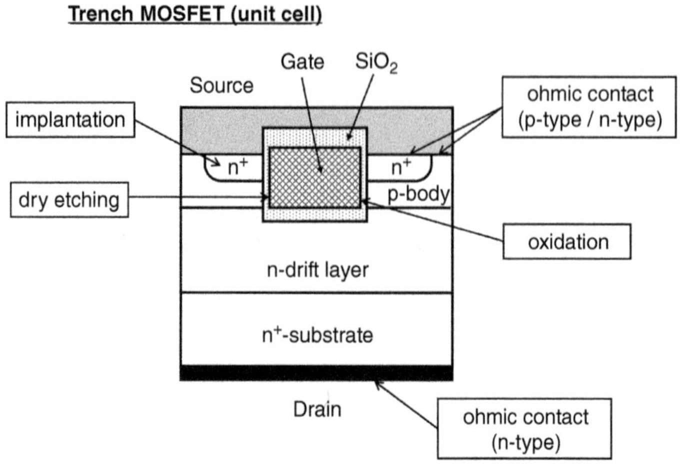

Figure 6.1 Schematic of a trench-type vertical metal-oxide-semiconductor field-effect transistor. Examples of device processing required to fabricate this device are shown.

6.1.1 Selective Doping Techniques

In Si and most III–V semiconductors both diffusion and ion implantation techniques are employed for selective doping. In general, a diffusion process introduces less lattice damage than ion implantation, in which bombardment with high-energy ions creates a variety of point and extended defects in the host material. Formation of deep junctions is easier in a diffusion process when the diffusion constant of a dopant is relatively large. On the other hand, a more flexible and accurate doping profile can be formed by ion implantation. The thermal budget is generally smaller in ion implantation because post-implantation annealing can be conducted in a rapid thermal process (RTP) [7].

Examples of selective doping in the fabrication of SiC devices are summarized in Table 6.1, where the typical doping densities and junction depths are indicated. In SiC, because of its very strong chemical bonding, diffusion constants of dopant impurities are extremely small, even above 1600 °C. Figure 6.2 shows the Arrhenius plot of diffusion constants for major dopants in SiC and Si [8, 9]. Diffusion constants for atoms other than boron are very small in SiC, ranging from 10^{-17} to 10^{-15} cm^2 s^{-1} at temperatures as high as 1800 °C. Extremely high temperature, above 2000 °C, is required to obtain reasonable diffusion constants. This fact makes diffusion processing of SiC unrealistic because such a high-temperature process causes generation of a high density of deep levels [10] and glide of basal-plane dislocations [11]. There exist no appropriate masking materials for a diffusion process at such high temperature. Therefore, selective doping of SiC has almost exclusively been performed using ion implantation. However,

Table 6.1 Examples of selective doping in fabrication of SiC devices.

Region	Depth (μm)	Doping density (cm^{-3})
Source/drain	0.2–0.3	10^{19}–10^{20}
p-body of FET	0.4–0.7	10^{17}–10^{18}
p$^+$ contact	0.2–0.3	10^{19}–10^{20}
Junction termination	0.4–0.8	10^{16}–10^{17}
Channel doping (optional)	0.1–0.2	10^{16}–10^{17}

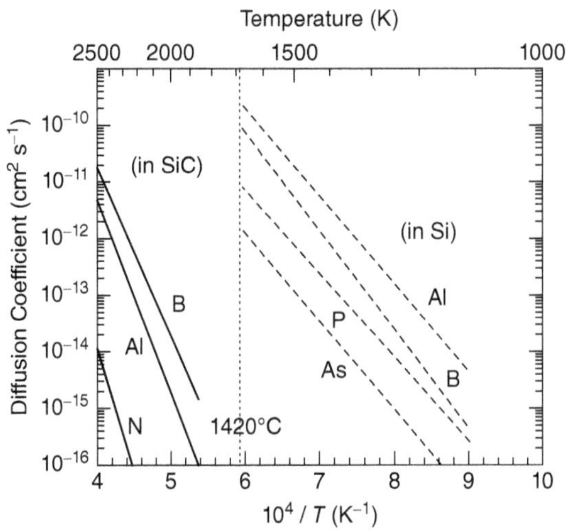

Figure 6.2 Arrhenius plot of diffusion constants for major dopants in SiC and Si [8, 9].

diffusion mechanisms of small impurities like hydrogen should be investigated [12] because SiC epitaxial layers grown by chemical vapor deposition (CVD) contain some hydrogen atoms. Furthermore, it is known that several transition metals, such as chromium (Cr) and nickel (Ni), exhibit significant diffusion above 1500 °C [13]. Although boron diffusion for p-type doping has been investigated to some extent [14, 15], a high density of boron-related deep levels (D center [16]) is generated, which hampers good electrical activation of boron acceptors. In fact, the tail of the boron-diffused region is highly resistive, which is not suitable for pn junctions of real devices. Abnormal diffusion of boron atoms, which is also encountered in boron ion implantation [6], is another problem.

6.1.2 *Formation of an* n-*Type Region by Ion Implantation*

Nitrogen or phosphorus ions are implanted to selectively form n-type regions in SiC. The implant profiles can be predicted well by Monte Carlo simulations, such as SRIM (stopping and range of ions in matter) [17]. Figure 6.3 shows depth profiles of the density of nitrogen atoms implanted into SiC(0001) with various implant angle [18]. In this particular case, the implant energy and dose are 100 keV and 1×10^{15} cm^{-2}, respectively. The open circles denote the nitrogen depth profile simulated by a SRIM code, assuming a mass density of 3.21 g cm^{-3} for SiC. The dashed and dotted lines represent the depth profiles determined by secondary ion mass spectrometry (SIMS) in samples implanted at various angles (with respect to <0001>). Here, a channeling effect can be suppressed by increasing the implantation angle to 7°. Ideally, the implantation angle should be larger than 5° from the <0001> axis (not normal to the wafer surface). Note that SRIM does not simulate channeling of the implanted ions accurately, leading to deviation in the implant-tail region. Implant profiles can also be designed using the Pearson distribution [19]. In the Pearson distribution an asymmetric Gaussian distribution is considered, and characterized by the projected range (mean), the straggle (standard deviation), the skewness, and kurtosis [19]. Figure 6.4 shows the projected mean range and straggle versus the implant energy in nitrogen and phosphorus ion implantation into SiC [20–23]. Because of the higher atom density of SiC ([Si]: 4.8×10^{22} cm^{-3}, [C]: 4.8×10^{22} cm^{-3}, total: 9.6×10^{22} cm^{-3}), a higher implant energy is required to obtain a given depth or peak position compared with the energy required for ion implantation into Si.

Figure 6.5 shows the depth profiles of phosphorus atoms which were implanted with multistep energies to form a 200-nm-deep box profile. The depth profiles before (as-implanted) and after annealing

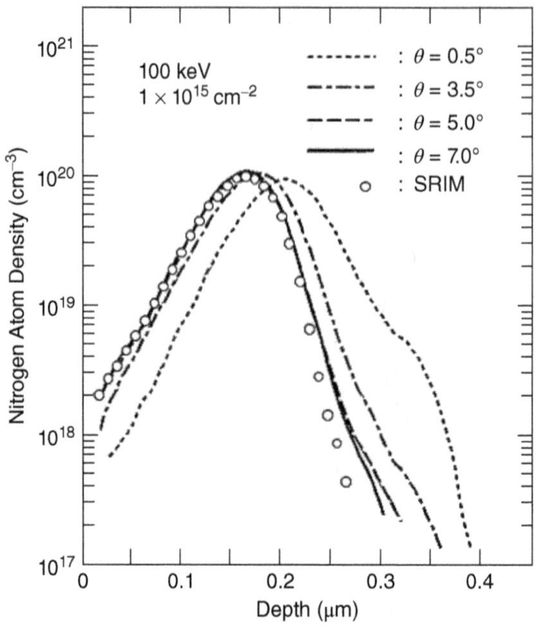

Figure 6.3 Depth profiles of the density of nitrogen atoms implanted into SiC(0001) with various implant angles ([18] reproduced with permission from The Japan Society of Applied Physics). The implant energy and dose are 100 keV and 1×10^{15} cm^{-2}, respectively.

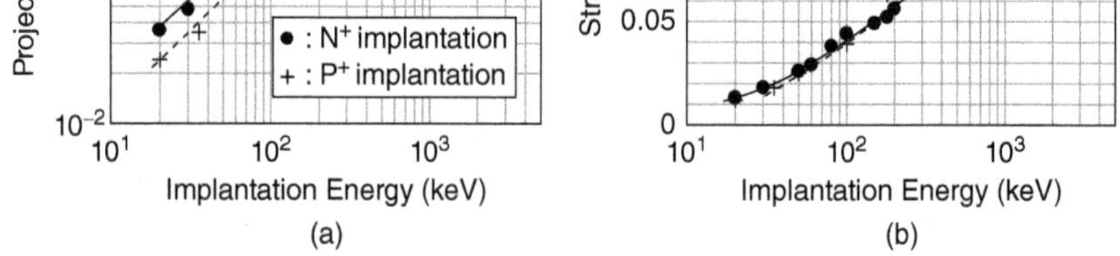

Figure 6.4 (a) Projected mean range (R_p) and (b) straggle (ΔR_p) versus the implant energy in nitrogen and phosphorus ion implantation into SiC [20–23].

in Ar at 1600 °C for 30 min are presented. The dopant profile exhibits very little diffusion during such high-temperature annealing, retaining the as-implanted profile. This is expected from the very small diffusion constants of phosphorus (also for nitrogen as shown in Figure 6.2), and any enhancement of impurity diffusion via implantation-induced defects is negligible. The lack of dopant diffusion makes it relatively easy to form a shallow junction, but difficult to form a very deep junction, in SiC.

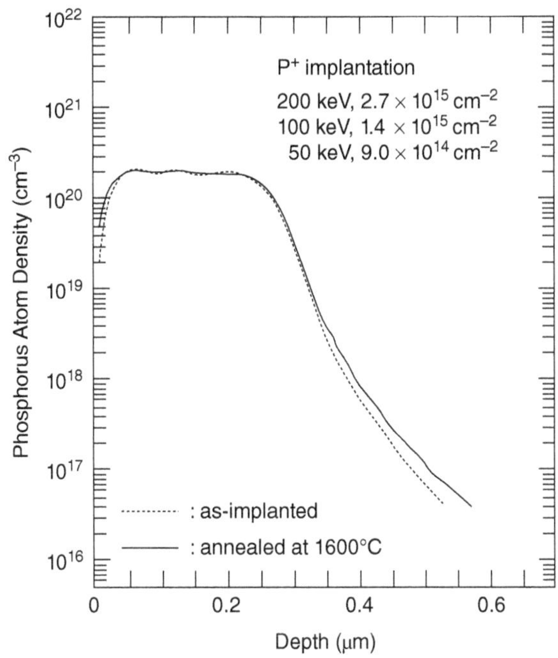

Figure 6.5 Depth profiles of phosphorus atoms, which were implanted with multistep energies to form a 200-nm-deep box profile, before (as-implanted) and after annealing in Ar at 1600 °C for 30 min.

Figure 6.6 shows typical sample structures employed for electrical characterization of implanted regions. In Figure 6.6a, the cross-section of a sample for van der Pauw and Hall effect measurements is illustrated. The implanted n-type layer is electrically isolated from the substrate by the np junction (n-type implanted layer/lightly-doped p-type epitaxial layer). The implanted region is processed into a clover-leaf pattern to improve the accuracy. By using this kind of sample the sheet resistance, the carrier density, and the mobility can be determined. In another type of test sample, a Schottky barrier is formed on the top of the implanted n-type layer, as shown in Figure 6.6b. In this case, electrical isolation is not necessary, and the implantation is conducted into a lightly-doped n-type epitaxial layer grown on an n-type substrate. From the capacitance–voltage characteristics, the depth profile of the net doping density (not the carrier density) is determined. This method is powerful when the doping density is relatively low because analysis of the deep levels can be performed on the same samples. For characterization of high-dose implanted samples, however, this technique is not very useful because formation of good Schottky barriers becomes difficult. For characterization of aluminum- or boron-implanted samples, the n- and p-type regions in Figure 6.6 should be inverted.

Figure 6.7 shows the annealing-temperature dependence of the electrical activation ratio in nitrogen- or phosphorus-implanted SiC. In this case, a box profile with a depth of about 400 nm was formed by multi-step ion implantation, and the total implant dose was 1×10^{14} cm^{-2}, which corresponds to a dopant atom density of 2×10^{18} cm^{-3} inside the box profile. The implantation was conducted into lightly-doped n-type epitaxial layers without intentional sample heating, and the samples were annealed at various temperatures in Ar for 30 min to promote electrical activation. The active donor density was determined by *capacitance–voltage* (*C–V*) characteristics of Ni/SiC Schottky structures. It should be noted that the *C–V* measurements do not give the *carrier density* but rather the *net doping density,* as further described in Section 6.4.1. As shown in Figure 6.7, the activation ratio is very low (<10%) after annealing

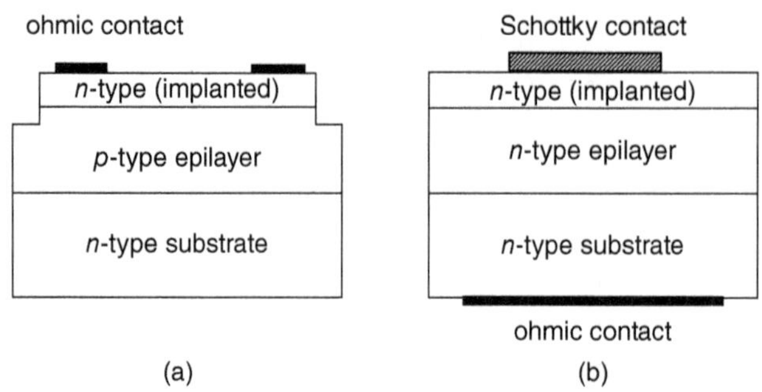

Figure 6.6 Typical sample structures employed for electrical characterization of implanted regions for: (a) Hall effect measurement and (b) $C-V$ measurement.

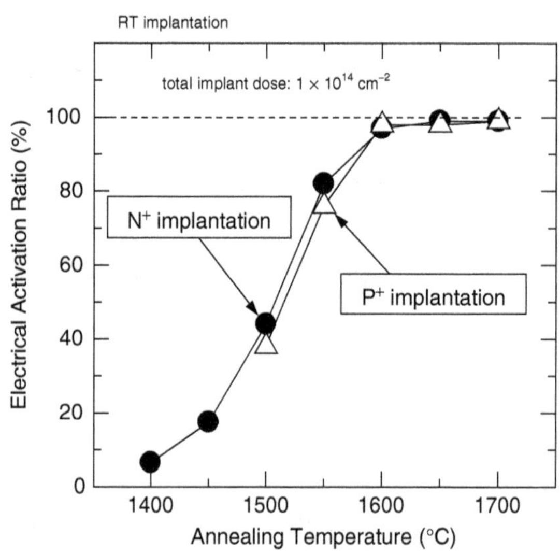

Figure 6.7 Annealing-temperature dependence of electrical activation ratio in nitrogen- or phosphorus-implanted SiC. The implantations were performed at room temperature (total implant dose: 1×10^{14} cm^{-2}).

at 1400 °C, and a high-temperature annealing at 1600 °C or higher is required to obtain nearly perfect (>95%) activation ratios. Note that the annealing temperature cannot be reduced very much, even if the implantation is performed at elevated temperatures (300–800 °C). In other words, hot implantation is not very effective to reduce the temperature required for post-implantation annealing. This can be understood by considering the thermal stability of the compensating deep levels generated by ion implantation, as described in Section 6.1.6.

When the implant dose becomes high (>10^{15} cm^{-2}), the situation is significantly changed. Striking differences are observed between room-temperature implantation and hot implantation, and also between nitrogen implantation and phosphorus implantation. This is especially true when, to improve contact characteristics, heavily-doped n$^+$-regions are formed by ion implantation, as described below.

Figure 6.8 shows the χ_{min} value determined by Rutherford backscattering spectrometry (RBS) channeling measurements as a function of the implant dose. The results are obtained from SiC(0001)

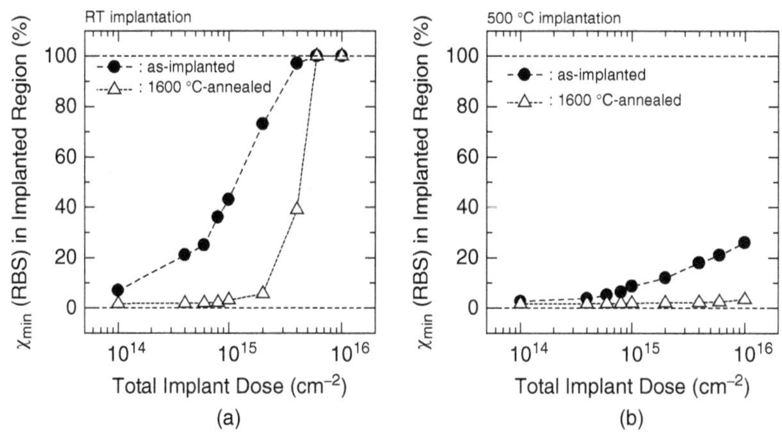

Figure 6.8 χ_{min} value determined by Rutherford backscattering spectrometry (RBS) channeling measurements as a function of the implant dose. The results are obtained from SiC(0001) samples implanted with nitrogen ions at (a) room temperature and (b) 500 °C. The χ_{min} values for both as-implanted and 1600 °C-annealed samples are shown (200-nm-deep box profile).

samples implanted with nitrogen ions at (a) room temperature and (b) 500 °C. The χ_{min} values for both as-implanted and 1600 °C-annealed samples are shown (200-nm-deep box profile) [5]. In these RBS measurements, a 2.0 MeV He^{2+} beam was employed with a backscattering angle of 170°. The lattice damage is monitored by the so-called χ_{min} value, which is the integrated scattering yield in the damaged region of a channeling spectrum divided by the integrated yield of a random spectrum. In the case of implantation at room temperature, the χ_{min} value from the implanted region reaches almost 100%, indicating complete amorphization by implantation-induced damage, when the implant dose is higher than $(3-4) \times 10^{15}$ cm^{-2} (this is called the *critical implant dose* for amorphization). The high χ_{min} value is not decreased very much after annealing at 1600 °C ($\chi_{min} > 90\%$ in Figure 6.8a). In contrast, the χ_{min} value remains rather low for an as-grown sample implanted at elevated temperature, owing to the *in situ annealing effect,* and it decreases significantly with further annealing ($\chi_{min} < 3\%$ in Figure 6.8b). Transmission electron microscope (TEM) observations and other structural analyses demonstrate that the surface region is indeed amorphous for as-implanted samples after room-temperature implantation, and the implanted region contains polycrystalline 3C-SiC grains after annealing [24–28].

The damage accumulation during ion implantation is approximately proportional to the implant dose until complete amorphization occurs [29]. When the implant dose is low, disordering of the carbon sub-lattice proceeds at a higher rate than for the silicon sub-lattice, as a result of the lower threshold-displacement energies for carbon atoms. However, amorphization of the carbon and silicon sub-lattices occurs at almost the same implant dose [30]. In the case of hot implantation, the implanted region can retain the original polytype structure and can be easily recovered. Very similar results are obtained for phosphorus, aluminum, and boron ion implantation. Therefore, this is a consistent and unique feature of ion implantation into hexagonal SiC polytypes. Because of the complicated polytypism and low stacking-fault energy in SiC, it is important to maintain the original crystalline structure during ion implantation [5, 6, 31–34]. Once the implanted region becomes completely amorphous, as a result of high-dose implantation, lattice recovery to the original polytype is not guaranteed. This is the main reason why hot implantation of SiC is employed, especially when the implant dose is high. The critical implant dose, above which the implanted region becomes amorphous, is approximately low to mid 10^{15} cm^{-2} for room-temperature implantation, though it depends on the implanted species and implant energy. It should be noted that the benefits of hot implantation are small when the implant dose is

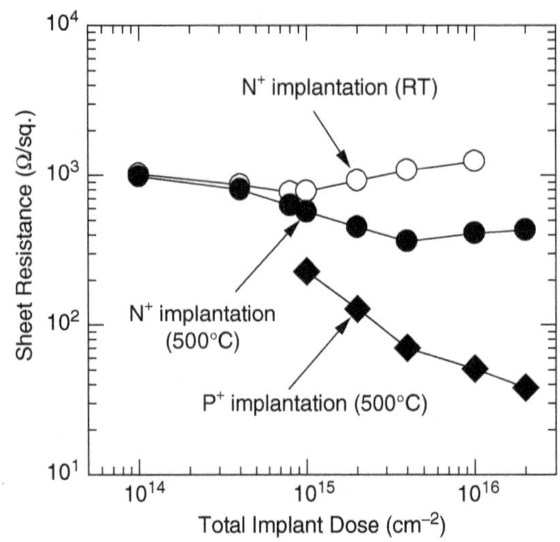

Figure 6.9 Sheet resistance versus the total implant dose for nitrogen- or phosphorus-implanted 4H-SiC(0001) annealed at 1700 °C for 30 min.

relatively low and thus the implanted region is not severely damaged. For example, hot implantation is not always necessary when the implant dose is in the range of $10^{13}-10^{14}$ cm^{-2} (and below), which is further supported by the data on electrical properties.

Figure 6.9 shows the sheet resistance versus the total implant dose for nitrogen- or phosphorus-implanted 4H-SiC(0001) annealed at 1700 °C for 30 min [35]. Multistep implantation at room temperature or 500 °C was performed to form a 200-nm-deep box profile. Here, high-dose implantation to form an n$^+$-region is considered. The sheet resistance is usually measured by the van der Pauw method. When the implant dose is relatively low ($<3 \times 10^{14}$ cm^{-2}), there exist no striking differences in sheet resistances, irrespective of implant species (N$^+$ or P$^+$) or implantation temperature (RT or 500 °C). In the case of room temperature implantation, the sheet resistance exhibits a minimum value at an implant dose of approximately $0.7-1 \times 10^{15}$ cm^{-2}, and increases when the implant dose is further increased. In this high-dose region, the lattice damage caused by room temperature implantation is so severe that the implanted region contains a high density of stacking faults and 3C-SiC grains after activation annealing, as described above. On the other hand, a continuous decrease in the sheet resistance is observed for hot implantation. The sheet resistance of the nitrogen-implanted region is almost saturated at 300 Ω/□, which may be limited by the relatively low solubility limit of nitrogen atoms in SiC. A much lower sheet resistance of 30–50 Ω/□ can be obtained by hot implantation of phosphorus owing to its higher solubility limit [35–37]. It is reported that high-dose arsenic ion implantation also gives a low sheet resistance, below 200 Ω/□ [38]. Systematic data on nitrogen or phosphorus implantation into SiC are found in a number of papers [39–46]. The obtained sheet resistance is low enough for most device fabrication. It is noteworthy that very good lattice recovery and high activation ratio are obtained, even in high-dose implantation at room temperature, when SiC(11$\bar{2}$0) is employed [47]. To increase the process throughput and to minimize formation of extended defects, *rapid thermal annealing* using a high-power infrared lamp [48] or microwave heating [49] has been investigated.

Based on these results, phosphorus implantation at elevated temperature is preferred when a heavily-doped n$^+$-region ($\gg 10^{19}$ cm^{-3}) must be formed. For formation of a moderately-doped n-region, selection of the implant species and implantation temperature is not critical.

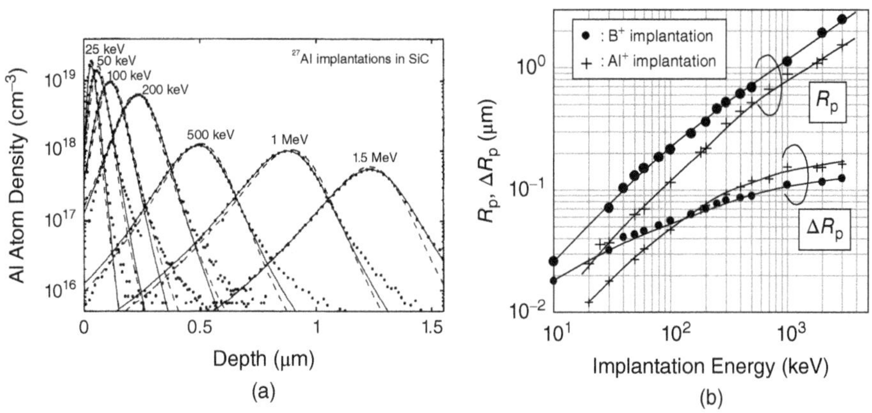

Figure 6.10 (a) Implant profiles of aluminum ion implantation ([22] reproduced with permission from AIP Publishing LLC) and (b) the projected range (R_p) and straggle (ΔR_p) versus the implant energy for aluminum and boron ion implantation into SiC [20–23].

6.1.3 Formation of a p*-Type Region by Ion Implantation*

Aluminum is the main acceptor used in SiC, and the situation is the same in ion implantation. As described below, boron implantation can cause several unwanted phenomena and is not usually employed for device fabrication in industry. As in the case of n-type doping, the implant profiles (apart from the tail region) can be predicted well by a SRIM code. Figure 6.10 presents (a) the implant profiles of aluminum ion implantation and (b) the projected range (R_p) and straggle (ΔR_p) versus the implant energy for aluminum and boron ion implantation into SiC [20–23]. As was found for nitrogen or phosphorus ion implantation, implanted aluminum atoms exhibit very little diffusion, even after high-temperature annealing at 1600–1700 °C. Implanted boron atoms, however, show significant out-diffusion and in-diffusion during the activation annealing [6, 50]. As a result of the out-diffusion, some portion of the implanted boron atoms is lost; the in-diffusion makes the junction depth extremely large compared with the designed depth. It has been accepted that diffusion of boron atoms is enhanced by implantation-induced damage via a kick-out mechanism [51, 52]. Boron interstitials created by implantation have a large diffusion constant, and can diffuse already at 1400–1500 °C. Damage-enhanced diffusion is also observed in boron-doped epitaxial layers. When any kind of ion is implanted into boron-doped epilayers, the boron atoms show abnormal diffusion during subsequent annealing at 1600–1800 °C [53], this diffusion means that the designed depth profile of acceptor density is completely destroyed. Thus, boron is not a good choice, even when producing p-type epitaxial layers for device development.

Figure 6.11 depicts the annealing-temperature dependence of the electrical activation ratio in aluminum- or boron-implanted SiC [54]. A 400-nm-deep box profile was formed by multistep ion implantation, and the total implant dose was 1×10^{14} cm^{-2}, which corresponds to a dopant atom density of 2×10^{18} cm^{-3} inside the box profile. The implantation was conducted into lightly-doped p-type epitaxial layers at room temperature, followed by activation annealing at various temperatures in Ar for 30 min. The electrically active acceptor density was determined from the C–V characteristics of Ti/SiC Schottky structures. The implanted region is highly resistive after annealing at temperatures lower than 1400 °C. High-temperature annealing at 1600 °C or higher is required to obtain nearly perfect (>90%) activation ratios, in agreement with the results of nitrogen or phosphorus ion implantation (Figure 6.7). Rapid thermal annealing has also been investigated [55].

High-temperature annealing is indeed required, not only for high electrical activation, but also for good junction characteristics in device fabrication [56–58]. Figure 6.12 shows the histograms of

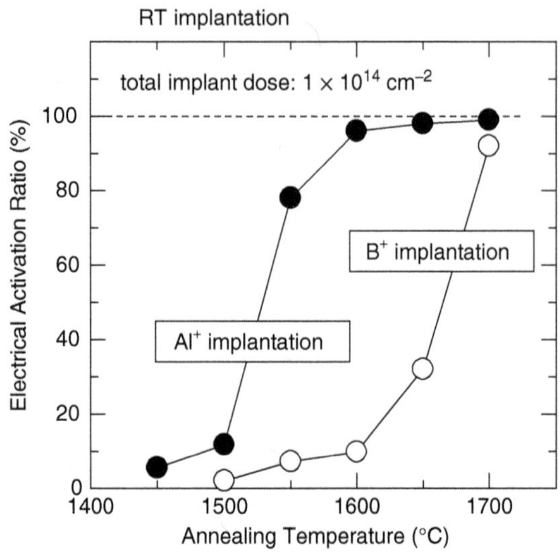

Figure 6.11 Annealing-temperature dependence of the electrical activation ratio in aluminum- or boron-implanted SiC [54]. The implantations were performed at room temperature (total implant dose: 1×10^{14} cm^{-2}).

leakage current density in Al^+-implanted 4H-SiC pn junction diodes annealed at various temperatures. Lightly-doped n-type epitaxial layers, 8 μm thick, were prepared, and pn junctions were formed by Al^+ implantation. The leakage current was monitored at a reverse voltage of 500 V. Though the breakdown voltage of these diodes was about 1400 V, exhibiting very little dependence on the annealing temperature, the distribution of leakage current is clearly dependent on the annealing temperature, as shown in Figure 6.12. The leakage current is rather high for the diodes annealed at 1550 °C, and it is significantly reduced in diodes which were formed by activation annealing at 1700 °C or higher.

Figure 6.13 shows the depth profiles of (a) aluminum and (b) boron atoms, which were implanted with multistep energies to form a 200-nm-deep box profile [59]. The depth profiles before (as-implanted) and after annealing in Ar at 1700 °C for 1 or 30 min are presented in both cases. In the figure, the depth profile of electrically active acceptor density is also plotted (the symbols). The depth profile of acceptor density was determined from the C–V characteristics of the Ti/SiC Schottky structure measured at different bias voltages. In this experiment, lightly-doped p-type epilayers ($N_A = 7 \times 10^{15}$ cm^{-3}) were employed to investigate the activation ratio in the implant tail region [59]. In aluminum ion implantation (Figure 6.13a), the implanted aluminum profile does not change during annealing (that is, the aluminum atoms exhibit very little diffusion) and the depth profile of acceptor density completely matches that of implanted atom density determined by SIMS. This result indicates that the electrical activation of implanted aluminum atoms is nearly perfect (>95%) inside the box-profile region as well as in the tail region. In boron ion implantation (Figure 6.13b), however, substantial out- and in-diffusion are observed, as mentioned above. The depth profile of active acceptor density shows a clear dip near the implant tail region (~0.3 μm), though the activation ratio is relatively high in the deeper (>0.5 μm) region. Analysis of the defects using deep level transient spectroscopy (DLTS) measurements revealed that a high density (>10^{15} cm^{-3}) of D centers ($E_v + 0.45$ eV) is generated near the tail region [59]. Although carbon co-implantation is effective to reduce the abnormal diffusion of boron atoms and generation of D centers [6, 50, 60], the suppression is not complete. Furthermore, the high ionization energy of boron acceptors hampers the formation of low-resistance p-type SiC by boron implantation [6, 60, 61].

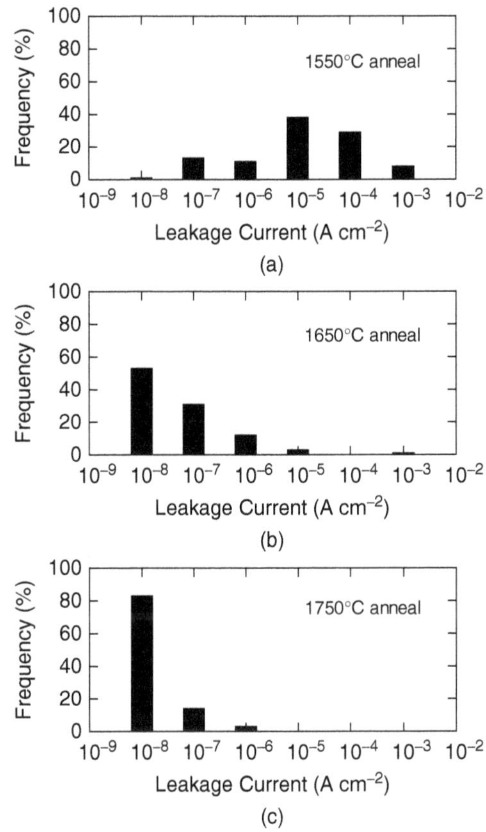

Figure 6.12 Histograms of leakage current density in Al^+-implanted 4H-SiC pn junction diodes annealed at (a) 1550, (b) 1650 and (c) 1750 °C. The Al^+-implantation was conducted at room temperature.

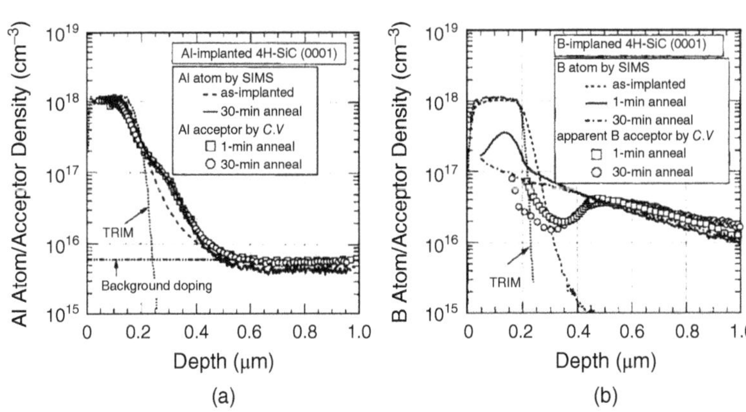

Figure 6.13 Depth profiles of (a) aluminum and (b) boron atoms, which were implanted with multistep energies to form a 200-nm-deep box profile ([59] reproduced with permission AIP Publishing LLC). The depth profiles before (as-implanted) and after annealing in Ar at 1700 °C for 1 or 30 min are presented.

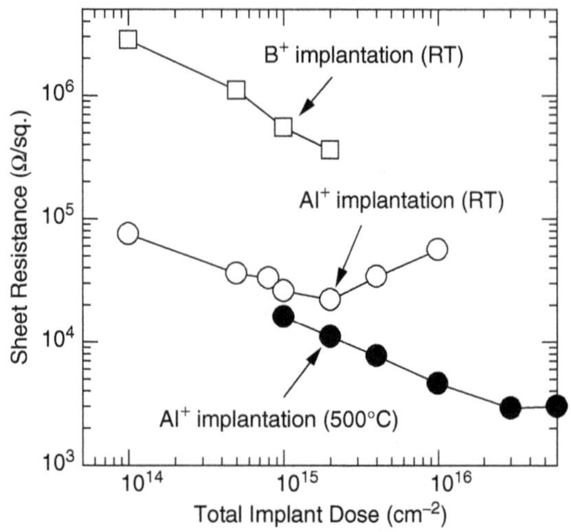

Figure 6.14 Sheet resistance versus the total implant dose for aluminum-implanted 4H-SiC(0001) annealed at 1800 °C for 30 min [62].

Figure 6.14 shows the sheet resistance versus the total implant dose for aluminum-implanted 4H-SiC(0001) annealed at 1800 °C for 30 min [62]. Multistep implantation at room temperature or 500 °C was performed to form a 200-nm-deep box profile. Here, high-dose implantation to form a p^+-region is considered. As in the case of nitrogen or phosphorus implantation, effects of hot implantation become remarkable when the implant dose is higher than 1×10^{15} cm^{-2}. For hot implantation, the sheet resistance of the aluminum-implanted region is almost saturated at approximately 3 kΩ/□. Carbon co-implantation gives a slightly lower sheet resistance, probably because carbon enrichment may enhance substitution of implanted aluminum atoms at the Si sites. The relatively high sheet resistance can be partly ascribed to the low hole mobility in SiC. The intrinsic mobility is further decreased by implantation-induced defects. Detailed characterization is found in the literature [63–67]. Although ion implantation of gallium [68] and beryllium [69] into SiC has been investigated for obtaining p-type conductivity, the obtained sheet resistances are relatively high because of the high ionization energy of the dopants.

Thus, aluminum implantation at elevated temperature is required when a heavily-doped p^+-region (> > 10^{19} cm^{-3}) must be formed. However, the resistivity of aluminum-implanted regions is still relatively high, and this must be carefully considered in design and characterization of SiC devices. To form a moderately-doped p-region, aluminum implantation at room temperature followed by high-temperature annealing is sufficient.

6.1.4 Formation of a Semi-Insulating Region by Ion Implantation

There are two different approaches to form semi-insulating regions by ion implantation. The first approach is creation of *intrinsic defects,* which form a very deep level in the bandgap, by ion bombardment. The second approach is implantation of a special *impurity* which forms a very deep level. In the former case, either protons [70, 71] or inert atoms, such as helium, neon, or argon [72], or the host element (silicon or carbon) [73] ions have been investigated. When a high enough density of deep levels has been created, the dopants are completely compensated and the Fermi level is pinned near a dominant deep level. A high resistivity of over 10^5 Ω cm is obtained when these ions are implanted into

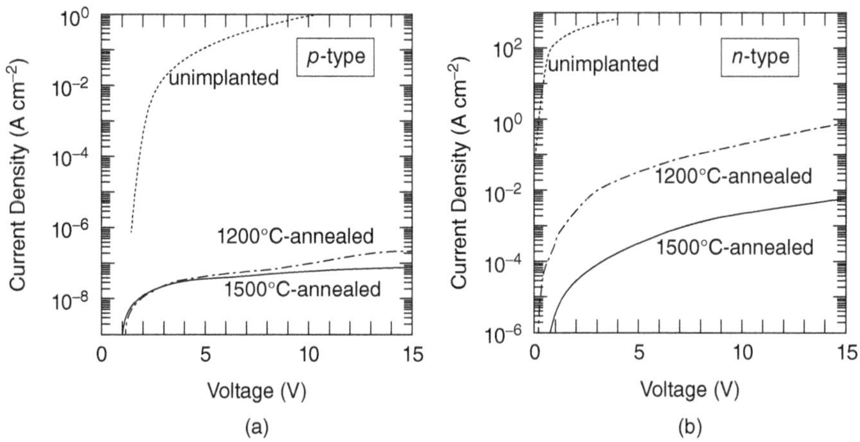

Figure 6.15 Current density–voltage characteristics of vanadium-implanted (a) $n^+/n^-/n^+$ and (b) $p^+/p^-/p^+$ 6H-SiC structures ([76] reproduced with permission from AIP Publishing LLC).

n-type 4H- or 6H-SiC. The semi-insulating property is maintained after high-temperature annealing when the implant dose is high. The dominant deep level(s) generated by ion implantation are described in Section 6.1.6.

It is known that a few metallic impurities form very deep levels in SiC and the most widely-used impurity for this purpose is vanadium [74]. Vanadium doping is also effective in producing semi-insulating SiC wafers in sublimation growth [75], as described in Section 3.4.4. Therefore, vanadium ion implantation is another approach to obtain semi-insulating SiC. Figure 6.15 shows the current density–voltage characteristics of vanadium-implanted (a) $n^+/n^-/n^+$ and (b) $p^+/p^-/p^+$ 6H-SiC structures [76]. In either case, vanadium ions were implanted into lightly-doped epitaxial layers to form a 300-nm-deep box profile. The vanadium density in the box profile is approximately $(2-3) \times 10^{17}$ cm^{-3}, which is higher than the doping density of the epitaxial layers (4×10^{16} cm^{-3}) and lower than the solubility limit of vanadium ($\sim 4 \times 10^{17}$ cm^{-3}). After post-implantation annealing at 1500 °C for 30 min, implanted vanadium atoms were electrically activated, and the resistivity of the implanted region increased to 10^6 Ω cm for n-type and over 10^{11} Ω cm for p-type SiC. Vanadium creates an acceptor-like level at $E_c - 0.7$ eV in n-type and a donor-like level at $E_v + 1.5$ eV in p-type 6H-SiC [74]. Because of the deeper energy level of vanadium in p-type SiC, the Fermi level is pinned near the midgap in vanadium-implanted p-type SiC, leading to the higher resistivity.

Though ion implantation is useful to selectively form semi-insulating regions, the depth of the semi-insulating region formed in this way is restricted to less than about 1–2 μm. If a very thick semi-insulating region is required then electron irradiation is a more powerful technique. 200–400 keV electron irradiation can create semi-insulating regions that are several tens of microns thick [77]. The semi-insulating property is stable even after high-temperature annealing at 1700 °C when the electron fluence is high enough.

6.1.5 *High-Temperature Annealing and Surface Roughening*

As described in previous subsections, post-implantation annealing must be carried out at very high temperatures of 1600–1700 °C to achieve reasonable lattice recovery and high electrical activation. The annealing temperature cannot be reduced, even if the implantation is conducted at elevated temperatures of 500–1000 °C. One of the main reasons for this is the thermal stability of several deep levels which are generated by ion implantation and cause compensation of dopants.

During high-temperature annealing one can usually observe surface degradation of SiC. Without special precautions the mirror-like surface is completely lost during annealing [78]. There are at least two mechanisms which cause surface degradation:

1. *Si desorption from the SiC surface*
   Si desorption leads to surface graphitization as well as severe roughening. This is naturally enhanced when SiC is annealed in vacuum (significant Si desorption in vacuum can already start at 900–1000 °C, even for unimplanted SiC surfaces). In the early stage, supply of Si overpressure was investigated with some success. For example, addition of SiH_4 flow into pure Ar ambient or introduction of Si pieces near the SiC samples during annealing was tried, and improvement of surface roughness was demonstrated [79, 80]. It is, however, difficult to suppress the second mechanism mentioned below with this technique.
2. *Migration of surface atoms*
   Migration of surface atoms at high temperature can take place in Si [81] and any other materials. The main driving force for surface migration is minimization of the surface energy of the crystal. In the case of hexagonal SiC polytypes, a several-degree off-angle is chosen for the SiC(0001) wafers to enable high-quality homoepitaxy. Since the SiC(0001) has a relatively high surface energy, significant migration of surface atoms occurs at high temperature and results in formation of macrosteps to minimize the total surface energy. Capping the surface with an appropriate material effectively suppresses this phenomenon because the surface atoms lose the freedom to migrate. Proper capping is also useful to minimize Si desorption from the SiC surface. So far, several capping materials have been investigated, such as SiO_2, Si_3N_4, AlN [82], and carbon [35]. A SiC face-to-face configuration was also investigated. Among these materials, a carbon cap gives the most successful results [35] and is employed for production of SiC devices in most industries.

Figure 6.16 shows the atomic force microscopy (AFM) images of Al^+-implanted off-axis 4H SiC(0001) annealed at 1800 °C for 5 min (a) with and (b) without a carbon cap. The total implant dose is 1×10^{16} cm^{-2}. The surface roughness, as defined by the root mean square (r_{rms}) value of height deviation, is 1.0 nm in a 10×10 μm^2 area for the sample capped with carbon, while the unprotected sample exhibits a much higher surface roughness (r_{rms} = 16.4 nm), because of macrostep formation. Figure 6.17 depicts the implant-dose dependence of the surface roughness (10×10 μm^2 area) for

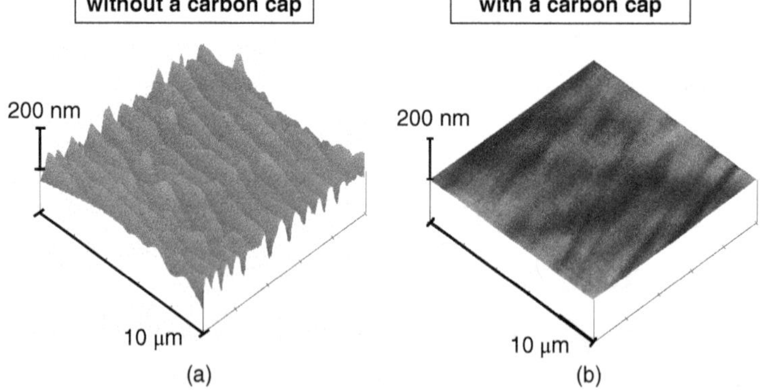

Figure 6.16 Atomic force microscopy (AFM) images of Al^+-implanted off-axis 4H-SiC(0001) annealed at 1800 °C for 5 min (a) without and (b) with a carbon cap. The total implant dose is 1×10^{16} cm^{-2}.

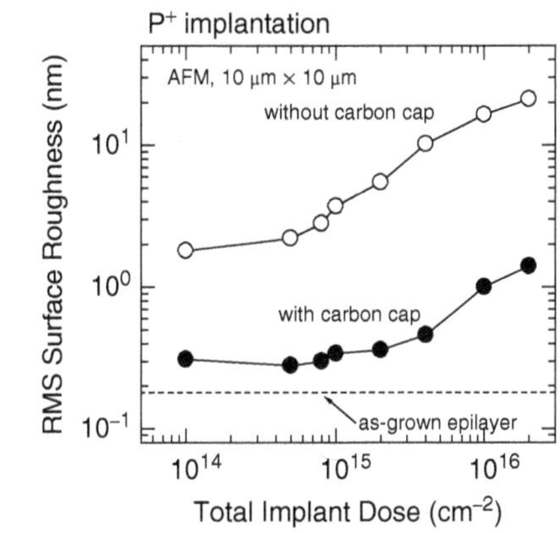

Figure 6.17 Implant-dose dependence of the surface roughness ($10 \times 10\ \mu m^2$ area) for P^+-implanted SiC(0001) annealed at 1700 °C for 20 min with and without a carbon cap.

P^+-implanted SiC(0001) annealed at 1700 °C for 20 min with and without a carbon cap. When the surface is not protected, the roughness increases significantly with increasing implant dose. Higher-dose implantation creates more broken bonds and more atoms can migrate at lower temperature (that is, the migration barrier is reduced by implantation-induced damage). In contrast, the surface roughness can be minimized by using a carbon cap, even for high-dose implantation. The carbon cap can be formed by RF sputtering or graphitization of a photoresist. During high-temperature annealing, no chemical reactions are observed at the carbon/SiC interface. After annealing, the carbon cap can be easily removed by O_2 plasma (ashing) or low-temperature (700–800 °C) oxidation; in these processes, the SiC is only minimally oxidized (less than a few nanometers).

6.1.6 *Defect Formation by Ion Implantation and Subsequent Annealing*

High-energy ion bombardment during ion implantation causes considerable displacement of Si and C atoms. Si and C atoms are kicked out from their lattice positions, and the kicked-out atoms reside to a large extent on interstitial sites, leaving vacancies behind. In SiC, C displacement occurs more easily than Si displacement. Created vacancies and interstitials become mobile during post-implantation annealing and can form complex point defects, such as vacancy clusters, interstitial clusters, antisite-vacancy pairs, and so forth. The profile of interstitial and vacancy densities generated by ion implantation can be estimated by an SRIM code. The vacancies, interstitials and antisites can also combine with implanted impurities or dopant atoms in the host material. All these point defects can create localized levels (shallow or deep levels) in the bandgap. Furthermore, the density of generated point defects is so high that they often segregate inside the implanted region during annealing, leading to formation of extended defects, such as dislocation loops and stacking faults. Since the activation annealing is carried out at high temperature, additional stress can be introduced in the material, leading to generation of new dislocations and/or movement of pre-existing dislocations.

The shallow and deep levels generated by ion-implantation processes have been investigated in detail [83–86]. Figure 6.18 shows the DLTS spectra obtained from Al^+-implanted (a) n-type and (b) p-type 4H-SiC epitaxial layers [86]. Here, Al^+ implantation was conducted at room temperature, followed by annealing at 1700 °C for 20 min. The total implant dose is low, $6 \times 10^{10}\ cm^{-2}$, which corresponds to an impurity density of about $1 \times 10^{15}\ cm^{-3}$. Therefore, the samples retain their original conduction type (n or p) after implantation. The major deep levels generated by the implantation are $Z_{1/2}$ (E_c – 0.63 eV)

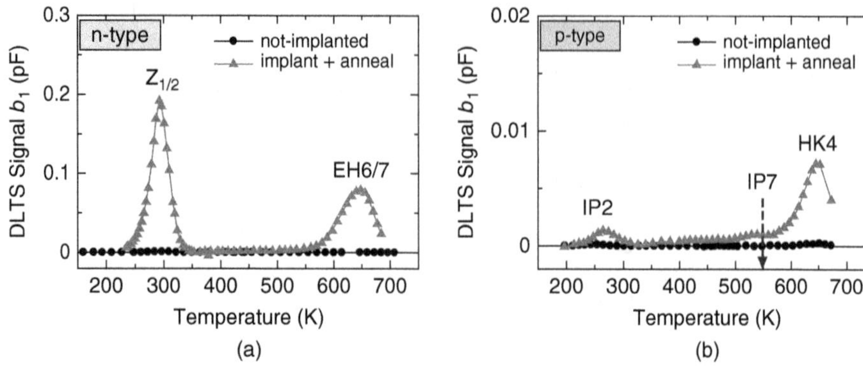

Figure 6.18 DLTS spectra obtained from Al^+-implanted (a) n-type and (b) p-type 4H-SiC epitaxial layers [86]. Here, Al^+ implantation was conducted at room temperature (total implant dose: 6×10^{10} cm^{-2}), followed by annealing at 1700 °C for 20 min.

[83], EH6/7 center ($E_c - 1.55$ eV) [87], and HK4 center ($E_v + 1.45$ eV) [88]. It should be noted that these deep levels are commonly observed in 4H-SiC samples implanted with any kinds of ions, such as Al^+, N^+, P^+, Ne^+, and Ar^+. Furthermore, the density of generated deep levels can exceed that of implanted ions. These results indicate that these major deep levels generated by low-dose ion implantation originate from intrinsic defects, which do not contain a specific impurity.

Figure 6.19 shows the depth profiles of the density of $Z_{1/2}$ centers in Al^+-, N^+-, P^+-, and Ne^+-implanted 4H-SiC [86]. In the ion implantation, an 800-nm-deep box profile, indicated by a dashed line in the figure,

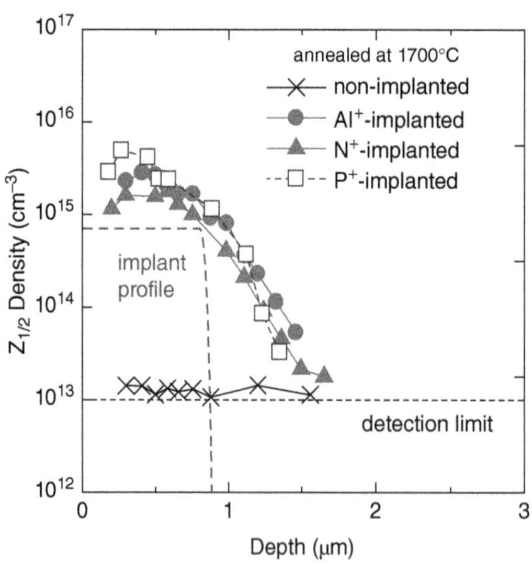

Figure 6.19 Depth profiles of the density of $Z_{1/2}$ centers in Al^+-, N^+-, P^+-, and Ne^+-implanted 4H-SiC ([86] reproduced with permission from AIP Publishing LLC). In the ion implantation, an 800-nm-deep box profile, indicated by a dashed line in the figure, was formed by multistep implantation of all ion species.

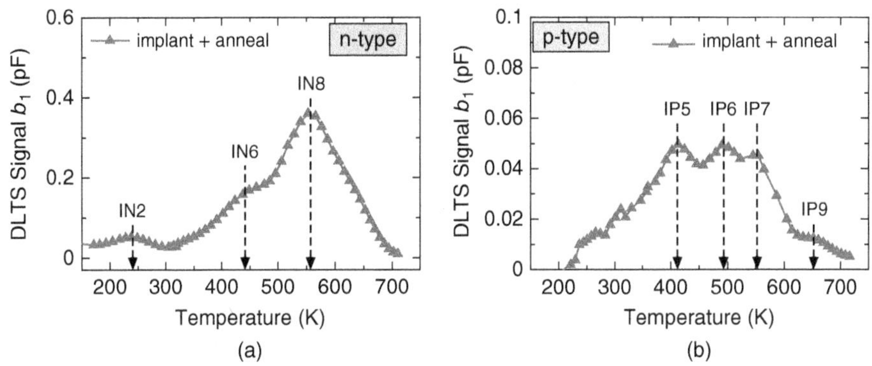

Figure 6.20 DLTS spectra obtained from (a) P^+-implanted n-type and (b) Al^+-implanted p-type 4H-SiC epitaxial layers [86]. The total implant dose is relatively high, 8×10^{13} cm^{-2}, which corresponds to an impurity density of about 1×10^{18} cm^{-3}. The implantation was carried out at room temperature and subsequent annealing was done at 1700 °C for 20 min.

was formed by multistep implantation of all ion species. The $Z_{1/2}$ center density inside the box profile is higher than the density of implanted ions and the depth profile extends much further into the sample than the implantation tail. Since the $Z_{1/2}$ center works as a carrier lifetime killer, as described in Section 5.3.2, this is one of the main reasons why pn junctions formed by ion implantation exhibit higher on-resistance and faster switching speed than epitaxial pn junctions [89, 90]. The high density of $Z_{1/2}$ centers can be reduced by a factor of 10–30 if appropriate thermal oxidation is performed after activation annealing [91].

Figure 6.20 depicts the DLTS spectra obtained from (a) P^+-implanted n-type and (b) Al^+-implanted p-type 4H-SiC epitaxial layers [86]. The total implant dose is relatively high in this case, 8×10^{13} cm^{-2}, which corresponds to an impurity density of about 1×10^{18} cm^{-3}. The implantation was carried out at room temperature, and subsequent annealing was done at 1700 °C for 20 min. DLTS spectra consist of several overlapping peaks, indicating that various deep levels with similar energies exist. The total density of observed deep levels reaches low 10^{16} cm^{-3}, which corresponds to about 3–10% of the implanted atom density. Thus, these levels work as carrier-trapping centers, increasing the resistance of the implanted regions. Thermal oxidation can also substantially reduce these deep levels, indicating that these defects contain a carbon vacancy [91].

Extended defects in implanted SiC have been investigated by TEM and X-ray topography [92–95]. Figure 6.21 shows a typical cross-sectional TEM image taken from implanted 4H-SiC [93]. Al^+ implantation (total implant dose: 7×10^{14} cm^{-2}) was performed at room temperature, followed by annealing at 1700 °C for 30 min. Inside the implanted region, one can observe a high density of small dark areas. Although the density and size of dark areas depend on the implant dose, annealing temperature, and also implanted species, these are common features of TEM observations of implanted SiC.

Figure 6.22 shows the high-resolution cross-sectional TEM image of B^+-implanted 6H-SiC [93]. Note that a similar structure is always observed when the dark areas shown in Figure 6.21 are magnified. There exists an extrinsic stacking fault (extra plane) in a {0001} basal plane, and a Frank-type partial dislocation is observed at the edge of the stacking fault. Furthermore, basal planes are distorted around both ends of the stacking fault. Figure 6.23 shows the implant-dose dependence of the number of atoms existing in the extra planes generated by ion implantation and annealing [93]. It is striking that the total number of atoms in the extra planes is proportional to the implant dose, and the absolute number is almost identical to that of the number of implanted ions. This result suggests that Si and C atoms kicked out by ion bombardment become mobile and segregate in the basal planes during annealing, leading to formation of extra planes. Furthermore, high-dose implanted regions exhibit a lattice tilt with respect to the original *c*-axis [96]. Figure 6.24 shows the reciprocal space mapping

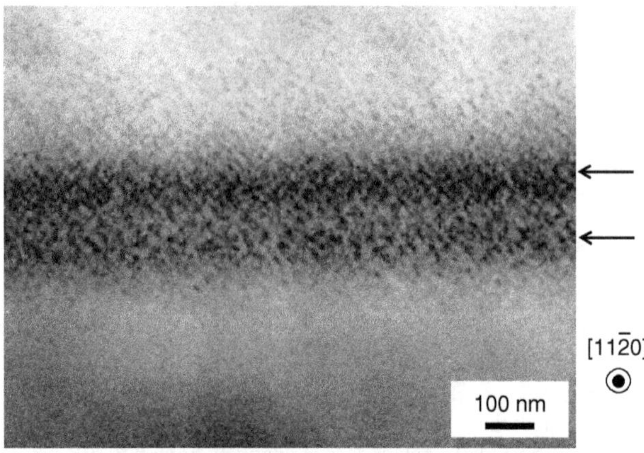

Figure 6.21 Typical cross-sectional TEM image taken from implanted 4H-SiC [93]. Al^{+} implantation (total implant dose: 7×10^{14} cm^{-2}) was performed at room temperature, followed by annealing at 1700 °C for 30 min.

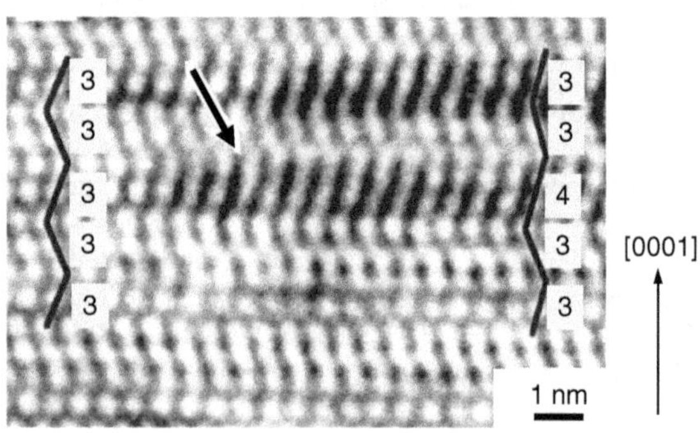

Figure 6.22 High-resolution cross-sectional TEM image of B^{+}-implanted 6H-SiC ([93] reproduced with permission from AIP Publishing LLC). Note that a similar structure is always observed when the dark areas shown in Figure 6.21 are magnified.

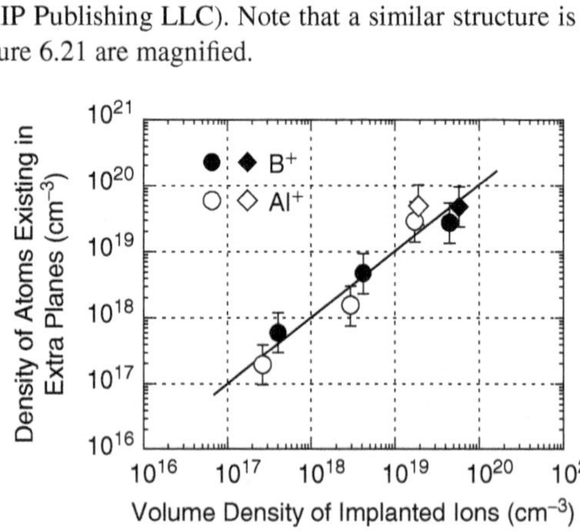

Figure 6.23 Implant-dose dependence of the number of atoms existing in the extra planes generated by ion implantation and annealing ([93] reproduced with permission from AIP Publishing LLC).

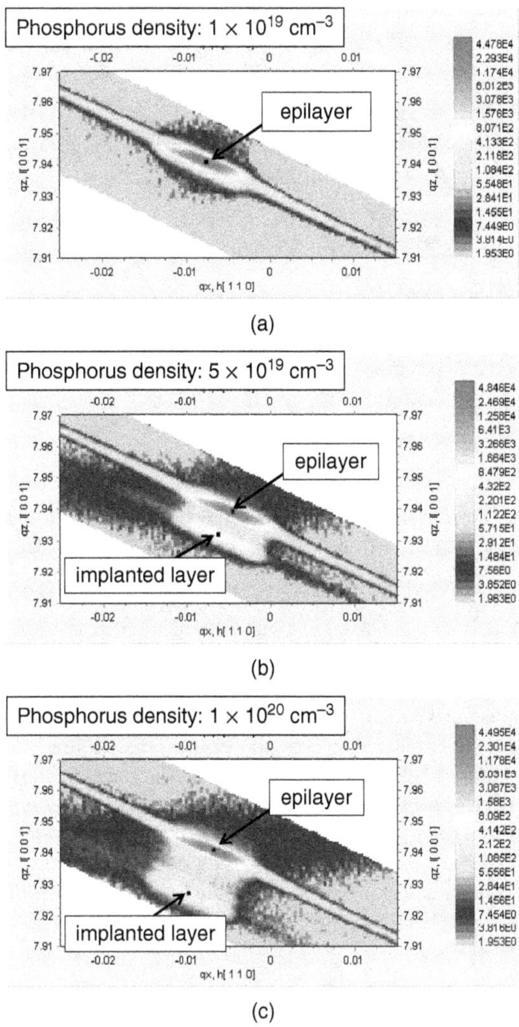

Figure 6.24 Reciprocal space mapping (RSM) images for the (0008) reflection of P$^+$-implanted 4H-SiC with a phosphorus density of (a) 1×10^{19}, (b) 5×10^{19}, and (c) 1×10^{20} atoms cm^{-3} ([96] reproduced with permission from AIP Publishing LLC).

(RSM) images for the (0008) reflection of P$^+$-implanted 4H-SiC with a phosphorus density of (a) 1×10^{19}, (b) 5×10^{19}, and (c) 1×10^{20} atoms cm^{-3}. Post-implantation annealing is performed at 1800 °C for 10 min. In the RSM images taken from high-dose implanted samples, two distinct reflection peaks, originating from the implanted layer and the epitaxial layer, respectively, are observed. These images reveal an increase in the c-lattice constant (decrease in q_z : [0001]) of the implanted layers. Furthermore, the peaks from the implanted layers are located at slightly different q_x values (q_x : $[11\bar{2}0]$), indicating the lattice tilt. The tilt angle increases with increasing implant dose, and the tilt direction is always along the upstream side of the off-direction, irrespective of the implanted species (N$^+$, P$^+$, and Al$^+$) [96]. These kinds of extended defects, of course, affect the carrier transport in the implanted regions [97, 98].

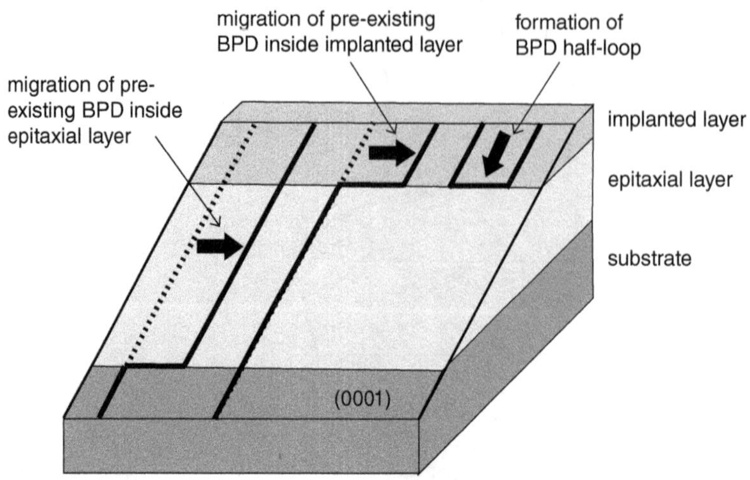

Figure 6.25 Schematic illustration of dislocation behaviors observed in implanted SiC annealed at 1600–1800 °C ([11, 99] reproduced with permission from Wiley-VCH Verlag KmbH).

During high-temperature activation annealing, new dislocations can be generated from the surface, and glide motion of pre-existing dislocations is also observed in SiC. Figure 6.25 shows a schematic illustration of typical phenomena observed in implanted SiC annealed at 1600–1800 °C [11, 99]. (i) A new dislocation half-loop is introduced in a basal plane from the surface. The bottom of this dislocation half-loop is often located at the implanted region/epilayer interface. (ii) Pre-existing basal-plane dislocations can migrate (glide motion) near the surface as well as near the epilayer/substrate interface, which results in formation of interface dislocations. Furthermore, (iii) Shockley-type stacking faults can be formed near the surface (not shown). Regarding (i) and (ii), misfit stress (caused by a large difference in dopant density) and thermal stress may be driving forces for dislocation nucleation or glide motion [100]. By improving the temperature gradient inside the annealing furnace, the glide of pre-existing basal-plane dislocations can be suppressed [101].

6.2 Etching

SiC is an extremely inert material against chemical solvents and wet etching of SiC is very difficult. SiC single crystals are attacked by neither acids nor alkali solutions at room temperature. Reactive ion etching (RIE) is widely employed to form mesa structures and trenches in SiC. A few review papers on dry etching of SiC have been published [102, 103]. Gas etching at high temperature can be useful for some applications.

6.2.1 Reactive Ion Etching

RIE of SiC is relatively easy, and commercial RIE systems developed for Si and other semiconductors can be used for etching SiC. Neither a very special etching gas nor high temperature is required for RIE of SiC, though the etching rates are relatively low compared with those of Si. In RIE, active radicals generated in a plasma diffuse toward the SiC surface and induce chemical etching. Positive ions are accelerated in a plasma sheath and ion bombardment onto the SiC surface induces physical etching. Capacitively-coupled plasma reactive ion etching (CCP-RIE) [104–111], inductively-coupled plasma reactive ion etching (ICP-RIE) [112–116], and electron cyclotron resonance (ECR) plasma [117–120] etching of SiC have been investigated.

Etching gas systems can be categorized into (i) fluorine-based, (ii) chlorine-based, and (iii) bromine-based gases as follows:

1. *Fluorine-based*: $SF_6, CF_4, NF_3, BF_3, CHF_3$
2. *Chlorine-based*: $Cl_2, SiCl_4, BCl_3$
3. *Bromine-based*: Br_2, IBr.

O_2 or Ar is often added to enhance removal of carbon atoms or to increase active species, especially in fluorine-based chemistry. The control of carbon removal is the main difference between the RIE of SiC and that of Si. Regarding the etching rate, fluorine-based chemistries generally give higher rates. In dry etching of SiC, the dominant mechanism is determined by the volatility of the reaction by-products and the energy of the ionized species. In practice, this translates into choices of the etching gases, the plasma pressure, and the bias voltage (or power) of the sample electrode [121]. Table 6.2 shows the boiling points of potential etch products in SiC RIE using fluorine- or chlorine-based chemistries. The fluorinated products are more volatile than the chlorinated ones, which may be one reason why higher etching rates are obtained with fluorine-based chemistries. Figure 6.26 shows the etching rate versus the gas composition in CCP-RIE with various fluorinated gas mixtures (a) without and (b) with H_2 additive [110]. Etching of SiC by NF_3 gives a high etching rate, though NF_3 is a toxic gas. Addition of O_2 promotes the etching, while the presence of H_2 results in decreased etching rate. Note that almost the same etching rates are obtained on (0001) and $(000\bar{1})$ faces.

Table 6.2 Boiling points of potential etch products in SiC RIE using fluorine- or chlorine-based chemistries.

Etching product	SiF_4	$SiCl_4$	CF_4	CCl_4	CO_2	CO
Boiling point (°C)	−86	58	−128	77	−79	−192

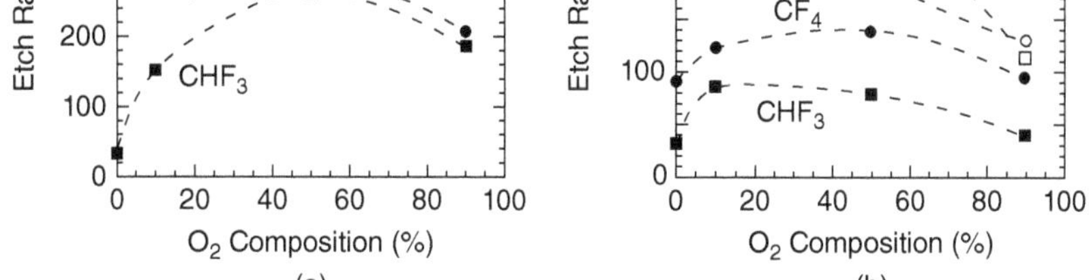

Figure 6.26 Etching rate of SiC versus the gas composition in CCP-RIE with various fluorinated gas mixtures (a) without and (b) with H_2 additive [110].

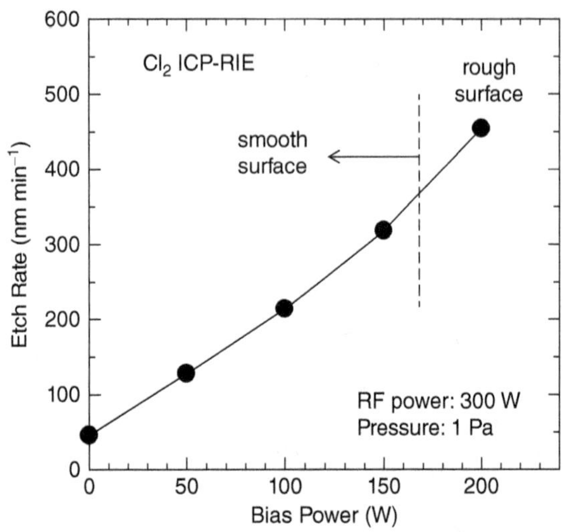

Figure 6.27 Etching rate of SiC as a function of self-bias power in ICP-RIE with a Cl_2–Ar system.

The etching rate of SiC can usually be increased with high-density plasma sources, such as ICP, helicon plasma, or ECR plasma. In particular, ICP-RIE offers an attractive etching technique, because the generation of high-density plasma and the RF bias to the samples (ion energy) can be independently controlled [121]. Figure 6.27 shows the etching rate of SiC as a function of self-bias power in ICP-RIE with a Cl_2–Ar system. The self-bias is a critical parameter, which determines the etching rate. By increasing the self-bias power, the etching is significantly enhanced, although it causes more surface damage. It has been reported that several deep levels are generated by RIE processes in n- and p-type SiC [122]. These defects can be generated at depths up to a few microns and affect the doping concentration, especially in p-type SiC. The densities of most of these deep levels can be reduced by more than 1 order of magnitude via thermal oxidation [122]. Exact surface reactions during RIE of SiC are not known and basic study of the etching mechanism is strongly desired.

Although dry etching of SiC itself is easy, obtaining high selectivity against a masking material is a challenge. Table 6.3 summarizes typical selectivities of SiC etching against that of various masking materials. Photoresist, which is often employed in RIE of Si, is not a good choice, because of very low selectivity. This is inevitable, because the etching gases and conditions for RIE of SiC have been adjusted to promote carbon etching. A high selectivity (over 10) is easily obtained by using metals, such as Al, Ni, or Cr, as the mask. However, nonvolatile by-products such as Al_2O_3 are formed during RIE, and small particles of these materials can be adsorbed on the surface. These particles work as a "micromask" and pillar-like hillocks are formed, leading to considerable surface roughening. This phenomenon is well known in the RIE of Si and other semiconductors [121]. By using a graphite plate on the cathode electrode, the micromasking effect can be greatly reduced [106]. The micromasking effect can also be reduced by adding a H_2 flow, because the H_2 supply promotes formation of volatile AlH_3, which effectively removes Al particles sputtered from the Al mask or the reactor walls [108, 110, 115]. One drawback of this process is degradation of the etching selectivity. In real device fabrication in industry, metallic contamination should be minimized throughout device processing. Therefore, metal masks are not usually employed in the fabrication of SiC devices. A common masking material in the RIE of SiC is SiO_2 deposited by CVD. By using slightly oxygen-rich conditions, or by increasing the self-bias, the selectivity of SiC against SiO_2 can be increased to 5–10 or even higher.

Control of etching profiles is as important as etching rates and selectivity. Figure 6.28 shows examples of etched SiC, as observed by a scanning electron microscope (SEM). In Figure 6.28a, a trench with an

Table 6.3 Typical selectivity of SiC etching against various masking materials.

Mask	SiO_2	ITO	Al	Ni	Photoresist
F-based RIE	0.8–3	10–20	5–30	>50	<0.5
Cl-based RIE	4–15	3–10	2–10	–	<0.8

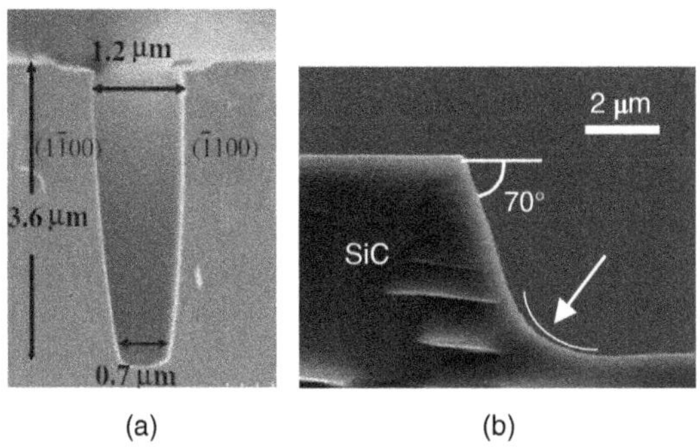

Figure 6.28 Examples of etched SiC, as observed by a scanning electron microscope (SEM). (a) Trench with an aspect ratio of 3 formed by ICP-RIE with a SiO_2 mask ([123] reproduced with permission from The Japan Society of Applied Physics), (b) mesa structure with a rounded bottom formed by CCP-RIE [124].

aspect ratio of 3 was formed by ICP-RIE with a SiO_2 mask [123], where the angle of the side walls is about 85°. In Figure 6.28b, a mesa structure with a rounded bottom was formed by CCP-RIE [124]. This structure was formed by using a SiO_2 mask, which was etched by a wet process. The rounded shape formed by the wet etching was transferred to SiC by RIE with a relatively low-selectivity condition. This mesa structure can be applied to an edge termination combined with implanted junction termination extension [124]. Furthermore, mesa structures with a very shallow bevel angle, which are also useful as termination structures, can be formed by using a thick photoresist baked at high temperature as an etching mask [125]. Dry etching is widely employed to fabricate SiC microelectromechanical systems (MEMSs) [126–128].

Thus, RIE processes of SiC have been developed to some extent but the etching mechanism is still not fully understood. The improvement of etching selectivity and reduction of surface roughness (of both the etched SiC surface and trench/mesa sidewalls) are remaining issues.

6.2.2 *High-Temperature Gas Etching*

Gas etching of SiC at high temperature has been investigated since the 1960s. Typical gases employed for etching are H_2 [129–131], $HCl + H_2$ [132–134] and $Cl_2 + O_2$ [135, 136]. Figure 6.29 shows Arrhenius plots of the etching rate of SiC by H_2 [131], $HCl + H_2$ [134], and $Cl_2 + O_2$ [136]. It is reported that ClF_3 is effective to obtain a very high etching rate [137]. Etching with H_2 or $HCl + H_2$ is routinely employed as *in situ* etching of substrates prior to epitaxial growth. At 1500 °C, the etching rate is about 0.05–0.1 μm min^{-1} for H_2 and 0.5–1 μm min^{-1} for $HCl(0.1\%)/H_2$, though the etching rate also depends on the process pressure. These etching processes give very flat surfaces, with periodic step-terrace structures formed on off-axis SiC{0001} [130, 134]. On the other hand, the etching rate by $Cl_2 + O_2$ is about

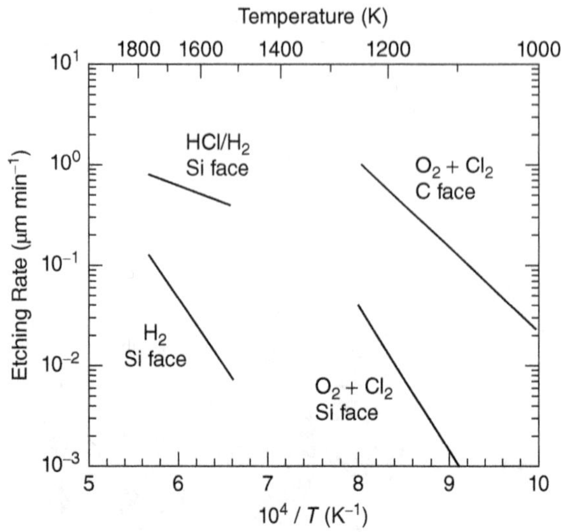

Figure 6.29 Arrhenius plots of etching rate of SiC by H_2 [131], HCl + H_2 [134], and $Cl_2 + O_2$ [136].

0.03 μm min^{-1} on (0001) and 1 μm min^{-1} on $(000\bar{1})$ at 1000 °C. However, this etching process leads to relatively large etch pits at dislocation sites on the (0001) face. Although high-temperature gas etching will not be employed as a main etching process (for formation of mesa or trench structures) in SiC device fabrication, it may be very effective in obtaining smooth side-walls of trenches [123, 136].

6.2.3 *Wet Etching*

SiC can be etched by molten KOH, NaOH, or Na_2O_2 at 450–600 °C. In these melts, SiC is oxidized and the formed oxide is subsequently removed by the melt [138]. However, this etching process usually creates dislocation pits or hillocks on the surface, as described in Section 5.1.5. Severe contamination from K or Na must be avoided in subsequent device processing, especially in oxidation.

SiC can be etched by electrochemical (or photoelectrochemical) etching. In these etching processes, holes must be supplied to the surface to induce oxidation. Therefore, p-type SiC can be selectively etched in electrochemical etching, when the sample is correctly biased, but etching of n-type SiC is suppressed because of the lack of holes [139, 140]. On the other hand, selective etching of n-type SiC is possible under illumination of above-bandgap photons. Under this illumination, photo-generated holes are accumulated at the electrolyte/n-type SiC interface as a result of surface band bending, leading to oxidation and subsequent etching of the oxide by the solution [141–145]. Since holes are depleted at the electrolyte/p-type SiC interface, p-type SiC is not etched in this process. Although this etching selectivity (n-type versus p-type) is of interest, these etching processes are not very suitable for fabrication of electronic devices. Major disadvantages include a fairly rough etched surface, the inability to pattern small-dimension features, and poor etching uniformity across entire wafers.

6.3 Oxidation and Oxide/SiC Interface Characteristics

A unique advantage of SiC is that it is the only compound semiconductor that can be thermally oxidized to give high-quality SiO_2. Therefore, thermal oxides of SiC are utilized as a gate dielectric in metal-oxide-semiconductor (MOS) devices as well as to passivate the SiC surface. However, the most striking difference from Si technology is, of course, carbon atoms, which are one of the host elements in SiC. A number of review papers on SiC MOS have been published [146–156]. In spite of continuous improvement of the SiC MOS interface, the quality, and the community's understanding of the factors which control this quality, is still at far from a satisfactory level. This subsection describes the common features, present understanding, and problems in SiC MOS technology.

6.3.1 Oxidation Rate

Thermal oxidation of SiC is expressed by the following simple equation:

$$SiC + 3/2\, O_2 \rightarrow SiO_2 + CO \tag{6.1}$$

Therefore, a thermal oxide of SiC is SiO_2, the formation of which can be confirmed by X-ray photoelectron spectroscopy (XPS), electron energy loss spectroscopy (EELS), and Auger electron spectroscopy (AES). Taking into account the Si density in SiC, the amount consumed during thermal oxidation of SiC can be calculated as 46%, which is close to the value for thermal oxidation of Si. For example, to grow 100 nm of SiO_2 on SiC, 46 nm of SiC are consumed. During thermal oxidation, most of the carbon atoms in SiC are removed and diffuse out as carbon monoxide (CO) molecules, and a small portion of carbon atoms diffuses into the SiC bulk region, leading to reduction of carbon-vacancy-related defects [157]. However, the thermal oxide is not completely free of carbon, and it is believed that carbon atoms remain near the oxide/SiC interface. More detail is given in Section 6.3.4.

Figure 6.30 shows the oxide thickness versus the oxidation time for thermal oxidation of SiC at various temperatures for (a) (0001) and (b) (000$\bar{1}$). The oxidation was performed in 100% dry O_2, and results on (0001) and (000$\bar{1}$) faces are plotted. Oxidation on the (000$\bar{1}$) face is about 8–15 times faster than that on the (0001) face. This phenomenon can be used to identify the polarity of unknown crystal faces [158]. The oxide thickness shown in Figure 6.30 can be reasonably explained by the *Deal–Grove model,* which was developed for Si technology [159, 160]:

$$d_{ox}^{\,2} + A\, d_{ox} = Bt \tag{6.2}$$

where d_{ox} is the oxide thickness and t the oxidation time. B and B/A are called the parabolic rate constant and linear rate constant, respectively. Note that a modified Deal–Grove model gives a better agreement

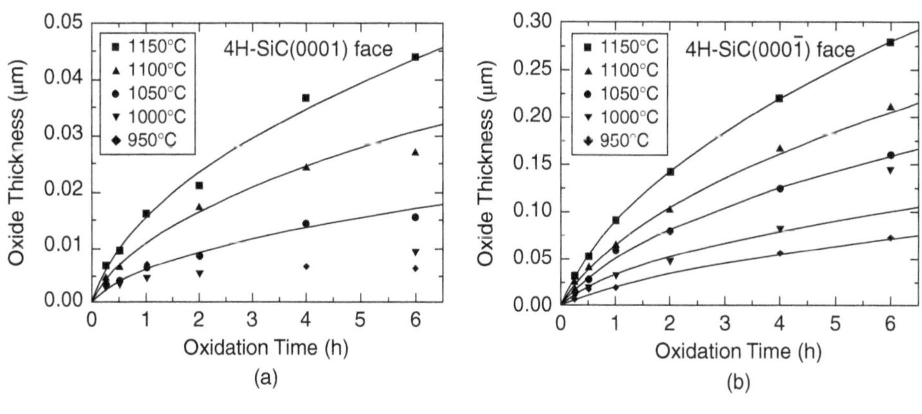

Figure 6.30 Oxide thickness versus the oxidation time for thermal oxidation of SiC at various temperatures for (a) (0001) and (b) (000$\bar{1}$) ([161] reproduced with permission from AIP Publishing LLC).

with experimental results [161]. The oxide thickness is almost proportional to the oxidation time when the oxide is very thin (*surface-reaction-limited regime*). The oxidation gradually exhibits slowdown, and the oxide thickness becomes almost proportional to the square root of the oxidation time when the oxide becomes thick (*diffusion-limited regime*) [161–164]. However, there is large scatter in the reported values of the activation energies of the linear rate constant (B/A) and the parabolic rate constant (B) in the literature. In recent years, a clear deviation of experimental results from this model has been pointed out and a modified model has been proposed [165]. In this model, emission of silicon and carbon atoms from the oxidation interface is considered and the abnormally fast oxidation rate in the initial stage can be qualitatively explained. This model explains well the oxidation rate of SiC(0001) and (000$\bar{1}$) under a wide range of oxidation conditions. However, the exact reactions during thermal oxidation of SiC at a microscopic level have not yet been fully clarified.

As is the case in Si technology, wet (including pyrogenic) oxidation results in a faster oxidation rate than dry oxidation. It is known that long wet oxidation can create surface pits at the locations of dislocations, because of enhanced oxidation near the dislocation cores [166, 167], while the pit formation is much smaller in dry oxidation. Enhanced oxidation is also observed at the implanted region. Even after post-implantation annealing at 1600–1700 °C, an implanted region contains weak Si-C bonds and can be oxidized at a faster rate. The oxides grown on the implanted regions are usually 10–40% thicker than those on the non-implanted regions; the enhancement factor depends on the implantation, annealing, and oxidation conditions.

As described above, the oxidation rate depends strongly on the crystal face in SiC [158, 168–170], and this anisotropy must be carefully taken into account in device fabrication. Under any oxidation conditions (dry/wet, any temperature), oxidation is the fastest on (000$\bar{1}$) and the slowest on (0001). The oxidation rates on the (11$\bar{2}$0) or (1$\bar{1}$00) faces are in between those on (0001) and (000$\bar{1}$). An example of this is shown in Figure 6.31, where the oxide thickness is plotted as a function of the angle from the (000$\bar{1}$) face. This strong anisotropy is particularly important in the fabrication of trench MOSFETs. When trench

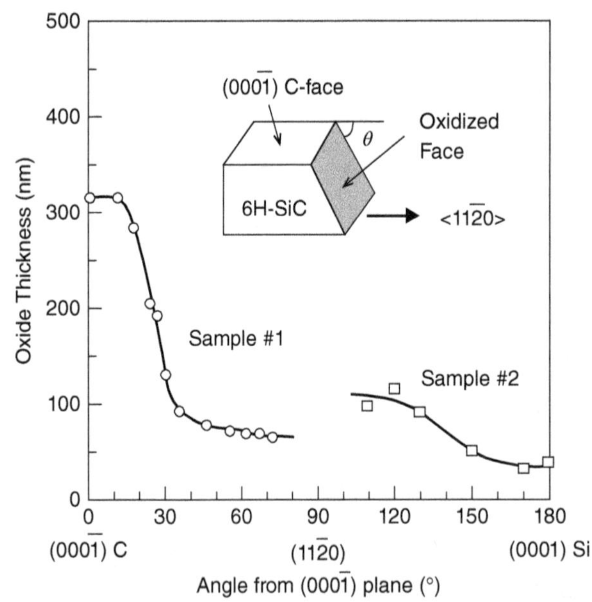

Figure 6.31 Anisotropy in the oxidation rate in hexagonal SiC, where the oxide thickness is plotted as a function of the angle from the (000$\bar{1}$) face ([169] reproduced with permission from Wiley-VCH Verlag GmbH).

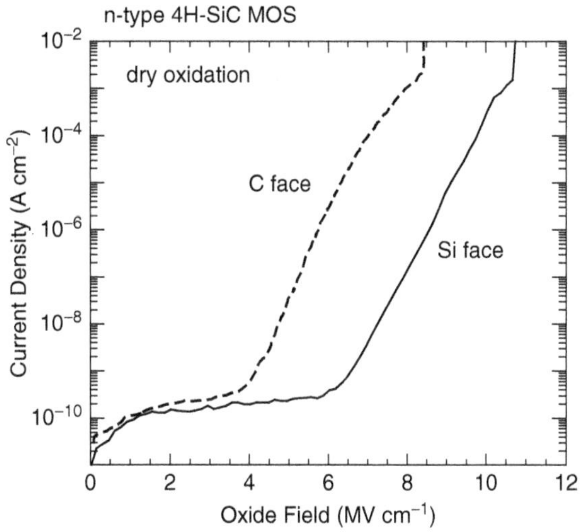

Figure 6.32 Current density–electric field (J–E) characteristics for 40-nm-thick thermal oxides grown on n-type 4H-SiC(0001) and (000$\bar{1}$) surfaces. The oxides were formed by dry oxidation at 1200 °C.

MOSFETs are processed on a SiC(0001) wafer, the oxide thickness is largest on the trench sidewalls and smallest near the bottom (and top) of a trench. Thus, a special design and structure must be employed to avoid breakdown of gate oxides near the trench bottom. In the case of trench MOSFETs on SiC(000$\bar{1}$), the profile of oxide thickness in the trench is ideal, the thickness is smallest on the sidewalls and largest near the trench bottom [170].

6.3.2 Dielectric Properties of Oxides

Figure 6.32 shows the current density–electric field (J–E) characteristics for 40-nm-thick thermal oxides grown on n-type 4H-SiC(0001) and (000$\bar{1}$) surfaces. The oxides were formed by dry oxidation at 1200 °C. The J–E characteristics were obtained by applying positive voltage to the gate of n-type MOS capacitors (accumulation state). Thermal oxides adequately formed on high-quality SiC exhibit very good dielectric properties, with a resistivity over 10^{16} Ω cm at low electric field (<3 MV cm^{-1}). The breakdown electric field of thermal oxides on SiC is approximately 9–13 MV cm^{-1}, which depends strongly on the oxidation condition, gate material, and surface roughness [171–174].

In the high-field region, the current through an oxide is governed by the *Fowler–Nordheim tunneling current* (J_{FN}), which is expressed by the following equation [8, 175].

$$J_{FN} = \frac{q^3E^2}{16\pi\hbar\phi_B}\exp\left(-\frac{4\sqrt{2m^*\phi_B{}^3}}{3q\hbar E}\right) \tag{6.3}$$

Here, E is the electric field strength in the oxide, m^* is the effective mass of an electron in the oxide, ϕ_B is the barrier height, or the conduction band offset (ΔE_c) between the oxide and semiconductor. The barrier height can be estimated from the slope of a Fowler–Nordheim (ln (J/E^2) – $1/E$) plot. The barrier height is also obtained from high-resolution XPS [176] or internal photoemission (IPE) measurements [177]. As shown in Figure 6.32, the onset electric fields for FN tunneling current are about 5.5–6.0 MV cm^{-1} for 4H-SiC(0001), 4.0–4.5 MV cm^{-1} for 4H-SiC(000$\bar{1}$), and 6.0 MV cm^{-1} for Si. This difference is caused

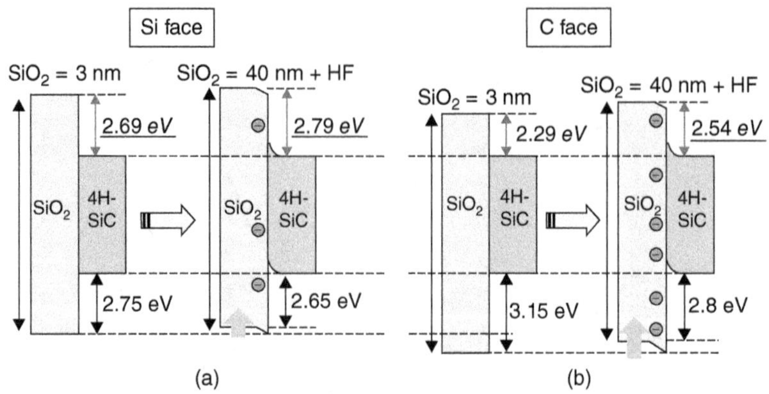

Figure 6.33 Band line-ups for dry oxide/n-type 4H-SiC (a) (0001) and (b) (000$\bar{1}$) structures, determined from synchrotron-radiation XPS ([176] reproduced with permission from Trans Tech Publications).

by different band structures, as shown in Figure 6.33 [176]. Figure 6.33 shows the band line-ups for dry oxide/n-type 4H-SiC(0001) and (000$\bar{1}$) structures which were determined from synchrotron-radiation XPS. Because of the wider bandgap and smaller electron affinity of SiC compared with those of Si, the barrier height (ϕ_B or ΔE_c) is inherently smaller in SiC. It has been argued that this smaller barrier height may limit the oxide reliability of SiC MOSFETs, especially at high electric field and high temperature [178]. Furthermore, the barrier height for SiC is dependent on the crystal face. Since SiC{0001} is a polar face, there exists a significant dipole at the oxide/SiC{0001} interface. As a result, the barrier height on 4H-SiC(0001) (2.7–2.8 eV) is higher than that on (000$\bar{1}$) (2.4–2.5 eV). This trend is valid for any SiO_2/SiC system, though the absolute values of barrier height vary for different processes (for example, the barrier height is increased on (000$\bar{1}$) in the case of wet oxidation).

Oxide reliability under high electric field has been extensively investigated [173, 179–182]. The oxide reliability in terms of dielectric properties is usually evaluated by *time-dependent dielectric breakdown* (TDDB) tests [183]. For example, a number of MOS capacitors are biased at very high electric field (>8 MV cm^{-1}) in the accumulation state, to force the Fowler–Nordheim tunneling current. The capacitors are either maintained at a fixed voltage (constant field stress) or constant current density (constant current stress). Either the time *time-to-breakdown,* t_{BD}, or the *charge-to-breakdown,* Q_{BD} (which is the integrated current until breakdown) is monitored for each capacitor. After collecting these data, the distribution of t_{BD} or Q_{BD} is analyzed by using a *Weibull plot* [183]:

$$\ln\{-\ln(1-F)\} = \beta \ln(t/\tau) \tag{6.4}$$

Here F is the cumulative failure probability, t the time-to-failure for each device. τ is the characteristic time-to-failure and β the shape parameter (or Weibull slope). When $F = 0.63212$ ($= 1 - e^{-1}$), the left-hand side of Equation 6.4 goes to zero, indicating that the characteristic time-to-failure (τ) is the time for 63.212% of the devices to fail. When a number of devices with different areas are compared, the Weibull plot scaled by the area is useful [183, 184]:

$$\ln\{-\ln(1-F_2)\} - \ln\{-\ln(1-F_1)\} = \ln(A_2/A_1) \tag{6.5}$$

Here A_1 and A_2 are the areas of the test devices, F_1 and F_2 are their cumulative failure probabilities. In TDDB tests on SiC MOS capacitors, dislocations and particles can affect the oxide reliability, and the t_{BD} and Q_{BD} tend to become smaller with increasing device area. When the area-scaled Weibull plot is employed, a more universal trend appears [184].

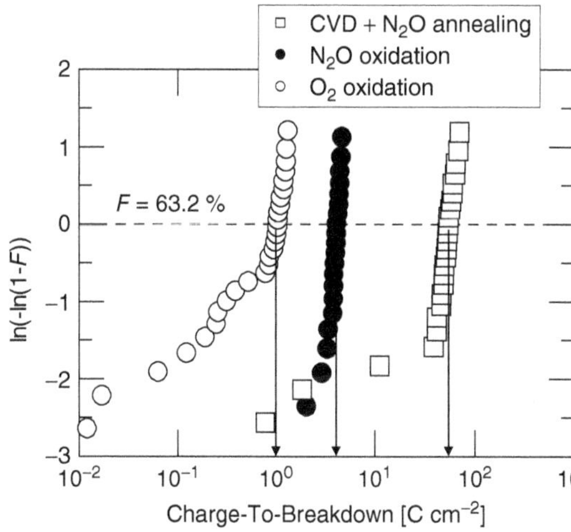

Figure 6.34 Weibull plots of Q_{BD} for several oxides formed on n-type 4H-SiC(0001) [185].

Figure 6.34 shows Weibull plots of Q_{BD} for several oxides formed on n-type 4H-SiC(0001) [185]. In this case, a high mean value of Q_{BD} and its steep slope (tight distribution) are desirable to ensure the oxide reliability. The oxide reliability depends strongly on the oxidation/annealing conditions, the quality (defect density) of SiC, the surface roughness, the device area, and the gate material, as well as other factors. In spite of several concerns pointed out in the early stages, the oxide reliability reported in the literature is rather promising, at least for temperatures below 300 °C. A few groups reported that screw and basal-plane dislocations in SiC epilayers cause severe reduction in t_{BD} and Q_{BD} of the thermal oxides formed on these areas [173, 181]. It was eventually discovered that these dislocations themselves are not always harmful to the oxide reliability. During the post-epitaxial growth (cooling down) process and other processing steps, surface pits can form at these dislocation sites. The surface pits cause electric field crowding of the oxides grown on them, leading to degraded oxide reliability. By suppressing the surface-pit formation or by surface planarization prior to thermal oxidation, very good oxide reliability has been achieved, even on the areas which contain dislocations [186]. Another approach to improve the oxide reliability is the usage of deposited oxides. By using an oxide–nitride–oxide (ONO) stack or a deposited oxide nitrided in NO or N_2O at high temperature, high breakdown fields over 12 MV cm^{-1} and very high Q_{BD} values of 50–150 C cm^{-2} have been attained [185, 187–189]. The oxide reliability at high temperature is also described in Section 8.2.11. Owing to rapid progress in SiC process technology, the data on oxide reliability is still being updated. Please refer to the latest papers and conference proceedings for the state-of-the-art in this field.

6.3.3 *Structural and Physical Characterization of Thermal Oxides*

Structure analyses of the SiO_2/SiC interface are not straightforward and several conflicting results have been reported. One main concern is the detection of carbon, which may remain near the interface as well as in the thermal oxide. The density of carbon atoms is usually below the detection limits of XPS and AES. In early SIMS measurements, carbon contamination in thermal oxides in the range of 10^{18}–10^{19} cm^{-3} was claimed. However, more recently, it has been confirmed that the residual carbon inside the oxides is close to the detection limit of SIMS when the oxides are grown under appropriate conditions, such as at high temperature and an appropriate nitridation process is conducted. The detection of carbon near

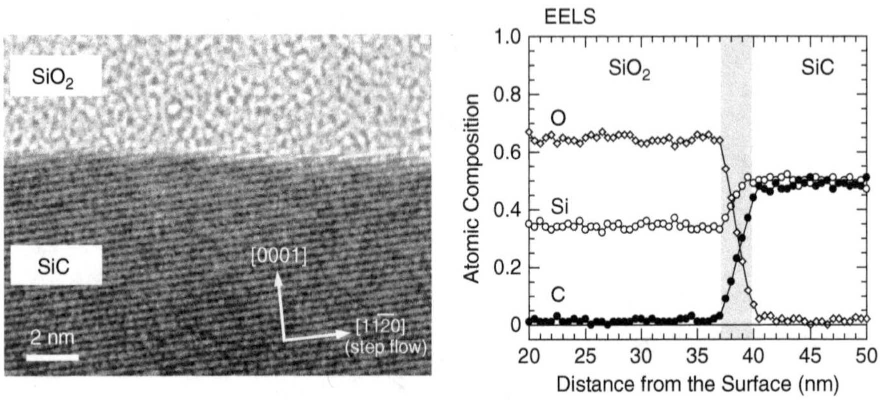

(a) (b)

Figure 6.35 (a) Typical cross-sectional TEM image and (b) intensity profiles of Si, C, and O signals in EELS measurements on a 4H-SiC(0001) MOS structure with a 40-nm-thick thermal oxide.

the interface by SIMS is difficult, because the interface is rather abrupt and the secondary ion yield is very sensitive to the host material. An early EELS analysis suggested that sp^2-like carbon exists near the interface [190], but this result has not always been reproduced in recent investigations. More recently, several groups tried careful TEM/EELS studies of SiO_2/SiC structures. One group showed a carbon-rich transition layer near the interface [191, 192], while another group claimed that the interface is very abrupt (the thickness of the transition layer is less than 1–2 nm) and that it is hard to detect it by EELS [193]. Figure 6.35 shows (a) a typical cross-sectional TEM image and (b) intensity profiles of Si, C, and O signals in EELS measurements on a 4H-SiC(0001) MOS structure with a 40-nm-thick thermal oxide. The oxide was grown by dry oxidation at 1300 °C. In the TEM image, neither significant disorder nor a thick transition layer is observed. A slightly darker contrast near the interface in the SiC lattice shown in literature originates from a slight "overfocus" in TEM observation, and this does not mean a transition layer. The contrast can be darker or brighter, depending on the focus condition of the TEM. In the intensity profile of EELS (Figure 6.35b), the measurement point was swept normal to the interface with a resolution below 1 nm. All the Si, C, and O signal intensities exhibit a reasonably abrupt change at the interface, and accumulation of C atoms near the interface is not observed within the resolution of EELS. The thickness of a transition layer, even if it exists, is smaller than 2 nm.

The interface structure has also been investigated by high-resolution XPS [194–196], including synchrotron XPS. In these studies, the interface is more abrupt than one expects, and the existence of only a few monolayer sub-oxides is indicated [196]. Spectroscopic ellipsometry measurements suggested that the interface is rather abrupt, with a very thin (< 2 nm) transition layer at the SiC side of the interface [197]. A high-resolution medium energy ion scattering (MEIS) analysis also revealed a fairly abrupt SiO_2/SiC interface [198]. It is important to reveal how the interface structure is changed when the oxidation/annealing conditions are varied. Cathodoluminescence (CL) and attenuated-total-reflectance Fourier transform infrared (ATR-FTIR) spectroscopy measurements have been applied to characterize SiO_2/SiC structures. In CL measurements, luminescence peaks, which are attributed to oxygen-vacancy centers, were detected at 460 and 490 nm, and some correlation with interface state density was reported [199]. Theoretical studies [200–205] should be linked to experimental investigations to reveal the real microscopic structure of the SiC MOS interface. Furthermore, one must be aware of the limitation of characterization techniques. In normal structural analyses such as XPS, AES, and EELS, the detection limit of foreign elements is not very good (0.3–1% of the host elements). On the other hand, imperfection of 0.1% of the host elements will cause a huge number of electronic defects, which are detectable in electrical characterization.

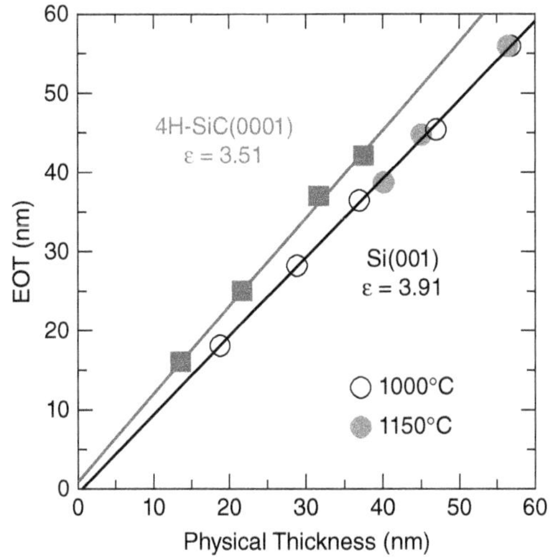

Figure 6.36 Relationship between the equivalent oxide thickness (EOT) and the physical oxide thickness for SiC MOS structures ([206] reproduced with permission from Trans Tech Publications).

Regarding the relative dielectric constant of thermal oxides, it has been assumed to be identical to that of thermal oxides grown on Si ($\varepsilon_{ox} = 3.9$). Figure 6.36 shows the relationship between the *equivalent oxide thickness* (EOT) and the physical oxide thickness for SiC MOS structures [206]. The EOT and physical thickness were determined by the accumulation capacitance and AFM, respectively. From this plot, the relative dielectric constant of thermal oxides on SiC is estimated to be 3.51. The physical reason for this low value is not clear at present.

6.3.4 Electrical Characterization Techniques and Their Limitations

6.3.4.1 Basic Phenomena Specific to SiC

To assess the quality of MOS interfaces, electrical characterization such as $C-V$ and conductance measurements of MOS capacitors is performed. In electrical characterization of SiC MOS capacitors, one has to bear the following points in mind:

1. Since the intrinsic carrier density and the carrier generation rate are extremely low at room temperature, no inversion layers are created in normal SiC MOS capacitors. Thus, $C-V$ curves exhibit "*deep depletion*", even in quasi-static (low-frequency) $C-V$ measurements, unless illuminated with appropriate light.
2. Because of its wide bandgap, the majority of interface states are energetically very deep. The emission time constant of electrons from interface states $\tau(E)$ follows the simple equation [146, 207]:

$$\tau(E) = \frac{1}{\sigma_n v_{th} N_C} \exp\left(\frac{E_C - E}{kT}\right) \tag{6.6}$$

where σ_n is the capture cross section, v_{th} the thermal velocity of electrons, N_c the effective density of states of the conduction band, and E_c the energy of the conduction band edge, k the Boltzmann constant, and T the absolute temperature. Figure 6.37 shows the emission time constant from the interface states as a function of the energy level, assuming a capture cross section of 1×10^{-15} cm^2.

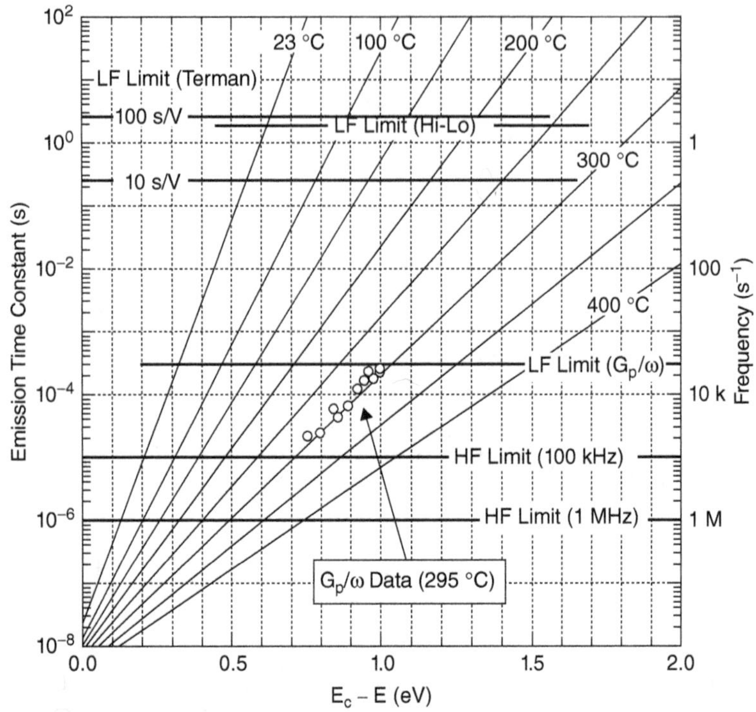

Figure 6.37 Emission time constant from the interface states as a function of the energy level, assuming a capture cross section of 1×10^{-15} cm^2 ([146] reproduced with permission from Wiley-VCH Verlag GmbH).

For example, the emission time constant of an interface state at $E_c - 1.0$ eV is as long as 6×10^5 s (about seven days) at room temperature. This means that electrons trapped at such deep states are never emitted to the conduction band and, unless appropriate illumination, or heating is used, stay trapped for the entire measurement period. In other words, such deep states are frozen and do not respond to any probe frequency or voltage sweep rate employed in the measurements. It should be noted that the capture cross section of interface states is strongly dependent on the energy level, and the emission time constant in actual MOS structures does not change as that shown in Figure 6.37. More accurate discussion is described in Section 6.3.4.7.

Note that these situations are similar to the case of Si MOS at very low temperatures (30–77 K) [208, 209].

Figure 6.38 shows the high-frequency (100 kHz) C–V curves of MOS capacitors fabricated on (a) n-type and (b) p-type 4H-SiC(0001). Gate oxides, about 40 nm thick, were formed by either dry or wet oxidation at 1200 °C. Post-oxidation annealing (POA) was carried out in Ar at the same temperature for 30 min. In general, a positive shift in C–V curves is observed for n-type SiC MOS capacitors, while a negative shift is seen for p-type MOS capacitors. From the voltage shift at the flat band (ΔV_{FB}), the effective fixed (or oxide) charge density (Q_{eff}) can be determined by the following equation [160]:

$$Q_{eff} = C_{ox}\Delta V_{FB} = C_{ox}(V_{FB,theory} - V_{FB,exp}) \tag{6.7}$$

where C_{ox} is the oxide capacitance, and $V_{FB,theory}$ and $V_{FB,exp}$ are the theoretical and experimental flat-band voltages, respectively. Since carriers trapped at deep interface states are frozen and act as fixed charges at

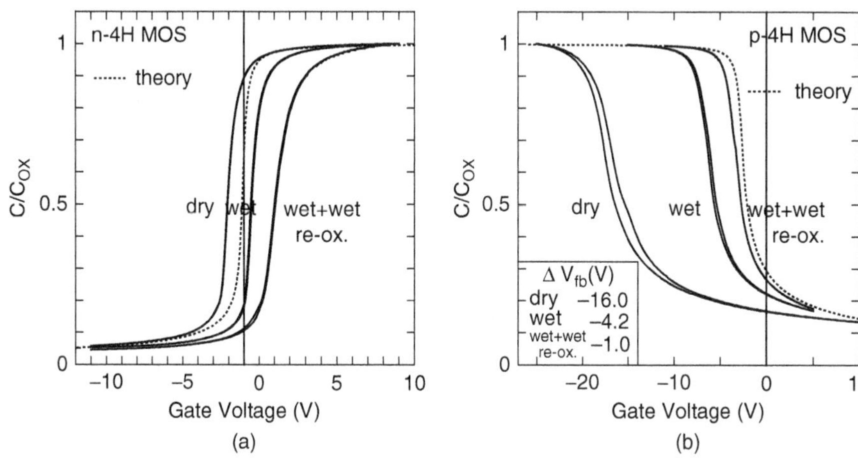

Figure 6.38 High-frequency (100 kHz) $C-V$ curves of MOS capacitors fabricated on (a) n-type and (b) p-type 4H-SiC(0001). About 40 nm-thick gate oxides were formed by either dry or wet oxidation at 1200 °C.

the interface, Q_{eff} is the sum of real fixed charges (positive or negative) and the charges of trapped carriers (negative (electrons) in n-type and positive (holes) in p-type MOS). This is the reason why this is called the *effective* fixed charge density. In this particular case (Figure 6.38), the effective fixed charge density is calculated to be 8×10^{11} cm^{-2} (negative) for the dry oxide and 2×10^{12} cm^{-2} (negative) for the wet oxide on n-type 4H-SiC(0001). In the case of p-type MOS capacitors, the effective fixed charge density is calculated to be 5×10^{12} cm^{-2} for the dry oxide and 2×10^{12} cm^{-2} (both positive) for the wet oxide.

Separation of real fixed charge (positive or negative) and charges of trapped carriers can be achieved by illuminating MOS capacitors with appropriate below-gap light (so-called "photo $C-V$"). Figure 6.39 shows the high-frequency $C-V$ curves of a MOS capacitor with a dry oxide on p-type 4H-SiC(0001) [146]. The voltage is first swept from accumulation to deep depletion in darkness. Light from a Xe lamp was then shone on the capacitor while it was held at a bias voltage of +10 V, leading to formation of an inversion layer. After stabilization the voltage sweep was started. The $C-V$ curve exhibits a large hysteresis in the deep depletion region: After light illumination, the $C-V$ curve from 0 V to about −3 V is shifted significantly and approaches the theoretical curve; this change can be attributed to the greatly reduced positive interface charge upon illumination. From the voltage shift in the deep depletion (ΔV_{DD}), the charge density of holes trapped at deep donor-like interface states can be estimated to be approximately 4×10^{12} cm^{-2}. Thus, the majority of interface charges in p-type MOS structures originate from holes trapped at deep interface states, rather than real fixed charges. Note that the density of effective fixed charges (mostly negative in n-type and positive in p-type) is still rather high, $(0.4-2) \times 10^{12}$ cm^{-2}, even after process optimization [153, 154], and can affect MOSFET characteristics. The effective charges at the oxide (passivation layer)/SiC interface also affect the charge balance in the junction termination region [210].

In general, ion motion shows a clockwise hysteresis in $C-V$ curves of MOS capacitors, while carrier injection (trapping) produces a counterclockwise hysteresis, regardless of the conduction type [160]. $C-V$ curves of SiC MOS capacitors, especially on n-type 4H-SiC(0001), often exhibit a carrier-injection-type hysteresis to some extent. The hysteresis depends on the probe frequency, the voltage sweep rate, and the maximum applied voltage during the accumulation [211, 212]. Figure 6.40a shows an example of high-frequency $C-V$ curves of a MOS capacitor with a dry oxide on n-type 4H-SiC(0001) [211]. As the maximum bias voltage is increased, the $C-V$ curve from accumulation to depletion is shifted toward the positive direction. This result originates from the increase in electron

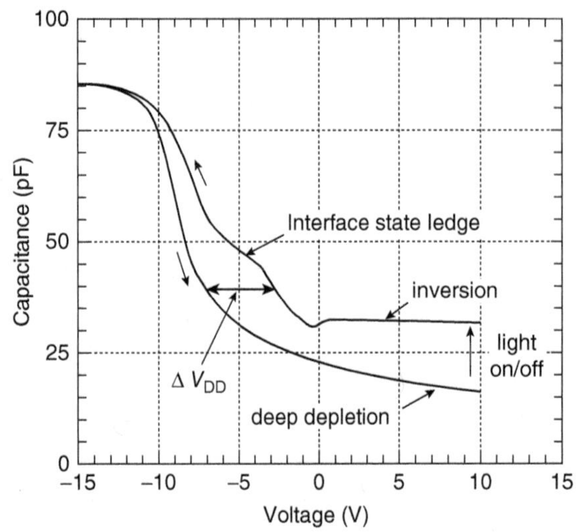

Figure 6.39 High-frequency $C-V$ curves of a MOS capacitor with a dry oxide on p-type 4H-SiC(0001) ([146] reproduced with permission from Wiley-VCH Verlag GmbH). After the voltage is first swept from accumulation to deep depletion in darkness, light illumination was performed at +10 V, leading to formation of an inversion layer. After stabilization, the voltage sweep was started toward the accumulation.

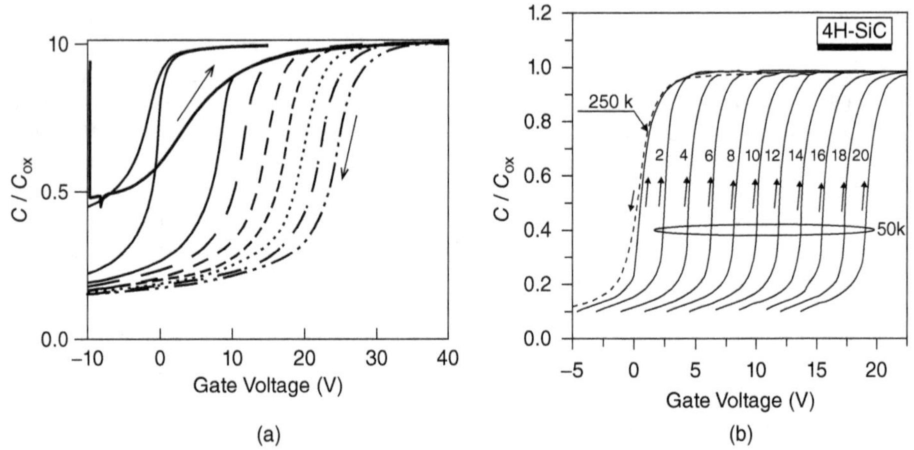

Figure 6.40 (a) High-frequency $C-V$ curves of a MOS capacitor with a dry oxide on n-type 4H-SiC(0001) ([211] reproduced with permission from Elsevier). As the maximum bias voltage is increased, the $C-V$ curve from accumulation to depletion is shifted toward the positive direction. (b) High-frequency $C-V$ curves measured sequentially at low temperature ([212] reproduced with permission fro Elsevier).

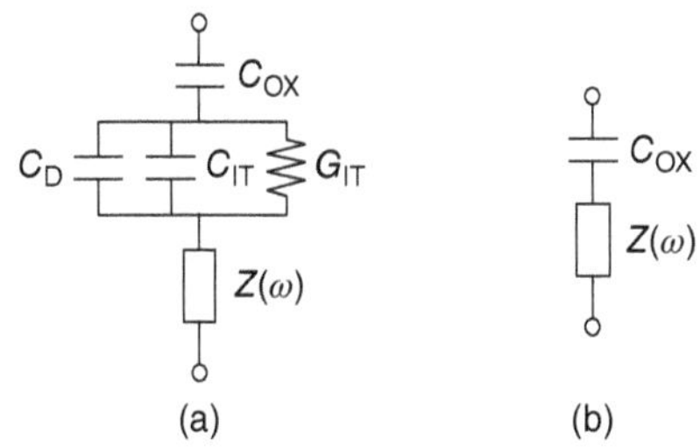

Figure 6.41 Equivalent circuits for (a) depletion to weak accumulation and (b) strong accumulation, where $C_{ox}, C_D, C_{IT}, G_{IT}$, and Z are the oxide capacitance, the semiconductor capacitance, the interface-state capacitance, the interface-state conductance, and the series parasitic impedance, respectively.

trapping at shallow (but slow) interface states and/or oxide traps near the interface when a higher positive voltage is applied. This phenomenon is markedly pronounced when C–V measurements are performed at low temperature, as shown in Figure 6.40b [212]. This is a useful technique to monitor such trapping but is an obstacle for accurate characterization when using C–V curves.

The interface state density can be characterized by several techniques, which possess their own merits and limitations [146, 160, 207]. Several characterization techniques are summarized below.

6.3.4.2 Equivalent Circuit of a MOS Capacitor

Before describing the characterization techniques, an equivalent circuit of a MOS capacitor must be understood. Figure 6.41 shows the equivalent circuits for (a) depletion to weak accumulation and (b) strong accumulation, where $C_{ox}, C_D, C_{IT}, G_{IT}$, and Z are the oxide capacitance, the semiconductor capacitance, the interface-state capacitance, the interface-state conductance, and the series parasitic impedance, respectively. The value of $Z(\omega)$ is determined for each frequency with an impedance analyzer. In strong accumulation, C_D, C_{IT}, and G_{IT} can be ignored because of the infinitely large C_D, and the measured capacitance, and conductance are almost independent of the gate voltage. Under moderately biased conditions, the effects of C_{IT} and G_{IT} become prominent.

6.3.4.3 Determination of the Surface Potential

It is important to accurately determine the surface potential ψ_s, because it determines the energy position of the interface states and is required in all of the characterization techniques described below. The determination of the surface potential is critical for SiC because the interface state density usually exhibits a sharp increase near the band edge, especially on SiC(0001). According to a theory in MOS physics, the surface potential (ψ_S) can be calculated from the low-frequency C–V curves using [160, 207]:

$$\psi_S(V_G) = \int (1 - C_{LF}/C_{ox})\, dV_G + A \tag{6.8}$$

where C_{LF} is the low-frequency (usually quasi-static) capacitance and V_G is the gate voltage. Here, a certain ambiguity exists in the determination of the integration constant (A). For example, this constant is often determined based on the flat-band capacitance in high-frequency measurements, by assuming that $\psi_s = 0$ (flat band) when the high-frequency capacitance C_{HF} equals the ideal flat-band capacitance (C_{FB}). This is correct only when the high-frequency capacitance does not include any contribution from the interface states. If the probe frequency is not high enough, the flatband capacitance contains a component

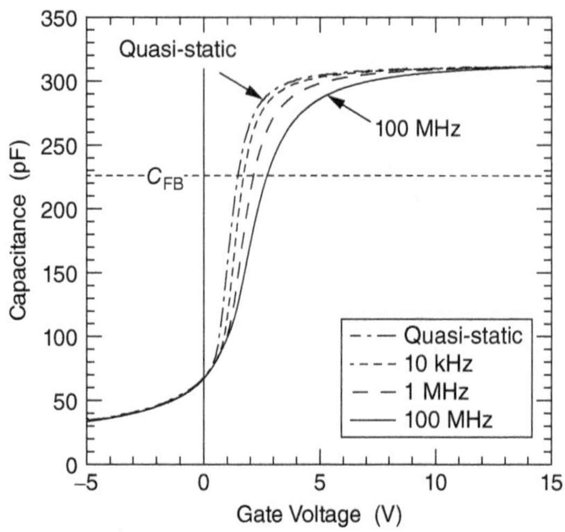

Figure 6.42 High-frequency $C-V$ curves of a MOS capacitor on n-type 4H-SiC(0001) at various probe frequencies (the influence of parasitic impedance at very high frequency was calibrated) [213].

from the fast interface states, leading to an error in the surface potential. This method results in a relatively large error of 0.06–0.15 eV when a high density of fast interface states exists, as is the case in SiC MOS structures. Some evidence for this is shown in Figure 6.42, where the high-frequency $C-V$ curves of a MOS capacitor on n-type 4H-SiC(0001) at various probe frequencies are shown (The influence of parasitic impedance at very high frequency was calibrated). There exists clear frequency dispersion in the $C-V$ curves, and the voltage at which the ideal flat-band capacitance C_{FB} is obtained clearly depends on the probe frequency. This result suggests that some interface states respond to such high frequency, and that the interface-state capacitance is involved in the high-frequency capacitance.

Taking account of the C_{ox} and Z values, $C_D + C_{IT}$ can be determined by using the equivalent circuit shown in Figure 6.41a. On the other hand, the surface potential $\psi_S(V_G)$ can be obtained using Equation 6.8, except for the integration constant A. The integration constant A can be uniquely determined, as shown in Figure 6.43, where $1/(C_D + C_{IT})^2$ is plotted against ψ_S. In Figure 6.43, a linear correlation is evident for sufficiently negative ψ_S (in the depletion region). At a sufficiently high frequency, and upon depletion, the interface states do not respond and no inversion carriers are generated at the SiC MOS interface. Therefore, $C_D + C_{IT}$ can be approximated as the depletion capacitance (C_{dep}), and a linear relationship can be established between $1/(C_D + C_{IT})^2$ and ψ_S [160, 207, 213]:

$$\frac{1}{(C_D + C_{IT})^2} \approx \frac{1}{{C_{dep}}^2} = -\frac{2\psi_s}{\varepsilon_s q N_D S^2} \tag{6.9}$$

where ε_s is the dielectric constant of the semiconductor, N_D the donor density, and S the area of the gate electrode. Based on Equation 6.9, the constant A can be determined so that extrapolation of the straight line should intersect the origin of the plot, as shown in Figure 6.43 [213]. The donor density is simultaneously determined from the slope of the plot.

6.3.4.4 Terman Method

In the Terman method, a voltage shift of a high-frequency $C-V$ curve from the ideal curve is extracted and then the interface state density is determined for each energy level (surface potential) [160, 207]. This

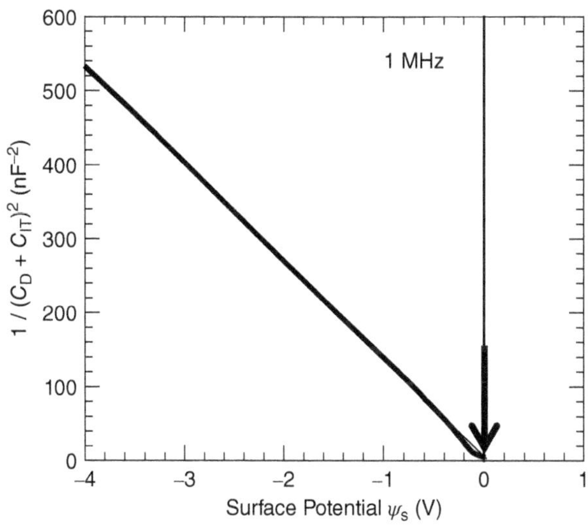

Figure 6.43 $1/(C_D + C_{IT})^2$ versus the surface potential ψ_S. The integration constant A can be uniquely determined so that extrapolation of the straight line should intersect the origin of the plot [213].

method is based on the assumption that the interface states do not respond to the high frequency and, therefore, the high-frequency capacitance does not include any contribution from the interface states. However, this assumption is usually not valid. Furthermore, the extracted interface state density includes a relatively large error because a slight deviation in the surface potential or doping density results in a large change in the extracted interface state density. Thus, this method has been employed only when the interface state density is extremely high (>10^{12} cm^{-2} eV^{-1}), and is not a preferred method to characterize the interface state density.

6.3.4.5 High–Low Method

In the high-low (or often hi–lo) method, the frequency response of interface states is utilized. The essential assumption is that the interface states fully respond to a frequency used for low-frequency C–V measurements and do not respond at all to a frequency used for high-frequency C–V. If this condition is satisfied, then the low-frequency capacitance includes contributions from all the interface states, while the high-frequency capacitance does not include any interface state contributions. Under this assumption, the interface state density D_{IT} is given by [160, 207]

$$D_{IT} = \frac{(C_D + C_{IT})_{LF} - (C_D + C_{IT})_{HF}}{q^2 S} \approx \frac{(C_D + C_{IT})_{LF} - (C_D)_{HF}}{q^2 S} \tag{6.10}$$

where $(C_D + C_{IT})_{HF}$ is the $C_D + C_{IT}$ measured at high frequency and is assumed to be C_D ($C_{IT} \approx 0$ at high frequency). The low- and high-frequency C–V measurements are usually performed in the quasi-static (QS) mode and at between 100 kHz and 1 MHz, respectively. To minimize the voltage shift in C–V curves caused by carrier trapping during the accumulation state, "simultaneous high–low" measurements are employed, where the high-frequency and quasi-static capacitances are measured at every bias point before changing the bias voltage. Figure 6.44 shows the typical high-frequency (1 MHz) and quasi-static C–V characteristics measured on a MOS capacitor with a dry oxide on n-type 4H-SiC(0001). The larger value of the quasi-static capacitance than the high-frequency capacitance reflects the interface state

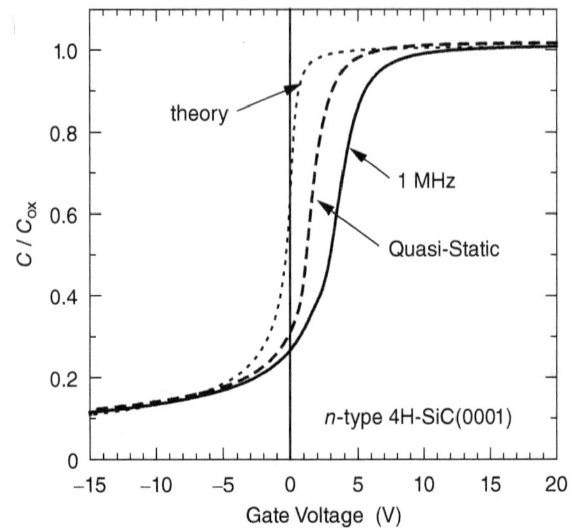

Figure 6.44 Typical high-frequency (1 MHz) and quasi-static $C-V$ characteristics measured on a MOS capacitor with a dry oxide on n-type 4H-SiC(0001).

density. This is a convenient technique to extract the interface state density, because a theoretical $C-V$ curve is not necessary and the technique is rather easy. Therefore, the high–low method has been a popular technique to determine the interface state density in many semiconductors, including SiC. However, it must be pointed out that the essential assumption described above is valid only in a very limited range of energy and measurement conditions. Problems that may occur are described below [146, 207, 214]:

1. The quasi-static capacitance does not include the capacitance of the interface states (C_{IT}) which possess slow emission rates. For example, the interface states located at energy levels deeper than $E_c - 0.5$ eV or $E_v + 0.5$ eV exhibit very long emission time constants (roughly >0.1–1 s) at room temperature and do not contribute to the quasi-static capacitance. Therefore, the interface state density in the energetically deep region is severely underestimated (or almost impossible to be monitored).
2. The high-frequency capacitance may include the capacitance of the interface states which possess fast emission rates. For example, the interface states located at energy levels shallower than $E_c - 0.2$ eV or $E_v + 0.2$ eV exhibit very short emission time constants (roughly $<10^{-5}$ to 10^{-6} s) at room temperature and thus contribute to the high-frequency capacitance. It is reported that even deep interface states can respond to a high frequency of 1 MHz at room temperature when the interface is nitrided [215]. Therefore, the interface state density in the energetically shallow region is severely underestimated. This is particularly detrimental in SiC, because a high density of shallow interface states is usually responsible for the low channel mobility.
3. Another important character of SiC MOS structures is that they can show a large *standard deviation of surface potential* σ_s and thereby a *large dispersion of the time constant* [146, 148, 216, 217]. This large dispersion can broaden the frequency range over which the transition from high-frequency to low-frequency behavior takes place. This leads to severe underestimation of the interface state density [214].

Figure 6.45 shows the limitations in the determination of the interface state density by the high–low method [214]. A typical distribution of interface state density and standard deviation of surface potential ($\sigma_s = 4$) are assumed. In Figure 6.45a, the interface state density deduced from the high(1 MHz)–low(0.5 Hz) method is plotted, together with the assumed distribution. The relative

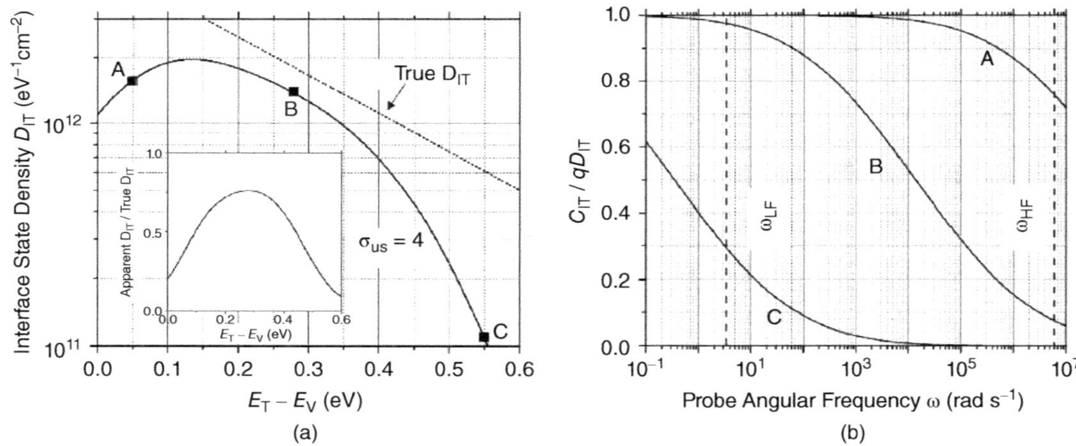

Figure 6.45 Limitations in the determination of the interface state density by the high-low method ([214] reproduced with permission from IEEE). A typical distribution of interface state density and standard deviation of surface potential ($\sigma_s = 4$) are assumed. (a) Interface state density deduced from the high(1 MHz)–low(0.5 Hz) method is plotted, together with the assumed distribution. (b) C_{IT} versus angular frequency ω at bias points A, B, and C denoted in (a).

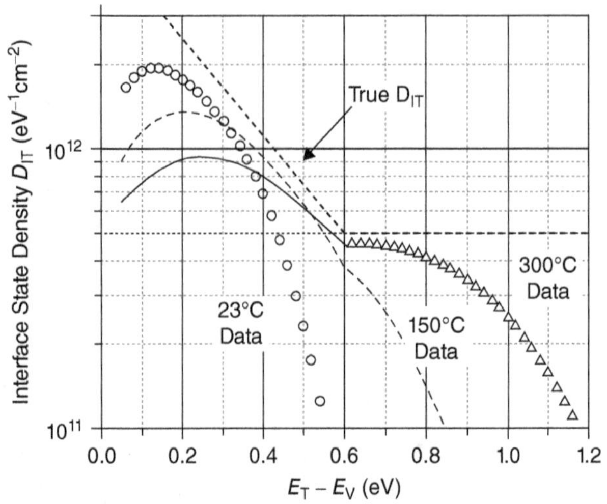

Figure 6.46 Extracted interface state density versus energy for a typical SiC MOS structure ($\sigma_s = 4$) at different temperatures ([214] reproduced with permission from IEEE).

accuracy of the deduced interface state density is shown in the inset. C_{IT} versus angular frequency ω at bias points A, B, and C in (a) are shown in Figure 6.45b. The large time-constant dispersion causes a very gradual transition from low- to high-frequency behavior. Thus, there is only a narrow range of energy where the interface state density can be reasonably estimated. The high–low method at room temperature monitors the interface states in a rather narrow energy range within the bandgap, typically between 0.2 and 0.45 eV from the majority carrier band edge; often the range can be even narrower.

To improve the accuracy in extraction of the interface state density, the measurements must be performed over a wider temperature range. Low-temperature C–V measurements should be used to monitor fast interface states and high-temperature C–V measurements to monitor slow states. This approach, however, still does not provide a very satisfactory result, as shown in Figure 6.46 [214]. To characterize shallow or fast interface states more accurately, the high-frequency measurements can be done at a much higher frequency, such as 100 MHz [213].

6.3.4.6 C–ψ_s Method

This is a modified version of the high–low method [213]. By using the theoretical capacitance, the high-frequency limit is considerably extended. Using the obtained surface potential, the theoretical semiconductor capacitance ($C_{D,theory}$) of n-type MOS capacitor can be calculated by [207, 213]:

$$C_{D,theory}(\psi_S) = \frac{SqN_D \left| \exp\left(\frac{q\psi_S}{kT}\right) - 1 \right|}{\sqrt{\frac{2kTN_D}{\varepsilon_S}\left\{\exp\left(\frac{q\psi_S}{kT}\right) - \frac{q\psi_S}{kT} - 1\right\}}} \tag{6.11}$$

assuming that there are no holes in the n-type semiconductor. Figure 6.47 shows the $C_D + C_{IT}$ values plotted against the surface potential ψ_S measured for an n-type 4H-SiC MOS capacitor at different frequencies, where $C_{D,\,theory}$ calculated from Equation 6.11 is also plotted. In this particular MOS capacitor, a 32-nm-thick dry oxide was formed on 4H-SiC(0001). The measured $C_D + C_{IT}$ approached $C_{D,theory}$ as the

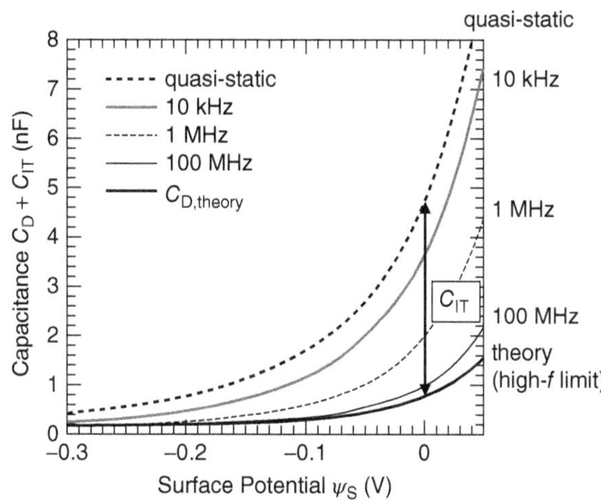

Figure 6.47 $C_D + C_{IT}$ values plotted against the surface potential ψ_S measured for an n-type 4H-SiC MOS capacitor at different frequencies, where $C_{D,theory}$ calculated from Equation 6.11 is also plotted [213].

frequency increased, because the carriers trapped at the interface states hardly respond once the frequency is sufficiently high ($C_{IT} \approx 0$). A definite difference exists between $C_D + C_{IT}$ at 1 MHz and $C_{D,theory}$, indicating that a significant portion of the fast interface states respond at 1 MHz. In contrast, 100 MHz seems to be almost sufficient for the interface carriers to not respond (in the case of dry oxides). The interface state density D_{IT} is given by:

$$D_{IT} = \frac{(C_D + C_{IT})_{LF} - C_{D,\,theory}}{q^2 S} \tag{6.12}$$

where $(C_D + C_{IT})_{LF}$ is $C_D + C_{IT}$ measured under quasi-static conditions. This method is superior to the other methods from two points of view. (i) The $C-\psi_S$ method can detect fast interface states almost without frequency limits. (ii) The $C-\psi_S$ method requires simple measurements (similar to those needed in the high–low method). An almost continuous D_{IT} distribution (as opposed to a point by point measurement) can be obtained by this method. However, it is difficult to monitor very slow states, which cannot respond to the voltage sweep; this problem also occurs with the other techniques. Furthermore, this method can result in an erroneous D_{IT} distribution when an accurate surface potential cannot be determined. For example, if the doping density varies significantly along the depth, it is difficult to obtain a good plot (such as the one shown in Figure 6.43). In such a case, erroneous surface potentials give a wrong $C_{D,theory}-\psi_S$ curve, leading to an incorrect D_{IT} distribution.

6.3.4.7 Conductance Method

As expected from the equivalent circuit of a MOS capacitor (Figure 6.41), the conductance–frequency (G–f) characteristics of the MOS capacitor should give a peak at a specific frequency, originating from the interface states. The interface state density is related to G/ω (ω: angular frequency) by [160, 207]:

$$\frac{G}{\omega} = q^2 S D_{IT} \int_{-\infty}^{+\infty} \frac{\ln(1 + (\omega\tau \exp(\eta))^2)}{2\omega\tau \exp(\eta)} \frac{1}{\sqrt{2\pi\sigma^2}} \exp\left(-\frac{\eta^2}{2\sigma^2}\right) \mathrm{d}\eta \tag{6.13}$$

where the interface state density (D_{IT}), the time constant of the interface states (τ), and the standard deviation (σ) are fitting parameters. And $\eta = u_s - \langle u_s \rangle$, where u_s is the surface potential normalized to

kT/q. This technique is regarded as the most sensitive method to determine the interface state density, and a D_{IT} of 10^9 cm^{-2} eV^{-1} can be measured in Si, though this method is time-consuming when compared with $C-V$ analyses [207]. All the interface states give definite peaks in $G/\omega - f$ curves and can be monitored, as long as the inverse of the response time is roughly within the range of the probe frequency (typically from 1 kHz to 10 MHz). However, it is difficult to monitor very slow states, which cannot respond to these frequencies, and very fast states, which show sufficient response to the highest probe frequency, because no conductance peaks appear in these cases. Since there are lower and upper limits of the probe frequency in the conductance measurements, it is important to perform the conductance measurements over a wide range of temperatures. In other words, however, this is one of the advantages of the conductance technique, because it does not provide data under conditions when that data would be invalid (i.e., the observed peaks give valid data).

It should also be noted that other very important information is obtained from the conductance method, namely, the time constant of the interface states and the standard deviation of the surface potential. The time constant is a useful indicator of the nature of the interface states, and the standard deviation reflects the microscopic fluctuations of the interface structure. Both are essential to understanding the low channel mobility in SiC MOSFETs. The time constant for transitions between interface states and the conduction band is given by

$$\tau(E) = \frac{1}{\sigma_N v_{th} n_s} \tag{6.14}$$

where σ_N is the capture cross section of the state for electrons, v_{th} the thermal velocity, and n_S the volume density of electrons at the interface. Given the position E of the Fermi level at the surface, the electron density is given by

$$n_s = N_C \exp\left(\frac{E - E_C}{kT}\right) \tag{6.15}$$

Inserting Equation 6.15 into Equation 6.14 shows that the interface state time constant increases exponentially as the Fermi level moves further from the conduction band. However, it is routinely observed for both Si and 4H-SiC that the capture cross section σ_N decreases exponentially with energy toward the majority carrier band edge. This exponential dependence can be characterized by a slope factor γ,

$$\sigma_N(E) = \sigma_{N0} \exp\{\gamma(E_C - E)\} \tag{6.16}$$

where σ_{N0} is a constant. In Si, the capture cross sections also fall exponentially toward the band edges, but are typically constant near the middle of the bandgap [218]. In 4H-SiC, no clear indication of a constant region been observed. Figure 6.48 shows capture cross sections from several studies, along with the associated γ values [219, 220]. The strong energy dependence in Equation 6.16 tends to cancel the energy dependence in Equation 6.15, making the time constant only weakly dependent on energy. If γ is close to unity, as observed in Figure 6.48, the energy dependence of the time constants becomes very small. In this situation the $G(\omega)/\omega$ curves shift very little in frequency as the bias is changed, making it possible to obtain data over a wider energy range in the bandgap. This also makes the $G(\omega)/\omega$ curves narrower [221], so that the *apparent* surface potential variation σ is less than the *true* surface potential variation. The apparent σ in 4H-SiC is typically between 4 and 5, so the *actual* σ is even larger. To date, the true standard deviation of surface potential has not been accurately determined.

Figure 6.49 shows the frequency dependence of G/ω at various surface potentials for a MOS capacitor with a dry oxide on n-type 4H-SiC(0001). Bell-shaped peaks originate from the interface states, and the bold lines in Figure 6.49 are the $G/\omega - f$ curves calculated from Equation 6.13 to fit the experimental results. It is noted that these data were acquired by conductance measurements at up to 100 MHz using a special probe. Figure 6.50 shows the τ and σ values obtained from fitting. The time constant is about 3×10^{-6} s at $E_c - 0.2$ eV and becomes longer for deeper states, 6×10^{-5} s at $E_c - 0.4$ eV. The standard deviation in SiC MOS structures (typically between 4 and 5) is much higher than that in Si MOS structures (~2), indicating inhomogeneity of the SiC MOS interface.

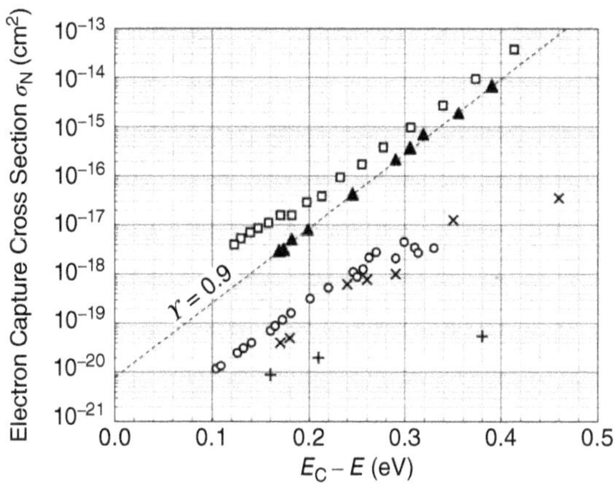

Figure 6.48 Capture cross sections of interface states obtained from several studies, along with the associated γ values [219, 220].

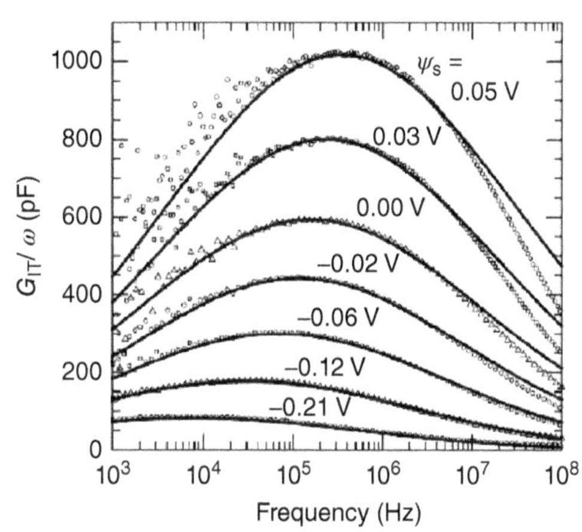

Figure 6.49 Frequency dependence of G/ω at various surface potentials for a MOS capacitor with a dry oxide on n-type 4H-SiC(0001) ([213] reproduced with permission from AIP Publishing LLC).

Figure 6.51a shows the interface state densities obtained by the high(1 MHz)–low method, the high(100 MHz)–low method, the C–ψ_S method, and the conductance method, where the $E_c - E_T$ (E_T: energy level of interface traps) of the horizontal axis is determined from the surface potential calculated by Equation 6.9 [213]. In the conductance measurements, the highest probe frequency was 100 MHz. All the measurements were performed on the same MOS capacitor with a 32-nm-thick dry oxide formed on n-type 4H-SiC(0001). The D_{IT} distribution obtained by the C–ψ_S method agrees very well with that obtained by the conductance method. The high(100 MHz)–low method also gave the same interface state density, because 100 MHz is almost sufficient for the fast states not to respond. However, the D_{IT} distribution obtained by the conventional high(1 MHz)–low method is two or three times lower than that from the other methods, because fast interface states that respond to higher than 1 MHz are not detected. The interface state density of SiC MOS structures is very high, about 1×10^{13} cm^{-2} eV^{-1} at E_c – 0.2 eV. It should be noted that this underestimation of interface state density by the conventional

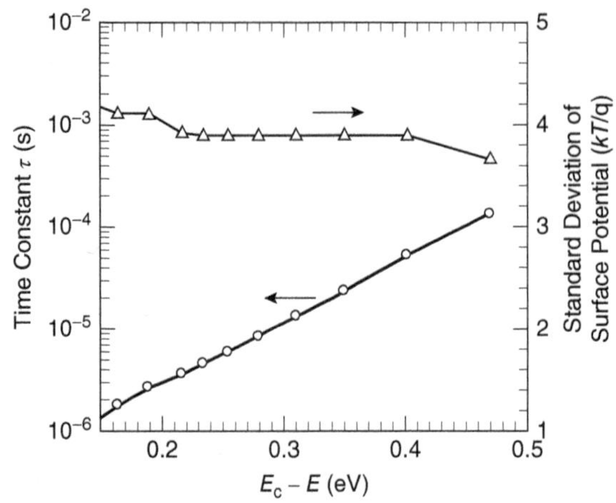

Figure 6.50 Emission time constant τ (at room temperature) and standard deviation σ of interface states observed for SiC MOS structures ([213] reproduced with permission from AIP Publishing LLC).

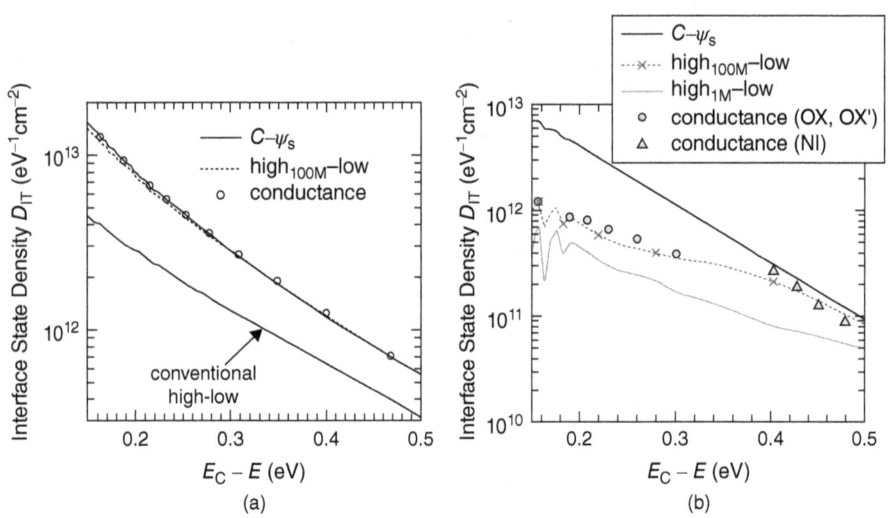

Figure 6.51 (a) Interface state density (D_{IT}) obtained by the high(1 MHz)–low method, the high(100 MHz)–low method, the C–ψ_S method, and the conductance method. All the measurements were performed on the same MOS capacitor with a 32-nm-thick dry oxide formed on n-type 4H-SiC(0001). (b) Comparison of D_{IT} distributions obtained by several methods for a 4H-SiC(0001) MOS structure annealed in NO ([213] reproduced with permission from AIP Publishing LLC).

high(1 MHz)–low method is more severe in nitrided SiC MOS structures [215]. Figure 6.51b shows comparison of D_{IT} distributions obtained by several methods for a 4H-SiC(0001) MOS structure annealed in NO. The D_{IT} values near the conduction band edge determined by the high(1 MHz)–low method are one-order-of-magnitude lower than those determined by the C–ψ_S method. Improvement of interface properties is described in Section 6.3.5.

6.3.4.8 Other Methods

Several other techniques have been applied to characterize the interface properties of SiC MOS structures. The *charge pumping method* monitors charge emission from the interface states by applying repetitive pulses to the gate [207, 222–224]. Although this technique gives total density of interface states inside the bandgap, it is difficult to accurately determine the energy distribution. A variety of DLTS measurements, such as constant-capacitance deep level transient spectroscopy (CC-DLTS) have been tried [225, 226]. In this technique, carrier emission from interface states, oxide traps, and bulk traps can be distinguished by carefully designed biasing. For example, two oxide traps, energetically located at $E_c - 0.15$ eV and $E_c - 0.39$ eV, are detected in n-type 4H-SiC(0001) MOS capacitors by CC-DLTS [226]. However, the energy position of interface states cannot be determined directly and is usually estimated by assuming a constant capture cross section of interface states, except for distinct peaks in CC-DLTS spectra. The situation is very similar in the *thermal dielectric relaxation current* (TDRC) method [212]. In the Zerbst method a voltage pulse is applied so that a MOS capacitor reaches deep depletion, and the capacitance transient to recover its steady state is monitored. The generation rate via the interface states can be obtained, but it is difficult to determine the D_{IT} distribution [160, 227, 228].

As described in Section 6.3.3, no inversion states appear in normal SiC MOS capacitors, even when large gate bias (negative for n-type, positive for p-type) is applied, because of the extremely low generation rate of minority carriers. A *gate-controlled diode* (GCD) is a test structure, where the inversion layer is induced and, thereby, the interface states near the minority carrier band edge can be characterized. Figure 6.52 shows the terminal connection of a GCD for $C-V$ measurements. The basic structure is a planar MOSFET and the gate capacitance is measured by sweeping the gate voltage. In these measurements, the source and drain terminals are connected to the grounded p-base. When sufficiently positive gate bias is applied, electrons are quickly injected from the source and drain, leading to the formation of an inversion layer underneath the gate oxide (in the case of an n-channel MOSFET). Figure 6.53 shows examples of low-frequency (20 Hz) $C-V$ characteristics obtained from a 4H-SiC(0001) GCD with a (a) dry oxide followed by wet-annealing and (b) dry oxide annealed in NO [151]. As in the case of Si MOS capacitors at room temperature, the low-frequency $C-V$ curve reaches the oxide capacitance C_{ox} when the gate voltage is sufficiently larger than the threshold voltage. The high-frequency $C-V$ curve exhibits a saturated capacitance determined by the maximum width of the space-charge region (not shown). From the $C-V$ curve, the D_{IT} distribution near the conduction band edge can be extracted [229]. It is very useful to determine the D_{IT} distribution using the MOSFET structure, because direct comparison between

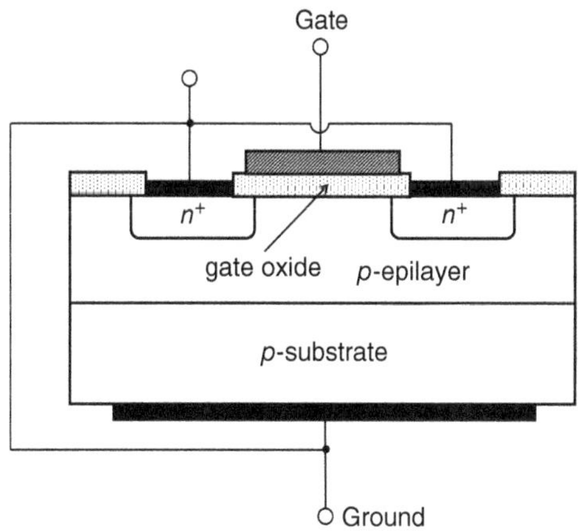

Figure 6.52 Terminal connection of a gate-controlled diode (GCD) for $C-V$ measurements.

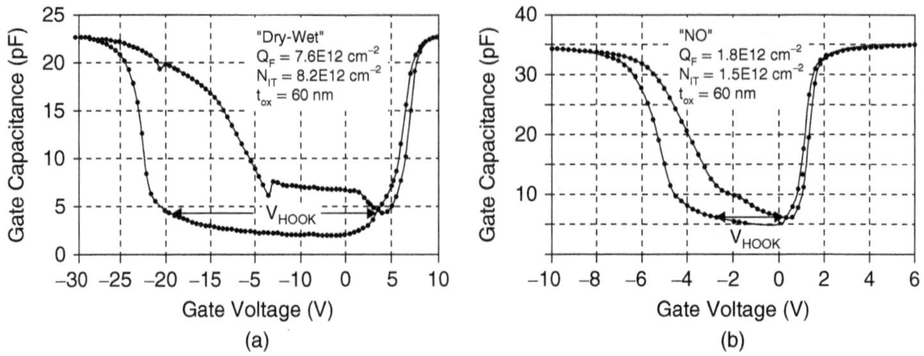

Figure 6.53 Low-frequency (20 Hz) $C-V$ characteristics obtained from a 4H-SiC(0001) GCD with a (a) dry oxide followed by wet-annealing and (b) dry oxide annealed in NO ([151] reproduced with permission from Springer-Verlag).

the interface properties and the channel mobility is possible on the same sample. However, this method possesses the same limitations as the conventional high–low method, in that both very fast states and very slow states cannot be detected.

6.3.5 Properties of the Oxide/SiC Interface and Their Improvement

6.3.5.1 Distribution of Interface States

In spite of the limited capability of characterization techniques described above, a rough picture of the interface state distribution inside the SiC bandgap has been revealed. Figure 6.54 shows schematic distributions of interface states inside the bandgaps of various SiC polytypes. It is known that the energy position of the valence band top is almost aligned for different SiC polytypes and that of the conduction band bottom is scaled according to the individual bandgaps [177].

When looking at the interface states in the lower half of the bandgap, most of these states are *donor-like*. These states are positively charged by trapped holes when the states are located above the Fermi level. These donor-like states, especially those located in the deep energy region, capture holes in p-type SiC and do not emit holes at room temperature. Thus, holes trapped at deep interface states behave as if they were "*positive fixed charges*". The density of these donor-like states is extremely high ($>5 \times 10^{12}$ cm^{-2} eV^{-1}) in MOS structures formed by dry oxidation, and can be remarkably reduced by wet oxidation [230]. These donor-like states do contribute to the negative shift of $C-V$ curves for p-type SiC MOS capacitors (and p-channel mobility). However, the donor-like states do not directly affect n-channel mobility, because these states become neutral when the Fermi level moves up [230, 231]. This behavior is common for the various SiC polytypes.

The interface state distribution seems to be intrinsic and common for all SiC polytypes. Thus, the distribution near the conduction band edge is determined by the intrinsic distribution and the location of the conduction band bottom for each polytype, as shown in Figure 6.54. Most of these states near the conduction band edge are acceptor-like. These states are negatively charged by trapped electrons when the states are located below the Fermi level. Electrons trapped at deep acceptor levels act as if they were "negative fixed charges". Note that the charge neutrality level in SiC MOS structures is not known at present. Furthermore, the existence of donor-like states near the conduction band edge in "nitrided" 4H-SiC MOS structures has been suggested. On α-SiC(0001), the interface state density increases almost exponentially toward the conduction band edge [232–234]. Because of this rapid increment with increasing energy level, the interface state density near the band edge is very high for 4H-SiC(0001)

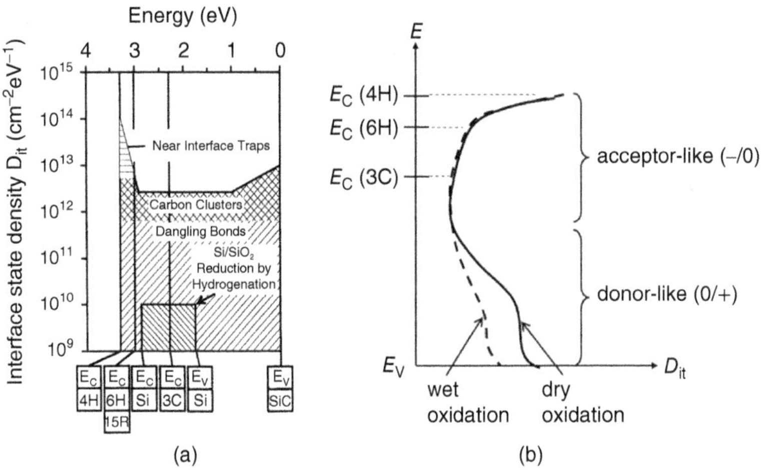

Figure 6.54 Schematic distributions of interface states inside the bandgaps of various SiC polytypes. (a) Model for explaining a high density of interface state density in SiC and its polytype dependence ([150] reproduced with permission from Springer-Verlag), (b) Schematic distribution of interface states of SiC MOS structures formed by dry or wet oxidation.

and relatively low for 3C-SiC(111). In n-channel MOSFETs, electrons in the inversion layer induced by the gate bias are trapped at these interface states and become almost immobile. The trapped electrons also act as Coulomb scattering centers. Therefore, the acceptor-like states near the conduction band edge are detrimental to n-channel mobility. This is why 4H-SiC(0001) MOSFETs usually exhibit poor channel mobility (5–8 $cm^2V^{-1}s^{-1}$ without appropriate annealing), while high channel mobility (over 100 $cm^2V^{-1}s^{-1}$) can be obtained easily for 3C-SiC MOSFETs [235–237].

The exact origins of high interface state density for SiC are not very well understood. In the case of Si MOS structures, dangling bonds at the interface are dominant defects (for example, the P_b center) [160, 175]. Since the interface state density of thermal oxide/Si structures is in the 10^9–10^{10} cm^{-2} eV^{-1} range, it is unlikely that the high interface state density (10^{12}–10^{13} cm^{-2} eV^{-1}) of SiC MOS structures can be attributed simply to dangling bonds. Afanas'ev *et al.* suggested that the donor-like interface states in the lower half of the bandgap may originate from carbon clusters (*carbon-cluster model*), based on IPE spectroscopy studies of SiC MOS structures and graphite [147, 238]. Although the role of residual carbon near the interface has been argued, a direct link between the carbon density near the interface and the interface state density has not been proven. The abnormally high interface state density near the conduction band edge of 4H-SiC(0001) is also a mystery. To explain this phenomenon, "*near-interface traps*" (NITs) were proposed as the origin of the shallow states [147]. NITs are traps existing inside the oxide near the interface and, because of their nature, their response can be very slow. It is also claimed that NITs are intrinsic to SiO_2 rather than only to the thermal oxide/SiC system. It is, however, not very easy to explain the experimental facts that a high density of shallow states is not observed in thermal oxide/4H-SiC(000$\bar{1}$) [239] and (11$\bar{2}$0) [233], and that such shallow states are observed even when other dielectric materials (Si_3N_4, Al_2O_3, and AlN) are employed. Systematic theoretical studies are required to clarify the origins of the interface states and NITs.

6.3.5.2 Post-Oxidation Annealing

POA and *post-metallization annealing* (PMA) are crucial to obtain high-quality MOS interfaces in any semiconductor materials. In Si technology, one major technique to reduce the interface state density is hydrogen passivation of dangling bonds near the interface; this can be achieved by PMA in a hydrogen-containing gas, like a N_2–H_2 mixture, at about 400–500 °C [8, 160, 175]. As a result, the interface state density can be reduced down to values as low as 10^9 cm^{-2} eV^{-1}. In the case of SiC, however, impacts of hydrogen annealing at 400 °C on the interface state density are very small, which suggests that the essential problem in SiC MOS is totally different from that in Si. It has been reported that hydrogen annealing at much higher temperatures, 800–1000 °C, is beneficial to reduce the interface state density [240], but the exact mechanism is not clear.

To reduce deep interface states, especially those lying in the lower half of the bandgap, "*re-oxidation*" annealing is effective [241]. The key step of this technique is POA at relatively low temperature (typically 950 °C) in a wet ambient in which additional oxidation is negligibly small. After this re-oxidation annealing, the final interface state density seems to be similar irrespective of the oxidation process (dry or wet). The interface properties are also dependent on the H_2O content in the vapor during re-oxidation annealing. By re-oxidation annealing with a high H_2O vapor content, the n-channel mobility is improved to 50 cm^{2}V^{-1}s^{-1} for 4H-SiC(0001) and 98 cm^{2}V^{-1}s^{-1} for 6H-SiC(0001) [242]. As mentioned above, this process also reduces the deep interface state density near the valence band edge, leading to enhancement of p-channel mobility of SiC MOSFETs [243]. The physical reasons for this improvement, however, are still unknown.

Annealing oxides in an inert gas (Ar or N_2) immediately after thermal oxidation is a popular process in SiC technology. So far, this inert POA is usually carried out at almost the same temperature as oxidation (1100–1200 °C). This step is believed to aid removal of excess carbon from the oxide or the interface, but no direct evidence for the out-diffusion of carbon has been given. Nevertheless, appropriate Ar POA improves the dielectric properties and reliability of oxides [216], and has been widely employed. In recent years, Ar POA at a very high temperature of 1300–1350 °C was tried to enhance diffusion of excess carbon interstitials based on a deep level study [244]. Although clear reduction in the interface state density, and enhancement of n-channel mobility are attained after high-temperature Ar POA, the improvement is still not satisfactory.

The oxidation condition itself has been investigated to improve the interface quality. As far as the SiC(0001) face is concerned, the oxidation atmosphere (dry versus wet) does not give striking differences in the interface state density near the conduction band edge and, thereby, the n-channel mobility. In general, wet oxidation results in a slightly lower interface state density near the conduction band edge, a significantly lower interface state density near the valence band edge, and improved p-channel mobility, as shown in Figure 6.54. Generation of negative charges near the interface in wet oxidation is also suggested. Note that the effects of wet oxidation become much greater when MOS structures are formed on SiC(000$\bar{1}$) [239] and non-basal planes like (11$\bar{2}$0) [233]. High-temperature dry oxidation of SiC(0001) at 1250–1300 °C is effective to reduce the interface state density near the conduction band edge and to improve the n-channel mobility [245, 246]. It has been reported that the interface quality is degraded once the thickness of the thermal oxide exceeds about 20 nm, as shown in Figure 6.55 [196]. In the figure, a sudden increase in interface state density and a change in the slope of the flatband-voltage are observed when the oxide thickness exceeds 20 nm. This degradation of the interface quality is correlated with an increase in the amount of intermediate (suboxide-like) structures, as determined by synchrotron XPS [196]. Despite the various studies described above, we do not at present understand the physical basis for the effects of oxidation.

6.3.5.3 Interface Nitridation

One promising process is *post-oxidation nitridation* in a nitrogen-containing gas such as nitric oxide (NO) [247–261], nitrous oxide (N_2O) [262–264], or ammonia (NH_3) [265, 266], or in the presence of nitrogen radicals [267]. Direct oxidation in N_2O or NO is also proposed. In particular, interface nitridation by NO or N_2O is widely employed in academic research, as well as in mass production of SiC power MOSFETs. Figure 6.56 shows the distribution of interface state density near the conduction and valence

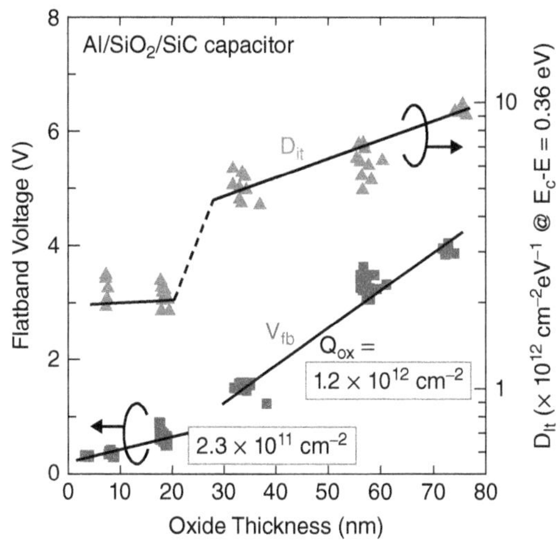

Figure 6.55 Interface state density and flatband-voltage of n-type 4H-SiC MOS capacitors with dry oxides as a function of the oxide thickness ([196] reproduced with permission from AIP Publishing LLC).

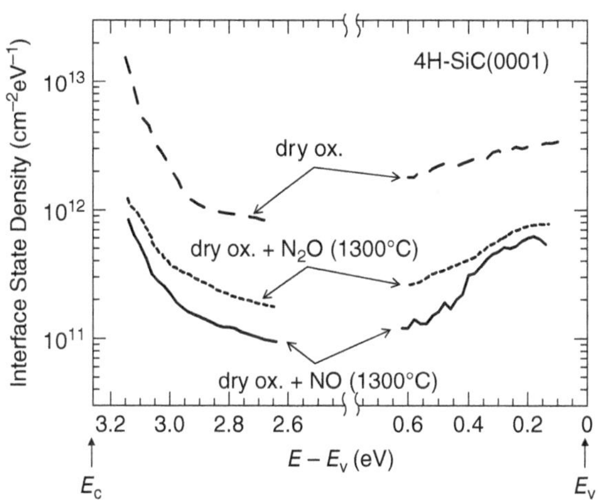

Figure 6.56 Distribution of interface state density near the conduction and valence band edges obtained from n- and p-type 4H-SiC(0001) MOS capacitors, respectively [154].

band edges obtained from n- and p-type 4H-SiC(0001) MOS capacitors, respectively [154]. The interface state densities of MOS structures formed by dry oxidation, and dry oxidation followed by nitridation in NO or N_2O are plotted. In this figure, the interface state density was evaluated by a conventional high (1 MHz)–low method. It is very clear that reduction in the interface state density over the entire range of energies within the bandgap is achieved by nitridation. As a result, the effective mobility of n-channel 4H-SiC(0001) MOSFETs fabricated on lightly-doped p-type epilayers has been enhanced from a single digit (4–8 $cm^2V^{-1}s^{-1}$) for dry oxides to 25–35 $cm^2\ V^{-1}s^{-1}$ for N_2O-nitrided oxides [262–264], and to 40–52 $cm^2V^{-1}s^{-1}$ for NO-nitrided oxides [250, 252, 253]. The effective mobility of p-channel

MOSFETs is also improved from 1–2 $cm^2V^{-1}s^{-1}$ for dry oxides to 7–12 $cm^2V^{-1}s^{-1}$ for nitrided oxides [268].

Annealing in NO or N_2O naturally results in a pile-up of nitrogen at the SiO_2/SiC interface. The nitrogen atom density at the interface depends strongly on the nitridation conditions, and can reach 5×10^{20} cm^{-3} or even higher. Correlation between the increase in the nitrogen density at the interface and reduction in the interface state density has been presented by several groups [153, 260, 263]. Furthermore, it has been reported that the nitridation process causes not only reduction of the interface state density near the conduction band edge but also increase in the number of hole traps [269]. Thus, excessive nitridation is not desirable because of pronounced hole-trapping effects. Again, it is unclear how the interface defects are passivated by nitridation. Passivation of dangling bonds with nitrogen or a simple shift of defect levels by nitridation is unlikely, based on the considerations mentioned above. More efficient removal of carbon from the interface by the nitridation annealing has been suggested; there is some experimental evidence supporting this [151, 192, 263].

Comparison between nitridation by NO and that by N_2O is of interest. Historically, the first reported success in nitridation used NO [247, 249]. Because of the highly toxic nature of NO, N_2O nitridation, or direct oxidation in N_2O was proposed as an alternative technique [262]. It seems that nitridation by NO gives a slightly better result than that by N_2O after process optimization. This result can be qualitatively interpreted by considering the chemical reactions in the gas phase. N_2O molecules are stable at temperatures below 1100 °C, and start to be decomposed at about 1200 °C according to the following reactions [270]:

$$N_2O \rightarrow N_2 + O \tag{6.17}$$

$$N_2O + O \rightarrow 2\,NO \tag{6.18}$$

Since N_2O molecules are larger, the diffusion coefficient of N_2O inside SiO_2 must be much lower than those of NO or O_2. Therefore, interface nitridation must proceed not by N_2O itself but by the NO which is created from N_2O in the gas phase above about 1200 °C. This is consistent with the observation that relatively high temperatures of 1250–1300 °C are required to achieve reasonable nitridation effects using N_2O. However, when the NO is generated from N_2O, atomic oxygen or oxygen molecules are also produced and thus must also be present in the gas phase. This means that simultaneous oxidation takes place during interface nitridation by N_2O. This is supported by the observation that noticeable increase in the oxide thickness is observed after nitridation in N_2O. On the other hand, NO molecules start to be decomposed at about 1300 °C, according to the following reaction (This decomposition is not desirable.) [271]:

$$2NO \rightarrow N_2 + O_2 \tag{6.19}$$

Therefore, significant nitridation can be achieved at a lower temperature of 1150–1250 °C when NO is used, and additional oxidation (increase of oxide thickness) during nitridation in NO is minimal.

It has been found that a high density of very fast interface states is generated near the conduction band edge by nitridation annealing [215]. The very fast states can respond to a probe frequency of 100 MHz or even higher (>GHz) at room temperature. Such very fast states cannot be monitored by the conventional high–low method at room temperature, and were characterized by the C–ψ_s method or conductance measurements at low temperature (40–150 K). Measurements confirmed that the density of such very fast states is low in MOS structures without nitridation annealing, and that the states are indeed created by nitridation in NO or N_2O. Figure 6.57 shows the density of interface states at $E_c - 0.3$ eV versus the area density of nitrogen atoms near the interface [215]. These data were acquired from 4H-SiC(0001) MOS structures with dry oxides, that had been annealed in NO at various temperatures (1150–1350 °C). Here, the densities of relatively slow (<1 MHz) states, very fast states (>100 MHz), and all the states are plotted. The density of slow states decreases with increasing nitrogen density at the interface. In contrast, the density of very fast states increases almost in proportion to the nitrogen area density. Therefore, the very fast states are linked to interface nitridation. As a result, the total interface state density exhibits

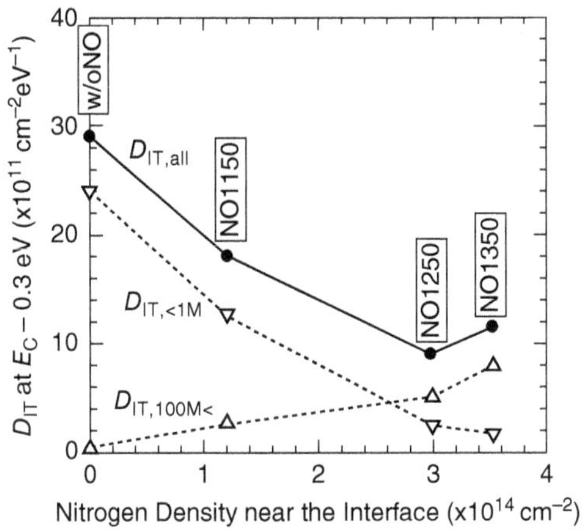

Figure 6.57 Density of interface states at $E_c - 0.3$ eV versus the area density of nitrogen atoms near the interface ([215] reproduced with permission from AIP Publishing LLC). These data were acquired from 4H-SiC(0001) MOS structures, with dry oxides, that had been annealed in NO at various temperatures (1150–1350 °C).

a minimum at a nitrogen area density of 3.0×10^{14} cm^{-2}, which was obtained by nitridation in NO at 1250 °C for 70 min. When n-channel MOSFETs were fabricated by the same process as that used for the samples characterized in Figure 6.57, the effective channel mobility showed a maximum when the sum of fast and slow interface state density takes a minimum (nitridation in NO at 1250 °C for 70 min). Thus, the very fast interface states are also responsible for limiting the channel mobility. More details are given in Section 6.3.7.

6.3.5.4 Other Approaches

Significant reduction of interface state density and improvement of n-channel mobility are achieved by POA in $POCl_3$ [272, 273]. By annealing in $POCl_3$ at 1000 °C for 10 min, a high channel mobility of 89 cm^{2}V^{-1}s^{-1} is attained. The mobility can be further improved to 101 cm^{2}V^{-1}s^{-1} by two-step annealing in $POCl_3$ at 1000 °C and forming gas at 700 °C [274]; this is one of the highest channel mobilities reported for 4H-SiC(0001) MOSFETs. The phosphorus atoms are almost uniformly distributed with a density in excess of 1×10^{21} cm^{-3} inside the gate oxide after the $POCl_3$ annealing. The interface state density and the subthreshold swing of MOSFETs are also significantly improved by the $POCl_3$ annealing.

Another striking technique to enhance the channel mobility is "sodium-contaminated" oxidation [275, 276]. In the initial reports on high channel mobility of 120–150 cm^{2}V^{-1}s^{-1} for 4H-SiC(0001) MOSFETs, an Al_2O_3 tube was used for gate oxidation. It turned out later that the furnace and oxides were highly contaminated with sodium (Na). An accelerated oxidation rate is reported for this Na-contaminated oxidation. Dipping SiC samples in Na-containing solutions prior to normal oxidation in a clean furnace gave very similar results (high oxidation rate and high channel mobility) [277]. This technique cannot be employed for device manufacturing because the processed MOSFETs exhibit clear instability in their threshold voltage, as is also well known in Si technology. It is, however, worth investigating this mechanism in detail to acquire important insights into mobility-limiting factors.

Deposited oxides, followed by appropriate processing, have been investigated with success. The oxide is typically formed by deposition of SiO_2 and subsequent annealing in NO or N_2O [185, 187, 278, 279]. The obtained interface state density is lower and the channel mobility is higher compared with those

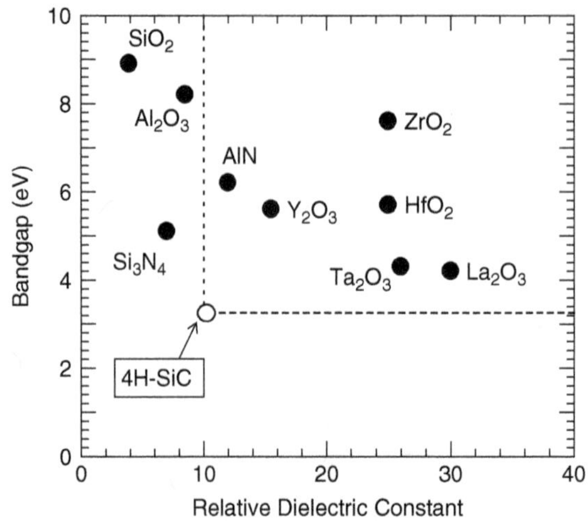

Figure 6.58 Relative dielectric constant and bandgap for major high-κ dielectrics.

of thermal oxides annealed in NO or N_2O. Since the barrier height at the SiO_2/4H-SiC interface is smaller than that of a SiO_2/Si system, the tunneling current at high electric field and high temperature is inherently higher in the case of SiC MOS structures, which can limit the oxide reliability at high temperature. In this sense, *high-κ dielectrics* show definite promise [150]. The physical thickness of gate insulators can be increased while keeping the same gate capacitance, because the material itself has a higher relative dielectric constant. Figure 6.58 shows the relative dielectric constant and bandgap for major high-κ dielectrics. Although HfO_2 has been intensively studied in advanced Si MOS, it may not be a good choice for SiC because of the small barrier height at the HfO_2/SiC interface. High-κ dielectrics which have relatively large bandgaps, such as Al_2O_3, AlN, and AlON are attractive for SiC MOSFETs [280–284]. A high channel mobility of 100–200 $cm^2V^{-1}s^{-1}$ [281, 283] or a high breakdown electric field of 12–15 MV cm^{-1} has been reported [284].

Buried channel structures have been investigated to alleviate the adverse influence of interface states. A thin (0.1–0.3 μm) n-type layer is formed by epitaxial growth [285, 286] or ion implantation [287–289] beneath the gate oxide. Since the buried channel is thicker than a normal inversion channel, high channel mobility is obtained, especially at low gate bias. The mobility significantly decreases at high gate bias because the energy band diagram of a MOSFET with a buried channel approaches that of a normal MOSFET at high gate bias, so once again the electron transport is severely affected by the interface states. The thickness and doping density of the channel must be carefully designed to ensure normally-off operation. Even though high peak mobilities of 80–170 $cm^2V^{-1}s^{-1}$ have been reported, it is not easy to retain a high threshold voltage at elevated temperature or with a short channel length.

So far, relatively high channel mobilities have been demonstrated for the peak mobility of MOSFETs on lightly-doped p-type SiC epitaxial layers. From a technological point of view, however, it is important to attain high channel mobility in MOSFETs fabricated on moderately-doped ($>10^{17}$ cm^{-3}) p-type bodies formed by aluminum implantation. Furthermore, high channel mobility should be obtained at high gate bias (oxide field > 2 MV cm^{-1}).

It should be noted that surface cleaning of SiC prior to thermal oxidation or oxide deposition has not been studied in great detail. A standard cleaning process employed in the SiC community is RCA cleaning followed by a dip in an HF-solution to remove the native oxide [290]. Ozone cleaning [291] and high-temperature hydrogen treatment [292] were also investigated. So far, the essential problems with SiC MOS interfaces are not solved by altering the cleaning process. Nevertheless, fundamental study on surface cleaning of SiC is required for full control of device fabrication.

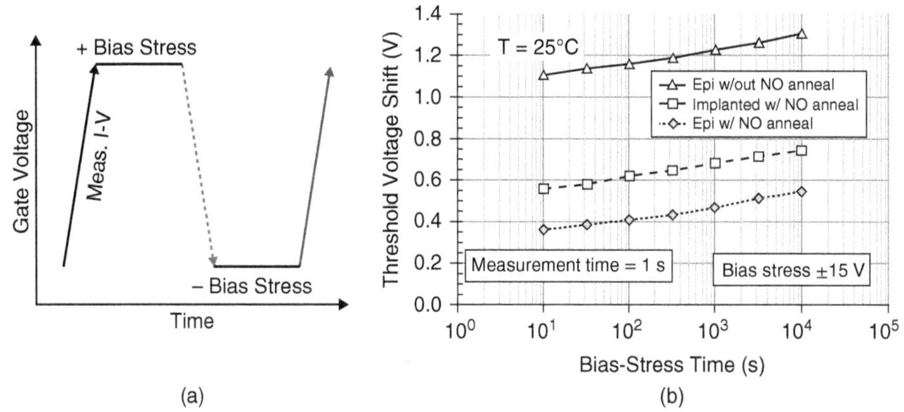

Figure 6.59 Threshold voltage shift of various 4H-SiC MOSFETs when stressed with a gate oxide field of ± 3 MV cm^{-1} ([294] reproduced with permission from IEEE). (a) Program of a bias-stress test, (b) threshold voltage shift as a function of bias-stress time for various MOSFETs.

6.3.5.5 Interface Instability

Though the channel mobility can be improved by several techniques, the instability of the threshold voltage remains a problem, as mentioned earlier. The instability is caused by, at least, two different phenomena; (i) *charge injection* into the oxide (or carrier trapping at slow states) and (ii) *mobile ions*. Since there exists a high density of interface states near the conduction band edge and oxide traps in SiC MOS structures, a large positive gate bias induces severe electron injection and trapping, leading to a large positive shift in threshold voltage (or flatband voltage) [212, 293]. Figure 6.59 shows the threshold voltage shift of various 4H-SiC MOSFETs when stressed with a gate oxide field of ± 3 MV cm^{-1} [294]. The threshold voltage shift increases with increasing the bias voltage during the stress. It is of interest that the 20-μs-long gate ramp used to measure the gate characteristics gives a much larger threshold voltage instability than the measurements with a 1-s-long gate ramp. These results can be interpreted by electron tunneling in and out of near-interface oxide traps [294]. As shown in Figure 6.59, MOSFETs with nitrided gate oxides exhibit an improved instability. More recently, another type of instability was found. When a negative gate bias is applied, the threshold voltage exhibits a significant negative shift and a stretch-out of the gate characteristics [295]. This shift is pronounced at elevated temperature. This instability is attributed to hole trapping near the MOS interface under a negative gate bias. Since interface nitridation creates more hole traps [269], optimization of the nitridation process is crucial.

Another instability is so-called *bias-temperature instability*, which is often observed in SiC MOS structures [296–298]. This phenomenon is caused by mobile ions; positive mobile ions accumulate at the SiO_2/SiC interface under positive bias-temperature (~200 °C) stress (PBTS), and result in a negative shift of the threshold voltage (or flatband voltage); conversely, positive mobile ions accumulate at the gate/SiO_2 interface under negative bias-temperature stress (NBTS), leading to a positive shift in threshold voltage. Thus, $C-V$ curves exhibit an ion-drift-type hysteresis. Annealing in a hydrogen-containing gas, such as forming gas, at high temperature (>700 °C) results in enhanced bias-temperature instability [298].

6.3.6 *Interface Properties of Oxide/SiC on Various Faces*

The densities and distributions of interface states are very different when different crystal faces (other than off-axis 4H-SiC(0001)) are employed. The major crystal faces described here are off-axis 4H-SiC(000$\bar{1}$), on-axis 4H-SiC(11$\bar{2}$0), and 4H-SiC(1$\bar{1}$00).

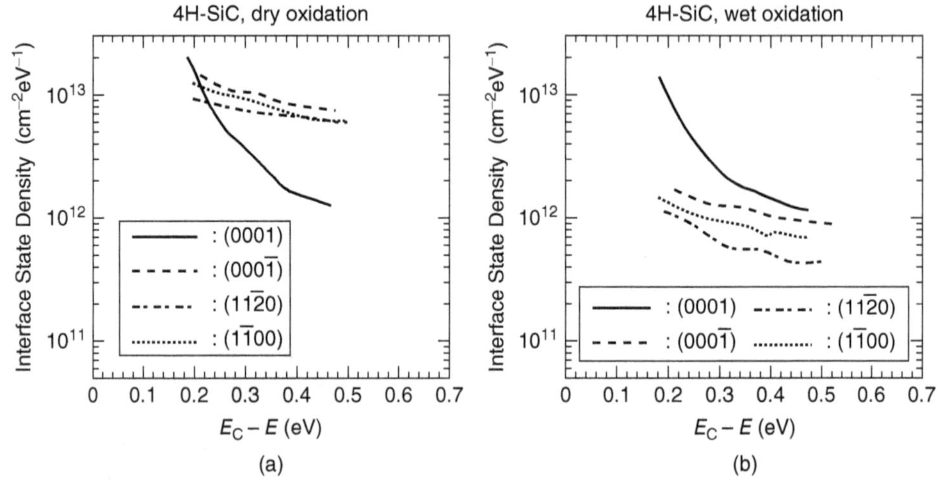

Figure 6.60 Interface state density distributions obtained from n-type 4H-SiC(0001), $(000\bar{1})$, $(11\bar{2}0)$, and $(1\bar{1}00)$ MOS structures prepared by (a) dry and (b) wet oxidation.

In SiC $(000\bar{1})$, $(11\bar{2}0)$, and $(1\bar{1}00)$ MOS structures, the atmosphere used for oxidation gives striking differences in the distribution of interface states near the conduction band edge. Figure 6.60 shows the interface state density distributions obtained from n-type 4H-SiC(0001), $(000\bar{1})$, $(11\bar{2}0)$, and $(1\bar{1}00)$ MOS structures prepared by (a) dry and (b) wet oxidation. Here, the interface state density is evaluated by the conventional high(1 MHz)–low method. On 4H-SiC(0001), the interface state density near the conduction band edge is not very sensitive to the oxidation conditions (dry/wet). On the other faces, dry oxidation results in very high interface state density, while much lower interface state density can be obtained using wet oxidation (or by POA in the presence of water vapor) [239, 299]. It is reported that the effects of hydrogen annealing at 800–900 °C are much larger on SiC$(000\bar{1})$ [239] and $(11\bar{2}0)$ [300]. Another important result in Figure 6.60 is the lack of sharp increase in interface state density toward the conduction band edge for these non-standard faces. In other words, the distributions of interface states are rather flat on SiC$(000\bar{1})$, $(11\bar{2}0)$, and $(1\bar{1}00)$. High mobilities of 120 cm^{2}V^{-1}s^{-1} on $(000\bar{1})$ [239] and 100–240 cm^{2}V^{-1}s^{-1} on $(11\bar{2}0)$ [300–302] have been achieved.

Post-oxidation nitridation or direct oxidation in a nitrogen-containing gas is also effective in reduction of interface state density on SiC$(000\bar{1})$, $(11\bar{2}0)$, and $(1\bar{1}00)$ [263, 303, 304]. Figure 6.61 shows the interface state density distributions obtained from n-type 4H-SiC(0001), $(000\bar{1})$, $(11\bar{2}0)$, and $(1\bar{1}00)$ MOS structures prepared by dry oxidation and subsequent nitridation in NO. In spite of dry oxidation, the interface state density is drastically reduced by nitridation in NO (or N_2O). Again, the distributions of interface states are rather flat on SiC$(000\bar{1})$, $(11\bar{2}0)$, and $(1\bar{1}00)$. Note that the interface state density near the valence band edge can be also decreased by nitridation of the interface [268].

Figure 6.62 shows the field-effect mobility as a function of the gate voltage of n-channel MOSFETs fabricated on 4H-SiC(0001), $(000\bar{1})$, $(11\bar{2}0)$, and $(1\bar{1}00)$ [304]. About 40-nm-thick oxides were grown by dry oxidation, and subsequent nitridation was performed in a NO(10%)/N_2(90%) atmosphere at 1250 °C. The obtained channel mobility is reasonably high, 46 cm^{2}V^{-1}s^{-1}, on $(000\bar{1})$, and very high, 95–115 cm^{2}V^{-1}s^{-1}, on $(11\bar{2}0)$ and $(1\bar{1}00)$. The latter result is promising for development of trench MOSFETs on SiC(0001) wafers [305, 306]. The effects of the tilt angle on the channel mobility of 4H-SiC MOSFETs with $(11\bar{2}0)$ and $(1\bar{1}00)$ sidewalls have also been investigated [307].

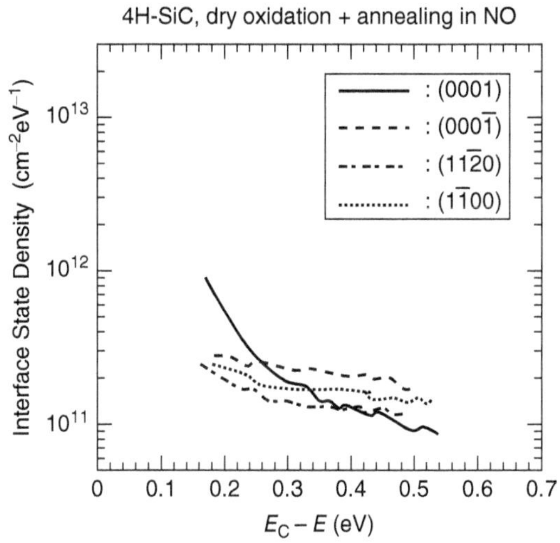

Figure 6.61 Interface state density distributions obtained from n-type 4H-SiC(0001), (000$\bar{1}$), (11$\bar{2}$0), and (1$\bar{1}$00) MOS structures prepared by dry oxidation and subsequent nitridation in NO.

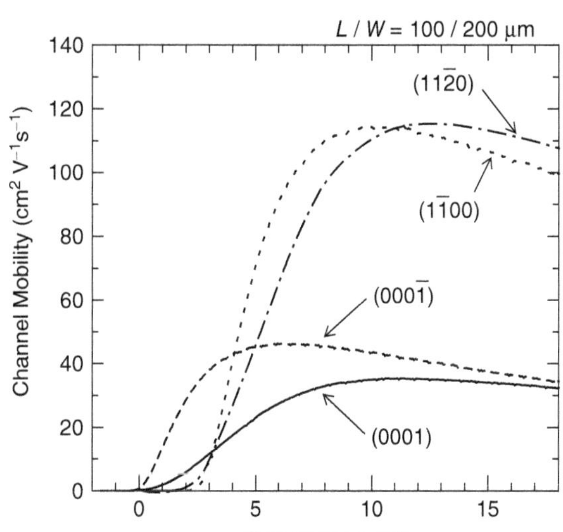

Figure 6.62 Field-effect mobility as a function of the gate voltage of n-channel MOSFETs fabricated on 4H-SiC(0001), (000$\bar{1}$), (11$\bar{2}$0), and (1$\bar{1}$00) [304]. About 40-nm-thick oxides were grown by dry oxidation, and subsequent nitridation was performed in a NO(10%)/N_2(90%) atmosphere at 1250 °C.

6.3.7 Mobility-Limiting Factors

The physical reason for low channel mobility of SiC MOSFETs has been debated. In Si MOSFETs, the main factors limiting the channel mobility are the fixed charge and surface roughness [308], because the interface state density in Si MOS structures formed by adequate processes is low enough not to limit channel mobility. In SiC MOSFETs, Coulomb scattering had often been proposed as the main limiting factor. This is because the channel mobility of 4H-SiC(0001) MOSFETs usually shows a positive temperature coefficient (increases at elevated temperature). However, this is a misleading interpretation, and thermally activated transport or electron localization in inversion layers [148, 309, 310] are more likely to be responsible. In recent years, deeper insights have been provided based on a few careful investigations. One can also see a detailed description of channel mobility in Section 8.2.10.

The channel mobility (n-channel) in the linear region is usually estimated from the following equations [311]:

$$\text{Field-effect mobility } \mu_{\mathrm{FE}}\text{: } \mu_{\mathrm{FE}} = \frac{L}{WC_{\mathrm{OX}}V_{\mathrm{D}}}\frac{\mathrm{d}I_{\mathrm{D}}}{\mathrm{d}V_{\mathrm{G}}} \tag{6.20}$$

$$\text{Effective mobility } \mu_{\mathrm{eff}}\text{: } \mu_{\mathrm{eff}} = \frac{L}{WC_{\mathrm{OX}}(V_{\mathrm{G}} - V_{\mathrm{T}})}\frac{\mathrm{d}I_{\mathrm{D}}}{\mathrm{d}V_{\mathrm{D}}} \tag{6.21}$$

Here I_{D} and V_{D} are the drain current and drain voltage, respectively. L and W are the channel length and the channel width, respectively. In these estimates, it is assumed that all the electrons in the inversion layer are mobile, traveling in the conduction band from the source to the drain. In other words, the sheet electron density in the inversion layer, n_{sheet}, is approximately given by $C_{\mathrm{ox}}(V_{\mathrm{G}} - V_{\mathrm{T}})$, where C_{ox} is the oxide capacitance, V_{G} the gate voltage, and V_{T} the threshold voltage. However, the interface state density in SiC MOS structures is so high that the integrated density of interface states can be of the same order as the induced sheet electron density ($\sim 10^{12}$ cm^{-2}). Electrons trapped at the interface states must be almost immobile. If 90% of induced electrons are trapped, for example, only 10% of induced electrons can travel and contribute to the drain current. Even if the mobile electrons drift with a mobility of 100 cm^{2}V^{-1}s^{-1}, the overall channel mobility is calculated as 10 cm^{2}V^{-1}s^{-1}, according to Equation 6.20 or 6.21. In this circumstance, more and more electrons are excited to the conduction band and become mobile with increasing temperature. As a result, the calculated channel mobility exhibits a positive temperature coefficient. In this case "Coulomb scattering" is a misleading explanation for the mobility-limiting factor. A more correct term is "*electron trapping effect*". Therefore, the temperature dependence of field-effect mobility or effective mobility calculated from Equation 6.20 or 6.21 does not give clear insight into the scattering mechanisms involved in carrier transport.

As described above, the total density of induced electrons (n_{total}) is given by:

$$n_{\mathrm{total}} = n_{\mathrm{mobile}} + n_{\mathrm{trap}} \tag{6.22}$$

where n_{mobile} and n_{trap} are the densities of mobile electrons and of electrons trapped at interface states, respectively. The real mobility of mobile electrons (μ_{real}) and the calculated mobility (μ_{ch}) are linked by the following equation:

$$\mu_{\mathrm{ch}} = \mu_{\mathrm{real}}\frac{n_{\mathrm{mobile}}}{n_{\mathrm{mobile}} + n_{\mathrm{trap}}} \tag{6.23}$$

The real mobility of mobile electrons in the conduction band can be obtained by MOS-Hall effect measurements [152, 260, 312, 313]. Figure 6.63 shows the typical pattern configuration for Hall effect measurements of a MOSFET. A long channel region is formed and three or four additional terminals are made for Hall effect measurements. In MOS-Hall effect measurements, not only the real mobility, but also the real mobile electron density (n_{mobile}), can be determined. Since the total density of induced electrons (n_{total}) is calculated as $C_{\mathrm{ox}}(V_{\mathrm{G}} - V_{\mathrm{T}})$, the degree of electron trapping ($(n_{\mathrm{total}} - n_{\mathrm{mobile}})/n_{\mathrm{total}}$) can

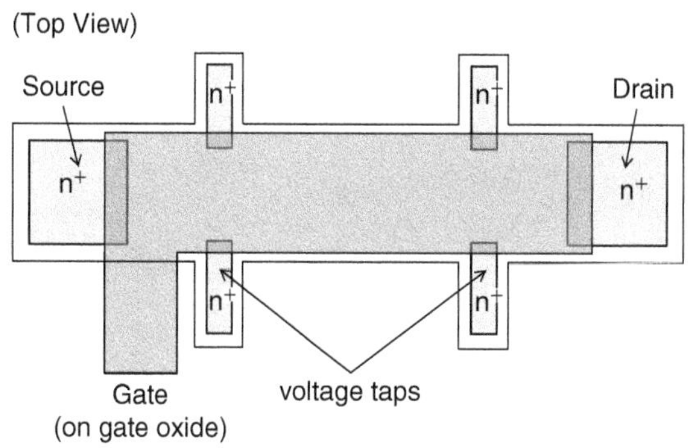

Figure 6.63 Typical pattern configuration for Hall effect measurements of a MOSFET.

be estimated. In a different manner, the field-effect mobility μ_{FE} can be expressed as [313]:

$$\mu_{FE} = \frac{\mu_{real}}{1 + \frac{dQ_T/dE_F}{dQ_{inv}/dE_F}} = \frac{\mu_{real}}{1 + \frac{q^2 D_{IT}(E_F)}{C_{inv}}} \tag{6.24}$$

Here Q_T is the trapped charge density, Q_{inv} the total charge in the inversion layer, $D_{IT}(E_F)$ the interface state density at a Fermi level (E_F), and C_{inv} the differential capacitance of the inversion layer. Thus, the estimated channel mobility is much smaller than the real value when $D_{IT}(E_F) > C_{inv}/q^2$ because of electron trapping. If the interface state density increases rapidly near the band edge, as is the case for SiC(0001), this trapping effect applies for almost the whole range of gate bias. It is reported, however, that the trapping effect is not very significant after recent optimization of the nitridation process [260].

Figure 6.64 shows examples of real mobility (as a function of gate bias) determined by MOS-Hall effect measurements on an n-channel (a) 6H- and (b) 4H-SiC(0001) MOSFETs [152, 312]. There exists a certain discrepancy between the Hall mobility and the calculated effective channel mobility, and the degree of electron trapping can be estimated to be about 30–50% for 6H-SiC and 70–85% for 4H-SiC MOSFETs. By MOS-Hall effect measurements, the sheet density of mobile carriers can be directly determined as a function of gate bias. Therefore, this is a powerful technique to investigate carrier transport in SiC MOSFETs.

In SiC MOSFETs, one often observes a negative correlation between the channel mobility and the threshold voltage determined by linear extrapolation of the transfer characteristics. The channel mobility decreases when the threshold voltage increases in MOSFETs, the oxides of which were formed by the same process but under slightly different conditions (e.g., nitridation at different temperatures or for different periods). This phenomenon is partly attributable to a problem in determining the threshold voltage and can be interpreted by considering the electron trapping, as shown schematically in Figure 6.65. As the interface state density near the conduction band increases, more and more of the electrons induced by the gate bias are trapped at the interface. This leads to a decrease in drain current as well as the gradual (not abrupt) increase near the threshold. Thus, the threshold voltage becomes artificially high when determined by linear extrapolation.

When the interface state density is very high, one can see a clear correlation between the n-channel mobility and the interface state density near the conduction band edge [151–154, 229, 314]. However, there is still debate about what the main mobility-limiting factor is once the interface state density has been reduced by nitridation or wet re-oxidation. For example, on SiC(0001), the interface state density at E_c − (0.2–0.3) eV can be reduced from mid 10^{12} cm^{-2} eV^{-1} to low 10^{11} cm^{-2} eV^{-1} (as evaluated by the high–low method) by appropriate nitridation, but the resulting n-channel mobility is 30–50 cm^{2}V^{-1}s^{-1},

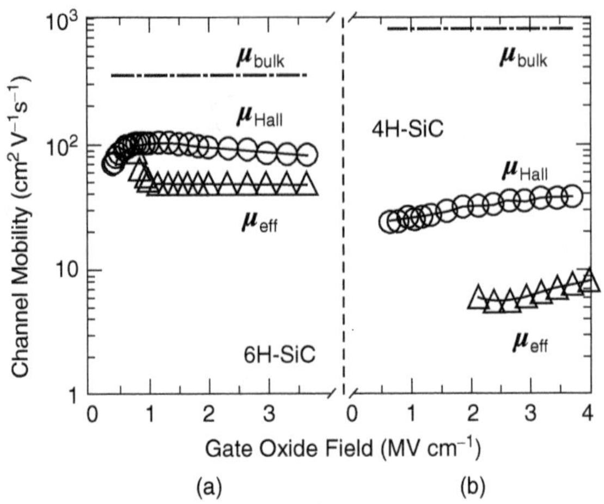

Figure 6.64 Examples of real mobility (as a function of gate bias) determined by MOS-Hall effect measurements on an n-channel (a) 6H- and (b) 4H-SiC(0001) MOSFETs ([152, 312] reproduced with permission from Springer Verlag). The effective mobility is also shown.

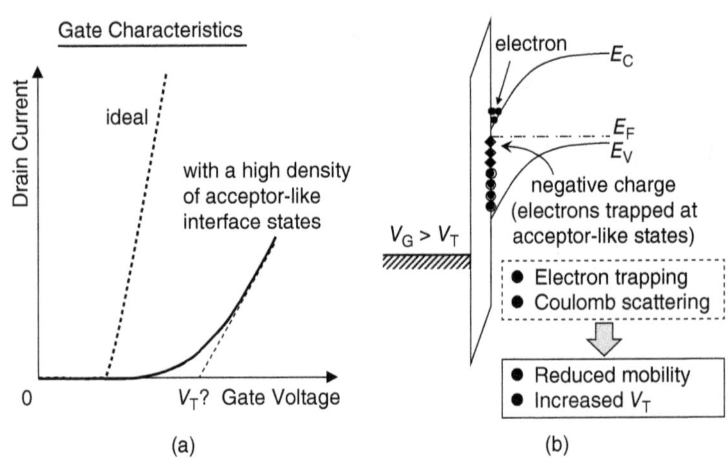

Figure 6.65 Schematic illustration of a problem of electron trapping in determining the threshold voltage of n-channel SiC MOSFETs. (a) Degradation of gate characteristics due to a high density of interface states, (b) major limiting factors of channel mobility in SiC MOSFETs.

which corresponds to only 3–5% of the bulk mobility. Figure 6.66 shows the n-channel mobility versus the interface state density at $E_c - 0.2$ eV determined by (a) the high (1 MHz)–low method and (b) the $C\text{–}\psi_s$ method. The data were obtained from 4H-SiC MOSFETs and MOS capacitors processed under various conditions. In Figure 6.66a, there exists large scatter, though a correlation is observed to some extent. For example, at an interface state density of about 1×10^{12} cm^{-2} eV^{-1}, the channel mobility of one MOSFET on (0001) is 8 cm^{2}V^{-1}s^{-1} and the other is 17 cm^{2}V^{-1}s^{-1}. Furthermore, the MOSFET on $(11\bar{2}0)$ with a similar interface state density is as high as 71 cm^{2}V^{-1}s^{-1}. It is hard to explain this mobility difference simply by the difference in surface roughness between two crystal faces. In addition,

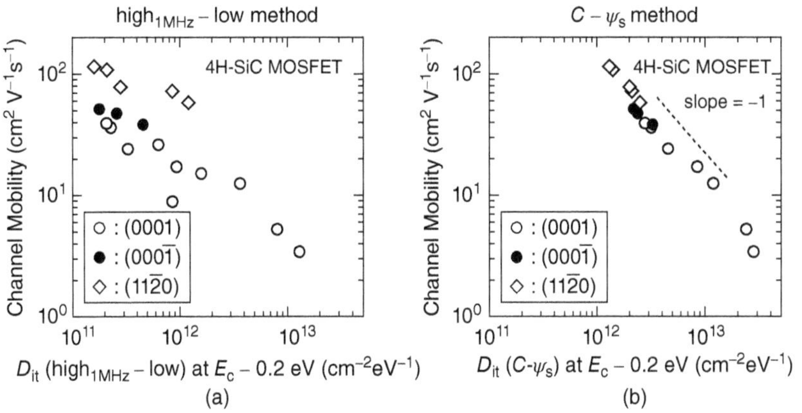

Figure 6.66 n-channel mobility versus the interface state density at $E_c - 0.2$ eV determined by (a) the high (1 MHz)–low method and (b) the $C-\psi_s$ method.

the improvement of channel mobility in MOSFETs on (0001) is not very significant (only a factor of 10), in spite of the remarkable reduction of interface state density by process optimization (a factor of 100). On the other hand, a very clear trend can be seen in Figure 6.66b, where the slope of the plot is nearly -1. This result indicates the following:

1. The n-channel mobility is still limited mainly by very high interface state density, even after recent process optimization.
2. Very fast interface states, which can be detected by the $C-\psi_s$ method (not detectable by the high(1 MHz)–low method) affect the channel mobility.

In general, SiC MOSFETs exhibit poor subthreshold slopes, 200–500 mV/decade, much larger than the ideal value, approximately 60 mV/decade at room temperature. The poor subthreshold slope can be reproduced well using the distribution of interface state density determined by the $C-\psi_s$ method. Figure 6.67 depicts the correlation between the interface state densities determined from $C-V$ data of MOS capacitors and those from the subthreshold slope of MOSFETs. The interface state densities determined from the subthreshold slope are much higher than those determined by the high(1 MHz)–low method (Figure 6.67a), which was a mystery in this field. However, the interface state densities determined by the $C-\psi_s$ method show good agreement with those obtained from the subthreshold slope, as shown in Figure 6.67b. This result indicates that fast interface states indeed affect the performance of MOSFETs.

However, the mechanism of channel conduction in SiC MOSFETs must be much more complicated. In SiC, the surface roughness is much larger compared with that of Si because of the use of off-axis {0001} wafers and immature surface preparation technology. The roughness scattering may be prominent when the interface state density is further reduced. This is especially important at high gate bias, where real MOS devices operate. The influence of real (not effective) fixed charge on the channel mobility has not yet been clarified.

Another possible problem is the large fluctuation of surface potential, as indicated by the conductance measurements [216]. Since the interface structure of SiC MOS structures can be highly inhomogeneous, the surface potential can fluctuate inside the MOS channel. Thus the total conduction in the inversion layer may be limited by microscopic regions where the conductivity is greatly suppressed by the fluctuation. Furthermore, recent defect studies indicate that excess carbon atoms are emitted from the oxidation interface into the SiC bulk region during thermal oxidation [157] and form defect levels near

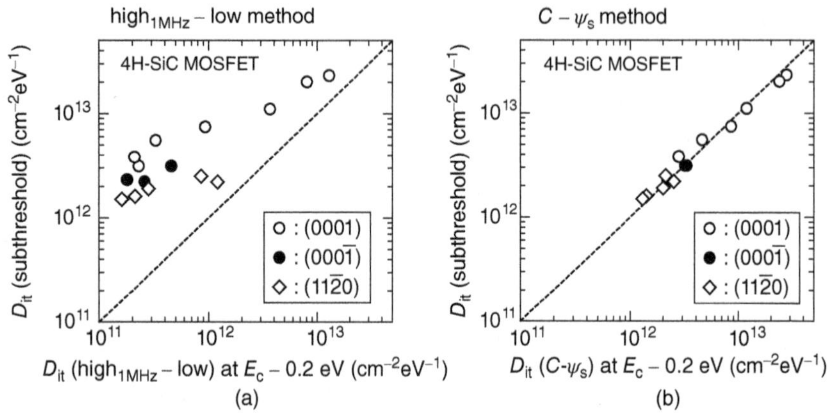

Figure 6.67 Interface state density estimated from the subthreshold slope of MOSFET versus The density at $E_c - 0.2$ eV determined by (a) the high (1 MHz)–low method and (b) the $C-\psi_s$ method.

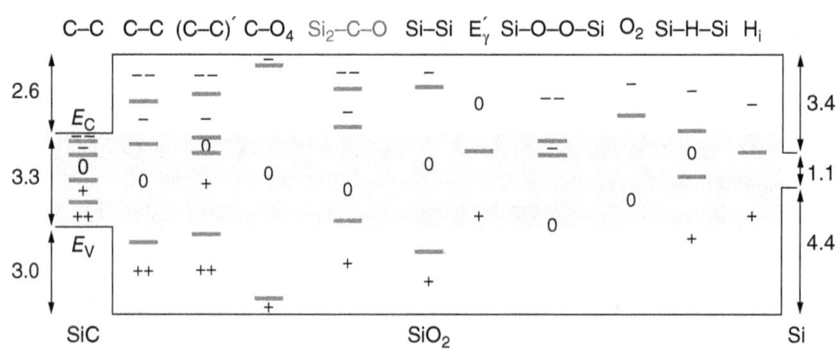

Figure 6.68 Charge transition levels of various possible defects near the SiO_2/SiC interface predicted by a hybrid-density function theory ([318] reproduced with permission from American Physical Society).

the conduction and valence band edges [315, 316]. *Ab initio* calculation has predicted that interstitial carbon dimers are formed in the presence of carbon interstitials in SiC, and that these carbon dimers form a shallow acceptor level at $E_c - 0.2$ eV [317]. Thus, the bulk mobility just near the interface may be severely affected by such defects. Figure 6.68 shows the charge transition levels of various possible defects near the SiO_2/SiC interface predicted by a hybrid-density function theory [318]. According to this study, C-C and Si_2-C-O defects may be responsible for the levels near the conduction band edge of 4H-SiC. However, more fundamental studies are required to reveal the nature of interface defects and carrier transport properties in the inversion layers.

6.4 Metallization

Schottky contacts are the key component of Schottky barrier diodes (SBDs) and metal-semiconductor field effect transistors (MESFETs). The basic theory and knowledge of Schottky contacts are

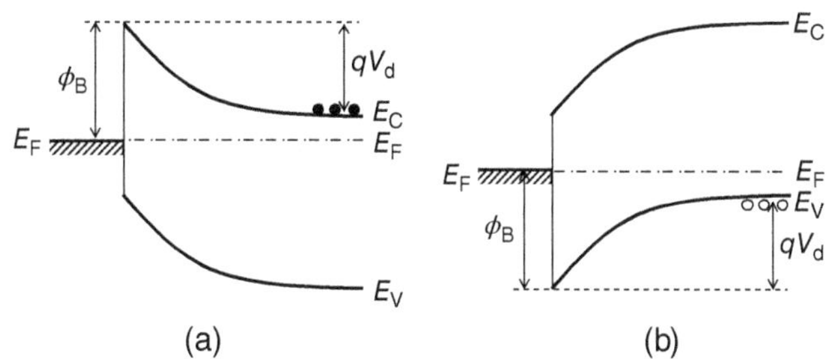

Figure 6.69 Energy band diagrams of a Schottky barrier on (a) n-type and (b) p-type semiconductor at zero bias.

also important in understanding ohmic contacts. Formation of ohmic contacts is essential in any semiconductor devices. Both types of contacts are required for a variety of electrical characterization techniques of materials, including $I-V, C-V$, DLTS, and Hall effect measurements. Several review papers on Schottky contacts [319–321] and ohmic contacts [320, 322–327] on SiC have been published.

6.4.1 Schottky Contacts on n-*Type and* p-*Type SiC*

6.4.1.1 Fundamentals

In the case of SiC, most metals deposited on SiC work as Schottky contacts, as long as the SiC material is not heavily doped and high-temperature (>700 °C) contact sintering is not performed. When a metal and a semiconductor are brought into contact, the Fermi levels of the two materials line up at equilibrium (zero bias). The energy band diagrams of a Schottky barrier on (a) n-type and (b) p-type semiconductor at zero bias are shown in Figure 6.69. Here, ϕ_B is the barrier height, V_d the built-in potential, and ΔE_{fn} (or ΔE_{fp}) the position of the Fermi level from the conduction band (or valence band) edge. From the band diagram, the following equation is satisfied:

$$\phi_B = qV_d + \Delta E_{fn}(n\text{-type}); \quad \phi_B = qV_d + \Delta E_{fp}(p\text{-type}) \tag{6.25}$$

In an ideal model, the barrier height is determined simply by the difference in the work function between the metal and the semiconductor. In actual semiconductors, however, the barrier height is affected by the surface states [8, 328]. The barrier height can be determined by several techniques, including $I-V, C-V$, IPE, and XPS.

Figure 6.70 shows the forward current density versus bias voltage for Ti, Ni, and Au/n-type 4H-SiC(0001) SBDs at room temperature [329]. The current density can be expressed by [8, 328]:

$$J = A^*T^2 \exp\left(-\frac{\phi_B}{kT}\right)\left\{\exp\left(\frac{qV}{nkT}\right) - 1\right\} \tag{6.26}$$

where n is the *ideality factor* and A^* is the *effective Richardson's constant* for the semiconductor, given by:

$$A^* = \frac{4\pi q m_n{}^* k^2}{h^3} \tag{6.27}$$

Here $m_n{}^*$ (or $m_p{}^*$) is the majority carrier effective mass and h is the Planck constant. In SBDs on n-type 4H-SiC(0001), A^* is determined as 146 A cm^{-2} K^{-2} by using an M_c of 3 and $m_n{}^*$ of 0.4 m_0 [329]. This value agrees with that obtained from a careful experiment [330]. Note that the effective Richardson's constant depends on the crystal face as well as on the conductivity type (n-type, p-type). Therefore, the

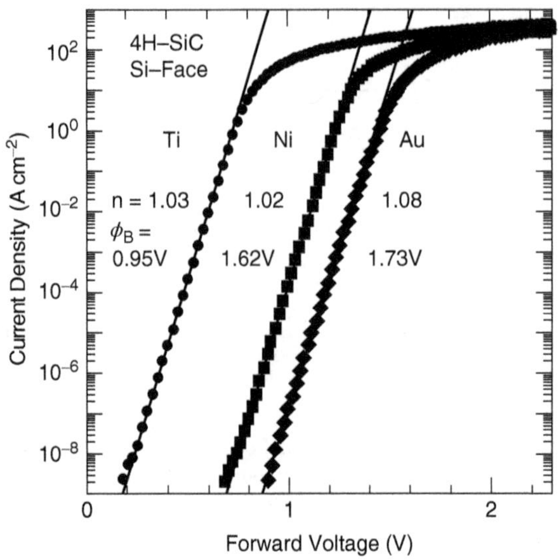

Figure 6.70 Forward current density versus bias voltage for Ti, Ni, and Au/n-type 4H-SiC(0001) Schottky barrier diodes (SBDs) at room temperature ([329] reproduced with permission from IEEE).

barrier height can be extracted from the J-intercept (J_0) of the semilogarithmic plot shown in Figure 6.70. Figure 6.71 shows the extracted barrier height versus the ideality factor for Ni/n-type 4H-SiC SBDs fabricated by the same process. It is clear that the barrier height is severely underestimated when the ideality factor is larger than 1.10–1.15. Thus, the barrier height extracted from the J-intercept (J_0) can be reliable only when the ideality factor is smaller than 1.10. Even when a relatively good ideality factor of 1.05 is obtained, the extracted barrier height is clearly smaller than the real value. The ideality factor is degraded by several reasons including inhomogeneity of the barrier height and imperfections (such as contamination) at the interface [331–333].

The capacitance of the space-charge region of SBDs (n-type) per unit area is given by:

$$C = \sqrt{\frac{\varepsilon_s q N_D}{2(V_d - V)}} \tag{6.28}$$

where N_D is the donor density and ε_s the dielectric constant of the semiconductor [8, 328]. Squaring and inverting this equation, one obtains:

$$\frac{1}{C^2} = \frac{2(V_d - V)}{\varepsilon_s q N_D} \tag{6.29}$$

By plotting $1/C^2$ versus the bias voltage, the built-in potential (V_d) can be determined from the V-intercept of the plot. Figure 6.72 shows $1/C^2$ versus bias voltage for Ti, Ni, and Au/n-type 4H-SiC(0001) SBDs. From the built-in potential and the calculated Fermi level (ΔE_{fn}), the barrier height can be extracted using $\phi_B = qV_d + \Delta E_{fn}$. The Schottky barrier heights determined from C–V measurements are slightly higher than those extracted from I–V measurements. This is natural – local lowering of Schottky barrier height makes a significant difference in I–V measurements, because the current depends exponentially on the barrier height while C–V measurements yield the value averaged over the entire contact area and are thus less affected by local variation.

Furthermore, the *net donor density* ($N_D - N_A$) can be determined from the inverse slope of the plot. Again, note that this is the net doping density, not "the carrier density." Inside the space-charge region

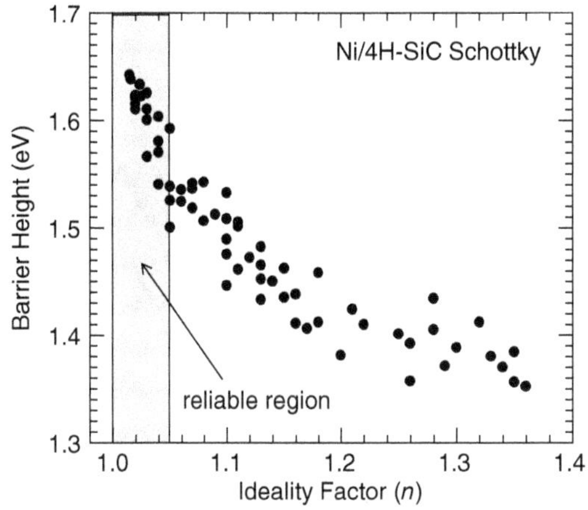

Figure 6.71 Extracted barrier height versus the ideality factor for Ni/n-type 4H-SiC SBDs fabricated by the same process.

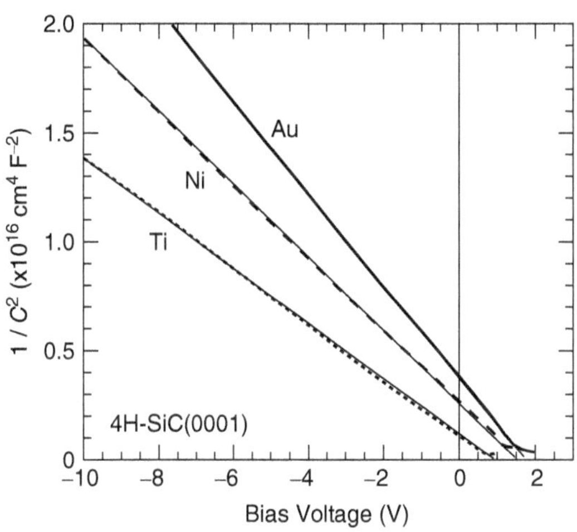

Figure 6.72 $1/C^2$ versus bias voltage for Ti, Ni, and Au/n-type 4H-SiC(0001) SBDs.

(except near the edge of the space-charge region) all the dopants are ionized, and the width of the space-charge region determines the capacitance. This is not very important in Si owing to nearly perfect ionization of dopants at room temperature. In SiC, however, the carrier density can be much lower than the doping density, especially for p-type materials at room temperature. Thus, one has to be aware of the fact that $C-V$ measurements always give the net doping density. This can be easily confirmed by very little change in the density obtained from $C-V$ measurements on p-type SBDs at different temperatures (e.g., from RT to 300 °C or from −100 °C to RT).

Another technique to determine the barrier height is IPE. In this technique, semi-transparent (<20 nm) Schottky contacts must be prepared. Monochromatic light is projected through the contact, and the photocurrent is monitored as the wavelength of the light is changed. When the photon energy ($h\nu$) is larger than the barrier height, the square root of the photocurrent yield (photocurrent divided by the photon

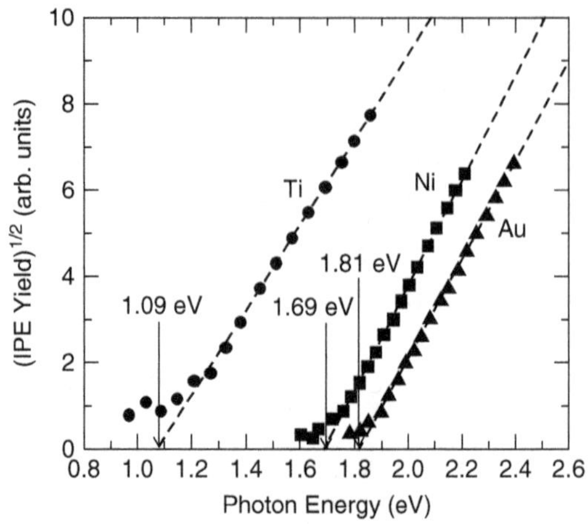

Figure 6.73 Square root of photocurrent yield versus the photon energy for Ti, Ni, and Au/n-type 4H-SiC(0001) SBDs (internal photoemission measurements) [334].

number) Y is linearly dependent on the photon energy as follows:

$$Y = A(h\nu - \phi_B)^2 \tag{6.30}$$

where A is a proportionality constant. The linearity in the square root of photocurrent comes from the energy dependence of the density of states in the energy band. Figure 6.73 shows the square root of photocurrent yield versus the photon energy for Ti, Ni, and Au/n-type 4H-SiC(0001) SBDs [334]. The barrier height can be directly determined from the $h\nu$ intercept of the plot. IPE gives the most reliable values of Schottky barrier heights, though it requires special equipment and semi-transparent Schottky contacts.

6.4.1.2 Schottky Contacts on SiC

Figure 6.74 shows the barrier height versus metal work function for n-type 4H-SiC SBDs with various metals [334–336]. Data for 4H-SiC(0001), (000$\bar{1}$), and (11$\bar{2}$0) are plotted. For a given metal, the barrier height is slightly higher on (000$\bar{1}$), slightly lower on (0001), and that on (11$\bar{2}$0) is in between. This difference may be attributed to the existence of polarity-dependent dipoles at the interface and/or different distribution of surface states. The slope of this plot is 0.8–0.9, indicating that the metal/SiC interface is free from Fermi-level pinning, and is close to the *Schottky–Mott limit* [328, 337]. To obtain this kind of data, of course, high-quality materials and careful surface cleaning are necessary. The barrier heights can be slightly changed by employing different processes, such as surface treatment, prior to metal deposition. It is noted that the ideality factor and reproducibility are improved by annealing at relatively low temperatures (200–500 °C) after metal deposition. When the Schottky barrier height on n-type 4H-SiC is compared with that on n-type 6H-SiC [338] for a given Schottky metal, the barrier height on 4H-SiC is about 0.2 eV higher, which corresponds to the difference in the bandgaps of the two polytypes. The barrier height is also discussed in Section 7.2.

Schottky barrier heights on p-type SiC have been much less studied [339]. Figure 6.75 shows the barrier height versus the metal work function for p-type 4H-SiC(0001) SBDs with various metals. The slope of the plot is about −0.8, and the sum of Schottky barrier heights on n- and p-type SiC for a Schottky metal material is close to the bandgap (E_g):

$$\phi_{Bn} + \phi_{Bp} \approx E_g \tag{6.31}$$

Therefore, a metal having a small work function, such as Ti, gives a high barrier height on p-type SiC.

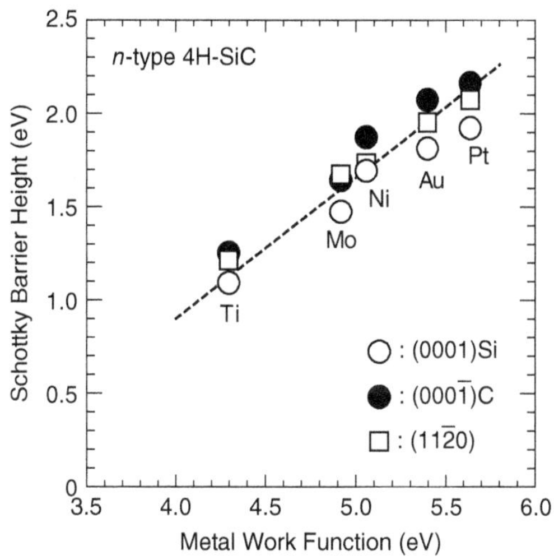

Figure 6.74 Barrier height versus metal work function for n-type 4H-SiC SBDs with various metals [334–336]. Data for 4H-SiC(0001), (000$\bar{1}$), and (11$\bar{2}$0) are plotted.

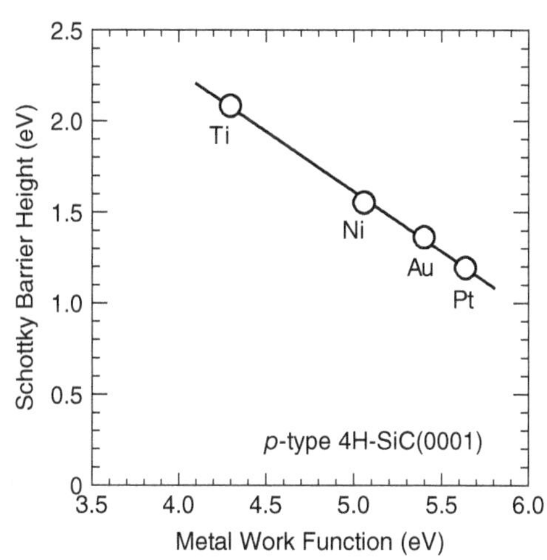

Figure 6.75 Barrier height versus the metal work function for p-type 4H-SiC(0001) SBDs with various metals.

Unique device physics is involved in the reverse leakage current of SiC SBDs. In SiC, the electric field strength in the space-charge region can be almost 10 times higher than in space-charge regions of Si-based devices. Therefore, the band bending can be so sharp that the potential barrier can be very thin. Figure 6.76 shows the band diagram of an n-type SiC Schottky barrier under high reverse bias. In Si SBDs, the reverse leakage current is well described by a thermionic emission model, taking into account the barrier height lowering by the image force [8, 328], unless the semiconductor is heavily doped. In SiC SBDs, however, the observed leakage current density is many orders of magnitude higher than the value calculated by the thermionic emission model (taking into account barrier height lowering).

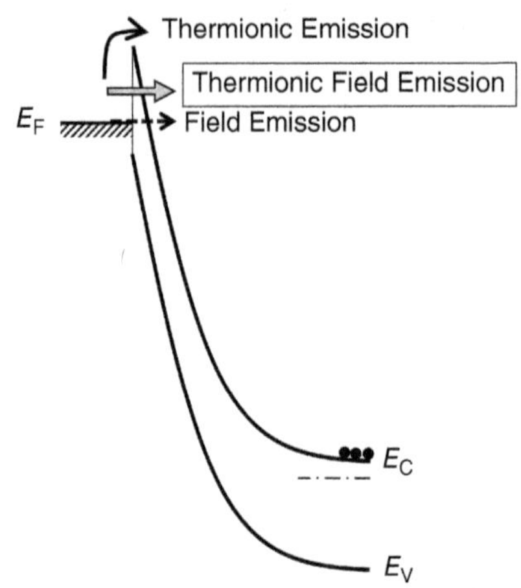

Figure 6.76 Band diagram of an n-type SiC Schottky barrier under high reverse bias.

At first, leakage current through crystalline defects or severe local lowering of Schottky barrier heights was suspected. However, it turns out that there is an essential reason for the relatively large leakage current in SiC SBDs. In SiC, the triangular-like potential barrier can be very thin because of the high electric field (slope of the band diagram), and the leakage current is governed by *thermionic field emission* (TFE) [340, 341], as shown in Figure 6.76. The leakage current density based on a TFE model can be expressed by the following equation [341]:

$$J_{\mathrm{TFE}} = \frac{A^{*}Tq\hbar E}{k}\sqrt{\frac{\pi}{2m^{*}kT}}\exp\left[-\frac{1}{kT}\left\{\phi_{\mathrm{B}} - \frac{(q\hbar E)^{2}}{24m^{*}(kT)^{2}}\right\}\right] \tag{6.32}$$

Here m^* and E are the tunneling mass of carriers and the electric field strength, respectively. Figure 6.77 shows an example of reverse $I-V$ characteristics of a Ti/4H-SiC(n-type) SBD at various temperatures [341]. The considerable increase in leakage current with increasing bias voltage, as well as the small temperature dependence, can be reproduced by the TFE model. It has been reported that the leakage current of SBDs formed on high-quality GaN(0001) can also be reproduced by the TFE model [342]. Thus, the TFE current will be dominant in SBDs on any wide-bandgap materials which exhibit high electric field strength, such as SiC, GaN, Ga_2O_3, and diamond.

When the barrier height is highly inhomogeneous, abnormal $I-V$ characteristics are observed [343, 344]. Such an example is shown in Figure 6.78, where forward $I-V$ characteristics of a Ti/SiC SBD at various temperatures are plotted [344]. At room temperature, a small shoulder-like feature is seen in the $I-V$ characteristics, suggesting non-ideal behavior and a non-unity ideality factor. The shoulder becomes distinct at low temperature, because the slope of the log I versus V plot (q/nkT) indicated in Equation 6.26 becomes sharp at low temperature. In this case, there exist small regions where the local barrier height ($\phi_{\mathrm{B,local}}$) is considerably lower than the main Schottky barrier (ϕ_{B}). The ratio of the low-ϕ_{B} region and the total contact region is estimated from the ratio of the almost-saturated current of the low-ϕ_{B} component and that of the main Schottky barrier (the "saturation" is caused by a series resistance). The ratio is approximately 10^{-5} in Figure 6.78. Note that the ideality factor is close to unity for both the low-ϕ_{B} region and the main diode. When the value of the low ϕ_{B} is estimated, the corresponding area (not the main area) should be employed for the calculation. Furthermore, such a low ϕ_{B} region can cause considerably increased leakage under reverse bias [345, 346]. The diode can be regarded as a parallel connection of the main diode and a diode with low ϕ_{B}. Thus, the reverse leakage current is governed by TFE through the low-ϕ_{B} region of the diode. Note that it is difficult to detect a low-ϕ_{B} region by $C-V$ or IPE measurements, since these techniques measure the average values.

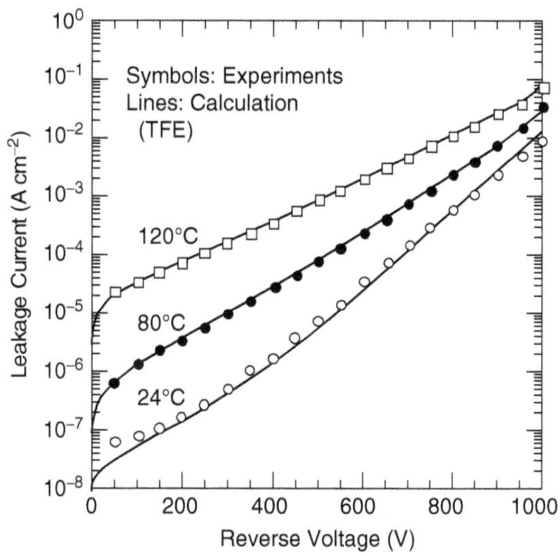

Figure 6.77 Reverse I–V characteristics of a Ti/4H-SiC(n-type) SBD at various temperatures ([341] reproduced with permission from Trans. Tech. Publication). The considerable increase in leakage current with increasing bias voltage, as well as the small temperature dependence, can be reproduced by the TFE model.

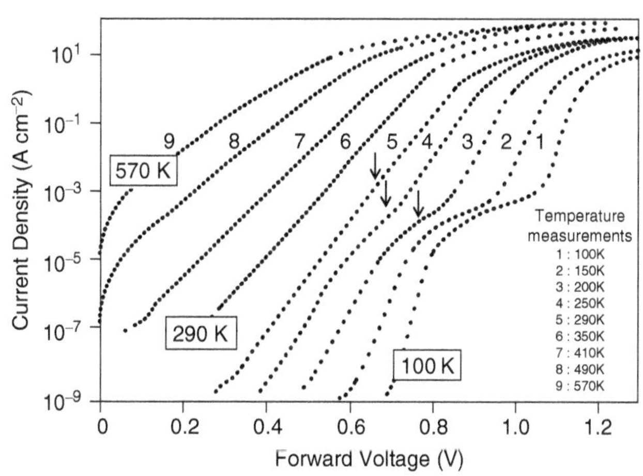

Figure 6.78 Forward I–V characteristics of a Ti/SiC SBD with an inhomogeneous barrier height measured at various temperatures ([344] reproduced with permission from IEEE).

6.4.2 Ohmic Contacts to n-*Type and* p-*Type SiC*

6.4.2.1 Fundamentals

The characteristics required for ohmic contacts include low *contact resistivity* (or *specific contact resistance*), surface flatness, and long-term stability. In particular, the contact resistivity is a key electrical parameter. This is the contact resistance per unit area, measured in Ω cm^2, and the voltage drop across

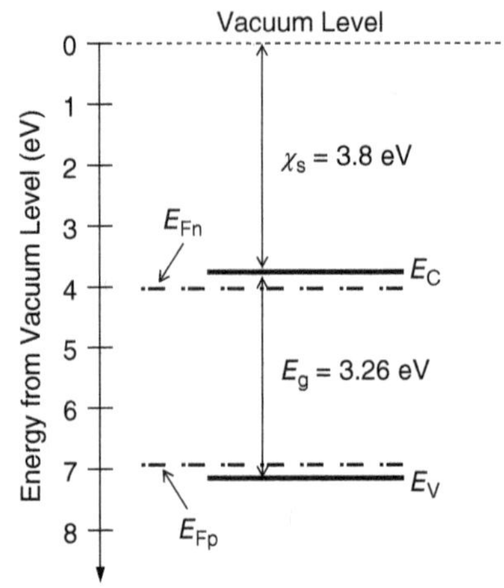

Figure 6.79 Band diagram of 4H-SiC, including the vacuum level.

the contact is thus calculated by multiplying the contact resistance by the current density (A cm^{-2}). The contact resistance should be negligibly small (<1%) compared with the total resistance between two main terminals of devices (e.g., anode-cathode and drain-source). In the cathode contact of a SBD and the drain contact of a vertical field-effect transistor (FET), the contact can cover almost all the device area (active area). In the source contact of a vertical FET and the emitter contact of a bipolar junction transistor, however, the contact area is usually much smaller (<20–30% of the device area), meaning that very low contact resistivity in the range of 10^{-6} Ω cm^2 is generally required.

Because of the wide bandgap of SiC, one can find very few contact metals with very low barrier heights. Figure 6.79 shows the band diagram of 4H-SiC, including the vacuum level. Since the electron affinity of 4H-SiC is about 3.8 eV [147, 176] and the bandgap 3.26 eV at room temperature, a contact metal with a work function lower than about 4 eV is ideal for an ohmic contact to n-type 4H-SiC, and that with a work function higher than about 7 eV is ideal for a contact to p-type. This rough sketch clearly shows the difficulty in formation of ohmic contacts, especially on p-type 4H-SiC.

Ohmic behavior is usually achieved by using the tunneling current through the thin potential barrier in SiC. When the tunneling current is caused primarily by electrons with energies near the Fermi level (that is, in the case of relatively low temperature or low barrier height), the tunneling is called *field emission* (FE). In the FE regime, the contact resistivity ρ_c for an n-type semiconductor is given approximately by the following equation [8, 347]:

$$\rho_C \approx \frac{k \sin(\pi c_1 kT)}{A^* \pi q T} \exp\left(\frac{\phi_B}{E_{00}}\right) \tag{6.33}$$

$$E_{00} = \frac{q\hbar}{2}\sqrt{\frac{N_D}{m^* \varepsilon_S}} \tag{6.34}$$

$$c_1 = \frac{1}{2E_{00}} \log\left\{-\frac{4(\phi_B - qV)}{\Delta E_{Fn}}\right\} \tag{6.35}$$

As the temperature increases, an appreciable number of electrons with energies above the Fermi level can contribute to the tunneling. This thermally-assisted tunneling is called *thermionic field emission*. In the TFE regime, the contact resistivity ρ_c for an n-type semiconductor is given approximately by the

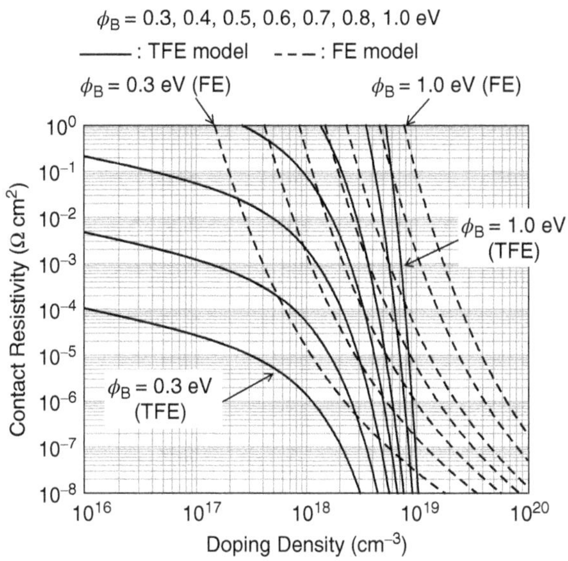

Figure 6.80 Contact resistivity versus dopant density for barrier heights from 0.3 to 1.0 eV, calculated for the contacts on n-type 4H-SiC, taking both FE (field emission) and TFE (thermionic field emission) currents into account.

following equation [8, 347]:

$$\rho_C \approx \frac{k\sqrt{E_{00}}\cosh\left(\frac{E_{00}}{kT}\right)\coth\left(\frac{E_{00}}{kT}\right)}{A^*qT\sqrt{\pi q(\phi_B - \Delta E_{Fn})}}\exp\left\{\frac{\phi_B - \Delta E_{Fn}}{E_{00}\coth\left(\frac{E_{00}}{kT}\right)} + \frac{\Delta E_{Fn}}{kT}\right\} \tag{6.36}$$

As seen in Equations 6.33 and 6.36, the contact resistivity is proportional to $\exp\{\phi_B/(N_D)^{1/2}\}$ in both cases. More detailed modeling in the case of SiC is found in the literature [348, 349]. In general, the pure FE mechanism is dominant when $E_{00} > kT$, and TFE becomes important when $E_{00} < kT$ [347].

Therefore, the strategy for obtaining low contact resistivity is, in principle, simple: (i) increase the surface doping density and (ii) select a metal which forms a low barrier height. In SiC, however, complicated interface reactions are involved to reduce the high barrier heights commonly observed. Figure 6.80 shows the contact resistivity versus the dopant density for barrier heights from 0.3 to 1.0 eV, calculated for the contacts on n-type 4H-SiC, taking both FE and TFE currents into account. In almost all the regime, the TFE mechanism dominates. It is clear that heavy doping (>5 × 10^{18} cm^{-3}) is necessary to obtain a low contact resistivity of 1 × 10^{-6} Ω cm^2 for a metal with a barrier height of 0.5 eV.

A special test structure must be prepared to measure the contact resistivity. The most common structure is shown in Figure 6.81. On the top of a rectangular mesa structure, multiple rectangular contacts are formed with varying spacing. The resistance between two terminals is measured with four probes (two for current, two for voltage), and the measured resistance is plotted against the contact spacing. Figure 6.82 shows an example of this plot for Ni/n-type 4H-SiC sintered at 1000 °C. Since the measured resistance (R_T) consists of the two contact resistances and the resistance of the semiconductor, the following equation is satisfied [207]:

$$R_T = R_{sh}\frac{d}{w} + 2R_C \tag{6.37}$$

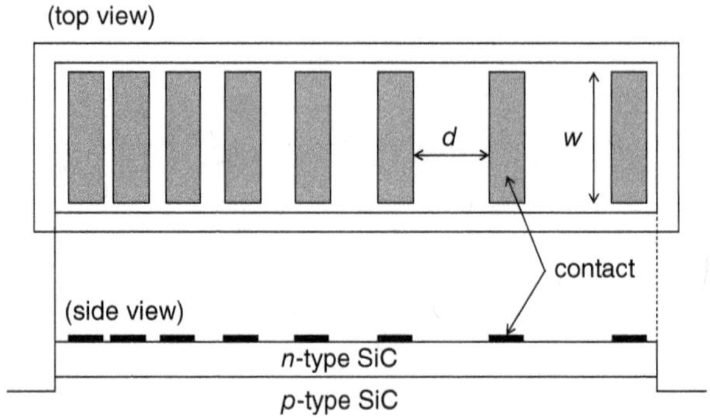

Figure 6.81 Test structure to measure the contact resistivity by the linear transfer length method (TLM).

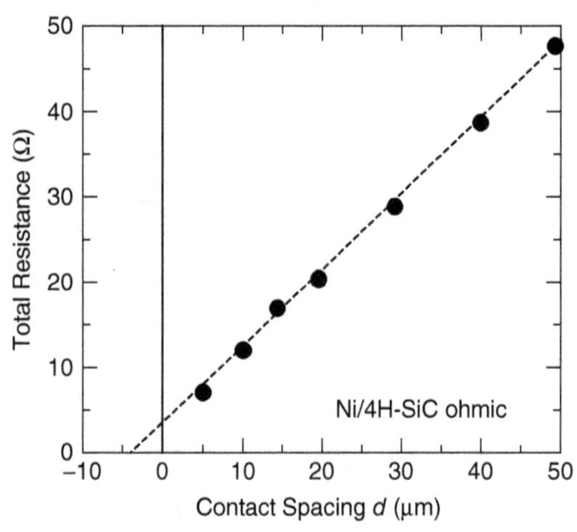

Figure 6.82 Measured resistance plotted against the contact spacing (linear transfer length method) for Ni/n-type 4H-SiC sintered at 1000 °C.

where R_{sh} is the sheet resistance of the semiconductor, d the spacing between contacts, w the contact width, and R_C the contact resistance given by the R_T-axis intercept. L_T, the linear transfer length, can be extracted from the d-axis intercept of the plot, as shown in Figure 6.81. The contact resistivity (ρ_c) is given by:

$$\rho_c = R_C L_T w \tag{6.38}$$

Note that the sheet resistance can also be determined from the slope of the plot. This measurement is called the *linear transfer length method* (TLM). This method is based on the *linear transmission line model,* which takes into account the voltage and current distribution beneath contacts [207]. If the contacts exhibit non-ohmic characteristics, some modification is required [350]. With the TLM method, contact resistivities down to about 10^{-6} Ω cm^2 can be evaluated. For more accurate measurements of the contact resistivity in the 10^{-7} to 10^{-8} Ω cm^2 range, so-called *Kelvin structures* must be formed [207].

In formation of good ohmic contacts, the metal chosen, and sintering after metal deposition are critical. In the case of SiC, one must be aware of reactions between the metal and Si or C as a function of temperature. In this sense, contact metals are classified into (i) metals which form only silicide(s) (no

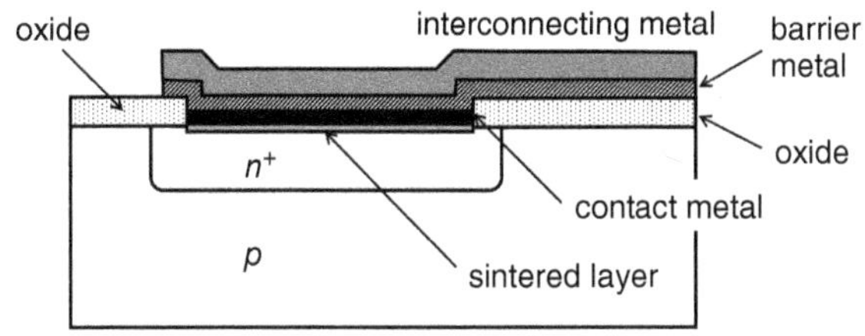

Figure 6.83 Multilayer contact structure used for practical device fabrication.

Table 6.4 Typical ohmic contacts for n- and p-type SiC reported in the literature.

	n-type	p-type
Ohmic contacts	Ni (sintered)	Al/Ti (sintered)
	Ti (sintered)	Al/Ni/Ti (sintered)
	Al (sintered)	Al/Ti/Al (sintered)
	Mo (sintered)	Al/Ti/Ge (sintered)
	W (sintered)	AlSi (sintered)
	Al/Ni (sintered)	Pt (sintered)
	Al/Ti (sintered)	Ni (sintered)
	Ni/Ti/Al (sintered)	Pd (sintered)
	TiC (sintered)	Ta (sintered)
	TiW (sintered)	Si/Co (sintered)
	NiCr (sintered)	–

carbides), (ii) metals which form only carbide(s) (no silicides), and (iii) metals which form both silicide(s) and carbide(s). Therefore, a metal-Si-C phase diagram is useful for selection of contact metals, and for obtaining insights into the ohmic behavior. However, formation of silicide(s) or carbide(s) is not always sufficient to ensure ohmic behavior in SiC, and more complicated phenomena, such as vacancy formation and associated diffusion of carbon (or silicon) must occur to attain good ohmic contacts. This is evidenced by the observation that non-sintered, as-deposited metal silicide or carbide on SiC usually exhibits Schottky characteristics.

Because of the chemical inertness of SiC, the sintering process to obtain ohmic contacts is typically performed at 900–1000 °C. It is a challenge to control the structure and thickness of the chemically reacted layer at the interface. Furthermore, the reaction between the metal and SiC can lead to surface roughening [351] which makes wire bonding difficult. Thus, the contact metal must be thin enough to minimize the surface roughening, and additional thick layers of metals (both a barrier metal and an interconnecting metal) are deposited as overlays after high-temperature sintering, as shown in Figure 6.83.

So far, a variety of metals and annealing processes have been investigated for both n- and p-type SiC. Table 6.4 shows typical ohmic contacts reported in the literature. Because of the limited space, the next sections mainly describe common ohmic contacts: Ni for n-type SiC and Al/Ti for p-type SiC. For a survey of ohmic contacts on SiC, please see review papers [322–327].

6.4.2.2 Ohmic Contacts to *n*-Type SiC

Ni is a good Schottky contact when the sintering temperature is lower than 500 °C on lightly doped SiC. When sintering is carried out at above 700–800 °C on relatively heavily doped n-type SiC, Ni forms a

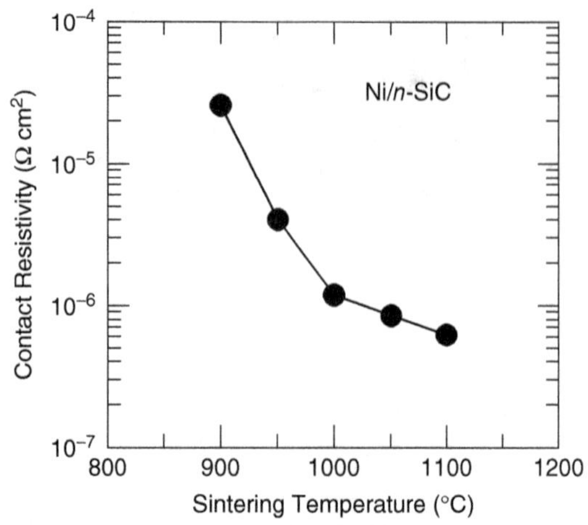

Figure 6.84 Contact resistivity as a function of sintering temperature for Ni/n-type 4H-SiC. The thickness of deposited Ni is 100 nm, and the donor density of SiC is about 2×10^{19} cm^{-3}.

good ohmic contact [322–327, 352–358]. Figure 6.84 shows the contact resistivity as a function of sintering temperature for Ni/n-type 4H-SiC. Sintering was done for 2 min by a RTP. The thickness of deposited Ni is 100 nm, and the donor density of SiC is about 2×10^{19} cm^{-3}. The contact resistivity decreases significantly with increasing sintering temperature, and is almost saturated at about 1×10^{-6} Ω cm^2 above 1000 °C. Since Ni does not form carbides, a carbon film or clusters can be formed near the interface as well as the metal surface. The carbon layer on the metal surface must be carefully removed (if formed) to ensure low contact resistivity and to improve adhesion of the subsequent layer of metals. When the Ni layer is too thin, the contact resistivity is increased. On the other hand, the surface is severely roughened when the Ni layer is too thick. Optimum Ni thickness is about 50–100 nm. Figure 6.85 shows the contact resistivity of Ni versus the donor density of 4H-SiC; sintering was performed at 1000 °C for 2 min by RTP. To obtain a low contact resistivity ($<1 \times 10^{-5}$ Ω cm^2), the doping concentration should be increased to $\geq 1 \times 10^{19}$ cm^{-3}.

In spite of its long history, the mechanism of ohmic behavior for Ni contacts to n-type SiC is not fully understood. Physical characterization by RBS and AES showed that Ni and SiC react to form Ni_2Si during sintering [353]. Ni_2Si formation already occurs at a sintering temperature of 600 °C, although the resultant contact is not ohmic. To achieve good ohmic characteristics, further sintering (typically at higher temperature) is required. During this sintering, some carbon is accumulated near the interface and some carbon moves up to the surface. It is suggested that the excess carbon formed near the interface by high-temperature sintering yields a low barrier height on n-type 4H-SiC, and a role for intermediate carbon in the ohmic behavior is argued [359]. However, further basic studies are required to elucidate the ohmic mechanism.

Since Ni does not form carbides, the control of excess carbon is a critical issue. To overcome this issue, Ni-Cr [360], Ti-based alloys [361, 362], Ta-based alloys [363], W-based alloys [364, 365], Co-based alloys [366], and several multilayer structures have been investigated. At present, however, Ni is the most popular metal for ohmic contacts employed in real device fabrication, though modifications are made in different groups.

As predicted from the contact theory, a metal with a very low barrier height can work as an ohmic contact without sintering (*as-deposited ohmic contact*) if the doping density of the semiconductor is sufficiently high. This is true even in n-type SiC. Al and Ti are good ohmic contacts without a sintering process if the donor density is higher than $(1–2) \times 10^{19}$ cm^{-3}. The contact resistivity is approximately 10^{-3} to 10^{-4} Ω cm^2 for as-deposited contacts and it can be reduced by high-temperature sintering.

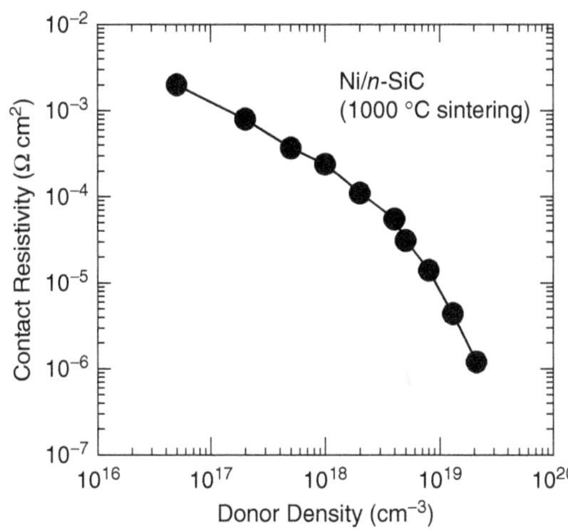

Figure 6.85 Contact resistivity of Ni versus the donor density of 4H-SiC. The sintering was performed at 1000 °C for 2 min by a rapid thermal process.

In power MOSFETs and junction field-effect transistors (JFETs), a contact, which exhibits ohmic behavior to both n-type and p-type SiC, is required to simplify the fabrication process. Such a simultaneous ohmic contact can be formed by using Ni when the p-type SiC is heavily doped [357, 358]. TiW [367] or Al/Ti/Ni [368], sintered at 900–950 °C, also form simultaneous ohmic contacts.

6.4.2.3 Ohmic Contacts to *p*-Type SiC

An Al-based metal is a good ohmic contact to p-type SiC after sintering at 900–1000 °C. Although aluminum is an effective acceptor in SiC, no direct evidence on Al doping after sintering has been reported. An obvious problem in this system is the low melting point of Al (about 630 °C). Because of severe segregation of Al during high-temperature sintering, it is difficult to form uniform ohmic contacts using a pure Al metal sintered at high temperature. To solve this problem, AlSi alloy or Al/Ti stacks have been commonly employed [369–371]. In particular, Al/Ti and its modification (e.g., Al/Ni/Ti) are the standard ohmic contacts to p-type SiC [369–376]. An optimized stack structure is Al (300 nm)/Ti (80 nm)/SiC [325, 357]. Figure 6.86 shows the contact resistivity of Al/Ti versus the acceptor density of 4H-SiC; sintering was performed at 1000 °C for 2 min by RTP. To obtain a contact resistivity $<1 \times 10^{-5}\ \Omega\ \mathrm{cm}^2$, the doping concentration should be increased to $\geq 3 \times 10^{19}\ \mathrm{cm}^{-3}$. TEM observation revealed that Ti_3SiC_2 is the main phase in contact with SiC after the sintering process [377, 378]. It is, however, not very clear how this compound contributes to the ohmic behavior. On p-type SiC, as-deposited ohmic contacts can be obtained if the acceptor concentration is extremely high ($>2 \times 10^{20}\ \mathrm{cm}^{-3}$). However, the contact resistivity on heavily doped p-type SiC formed by Al^+ implantation is always much higher than that on heavily Al-doped epitaxial layers with the same acceptor density. Other than Al/Ti, Pd-based metals [379], Ni-based metals [380, 381], Ti-based metals [362], and an Al/Ti/Ge stack [377] have been investigated.

The long-term stability of ohmic contacts has been assessed. The contact resistivity and the surface roughness are monitored after high-temperature aging in an inert atmosphere and promising results are reported. For example, no degradation is observed after aging at 300 °C for 5000 h nor at 500 °C for 500 h for both Ni/n-SiC and Al/Ti/p-SiC contacts [323, 325, 357]. Therefore, the ohmic contacts are

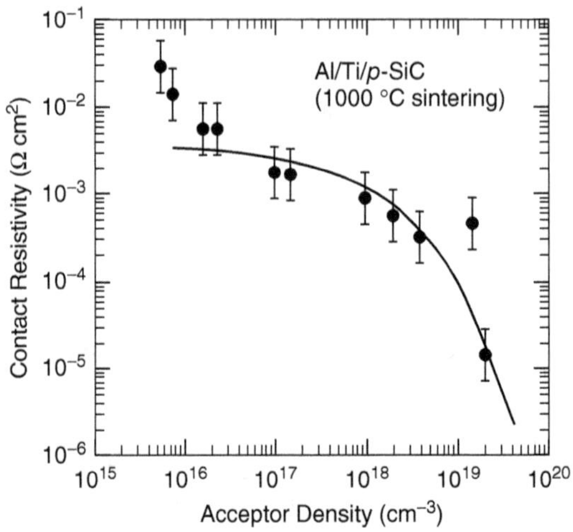

Figure 6.86 Contact resistivity of Al/Ti versus the acceptor density of 4H-SiC. The sintering was performed at 1000 °C for 2 min by rapid thermal process ([371] reproduced with permission from AIP Publishing LLC).

rather stable, and the contact reliability will not be limited by degradation of ohmic contacts themselves. Reactions between the interconnecting metal and a dielectric layer are of more concern.

6.5 Summary

Basic issues of device process technologies in SiC were described. Selective doping of donors and acceptors by ion implantation is feasible, and relatively high activation ratio values over 90% are obtained unless the implant dose is very high ($>5 \times 10^{15}$ cm^{-2}). Formation of heavily-doped n-type SiC is successful (sheet resistance < 100 Ω/□ by high-dose phosphorus implantation), while the resistivity of heavily-doped p-type SiC formed by aluminum implantation needs further improvement. The surface roughening during high-temperature (~1700 °C) activation annealing has been greatly reduced by employing a carbon cap. The pn junctions formed by aluminum implantation into n-type SiC and np junctions formed by nitrogen or phosphorus implantation into p-type SiC exhibit good characteristics. However, a high density of extended and point defects are generated inside the implanted region as well as the tail region. Impacts of these defects on SiC device performance should be carefully investigated. To produce lightly-doped n-type SiC with high uniformity, neutron transmutation doping (NTD), which utilizes conversion from ^{29}Si to ^{31}P by neutron irradiation, is promising for the future [382].

Dry etching is relatively easy in SiC, while wet etching is not a choice for device fabrication. Both fluorine and chlorine chemistries give reasonable results for SiC etching. SiO_2 deposited by CVD is the preferred masking material. Remaining issues include increasing the etching rate and improving the etching selectivity against the mask material. Control of etching profiles has been tried.

Control of MOS interfaces and their accurate characterization are still big challenges. In spite of extensive efforts to improve interface quality, the interface state density of 4H-SiC is still very high. Although the n-channel mobility has been improved to about 40–80 cm^{2}V^{-1}s^{-1} on 4H-SiC(0001), 60–120 cm^{2}V^{-1}s^{-1} on 4H-SiC(000$\bar{1}$), and 100–240 cm^{2}V^{-1}s^{-1} on 4H-SiC(11$\bar{2}$0) and (1$\bar{1}$00), the mobility is significantly lower in processed power MOSFETs. Both thermal oxides and deposited oxides have individual advantages and disadvantages. Importantly, the improvement of interface properties has been achieved by process optimization without solid understanding of the underlying physics. Insights into the origin of interface states as well as the physical/chemical structure of the SiC MOS interface are very limited. Therefore, many more fundamental experimental and theoretical studies are required in the future. Regarding the oxide reliability, continuous progress has been made, and a sufficiently long life (>100 years) at elevated temperature of 200–250 °C has been reported by several groups. However, this still needs to be proven in real MOS devices with a large chip size.

A basic process to form Schottky and ohmic contacts has been established. Owing to the lack of surface Fermi-level pinning, the Schottky barrier height can be well controlled over a wide range for both n-type and p-type SiC. Ni and Al/Ti are the standard ohmic contacts to n-type and p-type SiC, respectively. To obtain a low contact resistivity ($10^{-6}\ \Omega\, cm^2$), however, sintering at 950–1000 °C and high doping density are required. The physical/chemical mechanism of ohmic behaviors is not fully understood at present.

References

[1] A. Schöner, *Process Technology for Silicon Carbide Devices* , Chapter 3, ed. C.M. Zetterling (INSPEC, 2002) 51–84 Ion implantation and diffusion in SiC.

[2] Rao, M.V. (2004) Ion-implantation in SiC, in *Silicon Carbide, Materials, Processing, and Devices* (eds Z.C. Feng and J.H. Zhao), Taylor & Francis Group, p. 165.

[3] Laube, M., Schmid, F., Semmelroth, K. *et al.* (2004) Phosphorus-related centers in SiC, in *Silicon Carbide – Recent Major Advances* (eds W.J. Choyke, H. Matsunami and G. Pensl), Springer, p. 493.

[4] Rambach, M., Bauer, A.J. and Ryssel, H. (2008) Electrical and topographical characterization of aluminum implanted layers in 4H silicon carbide. *Phys. Status Solidi B*, **245**, 1315.

[5] Kimoto, T., Inoue, N. and Matsunami, H. (1997) Nitrogen ion implantation into α-SiC epitaxial layers. *Phys. Status Solidi A*, **162**, 263.

[6] Troffer, T., Schadt, M., Frank, T. *et al.* (1997) Doping of SiC by implantation of boron and aluminum. *Phys. Status Solidi A*, **162**, 277.

[7] Ryssel, H. and Ruge, I. (1986) *Ion Implantation*, John Wiley & Sons, Ltd, Chichester.

[8] Sze, S.M. (2002) *Semiconductor Devices, Physics and Technology*, 2nd edn, John Wiley & Sons, Inc..

[9] Vodakov, Y.A. and Mokhov, E.N. (1974) Diffusion and solubility of impurities in silicon carbide, in *Silicon Carbide* (eds R.C. Marshall, J.W. Faust Jr., and C.E. Ryan), University of South Carolina Press, p. 508.

[10] Zippelius, B., Suda, J. and Kimoto, T. (2012) High temperature annealing of n-type 4H-SiC: Impact on intrinsic defects and carrier lifetime. *J. Appl. Phys.*, **111**, 033515.

[11] Nagano, M., Tsuchida, H., Suzuki, T. *et al.* (2010) Annealing induced extended defects in as-grown and ion-implanted 4H–SiC epitaxial layers. *J. Appl. Phys.*, **108**, 013511.

[12] Linnarsson, M.K., Janson, M.S., Karlsson, S. *et al.* (1999) Diffusion of light elements in 4H– and 6H–SiC. *Mater. Sci. Eng., B*, **61–62**, 275.

[13] Danno, K., Saitoh, H., Seki, A. *et al.* (2012) Diffusion of transition metals in 4H-SiC and trials of impurity gettering. *Appl. Phys Express*, **5**, 031301.

[14] Gao, Y., Soloviev, S.I., Sudarshan, T.S. and Tin, C.C. (2001) Selective doping of 4H-SiC by codiffusion of aluminum and boron. *J. Appl. Phys.*, **90**, 5647.

[15] Gao, Y., Soloviev, S.I. and Sudarshan, T.S. (2003) Investigation of boron diffusion in 6H-SiC. *Appl. Phys. Lett.*, **83**, 905.

[16] Sridhara, S.G., Clemen, L.L., Devaty, R.P. *et al.* (1998) Photoluminescence and transport studies of boron in 4H SiC. *J. Appl. Phys.*, **83**, 7909.

[17] Ziegler, J.J.F., Biersack, P. and Littmark, U. (1985) *The Stopping and Range of Ions in Solids*, Pergamon Press, New York.

[18] Yaguchi, S., Kimoto, T., Ohyama, N. and Matsunami, H. (1995) Nitrogen ion implantation into 6H-SiC and application to high-temperature, radiation-hard diodes. *Jpn. J. Appl. Phys.*, **34**, 3036.

[19] Wilson, R.G. (1980) The Pearson IV distribution and its application to ion implanted depth profiles. *Radiat. Eff.*, **46**, 141.

[20] Ahmed, S., Barber, C.J., Sigmon, T.W. and Erickson, J.W. (1995) Empirical depth profile simulator for ion implantation in 6Hα-SiC. *J. Appl. Phys.*, **77**, 6194.

[21] Rao, M.V., Tucker, J., Holland, O.W. *et al.* (1999) Donor ion-implantation doping into SiC. *J. Electron. Mater.*, **28**, 334.

[22] Janson, M.S., Hallen, A., Linnarsson, M.K. and Svensson, B.G. (2003) Ion implantation range distributions in silicon carbide. *J. Appl. Phys.*, **93**, 8903.

[23] Janson, M.S., Linnarsson, M.K., Hallen, A. and Svensson, B.G. (2002) Range distributions of implanted ions in silicon carbide. *Mater. Sci. Forum*, **389–393**, 779.

[24] Edmond, J.A., Withrow, S.P., Wadlin, W. and Davis, R.F. (1987) High temperature implantation of single crystal beta silicon carbide thin films. *Mater. Res. Soc. Symp. Proc.*, **77**, 193.

[25] Suttrop, W., Zhang, H., Schadt, M. *et al.* (1992) Recrystallization and electrical properties of high-temperature implanted (N, Al) 6H-SiC layers, in *Amorphous and Crystalline Silicon Carbide IV* (eds C.Y. Yang, M.M. Rahman and G.L. Harris), Springer-Verlag, p. 143.

[26] Heera, V., Stoemenos, J., Kögler, R. and Skorupa, W. (1995) Amorphization and recrystallization of 6H-SiC by ion-beam irradiation. *J. Appl. Phys.*, **77**, 2999.

[27] Satoh, M., Nakaike, Y. and Nakamura, T. (2001) Solid phase epitaxy of implantation-induced amorphous layer in $(1\bar{1}00)$-and $(11\bar{2}0)$ -oriented 6H-SiC. *J. Appl. Phys.*, **89**, 1986.

[28] Okada, T., Negoro, Y., Kimoto, T. *et al.* (2004) Defect formation in (0001)- and $(11\bar{2}0)$-oriented 4H-SiC crystals P^+-implanted at room temperature. *Jpn. J. Appl. Phys.*, **43**, 6884.

[29] Henkel, T., Heera, V., Kogler, R. and Skorupa, W. (1998) *In situ* laser reflectometry study of the amorphization of silicon carbide by MeV ion implantation. *J. Appl. Phys.*, **84**, 3090.

[30] Zhang, Y., Weber, W.J., Jiang, W. *et al.* (2002) Damage evolution and recovery in Al-implanted 4H-SiC. *Mater. Sci. Forum*, **389–393**, 815.

[31] Edmond, J.A., Davis, R.F. and Withrow, S.P. (1989) Structural characterization of ion implanted beta-SiC thin films. *Ceram. Trans.*, **2**, 479.

[32] Ghezzo, M., Brown, D.M., Downey, E. *et al.* (1992) Nitrogen-implanted SiC diodes using high-temperature implantation. *IEEE Electron Device Lett.*, **13**, 639.

[33] Kawase, D., Ohno, T., Iwasaki, T. and Yatsuo, T. (1996) Amorphization and re-crystallization of Al-implanted 6H-SiC. *Inst. Phys. Conf. Ser.*, **142**, 513.

[34] Itoh, H., Ohshima, T., Aoki, Y. *et al.* (1997) Characterization of residual defects in cubic silicon carbide subjected to hot-implantation and subsequent annealing. *J. Appl. Phys.*, **82**, 5339.

[35] Negoro, Y., Katsumoto, K., Kimoto, T. and Matsunami, H. (2004) Electronic behaviors of high-dose phosphorus-ion implanted 4H–SiC (0001). *J. Appl. Phys.*, **96**, 224.

[36] Capano, M.A., Santhakumar, R., Venugopal, R. *et al.* (2000) Nitrogen and phosphorus implantation into 4H-silicon carbide. *J. Electron. Mater.*, **29**, 210.

[37] Schmid, F., Laube, M., Pensl, G. *et al.* (2002) Electrical activation of implanted phosphorus ions in [0001]- and $[11\bar{2}0]$-oriented 4H-SiC. *J. Appl. Phys.*, **91**, 9182.

[38] Senzaki, J., Harada, S., Kosugi, R. *et al.* (2002) Electrical characteristics and surface morphology for arsenic ion-implanted 4H-SiC at high temperature. *Mater. Sci. Forum*, **389–393**, 799.

[39] Gardner, J.A., Rao, M.V., Tian, Y.L. *et al.* (1997) Rapid thermal annealing of ion implanted 6H-SiC by microwave processing. *J. Electron. Mater.*, **26**, 144.

[40] Pan, J.N., Cooper, J.A. Jr., and Melloch, M.R. (1997) Activation of nitrogen implants in 6H-SiC. *J. Electron. Mater.*, **26**, 208.

[41] Gardner, J.A., Edwards, A., Rao, M.V. *et al.* (1998) Material and n-p junction properties of N-, P-, and N/P-implanted SiC. *J. Appl. Phys.*, **83**, 5118.

[42] Khemka, V., Patel, R., Ramungul, N. *et al.* (1999) Characterization of phosphorus implantation in 4H-SiC. *J. Electron. Mater.*, **28**, 752.

[43] Saks, N.S., Agarwal, A.K., Mani, S.S. and Hegde, V.S. (2000) Low-dose nitrogen implants in 6H-silicon carbide. *Appl. Phys. Lett.*, **76**, 1896.

[44] Capano, M.A., Cooper, J.A. Jr.,, Melloch, M.R. *et al.* (2000) Ionization energies and electron mobilities in phosphorus- and nitrogen-implanted 4H-silicon carbide. *J. Appl. Phys.*, **87**, 8773.

[45] Handy, E.M., Rao, M.V., Holland, O.W. *et al.* (2000) Variable-dose ($10^{17}-10^{20}$ cm^{-3}) phosphorus ion implantation into 4H–SiC. *J. Appl. Phys.*, **88**, 5630.

[46] Imai, S., Kobayashi, S., Shinohe, T. *et al.* (2000) Hot-implantation of phosphorus ions into 4H-SiC. *Mater. Sci. Forum*, **338–342**, 861.

[47] Negoro, Y., Miyamoto, N., Kimoto, T. and Matsunami, H. (2002) Remarkable lattice recovery and low sheet resistance of phosphorus-implanted 4H–SiC $(11\bar{2}0)$. *Appl. Phys. Lett.*, **80**, 240.

[48] Senzaki, J., Harada, S., Kosugi, R. *et al.* (2002) Improvements in electrical properties of n-type-implanted 4H-SiC substrates using high-temperature rapid thermal annealing. *Mater. Sci. Forum*, **389–393**, 795.

[49] Nipoti, R., Nath, A., Cristiani, S. *et al.* (2011) Improving doping efficiency of P^+ implanted ions in 4H-SiC. *Mater. Sci. Forum*, **679–680**, 393.

[50] Kumar, R., Kojima, J. and Yamamoto, T. (2000) Novel diffusion resistant p-base region implantation for accumulation mode 4H–SiC epi-channel field effect transistor. *Jpn. J. Appl. Phys.*, **39**, 2001.

[51] Bracht, H., Stolwijk, N.A., Laube, M. and Pensl, G. (2000) Diffusion of boron in silicon carbide: Evidence for the kick-out mechanism. *Appl. Phys. Lett.*, **77**, 3188.

[52] Bockstedte, M., Mattausch, A. and Pankratov, O. (2004) Different roles of carbon and silicon interstitials in the interstitial-mediated boron diffusion in SiC. *Phys. Rev. B*, **70**, 115203.

[53] Negoro, Y., Kimoto, T., Matsunami, H. and Pensl, G. (2007) Abnormal out-diffusion of epitaxially doped boron in 4H–SiC caused by implantation and annealing. *Jpn. J. Appl. Phys.*, **46**, 5053.

[54] Kimoto, T., Takemura, O., Matsunami, H. *et al.* (1998) Al^+ and B^+ implantations into 6H-SiC epilayers and application to pn junction diodes. *J. Electron. Mater.*, **27**, 358.

[55] Poggi, A., Bergamini, F., Nipoti, R. *et al.* (2006) Effects of heating ramp rates on the characteristics of Al implanted 4H–SiC junctions. *Appl. Phys. Lett.*, **88**, 162106.

[56] Hölzlein, K., Mitlehner, H., Rupp, R. *et al.* (1996) Annealing behavior and electrical properties of boron implanted 4H-SiC-layers. *Inst. Phys. Conf. Ser.*, **142**, 561.

[57] Peters, D., Schörner, R., Hölzlein, K.H. and Friedrichs, P. (1997) Planar aluminum-implanted 1400 V 4H silicon carbide p-n diodes with low on resistance. *Appl. Phys. Lett.*, **71**, 2996.

[58] Kimoto, T., Miyamoto, N., Saitoh, A. and Matsunami, H. (2002) High-energy (MeV) Al and B ion implantations into 4H-SiC and fabrication of pin diodes. *J. Appl. Phys.*, **91**, 4242.

[59] Negoro, Y., Kimoto, T. and Matsunami, H. (2005) Carrier compensation near tail region in aluminum- or boron-implanted 4H–SiC (0001). *J. Appl. Phys.*, **98**, 043709.

[60] Itoh, H., Troffer, T., Peppermüller, C. and Pensl, G. (1998) Effects of C or Si co-implantation on the electrical activation of B atoms implanted in 4H–SiC. *Appl. Phys. Lett.*, **73**, 1427.

[61] Kimoto, T., Itoh, A., Matsunami, H. *et al.* (1996) Aluminum and boron ion implantations into 6H-SiC epilayers. *J. Electron. Mater.*, **26**, 879.

[62] Negoro, Y., Kimoto, T., Matsunami, H. *et al.* (2004) Electrical activation of high-concentration aluminum implanted in 4H-SiC. *J. Appl. Phys.*, **96**, 4916.

[63] Rao, M.V., Griffiths, P., Gardner, J. *et al.* (1996) Al, Al/C and Al/Si implantations in 6H-SiC. *J. Electron. Mater.*, **25**, 75.

[64] Tone, K. and Zhao, J.H. (1999) A comparative study of C plus Al coimplantation and Al implantation in 4H- and 6H-SiC. *IEEE Trans. Electron Devices*, **46**, 612.

[65] Wirth, H., Panknin, D., Skorupa, W. and Niemann, E. (1999) Efficient p-type doping of 6H-SiC: Flash-lamp annealing after aluminum implantation. *Appl. Phys. Lett.*, **74**, 979.

[66] Bluet, J.M., Pernot, J., Camassel, J. *et al.* (2000) Activation of aluminum implanted at high doses in 4H–SiC. *J. Appl. Phys.*, **88**, 1971.

[67] Tanaka, H., Tanimoto, S., Yamanaka, M. and Hoshi, M. (2002) Electrical characteristics of Al^+ ion-implanted 4H-SiC. *Mater. Sci. Forum*, **389–393**, 803.

[68] Tanaka, Y., Kobayashi, N., Hasegawa, M. *et al.* (2000) Coimplantation effects of (C and Si)/Ga in 6H-SiC. *Mater. Sci. Forum*, **338–342**, 917.

[69] Ramungul, N., Khemka, V., Zheng, Y. *et al.* (1999) 6H-SiC p^+-n junctions fabricated by beryllium implantation. *IEEE Trans. Electron Devices*, **46**, 465.

[70] Nadella, R.K. and Capano, M.A. (1997) High-resistance layers in *n*-type 4H-silicon carbide by hydrogen ion implantation. *Appl. Phys. Lett.*, **70**, 886.

[71] Achtziger, N., Grillenberger, J., Witthuhn, W. *et al.* (1998) Hydrogen passivation of silicon carbide by low-energy ion implantation. *Appl. Phys. Lett.*, **73**, 945.

[72] Alok, D., Raghunathan, R. and Baliga, B.J. (1996) Planar edge termination for 4H-silicon carbide devices. *IEEE Trans. Electron Devices*, **43**, 1315.

[73] Edwards, A., Dwight, D.N., Rao, M.V. *et al.* (1997) Compensation implants in 6H–SiC. *J. Appl. Phys.*, **82**, 4223.

[74] Kunzer, M., Kaufmann, U., Maier, K. and Schneider, J. (1995) Magnetic circular dichroism and electron spin resonance of the A^- acceptor state of vanadium, V^{3+}, in 6H-SiC. *Mater. Sci. Eng., B*, **29**, 118.

[75] Hobgood, H.M., Glass, R.C., Augustine, G. *et al.* (1995) Semi-insulating 6H–SiC grown by physical vapor transport. *Appl. Phys. Lett.*, **66**, 1364.

[76] Kimoto, T., Nakajima, T., Matsunami, H. *et al.* (1996) Formation of semi-insulating 6H-SiC layers by vanadium ion implantations. *Appl. Phys. Lett.*, **69**, 1113.

[77] Kaneko, H. and Kimoto, T. (2011) Formation of a semi-insulating layer in n-type 4H-SiC by electron irradiation. *Appl. Phys. Lett.*, **98**, 262106.

[78] Capano, M.A., Ryu, S.-H., Melloch, M.R. *et al.* (1998) Dopant activation and surface morphology of ion implanted 4H- and 6H-silicon carbide. *J. Electron. Mater.*, **27**, 370.
[79] Capano, M.A., Ryu, S., Cooper, J.A. Jr., *et al.* (1999) Surface roughening in ion implanted 4H-silicon carbide. *J. Electron. Mater.*, **28**, 214.
[80] Saddow, S.E., Williams, J., Isaacs-Smith, T. *et al.* (2000) High temperature implant activation in 4H and 6H-SiC in a silane ambient to reduce step bunching. *Mater. Sci. Forum*, **338–342**, 901.
[81] Sato, T., Mitsutake, K., Mizushima, I. and Tsunashima, Y. (2000) Micro-structure transformation of silicon: A newly developed transformation technology for patterning silicon surfaces using the surface migration of silicon atoms by hydrogen annealing. *Jpn. J. Appl. Phys.*, **39**, 5033.
[82] Handy, E.M., Rao, M.V., Jones, K.A. *et al.* (1999) Effectiveness of AlN encapsulant in annealing ion-implanted SiC. *J. Appl. Phys.*, **86**, 746.
[83] Dalibor, T., Pensl, G., Matsunami, H. *et al.* (1997) Deep defect centers in silicon carbide monitored with deep level transient spectroscopy. *Phys. Status Solidi A*, **162**, 199.
[84] David, M.L., Alfieri, G., Monakhov, E.N. *et al.* (2004) Electrically active defects in irradiated 4H-SiC. *J. Appl. Phys.*, **95**, 4728.
[85] Lebedev, A.A., Veinger, A.I., Davydov, D.V. *et al.* (2000) Doping of n-type 6H–SiC and 4H–SiC with defects created with a proton beam. *J. Appl. Phys.*, **88**, 6265.
[86] Kawahara, K., Alfieri, G. and Kimoto, T. (2009) Detection and depth analyses of deep levels generated by ion implantation in n- and p-type 4H-SiC. *J. Appl. Phys.*, **106**, 013719.
[87] Hemmingsson, C., Son, N.T., Kordina, O. *et al.* (1997) Deep level defects in electron-irradiated 4H SiC epitaxial layers. *J. Appl. Phys.*, **81**, 6155.
[88] Danno, K. and Kimoto, T. (2007) Deep level transient spectroscopy on as-grown and electron-irradiated p-type 4H-SiC epilayers. *J. Appl. Phys.*, **101**, 103704.
[89] Mitlehner, H., Friedrichs, P., Peters, D. *et al.* (1998) Proceedings of 1998 International Symposium on Power Semiconductor Devices & ICs, p. 127 Switching behaviour of fast high voltage SiC pn-diodes.
[90] Lendenmann, H., Mukhitdinov, A., Dahlquist, F. *et al.* (2001) Proceedings of 2001 International Symposium on Power Semiconductor Devices & ICs, p. 31 4.5 kV 4H-SiC diodes with ideal forward characteristic.
[91] Kawahara, K., Suda, J., Pensl, G. and Kimoto, T. (2010) Reduction of deep levels generated by ion implantation into n- and p-type 4H–SiC. *J. Appl. Phys.*, **108**, 033706.
[92] Ohno, T. and Kobayashi, N. (2001) Structure and distribution of secondary defects in high energy ion implanted 4H-SiC. *J. Appl. Phys.*, **89**, 933.
[93] Ohno, T. and Kobayashi, N. (2002) Difference of secondary defect formation by high energy B^+ and Al^+ implantation into 4H–SiC. *J. Appl. Phys.*, **91**, 4136.
[94] Ishimaru, M., Dickerson, R.M. and Sickafus, K.E. (1999) High-dose oxygen ion implantation into 6H-SiC. *Appl. Phys. Lett.*, **75**, 352.
[95] Persson, P.O.Å., Hultman, L., Janson, M.S. *et al.* (2002) On the nature of ion implantation induced dislocation loops in 4H-silicon carbide. *J. Appl. Phys.*, **92**, 2501.
[96] Sasaki, S., Suda, J. and Kimoto, T. (2012) Lattice mismatch and crystallographic tilt induced by high-dose ion-implantation into 4H-SiC. *J. Appl. Phys.*, **111**, 103715.
[97] Åberg, D., Hallen, A., Pellegrino, P. and Svensson, B.G. (2001) Nitrogen deactivation by implantation-induced defects in 4H–SiC epitaxial layers. *Appl. Phys. Lett.*, **78**, 2908.
[98] Mitra, S., Rao, M.V., Papanicolaou, N. *et al.* (2004) Deep-level transient spectroscopy study on double implanted $n^+ - p$ and $p^+ - n$ 4H-SiC diodes. *J. Appl. Phys.*, **95**, 69.
[99] Tsuchida, H., Kamata, I., Nagano, M. *et al.* (2007) Migration of dislocations in 4H-SiC epilayers during the ion implantation process. *Mater. Sci. Forum*, **556–557**, 271.
[100] Zhang, X., Miyazawa, T. and Tsuchida, H. (2012) Critical conditions of misfit dislocation formation in 4H-SiC epilayers. *Mater. Sci. Forum*, **717–720**, 313.
[101] Zhang, X., Nagano, M. and Tsuchida, H. (2011) Correlation between thermal stress and formation of interfacial dislocations during 4H-SiC epitaxy and thermal annealing. *Mater. Sci. Forum*, **679–680**, 306.
[102] Yih, P.H., Saxena, V. and Steckl, A.J. (1997) A review of SiC reactive ion etching in fluorinated plasmas. *Phys. Status Solidi B*, **202**, 605.
[103] S.J. Pearton, *Process Technology for Silicon Carbide Devices*, Chapter 4, ed. C.M. Zetterling (INSPEC, 2002) 85–92 Wet and dry etching of SiC.
[104] Dohmae, S., Shibahara, K., Nishino, S. and Matsunami, H. (1985) Plasma etching of CVD grown cubic SiC single crystals. *Jpn. J. Appl. Phys.*, **24**, L873.

[105] Palmour, J.W., Davis, R.F., Wallett, T.W. and Bhasin, K.B. (1986) Dry etching of β-SiC in CF_4 and CF_4+O_2 mixtures. *J. Vac. Sci. Technol., A*, **4**, 590.

[106] Palmour, J.W., Davis, R.F., Astell-Burt, P. and Blackborow, P. (1987) Surface characteristics of monocrystalline ß-SiC dry etched in fluorinated gases. *Mater. Res. Soc. Symp. Proc.*, **76**, 185.

[107] Kelner, G., Binari, S.C. and Klein, P.H. (1987) Plasma etching of β-SiC. *J. Electrochem. Soc.*, **134**, 253.

[108] Yih, P.H. and Steckl, A.J. (1993) Effects of hydrogen additive on obtaining residue-free reactive ion etching of β-SiC in fluorinated plasmas. *J. Electrochem. Soc.*, **140**, 1813.

[109] Wu, J., Parsons, J.D. and Evans, D.R. (1995) Sulfur hexafluoride reactive ion etching of (111) beta-SiC epitaxial layers grown on (111) TiC substrates. *J. Electrochem. Soc.*, **142**, 669.

[110] Yih, P.H. and Steckl, A.J. (1995) Residue-free reactive ion etching of silicon carbide in fluorinated plasmas: II 6H SiC. *J. Electrochem. Soc.*, **142**, 312.

[111] Casady, J., Luckowski, E.D., Bozack, M. *et al.* (1996) Etching of 6H-SiC and 4H-SiC using NF_3 in a reactive ion etching system. *J. Electrochem. Soc.*, **143**, 1750.

[112] Cao, L.H., Li, B.H. and Zhao, J.H. (1998) Etching of SiC using inductively coupled plasma. *J. Electrochem. Soc.*, **145**, 3609.

[113] Wang, J.J., Lambers, E.S., Pearton, S.J. *et al.* (1998) Inductively coupled plasma etching of bulk 6H-SiC and thin-film SiCN in NF_3 chemistries. *J. Vac. Sci. Technol., B*, **16**, 2204.

[114] Khan, F.A. and Adesida, I. (1999) High rate etching of SiC using inductively coupled plasma reactive ion etching in SF_6-based gas mixtures. *Appl. Phys. Lett.*, **75**, 2268.

[115] Jiang, L.D., Cheung, R., Brown, R. and Mount, A. (2003) Inductively coupled plasma etching of SiC in SF_6/O_2 and etch-induced surface chemical bonding modifications. *J. Appl. Phys.*, **93**, 1376.

[116] Mikami, H., Hatayama, T., Yano, H. *et al.* (2005) Role of hydrogen in dry etching of silicon carbide using inductively and capacitively coupled plasma. *Jpn. J. Appl. Phys.*, **44**, 3817.

[117] Flemish, J.R., Xie, K. and Zhao, J.H. (1994) Smooth etching of single crystal 6H-SiC in an electron cyclotron resonance plasma reactor. *Appl. Phys. Lett.*, **64**, 2315.

[118] Flemish, J.R. and Xie, K. (1996) Profile and morphology control during etching of SiC using electron cyclotron resonant plasmas. *J. Electrochem. Soc.*, **143**, 2620.

[119] Lanois, F., Lassagne, P., Planson, D. and Locatelli, M.L. (1996) Angle etch control for silicon carbide power devices. *Appl. Phys. Lett.*, **69**, 236.

[120] Wang, J.J., Lambers, E.S., Pearton, S.J. *et al.* (1998) High rate etching of SiC and SiCN in NF_3 inductively coupled plasmas. *Solid State Electron.*, **42**, 743.

[121] Manos, D.M. and Flamm, D.J. (1989) *Plasma Etching*, Academic Press.

[122] Kawahara, K., Krieger, M., Suda, J. and Kimoto, T. (2010) Deep levels induced by reactive ion etching in n- and p-type 4H–SiC. *J. Appl. Phys.*, **108**, 023706.

[123] Kawada, Y., Tawara, T., Nakamura, S. *et al.* (2009) Shape control and roughness reduction of SiC trenches by high-temperature annealing. *Jpn. J. Appl. Phys.*, **48**, 116508.

[124] Hiyoshi, T., Hori, T., Suda, J. and Kimoto, T. (2008) Simulation and experimental study on the junction termination structure for high-voltage 4H-SiC PiN diodes. *IEEE Trans. Electron Devices*, **55**, 1841.

[125] Yan, F., Qin, C., Zhao, J.H. and Weiner, M. (2002) A novel technology for the formation of a very small bevel angle for edge termination. *Mater. Sci. Forum*, **389–393**, 1305.

[126] Zorman, C. and Mehregany, M. (2004) Micromachining of SiC, in *Silicon Carbide – Recent Major Advances* (eds W.J. Choyke, H. Matsunami and G. Pensl), Springer, p. 671.

[127] Cheung, R. (2006) *Silicon Carbide Microelectromechanical Systems for Harsh Environments*, Imperial College Press.

[128] Zorman, C.A. and Parro, R.J. (2008) Micro- and nanomechanical structures for silicon carbide MEMS and NEMS. *Phys. Status Solidi B*, **245**, 1404.

[129] Kong, H.S., Glass, J.T. and Davis, R.F. (1988) Chemical vapor deposition and characterization of 6H-SiC thin films on off-axis 6H-SiC substrates. *J. Appl. Phys.*, **64**, 2672.

[130] Hallin, C., Owman, F., Mårtensson, P. *et al.* (1997) In situ substrate preparation for high-quality SiC chemical vapour deposition. *J. Cryst. Growth*, **181**, 241.

[131] Akiyama, K., Ishii, Y., Abe, S. *et al.* (2009) *In situ* Gravimetric monitoring of thermal decomposition and hydrogen etching rates of 6H-SiC(0001) Si face. *Jpn. J. Appl. Phys.*, **48**, 095505.

[132] Powell, J.A., Petit, J.B., Edgar, J.H. *et al.* (1991) Controlled growth of 3C-SiC and 6H-SiC films on low-tilt-angle vicinal (0001) 6H-SiC wafers. *Appl. Phys. Lett.*, **59**, 333.

[133] Burk, A.A. and Rowland, L.B. (1996) Reduction of unintentional aluminum spikes at SiC vapor phase epitaxial layer/substrate interfaces. *Appl. Phys. Lett.*, **68**, 382.

[134] Nakamura, S., Kimoto, T., Matsunami, H. *et al.* (2000) Formation of periodic steps with a unit-cell height on 6H–SiC (0001) surface by HCl etching. *Appl. Phys. Lett.*, **76**, 3412.

[135] Jennings, V.J. (1969) The etching of silicon carbide. *Mater. Res. Bull.*, **4**, 5199.

[136] Hatayama, T., Shimizu, T., Koketsu, H. *et al.* (2009) Thermal etching of 4H-SiC(0001) Si faces in the mixed gas of chlorine and oxygen. *Jpn. J. Appl. Phys.*, **48**, 066516.

[137] Miura, Y., Habuka, H., Katsumi, Y. *et al.* (2007) Determination of etch rate behavior of 4H–SiC using chlorine trifluoride gas. *Jpn. J. Appl. Phys.*, **46**, 7875.

[138] Katsuno, M., Ohtani, N., Takahashi, J. *et al.* (1999) Mechanism of molten KOH etching of SiC single crystals: comparative study with thermal oxidation. *Jpn. J. Appl. Phys.*, **38**, 4661.

[139] Chang, W.H. (2004) Micromachining of p-type 6H-SiC by electrochemical etching. *Sens. Actuators, A*, **112**, 36.

[140] Ke, Y., Yan, F., Devaty, R.P. and Choyke, W.J. (2009) Surface polishing by electrochemical etching of p-type 4*H* SiC. *J. Appl. Phys.*, **106**, 064901.

[141] Shor, J.S., Bemis, L., Kutz, A.D. *et al.* (1994) Characterization of nanocrystallites in porous p-type 6H-SiC. *J. Appl. Phys.*, **76**, 4045.

[142] Shor, J.S., Osgood, R.M. and Kutz, A.D. (1992) Photoelectrochemical conductivity selective etch stops for SiC. *Appl. Phys. Lett.*, **60**, 1001.

[143] Shishkin, Y., Choyke, W. and Devaty, R.P. (2004) Photoelectrochemical etching of *n* -type 4H silicon carbide. *J. Appl. Phys.*, **96**, 2311.

[144] Mikami, H., Hatayama, T., Yano, H. *et al.* (2005) Analysis of photoelectrochemical processes in α-SiC substrates with atomically flat surfaces. *Jpn. J. Appl. Phys.*, **44**, 8329.

[145] Ke, Y., Devaty, R.P. and Choyke, W.J. (2008) Comparative columnar porous etching studies on n-type 6H SiC crystalline faces. *Phys. Status Solidi B*, **245**, 1396.

[146] Cooper, J.A. Jr., (1997) Advances in SiC MOS technology. *Phys. Status Solidi A*, **162**, 305.

[147] Afanas'ev, V.V., Bassler, M., Pensl, G. and Schulz, M. (1997) Intrinsic SiC/SiO_2 interface states. *Phys. Status Solidi A*, **162**, 321.

[148] Ouisse, T. (1997) Electron transport at the SiC/SiO_2 interface. *Phys. Status Solidi A*, **162**, 339.

[149] E.Ö. Sveinbjörnsson and C.M. Zetterling, *Process Technology for Silicon Carbide Devices*, Chapter 5, ed. C.M. Zetterling (INSPEC, 2002) 93–110 Thermally grown and deposited dielectrics on SiC.

[150] Afanas'ev, V.V., Ciobanu, F., Pensl, G. and Stesmans, A. (2004) Contributions to the density of interface states in SiC MOS structures, in *Silicon Carbide – Recent Major Advances* (eds W.J. Choyke, H. Matsunami and G. Pensl), Springer, p. 343.

[151] Dimitrijev, S., Harrison, H.B., Tanner, P. *et al.* (2004) Properties of nitride oxides on SiC, in *Silicon Carbide – Recent Major Advances* (eds W.J. Choyke, H. Matsunami and G. Pensl), Springer, p. 373.

[152] Saks, N.S. (2004) Hall effect studies of electron mobility and trapping at the SiC/SiO_2 interface, in *Silicon Carbide – Recent Major Advances* (eds W.J. Choyke, H. Matsunami and G. Pensl), Springer, p. 387.

[153] Dhar, S., Pantelides, S.T., Williams, J.R. and Feldman, L.C. (2009) Inversion layer electron transport in 4H-SiC metal–oxide–semiconductor field-effect transistors, in *Defects in Microelectronic Materials and Devices* (eds D.M. Fleetwood, S.T. Pantelides and R.D. Schrimpf), CRC Press, p. 575.

[154] Noborio, M., Suda, J., Beljakowa, S. *et al.* (2009) 4H-SiC MISFETs with nitrogen-containing insulators. *Phys. Status Solidi A*, **206**, 2374.

[155] Tilak, V. (2009) Inversion layer electron transport in 4H-SiC metal–oxide–semiconductor field-effect transistors. *Phys. Status Solidi A*, **206**, 2391.

[156] Chow, T.P., Naik, H. and Li, Z. (2009) Comparative study of 4H-SiC and 2H-GaN MOS capacitors and FETs. *Phys. Status Solidi A*, **206**, 2478.

[157] Hiyoshi, T. and Kimoto, T. (2009) Reduction of deep levels and improvement of carrier lifetime in n-type 4H-SiC by thermal oxidation. *Appl. Phys. Express*, **2**, 041101.

[158] von Münch, W. and Pfaffeneder, I. (1975) Thermal oxidation and electrolytic etching of silicon carbide. *J. Electrochem. Soc.*, **122**, 642.

[159] Deal, B.E. and Grove, A.S. (1965) General relationship for the thermal oxidation of silicon. *J. Appl. Phys.*, **36**, 3770.

[160] Nicollian, E.H. and Brews, J.R. (1982) *MOS Physics and Technology*, John Wiley & Sons, Inc., New York.

[161] Song, Y., Dhar, S., Feldman, L.C. *et al.* (2004) Modified Deal-Grove model for the thermal oxidation of silicon carbide. *J. Appl. Phys.*, **95**, 4953.

[162] Suzuki, A., Ashida, H., Furui, N. *et al.* (1982) Thermal oxidation of SiC and electrical properties of Al–SiO_2–SiC MOS structure. *Jpn. J. Appl. Phys.*, **21**, 579.

[163] Zheng, Z., Tressler, R.E. and Spear, K.E. (1990) Oxidation of single-crystal silicon carbide: Part I. Experimental studies. *J. Electrochem. Soc.*, **137**, 854.
[164] Raynaud, J. (2001) Silica films on silicon carbide: a review of electrical properties and device applications. *J. Non-Cryst. Solids*, **280**, 1.
[165] Hijikata, Y., Yaguchi, H. and Yoshida, S. (2009) A kinetic model of silicon carbide oxidation based on the interfacial silicon and carbon emission phenomenon. *Appl. Phys. Express*, **2**, 021203.
[166] Powell, J.A., Petit, J.B., Edgar, J.H. *et al.* (1991) Application of oxidation to the structural characterization of SiC epitaxial films. *Appl. Phys. Lett.*, **59**, 183.
[167] Nakano, Y., Nakamura, T., Kamisawa, A. and Takasu, H. (2009) Investigation of pits formed at oxidation on 4H-SiC. *Mater. Sci. Forum*, **600–603**, 377.
[168] Christiansen, K. and Helbig, R. (1996) Anisotropic oxidation of 6H-SiC. *J. Appl. Phys.*, **79**, 3276.
[169] Ueno, K. (1997) Orientation dependence of the oxidation of SiC surfaces. *Phys. Status Solidi A*, **162**, 299.
[170] Onda, S., Kumar, R. and Hara, K. (1997) SiC integrated MOSFETs. *Phys. Status Solidi A*, **162**, 369.
[171] Lipkin, L.A. and Palmour, J.W. (1999) Insulator investigation on SiC for improved reliability. *IEEE Trans. Electron Devices*, **46**, 525.
[172] Friedrichs, P., Burte, E.P. and Schörner, R. (1994) Dielectric strength of thermal oxides on 6H-SiC and 4H-SiC. *Appl. Phys. Lett.*, **65**, 1665.
[173] Tanimoto, S. (2006) Impact of dislocations on gate oxide in SiC MOS devices and high reliability ONO dielectrics. *Mater. Sci. Forum*, **527–529**, 955.
[174] Grieb, M., Peters, D., Bauer, A.J. *et al.* (2009) Influence of the oxidation temperature and atmosphere on the reliability of thick gate oxides on the 4H-SiC C(000$\bar{1}$) face. *Mater. Sci. Forum*, **600–603**, 597.
[175] Balk, P. (1988) *The Si-SiO_2 System*, Elsevier.
[176] Watanabe, H., Kirino, T., Kagei, Y. *et al.* (2011) Energy band structure of SiO_2/4H-SiC interfaces and its modulation induced by intrinsic and extrinsic interface charge transfer. *Mater. Sci. Forum*, **679–680**, 386.
[177] Afanas'ev, V.V., Bassler, M., Pensl, G. *et al.* (1996) Band offsets and electronic structure of SiC/SiO_2 interfaces. *J. Appl. Phys.*, **79**, 3108.
[178] Agarwal, A.K., Seshadri, S. and Rowland, L.B. (1997) Temperature dependence of Fowler-Nordheim current in 6H- and 4H-SiC MOS capacitors. *IEEE Electron Device Lett.*, **18**, 592.
[179] Maranowski, M.M. and Cooper, J.A. Jr., (1999) Time-dependent-dielectric-breakdown measurements of thermal oxides on n-Type 6H-SiC. *Trans. Electron Devices*, **46**, 520.
[180] Senzaki, J., Kojima, K. and Fukuda, K. (2004) Effects of n-type 4H-SiC epitaxial wafer quality on reliability of thermal oxides. *Appl. Phys. Lett.*, **85**, 6182.
[181] Senzaki, J., Kojima, K., Kato, T. *et al.* (2006) Correlation between reliability of thermal oxides and dislocations in *n*-type 4H-Si*C* epitaxial wafers. *Appl. Phys. Lett.*, **89**, 022909.
[182] Das, M., Haney, S., Richmond, J. *et al.* (2012) SiC MOSFET reliability update. *Mater. Sci. Forum*, **717–720**, 1073.
[183] McPherson, J.W. (2010) *Reliability Physics and Engineering*, Springer.
[184] Senzaki, J., Shimozato, A., Okamoto, M. *et al.* (2009) Evaluation of 4H-SiC thermal oxide reliability using area-scaling method. *Jpn. J. Appl. Phys.*, **48**, 081404.
[185] Noborio, M., Grieb, M., Bauer, A.J. *et al.* (2011) Reliability of nitrided gate oxides for n- and p-type 4H-SiC(0001) metal–oxide–semiconductor devices. *Jpn. J. Appl. Phys.*, **50**, 090201.
[186] Senzaki, J., Shimozato, A., Kojima, K. *et al.* (2012) Challenges of high-performance and high-reliablity in SiC MOS structures. *Mater. Sci. Forum*, **717–720**, 703.
[187] Tanimoto, S., Tanaka, H., Hayashi, T. *et al.* (2005) High-reliability ONO gate dielectric for power MOSFETs. *Mater. Sci. Forum*, **483–485**, 685.
[188] Wang, X.W., Luo, Z.J. and Ma, T.P. (2000) High-temperature characteristics of high-quality SiC MIS capacitors with O/N/O gate dielectric. *IEEE Trans. Electron Devices*, **47**, 458.
[189] Fujihira, K., Yoshida, S., Miura, N. *et al.* (2009) TDDB measurement of gate SiO_2 on 4H-SiC formed by chemical vapor deposition. *Mater. Sci. Forum*, **600–603**, 799.
[190] Chang, K.C., Nuhfer, N.T., Porter, L.M. and Wahab, Q. (2000) High-carbon concentrations at the silicon dioxide–silicon carbide interface identified by electron energy loss spectroscopy. *Appl. Phys. Lett.*, **77**, 2186.
[191] Zheleva, T., Lelis, A., Duscher, G. *et al.* (2008) Transition layers at the SiO_2/SiC interface. *Appl. Phys. Lett.*, **93**, 022108.
[192] Biggerstaff, T.L., Reynolds, C.L. Jr.,, Zheleva, T. *et al.* (2009) Relationship between 4H-SiC/SiO_2 transition layer thickness and mobility. *Appl. Phys. Lett.*, **95**, 032108.
[193] Hatakeyama, T., Matsuhata, H., Suzuki, T. *et al.* (2011) Microscopic examination of SiO_2/4H-SiC interfaces. *Mater. Sci. Forum*, **679–680**, 330.

[194] Hornetz, B., Michel, H.J. and Halbritter, J. (1994) ARXPS studies of SiO_2-SiC interfaces and oxidation of 6H SiC single crystal Si-(001) and C-(00$\bar{1}$) surfaces. *J. Mater. Res.*, **9**, 3088.

[195] Hijikata, Y., Yaguchi, H., Yoshida, S. *et al.* (2006) Characterization of oxide films on 4H-SiC epitaxial (0001) faces by high-energy-resolution photoemission spectroscopy: Comparison between wet and dry oxidation. *J. Appl. Phys.*, **100**, 053710.

[196] Watanabe, H., Hosoi, T., Kirino, T. *et al.* (2011) Synchrotron x-ray photoelectron spectroscopy study on thermally grown SiO_2/4H-SiC(0001) interface and its correlation with electrical properties. *Appl. Phys. Lett.*, **99**, 021907.

[197] Seki, H., Wakabayashi, T., Hijikata, Y. *et al.* (2009) Characterization of 4H-SiC-SiO_2 interfaces by a deep ultraviolet spectroscopic ellipsometer. *Mater. Sci. Forum*, **615–617**, 505.

[198] Zhu, X., Lee, H.D., Feng, T. *et al.* (2010) Structure and stoichiometry of (0001) 4H-SiC/oxide interface. *Appl. Phys. Lett.*, **97**, 071908.

[199] Yoshikawa, M., Ogawa, S., Inoue, K. *et al.* (2012) Characterization of silicon dioxide films on 4H-SiC Si (0001) face by cathodoluminescence spectroscopy and x-ray photoelectron spectroscopy. *Appl. Phys. Lett.*, **100**, 082105.

[200] Wang, S., Di Ventra, M., Kim, S.G. and Pantelides, S.T. (2001) Atomic-scale dynamics of the formation and dissolution of carbon clusters in SiO_2. *Phys. Rev. Lett.*, **86**, 5946.

[201] Knaup, J.M., Deak, P., Frauenheim, T. *et al.* (2005) Defects in SiO_2 as the possible origin of near interface traps in the SiC/SiO_2 system: A systematic theoretical study. *Phys. Rev. B*, **72**, 115323.

[202] Gavrikov, A., Knizhnik, A., Safonov, A. *et al.* (2008) First-principles-based investigation of kinetic mechanism of SiC(0001) dry oxidation including defect generation and passivation. *J. Appl. Phys.*, **104**, 093508.

[203] Ebihara, Y., Chokawa, K., Kato, S. *et al.* (2012) Intrinsic origin of negative fixed charge in wet oxidation for silicon carbide. *Appl. Phys. Lett.*, **100**, 212110.

[204] Wang, S.W., Dhar, S., Wang, S.R. *et al.* (2007) Bonding at the SiC-SiO_2 interface and the effects of nitrogen and hydrogen. *Phys. Rev. Lett.*, **98**, 026101.

[205] Tuttle, B.R., Shen, X. and Pantelides, S.T. (2013) Theory of near-interface trap quenching by impurities in SiC-based metal-oxide-semiconductor devices. *Appl. Phys. Lett.*, **102**, 123505.

[206] Hosoi, T., Uenishi, Y., Mitani, S. *et al.* (2013) Dielectric properties of thermally grown SiO_2 on 4H-SiC(0001) substrates. *Mater. Sci. Forum*, **740–742**, 605.

[207] Schroder, D.K. (2006) *Semiconductor Material and Device Characterization*, 3rd edn, Wiley-IEEE.

[208] Goetzberger, A. and Irvin, J.C. (1968) Low-temperature hysteresis effects in metal-oxide-silicon capacitors caused by surface-state trapping. *IEEE Trans. Electron Devices*, **15**, 1009.

[209] Sheppard, S.T., Cooper, J.A. Jr., and Melloch, M.R. (1994) Nonequilibrium characteristics of the gate-controlled diode in 6H-SiC. *J. Appl. Phys.*, **75**, 3205.

[210] Niwa, H., Feng, G., Suda, J. and Kimoto, T. (2012) Breakdown characteristics of 15-kV-class 4H-SiC PiN diodes with various junction termination structures. *IEEE Trans. Electron Devices*, **59**, 2748.

[211] Afanas'ev, V.V. and Stesmans, A. (2000) Generation of interface states in α-SiC/SiO_2 by electron injection. *Mater. Sci. Eng., B*, **71**, 309.

[212] Rudenko, T.E., Osiyuk, I.N., Tyagulski, I.P. *et al.* (2005) Interface trap properties of thermally oxidized n-type 4H–SiC and 6H–SiC. *Solid-State Electron.*, **49**, 545.

[213] Yoshioka, H., Nakamura, T. and Kimoto, T. (2012) Accurate evaluation of interface state density in SiC metal-oxide-semiconductor structures using surface potential based on depletion capacitance. *J. Appl. Phys.*, **111**, 014502.

[214] Penumatcha, A.V., Swandono, S. and Cooper, J.A. (2013) Limitations of the MOS Hi-Lo CV technique for MOS interfaces with large time constant dispersion. *IEEE Trans. Electron Devices*, **60**, 923.

[215] Yoshioka, H., Nakamura, T. and Kimoto, T. (2012) Generation of very fast states by nitridation of the SiO_2/SiC interface. *J. Appl. Phys.*, **112**, 024520.

[216] Shenoy, J.N., Chindalore, G.L., Melloch, M.R. *et al.* (1995) Characterization and optimization of the SiO_2/SiC MOS interface. *J. Electron. Mater.*, **24**, 303.

[217] Bano, E., Ouisse, T., Cioccio, L.D. and Karmann, S. (1994) Surface potential fluctuations in metal–oxide–semiconductor capacitors fabricated on different silicon carbide polytypes. *Appl. Phys. Lett.*, **65**, 2723.

[218] Fahrner, W. and Goetzberger, A. (1970) Energy dependence of electrical properties of interface states in Si-SiO_2 interfaces. *Appl. Phys. Lett.*, **17**, 16.

[219] Das, M.K. (1999) PhD thesis. Purdue University, West Lafayette, IN, Fundamental studies of the silicon carbide MOS structure.

[220] Chen, X.D., Dhar, S., Isaacs-Smith, T. *et al.* (2008) Electron capture and emission properties of interface states in thermally oxidized and NO-annealed SiO_2/4H-SiC. *J. Appl. Phys.*, **103**, 33701.
[221] Cooper, J.A. and Schwartz, R.J. (1973) The effect of an energy-dependent capture cross section on data interpretation using the MOS conductance technique. *J. Appl. Phys.*, **44**, 5613.
[222] Scozzie, C.J. and McGarrity, J.M. (1998) Charge pumping measurements on SiC MOSFETs. *Mater. Sci. Forum*, **264–268**, 985.
[223] Okamoto, D., Yano, H., Hatayama, T. *et al.* (2008) Analysis of anomalous charge-pumping characteristics on 4H-SiC MOSFETs. *IEEE Trans. Electron Devices*, **55**, 2013.
[224] Gurfinkel, M., Xiong, H.D., Cheung, K.P. *et al.* (2008) Characterization of transient gate oxide trapping in SiC MOSFETs using fast I–V techniques. *IEEE Trans. Electron Devices*, **55**, 2004.
[225] Johnson, N.M. (1982) Measurement of semiconductor–insulator interface states by constant-capacitance deep-level transient spectroscopy. *J. Vac. Sci. Technol.*, **21**, 303.
[226] Basile, A.F., Rozen, J., Williams, J.R. *et al.* (2011) Capacitance-voltage and deep-level-transient spectroscopy characterization of defects near SiO_2/SiC interfaces. *J. Appl. Phys.*, **109**, 064514.
[227] Pan, J.N., Cooper, J.A. Jr., and Melloch, M.R. (1995) Extremely long capacitance transients in 6H-SiC metal-oxide-semiconductor capacitors. *J. Appl. Phys.*, **78**, 572.
[228] Neudeck, P.G., Kang, S., Petit, J. and Tabib-Azar, M. (1994) Measurement of n-type dry thermally oxidized 6H-SiC metal-oxide-semiconductor diodes by quasistatic and high-frequency capacitance versus voltage and capacitance transient techniques. *J. Appl. Phys.*, **75**, 7949.
[229] Suzuki, S., Harada, S., Kosugi, R. *et al.* (2002) Correlation between channel mobility and shallow interface traps in SiC metal–oxide–semiconductor field-effect transistors. *J. Appl. Phys.*, **92**, 6230.
[230] Yano, H., Katafuchi, F., Kimoto, T. and Matsunami, H. (1999) Effects of wet oxidation/anneal on interface properties of thermally oxidized SiO_2/SiC MOS system and MOSFETs. *IEEE Trans. Electron Devices*, **46**, 504.
[231] Stein von Kamienski, E.G., Leonhard, C., Scharnholz, S. *et al.* (1997) Passivation of interface traps in MOS-devices on n- and p-type 6H-SiC. *Diamond Relat. Mater.*, **6**, 1497.
[232] Das, M.K., Um, B.S. and Cooper, J.A. Jr., (2000) Anomalously high density of interface states near the conduction band in SiO_2/4H-SiC MOS devices. *Mater. Sci. Forum*, **338–342**, 1069.
[233] Yano, H., Kimoto, T. and Matsunami, H. (2002) Shallow states at SiO_2/4H-SiC interface on $(11\bar{2}0)$ and (0001) faces. *Appl. Phys. Lett.*, **81**, 301.
[234] Saks, N.S., Mani, S.S. and Agarwal, A.K. (2000) Interface trap profile near the band edges at the 4H-SiC/SiO_2 interface. *Appl. Phys. Lett.*, **76**, 2250.
[235] Shibahara, K., Saito, T., Nishino, S. and Matsunami, H. (1986) Fabrication of inversion-type n-channel MOSFET's using cubic-SiC on Si. *IEEE Electron Device Lett.*, **13**, 226.
[236] Wan, J.W., Capano, M.A., Melloch, M.R. and Cooper, J.A. (2002) N-channel 3C-SiC MOSFETs on silicon substrate. *IEEE Electron Device Lett.*, **23**, 482.
[237] Uchida, H., Minami, A., Sakata, T. *et al.* (2012) High temperature performance of 3C-SiC MOSFETs with high channel mobility. *Mater. Sci. Forum*, **717–720**, 1109.
[238] Bassler, M., Pensl, G. and Afanas'ev, V. (1997) "Carbon cluster model" for electronic states at SiC/SiO_2 interfaces. *Diamond Relat. Mater.*, **6**, 1472.
[239] Fukuda, K., Kato, M., Kojima, K. and Senzaki, J. (2004) Effect of gate oxidation method on electrical properties of metal-oxide-semiconductor field-effect transistors fabricated on 4H-SiC $C(000\bar{1})$ face. *Appl. Phys. Lett.*, **84**, 2088.
[240] Fukuda, K., Suzuki, S., Tanaka, T. and Arai, K. (2000) Reduction of interface-state density in 4H–SiC n-type metal–oxide–semiconductor structures using high-temperature hydrogen annealing. *Appl. Phys. Lett.*, **76**, 1585.
[241] Lipkin, L.A. and Palmour, J.W. (1996) Improved oxidation procedures for reduced SiO_2/SiC defects. *J. Electron. Mater.*, **25**, 909.
[242] Kosugi, R., Suzuki, S., Okamoto, M. *et al.* (2002) Strong dependence of the inversion mobility of 4H and 6H SiC(0001) MOSFETs on the water content in pyrogenic re-oxidation annealing. *IEEE Electron Device Lett.*, **23**, 136.
[243] Zhang, Q., Wang, J., Jonas, C. *et al.* (2008) Design and characterization of high-voltage 4H-SiC p-IGBTs. *IEEE Trans. Electron Devices*, **55**, 1912.
[244] Kato, M., Nanen, Y., Suda, J. and Kimoto, T. (2011) Improved characteristics of SiC MOSFETs by post-oxidation annealing in Ar at high temperature. *Mater. Sci. Forum*, **679–680**, 445.
[245] Okuno, E. and Amano, S. (2002) Reduction of interface trap density in 4H-SiC MOS by high-temperature oxidation. *Mater. Sci. Forum*, **389–393**, 989.

[246] Kimoto, T., Kosugi, H., Suda, J. *et al.* (2005) Design and fabrication of RESURF MOSFETs on 4H-SiC(0001), $(11\bar{2}0)$, and 6H-SiC(0001). *IEEE Trans. Electron Devices*, **52**, 112.

[247] Li, H., Dimitrijev, S., Harrison, H.B. and Sweatman, D. (1997) Interfacial characteristics of N_2O and NO nitrided SiO_2 grown on SiC by rapid thermal processing. *Appl. Phys. Lett.*, **70**, 2028.

[248] Dimitrijev, S., Li, H.F., Harrison, H.B. and Sweatman, D. (1997) Nitridation of silicon-dioxide films grown on 6H silicon carbide. *IEEE Trans. Electron Devices Lett.*, **18**, 175.

[249] Chung, G.Y., Tin, C.C., Williams, J.R. *et al.* (2000) Effect of nitric oxide annealing on the interface trap densities near the band edges in the 4H polytype of silicon carbide. *Appl. Phys. Lett.*, **76**, 1713.

[250] Chung, G.Y., Tin, C.C., Williams, J.R. *et al.* (2001) Improved inversion channel mobility for 4H-SiC MOSFETs following high temperature anneals in nitric oxide. *IEEE Electron Device Lett.*, **22**, 176.

[251] Jamet, P., Dimitrijev, S. and Tanner, P. (2001) Effects of nitridation in gate oxides grown on 4H-SiC. *J. Appl. Phys.*, **90**, 5058.

[252] Schörner, R., Friedrichs, P., Peters, D. *et al.* (2002) Enhanced channel mobility of 4H–SiC metal–oxide–semiconductor transistors fabricated with standard polycrystalline silicon technology and gate-oxide nitridation. *Appl. Phys. Lett.*, **80**, 4253.

[253] Lu, C.-Y., Cooper, J.A. Jr.,, Tsuji, T. *et al.* (2003) Effect of process variations and ambient temperature on electron mobility at the SiO_2/4H-SiC interface. *IEEE Trans. Electron Devices*, **50**, 1582.

[254] McDonald, K., Feldman, L.C., Weller, R.A. *et al.* (2003) Kinetics of NO nitridation in SiO_2/4H–SiC. *J. Appl. Phys.*, **93**, 2257.

[255] Afanas'ev, V.V., Stesmans, A., Ciobanu, F. *et al.* (2003) Mechanisms responsible for improvement of 4H–SiC/SiO_2 interface properties by nitridation. *Appl. Phys. Lett.*, **82**, 568.

[256] Dhar, S., Wang, S., Williams, J.R. *et al.* (2005) Interface passivation for silicon dioxide layers on silicon carbide. *Mater. Res. Soc. Bull.*, **30**, 288.

[257] Tilak, V., Matocha, K. and Dunne, G. (2007) Electron-scattering mechanisms in heavily doped silicon carbide MOSFET inversion layers. *IEEE Trans. Electron Devices*, **54**, 2823.

[258] Dhar, S., Chen, X.D., Mooney, P.M. *et al.* (2008) Ultrashallow defect states at SiO_2/4H–SiC interfaces. *Appl. Phys. Lett.*, **92**, 102112.

[259] Chen, X.D., Dhar, S., Isaacs-Smith, T. *et al.* (2008) Electron capture and emission properties of interface states in thermally oxidized and NO-annealed SiO_2/4H-SiC. *J. Appl. Phys.*, **103**, 033701.

[260] Dhar, S., Haney, S., Cheng, L. *et al.* (2010) Inversion layer carrier concentration and mobility in 4H–SiC metal-oxide-semiconductor field-effect transistors. *J. Appl. Phys.*, **108**, 054509.

[261] Zhu, X., Ahyi, A.C., Li, M. *et al.* (2011) The effect of nitrogen plasma anneals on interface trap density and channel mobility for 4H–SiC MOS devices. *Solid-State Electron.*, **57**, 76.

[262] Lipkin, L.A., Das, M.K. and Palmour, J.W. (2002) N_2O processing improves the 4H-SiC: SiO_2 interface. *Mater. Sci. Forum*, **389–393**, 985.

[263] Kimoto, T., Kanzaki, Y., Noborio, M. *et al.* (2005) Interface properties of metal-oxide-semiconductor structures on 4H-SiC{0001} and $(11\bar{2}0)$ formed by N_2O oxidation. *Jpn. J. Appl. Phys.*, **44**, 1213.

[264] Fujihira, K., Tarui, Y., Imaizumi, M. *et al.* (2005) Characteristics of 4H-SiC MOS interface annealed in N_2O. *Solid-State Electron.*, **49**, 896.

[265] Senzaki, J., Suzuki, T., Shimozato, A. *et al.* (2010) Significant improvement in reliability of thermal oxide on 4H-SiC (0001) face using ammonia post-oxidation annealing. *Mater. Sci. Forum*, **645–648**, 685.

[266] Soejima, N., Kimura, T., Ishikawa, T. and Sugiyama, T. (2013) Effect of NH_3 post-oxidation annealing on flatness of SiO_2/SiC interface. *Mater. Sci. Forum*, **740–742**, 723.

[267] Maeyama, Y., Yano, H., Furumoto, Y. *et al.* (2003) Improvement of SiO_2/SiC interface properties by nitrogen radical irradiation. *Jpn. J. Appl. Phys.*, **42**, L575.

[268] Noborio, M., Suda, J. and Kimoto, T. (2009) P-channel MOSFETs on 4H-SiC {0001} and nonbasal faces fabricated by oxide deposition and N_2O annealing. *IEEE Trans. Electron Devices*, **56**, 1953.

[269] Rozen, J., Dhar, S., Zvanut, M.E. *et al.* (2009) Density of interface states, electron traps, and hole traps as a function of the nitrogen density in SiO_2 on SiC. *J. Appl. Phys.*, **105**, 124506.

[270] Acevedo, A.M., Guillermo, C. and Lopez, J.C. (2001) Thermal oxidation of silicon in nitrous oxide at high pressures. *J. Electrochem. Soc.*, **148**, F 200.

[271] Laurendeau, N.M. (1975) Fast nitrogen dioxide reactions: Significance during NO decomposition and NO_2 formation. *Combust. Sci. Technol.*, **11**, 89.

[272] Okamoto, D., Yano, H., Hirata, K. *et al.* (2010) Improved inversion channel mobility in 4H-SiC MOSFETs on Si face utilizing phosphorus-doped gate oxide. *IEEE Electron Device Lett.*, **31**, 710.

[273] Okamoto, D., Yano, H., Hatayama, T. and Fuyuki, T. (2010) Removal of near-interface traps at SiO_2/4H–SiC (0001) interfaces by phosphorus incorporation. *Appl. Phys. Lett.*, **96**, 203508.

[274] Okamoto, D., Yano, H., Hatayama, T. and Fuyuki, T. (2012) Development of 4H-SiC MOSFETs with phosphorus-doped gate oxide. *Mater. Sci. Forum*, **717–720**, 733.
[275] Olafsson, H.Ö., Gudjonsson, G., Nilsson, P.Å. *et al.* (2004) High field effect mobility in Si face 4H-SiC MOSFET transistors. *Electron. Lett.*, **40**, 508.
[276] Gudjonsson, G., Olafsson, H.Ö., Allerstam, F. *et al.* (2005) High field-effect mobility in n-channel Si face 4H-SiC MOSFETs with gate oxide grown on aluminum ion-implanted material. *IEEE Electron Device Lett.*, **26**, 96.
[277] Das, M.K., Hull, B.A., Krishnaswami, S. *et al.* (2006) Improved 4H-SiC MOS interfaces produced via two independent processes: Metal enhanced oxidation and 1300 °C NO anneal. *Mater. Sci. Forum*, **527–529**, 967.
[278] Yano, H., Hatayama, T., Uraoka, Y. and Fuyuki, T. (2005) High temperature NO annealing of deposited SiO_2 and SiON films on n-type 4H-SiC. *Mater. Sci. Forum*, **483–485**, 685.
[279] Kimoto, T., Kawano, H., Noborio, M. *et al.* (2006) Improved dielectric and interface properties of 4H-SiC MOS structures processed by oxide deposition and N_2O annealing. *Mater. Sci. Forum*, **527–529**, 987.
[280] Aboelfotoh, M.O., Kern, R.S., Tanaka, S. *et al.* (1996) Electrical characteristics of metal/AlN/n-type 6H–SiC(0001) heterostructures. *Appl. Phys. Lett.*, **69**, 2873.
[281] Hino, S., Hatayama, S.T., Kato, J. *et al.* (2009) Anomalously high channel mobility in SiC-MOSFETs with Al_2O_3/SiO_x/SiC gate structure. *Mater. Sci. Forum*, **600–603**, 683.
[282] Hosoi, T., Kagei, Y., Kirino, T. *et al.* (2009) Improved characteristics of 4H-SiC MISFET with AlON/nitrided SiO_2 stacked gate dielectrics. *Mater. Sci. Forum*, **645–648**, 991.
[283] Lichtenwalner, D.J., Misra, V., Dhar, S. *et al.* (2009) High-mobility enhancement-mode 4H-SiC lateral field-effect transistors utilizing atomic layer deposited Al_2O_3 gate dielectric. *Appl. Phys. Lett.*, **95**, 152113.
[284] Hosoi, T., Azumo, S., Kashiwagi, Y. *et al.* (2012) Technical Digest of 2012 International Electron Devices Meeting, p. 7.4 Performance and reliability improvement in SiC power MOSFETs by implementing AlON high-k gate dielectrics.
[285] Okuno, E., Endo, T., Matsuki, H. *et al.* (2005) Low on-resistance in normally-off 4H-SiC accumulation MOSFET. *Mater. Sci. Forum*, **483**, 817.
[286] Tarui, Y., Watanabe, T., Fujihira, K. *et al.* (2006) Fabrication and performance of 1.2 kV, 12.9 mΩcm^2 4H-SiC epilayer channel MOSFET. *Mater. Sci. Forum*, **527–529**, 1285.
[287] Shenoy, P.M. and Baliga, B.J. (1997) The planar 6H-SiC ACCUFET: a new high-voltage power MOSFET structure. *IEEE Electron Device Lett.*, **18**, 589.
[288] Ueno, K. and Oikawa, T. (1999) Counter-doped MOSFETs of 4H-SiC. *IEEE Electron Device Lett.*, **20**, 624.
[289] Harada, S., Suzuki, S., Senzaki, J. *et al.* (2001) High channel mobility in normally-off 4H-SiC buried channel MOSFETs. *IEEE Electron Device Lett.*, **22**, 272.
[290] Shenoy, J.N., Das, M.K., Cooper, J.A. Jr., *et al.* (1996) Effect of substrate orientation and crystal anisotropy on the thermally oxidized SiO_2/SiC interface. *J. Appl. Phys.*, **79**, 3042.
[291] Afanas'ev, V.V., Stesmans, A., Bassler, M. *et al.* (1996) Elimination of SiC/SiO_2 interface states by preoxidation ultraviolet-ozone cleaning. *Appl. Phys. Lett.*, **68**, 2141.
[292] Ueno, K., Asai, R. and Tsuji, T. (1998) 4H-SiC MOSFETs utilizing the H_2 surface cleaning technique. *IEEE Electron Device Lett.*, **19**, 244.
[293] Bassler, M., Afanas'ev, V.V., Pensl, G. and Schulz, M. (1999) Degradation of 6H-SiC MOS capacitors operated at high temperatures. *Microelectron. Eng.*, **48**, 257.
[294] Lelis, A.J., Habersat, D., Green, R. *et al.* (2008) Time dependence of bias-stress-induced SiC MOSFET threshold-voltage instability measurements. *IEEE Trans. Electron Devices*, **55**, 1835.
[295] Green, R., Lelis, A. and Habersat, D. (2012) Charge trapping in SiC Power MOSFETs and its consequences for robust reliability testing. *Mater. Sci. Forum*, **717–720**, 1085.
[296] Marinella, M.J., Schroder, D.K., Isaacs-Smith, T. *et al.* (2007) Evidence of negative bias temperature instability in 4H-SiC metal oxide semiconductor capacitors. *Appl. Phys. Lett.*, **90**, 253508.
[297] Okayama, T., Arthur, S.D., Garrett, J.L. and Rao, M.V. (2008) Bias-stress induced threshold voltage and drain current instability in 4H–SiC DMOSFETs. *Solid State Electron.*, **52**, 164.
[298] Chanthaphan, A., Hosoi, T., Mitani, S. *et al.* (2012) Investigation of unusual mobile ion effects in thermally grown SiO_2 on 4H-SiC(0001) at high temperatures. *Appl. Phys. Lett.*, **100**, 252103.
[299] Yano, H., Kimoto, T. and Matsunami, H. (2001) Interface States of SiO_2/SiC on $(11\bar{2}0)$ and (0001) Si Faces. *Mater. Sci. Forum*, **353–356**, 627.
[300] Senzaki, J., Kojima, K., Harada, S. *et al.* (2002) Strong dependence of the inversion mobility of 4H and 6H SiC(0001) MOSFETs on the water content in pyrogenic re-oxidation annealing. *IEEE Electron Device Lett.*, **23**, 136.

[301] Yano, H., Hirao, T., Kimoto, T. *et al.* (1999) High channel mobility in inversion layers of 4H-SiC MOSFETs by utilizing $(11\bar{2}0)$ face. *IEEE Electron Device Lett.*, **20**, 611.

[302] Endo, T., Okuno, E., Sakakibara, T. and Onda, S. (2009) High channel mobility of MOSFET fabricated on 4H-SiC $(11\bar{2}0)$ face using wet annealing. *Mater. Sci. Forum*, **600–603**, 691.

[303] Dhar, S., Song, Y.W., Feldman, L.C. *et al.* (2004) Effect of nitric oxide annealing on the interface trap density near the conduction bandedge of 4H-SiC at the oxide/$(11\bar{2}0)$ 4H-SiC interface. *Appl. Phys. Lett.*, **84**, 1498.

[304] Nanen, Y., Kato, M., Suda, J. and Kimoto, T. (2013) Effects of nitridation on 4H-SiC MOSFETs fabricated on various crystal faces. *IEEE Trans. Electron Devices*, **60**, 1260.

[305] Tan, J., Cooper, J.A. Jr., and Melloch, M.R. (1998) High-voltage accumulation-layer UMOSFETs in 4H-SiC. *IEEE Electron Device Lett.*, **19**, 467.

[306] Nakamura, T., Miura, M., Kawamoto, N. *et al.* (2009) Development of SiC diodes, power MOSFETs and intelligent power modules. *Phys. Status Solidi A*, **206**, 2403.

[307] Yano, H., Nakao, H., Mikami, H. *et al.* (2007) Anomalously anisotropic channel mobility on trench sidewalls in 4H-SiC trench-gate metal-oxide-semiconductor field-effect transistors fabricated on 8° off substrates. *Appl. Phys. Lett.*, **90**, 042102.

[308] Sun, S.C. and Plummer, J.D. (1980) Electron mobility in inversion and accumulation layers on thermally oxidized silicon surfaces. *IEEE Trans. Electron Devices*, **27**, 1497.

[309] Ouisse, T. (1996) Electron localization and noise in silicon carbide inversion layers. *Philos. Mag. B*, **73**, 325.

[310] Bano, E., Ouisse, T., Sharnholz, S.P. *et al.* (1997) Analytical modelling of thermally-activated transport in SiC inversion layers. *Electron. Lett.*, **33**, 243.

[311] Taur, Y. and Ning, T.H. (2009) *Fundamentals of Modern VLSI Devices*, 2nd edn, Cambridge University Press.

[312] Saks, N.S. and Agarwal, A.K. (2000) Hall mobility and free electron density at the SiC/SiO_2 interface in 4H–SiC. *Appl. Phys. Lett.*, **77**, 3281.

[313] Arnold, E. and Alok, D. (2001) Effect of interface states on electron transport in 4H-SiC inversion layers. *IEEE Trans. Electron Devices*, **48**, 1870.

[314] Harada, S., Kosugi, R., Senzaki, J. *et al.* (2002) Relationship between channel mobility and interface state density in SiC metal-oxide-semiconductor field-effect transistor. *J. Appl. Phys.*, **91**, 1568.

[315] Kawahara, K., Suda, J. and Kimoto, T. (2013) Deep levels generated by thermal oxidation in p-type 4H-SiC. *J. Appl. Phys.*, **113**, 033705.

[316] Kawahara, K., Suda, J. and Kimoto, T. (2013) Deep levels generated by thermal oxidation in n-type 4H-SiC. *Appl. Phys. Express*, **6**, 051301.

[317] Shen, X. and Pantelides, S.T. (2011) Identification of a major cause of endemically poor mobilities in SiC/SiO_2 structures. *Appl. Phys. Lett.*, **98**, 053507.

[318] Devynck, F., Alkauskas, A., Broqvist, P. and Pasquarello, A. (2011) Charge transition levels of carbon-, oxygen-, and hydrogen-related defects at the SiC/SiO_2 interface through hybrid functionals. *Phys. Rev. B*, **84**, 235320.

[319] Itoh, A. and Matsunami, H. (1997) Analysis of Schottky barrier heights of metal/SiC contacts and its possible application to high-voltage rectifying devices. *Phys. Status Solidi A*, **162**, 389.

[320] C.M. Zetterling, S.K. Lee, and M. Östling, *Process Technology for Silicon Carbide Devices*, Chapter 6, ed. C.M. Zetterling (INSPEC, 2002) Schottky and ohmic contacts to SiC.

[321] Mönch, W. (2004) The continuum of interface-induced gap states – The unifying concept of the band lineup at semiconductor interfaces – Application to silicon carbide, in *Silicon Carbide – Recent Major Advances* (eds W.J. Choyke, H. Matsunami and G. Pensl), Springer, p. 317.

[322] Porter, L.M. and Davis, R.F. (1995) A critical review of ohmic and rectifying contacts for silicon carbide. *Mater. Sci. Eng., B*, **34**, 83.

[323] Crofton, J., Porter, L.M. and Williams, J.R. (1997) The physics of ohmic contacts to SiC. *Phys. Status Solidi B*, **202**, 581.

[324] Cole, M.W. and Joshi, P.C. (2004) in *Silicon Carbide, Materials, Processing, and Devices* (eds Z.C. Feng and J.H. Zhao), Taylor & Francis Group, p. 203 The physics of ohmic contacts to SiC.

[325] Tanimoto, S., Okushi, H. and Arai, K. (2004) in *Silicon Carbide – Recent Major Advances* (eds W.J. Choyke, H. Matsunami and G. Pensl), Springer, p. 651.

[326] Roccaforte, F., La Via, F. and Raineri, V. (2005) Ohmic contacts for power devices on SiC. *Int. J. High Speed Electron. Syst.*, **15**, 781.

[327] Tanimoto, S. and Ohashi, H. (2009) Reliability issues of SiC power MOSFETs toward high junction temperature operation. *Phys. Status Solidi A*, **206**, 2417.

[328] Rhoderick, E.H. and Williams, R.H. (1988) *Metal-Semiconductor Contacts*, 2nd edn, Clarendon Press.

[329] Itoh, A., Kimoto, T. and Matsunami, H. (1995) High performance of high-voltage 4H-SiC Schottky barrier diodes. *IEEE Electron Device Lett.*, **16**, 280.
[330] Toumi, S., Ferhat-Hamida, A., Boussouar, L. *et al. et al.* (2009) Gaussian distribution of inhomogeneous barrier height in tungsten/4H-SiC (000$\bar{1}$) Schottky diodes. *Microelectron. Eng.*, **86**, 303.
[331] Shigiltchoff, O., Bai, S., Devaty, R.P. *et al.* (2003) Schottky barriers for Pt, Mo and Ti on 6H and 4H SiC (0001), (000$\bar{1}$), (1$\bar{1}$00) and (1$\bar{2}$10) faces measured by I–V, C–V and internal photoemission. *Mater. Sci. Forum*, **433–436**, 705.
[332] Im, H.J., Ding, Y., Pelz, J.P. and Choyke, W.J. (2001) Nanometer-scale test of the Tung model of Schottky-barrier height inhomogeneity. *Phys. Rev. B*, **64**, 075310.
[333] Skromme, B.J., Luckowski, E., Moore, K. *et al.* (2000) Electrical characteristics of Schottky barriers on 4H-SiC: The effects of barrier height nonuniformity. *J. Electron. Mater.*, **29**, 376.
[334] Itoh, A., Takemura, O., Kimoto, T. and Matsunami, H. (1996) Low power-loss 4H-SiC Schottky rectifiers with high blocking voltage. *Inst. Phys. Conf. Ser.*, **142**, 685.
[335] Crofton, J. and Sriram, S. (1996) Reverse leakage current calculations for SiC Schottky contacts. *IEEE Trans. Electron Devices*, **43**, 2305.
[336] Schoen, K.J., Woodall, J.M., Cooper, J.A. Jr., and Melloch, M.R. (1998) Design considerations and experimental characterization of high voltage SiC Schottky barrier rectifiers. *IEEE Trans. Electron Devices*, **45**, 1595.
[337] Teraji, T. and Hara, S. (2004) Control of interface states at metal/6H-SiC(0001) interfaces. *Phys. Rev. B*, **70**, 035312.
[338] Wardrop, J.R., Grant, R.E., Wang, Y.C. and Davis, R.F. (1992) Metal Schottky barrier contacts to alpha 6H-SiC. *J. Appl. Phys.*, **72**, 4757.
[339] Wardrop, J.R. and Grant, R.E. (1993) Schottky barrier height and interface chemistry of annealed metal contacts to alpha 6H-SiC: Crystal face dependence. *Appl. Phys. Lett.*, **62**, 2685.
[340] Treu, M., Rupp, R., Kapels, H. and Bartsch, W. (2001) Temperature dependence of forward and reverse characteristics of Ti, W, Ta and Ni Schottky diodes on 4H-SiC. *Mater. Sci. Forum*, **353–356**, 679.
[341] Hatakeyama, T. and Shinohe, T. (2002) Reverse characteristics of a 4H-SiC Schottky barrier diode. *Mater. Sci. Forum*, **389–393**, 1169.
[342] Suda, J., Yamaji, K., Hayashi, Y. *et al.* (2010) Nearly ideal current–voltage characteristics of Schottky barrier diodes formed on hydride-vapor-phase-epitaxy-grown GaN free-standing substrates. *Appl. Phys. Express*, **3**, 101003.
[343] Bhatnagar, M., Baliga, B.J., Kirk, H.R. and Rozgonyi, G.A. (1996) Effect of surface inhomogeneities on the electrical characteristics of SiC Schottky contacts. *IEEE Trans. Electron Devices*, **43**, 150.
[344] Defives, D., Noblanc, O., Dua, C. *et al.* (1999) Barrier inhomogeneities and electrical characteristics of Ti/4H-SiC Schottky rectifiers. *IEEE Trans. Electron Devices*, **46**, 449.
[345] Zheng, L., Joshi, R.P. and Fazi, C. (1999) Effects of barrier height fluctuations and electron tunneling on the reverse characteristics of 6H–SiC Schottky contacts. *J. Appl. Phys.*, **85**, 3701.
[346] Saitoh, H., Kimoto, T. and Matsunami, H. (2004) Origin of leakage current in SiC Schottky barrier diodes at high temperature. *Mater. Sci. Forum*, **457–460**, 997.
[347] Tiwari, S. (1991) *Compound Semiconductor Device Physics*, Academic Press.
[348] Crofton, J., Williams, J.R., Bozack, M.J. and Barnes, P.A. (1994) A TiW high-temperature ohmic contacts to n-type 6H-SiC. *Inst. Phys. Conf. Ser.*, **137**, 719.
[349] Crofton, J., Luckowski, E.D., Williams, J.R. *et al.* (1996) Specific contact resistance as a function of doping for n-type 4H and 6H-SiC. *Inst. Phys. Conf. Ser.*, **142**, 569.
[350] Piotrzkowski, R., Litwin-Stanszewska, E. and Grzanka, S. (2011) Towards proper characterization of nonlinear metal-semiconductor contacts: Generalization of the transmission line method. *Appl. Phys. Lett.*, **99**, 052101.
[351] Mohney, S.E., Hull, B.A., Lin, J.Y. and Crofton, J. (2002) Morphological study of the Al–Ti ohmic contact to p-type SiC. *Solid State Electron.*, **46**, 689.
[352] Kelner, G., Binari, S., Shur, M. and Palmour, J.W. (1991) High temperature operation of alpha -silicon carbide buried-gate junction field-effect transistors. *Electron. Lett.*, **27**, 1038.
[353] Crofton, J., McMullin, P.G., Williams, J.R. and Bozack, M.J. (1995) High-temperature ohmic contact to n-type 6H-SiC using nickel. *J. Appl. Phys.*, **77**, 1317.
[354] Uemoto, T. (1995) Reduction of ohmic contact resistance on n-type 6H-SiC by heavy doping. *Jpn. J. Appl. Phys.*, **34**, L7.
[355] Roccaforte, F., La Via, F., Raineri, V. *et al.* (2001) Improvement of high temperature stability of nickel contacts on n-type 6H-SiC. *Appl. Surf. Sci.*, **184**, 295.
[356] Han, S.Y., Kim, K.H., Kim, J.K. *et al.* (2001) Ohmic contact formation mechanism of Ni on n-type 4H-SiC. *Appl. Phys. Lett.*, **79**, 1816.

[357] Tanimoto, S., Kiritani, N., Hoshi, M. and Okushi, H. (2002) Ohmic contact structure and fabrication process applicable to practical SiC devices. *Mater. Sci. Forum*, **389–393**, 879.

[358] Kiritani, N., Hoshi, M., Tanimoto, S. *et al.* (2003) Single material ohmic contacts simultaneously formed on the source/p-well/gate of 4H-SiC vertical MOSFETs. *Mater. Sci. Forum*, **433–436**, 669.

[359] Reshanov, S.A., Emtsev, K.V., Konstantin, V. *et al.* (2008) Effect of an intermediate graphite layer on the electronic properties of metal/SiC contacts. *Phys. Status Solidi B*, **245**, 1369.

[360] Luckowski, E.D., Williams, J.R., Bozack, M.J. *et al.* (1996) Improved nickel silicide ohmic contacts to n-type 4H and 6H-SiC using nichrome. *Mater. Res. Soc. Symp. Proc.*, **423**, 119.

[361] Moki, A., Shenoy, P., Alok, D. *et al.* (1995) Low resistivity as-deposited ohmic contacts to 3C-SiC. *J. Electron. Mater.*, **24**, 315.

[362] Lee, S.K., Zetterling, C.M. and Östling, M. (2000) Low resistivity ohmic titanium carbide contacts to n- and p-type 4H-silicon carbide. *Solid State Electron.*, **44**, 1179.

[363] Jang, T., Porter, L.M., Rutsch, G.W.M. and Odekirk, B. (1999) Tantalum carbide ohmic contacts to n-type silicon carbide. *Appl. Phys. Lett.*, **75**, 3956.

[364] Baud, L., Billon, T., Lassagne, P. *et al.* (1996) Low contact resistivity W ohmic contacts to n-type 6H-SiC. *Inst. Phys. Conf. Ser.*, **142**, 597.

[365] Kriz, J., Scholz, T., Gottfried, K. *et al.* (1998) Metal disilicide contacts to 6H-SiC. *Mater. Sci. Forum*, **264–268**, 775.

[366] Nakatsuka, O., Koide, Y. and Murakami, M. (2002) CoAl ohmic contact materials with improved surface morphology for p-type 4H-SiC. *Mater. Sci. Forum*, **389–393**, 885.

[367] Lee, S.K., Zetterling, C.M., Östling, M. *et al.* (2002) Low resistivity ohmic contacts on 4H-silicon carbide for high power and high temperature device applications. *Microelectron. Eng.*, **60**, 261.

[368] Tsukimoto, S., Sakai, T., Onishi, T. *et al.* (2005) Simultaneous formation of p- and n-type ohmic contacts to 4H-SiC using the ternary Ni/Ti/Al system. *J. Electron. Mater.*, **34**, 1310.

[369] Nakata, T., Koga, K., Matsushita, Y. *et al.* (1988) Single crystal growth of 6H-SiC by a vacuum sublimation method, and blue LEDs, in *Amorphous and Crystalline Silicon Carbide and Related Materials II*, Springer, p. 26.

[370] Suzuki, A., Fujii, Y., Saito, H. *et al.* (1991) Effect of the junction interface properties on blue emission of SiC blue LEDs grown by step-controlled CVD. *J. Cryst. Growth*, **115**, 623.

[371] Crofton, J., Barnes, P.A., Williams, J.R. and Edmond, J.A. (1993) Contact resistance measurements on *p*-type 6H-SiC. *Appl. Phys. Lett.*, **62**, 384.

[372] Crofton, J., Beyer, L., Williams, J.R. *et al.* (1997) Titanium and aluminum-titanium ohmic contacts to p-type SiC. *Solid State Electron.*, **41**, 1725.

[373] Johnson, B.J. and Capano, M.A. (2003) The effect of titanium on Al–Ti contacts to p-type 4H-SiC. *Solid State Electron.*, **47**, 1437.

[374] Vassilevski, K., Zekentes, K., Tsagaraki, K. *et al.* (2001) Phase formation at rapid thermal annealing of Al/Ti/Ni ohmic contacts on 4H-SiC. *Mater. Sci. Eng., B*, **80**, 370.

[375] Crofton, J., Mohney, S.E., Williams, J.R. and Isaacs-Smith, T. (2002) Finding the optimum Al–Ti alloy composition for use as an ohmic contact to p-type SiC. *Solid State Electron.*, **46**, 109.

[376] Moscatelli, F., Scorzoni, A., Poggi, A. *et al.* (2003) Improved electrical characterization of Al–Ti ohmic contacts on p-type ion implanted 6H-SiC. *Semicond. Sci. Technol.*, **18**, 554.

[377] Tsukimoto, S., Sakai, T. and Murakami, M. (2004) Electrical properties and microstructure of ternary Ge/Ti/Al ohmic contacts to p-type 4H–SiC. *J. Appl. Phys.*, **96**, 4976.

[378] Tsukimoto, S., Ito, K., Wang, Z. *et al.* (2009) Growth and microstructure of epitaxial Ti_3SiC_2 contact layers on SiC. *Mater. Trans.*, **50**, 1071.

[379] Kassamakova, L., Kakanakov, R., Nordell, N. and Savage, S. (1998) Thermostable ohmic contacts on p-type SiC. *Mater. Sci. Forum*, **264–268**, 787.

[380] Konishi, R., Yasukochi, R., Nakatsuka, O. *et al.* (2003) Development of Ni/Al and Ni/Ti/Al ohmic contact materials for p-type 4H-SiC. *Mater. Sci. Eng., B*, **98**, 286.

[381] Fursin, L.G., Zhao, J.H. and Weiner, M. (2001) Nickel ohmic contacts to p and n-type 4H-SiC. *Electron. Lett.*, **37**, 1092.

[382] Tamura, S., Kimoto, T., Matsunami, H. *et al.* (2000) Nuclear transmutation doping of phosphorus into 6H-SiC. *Mater. Sci. Forum*, **338–342**, 849–852.

7

Unipolar and Bipolar Power Diodes

7.1 Introduction to SiC Power Switching Devices

The majority of silicon carbide devices developed to date can be grouped into three general classes: power switching devices, microwave devices, and specialty devices (sensors, high-temperature integrated circuits, etc.). Of these, by far the most important and most fully developed are the power switching devices. Accordingly, Chapters 7–11 will be devoted to SiC power devices, and other SiC device applications will be considered in Chapter 12.

Power switching devices attempt to emulate an ideal switch. An ideal switch carries infinite current in the on state with zero voltage drop, and hence zero power dissipation. In the off state it blocks infinite voltage with zero leakage current, and it switches instantly between states using zero switching energy. Of course, real semiconductor switches only approach these ideals, and the degree to which they achieve these goals is indicated by the performance specifications of the device. Most critical of these are blocking voltage, maximum on-state current, on-state and off-state power dissipation, and switching loss. Since maximizing performance involves trade-offs between parameters, researchers have developed *figures of merit* (FOMs) that define the theoretical envelope of maximum performance and quantify the degree to which actual devices approach these theoretical limits. Device performance depends on fundamental material parameters, as well as on device parameters such as dopings, physical dimensions, and so on. SiC is of interest because its fundamental material parameters, especially its high breakdown field, result in theoretical performance that is orders of magnitude higher than silicon.

To begin, let us consider blocking voltage, and the relationship between blocking voltage and on-state power dissipation. Since all power devices support their terminal voltage in the off-state by means of a reverse-biased pn junction (or a reverse-biased metal–semiconductor junction), we will first consider the voltage limits of reverse-biased pn junctions.

7.1.1 Blocking Voltage

Consider a p+/n− one-sided step junction under reverse bias. If the doping is highly asymmetrical, we can assume all the depletion occurs on the n− side, and the electric field profile is as shown in Figure 7.1. The reverse voltage can be increased until the maximum field E_M equals the critical field for avalanche breakdown E_C, at which point breakdown occurs. The blocking voltage is the integral of the electric field at this point, or

$$V_B = E_C x_{DB}/2 \tag{7.1}$$

Fundamentals of Silicon Carbide Technology: Growth, Characterization, Devices, and Applications, First Edition. Tsunenobu Kimoto and James A. Cooper.

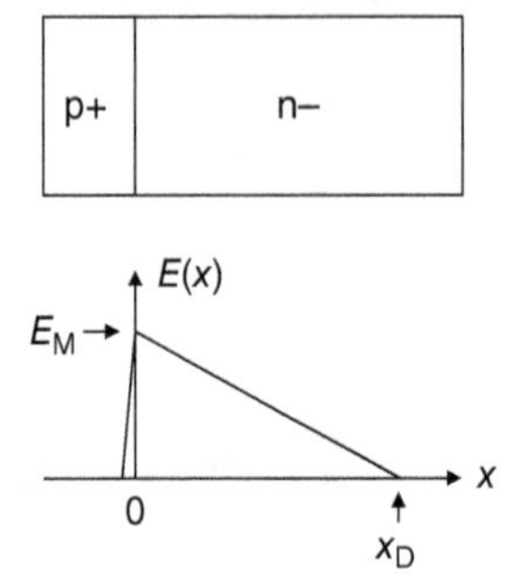

Figure 7.1 Electric field profile in a p+/n− one-sided step junction. The maximum field E_M occurs at the junction, and the depletion width into the n− region is x_D.

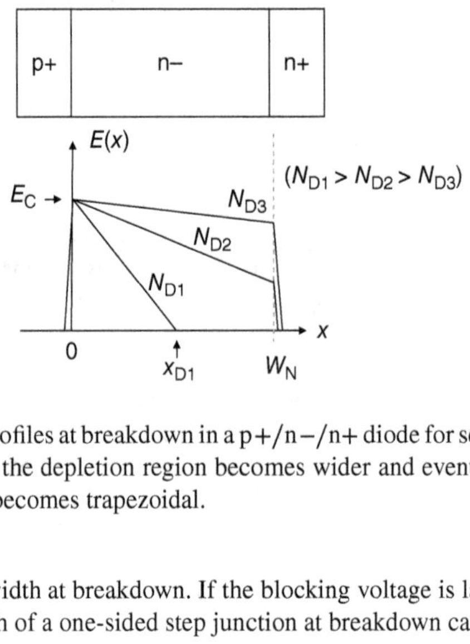

Figure 7.2 Electric field profiles at breakdown in a p+/n−/n+ diode for several dopings. As the doping in the n− region is reduced, the depletion region becomes wider and eventually reaches the n+ region, whereupon the field profile becomes trapezoidal.

where x_{DB} is the depletion width at breakdown. If the blocking voltage is large compared to the built-in potential, the depletion width of a one-sided step junction at breakdown can be written

$$x_{DB} = \sqrt{2\varepsilon_S V_B/(qN_D)} \tag{7.2}$$

where ε_S is the semiconductor dielectric constant and N_D is the doping on the lightly-doped side. Using Equation 7.1 to eliminate x_{DB} in Equation 7.2 allows us to solve for V_B,

$$V_B = (\varepsilon_S E_C^2)/(2qN_D) \tag{7.3}$$

Equation 7.3 tells us that the blocking voltage of a p+/n− junction of a given doping is proportional to the *square* of the critical field. The critical field in SiC is almost an order-of-magnitude higher than in silicon, so the blocking voltage for a given doping will be almost *two* orders-of-magnitude higher than in silicon.

Suppose the n− region is designed with a finite thickness W_N. Using Equation 7.3 we can get a relationship between blocking voltage and doping as a function of W_N. To see the basic features, let's first assume that the critical field is independent of doping, that is, a known constant. In this case, Equation 7.3 tells us that blocking voltage is inversely proportional to doping, that is, we can double the blocking voltage by halving the doping. However, each time we do this, Equation 7.2 tells us we also double the

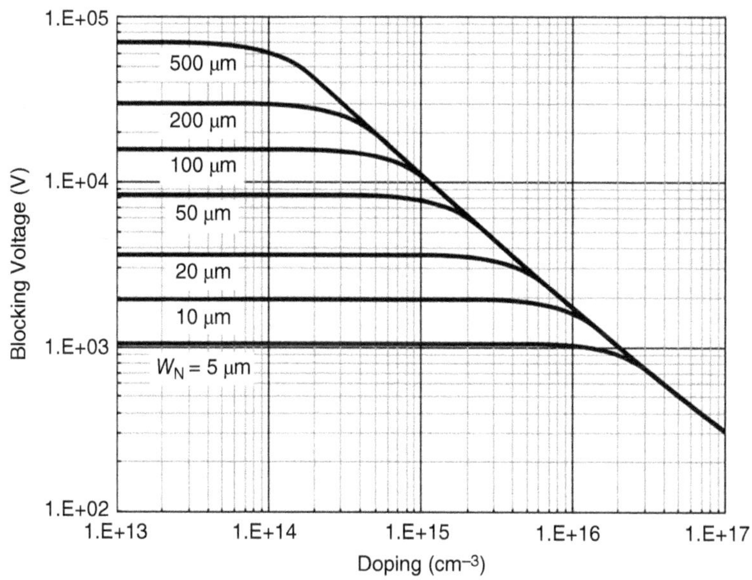

Figure 7.3 Blocking voltage as a function of doping with epilayer width as a parameter ([1] reproduced with permission from IEEE).

depletion width at breakdown x_{DB}. Eventually x_{DB} will exceed W_N, at which point our simple picture has to be revised. If the n− region is terminated in a heavily-doped n+ region, the depletion region will not extend very far into the n+ region, and the field profile will go from triangular to trapezoidal, as shown in Figure 7.2. As we further reduce the doping, the field trapezoid will approach a rectangle of height E_C and width W_N, whereupon the blocking voltage will become

$$V_{B,MAX} = E_C W_N \tag{7.4}$$

Figure 7.3 shows a plot of V_B versus doping in 4H-SiC for several fixed values of W_N. These curves are calculated using the ionization integral (to be discussed in Section 10.1), and include the actual doping dependence of E_C. The maximum possible blocking voltage for a fixed W_N is achieved by making the doping small. However, this has an undesirable effect on the on-state power dissipation, as explained next.

For a unipolar device, where current flow is due solely to majority carriers, the analysis is particularly simple. Assuming negligible contribution from the junction voltage drop, the on-state power dissipation per unit area within the device can be written

$$P_{ON} = R_{ON,SP} J_{ON}^2 \tag{7.5}$$

where J_{ON} is the on-state current density and $R_{ON,SP}$ is the specific on-resistance, defined as the resistance-area product, in units of $\Omega\,cm^2$. For the p+/n− one-sided step-junction, the on-resistance arises primarily from the resistance of the lightly-doped n− region, and we can write

$$R_{ON,SP} = \rho W_N = W_N/(q\mu_N N_D^+) \tag{7.6}$$

where ρ is the resistivity, μ_N the electron mobility in the direction of current flow, and $N_D{}^+$ the ionized dopant concentration in the n− region. Clearly, reducing N_D to achieve the maximum blocking voltage increases the specific on-resistance and the on-state power dissipation. As designers, how do we decide where the optimum lies? The answer is to maximize the unipolar power device figure of merit (*FOM*).

7.1.2 Unipolar Power Device Figure of Merit

A useful figure of merit for all power devices is the product of blocking voltage and on-state current, since an "ideal" power switch would maximize both. Thus we can write

$$\mathrm{FOM} = I_{ON}V_B = AJ_{ON}V_B \tag{7.7}$$

where A is the area of the device and J_{ON} is the on-state current density. The maximum allowable power dissipation P_{MAX} is determined by the thermal capability of the package, the heat sink temperature, and the maximum allowable junction temperature of the device. For a unipolar device we can attribute the majority of power dissipation to the on-state power P_{ON}. In this case, the maximum J_{ON} is obtained from Equation 7.5 as

$$J_{ON} = \sqrt{P_{MAX}/R_{ON,SP}} \tag{7.8}$$

and our FOM can be written

$$\boxed{\mathrm{FOM} = A\sqrt{P_{MAX}\left(V_B^2/R_{ON,SP}\right)}} \tag{7.9}$$

There are three factors in this FOM. The device area A is limited by material quality and fabrication technology, and ultimately by manufacturing yield. The maximum power dissipation P_{MAX} is limited by the thermal capability of the package and the maximum junction temperature of the device; using a package with a lower thermal resistance would increase the FOM. The remaining factor, $V_B^2/R_{ON,SP}$, represents the *unipolar device FOM,* and it is the goal of the designer to maximize this ratio.

The device FOM for a unipolar device can be calculated as follows. From Equation 7.3 we can write

$$N_D = \left(\varepsilon_S E_C^2\right)/\left(2qV_B\right) \tag{7.10}$$

We assume a *non-punch-through* design, such as shown in Figure 7.1, where the electric field profile is triangular. To minimize on-resistance, we reduce the n− region thickness W_N so that it just equals x_{DB}. In this case, Equation 7.1 yields

$$W_N = 2V_B/E_C \tag{7.11}$$

Now inserting Equations 7.10 and 7.11 into Equation 7.6, and assuming complete ionization of the donor impurities, we have

$$R_{ON,SP} = \left(4V_B^2\right)/\left(\mu_N \varepsilon_S E_C^3\right) \tag{7.12}$$

Equation 7.12 tells us that the specific on-resistance of an optimally designed non-punch-through unipolar device increases as the square of the desired blocking voltage, and is inversely proportional to the *cube* of the critical field. Since the critical field in 4H-SiC is almost an order-of-magnitude higher than in silicon, the on-resistance for a given blocking voltage will be almost 1000 times lower. This accounts for the great interest in developing power devices in this material (although the actual reduction in $R_{ON,SP}$ is closer to 400× due to the lower μ_N in SiC.) Finally, to write the desired unipolar device FOM, we can rearrange Equation 7.12 to yield

$$V_B^2/R_{ON,SP} = \left(\mu_N \varepsilon_S E_C^3\right)/4 \tag{7.13}$$

Equation 7.13 represents the maximum possible FOM for an optimally designed non-punch-through unipolar device [for a *punch-through* design, replace the 4 by $(3/2)^3$]. Real devices can only approach this theoretical limit, and comparing the actual measured $V_B^2/R_{ON,SP}$ to the theoretical limit given by Equation 7.13 is a useful indication of design and processing optimization. Note that the theoretical FOM depends on fundamental material constants, and not on specific device parameters. We will discuss how to achieve an optimal design in Section 10.2.

7.1.3 Bipolar Power Device Figure of Merit

The above discussion provides guidelines for optimizing the performance of unipolar devices, but how do we modify our thinking when dealing with bipolar devices, and how do we compare unipolar and bipolar devices for the same application? In bipolar devices, current flow involves both majority and minority carriers. A simple example is a p+/n −/n+ diode. If the n− region is very lightly doped, we can think of this as a p +/"intrinsic"/n+ or "pin" diode. In the previous discussion we assumed the current in the n− region consisted only of majority carrier electrons, but in a pin diode the current in the middle region includes both holes injected from the p+ region and electrons injected from the n+ region. The presence of both types of carriers in this region significantly reduces the resistivity ρ given by

$$\rho = 1/(q\mu_N n + q\mu_P p) \tag{7.14}$$

This "conductivity modulation" can lead to a large reduction in the on-state power dissipation, as shown by inserting Equation 7.14 into Equation 7.6, and Equation 7.6 into Equation 7.5. However, in a bipolar device we also have to consider power dissipation due to switching transients. This is because the minority carriers stored in the lightly-doped region need to be removed when the device turns off. This removal process involves recombination, drift, and diffusion, and as long as the carrier density remains high, the resistivity remains low. The low resistivity allows a significant reverse current to flow until all carriers have been removed. This reverse current, in turn, leads to a significant transient power dissipation. A quantitative analysis typically requires transient computer simulations that include the power dissipation in the external circuit elements, but the main point here is that power is dissipated during each switching event. It is convenient to consider the integral of the transient power over the switching event, or the *switching energy* E_{SW}. It is important to include both turn-on and turn-off events, although the turn-off energy is usually much larger than the turn-on energy. The switching power dissipation is then proportional to switching frequency f,

$$P_{SW} = E_{SW} f \tag{7.15}$$

We can now generalize our power device FOM given by Equation 7.7. The switching energy is a function of the carrier densities in the n− region established during the on state, and these densities depend on the on-state current J_{ON}, so we can write

$$E_{SW} = E_{OFF}\left(J_{ON}\right) + E_{ON}\left(J_{ON}\right) \tag{7.16}$$

where the exact dependence of E_{OFF} and E_{ON} on the on-state current J_{ON} is best established by transient computer simulations. The total power dissipation is the sum of the on-state power, off-state power, and switching power, so we can write

$$P_{TOTAL} = P_{ON}\left(J_{ON}\right)\delta + P_{OFF}(1-\delta) + E_{SW}\left(J_{ON}\right)f \tag{7.17}$$

where δ is the duty cycle. In the case of a unipolar device, P_{ON} is given by Equation 7.5, P_{OFF} is usually negligible, and $E_{SW}(J_{ON})$ is established by computer simulations. In the case of a bipolar device, P_{ON} may be a nonlinear function of J_{ON} (such as in a pn diode), and $E_{SW}(J_{ON})$ is established by computer simulations. Again, P_{OFF} is usually negligible. In any event, setting P_{TOTAL} in Equation 7.17 equal to the power dissipation limit of the package establishes an upper limit for J_{ON} at a given switching frequency f and duty cycle δ. At very low frequencies the switching loss can be ignored, but at high frequencies the switching loss can become the dominant loss.

Although the analysis for a bipolar device is more complicated than for a unipolar device, the basic procedure is similar. For a given desired blocking voltage, the bipolar device that meets that blocking voltage is evaluated to determine how P_{ON} and E_{SW} depend on J_{ON}. Given the switching frequency, we can calculate $P_{TOTAL}(J_{ON})$ using Equation 7.17 and adjust J_{ON} so that P_{TOTAL} equals the power dissipation limit of the package. The resulting current density J_{ON} then becomes our FOM; the device with the highest current density at the required switching frequency and blocking voltage wins. This technique can be used to compare bipolar devices to other bipolar devices, or to make a fair comparison between bipolar and unipolar devices for the same intended application.

A critical aspect of device design is to achieve a blocking voltage as close as possible to the theoretical plane-junction blocking voltage in Figure 7.3. Every real device has edges, and field crowding at the edges can reduce the actual blocking voltage to a small fraction of the value in Figure 7.3. This problem is mitigated by various types of *edge terminations,* as will be discussed in Section 10.1.

Having presented the general considerations in the design and optimization of unipolar and bipolar devices, we now turn to a discussion of specific devices, beginning with the Schottky diode.

7.2 Schottky Barrier Diodes (SBDs)

The Schottky barrier diode (SBD) is a rectifying metal–semiconductor contact whose band diagram is illustrated in Figure 7.4. The main features that distinguish a Schottky diode from an ohmic contact are the work function Φ_M of the metal and the doping of the semiconductor. In a Schottky contact, the work function places the metal Fermi level near the middle of the semiconductor bandgap. This creates a barrier to carrier injection from the metal into either band of the semiconductor, so current can only flow by the injection of majority carriers from the semiconductor into the metal. This insures unidirectionality of current and produces a rectifying current–voltage relationship. Likewise, the doping of the semiconductor must not be too high. If the doping is high, the depletion width in the semiconductor will be very small, and electrons can tunnel between the majority carrier band of the semiconductor and states at the same energy in the metal, leading to ohmic, or non-rectifying, behavior.

The theory of current flow in the SBD is developed in many textbooks, and the derivation will not be repeated here. The current density can be written

$$J = A^* T^2 \exp(-\Phi_B/kT)[\exp(qV_J/kT) - 1] \tag{7.18}$$

where A^* is the modified Richardson's constant for the semiconductor given by

$$A^* = 4\pi q m_N^* k^2/h^3 \tag{7.19}$$

Here Φ_B is the metal–semiconductor barrier height shown in Figure 7.4, k is Boltzmann's constant, T is absolute temperature, $m_N{}^*$ (or $m_P{}^*$) is the majority carrier effective mass, h is Planck's constant, and V_J is the voltage drop (or the offset in Fermi levels) across the junction. A^* is usually determined by experiment, and for 4H-SiC is approximately 145 A cm^{-2} K^{-2} [2]. The factors preceding the term in square brackets in Equation 7.18 constitute the saturation current, and it is notable that the current scales

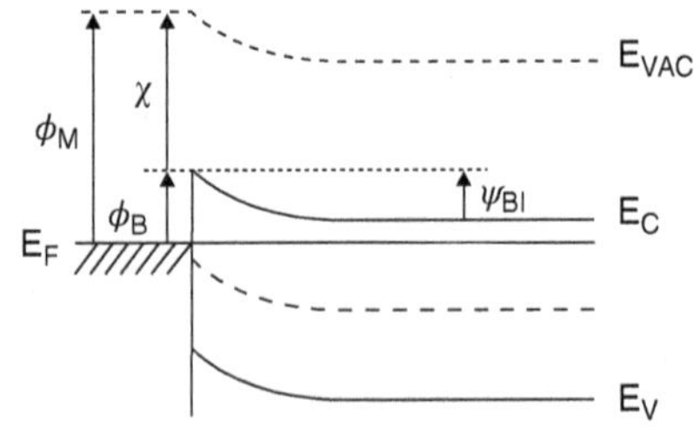

Figure 7.4 Band diagram of a Schottky diode on an n-type semiconductor. E_{VAC} is the vacuum level, Φ_M is the metal work function, χ is the electron affinity of the semiconductor, Φ_B is the barrier height for electrons, and ψ_{BI} is the built-in potential of the junction.

Table 7.1 Experimentally measured Φ_{B0} of metals on 4H-SiC.

Metal	Face	Φ_{B0}(C − V)	Φ_{B0}(I − V)	References
Ni	Si	1.70	1.60	[3]
Ni	Si	–	1.30	[1]
Ni	Si	–	1.4–1.5	[4]
Au	Si	1.80	1.73	[3]
Ti	Si	1.15	1.10	[3]
Ti	Si	–	0.80	[1]
Ni	C	–	1.55	[5]
Au	C	–	1.88	[5]
Ti	C	–	1.20	[5]

exponentially with barrier height Φ_B. Clearly, reducing the barrier height increases both forward and reverse current dramatically.

An important correction needs to be made to Equation 7.18 to account for *Schottky barrier lowering,* in which the *effective* barrier height Φ_B is reduced by the electric field at the surface of the semiconductor. This is especially important when the electric field is high, as under reverse-bias conditions. The effective barrier height can be written as

$$\Phi_B = \Phi_{B0} - q\sqrt{q|E_S|/(4\pi\varepsilon_S)} \tag{7.20}$$

where Φ_{B0} is the barrier height at zero field, E_S is the electric field in the semiconductor at the surface, and ε_S is the dielectric constant of the semiconductor. With this correction, the reverse current in the SBD does not saturate, but continues to increase gradually until breakdown. Since the second term in Equation 7.20 is independent of barrier height, the effect is more pronounced when the zero-field barrier height Φ_{B0} is low. For this reason it is important to ensure adequate Φ_{B0}. Table 7.1 lists the barrier height of various metals on 4H-SiC as measured by both $C-V$ and $I-V$ techniques.

The on-state voltage drop of a Schottky diode consists of the junction drop V_J plus the voltage drop across the lightly-doped drift region and the heavily-doped substrate. V_J for a given J_{ON} is given by Equation 7.18, and the voltage drop across the drift region and substrate is

$$(V_{DR} + V_{SUB}) = J_{ON}(R_{DR,SP} + R_{SUB,SP}) \tag{7.21}$$

where $R_{DR,SP}$ is given by Equation 7.6. Figure 7.5 shows measured forward current–voltage characteristics of Ni and Ti Schottky diodes as a function of temperature [1]. The higher current of the Ti diode is due to its lower barrier height, and the current increases with temperature, as predicted by Equation 7.18. The current saturation at high forward currents is due to the resistance of the drift region. This occurs because V_{DR} in Equation 7.21 increases linearly with current, while the junction voltage V_J in Equation 7.18 increases as the logarithm of current. At sufficiently high current the drift region voltage dominates, and the current–voltage characteristic becomes linear. In this region the specific on-resistance becomes the limiting factor, and must be minimized. Minimization of $R_{ON,SP}$ in unipolar devices for a given blocking voltage will be discussed in Section 10.2.

A major advantage of the Schottky diode over the pin diode is the lower V_J at a given J_{ON}. Roughly speaking, the junction voltage drop in a SiC pin diode at high currents is close to the bandgap energy, or about 3 V, whereas in a Schottky diode it is close to the barrier height, or around 0.5–1 V. Figure 7.6 shows the current–voltage characteristics of a SiC SBD and a SiC pin diode for a designed blocking voltage of 2400 V. The bend-over in the SBD characteristics at high currents is due to the resistance of the lightly-doped drift region. In the case of the pin diode, the drift region resistance is reduced by conductivity modulation, as will be discussed in the next section. We see that for currents below 1000 A cm^{-2}, the SBD has a lower voltage drop, and hence a lower on-state power dissipation at a given current.

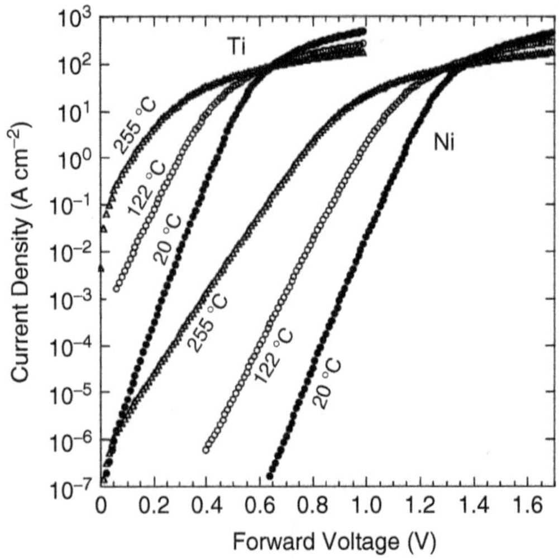

Figure 7.5 Forward current–voltage relationships of Ni and Ti Schottky diodes as a function of temperature ([1] reproduced with permission from IEEE). The lower turn-on voltage of the titanium diode is due to its lower barrier height, and the current saturation is due to the series resistance of the drift region.

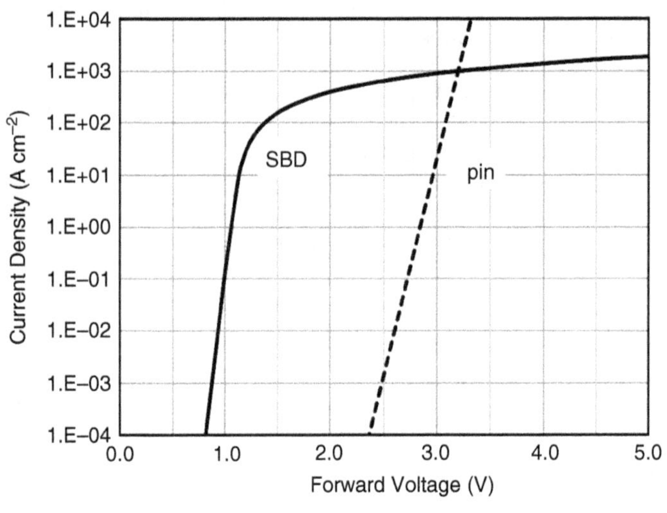

Figure 7.6 Calculated forward current–voltage relationship of SiC Schottky and pin diodes designed for a plane-junction blocking voltage of 2400 V. The SBD characteristics are obtained using Equations 7.18 and 7.21 with a Ni contact and $R_{ON,SP} = 2\ m\Omega\ cm^2$, and the pin characteristics are calculated using Equations 7.51 and 7.57–7.59.

The most important advantage of the SBD over the pin diode, and the main reason for using the Schottky diode, is the lack of minority carrier injection, leading to a very fast turn-off and low switching energy E_{SW}. For this reason, SiC SBDs are now being used in place of silicon pin diodes in high-frequency applications such as switched-mode power supplies, where switching loss is an issue. This is illustrated in Figure 7.7, which shows reverse recovery transients for a silicon pin diode and a Ni/4H-SiC

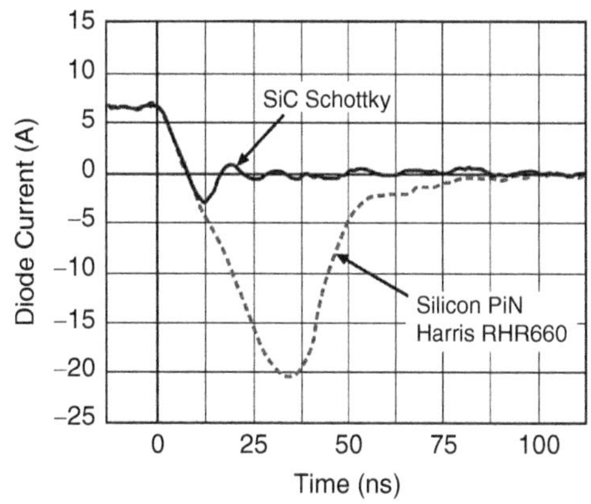

Figure 7.7 Reverse recovery transients of a Ni/4H-SiC SBD and a silicon pin diode (Harris RHR660) at 150 °C ([6] reproduced with permission from IEEE). The forward current is 6 A at a supply voltage of 300 V, with a dI/dt of 1000 A μs^{-1}. The recovery of the Schottky diode shows almost no overshoot.

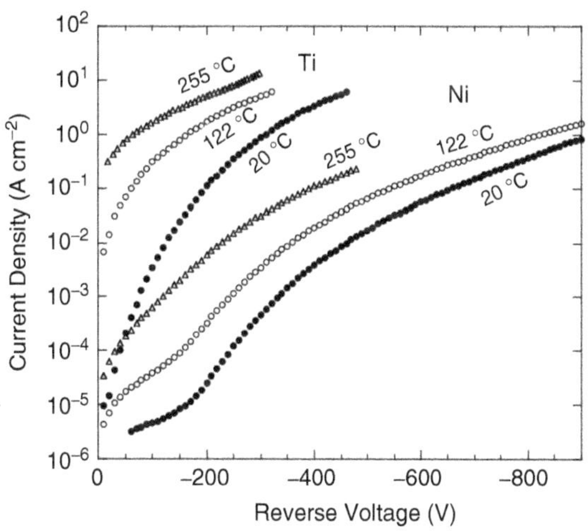

Figure 7.8 Reverse current–voltage relationships of Ni and Ti Schottky diodes as a function of temperature ([1] reproduced with permission from IEEE). The higher reverse current in the Ti diodes results from the lower barrier height Φ_{B0}.

Schottky diode at 150 °C in an inductively-loaded test circuit driven by a silicon IGBT (insulated gate bipolar transistor) [6]. The recovery transient of the Schottky diode is very short due to its lack of stored charge.

The very small switching loss of the SBD has to be balanced against the larger reverse leakage current that can lead to non-negligible off-state power dissipation. Reverse leakage in a Schottky diode is primarily due to thermionic field emission of carriers from the metal into the semiconductor, and is exacerbated by the barrier-lowering effect. The reverse currents of Ni and Ti Schottky diodes on 4H-SiC are shown in Figure 7.8 at several temperatures [1]. The reverse current is higher for the Ti diode due to its lower barrier height Φ_{B0}, and the current increases with temperature, as predicted by Equation 7.18. By comparison, the reverse leakage in a pin diode is due primarily to thermal generation, and is extremely small in SiC due to the wide bandgap.

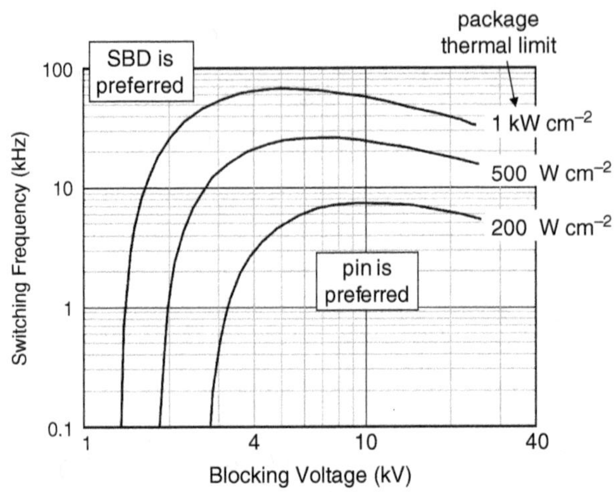

Figure 7.9 Loci of equal performance for Schottky and pin diodes as a function of the power dissipation limit of the package ([7] reproduced with permission from IEEE). For areas to the left and above the loci, the Schottky diode provides the highest current density, and hence is the preferred device. For areas below the loci, the pin diode is preferred.

Taking all these factors into account, the performance of SiC SBDs can be quantitatively compared to SiC pin diodes as follows: First select a blocking voltage V_B, and design both SBD and pin diode to meet this V_B specification. Then calculate the on-state power dissipation P_{ON} as a function of on-state current density J_{ON}. Next calculate the switching energy E_{SW} as a function of J_{ON} using computer simulations. Then use Equation 7.17 to calculate the maximum current density J_{ON} at each frequency such that the total power dissipation P_{TOTAL} equals the specified package limit, for example, 300 W cm^{-2}. At any given switching frequency and blocking voltage, the device with the higher current density is the preferred device. Using this procedure we can construct a map of relative performance of the SiC SBD and SiC pin diode in a two-dimensional parameter space of blocking voltage and switching frequency. Such a comparison is shown in Figure 7.9 [7]. Here we compare 4H-SiC Schottky diodes and pin diodes at several package thermal limits, assuming a 50% duty cycle and a 50% derating factor on blocking voltage. In regions above the lines, the SBD meets the blocking voltage and switching frequency specifications with a higher J_{ON} than the pin diode, and hence is the preferred device. The SBD is the better device at low blocking voltages and high switching frequencies, whereas the pin diode is better at high blocking voltages and low frequencies. For a package thermal limit of 200 W/cm^2, the SBD is preferred at *any* blocking voltage if the switching frequency is above about 8 kHz.

It is reasonable to ask, "If Schottky diodes are preferred over pin diodes at high switching frequencies, why not simply use *silicon* Schottky diodes?" It turns out that silicon SBDs are not suitable for high-voltage applications because of their high reverse leakage currents. SiC Schottky diodes have a significant advantage over silicon SBDs in terms of reverse leakage current due to the higher barrier heights in SiC. In 4H-SiC, Φ_{B0} can theoretically be as large as half the bandgap, or 1.6 V, whereas in silicon Φ_{B0} is limited to 0.56 V. Since reverse leakage scales exponentially with Φ_B, the additional ~1 V of barrier height in SiC reduces the reverse current by *17 orders of magnitude* at room temperature. The limited barrier height effectively restricts silicon SBDs to applications where the required blocking voltage is very low.

7.3 pn and pin Junction Diodes

PN and pin diodes are junction diodes whose doping profiles and band diagrams are shown in Figure 7.10. PN diode theory is covered in elementary semiconductor device textbooks and the equations will not

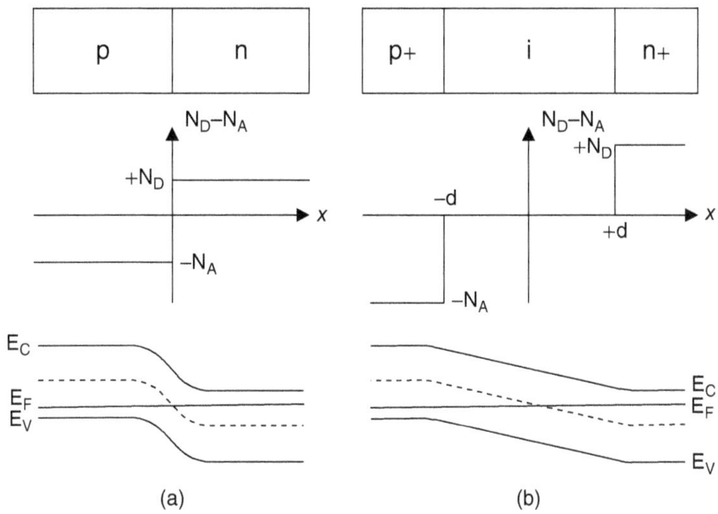

Figure 7.10 Structure, doping profile, and band diagrams of (a) pn and (b) pin diodes in equilibrium.

be derived here. Because of their importance in power switching applications, our discussion will concentrate on pin diodes. PiN diodes are used for power switching because a thick, lightly-doped region is needed to produce a high blocking voltage, as can be seen in Figure 7.3. As an illustrative example, Figure 7.3 shows that to achieve a plane-junction breakdown of 3.5 kV requires a 20 μm thick drift region doped below $\sim 10^{15}$ cm^{-3}. In practical terms, we will refer to such a lightly-doped region as an "i" region, even though it is not truly intrinsic.

In a pn diode under forward bias, the applied terminal voltage lowers the potential barriers confining electrons to the n region and holes to the p region. As a result, electrons from the n region flow into the p region where they are minority carriers, while holes from the p region flow into the n region where they are minority carriers. The current is determined by the rate at which electrons injected into the p region and holes injected into the n region can diffuse away from the junction. Assuming the minority carrier densities are low compared to the majority carrier densities ("low-level injection"), these diffusion rates are calculated by solving the minority carrier diffusion equations in the respective n and p neutral regions. To do this, it is convenient to express the electron and hole densities in terms of their equilibrium values plus their deviations from equilibrium,

$$n = n_0 + \Delta n \qquad p = p_0 + \Delta p \tag{7.22}$$

where n, p are the total carrier densities, n_0, p_0 are the carrier densities in equilibrium, and $\Delta n, \Delta p$ are the *excess carrier densities,* that is, the deviations from equilibrium. The minority carrier diffusion equations in one dimension can then be written

$$\begin{aligned} \frac{\partial \Delta n}{\partial t} &= D_N \frac{\partial^2 \Delta n}{\partial x^2} - \frac{\Delta n}{\tau_N} + G_L, \quad \text{p-type material} \\ \frac{\partial \Delta p}{\partial t} &= D_P \frac{\partial^2 \Delta p}{\partial x^2} - \frac{\Delta p}{\tau_P} + G_L, \quad \text{n-type material} \end{aligned} \tag{7.23}$$

where $D_{N,P}$ are the diffusion coefficients for electrons and holes, $\tau_{N,P}$ are the lifetimes of electrons and holes as minority carriers, and G_L is the generation rate due to light, if the sample is illuminated. The first term on the right-hand side is the rate of change of the carrier density due to diffusion, the second term is the rate of change due to recombination/generation, and the third term is the rate of change due to photogeneration. In steady-state, the time derivatives on the left-hand side are zero, and the general solutions can be written:

$$\begin{aligned}\Delta n(x) &= A_1 \exp\left(x/L_N\right) + A_2 \exp\left(-x/L_N\right) + \tau_N G_L, \quad L_N = \sqrt{D_N \tau_N} \\ \Delta p(x) &= B_1 \exp\left(x/L_P\right) + B_2 \exp\left(-x/L_P\right) + \tau_P G_L, \quad L_P = \sqrt{D_P \tau_P}\end{aligned} \tag{7.24}$$

where $L_{N,P}$ are defined as the minority carrier diffusion lengths. The constants $A_{1,2}$ and $B_{1,2}$ are chosen so that the solutions satisfy the boundary condition imposed by the "law of the junction", namely that the *pn* product within the depletion region be given by

$$pn = n_i^2 \exp\left(qV_J/kT\right) \tag{7.25}$$

where n_i is the intrinsic carrier concentration and V_J is the applied voltage, or equivalently, the splitting of the quasi-Fermi levels across the junction. Solution of the minority carrier diffusion equations subject to these boundary conditions in the dark ($G_L = 0$) leads directly to the Shockley diode equation

$$\boxed{J = qn_i^2 \left(\frac{D_N}{L_N N_A^-} + \frac{D_P}{L_P N_D^+}\right) \left[\exp\left(qV_J/kT\right) - 1\right]} \tag{7.26}$$

Here, N_A^- and N_D^+ are the *ionized* doping concentrations of the p- and n-sides of the junction, respectively. In 4H-SiC, the p-type dopants have relatively high ionization energies, and are not fully ionized at room temperature. The n-type dopants are shallow levels, and are nearly 100% ionized at room temperature and above. Incomplete ionization in 4H-SiC is discussed more fully in Appendix A. The diffusion lengths in Equation 7.26 are given by

$$L_{N,P} = \sqrt{D_{N,P} \tau_{N,P}} \tag{7.27}$$

where $\tau_{N,P}$ are the electron and hole minority carrier lifetimes. Equation 7.26 accounts for diffusion currents under both forward- and reverse-bias conditions, provided low-level injection can be assumed, that is, the forward voltage is not too large. An additional term needs to be added to account for generation–recombination in the depletion region, but this term is only important at low currents and will not be needed for this discussion.

Similarly, in a pin diode, forward current consists of electrons that flow from the n+ region into the "i" layer and holes that flow from the p+ region into the "i" layer. Here the similarity to the pn diode ends, because the minority carrier densities in the "i" layer immediately exceed the doping density, resulting in *high-level injection* conditions that preclude the use of the minority carrier diffusion equations. Instead, we must solve the electron and hole *continuity equations* in a self-consistent manner. Fortunately, by making some simple assumptions we can arrive at the high-level equivalent of the minority carrier diffusion equations, namely, the *ambipolar diffusion equation*. This equation can then be solved in the "i" region to obtain information on carrier densities, electrostatic potential, and current. The development is straightforward, but the reader who is more interested in the "bottom line" can skip ahead to Equation 7.39 without loss of understanding.

7.3.1 High-Level Injection and the Ambipolar Diffusion Equation

Assuming uniform doping and one-dimensional current flow, the continuity equations for electrons and holes can be written

$$\begin{aligned}\frac{\partial \Delta n}{\partial t} &= \frac{1}{q}\frac{\partial J_N}{\partial x} + G \\ \frac{\partial \Delta p}{\partial t} &= -\frac{1}{q}\frac{\partial J_P}{\partial x} + G\end{aligned} \tag{7.28}$$

where G is the local electron-hole net generation rate (or recombination rate, if negative), and $J_{N,P}$ are the electron and hole current densities given by

$$J_N = q\mu_N nE + qD_N \frac{\partial \Delta n}{\partial x}$$
$$J_P = q\mu_P pE - qD_P \frac{\partial \Delta p}{\partial x} \tag{7.29}$$

Inserting Equation 7.29 into Equation 7.28 yields

$$\frac{\partial \Delta n}{\partial t} = \mu_N \frac{\partial}{\partial x}(nE) + D_N \frac{\partial^2 \Delta n}{\partial x^2} + G$$
$$\frac{\partial \Delta p}{\partial t} = -\mu_P \frac{\partial}{\partial x}(pE) + D_P \frac{\partial^2 \Delta p}{\partial x^2} + G \tag{7.30}$$

We now assume that charge neutrality holds at every point, that is, that the electron and hole densities adjust so that at every point $\Delta n(x) \approx \Delta p(x)$. This is equivalent to assuming the electric field is uniform with position. For simplicity, we also assume the sample is in the dark, so G only consists of thermal generation/recombination. We then multiply Equations 7.30 by $(\mu_P p)$ and $(\mu_N n)$ respectively, add the two equations, and use the Einstein relation: $D_{N,P} = kT/q\ \mu_{N,P}$. The result is the desired *ambipolar diffusion equation,*

$$\frac{\partial \Delta p}{\partial t} = D_A \frac{\partial^2 \Delta p}{\partial x^2} - \frac{\Delta p}{\tau_A} - \left(\frac{n-p}{n/\mu_p + p/\mu_N}\right) E \frac{\partial \Delta p}{\partial x} \tag{7.31}$$

Here, D_A is the *ambipolar diffusion coefficient* given by

$$D_A = \frac{n+p}{n/D_P + p/D_N} \tag{7.32}$$

and τ_A is the *ambipolar lifetime* given by

$$\tau_A = -\Delta p/G \tag{7.33}$$

Equation 7.31 is written in terms of Δp, but since we have assumed $\Delta n = \Delta p$, it applies to Δn as well.

To apply Equation 7.31 to the "i" region of a pin diode under forward bias in the dark, we assume high-level injection exists so that the excess carrier densities are much larger than their equilibrium values, that is, $\Delta n \gg n_0$ and $\Delta p \gg p_0$. This allows us to set $n(x) \approx \Delta n(x)$ and $p(x) \approx \Delta p(x)$. Assuming charge neutrality, $\Delta n(x) \approx \Delta p(x)$, and we can write

$$\boxed{D_A = \left(\frac{\Delta n + \Delta p}{\Delta n/D_P + \Delta p/D_N}\right) = \frac{2D_N D_P}{D_N + D_P}} \tag{7.34}$$

The Shockley–Read–Hall thermal generation/recombination rate is given by

$$G = \frac{n_i^2 - pn}{\tau_P(n+n_1) + \tau_N(p+p_1)} \tag{7.35}$$

where

$$n_1 = n_i \exp\left[\left(E_T - E_i\right)/kT\right]$$
$$p_1 = n_i \exp\left[\left(E_i - E_T\right)/kT\right] \tag{7.36}$$

for R–G centers at energy E_T in the bandgap. Under high-level injection conditions, Equation 7.35 can be simplified to

$$G = \frac{-\Delta p^2}{\tau_P \Delta p + \tau_N \Delta p} = -\frac{\Delta p}{\tau_N + \tau_p} = -\frac{\Delta p}{\tau_A} \tag{7.37}$$

where the ambipolar lifetime defined by Equation 7.33 becomes

$$\boxed{\tau_A = \tau_N + \tau_P} \tag{7.38}$$

The ambipolar diffusion equation given by Equation 7.31 can now be written

$$\boxed{\frac{\partial \Delta p}{\partial t} = D_A \frac{\partial^2 \Delta p}{\partial x^2} - \frac{\Delta p}{\tau_A}} \tag{7.39}$$

Equation 7.39 has the same form as the familiar minority-carrier diffusion equations used under low-level injection conditions, Equation 7.23, with the $D_{N,P}$ coefficients replaced by D_A, the $\tau_{N,P}$ coefficients replaced by τ_A, and G_L set to zero. D_A and τ_A are computable constants given by Equations 7.34 and 7.38, respectively.

7.3.2 Carrier Densities in the "i" Region

Having derived the ambipolar diffusion equation in a form to describe the "i" region of the pin diode under high-level injection conditions, we now wish to solve Equation 7.39 subject to the appropriate boundary conditions. The situation is illustrated in Figure 7.11, where an x coordinate system is defined. Assuming steady-state conditions so that $\partial \Delta p/\partial t = 0$, the general solution to Equation 7.39 can be written

$$\Delta n(x) = \Delta p(x) = C_1 \sinh\left(x/L_A\right) + C_2 \cosh\left(x/L_A\right) \tag{7.40}$$

where L_A is the ambipolar diffusion length given by

$$L_A = \sqrt{D_A \tau_A} \tag{7.41}$$

and C_1 and C_2 are constants to be determined by the boundary conditions. Equation 7.40 and many of the equations to be developed below are written using hyperbolic functions. These functions are a convenient shorthand for the sum or difference of two exponential terms. Properties of hyperbolic functions and identities involving hyperbolic functions can be found in Appendix B. That Equation 7.40 is a solution to Equation 7.39 can be confirmed by direct substitution. We now assume *unity injection efficiency* at the boundaries $x = \pm d$. This allows us to set $J_N(-d) = J_P(+d) = 0$. With this insight, we can set $J_{TOTAL} = J_N(+d) = J_P(-d)$ and note that J_{TOTAL} is uniform throughout the "i" region in steady state.

To establish the boundary condition at $x = +d$, we note that

$$\begin{aligned} J_N(+d) = J_{TOTAL} &= q\mu_N n(d) E(d) + qD_N \left.\frac{\partial n}{\partial x}\right|_{x=d} \\ J_P(+d) = 0 &= q\mu_P p(d) E(d) - qD_P \left.\frac{\partial p}{\partial x}\right|_{x=d} \end{aligned} \tag{7.42}$$

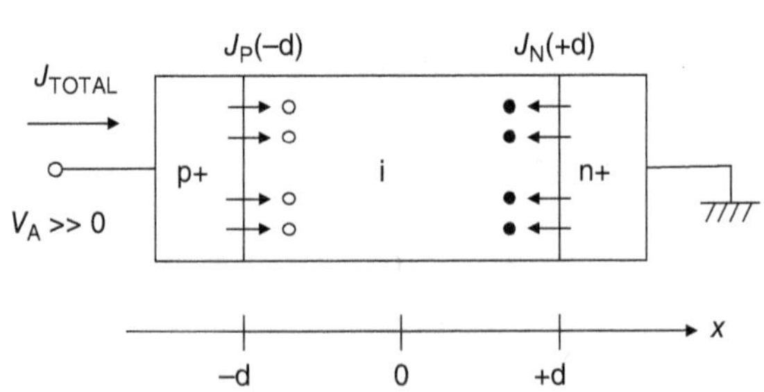

Figure 7.11 Structure of the pin diode with the x coordinate system and injected carriers depicted.

Making use of the fact that $n(x) \approx p(x)$ in the "i" region due to charge neutrality, we can solve the second equation for the electric field at $x = +d$:

$$E(d) = \frac{kT}{q} \frac{1}{n(d)} \frac{\partial n}{\partial x}\bigg|_{x=d} \tag{7.43}$$

Substituting Equation 7.43 into the first Equation 7.42 and solving for $\partial n/\partial x$ at $x = +d$ yields the desired boundary condition at $x = +d$,

$$\frac{\partial n}{\partial x}\bigg|_{x=d} = \frac{J_{\mathrm{TOTAL}}}{2qD_{\mathrm{N}}} \tag{7.44}$$

Using a similar approach at $x = -d$, we obtain the second boundary condition as

$$\frac{\partial n}{\partial x}\bigg|_{x=-d} = -\frac{J_{\mathrm{TOTAL}}}{2qD_{\mathrm{P}}} \tag{7.45}$$

Returning to the general solution, Equation 7.40, we apply the boundary conditions in Equations 7.44 and 7.45 to eliminate the constants C_1 and C_2, and after some algebra we find that

$$\Delta n(x) = \Delta p(x) = \frac{\tau_{\mathrm{A}} J_{\mathrm{TOTAL}}}{2qL_{\mathrm{A}}} \left[\frac{\cosh(x/L_{\mathrm{A}})}{\sinh(d/L_{\mathrm{A}})} - B \frac{\sinh(x/L_{\mathrm{A}})}{\cosh(d/L_{\mathrm{A}})} \right] \tag{7.46}$$

where

$$B = \frac{\mu_{\mathrm{N}} - \mu_{\mathrm{P}}}{\mu_{\mathrm{N}} + \mu_{\mathrm{P}}} \tag{7.47}$$

represents the fractional asymmetry in electron and hole mobilities. Figure 7.12 shows the carrier densities in the "i" region normalized to the pre-factor in Equation 7.46, for several ratios of $b = \mu_{\mathrm{N}}/\mu_{\mathrm{P}}$. The carrier distributions become asymmetrical when $\mu_{\mathrm{N}} \neq \mu_{\mathrm{P}}$ due to the second term in brackets in Equation 7.46.

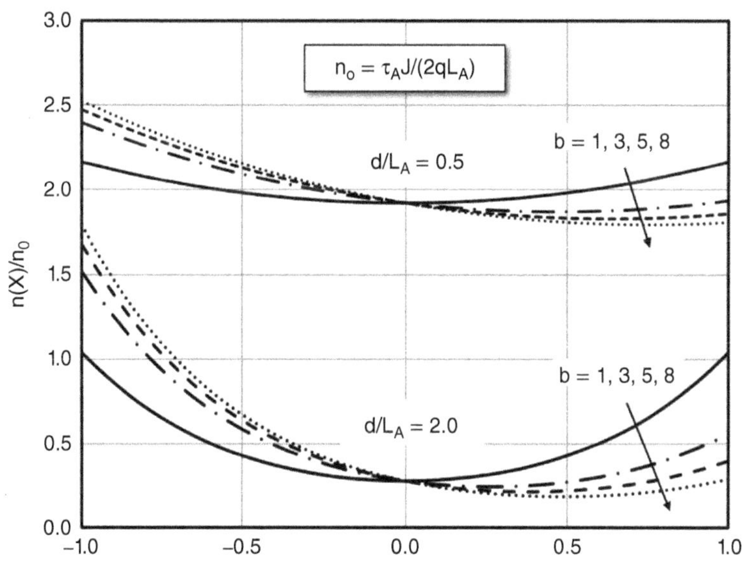

Figure 7.12 $\Delta n(x) \approx \Delta p(x)$ in the "i" region of a pin diode as a function of $b = \mu_{\mathrm{N}}/\mu_{\mathrm{P}}$ for $d/L_{\mathrm{A}} = 0.5$ and 2.0. In 4H-SiC, $b \approx 7.5$.

7.3.3 *Potential Drop across the "i" Region*

Having solved for the carrier densities in the "i" region as a function of current, we now wish to determine the total electrostatic potential drop $\Delta\psi_i$ across the "i" region. To do this, we first solve for the electric field $E(x)$ in the "i" region. Let's rewrite the current Equation 7.29 in the form

$$J_N(x) = q\mu_N\left(nE + \frac{kT}{q}\frac{\partial n}{\partial x}\right)$$
$$J_P(x) = q\mu_P\left(nE - \frac{kT}{q}\frac{\partial n}{\partial x}\right) \tag{7.48}$$

Adding the two equations to get the total current,

$$J_{TOTAL} = qn(\mu_N + \mu_P)E + kT(\mu_N - \mu_P)\frac{\partial n}{\partial x} \tag{7.49}$$

Solving Equation 7.49 for the electric field $E(x)$ yields

$$E(x) = \frac{J_{TOTAL}}{q(\mu_N + \mu_P)n} - \frac{kT}{q}B\frac{1}{n}\frac{\partial n}{\partial x} \tag{7.50}$$

The first term in Equation 7.50 is the ohmic voltage drop due to the resistivity of the "i" region with carrier densities $n(x) \approx p(x)$, and the second term represents the asymmetry due to unequal electron and hole mobilities. Note that since $n(x) \approx \Delta n(x)$ is proportional to current J_{TOTAL}, as shown by Equation 7.46, *the electric field E(x) is independent of current* so long as high-level injection conditions prevail. This is remarkable, because it suggests that *in high-level injection the voltage drop across the "i" region is independent of current.* The total electrostatic potential drop across the "i" region can be obtained by integrating Equation 7.50 with respect to distance from $x = -d$ to $x = +d$. Inserting Equation 7.46 into Equation 7.50 and integrating, we arrive at the desired result:

$$\Delta\psi_i = \frac{kT}{q}\left\{\frac{8b}{(b+1)^2}\frac{\sinh(d/L_A)}{\sqrt{1 - B^2\tanh^2(d/L_A)}}\tan^{-1}\left[\sqrt{1 - B^2\tanh^2(d/L_A)}\sinh(d/L_A)\right] + B\ln\left[\frac{1 + B\tanh^2(d/L_A)}{1 - B\tanh^2(d/L_A)}\right]\right\} \tag{7.51}$$

We use ψ for electrostatic potentials (or band banding) and V for voltages (electrochemical potentials, or equivalently, Fermi levels). Equation 7.51 is unwieldy, but it depends only upon device parameters and material constants, and is clearly independent of current. Figure 7.13 is a plot of Equation 7.51 for several values of the mobility ratio $b = \mu_N/\mu_P$. For lightly-doped 4H-SiC, $b \approx 7.5$.

Figure 7.14 is a plot of electric field and electrostatic potential versus position for $d/L_A = 2$, using Equation 7.50 and material parameters typical of lightly-doped 4H-SiC at room temperature. The asymmetry in $E(x)$ is due to the difference in electron and hole mobilities. Under high-level injection, the electric field in the "i" region is independent of current, and the magnitude is small ($\leq$26 V cm^{-1}). It is also *positive,* indicating the electric field sends electrons toward the p+ region and holes toward the n+ region. This is opposite to the field polarity under low-level injection. The total potential drop across the "i" layer is about 172 mV, as can be verified using Equation 7.51.

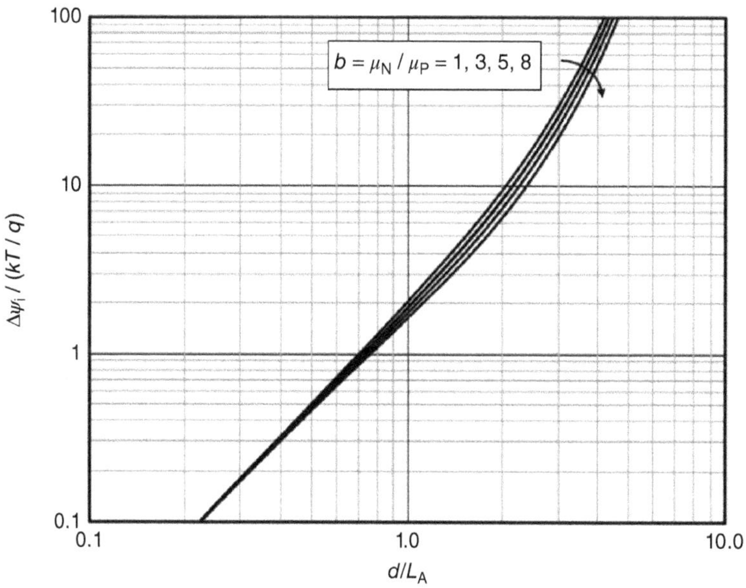

Figure 7.13 $\Delta\psi_i$ versus d/L_A for several values of b. In 4H-SiC, $b \approx 7.5$.

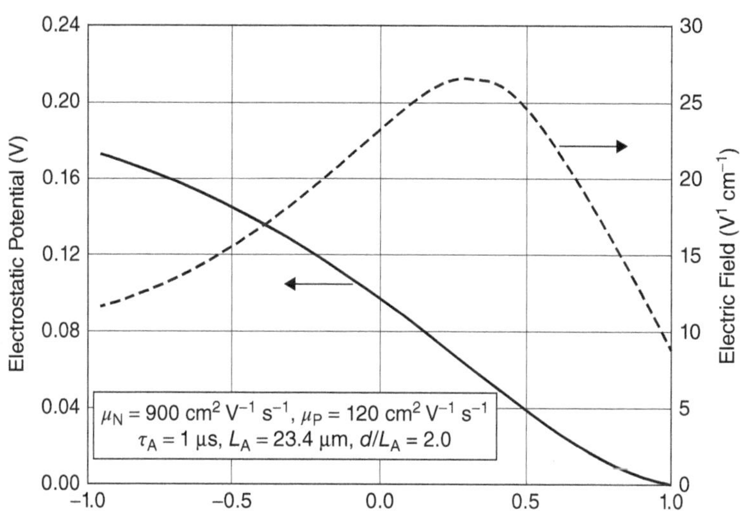

Figure 7.14 Electric field and electrostatic potential versus position for $d/L_A = 2$. The electric field is asymmetrical because of the difference in electron and hole mobilities.

7.3.4 *Current–Voltage Relationship*

To complete our analysis of the forward-bias current in the pin diode, we need to add to $\Delta\psi_i$ the potential drops across the p+/i and n+/i junctions. To aid in this analysis, we refer to the schematic band diagram of Figure 7.15 showing the electron and hole quasi-Fermi levels throughout the structure. The total

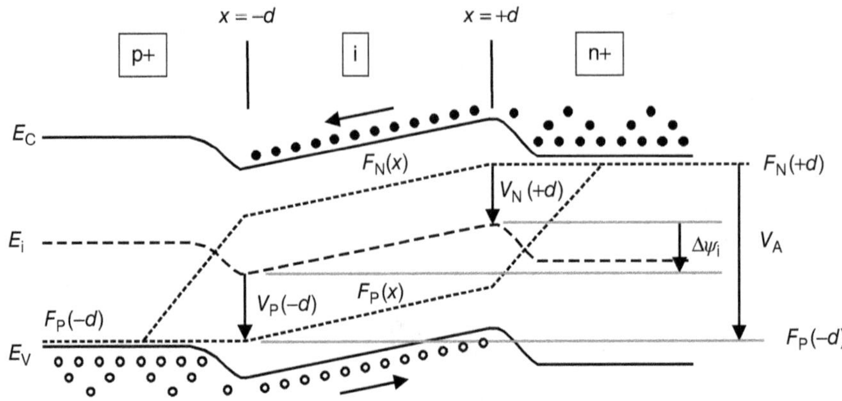

Figure 7.15 Band diagram of the pin diode under forward bias. The electric field in the "i" region is positive, taking electrons toward the p+ side and holes toward the n+ side. The terminal voltage V_A is the sum of $V_P(-d)$, $\Delta\psi_i$, and $V_N(+d)$.

voltage drop across the junctions is the offset between F_P on the left side and F_N on the right side. The electrostatic potential drop across the "i" region $\Delta\psi_i$ given by Equation 7.51 is the total band bending across the "i" region. We designate $V_P(-d)$ as the difference between F_P and E_i at $x = -d$, and $V_N(+d)$ as the difference between F_N and E_i at $x = +d$. These are known as the *chemical potentials* for holes and electrons, respectively. We can obtain the total voltage drop by adding $V_P(-d)$, $\Delta\psi_i$, and $V_N(+d)$. Working first on the p+/i junction, we note that

$$p(-d) = n_i \exp\left[qV_P(-d)/kT\right] \tag{7.52}$$

which can be rearranged to yield

$$V_P(-d) = \frac{kT}{q}\ln\left[\frac{n(-d)}{n_i}\right] \tag{7.53}$$

Similarly, at the n+/i junction we find that

$$V_N(+d) = \frac{kT}{q}\ln\left[\frac{n(+d)}{n_i}\right] \tag{7.54}$$

Clearly

$$V_P(-d) + V_N(+d) = \frac{kT}{q}\ln\left[\frac{n(-d)\,n(+d)}{n_i^2}\right] \tag{7.55}$$

where $n(-d)$ and $n(+d)$ can be obtained from Equation 7.46. When evaluated in equilibrium, Equation 7.55 represents the *built-in potential* of the junction. Equation 7.55 can be rearranged to yield

$$n(-d)n(+d) = n_i^2 \exp\left[q\left(V_A - \Delta\psi_i\right)/kT\right] \tag{7.56}$$

where V_A is the applied terminal voltage, and we have made use of the fact that $(V_A - \Delta\psi_i) = V_P(-d) + V_N(+d)$. Obtaining expressions for $n(-d)$ and $n(+d)$ from Equation 7.46 and inserting into Equation 7.56, we find that the left side of Equation 7.56 is proportional to $J_{TOTAL}{}^2$. Taking the square root and solving for J_{TOTAL} yields

$$\boxed{J_{TOTAL} = J_0 \exp(qV_A/2kT)} \tag{7.57}$$

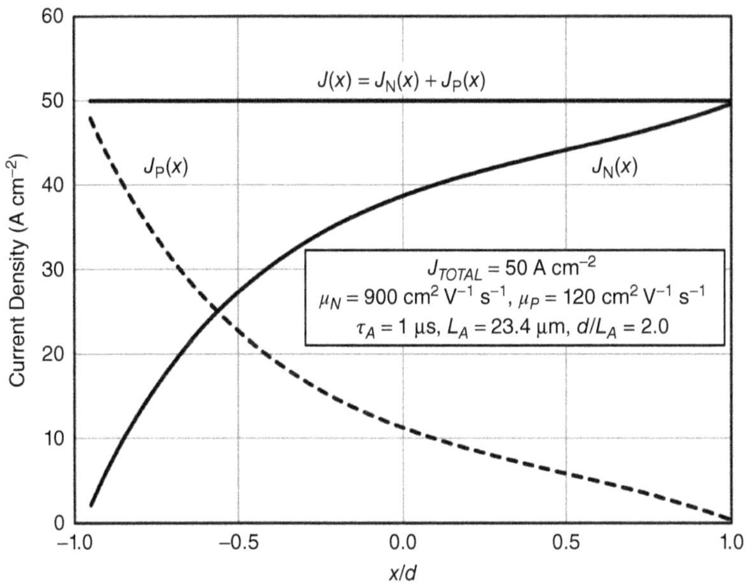

Figure 7.16 Electron, hole, and total currents versus position for $J_{TOTAL} = 50$ A cm^{-2} and $d/L_A = 2$. On the p+ end of the "i" region, all the current is carried by holes, while on the n+ end all the current is carried by electrons.

where J_0 is given by

$$J_0 = \left(2qn_i \frac{L_A}{\tau_A}\theta\right) \exp(-q\Delta\psi_i/2kT) \tag{7.58}$$

and θ is given by

$$\theta = \left\{\left[\coth\left(d/L_A\right) + B\tanh\left(d/L_A\right)\right]\left[\coth\left(d/L_A\right) - B\tanh\left(d/L_A\right)\right]\right\}^{-1/2} \tag{7.59}$$

Equation 7.59 evaluates to a constant that depends only on device and material parameters. Likewise, J_0 evaluates to a constant, using Equation 7.51 for $\Delta\psi_i$. The current–voltage relationship is then given by Equation 7.57 (to which must be added the voltage drop across the substrate). Note the ideality factor of 2 in the exponent of Equation 7.57. This is typical of conduction under high-level injection conditions.

Figure 7.16 shows electron current, hole current, and total current as a function of position for $d/L_A = 2.0$ at a total current density of 50 A cm^{-2}. The electron and hole currents are computed using Equation 7.48, with $n(x) \approx \Delta n(x) \approx p(x) \approx \Delta p(x)$ given by Equation 7.46. Note that the electron current goes to zero at the p+ end of the "i" region, $x = -d$, where the total current is carried by holes. Similarly, the hole current goes to zero at the n+ end of the "i" region, $x = +d$, where the total current is carried by electrons.

All our results depend on carrier lifetimes and mobilities in different regions of the device. In general, lifetimes increase and mobilities decrease with temperature. Mobility also decreases with doping concentration. Table 9.1 gives empirical equations for hole and electron mobilities and ambipolar lifetime as a function of doping and temperature in 4H-SiC.

7.4 Junction-Barrier Schottky (JBS) and Merged pin-Schottky (MPS) Diodes

The junction-barrier-controlled Schottky (JBS) diode and merged pin-Schottky (MPS) diode are structures that combine pin and Schottky diodes in a way that takes advantage of the best characteristics of both. At moderate forward currents the Schottky diode has a lower forward voltage drop than the pin diode, as shown in Figure 7.6. Since the SBD is a unipolar device, there is essentially no minority carrier charge storage, and the turn-off transient is very fast, as seen in Figure 7.7. These are desirable qualities, and make SBD the preferred device for blocking voltages below 2–3 kV or switching frequencies above 8–10 kHz, as shown in Figure 7.9. The main disadvantage of the SBD is the relatively large reverse leakage current caused by Schottky barrier lowering at high reverse biases, as seen in Figure 7.8. Even a small leakage current can lead to a large power dissipation in the off state, due to the high reverse voltage across the diode. This is not an issue in the pin diode, which has very low reverse leakage. With this in mind, the JBS diode is designed to behave like a Schottky diode under forward bias (to minimize on-state and switching losses) and like a pin diode under reverse bias (to minimize off-state losses). The MPS diode operates in a different mode under forward bias. We will discuss the JBS mode of operation first, then consider the MPS mode.

The structure of the JBS/MPS diode is illustrated in Figure 7.17. The metal layer on top forms ohmic contacts to the p+ regions and Schottky contacts to the n− regions, so the overall device consists of interdigitated Schottky and pin diodes connected in parallel. The p+ anode regions are spaced far enough apart that their depletion regions do not touch under zero or forward bias. This leaves a conductive path through the n− drift region between each Schottky contact and the n+ substrate. As forward bias is applied, the Schottky regions conduct first since, as shown in Figure 7.6, the current density of the SBD is orders of magnitude higher than the pin diode at the same forward voltage. The Schottky regions therefore effectively clamp the voltage drop across the pin regions, and the pin regions do not conduct. As a result, virtually all the forward current is due to electrons injected from the n− drift region through the Schottky contact into the metal. Since the p+ regions do not inject holes into the drift region, no minority carrier charge is stored and the turn-off transient is fast, minimizing switching loss. With no conductivity modulation, the series resistance of the drift region is determined by its thickness and doping, as given by Equation 7.6. This relatively high resistance leads to a voltage drop V_{DR} that dominates the total voltage drop at high currents, as illustrated by the SBD characteristics in Figure 7.6.

Before proceeding further, it is important to note that unlike the Schottky and pin diodes considered in the previous sections, the JBS/MPS diode has an interdigitated (or cellular) structure in which the

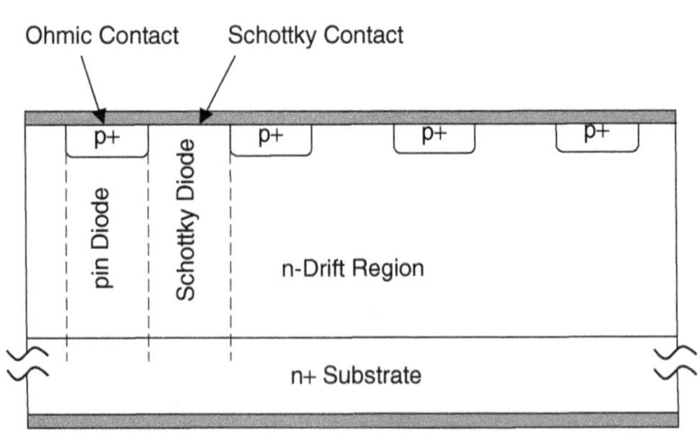

Figure 7.17 Structure of the JBS/MPS diode, consisting of interdigitated pin and Schottky diodes, electrically connected in parallel.

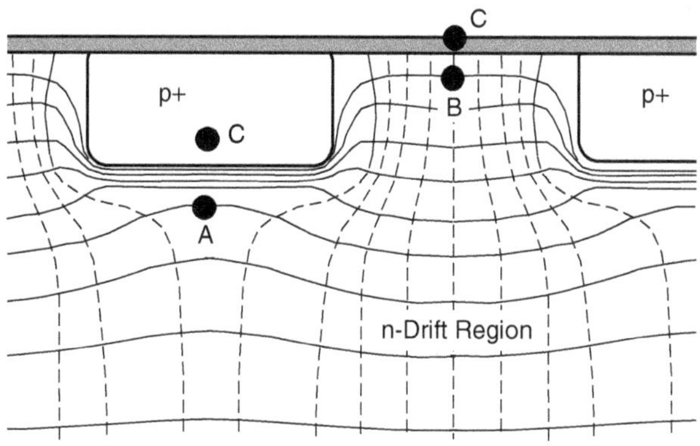

Figure 7.18 Current flow (dashed) and equipotential (solid) lines in the JBS/MPS diode under forward bias. As the current increases, a potential drop develops between points A and B, as illustrated, allowing the p+ diode to turn on. Note that the p+ anode and the Schottky metal, points C, are at the same potential.

lateral dimensions of the surface features are comparable to (or less than) the vertical thickness of the drift region. By comparison, Schottky and pin diodes are large-area devices, and within the interior of the diodes the current flow and electric field lines can be considered one-dimensional. Indeed, Figures 7.1 and 7.10 illustrate one-dimensional structures, and all our previous analyses were one-dimensional. The JBS/MPS diode is the first device we have considered where the one-dimensional assumption is not valid. This will also be true of the remaining power devices covered in later chapters, including the JFET (junction field-effect transistor), MOSFET (metal-oxide-semiconductor field-effect transistor), BJT (bipolar junction transistor), IGBT, and thyristor. Like the JBS/MPS diode, all these devices have interdigitated or cellular structures that require two-dimensional analysis, usually performed by computer simulation. As we study these devices, we will use one-dimensional approximations to obtain qualitative understanding, keeping in mind that a quantitative analysis must include the two-dimensional effects within the device.

With this in mind, let us now consider forward conduction in the JBS/MPS diode. Figure 7.18 illustrates the current flow lines and equipotential lines in the JBS/MPS diode under forward bias. As seen, the current spreads laterally under the p+ anodes and the flow is decidedly two-dimensional in this region. The true specific on-resistance is higher than Equation 7.6 because the surface area that conducts is less than the total area, but the area ratio cannot be used directly because of the current spreading under the p+ anodes. In practice, the on-resistance is best determined by computer analysis, but we can obtain a qualitative understanding by imagining (for the moment) that current flow in the pin and Schottky sections is one-dimensional in the vertical direction, and no carriers cross the dashed lines in Figure 7.17.

Proceeding under this assumption, for a given current density J we can use Equations 7.46, 7.51, 7.53, and 7.54 to calculate the voltage drop across the pin diode, assuming full conductivity modulation in the pin section and no hole flow into the Schottky section. Likewise, we can calculate the voltage drop across the Schottky contact using Equation 7.18, and across the (unmodulated) n− drift region of the SBD using Equation 7.21. These voltage drops are plotted as a function of forward current in Figure 7.19. In the pin diode, the potential drop $\Delta\psi_i$ is independent of current, and is negligible. The voltage drops V_P and V_N across the p+ and n+ junctions are almost equal, and have logarithmic slopes of q/kT. The total voltage V_{PIN} is the sum of $\Delta\psi_i$, V_P, and V_N, and has a logarithmic slope of $q/2kT$. In the SBD, the potential drop across the Schottky junction V_J has slope q/kT, and is smaller in magnitude than either V_P or V_N. The voltage drop across the un-modulated drift region V_{DR} is linear with current, and causes the total drop V_{SBD} (dashed line) to deviate from V_J for currents above about 100 A cm^{-2}.

If we think now about the real two-dimensional JBS/MPS structure in Figure 7.18, we note that the potential V_P across the p+ junction (points C to A) and the potential V_J across the Schottky junction

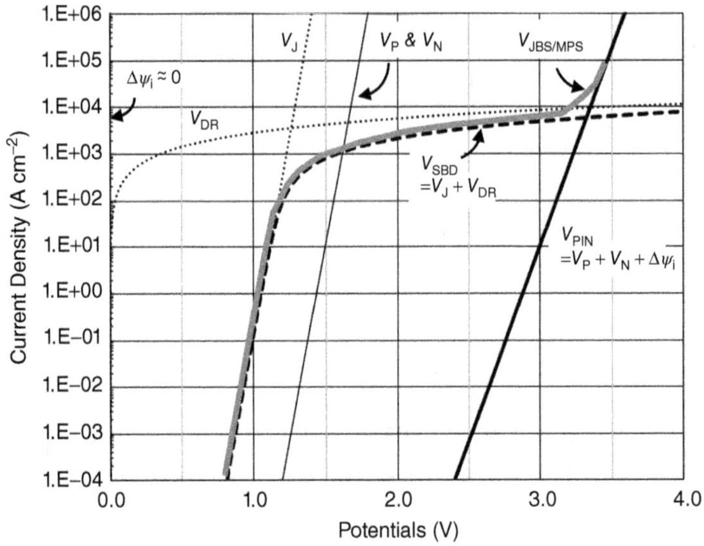

Figure 7.19 Voltage drops in unconnected pin and Schottky diodes as a function of forward current. Both diodes are designed for a plane-junction blocking voltage of 2.4 kV.

(points C to B) are the same, provided there is no potential drop in the n− drift region between points A and B. This will certainly be true when the current is very low. From Figure 7.19, assuming V_P and V_J are the same, the current crossing the p+ junction will be orders of magnitude lower than that crossing the Schottky junction. Thus there will be very little hole injection into the drift region, and negligible conductivity modulation. However, as the current increases, a lateral voltage drop will develop between points A and B. This is illustrated in Figure 7.18, where it is apparent by counting equipotential lines that the potential at point A is closer to ground (the substrate potential) than the potential at point B. Thus, the potential drop across the p+ junction (points C and A) is greater than across the Schottky junction (points C and B). This causes the p+ junction to begin injecting holes into the drift region, and these holes spread laterally throughout the drift region, modulating its conductivity. As current is increased, the p+ junction accounts for an increasing fraction of the total current, and the terminal current follows the heavy shaded line in Figure 7.19.

It is important to remember that the potentials shown in Figure 7.19 are calculated ignoring the two-dimensional nature of the device, and are not quantitatively correct. However, this description helps illustrate the processes occurring in the real device. Figure 7.19 is calculated for a device with a plane-junction blocking voltage of 2.4 kV, and for currents below about 1 kA cm^{-2} most of the current flows through the SBD section and negligible minority charge is injected into the drift region. At higher currents, the lateral voltage drops under the p+ regions are large enough that the p+ junction turns on, injecting holes into the drift region. This reduces $R_{ON,SP}$ and allows the current to follow the characteristics of a pin diode. The regime below the cross-over point is the JBS regime, where negligible minority charge storage takes place. The regime above the cross-over point is the MPS regime, where significant minority charge storage occurs. Operation in the MPS regime reduces the on-state loss at high current densities, but the stored charge increases the switching loss.

The transition between JBS and MPS regimes depends on the design blocking voltage, and can be understood as follows. When the drift region is unmodulated, $R_{ON,SP}$ increases as the square of the designed blocking voltage, as given by Equation 7.12. This happens because, to achieve a higher blocking voltage, the drift region must be thicker and more lightly doped, both of which increase the on-resistance.

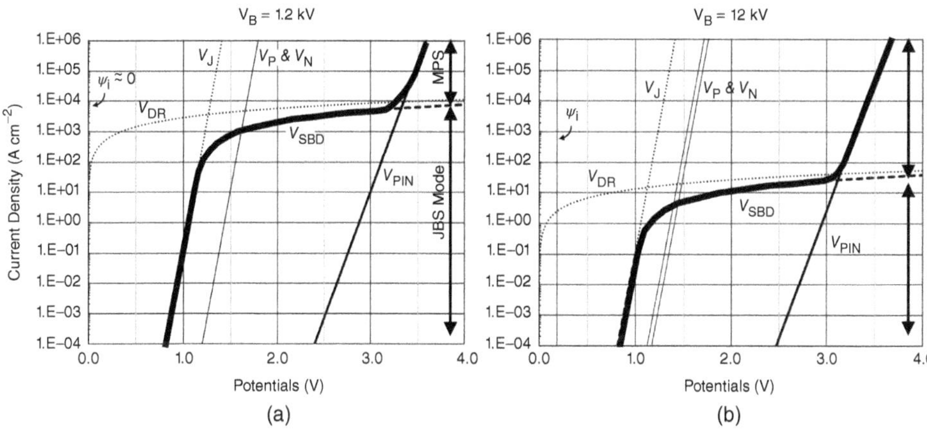

Figure 7.20 (a,b) Current–voltage characteristics of the JBS/MPS diode for two different blocking voltages: (a) 1.2 and (b) 12 kV.

For diodes with low blocking voltages, Equation 7.12 shows that $R_{ON,SP}$ is small, so a much higher current is needed before the p+ diode turns on. Therefore, the cross-over between JBS and MPS regimes occurs at high currents (say, above 1 kA cm^{-2}). On the other hand, when the blocking voltage is high, Equation 7.12 shows that $R_{ON,SP}$ is large, and the cross-over between JBS and MPS regimes occurs at lower currents (10–50 A cm^{-2}). For this reason, low-voltage JBS/MPS diodes typically operate in the JBS regime, whereas high-voltage JBS/MPS diodes operate in the MPS regime. This is illustrated in Figure 7.20, where the characteristics of a 1.2 and a 12 kV JBS/MPS diode are compared. At a current density of 200 A cm^{-2} for example, the 1.2 kV diode is in the JBS regime, whereas the 12 kV diode is in the MPS regime.

As reverse bias is applied to the JBS/MPS structure, the depletion regions of the p+ anodes quickly merge under the M–S contacts, then spread downward toward the n+ substrate. If the field lines were one-dimensional, the electric field would vary linearly with depth, as in Figure 7.1 or 7.2. In this case, the high surface field under the Schottky contact would lead to significant barrier lowering, as given by Equation 7.20. In the JBS/MPS structure, however, the electric field near the surface is two-dimensional, with many field lines terminating on the p+ anodes, as illustrated in Figure 7.21. This reduces the surface

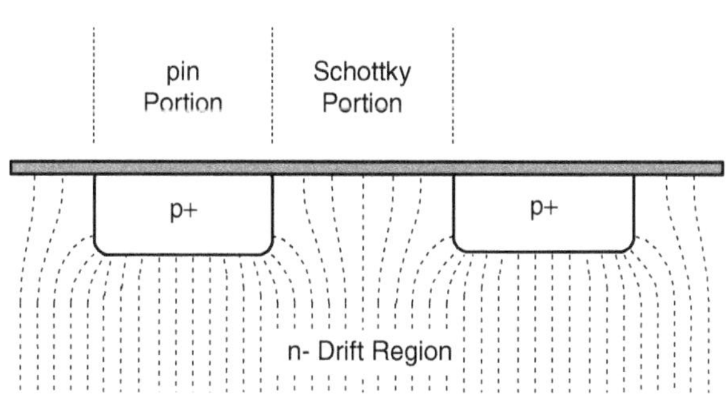

Figure 7.21 Illustration of field lines in the JBS diode under reverse bias. The width of the Schottky portion must be small enough that most field lines terminate on the surrounding p+ regions, and the pin regions should be small to minimize the total diode area.

field under the Schottky contacts, minimizing barrier lowering, and reducing the reverse leakage to levels approaching a pure pin diode. As a result, the JBS/MPS structure exhibits the desirable low reverse leakage current of a pin diode.

As with all diodes, proper edge terminations are critical to achieve a blocking voltage close to the theoretical plane-junction value. Techniques for edge terminations will be discussed in Section 10.1.

References

[1] Schoen, K.J., Woodall, J.M., Cooper, J.A. and Melloch, M.R. (1998) Design considerations and experimental characterization of high voltage SiC Schottky barrier rectifiers. *IEEE Trans. Electron. Devices*, **45** (7), 1595–1604.

[2] Toumi, S., Ferhat-Hanida, A., Boussouar, L. *et al.* (2009) Gaussian distribution of inhomogeneous barrier height in tungsten/4H-SiC (000-1) Schottky diodes. *Microelectron. Eng.*, **86** (3), 303–309.

[3] Itoh, A., Kimoto, T. and Matsunami, H. (1995) High performance of high-voltage 4H-SiC Schottky barrier diodes. *IEEE Electron. Device Lett.*, **16** (6), 280–282.

[4] Morisette, D.T. (2001) Development of robust power Schottky barrier diodes in silicon carbide. PhD thesis. Purdue University.

[5] Itoh, A., Kimoto, T., and Matsunami, H. (1995) Efficient power Schottky rectifiers of 4H-SiC. International Symposium on Power Semiconductor Devices & ICs, Yokohama, Japan.

[6] Morisette, D.T., Woodall, J.M., Cooper, J.A. *et al.* (2001) Static and dynamic characterization of large area high-current-density SiC Schottky diodes. *IEEE Trans. Electron. Devices*, **48** (2), 349–352.

[7] Morisette, D.T. and Cooper, J.A. (2002) Theoretical performance comparison of SiC pin and Schottky diodes. *IEEE Trans. Electron. Devices*, **49** (9), 1657–1664.

8

Unipolar Power Switching Devices

8.1 Junction Field-Effect Transistors (JFETs)

We turn now to the first of the three-terminal power devices, the junction field-effect transistor, or JFET. As discussed in Section 7.1, three-terminal power switching devices emulate ideal switches, that is, they attempt to carry high current with minimal voltage drop in the on state and to block high voltages with minimal leakage current in the off state.

The basic operation of the JFET can be understood by reference to the schematic cross-section in Figure 8.1. The prototype n-channel JFET consists of two p+ gate regions on either side of an n-type channel region. Each end of the channel is connected to an ohmic contact, designated the source and the drain. A depletion region exists at each gate-channel p+/n junction, and the width of this depletion region increases as the square root of the gate-to-channel voltage. Assuming a p+/n one-sided step junction, the depletion width can be written

$$x_{\mathrm{D}} = \sqrt{2\varepsilon_{\mathrm{S}}(\psi_{\mathrm{BI}} - V_{\mathrm{J}})/(qN_{\mathrm{D}})} \tag{8.1}$$

where V_{J} is the voltage difference (or Fermi level splitting) between the gate and the channel, N_{D} is the channel doping, and ψ_{BI} is the built-in potential (or band bending) of the gate-to-channel junction, given by

$$\psi_{\mathrm{BI}} = \frac{kT}{q}\ln\left(\frac{N_{\mathrm{A}}^{-}N_{\mathrm{D}}^{+}}{n_i^2}\right) \tag{8.2}$$

where N_{A}^{-} and N_{D}^{+} are the ionized doping concentrations of the p+ gate and the n-type channel, respectively (see Appendix A for a discussion of incomplete ionization in 4H-SiC. Note that N_{D} in Equation 8.1 is the *total* dopant concentration in the channel.) If a non-zero drain voltage $V_{\mathrm{D}} > 0$ is applied and the channel is not pinched-off by the depletion regions from the gate, an electron current will flow from the grounded source to the positive drain, and a voltage drop will exist along the channel from source to drain. Let us designate the voltage at point y in the channel as $V(y)$. In this case, V_{J} and x_{D} in Equation 8.1 become functions of y, and V_{J} can be written

$$V_{\mathrm{J}} = V_{\mathrm{G}} - V(y) \tag{8.3}$$

Clearly, at the grounded source $V(y) = 0$ and $V_{\mathrm{J}}(0) = V_{\mathrm{G}}$, while at the drain $V(L) = V_{\mathrm{D}}$ and $V_{\mathrm{J}}(L) = V_{\mathrm{G}} - V_{\mathrm{D}}$.

The JFET current–voltage relation is derived in many textbooks, but most treatments assume the top and bottom gates in Figure 8.1 are at the same potential. This is often not the case in practice, and we wish

Fundamentals of Silicon Carbide Technology: Growth, Characterization, Devices, and Applications, First Edition.
Tsunenobu Kimoto and James A. Cooper.

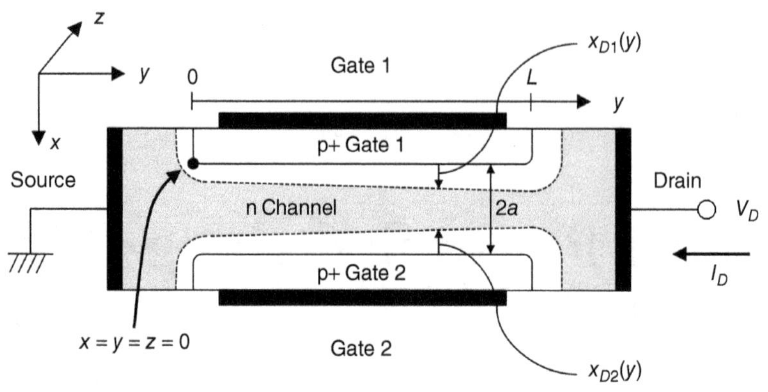

Figure 8.1 Schematic cross-section of a basic JFET. The width in the direction into the paper is W. For generality, we assume the two p+ gates may be at different potentials.

to allow the top and bottom gates to be at different potentials, say V_{G1} and V_{G2}. There are two practical configurations of interest for power JFETs: (i) the *double-gated* configuration where both gates are tied together, that is, $V_{G1} = V_{G2} = V_G$, and (ii) the *single-gated* configuration where one gate is grounded and the other gate modulates the channel, for example, $V_{G1} = V_G$ and $V_{G2} = 0$. We will consider both configurations in this chapter.

8.1.1 Pinch-Off Voltage

Before deriving the current–voltage relations, we should define the *pinch-off voltage* V_P as the gate voltage that just pinches off the channel at the source. V_P can be regarded as the *turn-on voltage* or *threshold voltage* of the JFET, since for gate voltages more negative than V_P there can be no current in the channel. V_P can be calculated using Equations 8.1 and 8.3 by setting $(x_{D1} + x_{D2}) = 2a$ with $V(y) = 0$ in Equation 8.3. For the double-gated configuration, we find

$$\boxed{V_P = \psi_{BI} - \frac{qN_D a^2}{2\varepsilon_s}} \tag{8.4}$$

and for the single-gated configuration,

$$V_P \begin{cases} = \dfrac{2qN_D a^2}{\varepsilon_S}\left(\sqrt{\dfrac{2\varepsilon_S \psi_{BI}}{qN_D a^2}} - 1\right), & N_D a^2 \geq \dfrac{\varepsilon_S \psi_{BI}}{2q} \\ > \psi_{BI}, & N_D a^2 < \dfrac{\varepsilon_S \psi_{BI}}{2q} \end{cases} \tag{8.5}$$

Caution is required in evaluating Equation 8.5. This is because whenever $N_D a^2 < \varepsilon_S \psi_{BI}/(2q)$ the actual V_P in the single-gated configuration exceeds ψ_{BI}, meaning it would require a positive gate bias in excess of ψ_{BI} to prevent the depletion regions from touching at the source. Such a gate bias would lead to extremely high gate current and is not permitted.

8.1.2 Current–Voltage Relationship

Turning now to the JFET current–voltage relationship, we can write the y-directed electron current density J_N at point y in the channel as

$$J_N(y) = q\mu_N N_D^+ E_y(y) = -q\mu_N N_D^+ \frac{dV}{dy} \tag{8.6}$$

where $E_y(y)$ is the y-directed electric field in the channel at point y. The total drain current I_D can be obtained by integrating $J_N(y)$ over the cross-sectional area of the channel at any point y,

$$I_D = -\iint J_N(y)\,dx\,dz = -W\int_{x_{D1}}^{2a-x_{D2}} J_N(y)\,dx = W\int_{x_{D1}}^{2a-x_{D2}} q\mu_N N_D^+ \frac{dV}{dy}\,dx \tag{8.7}$$

where x_{D1} and x_{D2} are given by Equations 8.1–8.3 with V_{G1} or V_{G2} used in Equation 8.3, as appropriate. A minus sign precedes the integral because I_D is defined as positive *into* the drain. Noting that the integrand in Equation 8.7 is independent of x, we may write

$$I_D = q\mu_N W N_D^+ \frac{dV}{dy}\left(2a - x_{D1} - x_{D2}\right) \tag{8.8}$$

Since I_D is uniform along the channel, independent of y, we can integrate Equation 8.8 with respect to y from source to drain,

$$\int_0^L I_D\,dy = I_D L = q\mu_N W N_D^+ \int_0^L \left(2a - x_{D1} - x_{D2}\right)\frac{dV}{dy}\,dy$$

$$I_D L = q\mu_N W N_D^+ \int_0^{V_D} \left(2a - x_{D1} - x_{D2}\right) dV \tag{8.9}$$

$$\boxed{I_D = q\mu_N N_D^+ \frac{W}{L}\int_0^{V_D} \left(2a - x_{D1} - x_{D2}\right) dV}$$

where in the second and third lines we have converted the integral over y to an integral over channel voltage V. Substituting for $x_{D1}(V)$ and $x_{D2}(V)$ using Equations 8.1–8.3 and performing the indicated integration, we obtain the desired current expression:

$$\boxed{\begin{aligned} I_D = q\mu_N\left(2aN_D^+\right)\frac{W}{L}\Bigg\{V_D - \frac{2}{3}\frac{\sqrt{2\varepsilon_S/qN_D}}{(2a)}\Big[\left(\psi_{BI} - V_{G2} + V_D\right)^{2/3} \\ - \left(\psi_{BI} - V_{G2}\right)^{3/2} + \left(\psi_{BI} - V_{G1} + V_D\right)^{3/2} - \left(\psi_{BI} - V_{G1}\right)^{3/2}\Big]\Bigg\}\end{aligned}} \tag{8.10}$$

We can simplify Equation 8.10 for the double-gated configuration where both gates are tied together, and the single-gated configuration where one gate is grounded and the other gate modulates the channel. For double-gated operation we can set $V_{G1} = V_{G2} = V_G$ in Equation 8.10 and obtain

$$\boxed{I_D = q\mu_N\left(2aN_D^+\right)\frac{W}{L}\left\{V_D - \frac{4}{3}\frac{\sqrt{2\varepsilon_S/qN_D}}{(2a)}\left[\left(\psi_{BI} - V_G + V_D\right)^{3/2} - \left(\psi_{BI} - V_G\right)^{3/2}\right]\right\}} \tag{8.11}$$

For single-gated operation we set $V_{G1} = V_G$ and $V_{G2} = 0$ in Equation 8.10 and obtain

$$\begin{aligned} I_D = q\mu_N \left(2aN_D^+\right) \frac{W}{L} \Bigg\{ V_D - \frac{2}{3} \frac{\sqrt{2\varepsilon_S/qN_D}}{(2a)} \Big[\left(\psi_{BI} + V_D\right)^{3/2} \\ - \left(\psi_{BI}\right)^{3/2} + \left(\psi_{BI} - V_G + V_D\right)^{3/2} - \left(\psi_{BI} - V_G\right)^{3/2} \Big] \Bigg\} \end{aligned} \tag{8.12}$$

Equation 8.11 is the form of the $I_D - V_D$ equation that appears in most device textbooks.

Let us now examine the $I_D - V_D$ relations in more detail. As V_D increases, the second terms in curly brackets in Equations 8.11 and 8.12 can no longer be neglected, and the current increases sub-linearly with V_D. The second terms in Equations 8.11 and 8.12 represent the widening of the gate-channel depletion regions x_{D1} and x_{D2} at the drain end of the channel as V_D increases, as illustrated in Figure 8.1. Eventually depletion regions x_{D1} and x_{D2} just touch at the drain end $y = L$. The drain voltage at which this occurs is called the *saturation* drain voltage $V_{D,SAT}$, because for $V_D > V_{D,SAT}$ the current saturates at the value it had at $V_D = V_{D,SAT}$. Further increases in V_D simply move the punch-through point slightly toward the source. Since the channel voltage at the punch-through point where $(x_{D1} + x_{D2}) = 2a$ is always $V_{D,SAT}$, the voltage drop between the source and the punch-through point remains constant at $V_{D,SAT}$. If the punch-through point does not move appreciably toward the source as V_D is increased, the channel current remains constant at the saturation current $I_D(V_{D,SAT}) = I_{D,SAT}$.

8.1.3 Saturation Drain Voltage

The saturation drain voltage $V_{D,SAT}$ can be derived for a given gate voltage by setting $x_{D1}(L) + x_{D2}(L) = 2a$. Using Equations 8.1 and 8.3 for the depletion widths, we can write

$$\sqrt{2\varepsilon_S(\psi_{BI} - V_{G1} + V_{D,SAT})/(qN_D)} + \sqrt{2\varepsilon_S(\psi_{BI} - V_{G2} + V_{D,SAT})/(qN_D)} = 2a \tag{8.13}$$

For the double-gated case where $V_{G1} = V_{G2} = V_G$, we can solve Equation 8.13 for $V_{D,SAT}$ as

$$V_{D,SAT} = \left(\frac{qN_D a^2}{2\varepsilon_S} + V_G - \psi_{BI}\right) \qquad (V_{G1} = V_{G2} = V_G) \tag{8.14}$$

For the single-gated case where $V_{G2} = 0$ and $V_{G1} = V_G$, we have

$$V_{D,SAT} = \frac{qN_D a^2}{2\varepsilon_S}\left(\frac{\varepsilon_S V_G}{2qN_D a^2} + 1\right) - \psi_{BI} \qquad (V_{G2} = 0, V_{G1} = V_G) \tag{8.15}$$

The saturation voltage for the double-gated case can also be written as $V_{D,SAT} = (V_G - V_P)$, as can be verified by substituting Equation 8.4 into Equation 8.14. For drain voltages $V_D > V_{D,SAT}$ the drain current saturates at the value $I_{D,SAT}$. $I_{D,SAT}$ can be calculated for the double-gated case by substituting Equation 8.14 for V_D into Equation 8.11, and for the single-gated case by substituting Equation 8.15 for V_D into Equation 8.12. Derivation of the resulting analytical equations for $I_{D,SAT}$ will be left to the reader.

The $I_D - V_D$ characteristics for the double-gated and single-gated configurations are illustrated in Figure 8.2. We note immediately that the currents are higher in the double-gated configuration for the same channel parameters N_D and a. A careful examination reveals that the saturation drain voltages $V_{D,SAT}$ are also higher in the double-gated configuration. This can be seen more clearly in Figure 8.3, where we plot $V_{D,SAT}$ and $I_{D,SAT}$ versus $(V_G - V_P)$. For the double-gated configuration, $V_{D,SAT} = (V_G - V_P)$, as pointed out above. For the single-gated configuration, $V_{D,SAT}$ is smaller and the corresponding $I_{D,SAT}$is smaller as well. This is because a smaller saturation voltage across the same channel will produce a smaller saturation current, as seen in the figure. All these differences arise

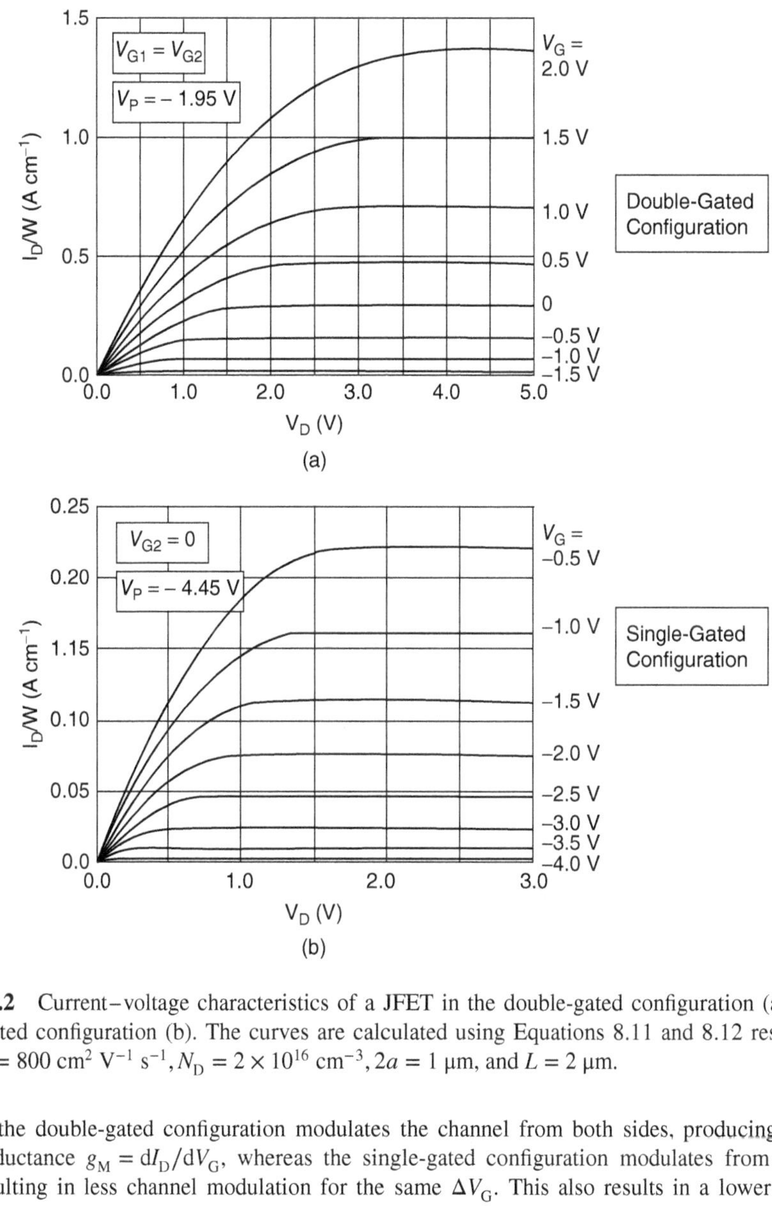

Figure 8.2 Current–voltage characteristics of a JFET in the double-gated configuration (a) and the single-gated configuration (b). The curves are calculated using Equations 8.11 and 8.12 respectively, with $\mu_N = 800$ cm^2 V^{-1} s^{-1}, $N_D = 2 \times 10^{16}$ cm^{-3}, $2a = 1$ μm, and $L = 2$ μm.

because the double-gated configuration modulates the channel from both sides, producing a higher transconductance $g_M = dI_D/dV_G$, whereas the single-gated configuration modulates from only one side, resulting in less channel modulation for the same ΔV_G. This also results in a lower pinch-off (or threshold) voltage for the double-gated configuration compared to the single-gated configuration. Pinch-off voltages for the two configurations are indicated on the individual plots in Figure 8.2.

8.1.4 Specific On-Resistance

An important parameter for power JFETs is the differential on-resistance, defined as the reciprocal slope of the $I_D - V_D$ relation at the origin,

$$\frac{1}{R_{ON}} = \left.\frac{dI_D}{dV_D}\right|_{V_D=0} \tag{8.16}$$

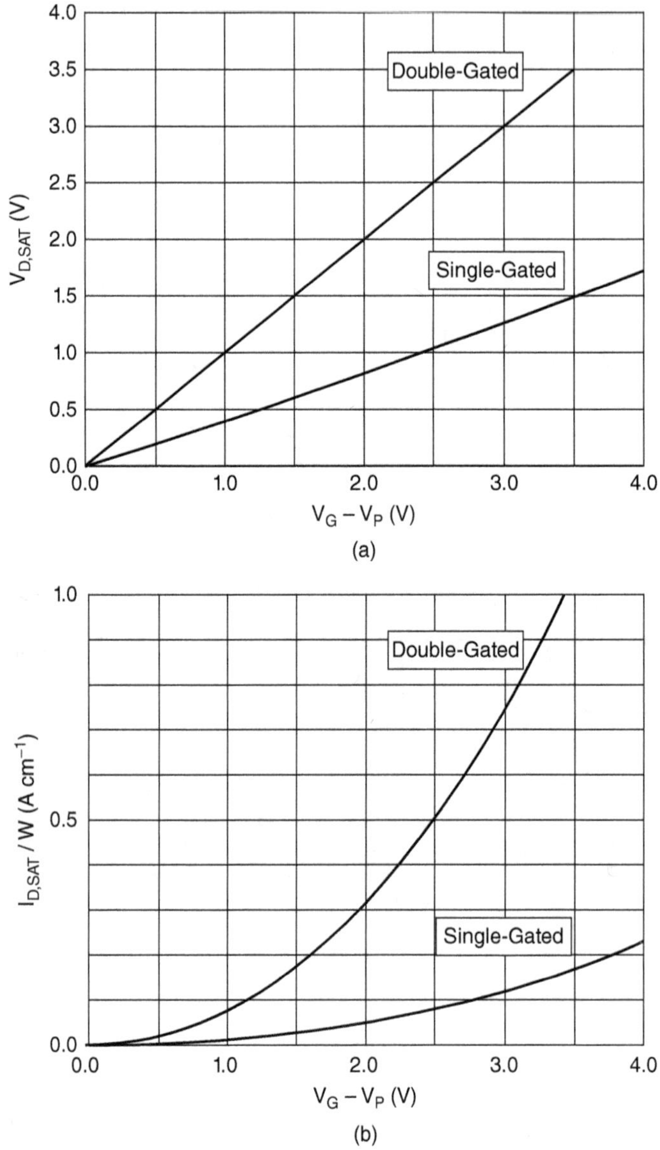

Figure 8.3 Saturation drain voltage (a) and saturation drain current (b) as a function of (V_G – V_P) for the double- and single-gated configurations, calculated using the parameters of Figure 8.2.

It is more useful to work in terms of the *specific on-resistance,* or resistance-area product given by $R_{ON,SP} = R_{ON} \cdot W \cdot S$, where R_{ON} is the resistance of one cell of the device, W is the width in the z direction (perpendicular to the page in Figure 8.1), and S is the width of the cell in the y direction (to be illustrated later). With this in mind, the specific on-resistance can be written

$$R_{ON,SP} = \frac{W \cdot S}{\left.\dfrac{dI_D}{dV_D}\right|_{V_D=0}} \tag{8.17}$$

We now wish to evaluate $R_{ON,SP}$ for the two cases. For the double-gated case, we take the derivative of Equation 8.11, evaluate at $V_D = 0$, and insert into the denominator of Equation 8.17 to obtain

$$R_{ON,SP} = \frac{L \cdot S}{q\mu_N(2aN_D^+)}\left(1 - \sqrt{\frac{2\varepsilon_S(\psi_{BI} - V_G)}{qN_D a^2}}\right)^{-1} \quad (V_{G1} = V_{G2} = V_G) \tag{8.18}$$

For the single-gated case, we take the derivative of Equation 8.12, evaluate at $V_D = 0$, and insert in the denominator of Equation 8.17, yielding

$$R_{ON,SP} = \frac{L \cdot S}{q\mu_N(2aN_D^+)}\left[1 - \frac{1}{2}\sqrt{\frac{2\varepsilon_S}{qN_D a^2}}\left(\sqrt{\psi_{BI}} + \sqrt{V_{BI} - V_G}\right)\right]^{-1} \quad (V_{G1} = V_G, V_2 = 0) \tag{8.19}$$

The $(2aN_D^+)$ term in the denominator is the total ionized dopant concentration per unit area in the channel, a quantity that in bipolar transistors is called the *Gummel number*. $R_{ON,SP}$ is directly proportional to channel length L and the cell pitch S, and inversely proportional to the Gummel number. It is obviously desirable to reduce the channel length L and to reduce the surface area of the cell by minimizing the cell pitch S.

Figure 8.4 plots $R_{ON,SP}$ as a function of $(\psi_{BI} - V_G)$ for double-gated and single-gated devices having the same channel doping and thickness. The viable operating range for gate voltage is between V_P where the drain current is zero and ψ_{BI} where the gate current becomes extremely large. We note that the double-gated device goes from pinch-off to full conduction over a smaller gate voltage swing, since the channel is gated from both sides. As V_G becomes more negative (moving to the right on the figure)

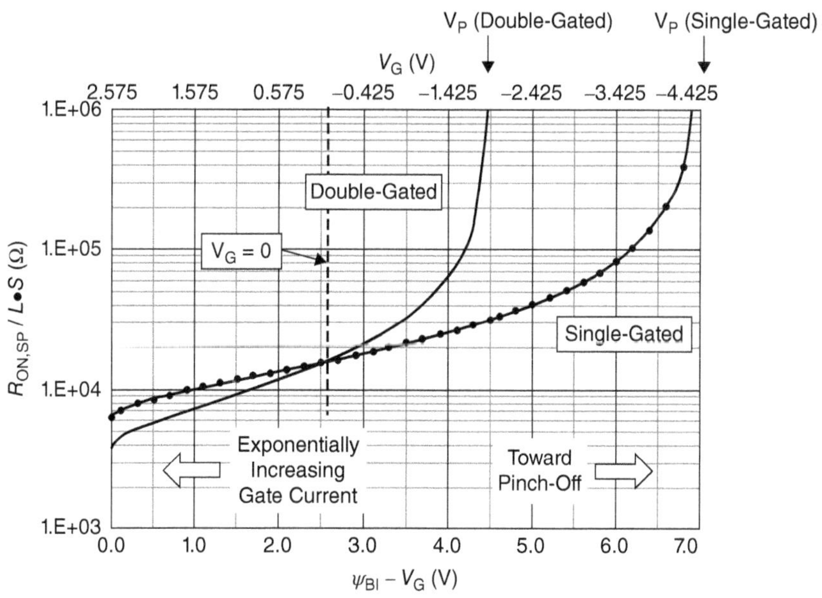

Figure 8.4 Specific on-resistance as a function of $(V_{BI} - V_G)$ using the parameters of Figure 8.2. The viable gate voltage range is between V_P and ψ_{BI}.

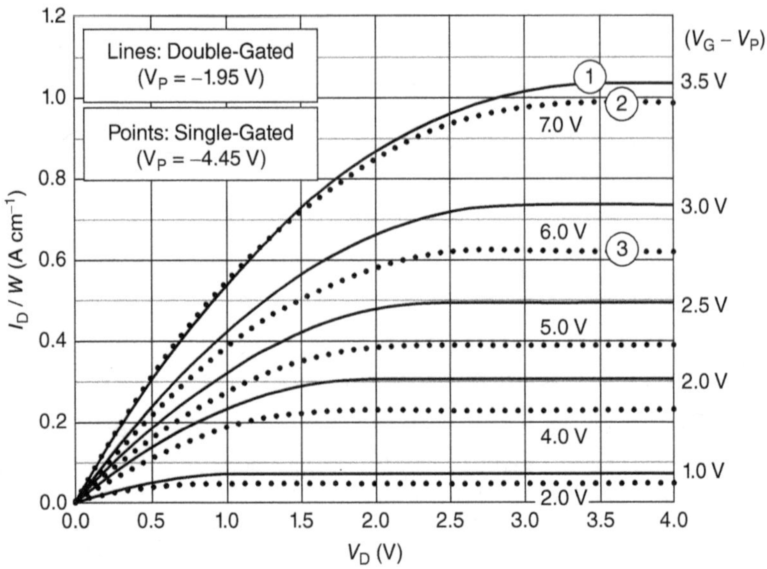

Figure 8.5 Current–voltage characteristics for JFETs with the parameters of Figure 8.2, operated in the double-gated and single-gated configurations. The gate voltage steps $\Delta(V_G - V_P)$ for the single-gated device are twice as large as for the double-gated device.

the on-resistance increases dramatically, since the channel is approaching pinch-off. This can also be seen from the slopes of the $I_D - V_D$ curves at the origin in Figure 8.2. On the other hand, as V_G becomes less negative (moving to the left in the figure), the channel opens up and $R_{ON,SP}$ decreases. For positive gate voltages corresponding to the region to the left of the $V_G = 0$ line, the double-gated device exhibits lower on-resistance than the single-gated device, but as V_G approaches ψ_{BI} the gate current increases exponentially, and this establishes an upper limit on V_G for either configuration.

In spite of its higher on-resistance, lower saturation current, and lower transconductance, the single-gated JFET is often used in practice because of its simpler fabrication. The reduced transconductance may be partially compensated by increasing the gate voltage steps. This is illustrated in Figure 8.5, where the voltage steps for the single-gated configuration are twice as large as for the double-gated configuration. However, we cannot continue this compensation indefinitely because we cannot allow V_G to approach ψ_{BI}. To compare gate currents for the two configurations, it is instructive to consider actual gate voltages rather than $(V_G - V_P)$. The gate voltages for the two upper curves (1 and 2) in Figure 8.5 are 1.55 V for the double-gated configuration (curve 1) and 2.55 V for the single-gated configuration (curve 2). Since ψ_{BI} of the gate-channel junction is 2.57 V in this example, the single-gated device at $V_G = 2.55$ V (curve 2) will draw substantial gate current. The second single-gated curve (curve 3) corresponds to an actual gate voltage of 1.55 V, and it is this curve that should be compared to curve 1 for the double-gated device. We can see that the double-gated device has a lower on-resistance and higher saturation current when each configuration is operating near the maximum practical gate voltage (curves 1 and 3).

8.1.5 Enhancement-Mode and Depletion-Mode Operation

Another important consideration in the operation of a power JFET is whether the device is depletion mode (normally on at $V_G = 0$) or enhancement mode (normally off at $V_G = 0$). Depletion-mode devices

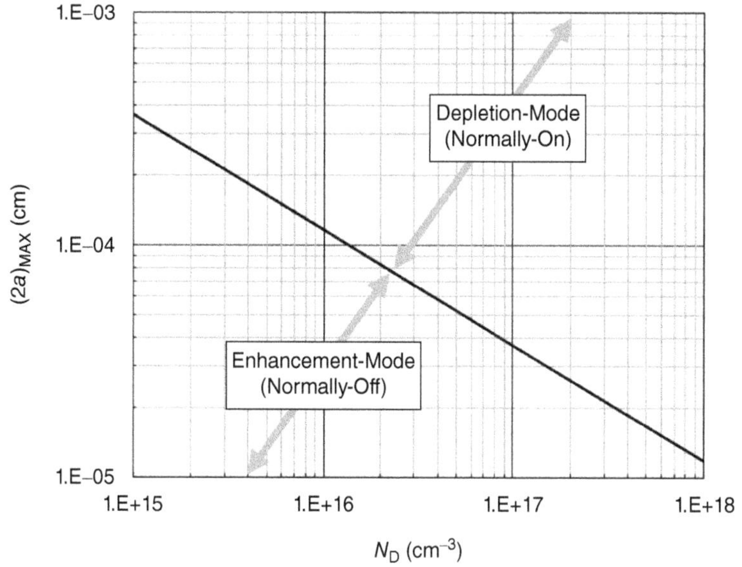

Figure 8.6 Channel thickness ($2a$) as a function of doping to achieve a V_P of zero, valid for both single-gated and double-gated configurations.

are problematic in power circuits, since a failure that removes gate voltage from the device would result in uncontrolled drain current. However, a depletion-mode JFET can be safely used if combined with an enhancement-mode metal-oxide-semiconductor field effect transistor (MOSFET) in a cascode arrangement, as will be discussed below. For a JFET to be enhancement mode (normally-off), the depletion regions of the two gates must touch at zero gate bias. This is equivalent to requiring that $V_P \geq 0$, so that when $V_G = 0$ the gate voltage is more negative than the pinch-off voltage and the channel is pinched off. For the double-gated device, setting $V_P \geq 0$ in Equation 8.4 leads to the requirement that

$$N_D a^2 \leq 2\varepsilon_S \psi_{BI}/q \tag{8.20}$$

Applying the same procedure for the single-gated device using Equation 8.5 leads to the same result, namely Equation 8.20. Taking account of the dependence of ψ_{BI} on N_D^+ given by Equation 8.2, we plot the maximum channel thickness $(2a)_{MAX}$ as a function of channel doping in Figure 8.6. For channel thicknesses above the line, the JFET is depletion mode (normally on), while for thicknesses below the line the JFET is enhancement mode (normally off).

Figure 8.7 shows the pinch-off voltage given by Equations 8.4 and 8.5 as a function of $N_D a^2$. Combinations of N_D and $2a$ that give positive V_P produce enhancement-mode devices. The borderline between enhancement- and depletion-mode operation ($V_P = 0$) is given by Equation 8.20 for both double-gated and single-gated configurations.

Enhancement-mode operation involves trade-offs. First, as seen in Figure 8.6, an enhancement-mode device has a thinner channel at the same doping, so the specific on-resistance will be higher than for a depletion-mode device at the same gate voltage, as can be seen from the prefactors in Equations 8.18 and 8.19. Second, the operating gate voltage range is from V_P to just below ψ_{BI}, and since V_P is positive in an enhancement-mode device, the working gate voltage range becomes smaller and control of V_P becomes more critical.

For the reasons given above, power circuits often utilize a depletion-mode SiC JFET and achieve normally-off behavior by connecting the JFET in cascode with a low-voltage enhancement-mode

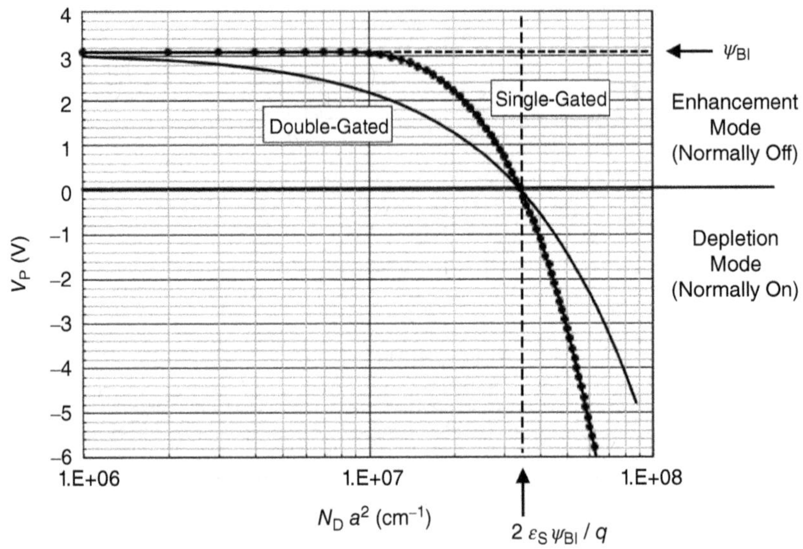

Figure 8.7 Pinch-off voltage as a function of $N_D\ a^2$, obtained using Equations 8.4 and 8.5.

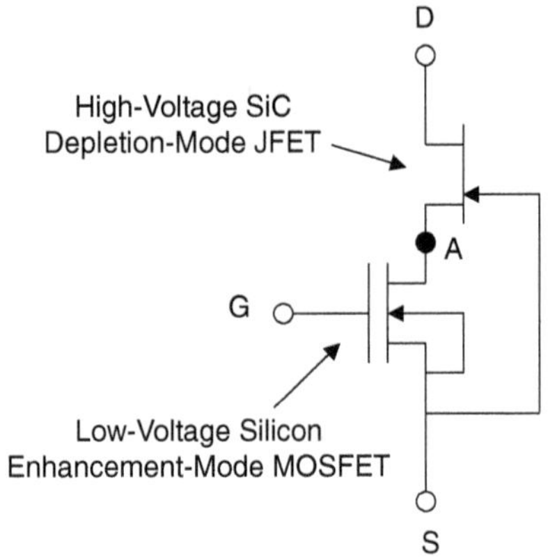

Figure 8.8 A cascode circuit incorporating a depletion-mode SiC JFET and a low-voltage enhancement-mode silicon power MOSFET.

silicon MOSFET, as shown in Figure 8.8. In this arrangement, the on-state current is controlled by the low-voltage silicon MOSFET, while the blocking voltage in the off state is supported by the high-voltage SiC JFET. The circuit behavior can be understood as follows: In operation, terminal S is grounded and terminal D is connected through a load to a positive voltage. When terminal G is above the MOSFET threshold, the MOSFET turns on, and node A is pulled toward ground by the low resistance of the silicon MOSFET. The depletion-mode JFET has a negative pinch-off voltage, and will be turned on whenever its gate-source voltage is more positive than this negative V_P. As node A (which was originally positive) approaches ground potential, the JFET gate-source voltage (which was originally negative) also approaches ground and the JFET turns on, allowing a large current flow. When terminal G is then taken to ground, the MOSFET turns off, and node A begins rising to increasingly positive voltages. This makes the gate-source voltage of the JFET more negative, and the JFET quickly turns off. The high voltage at terminal D is now held off by the gate-to-drain junction of the JFET, which is designed to have a high breakdown voltage, as will be discussed below. Thus, the cascode arrangement successfully achieves the normally-off behavior of a low-voltage silicon MOSFET with the high blocking voltage of a SiC depletion-mode JFET.

8.1.6 Power JFET Implementations

We have discussed the operation of the JFET structure of Figure 8.1 in detail, and we turn now to the implementation of the JFET as a power switch. To provide a high blocking voltage, the basic JFET is fabricated on a thick, lightly-doped n-type drift region located between the channel and the substrate, which acts as the drain terminal. Figure 8.9 shows three such implementations. The illustrated half-cells are reflected about their centerlines, and the cell pitch S is the width of the half-cells shown. In the first two implementations, the channels are horizontal and gate 2 is grounded to the source, forming a single-gated JFET. In the third implementation, the channel is vertical and is gated from both sides, forming a double-gated JFET.

In all implementations, the drift region is designed to withstand the desired blocking voltage while introducing the minimum possible specific on-resistance. Design of drift regions to achieve a desired blocking voltage was discussed in Section 7.1.1, and the blocking voltage is shown as a function of drift region doping and thickness in Figure 7.3. Since the JFET is a unipolar device, $R_{ON,SP}$ of the drift region is given by Equation 7.6. We will discuss how to maximize the drift region figure of merit, $V_B^2/R_{ON,SP}$, in Section 10.2.

It is clear from Figure 8.9 that the JFET current also flows through the drift region and the substrate, adding additional voltage drops in series with the JFET. The full on-state characteristics of the power JFET can be found by using Equation 8.11 or 8.12 to calculate $J = I_D/W$ as a function of the internal

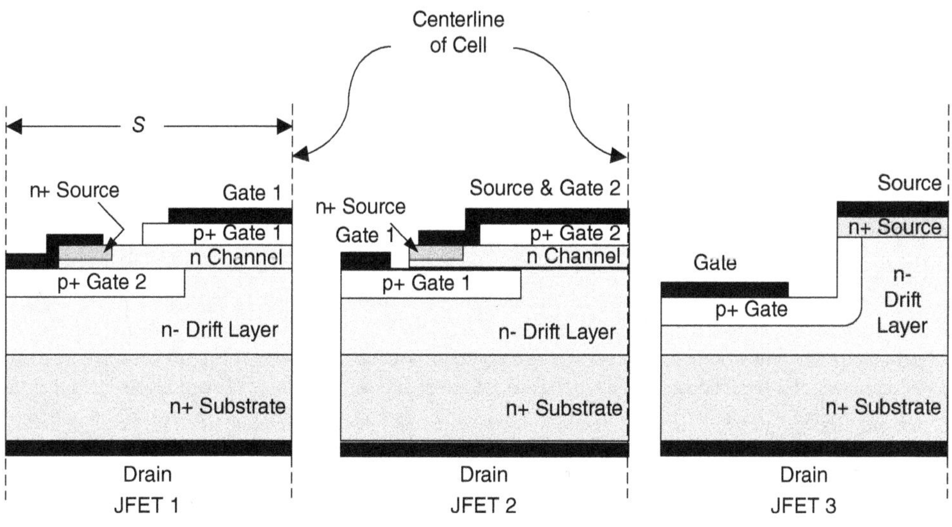

Figure 8.9 Three illustrative implementations of SiC power JFETs. In all cases, the current is controlled by a JFET on the surface of a thick, lightly-doped n-type epilayer that serves as the drift layer.

voltage V_D that develops across the JFET channel, then writing the terminal drain voltage $V_{D,TERM}$ as $V_D + J\ (R_{ON,DR} + R_{SUB})$, where $R_{ON,DR}$ is the specific resistance of the drift region given by Equation 7.6 and R_{SUB} is the specific resistance of the substrate. The on-resistance of the full JFET can be calculated by simply adding Equation 8.18 or 8.19, Equation 7.6, and R_{SUB}, noting that the $N_D{}^+$ in Equation 7.6 is that of the drift region, while the N_D in Equation 8.18 or 8.19 is the higher doping of the JFET channel. It is this overall on-resistance that must be minimized, as will be discussed in Section 10.2.

8.2 Metal-Oxide-Semiconductor Field-Effect Transistors (MOSFETs)

As discussed in Chapter 6, SiC can be thermally oxidized in the same manner as silicon, except that the chemical reaction produces both SiO_2 and CO. The CO escapes as a gas, while the SiO_2 forms a native passivation layer on the SiC. SiC is the only wide bandgap semiconductor whose native oxide is SiO_2, and this makes it possible to form the entire range of metal-oxide-semiconductor (MOS)-based devices in SiC. In this section we consider the simplest MOS-based power device, the power MOSFET. In Section 9.2 we will discuss another MOS-based power device, the insulated-gate bipolar transistor (IGBT).

8.2.1 *Review of MOS Electrostatics*

To begin our discussion of SiC MOSFETs, we will first review the concepts of MOS electrostatics. Since MOS electrostatics are covered in many semiconductor textbooks, we will simply present the outline here and refer the reader to the literature for detailed derivations. We initially assume an *ideal* MOS structure with no charges in the SiO_2, no interface states or fixed charges at the oxide/semiconductor interface, and zero metal-semiconductor work function difference ($\Phi_{MS} = \Phi_M - \Phi_S = 0$). Figure 8.10 shows energy band and block charge diagrams for an ideal MOS capacitor on p-type SiC at three bias points: flat band, depletion, and inversion. We define an x-coordinate system with the origin at the oxide/semiconductor interface and positive x into the semiconductor. The oxide thickness is t_{OX} and the gate is assumed to be metal, with work function Φ_M. The semiconductor substrate at $x = \infty$ is taken as our ground reference.

For future analyses, we now define two important parameters of the MOS structure. The *Fermi potential* ψ_F is the potential difference between midgap and the Fermi level far from the surface, as indicated in Figure 8.10. ψ_F is determined by doping and temperature, and is given by

$$\psi_F = \frac{kT}{q} \ln\left(\frac{N_A^-}{n_i}\right) \tag{8.21}$$

where N_A^- is the ionized dopant concentration. For acceptors in 4H-SiC, the ionized dopant concentration N_A^- at room temperature will be less than the total dopant concentration, as discussed in Appendix A. The *surface potential* ψ_S is the total band bending from the substrate to the surface at $x = 0$. ψ_S depends on the gate voltage, oxide thickness, and doping, and the exact relationship will be developed shortly.

We now wish to consider the three biasing conditions presented in Figure 8.10. Flat band is the special biasing condition where the bands in the semiconductor are flat ($\psi_S = 0$), and the flat-band voltage V_{FB} is defined as the gate voltage that produces this condition. In an ideal MOS structure such as illustrated in Figure 8.10, flat-band occurs at $V_G = 0$, since the metal and semiconductor Fermi levels line up at flat band. This is not always the case, and we will subsequently consider situations where V_{FB} is non-zero. The block-charge diagram for an ideal structure at flat band is particularly simple: no charges exist in the semiconductor, at the interface, or on the gate.

If the gate voltage is taken positive as in Figure 8.10b, holes are repelled from the surface of the semiconductor, forming a depletion region of width x_D. From elementary electrostatics, the depletion region width is given by

$$x_D = \sqrt{2\varepsilon_S \psi_S / (qN_A)} \tag{8.22}$$

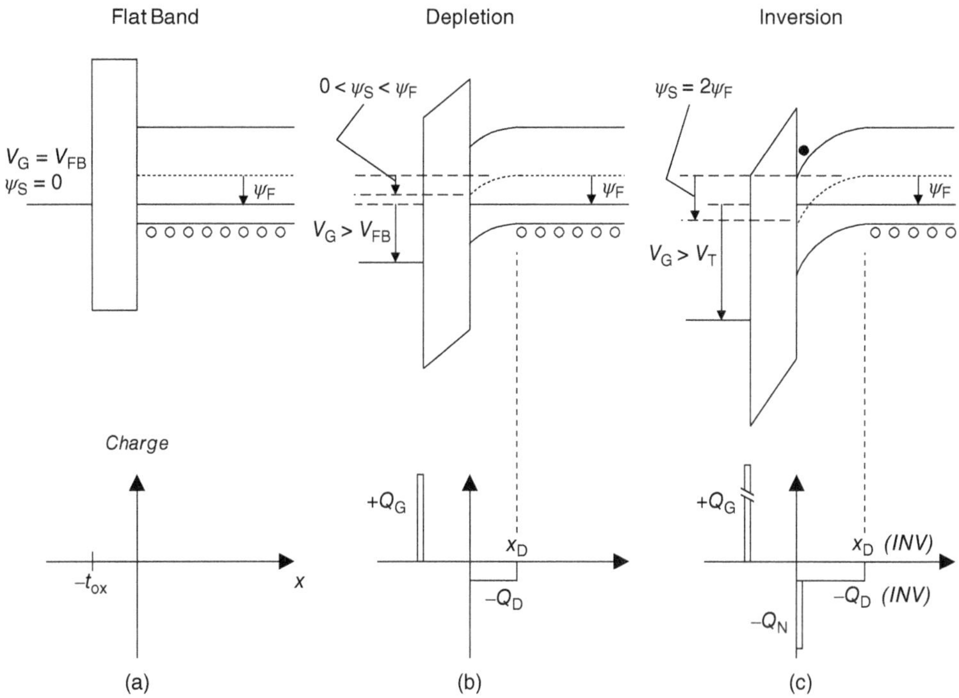

Figure 8.10 Band diagrams of a p-type MOS capacitor biased (a) at flat band, (b) in depletion, and (c) in inversion. In the delta-depletion approximation, the inversion charge Q_N is a sheet charge (delta function) at $x = 0$, and the surface potential ψ_S is defined on the *semiconductor* side of Q_N.

where ψ_S is the surface potential, or the band bending from the substrate to the surface at $x = 0$, as indicated in the figure. Since no holes are present in the depletion region, the depletion region has a net charge due to the negatively-charged acceptor dopant atoms, and the total charge in the depletion region can be written

$$Q_D = -qN_Ax_D = -\sqrt{2q\varepsilon_S N_A \psi_S} \tag{8.23}$$

The negative charge in the depletion region is balanced by an equal and opposite positive sheet charge on the surface of the gate, as shown.

Further increases in gate voltage from this point produce a wider depletion region and more band bending, until eventually the surface potential $\psi_S = 2\psi_F$. At this point the density of electrons per unit volume at the surface exactly equals the density of holes per unit volume in the bulk, and the surface is said to be *inverted* to n-type. The gate voltage at which this occurs is known as the threshold voltage V_T. If V_G is increased further, we enter the biasing regime known as inversion, as shown in Figure 8.10c.

As we develop the equations describing MOS electrostatics, we will utilize the "delta-depletion" approximation. In this approximation we represent the charge density in the depletion region as uniform with respect to x and equal to $-q\,N_A$ to a distance x_D from the surface, then zero beyond. We represent the charge in the inversion layer as an infinitely thin sheet, or delta function, at $x = 0$. We also assume that *any additional charges that develop in the semiconductor for $V_G > V_T$ lie entirely in the inversion layer*. This is equivalent to saying that every additional positive charge on the gate for $V_G > V_T$ induces one additional negative charge in the electron inversion layer, and the additional field lines from these positive gate charges all terminate on additional negative charges in the inversion layer. Under this assumption, none of these additional field lines penetrate the semiconductor, and the electric field in the semiconductor remains constant at the value it had when $V_G = V_T$. Since the electric field in the semiconductor does not increase, there is no increase in band bending, and the surface potential in inversion is the same as its value at threshold, namely $\psi_S(\text{inv}) = 2\psi_F$, where the surface potential is defined just on the *semiconductor* side of the inversion sheet charge, if present. It then follows from Equation 8.22 that the depletion width in inversion also remains fixed at

$$x_D(\text{inv}) = \sqrt{2\varepsilon_S \psi_S(\text{inv})/(qN_A)} \tag{8.24}$$

As illustrated in Figure 8.10c, when V_G is increased beyond V_T the Fermi level of the gate continues to move down in the band diagram, and the oxide electric field, given by the slope of the bands in the oxide, continues to increase. However, since the new charges on the gate are completely screened by the new charges in the inversion layer, the bands in the semiconductor remain unchanged.

We now return to the question we postponed earlier, namely how do we calculate the surface potential for a given gate voltage? We can write the total gate voltage, that is, the offset between the Fermi levels in the gate and semiconductor, as follows

$$\begin{aligned}(V_G - V_{FB}) &= \psi_S + \Delta\psi_{OX} = \psi_S + E_{OX} t_{OX} \\ (V_G - V_{FB}) &= \psi_S + \frac{\varepsilon_S}{\varepsilon_{OX}} E_S(0) t_{OX}\end{aligned} \tag{8.25}$$

In writing Equation 8.25, we are simply expressing $(V_G - V_{FB})$ as the sum of the potential drop across the semiconductor ψ_S plus the potential drop across the oxide $\Delta\psi_{OX}$. The electric field in the oxide E_{OX} is, by Gauss' law, equal to the surface electric field in the semiconductor $E_S(0)$ multiplied by the ratio of dielectric constants across the interface. Again using Gauss' law, $E_S(0)$ is simply the integral of the charge in the semiconductor divided by ε_S. For biases in depletion ($V_{FB} \leq V_G \leq V_T$), the only charge in the semiconductor is the depletion charge Q_D, so we can write

$$E_S(0) = -Q_D/\varepsilon_S = \sqrt{2qN_A\psi_S/\varepsilon_S} \tag{8.26}$$

where we have made use of Equation 8.23 for Q_D. The minus sign is inserted because the electric field of the negative Q_D forces holes in the $+x$ direction, and hence is a positive field. Inserting Equation 8.26 into Equation 8.25 yields the desired relationship between surface potential and gate voltage,

$$\boxed{(V_G - V_{FB}) = \psi_S + \sqrt{2V_0\psi_S}} \tag{8.27}$$

where the new constant V_0 is given by

$$V_0 = \frac{q\varepsilon_S N_A}{C_{OX}^2} \tag{8.28}$$

Here C_{OX} is the capacitance per unit area of the oxide, $C_{OX} = \varepsilon_{OX}/t_{OX}$. The term V_0 has no physical significance, and simply collects several parameters that often appear together in our equations. V_0 is determined by the doping and oxide thickness, and has the units of volts. Equation 8.27 can be put in a more useful form by solving for ψ_S in terms of V_G, yielding

$$\boxed{\psi_S = (V_G - V_{FB}) + V_0 - \sqrt{V_0^2 + 2V_0(V_G - V_{FB})}} \tag{8.29}$$

Under the delta-depletion approximation, we said that the surface potential in inversion remains fixed at its value at the threshold of inversion, namely ψ_S (inv). We can use this fact to obtain an equation for

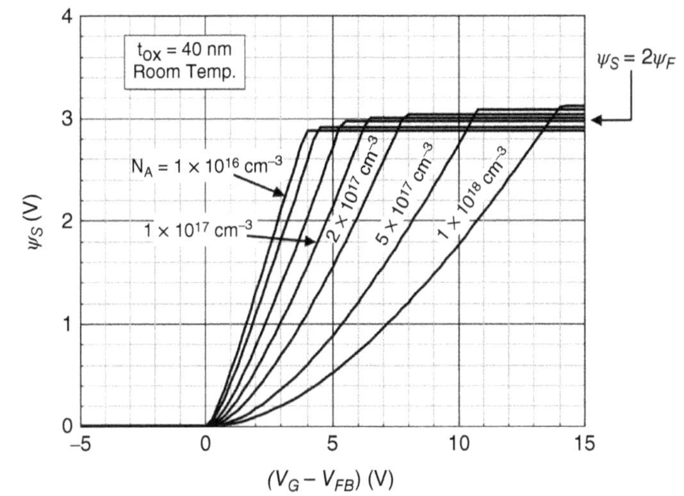

Figure 8.11 Surface potential–gate voltage relationship at room temperature for a 4H-SiC sample with an oxide thickness of 40 nm and several values of doping, calculated using Equation 8.29.

the threshold voltage V_T, the gate voltage at the onset of inversion. Setting $V_G = V_T$ and $\psi_S = \psi_S\,(\text{inv})$ in Equation 8.27, we can write

$$V_T = V_{FB} + \psi_S\,(\text{inv}) + \sqrt{2V_0\psi_S\,(\text{inv})} \tag{8.30}$$

where in the MOS capacitor, $\psi_S\,(\text{inv}) = 2\psi_F$.

Figure 8.11 plots the surface potential–gate voltage relation of Equation 8.29 for an oxide thickness of 40 nm and several values of doping. In this figure, we have set $\psi_S = 0$ for $V_G \leq V_{FB}$ and $\psi_S = 2\psi_F$ for $V_G \geq V_T$, in accordance with the assumptions in our delta-depletion model. We could have derived the exact $\psi_S - V_G$ relationship without invoking the delta-depletion approximation. The exact curves would follow the delta-depletion curves very closely for $V_{FB} < V_G < V_T$. For $V_G > V_T$ the exact curves would rise slightly above $\psi_S = 2\psi_F$, while for $V_G < V_{FB}$ the exact curves would lie slightly below $\psi_S = 0$. These differences are not essential for our analysis of the MOSFET, and will be ignored in this presentation.

8.2.2 *MOS Electrostatics with Split Quasi-Fermi Levels*

At this point we have developed all the equations needed for the MOS capacitor, and they will prove useful in analyzing the MOSFET. But before considering the MOSFET, we need to modify the above equations for the situation where the electron and hole quasi-Fermi levels in the semiconductor are unequal. This will allow us to analyze conditions in the MOSFET when a non-zero drain voltage is applied. Figure 8.12 shows the band diagram of an MOS capacitor biased in inversion when the electron quasi-Fermi level F_N lies below the hole quasi-Fermi level F_P by an amount qV_R. For the moment, we will not concern ourselves with *how* the quasi-Fermi levels become split, but this will become apparent when we discuss the MOSFET.

In examining Figure 8.12, we recall the definition of the onset of inversion as the bias condition where the electron density per unit volume at the surface is equal to the hole density per unit volume in the bulk.

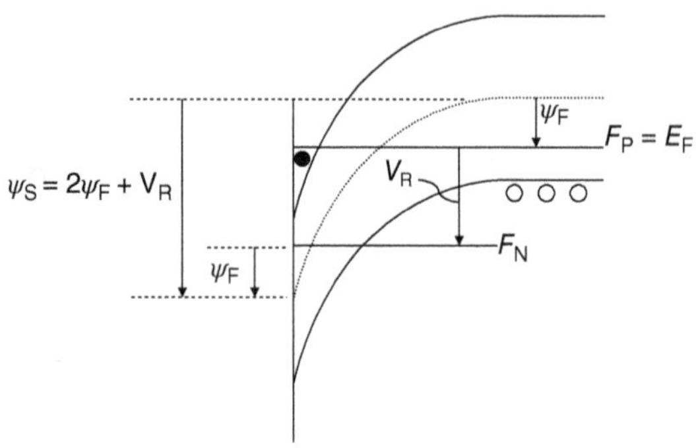

Figure 8.12 Band diagram of a p-type MOS capacitor in inversion with the electron and hole quasi-Fermi levels split by an amount qV_R.

The electron density at the surface depends on the electron quasi-Fermi level F_N as

$$n_S = n_i \exp\left[\frac{F_N - E_i(0)}{kT}\right] \tag{8.31}$$

while the hole density in the bulk is given by

$$P_B = n_i \exp\left[\frac{E_i(\infty) - F_P}{kT}\right] \tag{8.32}$$

Setting $n_S = p_B$ at the threshold of inversion, we can write

$$F_N - E_i(0) = E_i(\infty) - F_P = q\psi_F \tag{8.33}$$

Equation 8.33 tells us that at the onset of inversion, the electron quasi-Fermi level is as far *above* midgap at the surface as the hole quasi-Fermi level is *below* midgap in the bulk, where the latter quantity is $q\psi_F$. By reference to Figure 8.12, we see that the surface potential in this condition is

$$\boxed{\psi_S(\text{inv}) = 2\psi_F + V_R} \tag{8.34}$$

This is the surface potential in inversion for the general case where the quasi-Fermi levels are split by qV_R. Of course, if $V_R = 0$, Equation 8.34 reverts to our earlier definition: $\psi_S(\text{inv}) = 2\psi_F$.

8.2.3 *MOSFET Current–Voltage Relationship*

We are now ready to develop the current–voltage equations of the MOSFET. Figure 8.13 shows the structure of a basic MOSFET, along with relevant dimensions. The gate can be made of any high-conductivity material, such as doped polysilicon, but we will assume a metal gate in our discussions. The p-type body is grounded, and serves as our voltage reference. The source is usually also grounded, but we will derive our equations for the general case where the source may be at a voltage $V_S \geq 0$. The channel length L is measured from the edge of the source to the edge of the drain, and the oxide thickness is t_{OX}.

To begin, we apply a gate voltage V_G that is above threshold so that an inversion layer exists at the oxide/semiconductor interface. We will invoke the *gradual-channel approximation,* which assumes that the potential at any point y in the channel can be evaluated using one-dimensional electrostatics based on a vertical slice at point y. We assume no x-directed currents, and we neglect velocity saturation so that the electron drift velocity at point y is given by

$$\upsilon_y = \mu_N^* E_y(y) \tag{8.35}$$

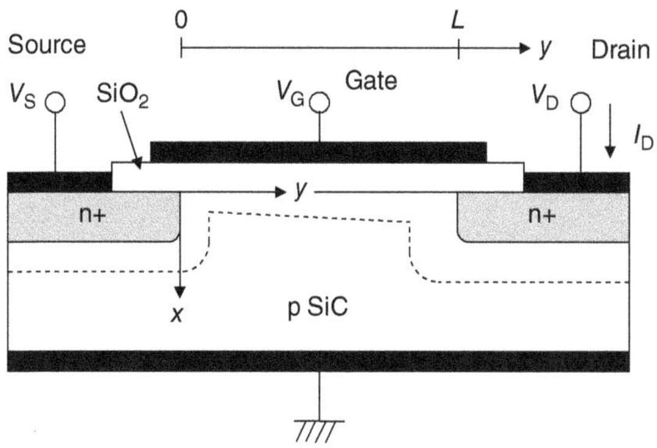

Figure 8.13 Structure of the basic MOSFET with important dimensions defined. The body contact is grounded, and all terminal voltages are referenced to the bulk.

where μ_N^* is the electron mobility at the oxide/semiconductor interface and $E_y(y)$ is the y-directed component of the electric field at point y in the channel. For now, we assume a continuous inversion layer exists between the source and drain. Considering a thin slice of the inversion layer of length dy at point y, we can write the electron current flowing through this slice as

$$I_D = -W\mu_N^* n_S \frac{dF_N}{dy} = W\mu_N^* Q_N \frac{d}{dy}\left(\frac{F_N}{q}\right) \tag{8.36}$$

where W is the width of the channel in the direction perpendicular to the page in Figure 8.13, n_S is the electron density in the inversion layer *per unit area,* and F_N is the electron quasi-Fermi level at point y. Q_N is the charge per unit area in the inversion layer, given by $Q_N = -q\, n_S$. We note that for a positive drain bias, dF_N/dy is negative (recall F_N is an energy, not a potential), and the minus sign in Equation 8.36 is chosen so that a positive I_D corresponds to a positive current *into* the drain terminal. The majority carrier holes are in equilibrium with the bulk, so $F_P = 0$ everywhere and we can write

$$V_R = -(F_N - F_P)/q = -F_N/q \tag{8.37}$$

Substituting Equation 8.37 into Equation 8.36 leads to

$$I_D = -W\mu_N^* Q_N \frac{dV_R}{dy} \tag{8.38}$$

This expression for the drain current at point y includes both drift and diffusion currents, since it is written in terms of the gradient of the electron quasi-Fermi level. We now multiply both sides of Equation 8.38 by dy and integrate from the source to the drain,

$$\int_0^L I_D dy = -W\mu_N^* \int_{V_S}^{V_D} Q_N dV_R \tag{8.39}$$

Since the current is uniform along the channel, I_D is not a function of y and can be taken out of the integral, yielding the desired result

$$\boxed{I_D = -\frac{W\mu_N^*}{L}\int_{V_S}^{V_D} Q_N\left(V_R\right) dV_R} \tag{8.40}$$

To obtain an analytical equation for the MOSFET drain current, we need to find an expression for the inversion charge Q_N as a function of the local quasi-Fermi splitting V_R, then evaluate the integral in Equation 8.40.

The relation between Q_N and V_R can be obtained by recalling that in the delta-depletion approximation, all the additional charges added to the semiconductor for $V_G > V_T$ reside in the inversion layer. Hence we can write

$$Q_N = -C_{OX}\left(V_G - V_T\right) \tag{8.41}$$

where the threshold voltage V_T is given by Equation 8.30. Substituting Equation 8.34 into Equation 8.30 yields

$$V_T = V_{FB} + \left(2\psi_F + V_R\right) + \sqrt{2V_0\left(2\psi_F + V_R\right)} \tag{8.42}$$

Equation 8.42 gives the *local* threshold voltage at a point in the channel where the quasi-Fermi splitting is qV_R. The threshold voltage of the transistor is given by Equation 8.42 evaluated at the source where $V_R = V_S$, and in most cases the source is grounded, that is, $V_S = 0$. Inserting Equation 8.42 into Equation 8.41 leads to

$$\boxed{Q_N = -C_{OX}\left[\left(V_G - V_{FB}\right) - \left(2\psi_F + V_R\right) - \sqrt{2V_0\left(2\psi_F + V_R\right)}\right]} \tag{8.43}$$

We next insert Equation 8.43 into Equation 8.40 and perform the indicated integral to obtain

$$\boxed{\begin{aligned} I_D = \mu_N^* C_{OX}\frac{W}{L}\Big\{&\left(V_G - V_{FB} - 2\psi_F - V_S\right)V_{DS} - \frac{1}{2}V_{DS}^2 - \frac{2}{3}\sqrt{2V_0}\Big[\left(2\psi_F + V_S + V_{DS}\right)^{3/2} \\ &-\left(2\psi_F + V_S\right)^{3/2}\Big]\Big\}\end{aligned}} \tag{8.44}$$

where we have made use of the fact that $V_{DS} = V_D - V_S$. This is the desired equation for MOSFET drain current.

The approach we have employed above is referred to as the "bulk-charge" theory, because we have accounted for the variation in bulk charge Q_D along the channel due to the varying V_R. The "bulk-charge" effect is represented by the term in square brackets that is multiplied by $(2V_0)^{1/2}$. This term depends on doping through the factor V_0, and hence makes Equation 8.44 doping-dependent. In the limit of light doping or thin oxide, $V_0 \to 0$ and the bulk charge term vanishes. In this case Equation 8.44 reduces to

$$I_D = \mu_N^* C_{OX}\frac{W}{L}\left[\left(V_G - V_{FB} - 2\psi_F - V_S\right)V_{DS} - \frac{1}{2}V_{DS}^2\right] \tag{8.45}$$

The MOSFET threshold voltage is given by Equation 8.42 with V_R evaluated at the source. If the source is grounded, then $V_R = V_S = 0$. If we also assume a light doping or a thin oxide, so that V_0 terms can be neglected, Equation 8.42 simplifies to

$$V_T \approx V_{FB} + 2\psi_F \tag{8.46}$$

Inserting Equation 8.46 into Equation 8.45 yields the familiar "square-law" equation for the MOSFET:

$$\boxed{I_D \approx \mu_N^* C_{OX}\frac{W}{L}\left[\left(V_G - V_T\right)V_{DS} - \frac{1}{2}V_{DS}^2\right]} \tag{8.47}$$

The square-law result is a special case of the more complete bulk-charge theory when the bulk doping is light or the oxide is thin. While simpler to use, the square-law equation overestimates the current, and the error increases as the doping increases.

8.2.4 Saturation Drain Voltage

Continuing now with the bulk-charge approach, we note that Equation 8.44 was derived assuming a continuous inversion layer from source to drain. As V_{DS} increases, the inversion layer narrows near the drain, and eventually pinches off. When this occurs, Equation 8.44 is no longer valid. The drain voltage that causes pinch-off can be found by setting the charge in the inversion layer at the drain to zero. Applying Equation 8.43 at $y = L$, we set $Q_N = 0$ and $V_R = V_S + V_{DS} = V_S + V_{DS,SAT}$. Solving for $V_{DS,SAT}$ yields

$$V_{D,SAT} = \left(V_G - V_{FB} - 2\psi_F - V_S\right) + V_0 - \sqrt{V_0^2 + 2V_0\left(V_G - V_{FB}\right)} \tag{8.48}$$

As with the JFET, for drain voltages beyond pinch-off the drain current saturates at the value it had at pinch-off, and the saturation drain current can be found by inserting Equation 8.48 in place of V_{DS} in Equation 8.44. In the case of light doping or thin oxide, we may neglect the V_0 terms, and $V_{D,SAT}$ becomes

$$\begin{aligned} V_{D,SAT} &\approx \left(V_G - V_{FB} - 2\psi_F - V_S\right) \\ V_{D,SAT} &\approx \left(V_G - V_T\right) \end{aligned} \tag{8.49}$$

Substituting Equation 8.49 for V_{DS} in Equation 8.47 gives the saturation current for the simple square-law theory.

8.2.5 Specific On-Resistance

The on-resistance at low drain voltages is of great interest, since this is the region of the $I - V$ characteristics where the power MOSFET operates in the on state. The specific on-resistance, or resistance–area product, is given by Equation 8.17, which is repeated below

$$R_{ON,SP} = \frac{W \cdot S}{\left.\dfrac{dI_D}{dV_D}\right|_{V_D=0}} \tag{8.17}$$

Since $V_D = V_S + V_{DS}$ and V_S is constant, the derivative in Equation 8.17 can be taken with respect to V_{DS}. Evaluating this derivative at the origin, $V_{DS} \approx 0$, we obtain

$$\left.\frac{dI_D}{dV_{DS}}\right|_{V_{DS}=0} = \mu_N^* C_{OX}\frac{W}{L}\left\{V_G - \left[V_{FB} + 2\psi_F + V_S + \sqrt{2V_0\left(2\psi_F + V_S\right)}\right]\right\} \tag{8.50}$$

The term in square brackets in Equation 8.50 is readily identified as the MOSFET threshold voltage, given by Equation 8.42 with $V_R = V_S$, so we can write

$$\left.\frac{dI_D}{dV_{DS}}\right|_{V_{DS}=0} = \mu_N^* C_{OX}\frac{W}{L}\left(V_G - V_T\right) \tag{8.51}$$

Inserting Equation 8.51 into Equation 8.17 gives the specific on-resistance of the MOSFET as

$$R_{ON,SP} = \frac{L \cdot S}{\mu_N^* C_{OX}\left(V_G - V_T\right)} \tag{8.52}$$

We should reiterate that in all our equations the gate voltage V_G is referred to the bulk rather than to the source. In power MOSFETs the source is normally grounded, so V_G is the same as V_{GS}. If for some reason the source is not grounded, $R_{ON,SP}$ will be modified through the dependence of V_T on V_S, given by Equation 8.42 with $V_R = V_S$.

8.2.6 *Power MOSFET Implementations: DMOSFETs and UMOSFETs*

We have now derived all the equations needed to describe the operation of the basic MOSFET, and we now turn to the implementation of a SiC MOSFET as a power switch. The power MOSFET incorporates the basic MOSFET structure of Figure 8.13 on a thick, lightly-doped n-type epilayer on an n+ substrate. The n+ drain is no longer at the surface, and instead the current exits the inversion channel and flows vertically through the lightly-doped n-drift region to the n+ substrate, which acts as the drain terminal. Consequently, the on-resistance of the power MOSFET is the sum of the on-resistance of the MOSFET channel plus the on-resistance of the drain drift region and the substrate. Depending on the actual geometry of the power MOSFET, other resistance components may also contribute, as will be discussed shortly.

Figure 8.14 shows two implementations of the power MOSFET in the form of a vertical, planar DMOSFET and a vertical, trench UMOSFET. The term DMOSFET derives from the silicon device of the same name, where the n+ source and p base regions are formed by diffusion of n-type and p-type impurities through the same mask opening (hence "double-diffused" MOSFET). In SiC the same structure is formed by double implantation, as will be discussed below. The term UMOSFET derives from the U-shaped geometry, but the term "trench MOSFET" is also used. Historically, the first SiC power MOSFETs were UMOSFETs [1], since they could be made without ion implantation, but these were soon joined by ion-implanted DMOSFETs [2] (sometimes called DIMOSFETs, or double-implanted MOSFETs).

As shown in Figure 8.14, both DMOSFETs and UMOSFETs consist of a MOSFET formed above a thick n-drift region, with the n+ substrate serving as the drain terminal. The blocking voltage in the off state is supported by the reverse-biased junction between the base and the drift region, but it is also supported by the MOS capacitor formed by the gate and the drift region.

Figure 8.15 illustrates the electric field lines in the DMOSFET in the blocking state. Also shown are electric field profiles along vertical cross-sections taken through the pn junction and the MOS capacitor. The field in the oxide of the MOS capacitor is higher than the peak field in the semiconductor by the

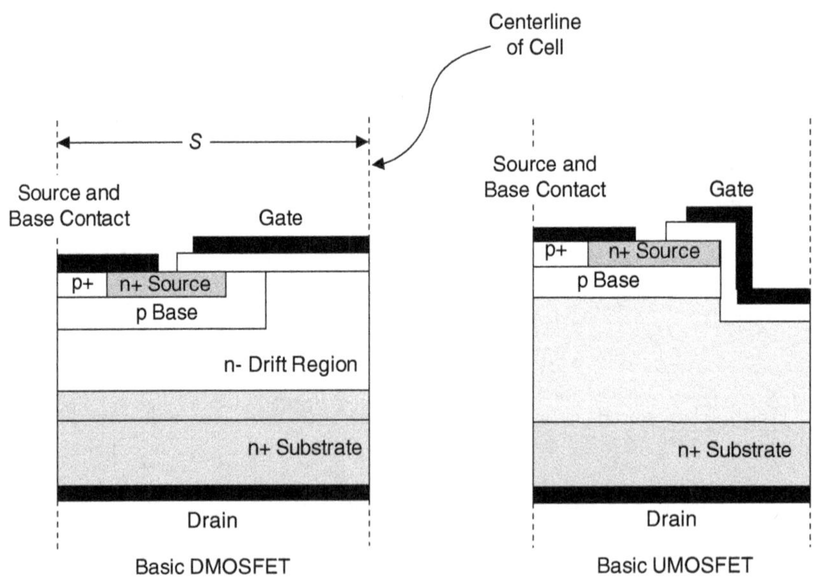

Figure 8.14 Two implementations of SiC power MOSFETs, the planar DMOSFET and the trench UMOSFET. Each cell is reflected left and right about the dashed centerlines.

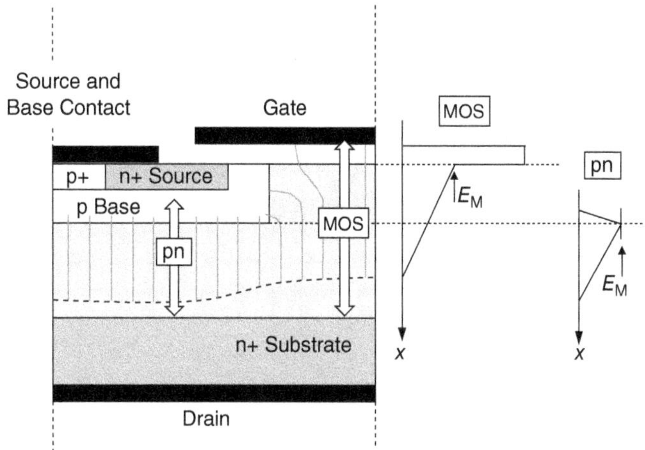

Figure 8.15 Electric fields in the DMOSFET in the blocking state. The blocking voltage must be supported by both the reverse-biased pn junction and the MOS capacitor.

ratio of dielectric constants, as required by Gauss' law, namely

$$E_{OX} = \frac{\varepsilon_S}{\varepsilon_{OX}} E_S(0) \tag{8.53}$$

For SiO_2/SiC (and SiO_2/silicon), $\varepsilon_S/\varepsilon_{OX} \approx 2.5$, meaning the oxide field is 2.5× higher than the peak field in the semiconductor. The breakdown field of SiO_2 is about 10 MV cm^{-1}, but for good long-term reliability it is advisable to keep the oxide field below about 4 MV cm^{-1}, as will be discussed in Section 8.2.11. In silicon this is not a problem, since $E_S(0)$ is limited to about 0.3 MV cm^{-1} by the critical field for avalanche breakdown. This corresponds to an oxide field of 0.75 MV cm^{-1}, well below the 4 MV cm^{-1} upper limit imposed by oxide reliability. However, in SiC the semiconductor field can reach 1.5 MV cm^{-1} due to the higher critical field. The corresponding oxide field is 3.75 MV cm^{-1}, close to the maximum allowable field, and even a small amount of field crowding can result in local fields above the acceptable value.

A similar situation exists in the UMOSFET geometry. Figure 8.16 illustrates the electric field lines in the UMOSFET in the blocking state, along with field profiles along two vertical slices. As in the DMOSFET, the oxide field is 2.5× higher than the peak semiconductor field, but there is significant field crowding at the trench corners, producing much higher local fields at these points. This is an important consideration in the design of SiC UMOSFETs, as will be discussed shortly.

Both DMOSFETs and UMOSFETs have undergone considerable development and optimization, and several features have been added to the basic structures of Figure 8.14 to improve their performance and reliability. In the following sections we will discuss these features and show how they improve the performance. Finally, we will consider limitations imposed by material properties and point out areas where further improvements may be realized.

8.2.7 Advanced DMOSFET Designs

In addition to the resistance of the MOSFET channel $R_{CH,SP}$ given by Equation 8.52 and the resistance of the drift region $R_{DR,SP}$ given by Equation 7.6, the DMOSFET includes several other resistive components, as illustrated in Figure 8.17. The current flow in the structure is two-dimensional, making it difficult to

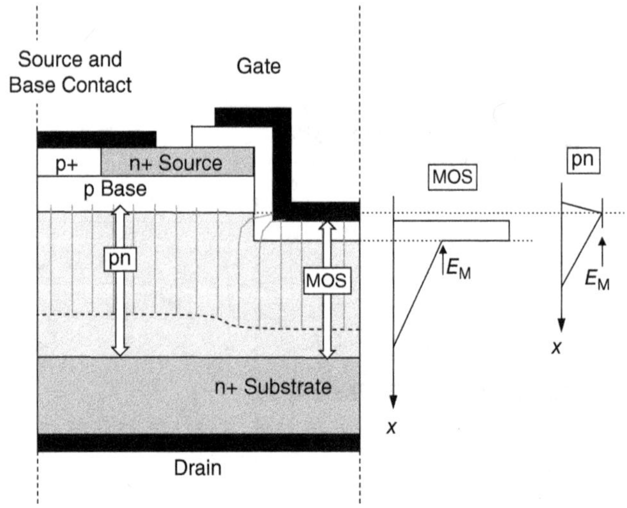

Figure 8.16 Electric fields in the UMOSFET in the blocking state. There is significant field crowding at the trench corners, making the oxide field higher at that point.

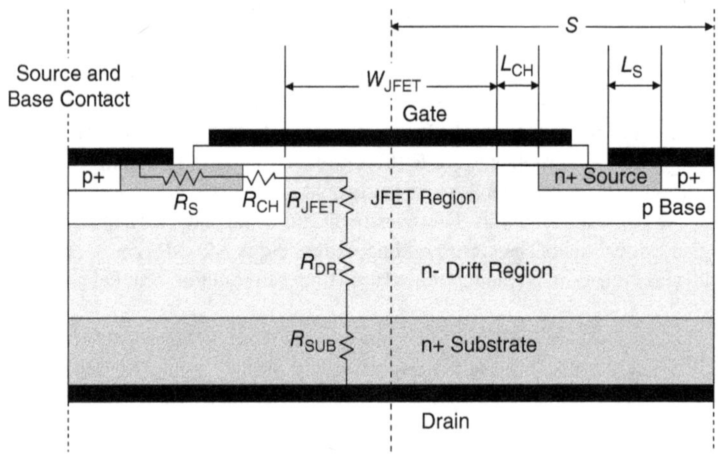

Figure 8.17 Cross-section of a full DMOSFET cell showing the important resistance components.

define individual "lumped" resistance elements, but it is conceptually useful to approximate the situation as shown. Here R_S represents the resistance of the source contact and the n+ source region, R_{JFET} is the resistance of the vertical JFET gated by the grounded p base regions, and R_{SUB} is the resistance of the substrate and its ohmic contact. We can express $R_{ON,SP}$ of the DMOSFET as

$$R_{ON,SP} = R_{CH,SP} + R_{DR,SP} + W \cdot S\left(R_S + R_{JFET} + R_{SUB}\right) \tag{8.54}$$

where W and S are the width and half-pitch of the cell in Figure 8.17.

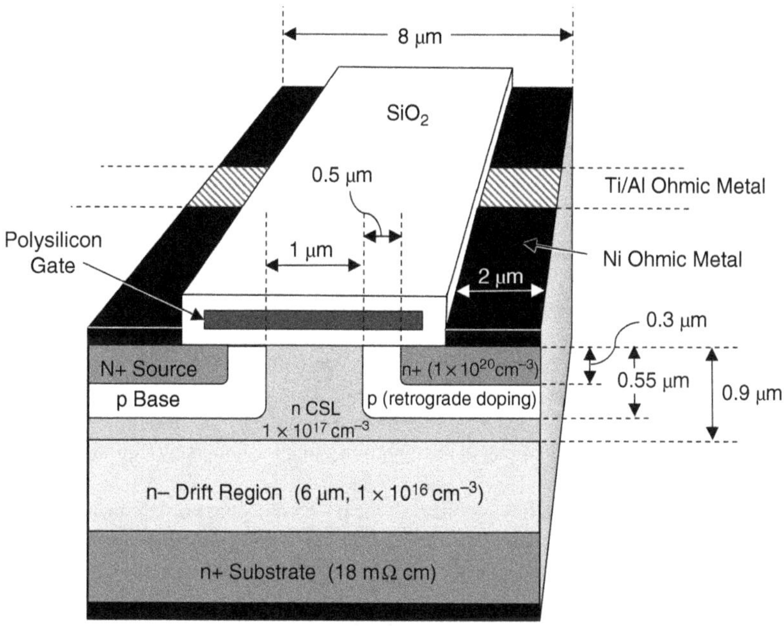

Figure 8.18 An advanced DMOSFET incorporating several features to enhance performance. In the actual device, the entire top surface is covered with Ni contact metal and thick Au (not shown).

As discussed in Section 7.1, the goal of the designer is to maximize the unipolar figure of merit $V_B{}^2/R_{ON,SP}$. Stated another way, for a given blocking voltage V_B our goal is to minimize the specific on-resistance given by Equation 8.54. This involves adjusting device parameters (dopings, thicknesses, and lateral dimensions) to minimize the total resistance and the surface area ($W \cdot S$), while maintaining the required blocking voltage. We must keep in mind that the blocking voltage may be limited by an oxide field that exceeds the limit for satisfactory oxide reliability, typically about 4 MV cm^{-1}. A comprehensive analysis must also include the two-dimensional effects due to current crowding and field crowding. Current crowding occurs at the end of the MOS channel as current spreads into the JFET region, and again at the bottom of the JFET region as current spreads into the wider drift region. In the blocking state, field crowding occurs at the corners of the p base regions within the cell and at the periphery of the full DMOSFET. These effects are usually analyzed by computer simulation.

The goals of minimizing resistance, minimizing area, and maintaining blocking voltage have resulted in several innovations. Figure 8.18 illustrates a modern SiC power DMOSFET [3] that incorporates many novel features, including (i) a self-aligned sub-micron MOS channel, (ii) a more heavily doped JFET region, (iii) a current spreading layer (CSL), (iv) source ohmic contacts that are self-aligned to the polysilicon gate, and (v) p base contacts that are segmentented along the length of the fingers. The parameters of this structure provide a blocking voltage of 1200 V with the minimum $R_{ON,SP}$. The new features and design considerations are discussed below.

The self-aligned sub-micron channel reduces the channel resistance by minimizing L in Equation 8.52. Simulations have shown that for typical base dopings, a channel length between 0.3 and 0.5 μm does not lead to punch-through of the channel at the maximum blocking voltage. The self-aligned process also reduces the cell pitch S. We will not describe the fabrication process here, but the reader is referred to the literature.

The CSL and more heavily doped JFET region are easily incorporated by specifying a higher doping for the top layer of epigrowth that forms the drift layer. Increasing the doping of the JFET region to about 1×10^{17} cm^{-3} allows the JFET width W_{JFET} to be reduced to about 1 μm. This has several beneficial effects: reduced cell pitch S, reduced field crowding at the corners of the base in the blocking state, and a lower field in the gate oxide in the blocking state. Of course, reducing W_{JFET} also increases R_{JFET}, so a trade-off is necessary, and this is done so as to maximize the overall figure of merit $V_B{}^2/R_{ON,SP}$ [3].

The CSL below the base regions reduces current crowding where electrons flow from the JFET region into the drift region, thereby reducing the resistance. In the blocking state, this thin layer of higher doping leads to a small increase in the peak electric field under the base, but the higher field is offset by an increase in the critical field for breakdown due to the higher doping, and the actual breakdown voltage is not reduced.

The use of self-aligned source contacts allows the source ohmic metal and top metal to be deposited directly over the polysilicon gate, isolated from the gate by a thick oxide. This reduces cell area by eliminating an alignment tolerance, reduces the source resistance by shortening the n+ source region, and reduces the resistance of the top metal in the direction along the fingers by extending the metal across the entire width of the cell.

The cell pitch S is further reduced by eliminating the p+ base contacts along the length of the fingers and inserting them only at isolated points, as shown in Figure 8.18. This can be done because very little current flows through the base regions. However, we must ensure that holes generated in the depletion region under reverse bias do not produce enough voltage drop as they flow laterally across the base to forward bias the source–base junction, as this would reduce the blocking voltage due to the gain of the parasitic bipolar transistor formed by the source, base, and drift regions.

Although not a new structural feature, the width of the source ohmic contact L_S is another relevant parameter. To minimize cell pitch, we would like to reduce L_S, but this increases the contact resistance. In general, we can reduce L_S until the source resistance begins to become a significant contributor to $R_{ON,SP}$ in Equation 8.54. A width of 1–2 μm is typically optimum.

The structure of Figure 8.18 is certainly not the end of SiC DMOSFET optimization, and the technology continues to advance. The reader is referred to the current literature for the latest developments.

8.2.8 *Advanced UMOS Designs*

Because of their trench geometry, UMOSFETs present both opportunities and challenges relative to planar devices such as the DMOSFET. The UMOSFET can be fabricated in a smaller surface area than the DMOSFET, since the MOS channel is oriented perpendicular to the surface. It is also easier to form a short (sub-micron) channel, since the channel length is determined by epigrowth. However, the MOS channel is formed on an etched non-polar face of the crystal, and the properties of the gate oxide and MOS interface are different from those on the (1000) plane. Critical parameters include interface state density, fixed charge density, inversion layer mobility, oxide reliability, and maximum allowable oxide field. We will discuss these parameters in more detail later, and for now we will confine ourselves to issues related to geometry and design optimization.

Figure 8.19 shows the major resistances in the vertical UMOSFET and it is apparent that the geometry of the device effectively eliminates the JFET resistance present in the DMOSFET. Since the current flow in the UMOSFET is two-dimensional, accurate analysis requires computer simulations.

Figure 8.16 illustrates the electric fields in the UMOSFET in the blocking state, and we noted earlier that the trench corners are locations of significant field crowding. Because the oxide field is ~ 2.5× higher than the peak field in the semiconductor, this is a serious problem. Fortunately, it is possible to reduce these oxide fields by modifications to the design. Figure 8.20 shows a modern SiC UMOSFET [4] that incorporates two features to reduce the oxide field and facilitate current spreading into the drift region, a p-type implant self-aligned to the bottom of the trench, and an n-type CSL incorporated below the base

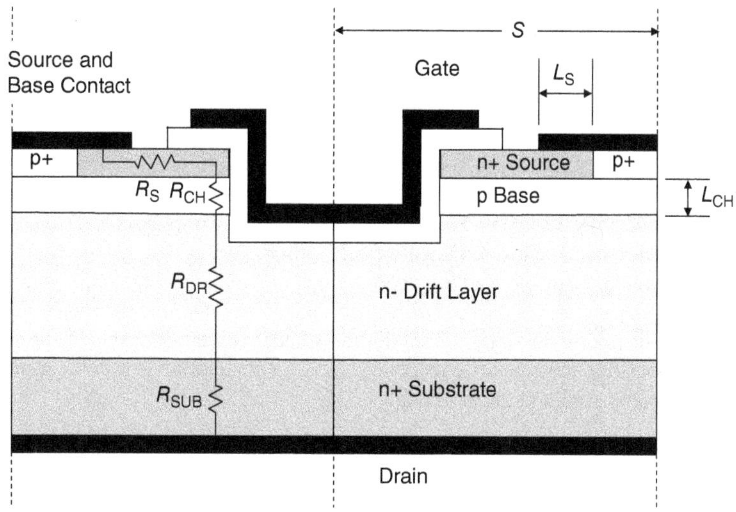

Figure 8.19 Cross-section of a full UMOSFET cell showing the important resistance components.

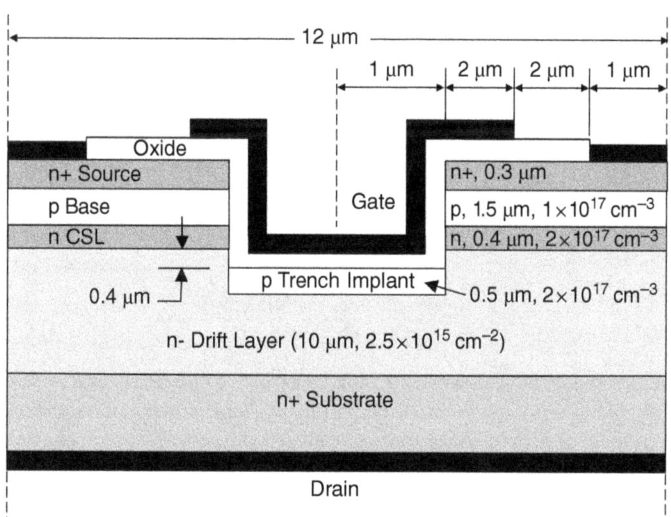

Figure 8.20 A UMOSFET design incorporating several features to enhance performance, including a p-type implant self-aligned to the bottom of the trench and a current spreading layer (CSL) below the base epilayer.

layer during epigrowth. In addition, the p+ base contact has been eliminated by placing the contact at the periphery of the cell array.

The p-type trench implant is performed immediately following trench etch using the same masking material as the etch. The implant is grounded by an ohmic contact at the periphery of the cell array (not shown). In the blocking state, the trench implant functions in the same manner as the p-type base epilayer. Just as field lines in the pn portion of Figure 8.16 terminate on the p base, field lines in the region under the trench now terminate on the trench implant and do not penetrate the oxide. As a result, the high oxide fields at the trench corners are drastically reduced, and do not limit the blocking voltage.

An undesirable side effect of the trench implant is to re-introduce a JFET constriction at the bottom end of the MOSFET channel. To alleviate this problem, a more heavily doped n-type CSL is incorporated under the base layer during epigrowth, as shown in the figure. The CSL minimizes the JFET resistance and facilitates lateral current spreading from the MOS channel into the drift region.

The reader will also notice that the UMOSFET channel length in Figure 8.20 is longer than the channel length in the DMOSFET of Figure 8.18. While the longer L_{CH} does not increase cell area, it does increase channel resistance, as shown by Equation 8.52. It would be easy to reduce L_{CH} by specifying a thinner base epilayer, but this could lead to punch-through of the base at high drain voltages. We could avoid punch-through by increasing the base doping, but this would raise the surface doping of the MOS channel, increasing the threshold voltage. This in turn would increase the channel resistance, as shown by Equation 8.52, negating the benefit of reducing L_{CH}. This conundrum is avoided in the DMOSFET by the retrograde doping profile of the base implant [3].

8.2.9 Threshold Voltage Control

Threshold voltage control during production and threshold stability during operation are critical issues for power MOSFETs. The threshold voltage of the MOSFET is given by Equation 8.42 evaluated at the source. Since the quasi-Fermi level splitting at the source $V_R(0) = V_S$, we can write

$$V_T = V_{FB} + (2\psi_F + V_S) + \sqrt{2V_0\,(2\psi_F + V_S)} \tag{8.55}$$

where ψ_F is given by Equation 8.21 and V_0 by Equation 8.28. In power MOSFETs the source is typically grounded, so $V_S = 0$. ψ_F is determined by doping and temperature, and V_0 is determined by doping and oxide thickness. Both doping and oxide thickness are well controlled in manufacture, and present no serious repeatability issues. The most critical term in Equation 8.55 is the flat-band voltage V_{FB}, defined as the gate voltage required to produce a flat-band condition in the semiconductor, that is, $\psi_S = 0$. The flat-band voltage is derived in most standard device textbooks, and can be written

$$V_{FB} = \frac{\Phi_{GS}}{q} - \frac{Q_F}{C_{OX}} - \frac{Q_{IT}\,(\psi_S = 0)}{C_{OX}} \tag{8.56}$$

where Φ_{GS} is the gate-semiconductor work function, given by the difference in work functions of the gate material and the semiconductor, Q_F is the density of fixed charge at the oxide/semiconductor interface, and $Q_{IT}(\psi_S = 0)$ is the charge in interface states at flat band. An additional term can be added to account for charges distributed within the gate oxide, if desired. In the ideal structure discussed earlier, Φ_{GS}, Q_F, and Q_{IT} were zero, resulting in $V_{FB} = 0$. We now consider the more general situation when these terms are not zero.

We begin with the gate-to-semiconductor work function Φ_{GS}. Figure 8.21 shows band offsets between several materials of importance in SiC MOS technology. The vacuum energy E_{VAC} represents the energy of an electron that has escaped the material into the vacuum with no remaining kinetic energy. The electron affinity χ of any material is the energy required to promote an electron from the conduction band to the vacuum, and the work function Φ of any material is the energy required to move an electron from the Fermi level to the vacuum. All the quantities shown in Figure 8.21 are fundamental properties of the materials, and are not affected by processing. The gate-to-semiconductor work function Φ_{GS} is the difference in work functions between the gate material and the semiconductor, that is,

$$\Phi_{GS} = \Phi_G - \Phi_S \tag{8.57}$$

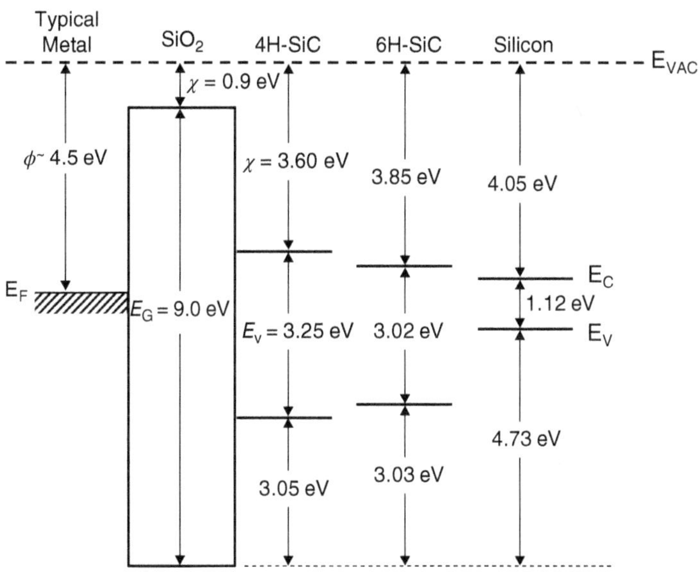

Figure 8.21 Band alignments of SiO_2, SiC, silicon, and a typical metal.

If the gate is metal, Φ_G is a fixed constant Φ_M that depends on the metal. If the gate is a semiconductor, Φ_G will depend upon the doping of the gate. The work function of a semiconductor can be written

$$\Phi_S = \chi + (E_C - E_F) = \chi + \frac{E_G}{2} \pm kT \ln\left(\frac{N_{A,D}^{-,+}}{n_i}\right) \tag{8.58}$$

where $N_{A,D}^{-,+}$ is the ionized dopant concentration in the semiconductor. Figure 8.22 shows the band diagram of an n-channel 4H-SiC MOSFET with a degenerately doped p-type polycrystalline silicon gate, placing the gate Fermi level at the valence band edge. For a degenerately doped p-type polysilicon gate, $\Phi_G = \chi + E_G = 5.17$ eV, and for 4H-SiC with an acceptor doping 2×10^{17} cm^{-3}, $\Phi_S = 6.64$ eV at room temperature, as given by Equation 8.58, allowing for the incomplete ionization of aluminum acceptors. Therefore the gate–semiconductor work function $\Phi_{GS} = -1.47$ eV.

We now turn to the fixed oxide charge and interface state charge terms in Equation 8.56. In thermally oxidized SiC, a fixed charge Q_F is present in a thin sheet at the oxide/semiconductor interface. This charge is typically positive and is determined solely by oxidation and anneal conditions, independent of oxide thickness and substrate doping. In addition, a thin sheet of interface states is also present at the interface, distributed with respect to energy across the bandgap. These states can exchange charge with the minority and majority carrier bands in the semiconductor, and the charge in interface states Q_{IT} depends on band bending. In equilibrium, states below the Fermi level are occupied and those above the Fermi level are empty. We represent the net charge in interface states at flat band by the value $Q_{IT}(\psi_S = 0)$. We should note that because both Q_F and Q_{IT} lie in a thin sheet at the interface, their effect on the electrostatics is the same, as seen from Equation 8.56. For this reason it is not possible to independently determine Q_F and $Q_{IT}(0)$ from electrical measurements, and it is common practice to represent the total interface charge at flat band by an *effective* fixed charge Q_F^*.

Since the charge in interface states depends on band bending, our expression for threshold voltage, Equation 8.55, needs to be modified to reflect the fact that Q_{IT} is larger at threshold than at flat band. The

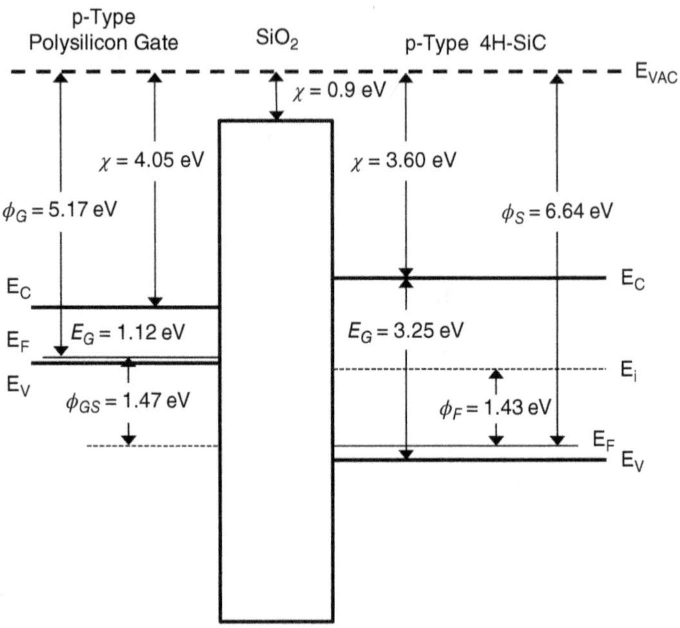

Figure 8.22 Band diagram of an n-channel 4H-SiC MOSFET with acceptor doping 2×10^{17} cm^{-3} with a degenerately doped p-type polysilicon gate. Here we assume no net charge at the interface, that is, $Q_F = 0$.

more complete expression can be written

$$V_T = \left[\frac{\Phi_{GS}}{q} - \frac{Q_F}{C_{OX}} - \frac{Q_{IT}(\psi_S = 2\psi_F + V_S)}{C_{OX}}\right] + (2\psi_F + V_S) + \sqrt{2V_0(2\psi_F + V_S)} \tag{8.59}$$

where the interface charge term is evaluated at a band bending corresponding to inversion at the surface.

The threshold voltage is a function of temperature. As temperature is raised, the band bending ($2\psi_F + V_S$) required to induce an inversion layer decreases due to the rapid increase in intrinsic carrier concentration in Equation 8.21. This is partially offset by the temperature pre-factor in Equation 8.21 and by the increased ionization of acceptors N_A^- as temperature is raised. Figure A.1 shows the ionization fraction of aluminum acceptors in 4H-SiC as a function of temperature for several acceptor doping levels, and Figure A.3 shows ψ_F computed using Equation 8.21 and the ionization fractions of Figure A.1. As temperature is increased, the Fermi level moves closer to midgap, and therefore the band bending ($2\psi_F + V_S$) needed to reach threshold decreases. This reduces the threshold voltage, as shown by Equation 8.59.

Two other factors influence the temperature dependence of V_T. Since the Fermi level lies closer to midgap at threshold, there will be less negative charge in interface states, so $Q_{IT}(\psi_S = 2\psi_F + V_S)$ decreases in magnitude. A less-negative Q_{IT} means a less positive threshold voltage. The semiconductor work function Φ_S given by Equation 8.58 is also a function of temperature. Φ_S decreases with temperature for a p-type substrate (n-channel MOSFET) and increases with temperature for an n-type substrate (p-channel MOSFET). The threshold variation with temperature can be minimized by reducing the density of interface states, but the decrease in band bending and the change in semiconductor work function cannot be mitigated by improved materials or processing techniques.

Control of threshold voltage is essential for a commercial process. One method for tailoring threshold voltage in production is to incorporate a thin n-type semiconductor layer immediately below the interface. This layer may be formed either by epitaxy or ion implantation, with the doping and thickness chosen to

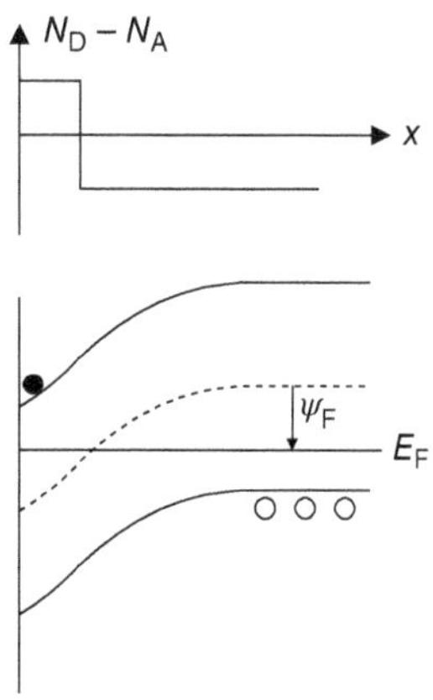

Figure 8.23 Doping profile and band diagram for MOSFETs with an n-type surface layer. The surface electric field is proportional to the slope of the bands at the surface, and is reduced in this structure.

ensure the layer is completely depleted at threshold. If the layer is sufficiently thin, the net charge in the layer can be regarded as a sheet charge at the interface. We can then add this donor charge to the effective oxide fixed charge, and the flat-band voltage can be written

$$V_{FB} = \frac{\Phi_{GS}}{q} - \frac{Q_F^*}{C_{OX}} - \frac{Q_{SURF}}{C_{OX}} \tag{8.60}$$

where Q_{SURF} is the sheet charge added by the surface n-layer. This allows the designer some latitude in adjusting the threshold voltage in Equation 8.55, provided the effective fixed charge Q_F^* is well controlled.

Although not shown in the figure, both the DMOSFET of Figure 8.18 and the UMOSFET of Figure 8.20 have thin n-type layers at the surface of the p-type base regions. In the UMOSFET, a 150 nm, 1×10^{17} cm^{-3} n-type epilayer is grown following the trench etch and implant activation, but prior to gate oxidation [4]. The main reason for this layer is to reduce the threshold voltage, but it also increases the mobility. This is because the surface n-layer is wide enough to affect the band diagram near the surface, as shown in Figure 8.23. Here we see that the electric field has been reduced near the surface (note the bands are concave upward near the surface). The lower field normal to the surface reduces interface scattering, increasing the inversion layer mobility. Devices with this feature are sometimes referred to as doped-channel FETs or accumulation-layer field-effect transistors (ACCUFETs). The DMOSFET of Figure 8.18 has a doped channel formed by implantation of nitrogen into the top 100 nm of the p-base at a dose of 5×10^{12} cm^{-2} [3]. In both the DMOSFET and UMOSFET, part of the surface layer is consumed during gate oxidation, so the remaining sheet concentration is somewhat lower than quoted above.

8.2.10 Inversion Layer Electron Mobility

8.2.10.1 Mechanisms Affecting Inversion Layer Mobility

In the development of the current–voltage characteristics of the MOSFET in Section 8.2.3, we denoted the mobility of electrons in the inversion layer by the symbol μ_N^*. At inversion biases, electrons are confined to the oxide/semiconductor interface by the strong band bending in the semiconductor. The mobility of electrons in the inversion layer is lower than in the bulk semiconductor due to increased scattering at the oxide/semiconductor interface. Moreover, the mobility decreases with increasing gate voltage, since higher gate voltages increase the electric field confining electrons to the interface, resulting in increased scattering. The gate voltage dependence can be described by the empirical equation

$$\mu_N^* \approx \frac{\mu_0^*}{1+\theta\left(V_G - V_T\right)} \tag{8.61}$$

where μ_0^* is the peak mobility at threshold and θ is a parameter characterizing the rate of decrease with field.

In silicon MOSFETs, the inversion layer mobility near threshold is about half the bulk mobility, but in 4H-SiC the inversion layer mobility on the silicon face is only about 5–10% of the bulk mobility. This impacts the performance of power MOSFETs by increasing the specific on-resistance, as shown by Equation 8.52. The low inversion layer mobility is a major limitation for MOSFETs and IGBTs in 4H-SiC, and inversion layer transport is the subject of intensive research.

Electron mobility in the inversion layer is limited by several scattering mechanisms, including *surface phonon scattering, Coulomb scattering* by fixed charges and charged interface states, and *surface roughness scattering* due to the structural and stoichiometric disorder at the interface. In addition, inversion electrons are also subject to the same scattering mechanisms as electrons in the bulk semiconductor. Mobility is inversely proportional to the total scattering rate, and assuming that scattering rates from the different processes add, the resultant mobility can be expressed using Matthiesson's rule as

$$\frac{1}{\mu_N^*} = \frac{1}{\mu_B} + \frac{1}{\mu_{PH}^*} + \frac{1}{\mu_C^*} + \frac{1}{\mu_{SR}^*} \tag{8.62}$$

where μ_B is the mobility of electrons in the bulk semiconductor, μ_{PH}^* is the mobility due to surface phonon scattering, μ_C^* is the mobility due to Coulomb scattering, and μ_{SR}^* is the mobility due to surface roughness scattering. The inversion layer mobility is therefore determined by the interplay of several scattering mechanisms whose magnitudes are affected by device processing and operating conditions. Many of these dependences can be related to the *effective normal field*, defined as the electric field in the semiconductor normal to the surface, evaluated at the centroid of the inversion layer [5]. Before considering the scattering mechanisms individually, we will examine how the effective normal field depends on parameters such as gate voltage, doping, and temperature.

In silicon, inversion layer mobility follows a universal curve when plotted against the effective normal field E_{EFF}, which can be written [5]

$$E_{EFF} = \left(Q_N/2 + Q_D\right)/\varepsilon_S \tag{8.63}$$

Here Q_N is the sheet charge density in the inversion layer and Q_D is the charge per unit area in the semiconductor depletion region. In an operating transistor, both Q_N and Q_D are functions of position y along the channel. Using Equations 8.23, 8.29, and 8.34, the depletion charge can be written

$$Q_D = -C_{OX}\sqrt{2V_0\left(2\psi_F + V_R\right)} \tag{8.64}$$

where ψ_F is given by Equation 8.21 and V_0 by Equation 8.28. The quasi-Fermi level splitting V_R is a function of position along the channel, going from $V_R = V_S$ at the source to $V_R = V_D$ at the drain. Under delta-depletion electrostatics, the charge in the semiconductor depletion region does not depend on gate voltage in inversion, as shown by Equation 8.64. A more exact analysis would reveal a weak dependence that we will neglect in the present discussion.

The inversion charge Q_N can be obtained from the gate voltage/surface potential relation of Equation 8.27 after appropriate modifications. Equation 8.27 is derived assuming depletion-region biasing with no inversion layer present. To apply Equation 8.27 in inversion, we must replace V_G with an "effective" gate voltage that includes the screening effect of the inversion charge. Performing this modification and setting the surface potential in inversion $\psi_S(\text{inv}) = (2\psi_F + V_R)$, we obtain

$$V_G - V_{FB} + Q_N/C_{OX} = (2\psi_F + V_R) + \sqrt{2V_0(2\psi_F + V_R)} \tag{8.65}$$

Solving for Q_N, and expanding the V_{FB} term to include charge in interface states Q_{IT}, we can write

$$Q_N = -C_{OX}\left[V_G - \frac{\Phi_{GS}}{q} + \frac{Q_F}{C_{OX}} + \frac{Q_{IT}\left(\psi_S = 2\psi_F + V_R\right)}{C_{OX}} - (2\psi_F + V_R) - \sqrt{2V_0(2\psi_F + V_R)}\right] \tag{8.66}$$

We can apply Equations 8.64 and 8.66 at the source by setting $V_R = V_S$. Then using Equation 8.59, the inversion charge at the source reduces to the simple expression $Q_N(y = 0) = -C_{OX}(V_G - V_T)$, showing that the inversion charge increases linearly with gate voltage once V_G exceeds V_T.

Q_N and V_T depend on temperature through two effects. As shown in Figure A.3, the Fermi potential ψ_F decreases with temperature, that is, the Fermi level moves closer to midgap as the temperature is raised. This means that less band bending is required to reach inversion, and this is represented in Equations 8.59 and 8.66 by the $2\psi_F$ terms. The Q_{IT} term in these equations is the charge in interface states, which depends on the position of the Fermi level at the surface. For an n-channel MOSFET in inversion, the Fermi level lies above midgap at the surface by an amount equal to ψ_F. As the temperature is raised and the Fermi level moves closer to midgap, the negative charge in interface states decreases and Q_{IT} becomes less negative. For an n-channel MOSFET, both these effects make Q_N more negative and V_T less positive as the temperature is raised. In contrast, Q_D becomes *less* negative with temperature, since ψ_F decreases with temperature (Figure A.3). Since Q_N and Q_D change in opposite directions as the temperature is raised, it often turns out that the effective normal field at a fixed gate voltage is only a weak function of temperature, as can be verified using the above equations.

Equations 8.63 and 8.64 indicate that samples with heavier substrate doping (larger V_0 and ψ_F) have a higher normal field at the same inversion charge density, and this is expected to increase scattering and reduce mobility by confining electrons closer to the interface. In addition, in heavier-doped samples the Fermi level lies closer to the conduction band in inversion, so there is more negative charge in interface states, and this can increase Coulomb scattering.

We will now consider each of the four scattering mechanisms in Equation 8.62. The mobility due to scattering in the bulk semiconductor depends on doping and temperature, and can be described by an equation of the form [6]

$$\mu_B = \frac{\mu_{MAX}(300/T)^{\eta}}{1 + \left(\dfrac{N_D^+ + N_A^-}{N_{REF}}\right)^{\gamma}} \tag{8.67}$$

where μ_{MAX} is the peak mobility at 300 K, T is absolute temperature, N_D^+ and N_A^- are the ionized dopant concentrations (which depend upon temperature), and η and γ are constants. Values for these parameters in 4H-SiC are given in Table 8.1, and Appendix A gives equations for N_D^+ and N_A^- as a function of temperature and doping. For normal operating conditions, the bulk mobility term is not the limiting factor in SiC inversion layer mobility.

Surface phonon scattering is the deflection of electrons by acoustic phonons at the surface. The phonon-limited mobility can be written [7]

$$\mu_{SP}^* = \frac{A}{E_{EFF}} + \frac{B}{TE_{EFF}^{1/3}} \tag{8.68}$$

where A and B are parameters evaluated by fitting theory to experiment and have typical values given in Table 8.1. The phonon-limited mobility decreases with increasing normal field (increasing gate voltage) and decreases with increasing temperature, since higher temperatures are associated with larger lattice vibrations. Although acoustic phonon scattering is important in silicon samples, this term is typically not the limiting factor in SiC MOSFETs fabricated to date.

Coulomb scattering is the deflection of electrons by charged centers at the interface, such as fixed charges Q_F and charged interface states Q_{IT}. The scattering effect is independent of the charge polarity, and depends on the *total* density of charged centers rather than on the *net* charge density. A number of

Table 8.1 Parameters of the inversion layer mobility model, Equations 8.67–8.70.

Parameter	Value	Units	References
μ_{MAX}	1141	$cm^2\ V^{-1}\ s^{-1}$	[8, 9]
N_{REF}	1.94×10^{17}	cm^{-3}	
η	2.8	–	
γ	0.61	–	
A	7.82×10^7	$cm\ s^{-1}$	[7]
B	9.92×10^6	$(V\ cm^{-1})^{-2/3}\ K\ cm\ s^{-1}$	
Γ_C	1.5×10^{11}	$eV^{-1}\ cm^{-2}$	[6]
n_C	1.5×10^{18}	cm^{-3}	
ζ_C	0.8	–	
Γ_{SR}	1.55×10^{13}	$V\ s^{-1}$	[10]

different formulations have been proposed for Coulomb scattering, but a commonly accepted form that captures the essential physics is [6]

$$\mu_C^* = \frac{\Gamma_C}{(N_F + N_{IT})} T \left(1 + \frac{n_S}{n_C}\right)^{\zeta_C} \tag{8.69}$$

where N_F is the total density of fixed charges (sum of positive and negative charges), N_{IT} is the total density of charged interface states (both positive and negative), n_S is the electron density *per unit volume* at the surface, and Γ_C, n_C, and ζ_C are parameters with typical values given in Table 8.1. Coulomb scattering limits inversion layer mobility at gate voltages near threshold where n_S is small, but drops rapidly at higher gate voltages (higher n_s values) due to carrier screening by inversion electrons. In samples with high fixed charge N_F or high interface state density N_{IT}, Coulomb scattering can limit μ_N^* to single digits, but improvements in oxidation and anneal processes have reduced both N_F and N_{IT} to the point that Coulomb scattering is comparable to surface roughness scattering in limiting inversion layer mobility.

Surface roughness scattering is the deflection of electrons by structural defects at the interface. In silicon, surface roughness scattering is attributed to surface steps and residual unoxidized silicon particles at the interface. However, in SiC the situation is more complex, and there is evidence for a *transition layer* at the interface where the chemical composition changes gradually from pure SiC to pure SiO_2 over a distance of several nanometers. The transition layer can be seen in high-resolution transmission electron microscope (TEM) images, where it has an apparent width of several nm [11]. Spatially-resolved electron energy loss spectroscopy (EELS) shows a gradual transition of the C/Si ratio, beginning about 3 nm inside the SiC and extending about 5 nm into the SiO_2. The surface layer of the SiC also exhibits some structural disorder resulting from the oxidation process, and the first monolayers of SiO_2 contain excess carbon [11]. The width of the transition layer is influenced by oxidation and anneal procedures, and samples with thinner transition layers have higher peak inversion mobilities [12].

Surface roughness scattering can be characterized by an expression borrowed from the silicon literature, having the general form [13]

$$\mu_{SR}^* = \frac{\Gamma_{SR}}{E_{EFF}^2} \tag{8.70}$$

where Γ_{SR} is a parameter that depends on the correlation length and mean height of the assumed roughness. This equation has been applied to 4H-SiC by several authors [7, 14], although the more complex nature of the SiC interface suggests the need for further studies. Nevertheless, the general

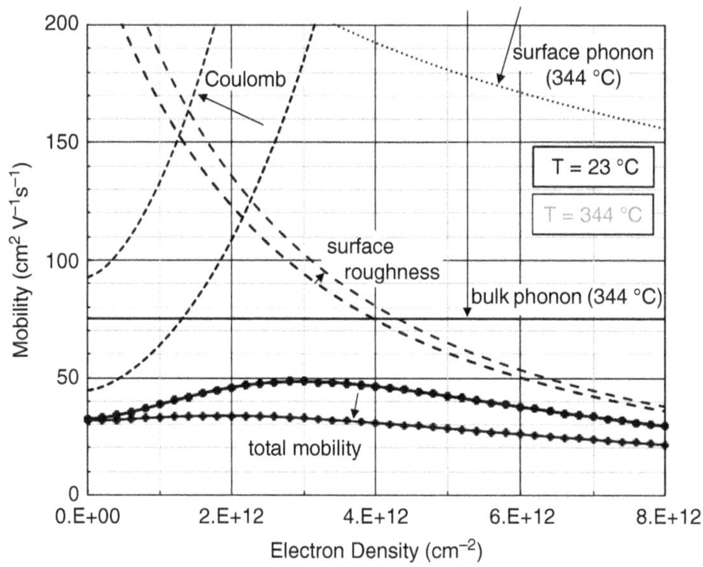

Figure 8.24 General dependence of surface scattering mechanisms on inversion charge density and temperature. The parameters used to generate these plots were modified slightly from those of Table 8.1 to produce results that are generally consistent with a broad range of reports in the literature.

form of Equation 8.70 indicates the strong dependence of surface roughness scattering on normal field. Roughness scattering limits the mobility of SiC MOSFETs at high normal fields and accounts for the decrease in mobility with gate voltage above threshold. This scattering is essentially independent of temperature, provided the normal field is held constant.

Figure 8.24 illustrates how the four scattering mechanisms vary with inversion charge density (or gate voltage) and temperature. Results are shown for two temperatures, 23 and 344 °C. The arrows indicate the trend of the particular scattering mechanism as temperature is increased. At room temperature, Coulomb scattering limits the mobility in most samples at low electron densities (gate voltages close to threshold) and surface roughness scattering limits at high electron densities (strong inversion). Bulk and surface phonon scattering do not play a significant role at room temperature. As temperature increases, Coulomb and surface roughness scattering decrease, while surface and bulk phonon scattering increase. The overall mobility decreases slightly with temperature, mainly due to the strong increase in bulk phonon scattering. This figure illustrates general trends in a way that is consistent with a broad range of reports in the literature, but it does not describe any particular device. As technology advances, interface state density and surface roughness scattering should continue to decrease, leading to higher mobilities and better MOSFET performance.

8.2.10.2 Device-Related Definitions of Inversion Layer Mobility

Having discussed the physical mechanisms governing electron transport in inversion layers, we now wish to consider three different device-related definitions of inversion mobility: effective (or conductivity) mobility μ_{EFF}^*, field-effect mobility μ_{FE}^*, and Hall mobility μ_H^*. While closely related, each definition provides a slightly different perspective on conditions and processes within the device.

Equation 8.47 gives the MOSFET current using the "square law" approximation with the source grounded (the effect of a non-zero source voltage is included in Equation 8.44). The *effective* (or conductivity) mobility is the value deduced by measuring the drain conductance g_D at small drain voltages where the V_{DS}^2 term in Equation 8.47 can be neglected. Neglecting the V_{DS}^2 term, we can write

$$g_D = \frac{\partial I_D}{\partial V_{DS}} = \mu_{EFF}^* C_{OX} \frac{W}{L}(V_G - V_T) \tag{8.71}$$

Thus,

$$\boxed{\mu_{EFF}^*(V_G) = \frac{L}{C_{OX} W (V_G - V_T)} g_D(V_G)} \tag{8.72}$$

In practice, μ_{EFF}^* is calculated as a function of gate voltage from the slope of the $I_D - V_{DS}$ relation at the origin, $V_{DS} = 0$. This mobility should be used for μ_N^* in Equations 8.44 and/or 8.47 to calculate the current. Since the mobility varies with gate voltage, μ_{EFF}^* is usually approximated using the empirical expression in Equation 8.61.

The mobility most often quoted in the literature is the *field-effect* mobility, obtained from the transconductance measured at a small drain voltage. From Equation 8.47 evaluated at low V_{DS} with $\mu_N^* = \mu_{EFF}^*$, the transconductance may be written

$$\begin{aligned} g_M &= \frac{\partial I_D}{\partial V_G} = C_{OX} \frac{W}{L} V_{DS} \frac{d}{dV_G}[\mu_{EFF}^*(V_G - V_T)] \\ &= C_{OX} \frac{W}{L} V_{DS} \left[\mu_{EFF}^* + (V_G - V_T) \frac{d\mu_{EFF}^*}{dV_G} \right] \\ &= \mu_{FE}^* C_{OX} \frac{W}{L} V_{DS} \end{aligned} \tag{8.73}$$

Thus,

$$\boxed{\mu_{FE}^*(V_G) = \frac{L}{C_{OX} W V_{DS}} g_M(V_G)} \tag{8.74}$$

From Equation 8.73, the field-effect mobility is related to the effective mobility by

$$\mu_{FE}^* = \mu_{EFF}^* + (V_G - V_T) \frac{d\mu_{EFF}^*}{dV_G} \tag{8.75}$$

Since μ_{EFF}^* decreases with V_G, the partial derivative is negative and $\mu_{FE}^* < \mu_{EFF}^*$ at any gate voltage. Figure 8.25 illustrates how μ_{EFF}^* and μ_{FE}^* vary with gate voltage. Near threshold the effective and field-effect mobilities are equal, but μ_{FE}^* decreases more rapidly than μ_{EFF}^* as gate voltage is increased.

It is important to note that using μ_{FE}^* instead of μ_{EFF}^* in Equation 8.44 or 8.47 would underestimate the current. Since μ_{FE}^* is a differential mobility, calculating the current based on μ_{FE}^* would require *integration* of the μ_{FE}^* versus V_G relation. From Equation 8.73, we can write

$$I_D = \int_0^{V_G} g_M dV_G' = C_{OX} \frac{W}{L} V_{DS} \int_0^{V_G} \mu_{FE}^* dV_G' \tag{8.76}$$

Thus the current at a given gate voltage is proportional to the *integral* of $\mu_{FE}^*(V_G)$ up to that gate voltage, and not to the *value* of μ_{FE}^* at that gate voltage. For this reason, any process that increases the peak field-effect mobility in the region just above threshold will increase the current at *all* gate voltages, even if μ_{FE}^* at high gate voltages is not improved.

Finally, we turn to the Hall mobility μ_H^*. Hall mobility is of interest because it allows us to account for the effect of charge sequestration by interface states. In deriving the current Equations 8.44 and 8.47 we assumed all the additional positive charge placed on the gate beyond threshold is balanced by an equal negative charge induced into the inversion layer, as evident from Equation 8.41. However, if interface

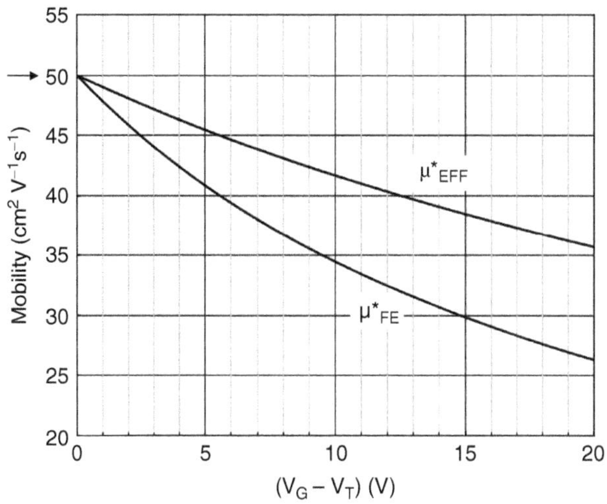

Figure 8.25 Relationship of effective mobility and field-effect mobility, with values typical of 4H-SiC MOSFETs on the (0001) face after thermal oxidation and post-oxidation anneal in nitric oxide.

states are present, some of the charge placed on the gate beyond threshold is balanced by additional charge in interface states, and this reduces the charge density in the inversion layer. Without knowing the distribution of interface states across the bandgap, it is not possible to calculate how much charge is trapped in interface states and how much remains in the inversion layer. If the actual inversion charge density is less than given by Equation 8.41, measurements of μ^*_{EFF} and μ^*_{FE} using Equations 8.72 and 8.74 will underestimate the true carrier mobility. The effective mobility is the correct mobility to use in the current Equations 8.44 and 8.47, since μ^*_{EFF} includes the combined effects of inversion layer mobility and charge sequestration by interface states, but the Hall mobility is the correct mobility to use if we are concerned with the true carrier mobility in the inversion layer.

The Hall technique is discussed extensively in the literature, and only an outline will be presented here. When used to study the inversion layer in an MOS transistor, the Hall technique imposes a magnetic field B_x perpendicular to the surface with a constant drain current J_y flowing along the channel. The magnetic field exerts a Lorentz force on the moving electrons given by

$$F_z = -qv_yB_x \tag{8.77}$$

where F_z is the force directed across the width of the channel (z direction) and v_y is the velocity of electrons along the channel in the y direction. The Lorentz force deflects electrons to one side of the channel, creating a lateral electric field E_z perpendicular to the current flow. In steady state the z-directed force due to the lateral electric field exactly balances the oppositely-directed z-directed Lorentz force, so $q\,E_z = -q\,v_yB_x$. The electric field E_z produces a lateral potential drop called the Hall voltage, given by

$$V_{\mathrm{H}} = -\left(\frac{J_yB_xW}{qn_{\mathrm{S}}}\right) \tag{8.78}$$

Here, W is the width of the channel, and we have used the fact that $J_y = -q\,n_{\mathrm{S}}v_y$. The Hall voltage can be measured by high-impedance probes that contact the inversion layer on opposite sides of the channel. Since we know the current density and magnetic field, we can determine the mobile electron density n_{S} using Equation 8.78. The mobility can then be deduced from the measured current, given knowledge of the true carrier density in the channel. The mobility thus determined is known at the Hall mobility μ^*_{H}.

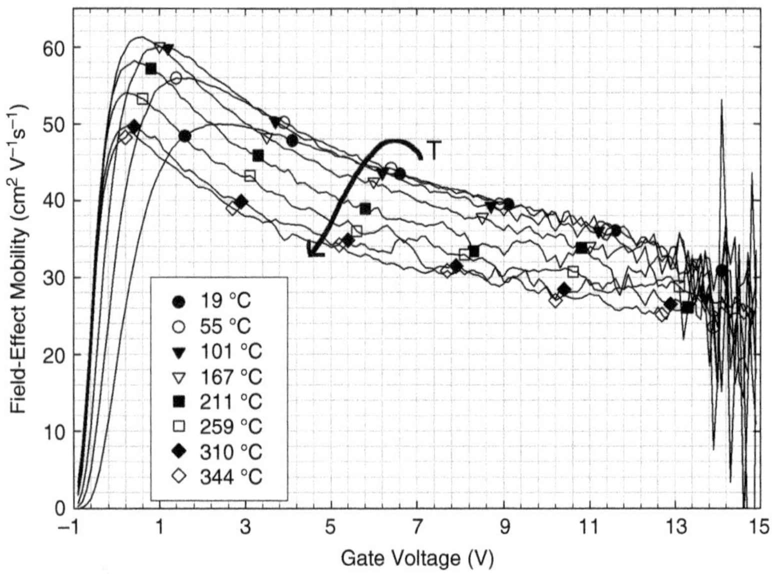

Figure 8.26 Field-effect mobility as a function of temperature for a 4H-SiC MOSFET on a p-type epilayer ([16] reproduced with permission from IEEE).

A more careful analysis that includes momentum relaxation effects would lead to a slightly modified form for Equation 8.78, namely

$$V_H = -r_H\left(\frac{J_y B_x W}{qn_S}\right) \tag{8.79}$$

Here r_H is the *Hall factor* given by $r_H =< \tau_m{}^2 > / < \tau_m >^2$, where τ_m is the mean time between electron scattering events. In 4H-SiC the Hall factor at room temperature typically lies in the range 0.96–0.99 [15], and in experimental work it is common to simply set $r_H = 1$.

Since Hall measurements are based on the actual mobile carrier density, the Hall mobility represents the true mobility of electrons in the inversion layer, and μ_H^* is invariably higher than μ_{EFF}^*.

8.2.10.3 Experimental Results on Inversion Layer Mobility in 4H-SiC

Figure 8.26 shows field-effect mobility measured as a function of temperature for a 4H-SiC MOSFET on a 3 μm p-type epilayer doped 2.5×10^{16} cm^{-3} with aluminum [16]. The p+ (0001) substrate is grown 8° off-axis. A 54 nm gate oxide was formed by pyrogenic oxidation at 1150 °C, followed by 30 min *in situ* argon anneal, a 950 °C re-oxidation anneal for 2 h, and a nitric oxide (NO) post-oxidation anneal at 1175 °C for 2 h. The final NO anneal is used to reduce the interface state density in the upper half of the bandgap [17]. Polysilicon gates were deposited at 625 °C and doped at 900 °C. At room temperature, the peak mobility is about 50 cm^{2}V^{-1}s^{-1}, decreasing to about 30 cm^{2}V^{-1}s^{-1} at 15 V. As temperature is increased, the peak low-field mobility increases to approximately 60 cm^{2}V^{-1}s^{-1} at 167 °C, consistent with Coulomb scattering, then decreases to about 50 cm^{2}V^{-1}s^{-1} at 344 °C. At higher fields the mobility decreases monotonically with temperature. The decrease in mobility with temperature is suggestive of bulk phonon scattering, as can be inferred from the curves in Figure 8.24. At 344 °C the field-effect mobility has dropped to about 80% of its room temperature value over much of the gate voltage range.

Figure 8.27 shows Hall mobility as a function of sheet carrier density for an n-channel MOSFET on a p-type epilayer doped 5×10^{15} cm^{-3} with aluminum [18]. The substrate is (0001) 4H-SiC, cut 4° off-axis.

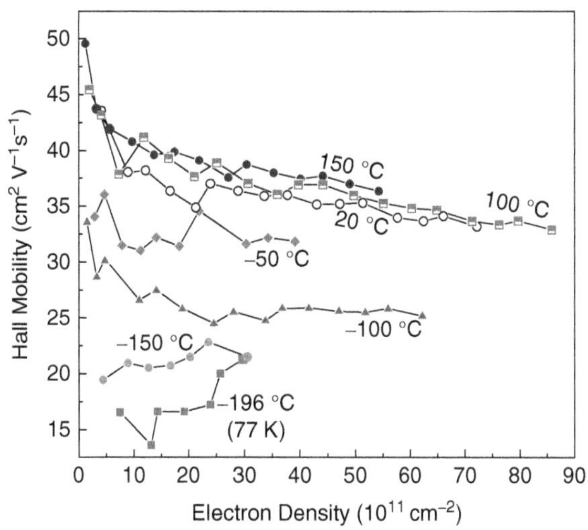

Figure 8.27 Hall mobility of inversion electrons as a function of effective normal field at several temperatures. The sample was prepared by a similar oxidation process to the MOSFET in Figure 8.26 ([18] reproduced with permission from Trans Tech Publications).

The oxidation was performed in dry oxygen at 1175 °C, followed by a wet re-oxidation anneal at 950 °C and an 1175 °C post-oxidation anneal in NO for 2 h, resulting in a 53 nm oxide. This is a similar oxidation process to the MOSFET of Figure 8.26, but with a slightly lower epilayer doping. Hall mobility increases with temperature below 20 °C, suggesting that transport at these temperatures is limited by Coulomb scattering. Above room temperature the mobility increases only slightly with temperature, suggesting that transport here is dominated by surface roughness scattering. These conclusions are consistent with the trends illustrated in Figure 8.24.

The MOSFET process described above has been investigated by several groups, all of which report increased mobilities as a result of the NO anneal. Recently, however, a number of alternative oxidation processes have been explored, some of which have produced even greater improvements in mobility. We will briefly discuss two of these processes next.

Post-oxidation annealing in phosphorus has been shown to reduce the interface state density and increase the mobility, even compared to the standard NO process. Figure 8.28 shows measured interface state density and field-effect mobility for several 4H-SiC MOSFETs [19]. The MOSFETs were formed on p-type epilayers doped 7×10^{15} cm^{-3} with aluminum, and interface state density was measured on MOS capacitors on n-type epilayers doped 8×10^{15} cm^{-3} with nitrogen. The substrate is n+ (0001), grown 4° off-axis. A 56 nm gate oxide was formed by dry oxidation at 1000 °C, followed by a post-oxidation anneal in $POCl_3$ at 900, 950, or 1000 °C for 10 min and a 30 min nitrogen anneal at the same temperature. Aluminum was evaporated to form the gate electrodes. Other samples were fabricated by dry oxidation at 1000 °C followed by a 1250 °C nitric oxide (NO) anneal for 90 min. As seen in the figure, the 900 °C phosphorus anneal did not reduce the interface state density, but anneals at 950 and 1000 °C decreased the interface state density below that of the NO sample in the upper half of the bandgap. The field-effect mobility of the 1000 °C $POCl_3$ sample is much higher than the 1250 °C NO sample, reaching a peak value of 89 cm^{2}V^{-1}s^{-1} before dropping to 58 cm^{2}V^{-1}s^{-1} at 20 V. If this result is compared to the MOSFET of Figure 8.26, the mobility improvement is not as great, but is still about a factor of 2.

The improved mobility is explained as a reduction in Coulomb scattering by charged interface states, but the phosphorus process may also introduce a shallow n-type layer at the surface of the SiC. This seems

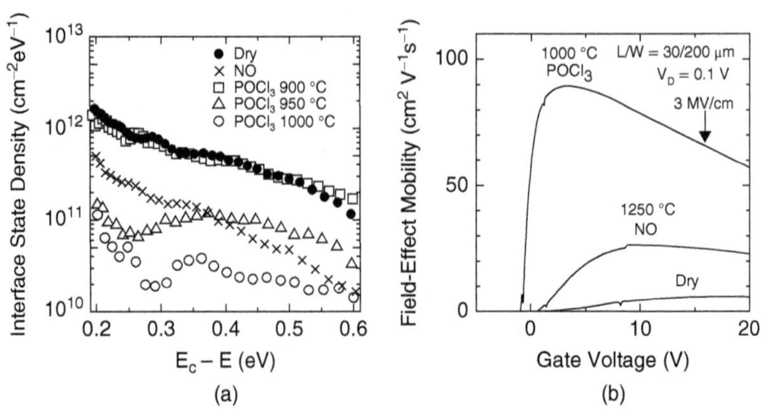

Figure 8.28 (a) Interface state density and (b) field-effect mobility for dry oxidized samples with no post-oxidation anneal, with a post-oxidation anneal in nitric oxide, and with a post-oxidation anneal in $POCl_3$. The large increase in mobility for the $POCl_3$ sample is correlated with a reduction in interface state density ([19] reproduced with permission from IEEE).

plausible, since phosphorus can be activated as a dopant in 4H-SiC at normal oxidation temperatures. This would produce a situation similar to the doped-channel FET discussed in Section 8.2.9. As shown in Figure 8.23, surface doping reduces the surface normal field at a given inversion electron density, which would lead to higher mobility. However, phosphorus annealing also converts the oxide to phosphosilicate glass (PSG), a polar material having piezoelectric properties that introduce an instability in the threshold voltage [20]. Phosphorus atoms are distributed throughout the gate insulator, and they act as electron traps that introduce a positive threshold shift at high electric fields [21]. Instabilities of this type would make it difficult to use this process in a production environment.

Several groups have explored inversion layer transport on alternative crystal faces such as the $(11\bar{2}0)$ a-face, where mobilities are typically higher than on the (0001) silicon face. These results are important for devices such as the UMOSFET (or UMOS IGBT) whose inversion layers are on the a-face. Figure 8.29 shows measured interface state density and field-effect mobility for 4H-SiC MOSFETs on both the silicon face and the $(11\bar{2}0)$ a-face [22]. Planar MOSFETs were formed on p-type epilayers doped 1×10^{16} cm^{-3} with aluminum, and interface state density was measured on MOS capacitors on n-type epilayers doped 1×10^{16} cm^{-3} with nitrogen. All samples were oxidized in dry oxygen at 1150 °C. Some samples received a post-oxidation anneal in NO at 1175 °C for 2 h, and others received a phosphorus post-oxidation anneal at 1000 °C for 4 h using a planar diffusion source. On a given crystal face, the field-effect mobility is 1.5–2 × higher with the phosphorus anneal compared to the NO anneal. For the same post-oxidation process, the mobility is also 1.5–2 × higher on the $(11\bar{2}0)$ face than on the silicon face. The highest mobility is obtained with the phosphorus anneal on the $(11\bar{2}0)$ face.

With the phosphorus anneal on the silicon face, the higher mobility is correlated with a lower interface state density, suggesting the mobility is limited by Coulomb scattering, at least at biases near threshold. However, on the $(11\bar{2}0)$ face the opposite is true: higher mobilities are obtained with the phosphorus anneal, but the interface state density is similar on both faces. Comparing phosphorus anneals on the two faces, a higher mobility is obtained on the $(11\bar{2}0)$ face even though that face has a higher interface state density. Two possibilities present themselves. Yoshioka *et al.* [23] have reported that certain annealing procedures create interface states with large capture cross sections whose frequency response is so fast that they are not detected with conventional CV methods, so the conclusion that the silicon face has lower total interface state density may be incorrect. On the other hand, if phosphorus achieves high mobility by creating a thin n-doped layer at the surface, this doping process could be more effective on the $(11\bar{2}0)$

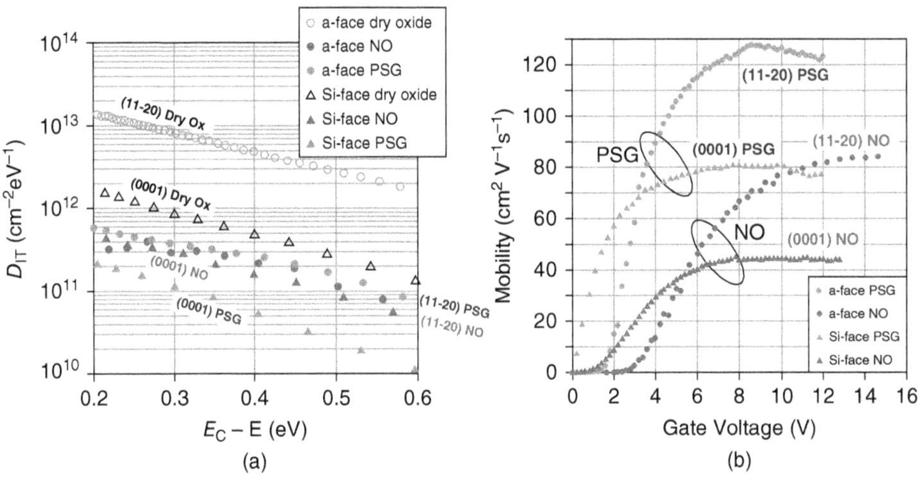

Figure 8.29 (a) Interface state density and (b) field-effect mobility for dry oxidized samples on two crystal orientations, the (0001) silicon face and the $(11\bar{2}0)$ a-face. Mobility is shown for both nitrogen and phosphorus post-oxidation anneals, and interface state density is shown for unannealed, nitrogen-annealed, and phosphorus-annealed samples ([22] reproduced with permission from IEEE).

face than on the silicon face. These issues can only be resolved by further experiments and the reader is advised to consult the literature for the latest information.

An intriguing result is the high mobility observed in samples oxidized in an alumina furnace tube [24], as shown in Figure 8.30 [25]. The high mobility is correlated with the presence and position of sodium ions in the oxide, introduced from impurities in the alumina tube. These ions are mobile in SiO_2 at elevated temperatures, and can be drifted toward the interface or toward the gate by application of a positive or negative gate voltage. The highest mobilities are observed when the sodium ions are close to the interface, the "initial" and "recovered" curves in Figure 8.30. When the sodium ions are drifted to the gate by negative bias-temperature stress, the threshold voltage increases, and the mobility decreases. Measurements of interface state density on n-type MOS capacitors indicate that the position of the sodium ions has no effect on interface state density [25], suggesting that a reduction of Coulomb scattering is not the mechanism responsible for the mobility improvement. It is also significant that the mobility at high gate fields is not affected by the ions. In other words, sodium enhances the mobility at low gate fields but not at high fields. The physical mechanism responsible for these observations is not understood. As might be expected, the instability in threshold voltage associated with sodium precludes the use of this process in practical devices.

Even though both phosphorus and sodium processes introduce threshold instabilities, the results are still encouraging because they show that significantly higher mobilities are possible in 4H-SiC MOS devices. The goal now is to understand the mechanisms for the mobility improvement and develop processes that deliver the improvement without introducing instabilities.

8.2.11 Oxide Reliability

Two of the main advantages of SiC compared to silicon are its higher bandgap energy and higher critical field for avalanche breakdown. The critical field in 4H-SiC is 6–7 times higher than in silicon (Figure 10.5) and, as discussed in Chapter 7, the on-resistance of SiC unipolar devices is about 400 times lower than silicon devices of the same blocking voltage. However, the higher critical field means

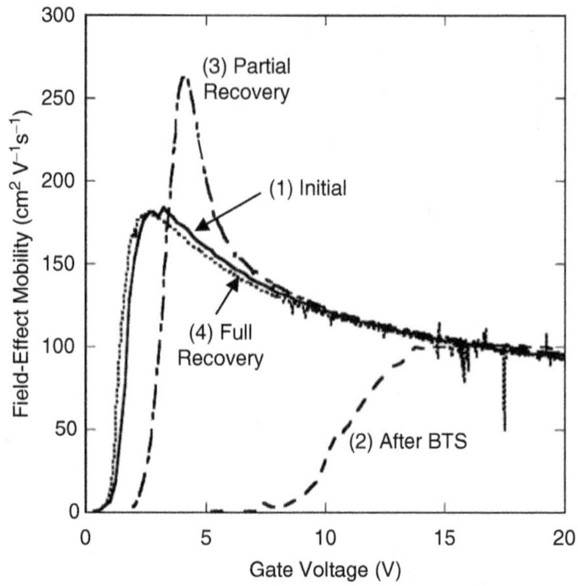

Figure 8.30 Field-effect mobility of a MOSFET whose oxide is contaminated with sodium ions. Negative bias-temperature stress (BTS) draws the positive sodium ions to the gate, where they have no effect on the electrostatics. This shifts the threshold positive and reduces the mobility. After the stress is removed, the threshold and mobility gradually recover to their pre-stress values ([25] reproduced with permission from Trans Tech Publications).

that gate oxides in SiC MOS devices may be subjected to higher fields than oxides in silicon devices. The oxide field is related to the surface field in the semiconductor by Gauss' law, so we can write

$$E_{OX} = \frac{\varepsilon_S}{\varepsilon_{OX}} E_S(0) \approx \frac{10}{3.9} E_S(0) \approx 2.6 E_S(0) \tag{8.80}$$

where $E_S(0)$ is the perpendicular component of the semiconductor field at the surface. In SiC, $E_S(0)$ (and therefore E_{OX}) can be 6–7 times higher than in silicon. The oxide field may be increased even further by fixed charges Q_F and interface trapped charges Q_{IT}. In addition, the barrier heights for electron and hole injection into SiO_2 are lower for SiC, as shown in Figure 8.21. The higher oxide field and lower barrier height raise concerns for the long term reliability of SiC MOS devices, particularly at high temperatures.

The literature contains very few careful studies of SiC MOS reliability. One reason is the large number of samples (20–50) and long measurement times (weeks to months) needed to obtain an adequate data set. Most studies make use of constant-voltage stress at elevated temperatures. The procedure is to subject a number of identical devices, either MOS capacitors or MOSFETs, to accelerated stress conditions (high oxide fields and high temperatures) and monitor the leakage currents through the gate oxides. A certain current density is chosen as indicative of oxide failure, and a set of devices is stressed simultaneously while the gate currents are monitored. As devices fail, their failure times are recorded and the test continues until the entire set of devices has failed. The failure times are plotted on a Weibull plot, where it is possible to distinguish between *extrinsic* and *intrinsic* failures. In this context, "extrinsic" refers to early failures due to oxide defects, and "intrinsic" refers to failure of an intrinsically good oxide. Since the extrinsic failures do not represent the fundamental properties of the oxide, they are eliminated from the distribution and the mean-time-before-failure (t_{50}) or the 63% failure time (t_{63}) of the remaining

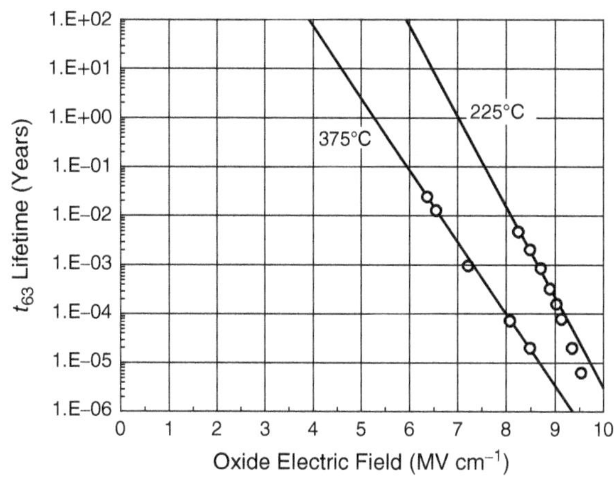

Figure 8.31 63% failure time as a function of oxide field for 4H-SiC MOS capacitors on n-type epilayers at two temperatures ([26] reproduced with permission from IEEE).

intrinsic set is determined. Constant-voltage stress is considered fairly representative of actual operating conditions. A quicker procedure, known as constant-current stress, forces a constant current through the oxide and measures the total charge-to-breakdown, or Q_{BD}. The discussion below deals only with constant-voltage stress.

Figure 8.31 shows 63% failure times as a function of oxide field for 4H-SiC MOS capacitors on n-type epilayers from two vendors [26]. As is the case in silicon [27], the time-to-failure increases exponentially as the oxide field is reduced, and a constant field acceleration factor γ is observed. At 225 °C, extrapolation to lower fields predicts a t_{63} of 100 years if the oxide field is kept below 5.9 MV cm^{-1}. The oxide field must be kept below 3.9 MV cm^{-1} if a t_{63} of 100 years is to be obtained at 375 °C. Although 100 year lifetimes may seem excessive, this is the time for 63% of the samples to fail. Failure times for lower failure percentages, say 10%, can be determined by examining the failure distributions in the Weibull plots at each field.

Figure 8.32 shows t_{63} failure time versus oxide field for 4H-SiC power DMOSFETs at three temperatures [28]. The n-channel DMOSFETs are formed on implanted p-type base regions, and a portion of the gate extends over implanted n+ source regions. Oxides over heavily-implanted regions are not as robust as those over epitaxial layers, and the t_{63} times for the DMOSFET are shorter than the MOS capacitors of Figure 8.31. Extrapolation of the high-field data indicates that a t_{63} of 100 years can be realized at 175 °C by keeping the oxide field below 6.5 MV cm^{-1}. The same lifetime can be obtained at 275 °C by keeping the oxide field below 4 MV cm^{-1}, and at 300 °C by keeping the oxide field below 1.5 MV cm^{-1}.

The data in Figure 8.32 also illustrate the danger in extrapolating high-field data to lower fields. At 275 and 300 °C the field acceleration factor γ is not independent of field, but exhibits a break at an oxide field around 8 MV cm^{-1}. If extrapolations were made using only data above 8 MV cm^{-1}, unrealistically long lifetimes would be predicted at normal operating fields. This brings into question the practice of predicting low-field reliability by extrapolation from high-field data. However, it seems safe to assume that the low-field reliability will not be *better* than predicted by extrapolating from high-field data. The severity of the change in field acceleration factor in Figure 8.32 decreases as temperature is reduced, and at 175 °C it appears that extrapolation using the high-field acceleration factor is reasonable.

Finally, we should point out that the operational lifetime of the MOSFET may be limited not by catastrophic oxide failure, but by a gradual drift in parameters such as threshold voltage or on-resistance.

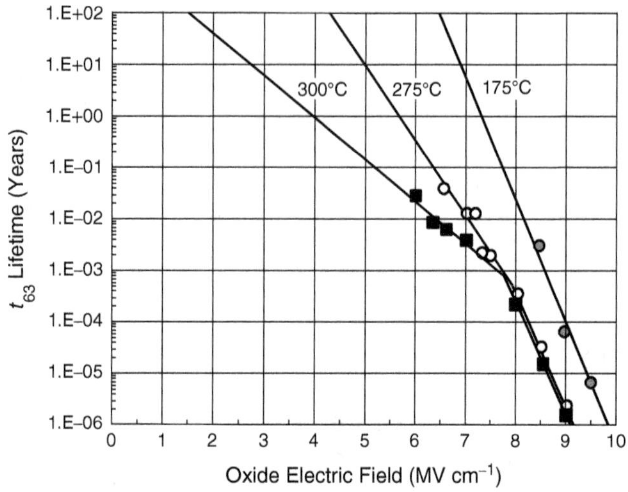

Figure 8.32 63% failure time as a function of oxide field for 4H-SiC power DMOSFETs at three temperatures ([28] reproduced with permission from Trans Tech Publications).

Unfortunately, no studies of long-term parameter drift have been published for SiC devices, leaving the question of parameter stability largely unanswered.

8.2.12 MOSFET Transient Response

One of the motivations for using MOSFETs in power switching systems is their high switching frequency and low switching loss. This allows the entire system to operate at a high frequency, leading to significant reductions in the volume and weight of passive components, such as transformers and filter capacitors, that often dominate system cost. In this section we will consider the transient response of the power MOSFET in some detail.

The switching transients of the power MOSFET can be analyzed using the clamped inductive load circuit of Figure 8.33. The inductor is representative of the windings of an electric motor, and its inductance is assumed large enough that the current I_L cannot change appreciably during the switching transient. R_G represents the internal resistance of the bond wires and the distributed gate resistance of the MOSFET. C_{GS} and C_{GD} are the gate-source and gate-drain capacitances of the MOSFET, R_S is the source resistance, and $R_D = R_{JFET} + R_{DR} + R_{SUB}$ is the lumped drain resistance of the MOSFET.

Since the p-type body regions of power MOSFETs are not lightly doped, the V_0 terms in Equations 8.44 and 8.48 cannot be ignored, and the simplified Equations 8.47 and 8.49 will overestimate the current and underestimate the switching time. However, retaining all the V_0 terms would make the mathematics very cumbersome. Therefore, to illustrate the main features of the switching transient, we will utilize the approximate Equations 8.47 and 8.49 and assume the source resistance R_S is negligible. This will not provide numerically precise answers, but will capture the qualitative features of the transient. A more quantitative analysis can be performed using the full MOSFET equations and a nonlinear circuit simulator such as SPICE™.

The turn-on transient will be considered first. Prior to $t = 0$ the MOSFET is off, and the inductor current circulates through the clamping diode. If we neglect the forward drop of the diode, which is likely to be a SiC junction-barrier Schottky (JBS) or merged pin-Schottky (MPS) diode, the drain voltage of the MOSFET is the supply voltage V_{DD}. The turn-on process can be divided into four phases, as illustrated in Figure 8.34.

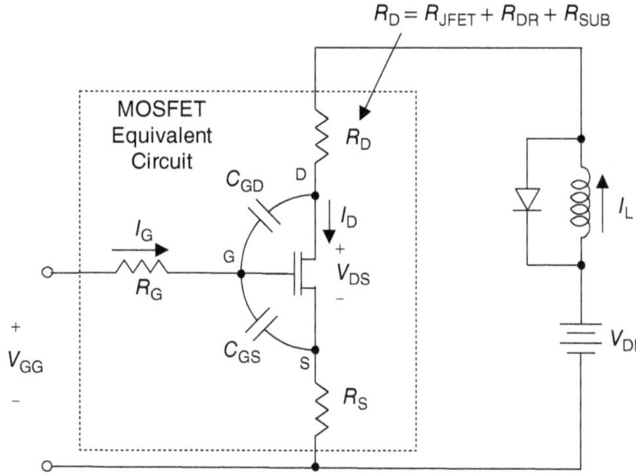

Figure 8.33 Equivalent circuit used for the transient analysis of a MOSFET driving a clamped inductive load. The dashed box encloses the MOSFET and its internal parasitic elements.

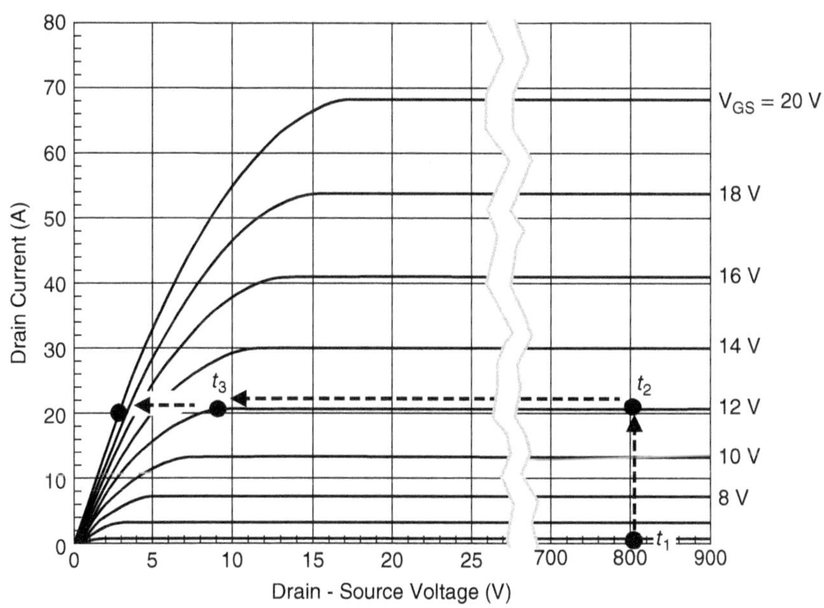

Figure 8.34 Drain characteristics of the MOSFET to be analyzed, with $\mu_N{}^* = 25\ \text{cm}^2\text{V}^{-1}\text{s}^{-1}$, C_{OX} = 69 nF, W = 125 cm, L = 0.5 μm, V_T = 2.2 V, area = 0.1 cm^2, and $R_S = R_D = 0$. In these calculations, V_{DD} = 800 V, I_L = 20 A, and V_{GG} switches between 0 and 20 V.

8.2.12.1 Turn-On, $0 < t < t_1$

At $t = 0$ the voltage at the gate terminal is stepped to $V_{GG} > V_T$, but the MOSFET does not turn on until V_{GS} has charged to V_T through the RC input circuit. The gate voltage during the charging process can be written

$$V_{GS}(t) = V_{GG}\left[1 - \exp(-t/\tau)\right] \tag{8.81}$$

where τ is given by

$$\tau = R_G\left(C_{GS} + C_{GD}\right) \tag{8.82}$$

Setting $V_{GS}(t_1) = V_T$ allows us to calculate t_1,

$$\boxed{t_1 = \tau \ln\left(\frac{V_{GG}}{V_{GG} - V_T}\right)} \tag{8.83}$$

The gate current during this phase is of interest, and can be written

$$I_G(t) = \frac{V_{GG} - V_{GS}(t)}{R_G} \tag{8.84}$$

8.2.12.2 Turn-On, $t_1 < t < t_2$

At time t_1 the MOSFET turns on and its drain current increases, but the clamping diode remains forward biased until the drain current through the MOSFET equals the inductive load current I_L. As long as the diode is forward biased, the drain-source voltage of the MOSFET remains at V_{DD} and the gate-source voltage continues to increase according to Equation 8.81. Since $V_{DS} \gg V_{D,SAT} = (V_G - V_T)$, the MOSFET is in its saturation region and the drain current can be described by Equation 8.47 with V_{DS} set equal to $V_{D,SAT}$, namely

$$I_D(t) = \mu_N^* C_{OX} \frac{W}{2L}\left[V_{GS}(t) - V_T\right]^2 \tag{8.85}$$

The drain current will increase according to Equations 8.85 and 8.81 until time t_2 when $I_D = I_L$, at which point the MOSFET is carrying all the load current and the diode becomes reverse biased. The trajectory between times t_1 and t_2 is illustrated on the $I_D - V_{DS}$ characteristics of the MOSFET in Figure 8.34. Time t_2 can be calculated by setting $I_D(t_2) = I_L$ in Equation 8.85 to obtain $V_{GS}(t_2)$, and inserting this into Equation 8.81 to solve for t_2. The result is

$$\boxed{t_2 = \tau \ln\left[\frac{V_{GG}}{V_{GG} - V_T - \sqrt{(2I_L L)/(\mu_N^* C_{OX} W)}}\right]} \tag{8.86}$$

8.2.12.3 Turn-On, $t_2 < t < t_3$

When the diode becomes reverse biased, the drain voltage of the MOSFET is no longer clamped at V_{DD}, and is free to decrease. However, the drain current is now held fixed at I_L by the inductor. With the drain current constant, the gate-source voltage cannot change, as shown by Equation 8.85. Since V_{GS} has not yet reached V_{GG}, gate current continues to flow through R_G and passes as displacement current through C_{GD}. (Although V_{GS} is constant, V_{DS} is now falling, and this draws displacement current through C_{GD}. In claiming that I_D is constant, we are assuming this displacement current is small compared to I_L.) The gate current can now be written

$$I_G = \frac{V_{GG} - V_{GS}(t_2)}{R_G} = C_{GS}\frac{dV_{GS}}{dt} + C_{GD}\frac{dV_{GD}}{dt} \tag{8.87}$$

Writing $V_{GD} = V_{GS} - V_{DS}$ and setting $dV_{GS}/dt = 0$ yields

$$I_G = \frac{V_{GG} - V_{GS}(t_2)}{R_G} = -C_{GD}\frac{dV_{DS}}{dt} \tag{8.88}$$

Since V_{GS} and I_G are constant during this period, Equation 8.88 tells us that the drain voltage decreases linearly with time. The drain voltage transient can be calculated by integrating Equation 8.88,

$$V_{DS}(t) = V_{DD} - I_L R_D - \left[\frac{V_{GG} - V_{GS}(t_2)}{R_G C_{GD}}\right](t - t_2) \tag{8.89}$$

The above analysis holds so long as the MOSFET remains in its saturation region, where the drain current is given by Equation 8.85. When V_{DS} falls below $V_{D,SAT}$, the MOSFET enters its quasi-linear region and I_D must be calculated using Equation 8.47. This marks the end of the third portion of the transient, and the time t_3 can be found by setting $V_{DS}(t_3)$ equal to $V_{D,SAT}$,

$$\boxed{t_3 = t_2 + R_G C_{GD}\left[\frac{V_{DD} - I_L R_D + V_T - V_{GS}(t_2)}{V_{GG} - V_{GS}(t_2)}\right]} \tag{8.90}$$

8.2.12.4 Turn-On, $t > t_3$

For $t > t_3$ the MOSFET is in its quasi-linear region. The gate current can be written

$$\begin{aligned} I_G = \frac{V_{GG} - V_{GS}(t)}{R_G} &= C_{GS}\frac{dV_{GS}}{dt} + C_{GD}\frac{dV_{GD}}{dt} \\ &= C_{GS}\frac{dV_{GS}}{dt} + C_{GD}\left(\frac{dV_{GS}}{dt} - \frac{dV_{DS}}{dt}\right) \end{aligned} \tag{8.91}$$

Since V_{DS} is now $\ll V_{DD}$, we can set $dV_{DS}/dt \approx 0$, so

$$I_G = \frac{V_{GG} - V_{GS}(t)}{R_G} \approx (C_{GS} + C_{GD})\frac{dV_{GS}}{dt} \tag{8.92}$$

Cross-multiplying,

$$\frac{dt}{R_G(C_{GS} + C_{GD})} \approx \frac{dV_{GS}}{(V_{GG} - V_{GS})} \tag{8.93}$$

Integrating Equation 8.93 from t_3 to t yields

$$V_{GS}(t) = V_{GG} - [V_{GG} - V_{GS}(t_3)]\exp[-(t - t_3)/\tau] \tag{8.94}$$

where we are reminded that $V_{GS}(t_3) = V_{GS}(t_2)$. During the period $t > t_3$ the drain voltage decreases toward the steady-state drain–source voltage of the MOSFET. Setting I_D in Equation 8.47 equal to I_L and solving for V_{DS} yields

$$V_{DS}(t) = (V_{GS} - V_T) - \sqrt{(V_{GS} - V_T)^2 - 2I_L L/(\mu_N^* C_{OX} W)} \tag{8.95}$$

As the device approaches steady state, V_{GS} approaches V_{GG}. Inserting V_{GG} into Equation 8.95 yields

$$V_{DS}(\infty) = (V_{GG} - V_T) - \sqrt{(V_{GG} - V_T)^2 - 2I_L L/(\mu_N^* C_{OX} W)} \tag{8.96}$$

The total gate charge during turn-on is found by setting $Q_G = C_{GS}\Delta V_{GS} + C_{GD}\Delta V_{GD}$. V_{GS} goes from zero to V_{GG} during the transient, while V_{GD} goes from $-V_{DD}$ at $t = 0$ to $V_{GG} - V_{DS}(\infty)$ at $t = \infty$. Using Equation 8.96 for $V_{DS}(\infty)$ yields

$$Q_G = C_{GS}V_{GG} + C_{GD}\left[V_{DD} + V_T + \sqrt{(V_{GG} - V_T)^2 - 2I_L L/(\mu_N^* C_{OX} W)}\right] \tag{8.97}$$

Figure 8.35a–c show gate voltage, drain current, and drain voltage for the MOSFET of Figure 8.34 as calculated using the above equations with parameters typical of a modern commercial power DMOSFET. In these calculations, the supply voltage $V_{DD} = 800$ V, the terminal gate voltage $V_{GG} = 20$ V, and the inductive load current $I_L = 20$ A. The time t_1 for V_{GS} to reach threshold is short, only 0.52 ns. The MOSFET drain current then rises according to Equation 8.85 until reaching the 20 A load current at time $t_2 = 3.97$ ns. At this point the diode becomes reverse biased and the MOSFET drain voltage decreases linearly according to Equation 8.89 until time $t_3 = 13.6$ ns. During this period V_{GS} remains constant at 11.8 V, I_G is constant at 1.78 A, and I_D is constant at 20 A. At time t_3 the drain voltage has reached

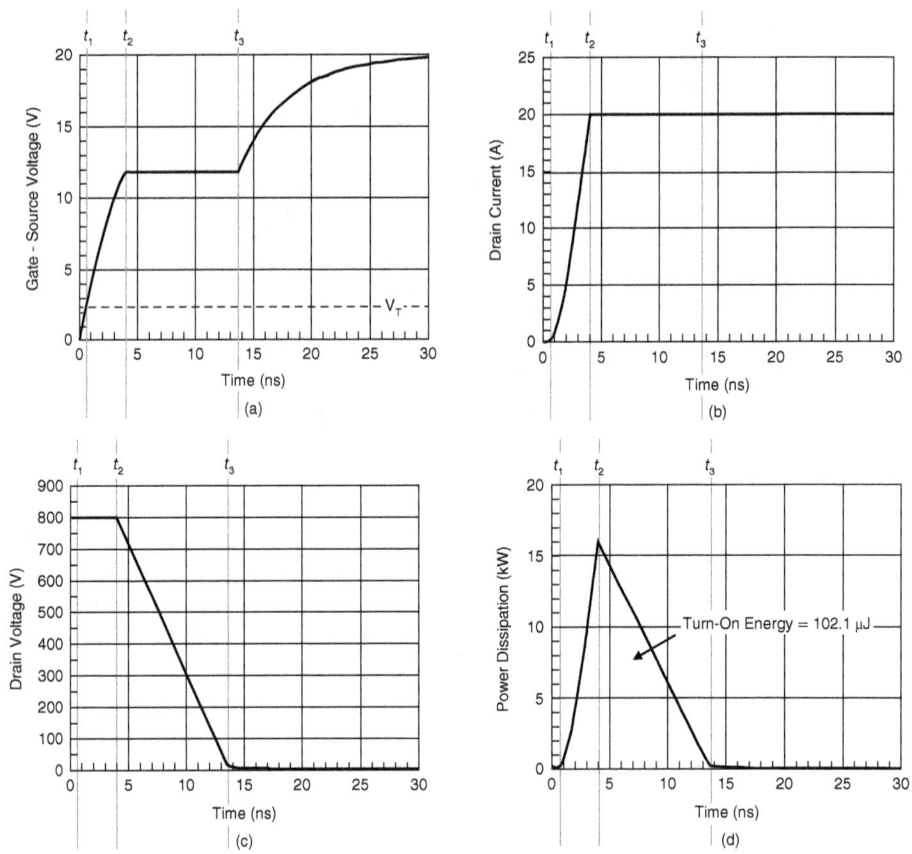

Figure 8.35 Transient waveforms during the turn-on transient, assuming $R_G = 4.6\ \Omega$, $C_{GS} = 0.944$ nF, $C_{GD} = 21.6$ pF. (a) Internal gate-source voltage V_{GS}, (b) drain current I_D, (c) drain-source voltage V_{DS}, and (d) instantaneous power dissipation within the MOSFET.

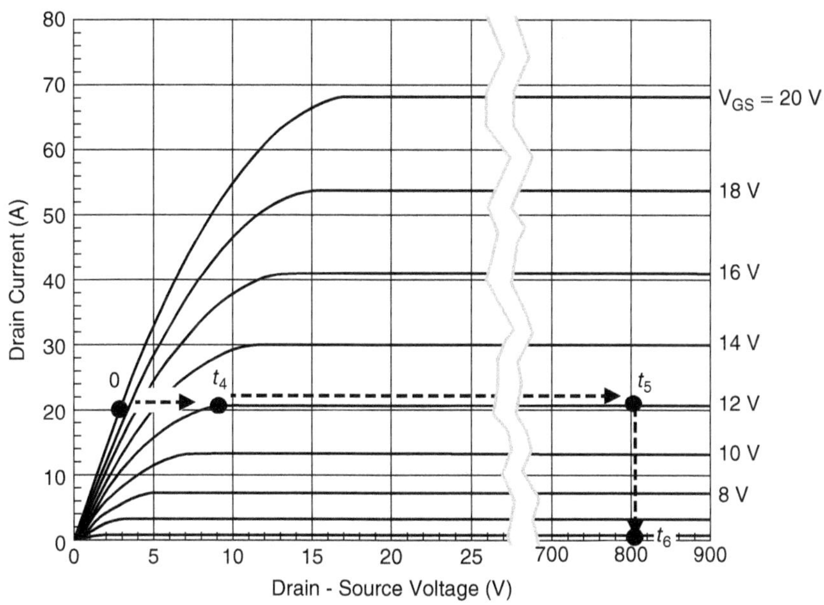

Figure 8.36 Idealized switching trajectory of the MOSFET during turn-off.

$V_{D,SAT} = V_{GS}(t_2) - V_T = 9.6$ V and the MOSFET enters its quasi-linear region. From this point, V_{DS} declines according to Equation 8.95 to a final value of $V_{DS}(\infty) = 0.26$ V.

The instantaneous power dissipated during the turn-on transient is given by

$$P(t) = I_D(t)^2(R_S + R_D) + I_D(t)V_{DS}(t) + I_G(t)^2R_G \tag{8.98}$$

The instantaneous power dissipation is plotted in Figure 8.35d, and the integral of this waveform gives the turn-on energy, 102 μJ in this example. From Equation 8.97, the gate charge during the turn-on transient is 36.6 nC.

The turn-off transient follows the same trajectory as the turn-on transient, but in the opposite direction, as illustrated in Figure 8.36. Prior to the start of the transient at $t = 0$, the MOSFET is in its conducting state with a very low forward voltage drop. The clamping diode is reverse biased, and all the load current flows through the MOSFET.

8.2.12.5 Turn-Off, $0 < t < t_4$

At $t = 0$ the voltage at the gate terminal is switched to zero, but the MOSFET does not turn off immediately because the RC input circuit must discharge. As V_{GS} decreases, V_{DS} must increase to keep the drain current at a constant value equal to I_L, but the change in V_{DS} during this period is small compared to the change in V_{GS}, so we can write the gate voltage as

$$V_{GS}(t) \approx V_{GG} \exp(-t/\tau) \tag{8.99}$$

where τ is the time constant of the gate circuit, given by Equation 8.82. However, as V_{DS} increases, it eventually reaches $V_{D,SAT}$ and the MOSFET enters saturation. At this point we can write

$$I_D = I_{D,SAT} = \mu_N^* C_{OX} \frac{W}{2L}(V_{GS} - V_T)^2 \tag{8.100}$$

Defining this point as $t = t_4$ and solving for $V_{GS}(t_4)$,

$$V_{GS}(t_4) = V_T + \sqrt{2I_L L/(\mu_N^* C_{OX} W)} \tag{8.101}$$

Inserting Equation 8.101 into Equation 8.99 gives

$$\boxed{t_4 = \tau \ln\left[\frac{V_{GG}}{V_T + \sqrt{2I_L L/\left(\mu_N^* C_{OX} W\right)}}\right]} \tag{8.102}$$

While the MOSFET is in its quasi-linear region, the drain current is given by

$$I_D = I_L = \mu_N^* C_{OX} \frac{W}{L}\left[\left(V_{GS} - V_T\right) V_{DS} - \frac{1}{2} V_{DS}^2\right] \tag{8.103}$$

Solving for V_{DS}, we can write

$$V_{DS}(t) = [V_{GS}(t) - V_T] - \sqrt{[V_{GS}(t) - V_T]^2 - 2I_L L/(\mu_N^* C_{OX} W)} \tag{8.104}$$

with $V_{GS}(t)$ given by Equation 8.99.

8.2.12.6 Turn-Off, $t_4 < t < t_5$

Between times t_4 and t_5 the MOSFET is in saturation and the drain current is fixed at I_L. Since the drain current remains constant, the gate voltage cannot change, and remains at the value given by Equation 8.101. The gate current can be written

$$I_G = \frac{0 - V_{GS}(t_4)}{R_G} = C_{GS}\frac{dV_{GS}}{dt} + C_{GD}\frac{dV_{GD}}{dt} \tag{8.105}$$

Writing $V_{GD} = V_{GS} - V_{DS}$ and setting $dV_{GS}/dt \approx 0$ yields

$$I_G = -\frac{V_{GS}(t_4)}{R_G} = -C_{GD}\frac{dV_{DS}}{dt} \tag{8.106}$$

Equation 8.106 tells us that the drain voltage increases linearly with time,

$$V_{DS}(t) = V_{DS}(t_4) + \frac{V_{GS}(t_4)}{R_G C_{GD}}(t - t_4) \tag{8.107}$$

At $t = t_5$, $V_{DS} = V_{DD} - I_L R_D$, so from Equation 8.107 we have

$$\boxed{t_5 = t_4 + R_G C_{GD}\left[\frac{V_{DD} - I_L R_D + V_T - V_{GS}\left(t_4\right)}{V_{GS}(t_4)}\right]} \tag{8.108}$$

where we have made use of the fact that $V_{DS}(t_4) = V_{D,SAT} = V_{GS}(t_4) - V_T$.

8.2.12.7 Turn-Off, $t_5 < t < t_6$

As V_{DS} tries to rise above V_{DD}, the clamping diode becomes forward biased and holds V_{DS} at V_{DD} plus a diode drop (which we will neglect). The diode carries an increasing fraction of the load current, and the

MOSFET drain current decreases toward zero. Since I_D decreases with V_{DS} held constant, V_{GS} must also decrease. Writing the equation for the gate current,

$$\begin{aligned} I_G = \frac{0 - V_{GS}(t)}{R_G} &= C_{GS}\frac{dV_{GS}}{dt} + C_{GD}\frac{dV_{GD}}{dt} \\ &= C_{GS}\frac{dV_{GS}}{dt} + C_{GD}\left(\frac{dV_{GS}}{dt} - \frac{dV_{DS}}{dt}\right) \\ &= \left(C_{GS} + C_{GD}\right)\frac{dV_{GS}}{dt} \end{aligned} \quad (8.109)$$

Solving Equation 8.109 for $V_{GS}(t)$,

$$V_{GS}(t) = V_{GS}(t_4)\exp\left[-\left(t - t_5\right)/\tau\right] \quad (8.110)$$

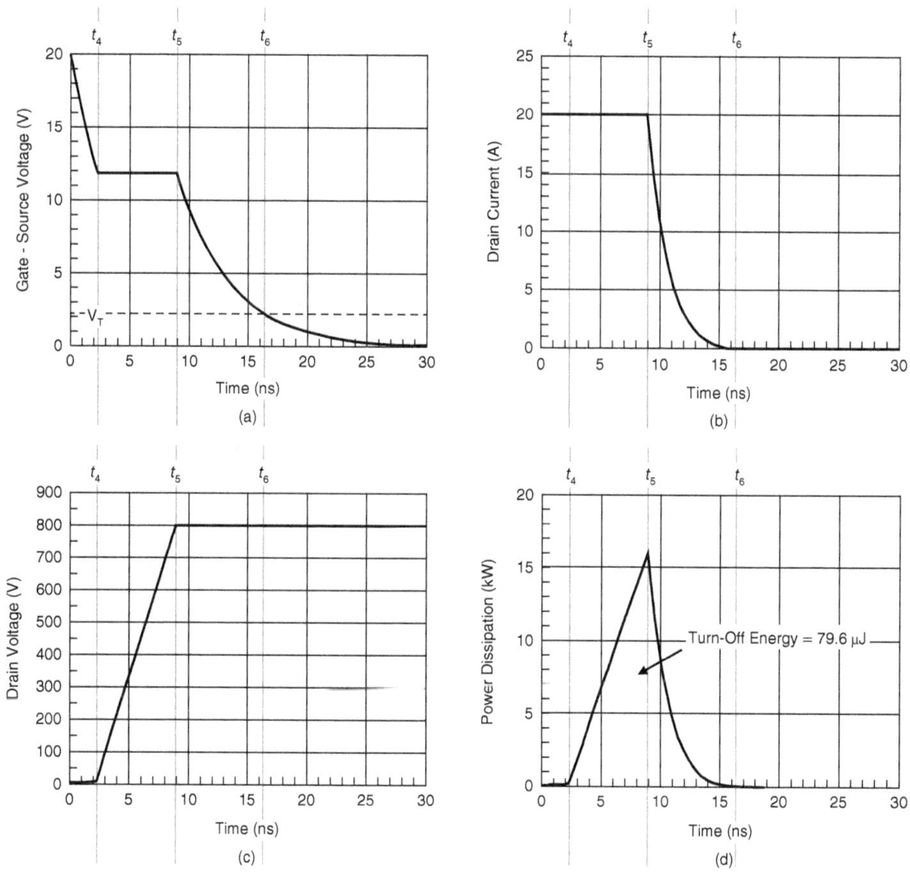

Figure 8.37 Transient waveforms during the turn-off transient. (a) Internal gate-source voltage V_{GS}, (b) drain current I_D, (c) drain-source voltage V_{DS}, and (d) instantaneous power dissipation within the MOSFET.

At $t = t_6$, V_{GS} reaches V_T and the MOSFET turns off. Setting $V_{GS}(t_6) = V_T$ in Equation 8.110, we find that

$$t_6 = t_5 + \tau \ln\left[\frac{V_T + \sqrt{2I_L L/\left(\mu_N^* C_{OX} W\right)}}{V_T}\right] \quad (8.111)$$

During the period from t_5 to t_6, the MOSFET is in its saturation region and I_D is given by Equation 8.85, with $V_{GS}(t)$ given by Equation 8.110.

8.2.12.8 Turn-Off, $t > t_6$

For times $t > t_6$ the MOSFET is off, and the gate voltage V_{GS} decreases toward zero according to Equation 8.110.

The gate charge Q_G removed by the input circuit during the turn-off transient is precisely the same as the charge supplied during the turn-on transient, since at the end of the turn-off process both C_{GS} and C_{GD} have returned to their original charge states.

Figure 8.37a–c show gate voltage, drain current, and drain voltage for the MOSFET of Figure 8.36. The time t_4 for V_{DS} to rise to $V_{D,SAT}$ is 2.33 ns. During this time the change in V_{DS}, 9.63 V, is almost imperceptible on the scale of Figure 8.37c. The MOSFET enters saturation at t_4 and the drain current remains fixed at the load current, 20 A. According to Equation 8.85, V_{GS} must also remain constant. However, V_{DS} can rise because I_D is independent of V_{DS} in saturation. The increase in V_{DS} is just sufficient to create the displacement current in C_{GD} required to satisfy Equation 8.106. The drain voltage reaches the supply voltage at time $t_5 = 8.96$ ns. At this point the clamping diode becomes forward biased and holds V_{DS} constant at V_{DD}. After t_5 the diode carries an increasing fraction of the load current, and I_D decreases. The drain current reaches zero at $t_6 = 16.4$ ns, when V_{GS} has fallen to V_T. The gate voltage continues to decrease as the input circuit completes its discharge. Figure 8.37d shows the instantaneous power during the turn-off transient, and the integral of this waveform is the turn-off energy, 79.6 μJ.

References

[1] Palmour, J.W., Edmond, J.A., Kong, H.S., and Carter, C. (1993) 6H-SiC power devices for aerospace applications. Proceedings of the 28th Intersociety Energy Conversion Conference, pp. 1.249–1.254.

[2] Shenoy, J.N., Melloch, M.R., and Cooper, J.A. (1996) High-voltage double-implanted power MOS transistors in 6H-SiC. IEEE Device Research Conference, Santa Barbara, CA.

[3] Saha, A. and Cooper, J.A. (2007) A 1200 V 4H-SiC power DMOSFET optimized for low on-resistance. *IEEE Trans. Electron. Devices*, **54** (10), 2786–2791.

[4] Tan, J., Cooper, J.A. and Melloch, M.R. (1998) High-voltage accumulation-layer UMOSFETs on 4H-SiC. *IEEE Electron. Device Lett.*, **19** (12), 487–489.

[5] Sabnis, A.G. and Clemens, J.T. (1979) Characterization of the electron mobility in the inverted <100> Si surface. IEEE International Electron Devices Meeting Technical Digest, pp. 18–21.

[6] Powell, S.K., Goldsman, N., McGarrity, J.M. *et al.* (2002) Physics-based numerical modeling and characterization of 6H-silicon carbide metal-oxide-semiconductor field-effect transistors. *J. Appl. Phys.*, **92** (7), 4053–4061.

[7] Potbhare, S., Goldsman, N., Pennington, G. *et al.* (2006) Numerical and experimental characterization of 4H-silicon carbide lateral metal-oxide-semiconductor field-effect transistor. *J. Appl. Phys.*, **100** (4), 044515.1–044515.8.

[8] Schaffer, W.J., Negley, G.H., Irvine, K.J. and Palmour, J.W. (1994) Conductivity anisotropy in epitaxial 6H and 4H-SiC, in *Diamond, SiC, and Nitride Wide-Bandgap Semiconductors*, Materials Research Society Proceedings, (eds C.H. Carter Jr.,, G. Gildenblatt, S. Nakamura and R.J. Nemanich), MRS, vol. 399, pp. 595–600.

[9] Ruff, M., Mitlehner, H. and Helbig, R. (1994) SiC devices: physics and numerical simulaton. *IEEE Trans. Electron. Devices*, **41** (6), 1040–1054.

[10] Tilak, V. and Matocha, K. (2007) Electron-scattering mechanisms in heavily doped silicon carbide MOSFET inversion layers. *IEEE Trans. Electron. Devices*, **54** (11), 2823–2829.

[11] Taillon, J. A., Yang, J. H., Ahyi, C. A. *et al.* (2013) Systematic structural and chemical characterization of the transition layer at the interface of NO-annealed 4H-SiC/SiO_2 metal-oxide-semiconductor field-effect transistors. *J. Appl. Phys.*, **113** (4) 044517.1–044517.6.
[12] Biggerstaff, T.L., Reynolds, C.L. Jr.,, Zheleva, T. *et al.* (2009) Relationship between 4H-SiC/SiO_2 transition layer thickness and mobility. *Appl. Phys. Lett.*, **95** (3), 032108.1–032108.3.
[13] Zeng, Y.A., White, M.H. and Das, M.K. (2005) Electron transport modeling in the inversion layers of 4H and 6H-SiC MOSFETs on implanted regions. *Solid-State Electron.*, **49** (6), 1017–1028.
[14] Dhar, S., Haney, S., Cheng, L. *et al.* (2010) Inversion layer carrier concentration and mobility in 4H-SiC metal-oxide-semiconductor field-effect transistors. *J. Appl. Phys.*, **108** (5), 054509.1–054509.5.
[15] Schmid, F., Krieger, M., Laube, M. *et al.* (2004) Hall scattering factor of electrons and holes in SiC, in *Silicon Carbide – Recent Major Advances* (eds W.J. Choyke, H. Matsunami and G. Pensl), Springer, Berlin, pp. 517–536.
[16] Lu, C.Y., Cooper, J.A., Tsuji, T. *et al.* (2003) Effect of processing variations and ambient temperature on electron mobility at the SiO_2/4H-SiC interface. *IEEE Trans. Electron. Devices*, **50** (7), 1582–1588.
[17] Chung, G.Y., Tin, C.C., Williams, J.R. *et al.* (2000) Effect of nitric oxide annealing on the interface trap densities near the band edges in the 4H polytype of silicon carbide. *Appl. Phys. Lett.*, **76** (13), 1713–1715.
[18] Dhar, S., Ahyi, A.C., Williams, J.R. *et al.* (2012) Temperature dependence of inversion layer carrier concentration and Hall mobility in 4H-SiC MOSFETs, in *Mater. Sci. Forum*, vol. 717–720, pp. 713–716.
[19] Okamoto, D., Yano, H., Harita, K. *et al.* (2010) Improved inversion channel mobility in 4H-SiC MOSFETs on Si face utilizing phosphorus-doped gate oxide. *IEEE Electron. Device Lett.*, **31** (7), 710–712.
[20] Sharma, Y.K., Ahyi, A.C., Isaacs–Smith, T. *et al.* (2013) High-mobility stable 4H-SiC MOSFETs using a thin PSG interfacial passivation layer. *IEEE Electron. Device Lett.*, **34** (2), 175–177.
[21] Yano, H., Hatayama, T. and Fuyuki, T. (2012) $POCl_3$ annealing as a new method for improving 4H-SiC MOS device performance. *Trans. Electrochem. Soc.*, **50** (3), 257–265.
[22] Liu, G., Ahyi, A.C., Xu, Y. *et al.* (2013) Enhanced inversion mobility on 4H-SiC ($11\bar{2}0$) using phosphorus and nitrogen interface passivation. *IEEE Electron. Device Lett.*, **34** (2), 181–183.
[23] Yoshioka, H., Nakamura, T. and Kimoto, T. (2012) Generation of very fast states by nitridation of the SiO_2/SiC interface. *J. Appl. Phys.*, **112** (1), 024520 (6 pages).
[24] Gudjonsson, G., Olafsson, O., Allerstam, F. *et al.* (2005) High field-effect mobility in n-channel Si-face 4H-SiC MOSFETs with gate oxide grown on aluminum ion-implanted material. *IEEE Electron. Device Lett.*, **26** (2), 96–98.
[25] Sveinbjörnsson, E.O., Gudjonsson, G., Allerstam, F. *et al.* (2006) High channel mobility 4H-SiC MOSFETs, in *Mater. Sci. Forum*, vol. **527–529**, pp. 961–966.
[26] Yu, L., Cheung, K.P., Campbell, J. *et al.* (2008) Oxide reliability of SiC MOS devices. 2008 IEEE International Integrated Reliability Workshop Final Report, S. Lake Tahoe, CA, October 12–16, pp. 141–144.
[27] Prendergast, J., Suehle, J., Chaparala, P. *et al.* (1995) TDDB characterization of thin SiO_2 films with bimodal failure populations. IEEE Reliability Physics Symposium, pp. 124–130.
[28] Das, M.K., Haney, S., Richmond, J. *et al.* (2012) SiC MOSFET reliability update, in *Mater. Sci. Forum*, vol. 717–720, pp. 1073–1076.

9

Bipolar Power Switching Devices

9.1 Bipolar Junction Transistors (BJTs)

Bipolar transistors were the first power transistors developed in silicon, but they have now been largely displaced by silicon insulated-gate bipolar transistors (IGBTs) and thyristors. This occurred because silicon bipolar junction transistors (BJTs) suffer from a phenomenon known as "second breakdown" that limits the safe operating area (SOA) of the device. As will be shown below, the higher critical field for avalanche breakdown in SiC essentially eliminates second breakdown, making high-performance power BJTs practical. BJTs are particularly attractive for high-temperature applications, since they are not dependent on a gate oxide for their operation, and are not subject to the oxide reliability limitations of metal-oxide-semiconductor field effect transistors (MOSFETs) and IGBTs.

As is our custom when considering a new device, we begin our discussion with a review of the basics of BJT operation. We then consider a number of special effects that occur during high-current operation of power BJTs. Figure 9.1 shows a basic n+/p/n− BJT, along with standard current and voltage definitions. In these devices the emitter is more heavily doped than the base, and the base is more heavily doped than the collector, so $N_{DE} \gg N_{AB} \gg N_{DC}$. We can identify four internal junction currents, the hole and electron currents crossing the emitter–base (EB) junction (I_{EP} and I_{EN}) and the hole and electron currents crossing the collector–base (CB) junction (I_{CP} and I_{CN}). The polarities in the figure correspond to positive current flow, and we should remember that electron flux is opposite to direction of electron current. Voltages V_{BE} and V_{BC} are developed across the junction depletion regions, shown cross-hatched in the figure. The widths of the emitter, base, and collector, measured from the metallurgical junctions, are W_E, W_B, and W_C, and the width of the neutral portion of the base, measured between the edges of the depletion regions, is simply W.

9.1.1 *Internal Currents*

Our first goal is to obtain equations for the four internal currents I_{EP}, I_{EN}, I_{CP}, and I_{CN} in terms of the device parameters and the terminal voltages V_{BE} and V_{BC}. We can then write expressions for the terminal currents I_E, I_B, and I_C, and from these we will obtain equations for important performance parameters such as current gain, specific on-resistance, and so on. We initially assume low-level injection in all regions, since this allows us to use the minority carrier diffusion equations (MCDEs) to obtain the desired internal currents. We will later remove this restriction in the base and collector.

The solution to the MCDE in the emitter and collector proceeds in the same way as in the neutral regions of a pn diode discussed in Section 7.3. We first write the general solution of the MCDE in the

Fundamentals of Silicon Carbide Technology: Growth, Characterization, Devices, and Applications, First Edition.
Tsunenobu Kimoto and James A. Cooper.

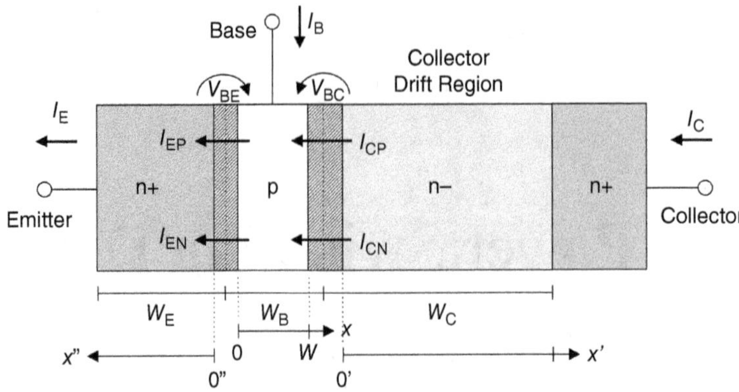

Figure 9.1 Basic structure of an n+/p/n– BJT showing internal current components I_{EN}, I_{EP}, I_{CN}, and I_{CP}, terminal currents I_E, I_C, and I_B, and x, x', and x'' coordinate systems. The arrows indicate assumed directions for positive current.

emitter and collector, subject to two boundary conditions in each region. One boundary condition, the carrier density at the edge of the depletion region, is provided by the law of the junction, Equation 7.25. The second boundary condition is based on the assumption that the minority carrier densities far from the junctions remain at their equilibrium values. However, in SiC BJTs it may not be accurate to assume that the emitter ohmic contact is infinitely far from the junction. In this case, we must set the minority carrier density at the position of the ohmic contact to its equilibrium value. The solution for $\Delta p_E(x)$ in the emitter involves hyperbolic functions (see Appendix B), and can be written

$$\Delta p_E(x) = \frac{n_{iE}^2}{N_{DE}^+}[\exp(qV_{BE}/kT) - 1]\frac{\sinh[(W_E - x)/L_{PE}]}{\sinh(W_E/L_{PE})} \tag{9.1}$$

Here n_{iE} is the intrinsic carrier concentration in the emitter, $N_{DE}{}^+$ is the ionized dopant concentration in the emitter, and $L_{PE} = (D_{PE}\tau_{PE})^{0.5}$ is the hole diffusion length in the emitter, where D_{PE} is the hole diffusion coefficient and τ_{PE} is the hole lifetime. We specify n_{iE} in the emitter separately from n_i in the other portions of the device, since the emitter is usually so heavily doped that bandgap narrowing occurs, resulting in a higher intrinsic carrier concentration. The hole current crossing the EB junction is obtained from the derivative of Equation 9.1 at the depletion edge $x = 0$,

$$\boxed{I_{EP} = \frac{qAD_{PE}}{L_{PE}}\frac{n_{iE}^2}{N_{DE}^+}\frac{\cosh(W_E/L_{PE})}{\sinh(W_E/L_{PE})}[\exp(qV_{BE}/kT) - 1]} \tag{9.2}$$

where A is the area of the junction. In the case of a "long" emitter where $W_E \gg L_{PE}$, Equation 9.2 reduces to

$$I_{EP} = \frac{qAD_{PE}}{L_{PE}}\frac{n_{iE}^2}{N_{DE}^+}[\exp(qV_{BE}/kT) - 1], \quad (W_E \gg L_{PE}) \tag{9.3}$$

The solution in the collector is similar to that in the emitter. By analogy with Equation 9.2 we can write the hole current crossing the CB junction as

$$\boxed{I_{CP} = \frac{qAD_{PC}}{L_{PC}}\frac{n_i^2}{N_{DC}^+}\frac{\cosh(W_C/L_{PC})}{\sinh(W_C/L_{PC})}[\exp(qV_{BC}/kT) - 1]} \tag{9.4}$$

which in the case of a "long" collector reduces to

$$I_{CP} = \frac{qAD_{PC}}{L_{PC}}\frac{n_i^2}{N_{DC}^+}[\exp(qV_{BC}/kT) - 1], \quad (W_C \gg L_{PC}) \tag{9.5}$$

Equations 9.3 and 9.5 are similar in form to the Shockley diode Equation 7.26.

The solution for the minority carrier density in the base is obtained in the same manner – we find the general solution of the MCDE, subject to the boundary conditions at the EB and CB junctions provided by the law of the junction. In a typical SiC BJT we cannot always assume $W \ll L_{NB}$, and the full solution for $\Delta n(x)$ will involve hyperbolic functions, namely

$$\Delta n_B(x) = \frac{n_i^2}{N_{AB}^-}[\exp(qV_{BE}/kT) - 1]\frac{\sinh[(W-x)/L_{NB}]}{\sinh(W/L_{NB})} + \frac{n_i^2}{N_{AB}^-}[\exp(qV_{BC}/kT) - 1]\frac{\sinh(x/L_{NB})}{\sinh(W/L_{NB})} \tag{9.6}$$

The electron current crossing the EB junction is obtained from the derivative of Equation 9.6 evaluated at $x = 0$, and the electron current crossing the CB junction is obtained from the derivative evaluated at $x = W$. After some algebra, we can write

$$\boxed{I_{EN} = \frac{qAD_{NB}}{L_{NB}}\frac{n_i^2}{N_{AB}^-}\left\{\frac{\cosh(W/L_{NB})}{\sinh(W/L_{NB})}[\exp(qV_{BE}/kT) - 1] - \frac{1}{\sinh(W/L_{NB})}[\exp(qV_{BC}/kT) - 1]\right\}} \tag{9.7}$$

and

$$\boxed{I_{CN} = \frac{qAD_{NB}}{L_{NB}}\frac{n_i^2}{N_{AB}^-}\left\{\frac{1}{\sinh(W/L_{NB})}[\exp(qV_{BE}/kT) - 1] - \frac{\cosh(W/L_{NB})}{\sinh(W/L_{NB})}[\exp(qV_{BC}/kT) - 1]\right\}} \tag{9.8}$$

If the base width is much shorter than the minority carrier diffusion length in the base, we can set $W \ll L_{NB}$, and the above equations reduce to

$$I_{EN} = I_{CN} = \frac{qAD_{NB}}{W}\frac{n_i^2}{N_{AB}^-}[\exp(qV_{BE}/kT) - \exp(qV_{BC}/kT)], \ (W \ll L_{NB}) \tag{9.9}$$

Equations 9.2, 9.4, 9.7, and 9.8 are the desired equations for the four internal currents in the BJT of Figure 9.1.

9.1.2 Gain Parameters

We can now write equations for some of the important performance parameters of the BJT. The common-base current gain α is defined as the ratio of collector current to emitter current. We assume the BJT is operated in the forward-active mode with the EB junction forward biased and the CB junction reverse biased, so that $V_{BE} \gg kT/q$ and $V_{BC} \ll -kT/q$. Under forward-active biasing the $[\exp(qV_{CB}/kT) - 1]$ terms in the above equations are small compared to the $[\exp(qV_{EB}/kT) - 1]$ terms, and the -1 terms are small compared to the $\exp(qV_{EB}/kT)$ terms. The current component I_{CP} represents the reverse-bias leakage current of the CB junction, and from Equation 9.4 we see it is very small and can be neglected. Therefore, the common-base current gain can be written

$$\alpha = \frac{I_C}{I_E} \approx \frac{I_{CN}}{I_{EN} + I_{EP}} = \left(\frac{I_{CN}}{I_{EN}}\right)\left(\frac{I_{EN}}{I_{EN} + I_{EP}}\right) = \alpha_T \gamma \tag{9.10}$$

where α_T is defined as the *base transport factor* and γ is the *emitter injection efficiency*. The base transport factor is the fraction of electrons injected from the emitter that diffuse across the base and are swept across the collector junction. The emitter injection efficiency is the fraction of current crossing the EB junction that is due to electrons injected from the emitter into the base. The base transport factor is obtained by dividing Equation 9.8 by Equation 9.7. Under forward-active biasing, α_T can be simply written

$$\alpha_T = \frac{1}{\cosh(W/L_{NB})} \tag{9.11}$$

Likewise, using Equations 9.7 and 9.2, the emitter injection efficiency is

$$\gamma = \frac{1}{1 + \dfrac{D_{PE}}{D_{NB}} \dfrac{L_{NB}}{L_{PE}} \dfrac{N_{AB}^-}{N_{DE}^+} \dfrac{n_{iE}^2}{n_i^2} \dfrac{\cosh(W_E/L_{PE})}{\sinh(W_E/L_{PE})} \dfrac{\sinh(W/L_{NB})}{\cosh(W/L_{NB})}} \tag{9.12}$$

Inserting Equations 9.11 and 9.12 into Equation 9.10, we can write the common-base current gain as

$$\boxed{\alpha = \frac{1}{\cosh\left(\dfrac{W}{L_{NB}}\right) + \dfrac{D_{PE}}{D_{NB}} \dfrac{L_{NB}}{L_{PE}} \dfrac{N_{AB}^-}{N_{DE}^+} \dfrac{n_{iE}^2}{n_i^2} \coth\left(\dfrac{W_E}{L_{PE}}\right) \sinh\left(\dfrac{W}{L_{NB}}\right)}} \tag{9.13}$$

The common-emitter current gain β is the ratio of collector current to base current,

$$\beta = \frac{I_C}{I_B} = \frac{I_C}{I_E - I_C} = \frac{I_C/I_E}{1 - I_C/I_E} = \frac{\alpha}{1 - \alpha} \tag{9.14}$$

Inserting Equation 9.13 into Equation 9.14, we can write

$$\boxed{\beta = \frac{1}{\cosh(W/L_{NB}) + \dfrac{D_{PE}}{D_{NB}} \dfrac{L_{NB}}{L_{PE}} \dfrac{N_{AB}^-}{N_{DE}^+} \dfrac{n_{iE}^2}{n_i^2} \dfrac{\cosh(W_E/L_{PE})}{\sinh(W_E/L_{PE})} \sinh(W/L_{NB}) - 1}} \tag{9.15}$$

We can simplify the above equations under certain conditions. If the base is narrow compared to a diffusion length, the hyperbolic functions involving W/L_{NB} can be replaced by the first term of their Taylor series expansions, namely,

$$\left.\begin{aligned} \sinh(\theta) &\approx \theta \\ \cosh(\theta) &\approx 1 \end{aligned}\right\} \quad (\theta \ll 1) \tag{9.16}$$

Moreover, if the emitter is long compared to a diffusion length, the hyperbolic functions involving W_E/L_{PE} can be simplified using

$$\frac{\cosh(\theta)}{\sinh(\theta)} \approx 1, \quad (\theta \gg 1) \tag{9.17}$$

Under these conditions, $\alpha_T \approx 1$ and we can write

$$\alpha \approx \gamma \approx \frac{1}{1 + \dfrac{D_{PE}}{D_{NB}} \dfrac{W}{L_{PE}} \dfrac{N_{AB}^-}{N_{DE}^+} \dfrac{n_{iE}^2}{n_i^2}} \tag{9.18}$$

From this, it follows that

$$\boxed{\beta \approx \frac{D_{NB}}{D_{PE}} \frac{L_{PE}}{W} \frac{N_{DE}^+}{N_{AB}^-} \frac{n_i^2}{n_{iE}^2}} \tag{9.19}$$

If the emitter is short compared to a diffusion length, $W_E \ll L_{PE}$, we simply replace L_{PE} by W_E in the above equations. Although these simplifications are appealing, a word of caution is in order. The "narrow base" and "long emitter" assumptions are often invoked for silicon BJTs, but they may not be accurate for all SiC power BJTs. When in doubt, it is advisable to use the full expressions that involve hyperbolic functions.

9.1.3 Terminal Currents

We next wish to obtain expressions for the terminal currents I_E, I_C, and I_B. This is easily done by recognizing that $I_E = I_{EN} + I_{EP}, I_C = I_{CN} + I_{CP}$, and $I_B = I_E - I_C$. Adding Equations 9.7 and 9.2 we obtain

$$\begin{aligned} I_E = qA &\left[\frac{D_{NB}}{L_{NB}} \frac{n_i^2}{N_{AB}^-} \frac{\cosh(W/L_{NB})}{\sinh(W/L_{NB})} + \frac{D_{PE}}{L_{PE}} \frac{n_{iE}^2}{N_{DE}^+} \frac{\cosh(W_E/L_{PE})}{\sinh(W_E/L_{PE})}\right] [\exp(qV_{BE}/kT) - 1] \\ - qA &\left[\frac{D_{NB}}{L_{NB}} \frac{n_i^2}{N_{AB}^-} \frac{1}{\sinh(W/L_{NB})}\right] [\exp(qV_{BC}/kT) - 1] \end{aligned} \tag{9.20}$$

Similarly, adding Equations 9.8 and 9.4 yields

$$\begin{aligned} I_C = qA &\left[\frac{D_{NB}}{L_{NB}} \frac{n_i^2}{N_{AB}^-} \frac{1}{\sinh(W/L_{NB})}\right] [\exp(qV_{BE}/kT) - 1] \\ - qA &\left[\frac{D_{NB}}{L_{NB}} \frac{n_i^2}{N_{AB}^-} \frac{\cosh(W/L_{NB})}{\sinh(W/L_{NB})} + \frac{D_{PC}}{L_{PC}} \frac{n_i^2}{N_{DC}^+} \frac{\cosh(W_C/L_{PC})}{\sinh(W_C/L_{PC})}\right] [\exp(qV_{BC}/kT) - 1] \end{aligned} \tag{9.21}$$

The equation for I_B follows by subtracting Equation 9.21 from Equation 9.20.

The above equations initially seem formidable, but a closer analysis reveals symmetries that lead to a simpler formulation. We note that each equation is the algebraic sum of two Shockley diode equations of the form given in Equation 7.26. We can exploit this symmetry by writing Equations 9.20 and 9.21 in the alternate forms

$$\boxed{I_E = I_{F0}[\exp(qV_{BE}/kT) - 1] - \alpha_R I_{R0}[\exp(qV_{BC}/kT) - 1]} \tag{9.22}$$

and

$$\boxed{I_C = \alpha_F I_{F0}[\exp(qV_{BE}/kT) - 1] - I_{R0}[\exp(qV_{BC}/kT) - 1]} \tag{9.23}$$

Equations 9.22 and 9.23 are known as the *Ebers–Moll equations* for the BJT. The four constants in the Ebers–Moll equations can be determined by comparing Equations 9.22 and 9.23 to Equations 9.20 and 9.21. For convenience, we summarize these parameters below.

$$\begin{aligned}
I_{F0} &= qA\left[\frac{D_{NB}}{L_{NB}}\frac{n_i^2}{N_{AB}^-}\frac{\cosh(W/L_{NB})}{\sinh(W/L_{NB})} + \frac{D_{PE}}{L_{PE}}\frac{n_{iE}^2}{N_{DE}^+}\frac{\cosh(W_E/L_{PE})}{\sinh(W_E/L_{PE})}\right]\\
I_{R0} &= qA\left[\frac{D_{NB}}{L_{NB}}\frac{n_i^2}{N_{AB}^-}\frac{\cosh(W/L_{NB})}{\sinh(W/L_{NB})} + \frac{D_{PC}}{L_{PC}}\frac{n_i^2}{N_{DC}^+}\frac{\cosh(W_C/L_{PC})}{\sinh(W_C/L_{PC})}\right]\\
\alpha_F &= \frac{1}{1 + \dfrac{D_{PE}}{D_{NB}}\dfrac{L_{NB}}{L_{PE}}\dfrac{N_{AB}^-}{N_{DE}^+}\dfrac{n_{iE}^2}{n_i^2}\dfrac{\sinh(W/L_{NB})}{\cosh(W/L_{NB})}\dfrac{\cosh(W_E/L_{PE})}{\sinh(W_E/L_{PE})}}\\
\alpha_R &= \frac{1}{1 + \dfrac{D_{PC}}{D_{NB}}\dfrac{L_{NB}}{L_{PC}}\dfrac{N_{AB}^-}{N_{DC}^+}\dfrac{\sinh(W/L_{NB})}{\cosh(W/L_{NB})}\dfrac{\cosh(W_C/L_{PC})}{\sinh(W_C/L_{PC})}}
\end{aligned} \tag{9.24}$$

As before, the approximations in Equations 9.16 and 9.17 can be applied to the above equations, if justified in a particular situation.

Equations 9.24 allow us to calculate all four constants in the Ebers–Moll equations for the BJT terminal currents. The Ebers–Moll equations are associated with a simple and easily-remembered equivalent circuit for the BJT known as the Ebers–Moll model, shown in Figure 9.2. The Ebers–Moll equations and model are valid in all biasing regimes of the BJT: forward-active, saturation, inverse-active, and cutoff, and they form the basis for the BJT model used in popular circuit analysis programs such as SPICE™.

Consider the Ebers–Moll model for a BJT in the forward-active mode, with the EB junction forward biased and the CB junction reverse biased. With the CB junction reverse biased, the diode current I_R is just the reverse leakage current $-I_{R0}$, which is small and can be neglected. Likewise, the dependent current generator $\alpha_R I_R$ can be neglected. Diode I_F then represents the hole and electron currents crossing the EB junction under forward bias, and the dependent current generator $\alpha_F I_F$ represents the fraction of electron current injected from the emitter that reaches the collector (keep in mind that the flow of electrons is opposite to the positive direction of electron current). If the BJT were operated in the inverse-active

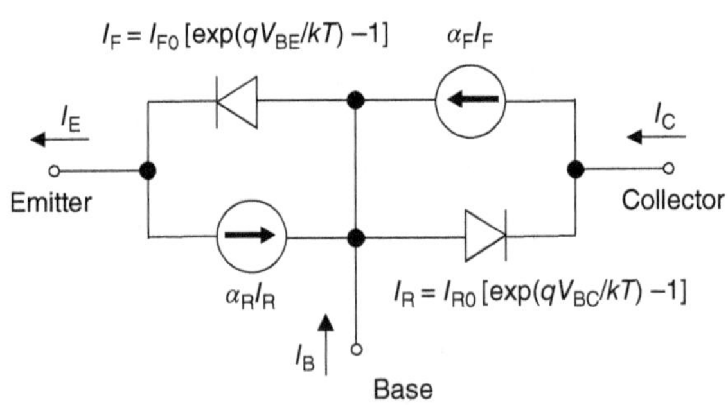

Figure 9.2 The Ebers–Moll equivalent circuit model of the npn BJT. This model can be used in all operating regimes: forward-active, saturation, inverse-active, and cutoff.

mode with the EB junction reverse biased and the CB junction forward biased, the roles of the I_F and I_R diodes would be reversed: the collector would inject electrons into the base, and the emitter would sweep them out. In saturation, both junctions are forward biased and all four elements of the Ebers–Moll model are active.

9.1.4 Current–Voltage Relationship

The Ebers–Moll model can be used to obtain a single equation for the I_C–V_{CE} characteristics with base current as a parameter. We first write I_B in terms of the Ebers–Moll circuit model as

$$I_B = (1 - \alpha_F)I_F + (1 - \alpha_R)I_R \tag{9.25}$$

where, as illustrated in Figure 9.2,

$$I_F = I_{F0}[\exp(qV_{BE}/kT) - 1] \tag{9.26}$$

and

$$I_R = I_{R0}[\exp(qV_{BC}/kT) - 1] \tag{9.27}$$

We can eliminate V_{BC} in Equation 9.27 by writing $V_{BC} = V_{BE} - V_{CE}$, resulting in

$$I_R = I_{R0}[\exp(qV_{BE}/kT)\exp(-qV_{CE}/kT) - 1] \tag{9.28}$$

Inserting Equations 9.26 and 9.28 into Equation 9.25 then provides an equation for I_B in terms of V_{BE} and V_{CE},

$$I_B = (1 - \alpha_F)I_{F0}[\exp(qV_{BE}/kT) - 1] + (1 - \alpha_R)I_{R0}[\exp(qV_{BE}/kT)\exp(-qV_{CE}/kT) - 1] \tag{9.29}$$

Equation 9.29 can be solved for $\exp(qV_{BE}/kT)$, yielding

$$\exp(qV_{BE}/kT) = \left[\frac{I_B + (1 - \alpha_F)I_{F0} + (1 - \alpha_R)I_{R0}}{(1 - \alpha_F)I_{F0} + (1 - \alpha_R)I_{R0}\exp(-qV_{CE}/kT)}\right] \tag{9.30}$$

From the Ebers–Moll model we can write the collector current as

$$I_C = \alpha_F I_F - I_R \tag{9.31}$$

We now substitute Equation 9.30 into Equations 9.26 and 9.28 to eliminate V_{BE}, then insert Equations 9.26 and 9.28 into Equation 9.31 to obtain the desired equation for I_C as a function of V_{CE} with I_B as a parameter. After some algebra we can write

$$\boxed{I_C = [\alpha_F I_{F0} - I_{R0}\exp(-qV_{CE}/kT)]\left[\frac{I_B + (1 - \alpha_F)I_{F0} + (1 - \alpha_R)I_{R0}}{(1 - \alpha_F)I_{F0} + (1 - \alpha_R)I_{R0}\exp(-qV_{CE}/kT)}\right] - \alpha_F I_{F0} + I_{R0}} \tag{9.32}$$

Equation 9.32 is valid in all biasing regimes: forward-active, saturation, inverse-active, and cut-off. Despite its apparent complexity, Equation 9.32 is a simple algebraic equation involving four constants (the four Ebers–Moll parameters) and the variables V_{CE} and I_B. Equation 9.32 can be used to generate a plot of I_C as a function of V_{CE} for various values of base current I_B, and such a plot is shown in Figure 9.3. The parameters used to generate this plot are representative of a typical 4H-SiC n+/p/n– BJT.

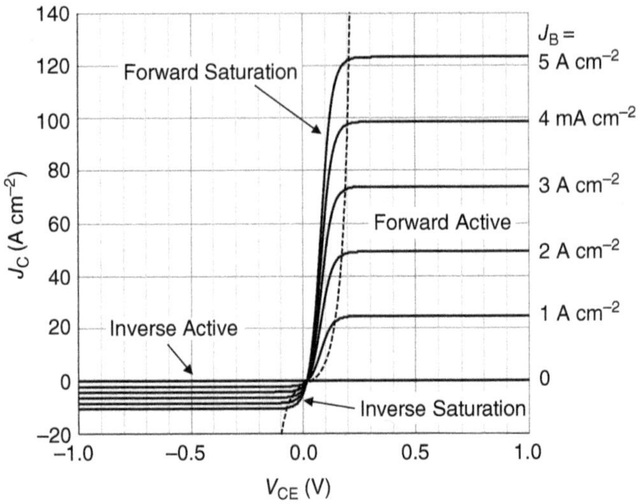

Figure 9.3 Current–voltage characteristics of an n+/p/n− BJT in 4H-SiC at room temperature, calculated using Equation 9.32. In this example, $N_{DE} = 1 \times 10^{19}$ cm^{-3}, $N_{AB} = 2 \times 10^{17}$ cm^{-3}, $N_{DC} = 2 \times 10^{15}$ cm^{-3}, and $W = 1$ μm. τ_{PE}, τ_{NB}, and τ_{PC} are 1 ns, 10 ns, and 2 μs, and μ_{PE}, μ_{PC}, μ_{NB}, and μ_{NC} are 78, 123, 565, and 1075 cm^{2}V^{-1}s^{-1}, respectively. The common-emitter current gain is 25 in the forward-active region and 1.1 in the inverse-active region.

The equations developed to this point involve a number of assumptions that may not be valid in all regions of operation. This is especially true of power BJTs, since they operate at high current densities where our assumptions of *low-level injection* fail, first in the collector and then in the base. Other effects occur at high voltages, and still others at high temperatures. In addition, there are important lateral effects that are not captured in our one-dimensional analysis. In the following sections we will discuss the most important of these effects, and how they influence practical SiC power BJTs.

9.1.5 High-Current Effects in the Collector: Saturation and Quasi-Saturation

To support a high blocking voltage in the off state, the power BJT incorporates a thick, lightly-doped collector drift region between the CB junction and the n+ substrate, as shown in Figure 9.4. As with the power junction field-effect transistor (JFET) and power MOSFET, this drift region is designed to support the desired blocking voltage, with doping and thickness given by Equations 7.10 and 7.11. In the absence of conductivity modulation, the drift region would add a series resistance given by Equation 7.12 to the basic BJT described above, but in the saturation regime the drift region is partially or totally conductivity modulated. The actual resistance and associated voltage drop added by the collector drift region will be considered next.

Figure 9.5 illustrates the I_C−V_{CE} relationships of an n+/p/n− power BJT with a thick lightly-doped collector drift region, and Figure 9.6 shows the corresponding minority carrier densities in the device at operating points (a), (c), and (e). The behavior in saturation consists of two distinct regimes: *saturation* and *quasi-saturation*. This can be understood as follows. In the forward-active region (a), the CB junction is reverse biased and minority carriers are not injected into the collector drift region. The drift region is not conductivity modulated, and the carrier densities are illustrated in Figure 9.6a. As V_{CE} is reduced,

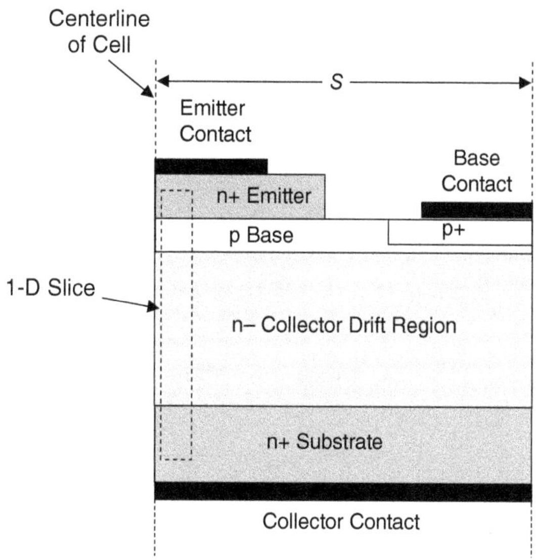

Figure 9.4 Implementation of a practical power BJT. All layers are epitaxially grown to avoid the reduced lifetimes associated with implanted regions.

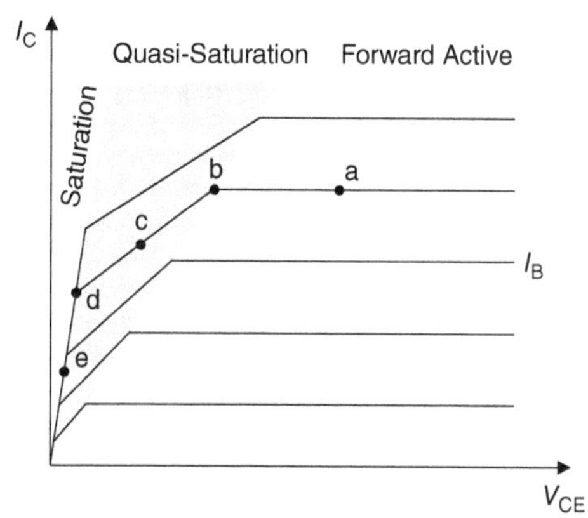

Figure 9.5 Illustration of the $I_C - V_{CE}$ characteristics of a power BJT with a thick drift region.

the reverse bias on the CB junction is reduced and V_{BC} eventually reaches zero at point (b) in Figure 9.5. This is the boundary between the forward-active and saturation regimes. As V_{CE} is reduced further, the CB junction becomes forward biased, injecting holes into the n− collector drift region. Since the collector is lightly doped, even a small injection of holes places the region near the junction into high-level injection, and the minority carrier density at bias point (c) is shown in Figure 9.6c. Here the drift region is conductivity modulated out to a distance x_M', but not beyond. The carrier densities vary linearly with distance from $x' = 0$ to $x' = x_M'$, as will be explained below. As we continue to reduce V_{CE}, the CB junction becomes even more forward biased, increasing the injection of holes into the drift region, and at bias point (d) the modulated portion extends across the entire drift region, that is, $x_M' = W_C$. The resistance of the drift region has now reached a very low value, and the $I_C - V_{CE}$ characteristic in Figure 9.5 follows a

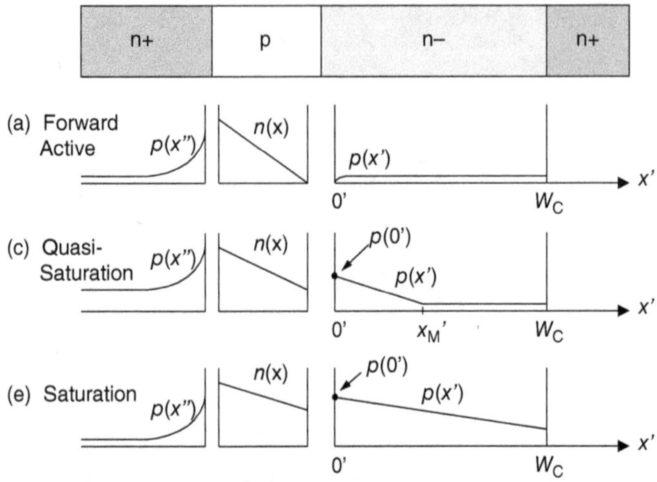

Figure 9.6 Minority carrier densities in the neutral regions of a power BJT in (a) forward-active, (c) quasi-saturation, and (e) saturation biasing regimes (see points a, c and e in Figure 9.5).

steep slope toward the origin. Thus, the term "saturation" in the power BJT refers to the condition where the drift region is fully conductivity-modulated, and the term "quasi-saturation" refers to the condition where the drift region is partially conductivity-modulated.

We will now develop equations describing the carrier density in the drift region and the voltage drop across the drift region in saturation and quasi-saturation. Since the region must remain charge neutral, we can write

$$n(x') = p(x') + N_{DC}^{+} \tag{9.33}$$

The hole concentration at $x' = 0$ is given by the law of the junction, Equation 7.25,

$$p(0) = p_{NC0} \exp(qV_{BC}/kT) \tag{9.34}$$

Since the drift region is lightly doped, $p(0)$ can exceed N_{DC} at even moderate forward biases V_{BC}, so high-level injection prevails, and in this region we can write

$$\frac{\partial p}{\partial x'} = \frac{\partial n}{\partial x'} \tag{9.35}$$

The electron and hole currents in the drift region are

$$J_N = q\mu_{NC} n(x')E(x') + qD_{NC}\frac{\partial n}{\partial x'} \tag{9.36}$$

$$J_P = q\mu_{PC} p(x')E(x') - qD_{PC}\frac{\partial p}{\partial x'} \tag{9.37}$$

We now assert that the injected hole current flowing into the drift region from the base, $J_P(0)$, is small compared to the electron current flowing across the base from the emitter, $J_N(0)$. The situation in the BJT is obviously different from the pin diode, where electrons are injected from the n+ region and holes from the p+ region. In the BJT drift region, the primary electron current comes from electrons injected from the emitter (I_{EN}) that diffuse across the base into the collector. The hole current injected from the base into the collector is much smaller, due to the gain of the transistor. Recall that $I_B = I_C/\beta$. Assuming $\beta \gg 1$, the base current is much smaller than the collector current. The base current in saturation consists of holes injected into the emitter (I_{EP}) plus holes injected into the collector (I_{CP}) plus holes recombining in the base, so we can be sure that $I_{CP} < I_B \ll I_C$. Thus the collector current is almost totally due to electrons, and we can set $J_P \approx 0$ in Equation 9.37. Solving for the electric field and employing the Einstein relationship, we find that

$$E(x') \approx \frac{kT}{q}\frac{1}{p}\frac{\partial p}{\partial x'} \tag{9.38}$$

The collector current, which is due primarily to electrons, can be obtained by combining Equation 9.36 with Equations 9.33, 9.35, and 9.38,

$$-J_C \approx J_{CN} = q\mu_{NC}(p + N_{DC}^+)\left(\frac{kT}{q}\frac{1}{p}\right)\frac{\partial p}{\partial x'} + qD_{NC}\frac{\partial p}{\partial x'} \tag{9.39}$$

Again using the Einstein relationship, Equation 9.39 can be written

$$-J_C \approx 2qD_{NC}\left(1 + \frac{N_{DC}^+}{2p}\right)\frac{\partial p}{\partial x'} \tag{9.40}$$

This equation can be solved by cross-multiplying both sides by $\partial x'$ and integrating with respect to x'. Performing the integration and solving for $p(x')$ yields

$$p(x') = p(0) - \frac{J_C x'}{2qD_{NC}} + \frac{N_{DC}^+}{2}\ln\left[\frac{p(0)}{p(x')}\right] \tag{9.41}$$

Since $p(x') \gg N_{DC}^+$ in saturation, the third term can be dropped, making the hole density a linearly decreasing function of position,

$$\boxed{p(x') \approx p(0) - \frac{J_C x'}{2qD_{NC}}} \tag{9.42}$$

The reader should recognize that the transport parameters μ_{NC} and D_{NC} in Equation 9.39 and the following equations are those of *majority carriers (electrons)* in the n-type collector drift region. This is different from our earlier development in Equations 9.1–9.32, where the transport parameters in each region were those of *minority carriers*.

As an aside, we note that Equation 9.42 could have been deduced directly from the ambipolar diffusion equation (ADE) Equation 7.39. If recombination in the drift region is negligible, we can set $\Delta p/\tau_A \approx 0$. In steady-state $\partial p/\partial t = 0$, so the ADE reduces to simply $\partial^2 p/\partial x^2 = 0$. This means $p(x')$ is a linear function of position, and we can write

$$p(x') = p(0) + \frac{\partial p}{\partial x'}x' \tag{9.43}$$

Since $p \gg N_{DC}^+$, Equation 9.40 can be rewritten in the form

$$\frac{\partial p}{\partial x'} = -\frac{J_C}{2qD_{NC}} \tag{9.44}$$

Inserting Equation 9.44 into Equation 9.43 leads directly to Equation 9.42.

Having obtained an expression for the electron and hole densities in the collector drift region as a function of current, Equation 9.42, we now wish to calculate the voltage drop across the drift region. This voltage drop can then be added to the V_{CE} in Equation 9.32 to obtain an I_C–V_{CE} relation for the power BJT. To begin, we note from Equation 9.42 that the x' value at which the hole density equals the background doping density is

$$x_M{}' = \frac{2qD_{NC}}{J_C}[p(0) - N_{DC}^+] \tag{9.45}$$

Inserting Equation 9.34 for $p(0)$ yields

$$x_M' = \frac{2qD_{NC}}{J_C}\left[\frac{n_i^2}{N_{DC}^+}\exp\left(qV_{BC}/kT\right) - N_{DC}^+\right] \tag{9.46}$$

Here, V_{BC} refers to the voltage drop across the *internal* CB junction, which is less than the voltage between the collector and base *terminals* because of the voltage developed across the collector drift region and the substrate. In Equation 9.46, V_{BC} is obtained from the internal V_{CE} by setting $V_{BC} = V_{BE} + V_{CE}$, with V_{BE} given by Equation 9.30.

We assume the drift region is conductivity modulated over the region $0 < x' < x_M'$ where the hole density exceeds the background doping density, but not in the region $x_M' < x' < W_C$. The voltage drop V_{UM} across the unmodulated portion of the drift region is then given by

$$V_{UM} = \begin{cases} \dfrac{J_C\left(W_C - x_M'\right)}{q\mu_{NC}N_{DC}^+}, & x_M' < W_C \\ 0, & x_M' \geq W_C \end{cases} \tag{9.47}$$

The voltage drop across the conductivity-modulated portion V_M is obtained by integrating the electric field given by Equation 9.38 over the region $0 < x' < x_M'$. Thus we can write

$$V_M = \begin{cases} -\displaystyle\int_0^{x_M'} E\left(x'\right)dx' = -\frac{kT}{q}\int_{p(0)}^{p(x_M')}\frac{dp}{p} = \frac{kT}{q}\ln\left[\frac{p(0)}{N_{DC}^+}\right], & x_M' < W_C \\ -\displaystyle\int_0^{W_C} E(x')dx' = -\frac{kT}{q}\int_{p(0)}^{p(W_C)}\frac{dp}{p} = \frac{kT}{q}\ln\left[\frac{p(0)}{p(0) - J_C W_C/(2qD_{NC})}\right], & x_M' \geq W_C \end{cases} \tag{9.48}$$

where $p(0)$ is given by Equation 9.34. In evaluating Equation 9.34, we set $V_{BC} = V_{BE} + V_{CE}$, with V_{BE} given by Equation 9.30. For SiC power BJTs, V_M is typically less than −15 mV and can be neglected compared to other voltage drops in the device.

Having obtained expressions for the voltage drop across the collector drift region, we are now in a position to modify the $I_C - V_{CE}$ relation given in Equation 9.32 to include these effects. The procedure is as follows. We chose a value for base current I_B and step the *internal* V_{CE} to calculate I_C using Equation 9.32. To obtain the *terminal* V_{CE}, we add Equations 9.47 and 9.48 to the internal V_{CE} using Equation 9.46 for x_M'. Figure 9.7 shows $I_C - V_{CE}$ curves for the transistor of Figure 9.3 with the drift region voltage included (solid lines). For reference, the dotted lines show the curves of Figure 9.3 that do not include the drift region voltage. Points (a) and (b) on the $J_B = 5$ A cm^{-2} curve denote boundaries between different modulation modes. From the origin to point (a), the drift region is fully conductivity modulated ($x_M' \geq W_C$) and its differential resistance is very small. Between (a) and (b), the drift region is partially conductivity modulated ($0 < x_M' < W_C$) and the differential resistance is higher. To the right of point (b), the injection level $p(0)$ is insufficient to produce any conductivity modulation ($x_M' \leq 0$), and the differential resistance is that of the full unmodulated drift region.

Characteristics like those in Figure 9.7 are often not seen in experimental devices, for several reasons. First, at the somewhat higher dopings needed to maximize $V_B^2/R_{ON,SP}$, the drift region often never reaches full conductivity modulation as $I_C \to 0$, and point (a) lies very close to the origin. Figure 9.8 shows the device of Figure 9.7 with a drift region doping of 1×10^{16} cm^{-3} instead of 2×10^{15} cm^{-3}. For the 10 μm drift region used here, this doping provides the optimum $V_B^2/R_{ON,SP}$ and a theoretical blocking voltage of 1580 V. At this higher doping, the device remains in quasi-saturation almost to the origin. A second reason is that lateral effects such as current spreading in the drift region and the voltage drop due

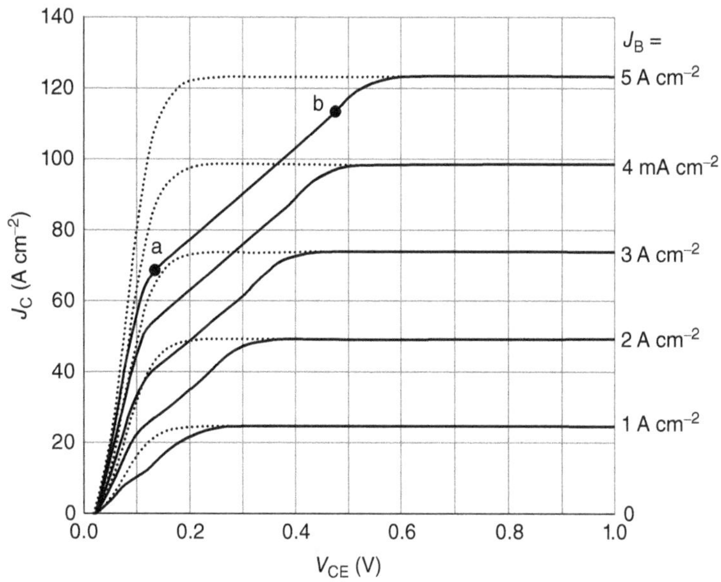

Figure 9.7 $I_C - V_{CE}$ characteristics including the voltage drop across the collector drift region in the BJT of Figure 9.3. The dotted curves are the characteristics of the BJT given by the basic Ebers–Moll model of Figure 9.3, without the voltage drop across the collector drift region.

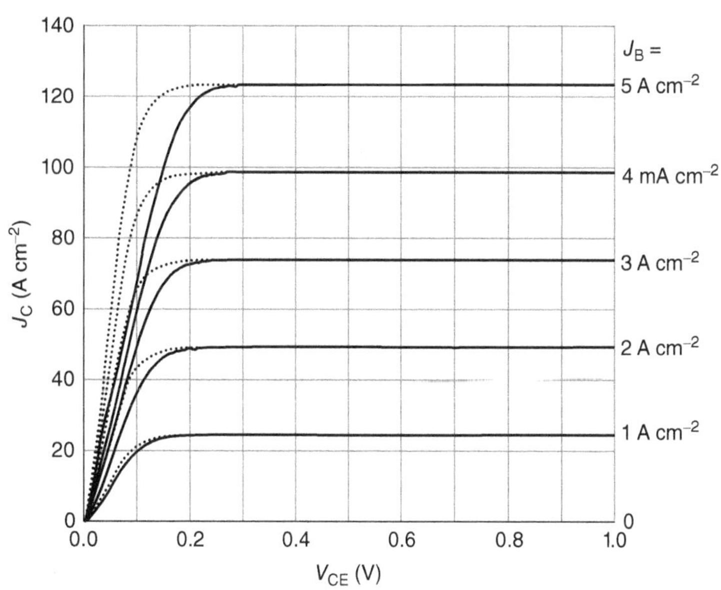

Figure 9.8 $I_C - V_{CE}$ of the BJT of Figure 9.7 with collector doping of 1×10^{16} cm^{-3} instead of 2×10^{15} cm^{-3}. The higher doping prevents complete modulation of the drift region, leaving the device in quasi-saturation until very near the origin.

to spreading resistance in the base often obscure the breakpoints in the idealized situation of Figure 9.7. Finally, we should caution that the development leading to Equation 9.32 assumes low-level injection in all regions, whereas in quasi-saturation and saturation the collector drift region is in high-level injection. This will alter the parameters I_{R0} and α_R in Equation 9.24 and subsequent equations, and this effect has not been included in Equation 9.32 or Figures 9.7 and 9.8.

9.1.6 High-Current Effects in the Base: the Rittner Effect

We have discussed high-level injection in the collector and examined how conductivity modulation can reduce the drift region voltage drop in saturation. We now turn to high-level injection in the base. When the BJT is operated at high currents, the injection of electrons from the emitter into the base can become large enough that the electron density in the base exceeds the ionized dopant density. When this occurs, charge neutrality demands that the hole concentration in the base also increase so that $\Delta p(x) \approx \Delta n(x)$. The higher hole concentration increases the back injection of holes from the base into the emitter. This is intuitively reasonable, since if we have more holes per unit volume in the base, we expect a greater flow of holes from base to emitter under forward bias conditions. The increased hole flow must be supplied by an increased base current, and this means a lower current gain, since $\beta = I_C/I_B$. This phenomenon is known as the *Rittner Effect*. We will now develop equations to describe this effect, culminating in an expression for β as a function of collector current.

To solve for β, we need to obtain expressions for the collector and base currents under conditions of high-level injection in the base. If we neglect recombination in the base, the base current in the forward-active region consists exclusively of holes injected from the base into the emitter, the component identified as I_{EP} in Figure 9.1. Since the emitter remains in low-level injection, this current can be found by solving the MCDE in the emitter, subject to the boundary condition imposed by the hole density at the depletion edge, $p_E(x'' = 0)$. Our first task is to find this boundary condition.

Invoking the law of the junction on the emitter side of the EB depletion region, we can write

$$p_E(0'')n_E(0'') = n_{iE}^2 \exp(qV_{BE}/kT) \tag{9.49}$$

Likewise, on the base side of the EB depletion region, we can write

$$p_B(0)n_B(0) = n_i^2 \exp(qV_{BE}/kT) \tag{9.50}$$

In these equations, $0''$ denotes the edge of the depletion region in the emitter and 0 denotes the edge of the depletion region in the base. The above equations can be combined to give the hole density in the emitter at the depletion edge,

$$p_E(0'') = p_B(0)\frac{n_B(0)n_{iE}^2}{n_E(0'')n_i^2} \tag{9.51}$$

Since the emitter remains in low-level injection, the electron density $n_E(0'')$ remains at its equilibrium value N_{DE}. However, if the base is in high-level injection, the hole density $p_B(0)$ will be greater than the doping density in the base. Charge neutrality in the base requires that

$$\Delta p_B(x) = p_B(x) - p_{B0} = \Delta n_B(x) = n_B(x) - n_{B0} \tag{9.52}$$

Solving for $p_B(x)$, we can write

$$p_B(x) = p_{B0} + n_B(x) - n_{B0} = N_{AB}^- + n_B(x) - \frac{n_i^2}{N_{AB}^-} \approx N_{AB}^- + n_B(x) \tag{9.53}$$

Evaluating Equation 9.53 at $x = 0$ and inserting into Equation 9.51 yields

$$p_E(0'') = [N_{AB}^- + n_B(0)]\frac{n_B(0)}{N_{DE}^+}\frac{n_{iE}^2}{n_i^2} \tag{9.54}$$

We see that the boundary condition for holes in the emitter $p_E(0'')$ now depends on the injection level in the base $n_B(0)$. From the law of the junction, we can write

$$n_B(0) = n_{B0}\exp(qV_{BE}/kT) = \frac{n_i^2}{N_{AB}^-}\exp(qV_{BE}/kT) \tag{9.55}$$

Inserting Equation 9.55 into Equation 9.54 yields

$$p_E(0'') = \left[1 + \frac{n_B(0)}{N_{AB}^-}\right]\frac{n_{iE}^2}{N_{DE}^+}\exp(qV_{BE}/kT) \tag{9.56}$$

Having obtained the boundary condition for the hole density in the emitter, we are now in a position to write expressions for the base and collector currents, and from their ratio we can find β. Assuming the emitter width W_E is long compared to the minority diffusion length L_{PE}, the solution of the MCDE for holes in the emitter is a decaying exponential,

$$\Delta p_E(x'') = \Delta p_E(0)\exp(-x''/L_{PE}) \tag{9.57}$$

The hole current flowing into the neutral emitter at $x'' = 0$ is given by

$$I_{EP} = -qAD_{PE}\left.\frac{\partial \Delta p_E}{\partial x''}\right|_{x''=0} = \frac{qAD_{PE}}{L_{PE}}\Delta p_E(0'') \approx \frac{qAD_{PE}}{L_{PE}}p_E(0'') \tag{9.58}$$

As stated earlier, with negligible recombination in the base, $I_B \approx I_{EP}$. The collector current I_C is the electron current flowing across the base from the emitter, namely, the components $I_{CN} \approx I_{EN}$ in Figure 9.1 (recall that electron particle flow is in the opposite direction to electron current). With negligible recombination in the base, the electron density decreases linearly from the emitter edge to the collector edge (see Figure 9.6a), and the collector current can be written

$$I_C = qAD_{NB}\frac{n_B(0)}{W} \tag{9.59}$$

This gives us another expression for $n_B(0)$, namely

$$n_B(0) = \frac{J_C W}{qD_{NB}} \tag{9.60}$$

We now divide Equation 9.59 by Equation 9.58 to obtain

$$\beta = \frac{I_C}{I_B} = \frac{D_{NB}}{D_{PE}}\frac{L_{PE}}{W}\frac{n_B(0)}{p_E(0'')} \tag{9.61}$$

Inserting Equation 9.55 for $n_B(0)$ and Equation 9.56 for $p_E(0'')$, and substituting Equation 9.60 into Equation 9.56 yields

$$\boxed{\beta = \left(\frac{D_{NB}}{D_{PE}}\right)\left(\frac{L_{PE}}{W}\right)\left(\frac{N_{DE}^+}{N_{AB}^-}\right)\left(\frac{n_i^2}{n_{iE}^2}\right)\left(1 + \frac{J_C W}{qD_{NB}N_{AB}^-}\right)^{-1}} \tag{9.62}$$

The last factor in Equation 9.62 describes the decrease in β due to high-level injection in the base. As the collector current density J_C is reduced, Equation 9.62 approaches the low-injection β_0 given in Equation 9.19, and we can rewrite Equation 9.62 in the form

$$\beta = \frac{\beta_0}{1 + J_C/J_R} \tag{9.63}$$

where J_R is the *Rittner current density* given by

$$J_R = qD_{NB}N_{AB}^-/W \tag{9.64}$$

The Rittner current can be regarded as the collector current density where β has fallen to half of its low-injection value. It is obviously desirable to have a high value of the Rittner current, which suggests increasing the base doping and/or reducing the base width. However, β_0 given in Equation 9.19 is inversely proportional to N_{AB}^-, so increasing the doping has the undesirable effect of reducing β_0. Reducing base width increases both β_0 and J_R, but to avoid punch-through in the base, the doping-thickness product needs to obey

$$N_{AB}W \geq \epsilon_S E_C/q \tag{9.65}$$

Taking the equality in Equation 9.65, the optimum β_0 and J_R can be written in terms of the critical field as

$$\beta_0 = \left(\frac{D_{NB}}{D_{PE}}\right)\left(\frac{qN_{DE}^+L_{PE}}{\epsilon_S E_C}\right)\left(\frac{N_{AB}}{N_{AB}^-}\right)\left(\frac{n_i^2}{n_{iE}^2}\right) \tag{9.66}$$

and

$$\boxed{J_R = \frac{\epsilon_S D_{NB}E_C}{W^2}\left(\frac{N_{AB}^-}{N_{AB}}\right)} \tag{9.67}$$

Thus, for highest gain and highest Rittner current, we should reduce W while increasing N_{AB} to satisfy the equality in Equation 9.65, thereby preventing punch-through. Equation 9.67 indicates another advantage of SiC for power BJTs, namely a higher Rittner current due to the higher critical field, but the Rittner current is reduced somewhat at room temperature by the incomplete ionization of acceptor dopants in the base.

9.1.7 High-Current Effects in the Collector: Second Breakdown and the Kirk Effect

In the forward-active region of operation, the n+ emitter injects electrons into the base, where they diffuse to the reverse-biased CB junction and are swept into the n− collector drift region. At high current densities, typical of power BJTs, a significant density of electrons exists in the base, throughout the CB depletion region, and in the collector drift region. As discussed in the previous section, when the density of electrons in the base exceeds the base doping, the base enters high-level injection, and this leads to a reduction in current gain β through the Rittner Effect. We now wish to consider conditions in the CB depletion region and the n− collector drift region.

Focusing first on the CB depletion region, the electric field at any point within the depletion region must obey Poisson's equation. On the collector side of the metallurgical junction we may write

$$\frac{\partial E(x)}{\partial x} = -\frac{q}{\epsilon_S}[N_{DC} - n(x)] \tag{9.68}$$

where $n(x)$ is the density of electrons drifting across the depletion region, and we have redefined our x coordinate system to place the origin at the CB metallurgical junction. N_{DC} here represents the total donor density, since donors atoms in the depletion region are fully ionized. The CB junction is strongly reverse biased and the electric field is high, so we may assume that electrons are moving at their saturated drift velocity $v_{SAT} \approx 2 \times 10^7$ cm s^{-1}. Under this assumption, the electron density in the depletion region is uniform with respect to position, and we can write

$$n(x) = J_C/(qv_{SAT}) \neq f(x) \tag{9.69}$$

We can calculate the electric field by inserting Equation 9.69 into Equation 9.68 and integrating with respect to x, resulting in

$$E(x) = E(0) - \frac{q}{\epsilon_S}\left(N_{DC} - \frac{J_C}{qv_{SAT}}\right)x \tag{9.70}$$

In equilibrium the collector current J_C is zero, and Equation 9.69 shows there are no electrons in the depletion region. The electric field in this case is illustrated by curve (a) in Figure 9.9. The depletion region spreads much further into the n− collector due to the asymmetry in doping between the base and collector. As the collector current is increased, the electron density given by Equation 9.69 becomes significant compared to the (very low) doping density N_{DC}, and the field taper given by Equation 9.68 is reduced, resulting in profile (b) in Figure 9.9. Here the depletion region extends through the entire collector drift region, and the field falls rapidly inside the n+ substrate. If the current is increased even more, the electron density given by Equation 9.69 will eventually equal the doping density N_{DC}, at which point the slope of the field profile becomes zero, curve (c). At even higher currents, the electron density will exceed the doping density and the net charge density changes sign. The electric field now *increases* with distance, resulting in profile (d). Here the peak field has shifted from the CB junction to the n−/n+ junction. If the current continues to increase, the field at the CB junction eventually goes to zero, profile (e). The current density corresponding to profile (e) is called the *Kirk current*, and can be calculated from Equation 9.70 by setting $E(0) = 0, x = W_C$, and $E(W_C) = 2\ V_{BC}/W_C$, resulting in

$$J_K = qv_{SAT}\left(N_{DC} + \frac{2\epsilon_S V_{BC}}{qW_C^2}\right) \tag{9.71}$$

If J_C exceeds the Kirk current J_K, the electric field shifts toward profile (f). When the peak field at the n−/n+ junction reaches the critical field for avalanche breakdown, as illustrated by profile (f), the device enters *second breakdown*. Holes generated by impact ionization in the high-field region at the n−/n+ junction flow toward the base and are injected into the emitter, acting as additional base current.

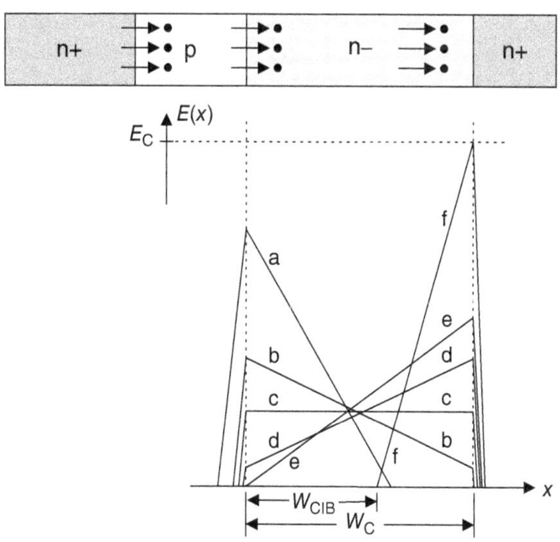

Figure 9.9 Electric field profiles in the collector when the BJT is carrying a high current density in the forward-active region. The device cross-section illustrates only the emitter, base, and collector doping regions, and does not distinguish between depletion and neutral regions.

This results in a much larger electron current from the emitter, further increasing the collector current and providing positive feedback to the breakdown process. It is important to note that this can occur at collector voltages *below* that required to cause avalanche breakdown at zero current. This is because at high current densities the field taper in the collector drift region given by Equation 9.70 and shown as profile (f) can exceed the normal field taper when the current is zero, profile (a). Since the area under the electric field remains equal to V_{BC}, which is held constant, the peak field in (f) can be higher than the peak field in (a), even though the voltage has not increased.

Now let us see how these concepts can be related to the critical field of the semiconductor. When the device is in the blocking state with $J_C = 0$, the electric field is shown in profile (a). As discussed in Section 7.1.1, in a non-punch-through (or NPT) design, the drift layer is totally depleted when the peak field just equals the critical field. Gauss' law and Poisson's equation lead to the relations

$$N_{DC} = \epsilon_S E_C^2/(2qV_B) \tag{9.72}$$

and

$$W_C = 2V_B/E_C \tag{9.73}$$

Inserting these expressions into Equation 9.71 yields

$$\boxed{J_K = \frac{\upsilon_{SAT}\epsilon_S E_C^2}{2V_B}\left(1 + \frac{V_{BC}}{V_B}\right)} \tag{9.74}$$

This indicates that for a given blocking voltage V_B, the Kirk current increases as the square of the critical field. The higher critical field of SiC leads to a much higher value for the Kirk current, effectively eliminating second breakdown as an issue in SiC BJTs.

We note that when the electric field has the form of profile (f), the field is zero over the portion of the collector drift region nearest the CB metallurgical junction. Poisson's equation requires that when the field is uniform (and in this case it is uniform at zero), the region must be charge neutral. This neutral region is referred to as the *current-induced base*, and has width W_{CIB} as indicated in Figure 9.9. Since the field is zero in this region, there can be no electron drift, and the collector current is supported entirely by electron diffusion. Since diffusion requires a concentration gradient, and since the region must remain charge neutral throughout, additional holes must be present in a concentration to exactly balance the additional electrons at each point. Electron diffusion in a charge-neutral region is the same process that takes place in the neutral base and, therefore, the current-induced base acts as an extension of the metallurgical base. This effect is referred to as *base push-out*, and results in a reduction in the current gain in the forward-active region at high current densities.

9.1.8 Common Emitter Current Gain: Temperature Dependence

The common emitter current gain of a narrow-base npn BJT under low-level injection was given in Equation 9.19, where the density of ionized acceptors in the base N_{AB}^- appears in the denominator. As discussed in Appendix A, aluminum acceptors in 4H-SiC have an ionization energy around 200 meV and not all the acceptors are ionized at room temperature. Figure A.1 shows that at a doping density of 2×10^{16} cm^{-3} only about 30% of the acceptors in the base are ionized and contribute holes at 23 °C. The low density of holes in the base increases both the emitter injection efficiency (Equation 9.18) and β (Equation 9.19). However, as temperature goes up, a larger fraction of the acceptors become ionized, and this causes β to decrease. For a BJT with base doping of 2×10^{16} cm^{-3}, the common emitter current gain will fall to one-third of its room temperature value at 300 °C. For this reason, it is important to evaluate the performance of the BJT, indeed all SiC power devices, at the highest junction temperature envisioned for the particular application.

The decrease in β with temperature is actually beneficial in one respect, since it helps prevent thermal runaway. This makes it possible to parallel multiple BJTs in high-current modules.

9.1.9 Common Emitter Current Gain: the Effect of Recombination

The current gain in SiC BJTs is profoundly affected by recombination taking place in the base and emitter regions, at the surfaces of the base and emitter, at the EB and CB junctions, and in the p+ implanted regions typically used to facilitate base contacts. We will discuss each of these effects below.

Recombination (and generation) is treated in most standard device textbooks. We are concerned here with recombination through deep levels in the bandgap associated with crystal defects, a process known as Shockley–Read–Hall (SRH) recombination. The net recombination rate per unit volume through single-level SRH centers in a bulk crystal can be written

$$R = \frac{\sigma_N \sigma_P v_T N_T (pn - n_i^2)}{\sigma_N (n + n_1) + \sigma_P (p + p_1)} = \frac{pn - n_i^2}{\tau_P (n + n_1) + \tau_N (p + p_1)} \tag{9.75}$$

where σ_P and σ_N are hole and electron capture cross sections, v_T is the thermal velocity (~10^7 cm s^{-1} at room temperature), N_T is the density of SRH centers per unit volume, τ_P and τ_N are the hole and electron minority carrier lifetimes, and n_1 and p_1 are given by

$$n_1 = n_i \exp[(E_T - E_i)/kT] \quad \text{and} \quad p_1 = n_i \exp[(E_i - E_T)/kT] \tag{9.76}$$

where E_T is the energy of the SRH center in the bandgap. For midgap centers, n_1 and p_1 are each equal to n_i.

With appropriate modifications, Equation 9.75 can also be applied to two-dimensional crystal surfaces or interfaces where we have a distribution of states with respect to energy across the bandgap. In these cases, the net recombination rate per unit area can be written

$$R_S = \frac{\sigma_N \sigma_P v_T N_{IT} (pn - n_i^2)}{\sigma_N (n + n_1) + \sigma_P (p + p_1)} = \frac{pn - n_i^2}{(n + n_1)/s_P + (p + p_1)/s_N} \tag{9.77}$$

where N_{IT} is the density of recombination centers per unit area at the surface or interface, and s_P and s_N are the *surface recombination velocities* for holes and electrons, respectively. In both Equations 9.75 and 9.77 the net recombination rate goes to zero in equilibrium ($pn = n_i^2$), and becomes negative (net generation) when $pn < n_i^2$. Both surface and bulk recombination rates are proportional to the density of SRH centers N_T or N_{IT}, so minimizing the density of such centers is essential to reducing recombination.

In SiC BJTs, recombination events remove electrons injected from the emitter into the base, preventing them from contributing to collector current. Recombination also removes holes injected from the base into the emitter. This increases the gradient of hole density in the emitter, which increases the base current. Both of these effects reduce β.

The dominant recombination sites in the BJT are illustrated schematically in Figure 9.10. Recombination can occur through defects in the neutral base and neutral emitter, indicated by (a) in the figure. We will discuss this recombination momentarily. Surface recombination can occur at the top surfaces of the base and emitter, and at the sidewalls of the emitter, labeled (b) in the figure. To minimize surface recombination, it is important to employ the best possible surface passivation, typically a thermal or deposited oxide followed by post-oxidation anneal in nitric oxide [1]. In addition, orienting the emitter fingers so their sidewalls lie on the $(1\bar{1}00)$ plane can reduce emitter sidewall recombination, since this plane has a lower surface recombination velocity [1].

Recombination can also occur in the implanted p+ regions under the base contacts, region (c) in the figure, due to crystal damage caused by the implantation [2]. An effective solution is to locate the p+ implants several diffusion lengths (or several base widths, whichever is shorter) away from

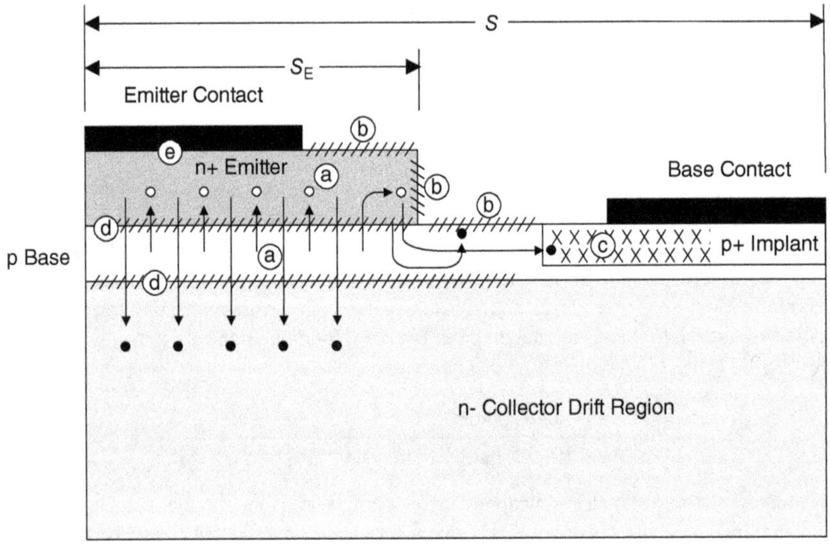

Figure 9.10 Illustration of the important recombination sites in a SiC power BJT.

the emitter edge. Recombination also occurs through defects at the pn junctions, shown as (d) in the figure, especially when the epigrowth is interrupted at these interfaces. This recombination can be reduced by growing the entire npn structure using continuous epigrowth [1]. Finally, we must consider recombination at the emitter ohmic contact, region (e) in the figure. This increases the hole current injected from the base and reduces β, but this degradation can be minimized by making the emitter thicker than the hole diffusion length L_{PE}.

Since surface recombination occurs along or close to the emitter edges, β can often be increased by using wide emitter fingers, thereby decreasing the perimeter-to-area ratio of the emitter [1]. However, this has to be balanced against the increase in specific on-resistance $R_{ON,SP}$. At first glance, increasing the emitter area relative to cell area would seem to make the cell more efficient, but we need to consider the effect of *base spreading resistance*. The lateral sheet resistivity of the base is given by

$$\rho_S = 1/(q\mu_P N_{AB}^- W) \tag{9.78}$$

This resistance can be significant in SiC BJTs due to the low hole mobility μ_P and low acceptor ionization percentage $N_{AB}{}^-$. Base current flowing from the base contact through the lateral resistance of the base produces a lateral voltage drop that reduces V_{EB} under the center of the emitter, reducing both electron and hole currents I_{EN} and I_{EP}. This does not affect β, since I_{EN} and I_{EP} are reduced by the same factor, but it makes the interior of the emitter inactive and increases the on-resistance. In choosing the width of the emitter, the designer must evaluate the trade-off between enhancing β and degrading $R_{ON,SP}$.

Process (a) in Figure 9.10 represents SRH recombination within the neutral base and neutral emitter, and this can obviously limit β. The dominant SRH centers in bulk 4H-SiC are Z_1/Z_2 centers associated with carbon–silicon divacancies, and EH_6/EH_7 centers associated with carbon vacancies. These deep levels can be minimized by a 5-h thermal oxidation at 1150 °C prior to the implant activation anneal, followed by another 5-h oxidation after the implant activation anneal [1]. SiC power BJTs receiving such anneals, along with the other measures discussed above, have demonstrated βs above 250 at room temperature [1].

9.1.10 Blocking Voltage

In the previous sections we considered the on-state performance of the BJT. We now turn to the blocking voltage. In the blocking state, the CB junction is reverse biased and supports the entire collector voltage. The maximum voltage that can be applied to the collector is limited either by punch-through of the neutral base or by avalanche breakdown of the CB junction.

Punch-through occurs when the depletion region of the CB junction extends across the neutral base and merges with the depletion region of the EB junction. When this occurs, the potential barrier confining electrons to the emitter is reduced and a large electron current flows from emitter to collector. The current is no longer controlled by the base, since the neutral portion of the base has vanished. As discussed in Section 10.1, punch-through can be prevented by insuring that the doping-thickness product of the base is large enough that the base cannot be completely depleted before the onset of avalanche breakdown at the CB junction. This condition is given by Equation 10.1. Since Equation 10.1 is satisfied in a well-designed BJT, the blocking voltage will normally be limited by avalanche breakdown.

Avalanche breakdown and edge termination techniques are discussed in Chapter 10, but there is one aspect of the BJT that requires special attention, namely the role of the internal current gain of the BJT. In the blocking state the base current is held at zero, which is equivalent to an open-circuit condition at the base terminal. As will be discussed in Section 10.1.1, avalanche breakdown is the result of impact ionization of carriers in the high-field region of the reverse-biased CB junction. Impact ionization generates electron–hole pairs in the depletion region, and the electric field separates the carriers, drawing the electrons into the collector and holes into the base. With I_B set to zero, these holes cannot flow out the base terminal and instead must flow into the emitter. This is equivalent to an *internal base current* that is amplified by the gain of the BJT. In other words, for every hole flowing into the emitter, approximately β electrons are injected from the emitter into the base. These electrons diffuse across the base and are swept into the CB depletion region where they initiate additional impact ionization. The holes generated by these new ionization events are themselves swept toward the emitter where they cause the injection of even more electrons, and so on. Once initiated, this process increases without bound, resulting in an uncontrolled increase in collector current *at a reverse voltage that would be insufficient to cause avalanche breakdown in an isolated CB junction.* As discussed in many textbooks, the blocking voltage of the BJT with base open-circuited $V_{B,CEO}$ is related to the breakdown voltage of an isolated CB junction $V_{B,CBO}$ by the equation

$$V_{B,CEO} = \frac{V_{B,CBO}}{(\beta + 1)^{1/m}} \tag{9.79}$$

where m is a constant, typically between 3 and 6. Since the denominator in Equation 9.79 is greater than unity, the open-base blocking voltage of the BJT is less than the breakdown voltage of the CB junction in isolation.

9.2 Insulated-Gate Bipolar Transistors (IGBTs)

We now turn our attention to the IGBT. The silicon IGBT was developed to reduce the drift region resistance R_{DR} of power MOSFETs through the use of conductivity modulation. Indeed, an early term for the IGBT was "COMFET", short for conductivity-modulated field effect transistor. Structurally, an n-channel IGBT can be viewed as a vertical n-channel power MOSFET, such as shown in Figure 8.17, where the n+ substrate has been replaced by a p+ substrate. In the conducting state, current flows through the channel of the MOSFET, vertically through the n− drift region, then through the forward-biased n−/p+ diode into the substrate. However, the physics of the IGBT is much richer than this. A more insightful interpretation would view the IGBT as an n-channel MOSFET merged with a pnp BJT. In the conducting state, the thick n-base of the pnp BJT is in high-level injection, resulting in conductivity modulation that reduces the voltage drop across this region. The penalty we pay for this benefit is the additional forward diode drop across the n−/p+ junction and a significant increase in turn-off time due to the minority carriers that must be removed from the base during turn-off.

To flesh out these concepts more thoroughly, we now consider a specific example. For our working example we chose a p-channel IGBT rather than an n-channel IGBT. This is done to avoid the use of a p+ substrate. p+ substrates in SiC are highly resistive due to the low mobility of holes and the low ionization fraction of acceptors at these doping levels, as shown in Figure A.1. Our p-channel IGBT is shown schematically in Figure 9.11. This structure can be viewed as a p-channel MOSFET that supplies base current to an npn BJT. In contrast to the narrow-base BJT discussed in the last section, this BJT has a thick, lightly-doped base that produces a current gain β in single digits. The IGBT also incorporates a thin p+ buffer layer between the drift region and the substrate to prevent punch-through of the drift region in the blocking state. The buffer layer has minimal effect on the on-state performance, but it becomes important in switching, as will be discussed shortly.

9.2.1 Current–Voltage Relationship

The current–voltage characteristics of the IGBT are illustrated in Figure 9.12. These characteristics resemble those of a power MOSFET except for the offset voltage near the origin, caused by the potential drop across the forward-biased p−/n+ substrate junction. Another difference that is not apparent from this figure is the steeper slope of the $I-V$ characteristics in the linear region. This is due to the lower on-resistance of the drift region caused by conductivity modulation, as will be discussed below.

To derive the operating equations of the IGBT, we will first consider conduction in the thick p− drift region. Since this region is lightly doped, we can assume high-level injection conditions prevail in the on state. The approach we use is similar to that employed in the pin diode discussion in Section 7.3, and we will emphasize comparisons with the pin diode as we go forward. Consider first a one-dimensional vertical slice through the drift region, as shown in Figure 9.11. We redraw this region in Figure 9.13, along with a coordinate system for analysis. In this representation we have neglected the p+ buffer layer, since it has negligible effect on the on-state operation.

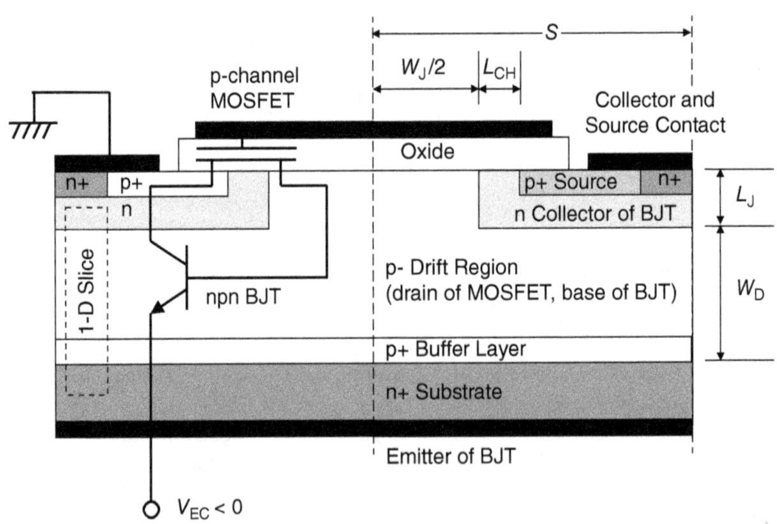

Figure 9.11 Cross-section of a p-channel IGBT in 4H-SiC. This device can be viewed as a p-channel MOSFET that supplies base current to an internal npn BJT. The dashed box indicates a one-dimensional slice that will be used for analysis.

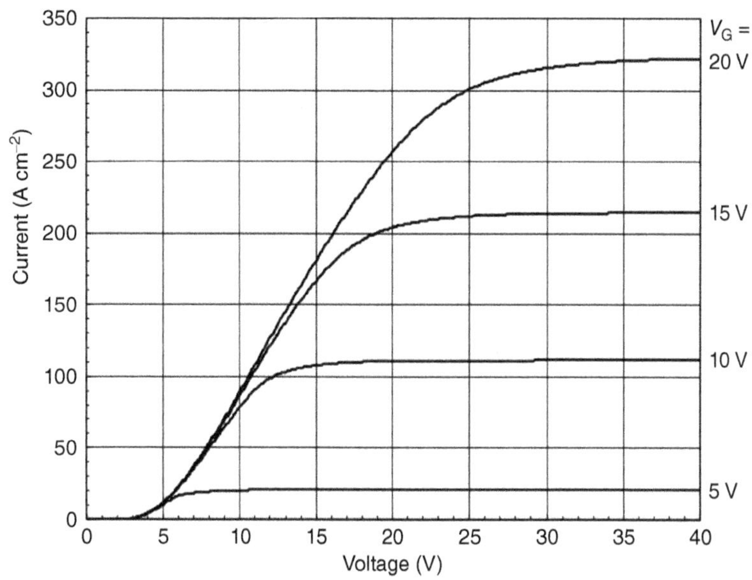

Figure 9.12 Current–voltage characteristics of a 4H-SiC IGBT. The offset voltage at the origin is due to the potential drop across the forward-biased substrate diode that serves as the emitter junction of the internal BJT.

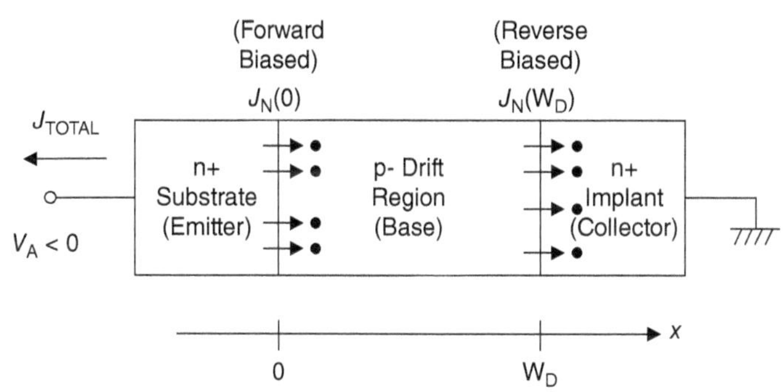

Figure 9.13 Simplified one-dimensional section of the IGBT used for analysis.

In the conducting state, the internal BJT is operating in the forward-active region, with the n+ substrate acting as emitter and the p− drift region acting as the base. Base current is supplied through the p-channel MOSFET, which is not shown in Figure 9.13, but we can assume that holes will be made available as needed to satisfy recombination in the base.

Our first objective is to obtain an expression for the potential drop across the p− base under high-level injection. The approach is the same as employed for the pin diode in Section 7.3, where we assumed $\Delta n \approx \Delta p$ and applied the ambipolar diffusion equation, Equation 7.39, subject to the boundary conditions at $x = 0$ and $x = d$. The general solution to the ambipolar diffusion equation is given by Equation 7.40, which is repeated here for convenience:

$$\Delta n(x) = \Delta p(x) = C_1 \sinh(x/L_A) + C_2 \cosh(x/L_A) \tag{7.40}$$

where L_A is the ambipolar diffusion length. All electrons that reach the reverse-biased CB junction are immediately swept across by the electric field, so the boundary condition here is $\Delta n(W_D) = 0$. For now, we will denote the boundary condition at the EB junction as $\Delta n(0)$. Inserting these boundary conditions into Equation 7.40 leads to two equations that can be solved for the constants C_1 and C_2. Performing the solution, we find

$$\begin{aligned} C_1 &= -\Delta n(0)\frac{\cosh(W_D/L_A)}{\sinh(W_D/L_A)} \\ C_2 &= \Delta n(0) \end{aligned} \tag{9.80}$$

Inserting C_1 and C_2 into the general solution Equation 7.40 yields

$$\Delta n(x) = \Delta n(0)\frac{\sinh[(W_D - x)/L_A]}{\sinh(W_D/L_A)} \tag{9.81}$$

To obtain the boundary condition $\Delta n(0)$, we proceed as we did with the pin diode and assume unity injection efficiency at the EB junction, that is, we assume hole injection into the emitter is negligible compared to the total current. Writing the expressions for hole and electron current at $x = 0$, we obtain results analogous to Equation 7.42, namely

$$\begin{aligned} J_N(0) = -J_{TOTAL} &= q\mu_N n(0)E(0) + qD_N \left.\frac{\partial n}{\partial x}\right|_{x=0} \\ J_P(0) = \quad 0 \quad &= q\mu_P p(0)E(0) - qD_P \left.\frac{\partial p}{\partial x}\right|_{x=0} \end{aligned} \tag{9.82}$$

where the minus sign in front of J_{TOTAL} arises because we define positive current to be in the negative x direction, as shown in Figure 9.13. We now set $n(x) \approx p(x)$ in the p− base to insure charge neutrality (since the doping is small compared to the injection level), and solve the second equation to obtain the electric field at $x = 0$,

$$E(0) = \frac{kT}{q}\frac{1}{n(0)}\left.\frac{\partial n}{\partial x}\right|_{x=0} \tag{9.83}$$

Substituting Equation 9.83 into the first Equation 9.82 and solving for $\partial n/\partial x$ at $x = 0$ yields the desired boundary condition at $x = 0$,

$$\left.\frac{\partial n}{\partial x}\right|_{x=0} = -\frac{J_{TOTAL}}{2qD_N} \tag{9.84}$$

Inserting Equation 9.81 into Equation 9.84 and noting that in high-level injection $\Delta n(x) \approx n(x)$, we find that

$$\Delta n(0) = \frac{J_{TOTAL}L_A}{2qD_N}\tanh(W_D/L_A) \tag{9.85}$$

Inserting this into Equation 9.81 leads to the desired equation for carrier density,

$$\boxed{\Delta n(x) = \Delta p(x) = \frac{J_{TOTAL}L_A}{2qD_N}\frac{\sinh[(W_D - x)/L_A]}{\cosh(W_D/L_A)}} \tag{9.86}$$

This result may be compared with Equation 7.46 for the pin diode. The differences arise from the nature of the boundary condition at $x = W_D$ (or $x = +d$ in the pin diode coordinate system). In the IGBT this boundary extracts electrons and injects no holes. In the pin diode this boundary extracts no electrons, and injects holes with unity injection efficiency. We should also point out that Equation 9.86 involves the *electron* diffusion coefficient, while Equation 7.46 involves only *ambipolar* factors. Figure 9.14 shows carrier densities in the pin diode and IGBT at the same total current, computed using Equations 7.46 and 9.86. Here we see the effect of the reverse-biased CB junction that quickly extracts all arriving electrons. As a result, the IGBT exhibits less conductivity modulation than the pin diode.

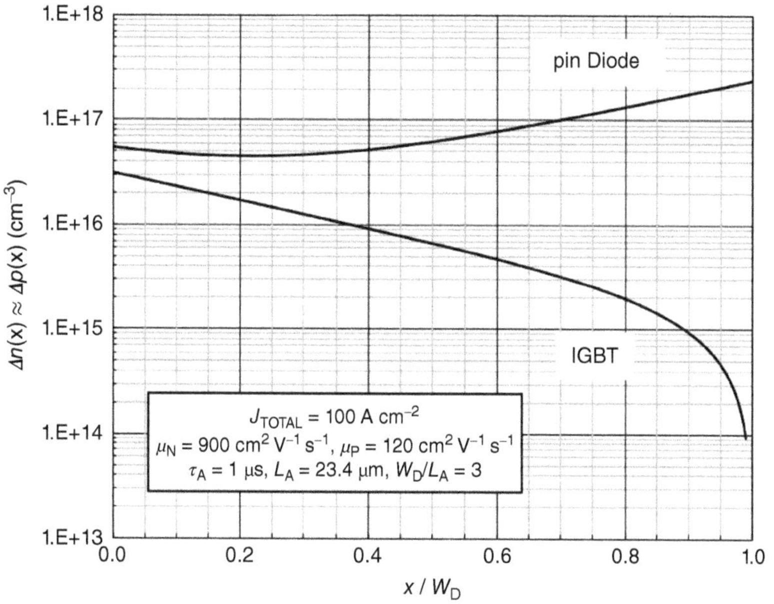

Figure 9.14 Excess electron and hole densities in the p-channel IGBT and the pin diode as a function of position.

Continuing to work toward an expression for the total potential drop across the p− layer, we next wish to find an expression for the electric field as a function of position. We can write the total current as

$$-J_{\mathrm{TOTAL}} = J_{\mathrm{N}}(x) + J_{\mathrm{P}}(x) = q\mu_{\mathrm{N}}\left[n(x)E(x) + \frac{kT}{q}\frac{\partial n}{\partial x}\right] + q\mu_{\mathrm{P}}\left[p(x)E(x) - \frac{kT}{q}\frac{\partial p}{\partial x}\right] \tag{9.87}$$

where the minus sign in front of J_{TOTAL} again enters because the terminal current is defined to be positive in the negative x direction. We now set $n(x) = \Delta n(x)$ and $p(x) = \Delta p(x) + N_{\mathrm{A}}^{-}$, with $\Delta p(x) = \Delta n(x)$ for charge neutrality. Solving for $E(x)$ yields

$$E(x) = \frac{-J_{\mathrm{TOTAL}} - kT(\mu_{\mathrm{N}} - \mu_{\mathrm{P}})(\partial n/\partial x)}{q(\mu_{\mathrm{N}} + \mu_{\mathrm{P}})\Delta n(x) + q\mu_{\mathrm{P}}N_{\mathrm{A}}^{-}} \tag{9.88}$$

Inserting Equation 9.86 into Equation 9.88 yields, after some algebra,

$$E(x) = \frac{kT}{qL_{\mathrm{A}}}\left\{\frac{\dfrac{-2\mu_{\mathrm{N}}}{(\mu_{\mathrm{N}} + \mu_{\mathrm{P}})}\cosh\left(\dfrac{W_{\mathrm{D}}}{L_{\mathrm{A}}}\right) + \dfrac{(\mu_{\mathrm{N}} - \mu_{\mathrm{P}})}{(\mu_{\mathrm{N}} + \mu_{\mathrm{P}})}\cosh\left(\dfrac{W_{\mathrm{D}} - x}{L_{\mathrm{A}}}\right)}{\sinh[(W_{\mathrm{D}} - x)/L_{\mathrm{A}}] + \theta}\right\} \tag{9.89}$$

where

$$\theta = \frac{qD_{\mathrm{A}}N_{\mathrm{A}}^{-}}{J_{\mathrm{TOTAL}}L_{\mathrm{A}}}\cosh\left(\frac{W_{\mathrm{D}}}{L_{\mathrm{A}}}\right) \tag{9.90}$$

The electric field within the p− drift layer is *independent of current* in the regions where high-level injection prevails, that is, where $\Delta p(x) \gg N_{\mathrm{A}}^{-}$. In these regions, θ can be neglected compared to the sinh

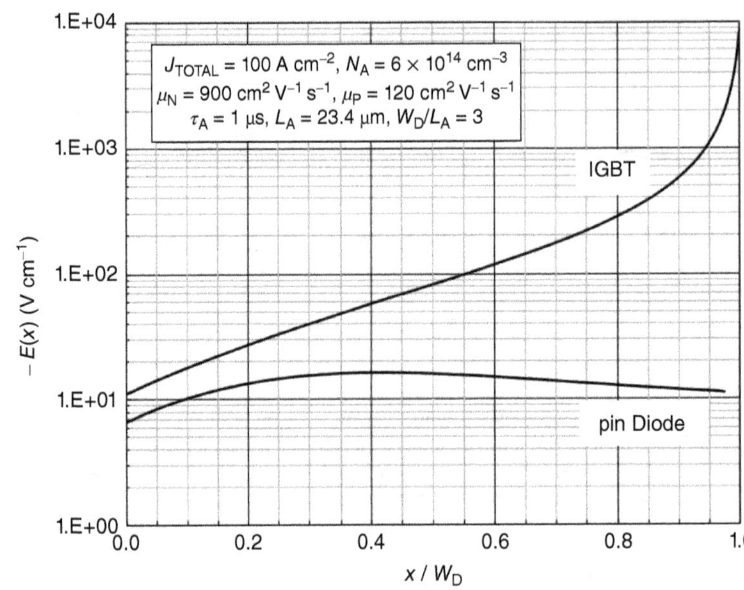

Figure 9.15 Electric field as a function of position in the IGBT and the pin diode. The electric field is negative, and tends to drive holes to the left and electrons to the right.

term in the denominator of Equation 9.89. The region very close to $x = W_D$ is not in high-level injection, since Δp approaches zero near the collector, but the θ term in the denominator prevents the electric field from becoming infinite at $x = W_D$.

The electric field calculated using Equation 9.89 is shown in Figure 9.15 for the same conditions as Figure 9.14. The electric field is negative, which drives holes to the left and electrons to the right. The diffusion process, on the other hand, carries holes and electrons in the same direction, since $\mathrm{d}\Delta p(x)/\mathrm{d}x \approx \mathrm{d}\Delta n(x)/\mathrm{d}x$. From Figure 9.14 we see that diffusion moves both electrons and holes to the right throughout the IGBT whereas, in the pin diode, diffusion brings carriers into the drift region from both edges.

Having an expression for the electric field, we simply need to integrate Equation 9.89 with respect to x to obtain the electrostatic potential as a function of position. We choose to write the potential in the form

$$\psi(x) = \psi(0) - \int_0^x E(x)\mathrm{d}x \tag{9.91}$$

Inserting Equation 9.89 and performing the integral, we obtain the rather intimidating expression

$$\boxed{\begin{aligned}\psi(x) = \psi(0) + \frac{kT}{q}\left\{\frac{(\mu_N - \mu_P)}{(\mu_N + \mu_P)}\ln\left[\frac{\sinh\left(\frac{W_D - x}{L_A}\right) + \theta}{\sinh(W_D/L_A) + \theta}\right] - \frac{4\mu_N}{(\mu_N + \mu_P)}\frac{\cosh(W_D/L_A)}{\sqrt{\theta^2 + 1}}\right. \\ \left.\times\left[\tanh^{-1}\left(\frac{\theta\tanh\left(\frac{W_D - x}{2L_A}\right) - 1}{\sqrt{\theta^2 + 1}}\right) - \tanh^{-1}\left(\frac{\theta\tanh\left(\frac{W_D}{2L_A}\right) - 1}{\sqrt{\theta^2 + 1}}\right)\right]\right\}\end{aligned}} \tag{9.92}$$

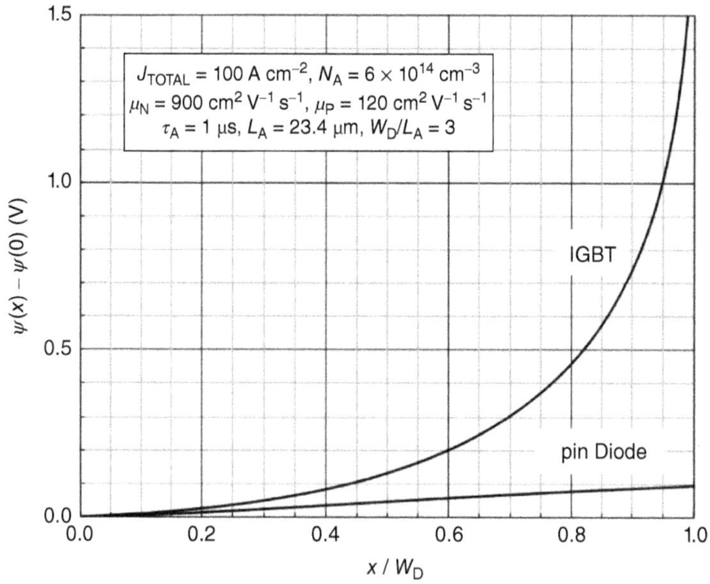

Figure 9.16 Potential drop as a function of position in the drift region of a p-channel IGBT and a comparable pin diode. The total drop in the IGBT is much higher than in the pin diode, since the drift region is only modulated from one side.

Figure 9.16 shows the electrostatic potential given by Equation 9.92, along with the potential in a comparable pin diode. The potential drop in the IGBT is much greater than in a pin diode because injection occurs at only one boundary of the drift region rather than two. Nevertheless, the potential drop is much less than would be expected in the unmodulated drift region of a power MOSFET.

The total electrostatic potential drop across the p− drift layer can be found by setting $x = W_D$ in Equation 9.92. However, this is not the *voltage* drop across the drift layer. The electrostatic potential drop in Equation 9.92 describes the band bending, or change in $E_i(x)$, across the drift layer, whereas the *voltage* drop is the change in the hole quasi-Fermi level F_P. The difference can be understood from the band diagram in Figure 9.17. The splitting in F_P across the n+/p− junction, designated V_{N+P}, will be calculated in a few moments. The additional change in F_P across the drift region is labeled V_P, and this is the voltage drop we desire.

To calculate V_P, we need an expression for the hole quasi-Fermi level, or the electrochemical potential for holes in the drift region. We note that the hole density at any point is related to F_P by

$$p(x) = \Delta p(x) + N_A^- = n_i \exp\left[\frac{E_i(x) - F_P(x)}{kT}\right] \quad (9.93)$$

From this it follows that

$$E_i(x) - F_P(x) = kT \ln\left[\frac{\Delta p(x) + N_A^-}{n_i}\right] \quad (9.94)$$

Defining the hole quasi-Fermi potential $\psi_P(x) = -F_P(x)/q$ and noting that the electrostatic potential $\psi(x) = -E_i(x)/q$, we can write

$$\psi_P(x) = \psi(x) + \frac{kT}{q} \ln\left[\frac{\Delta p(x) + N_A^-}{n_i}\right] \quad (9.95)$$

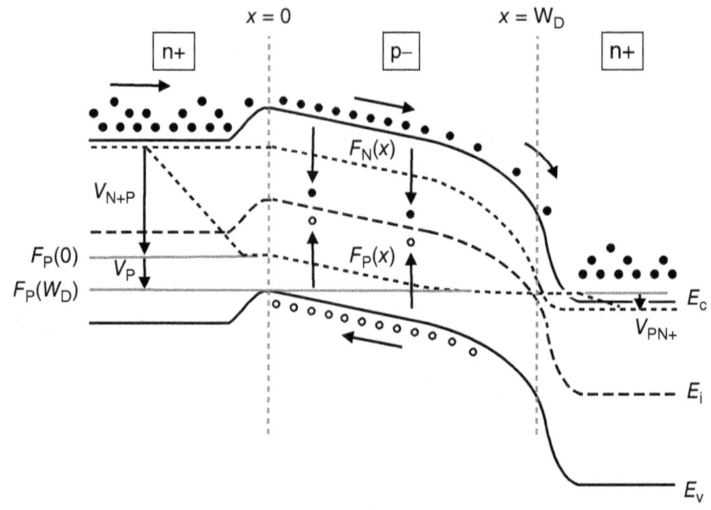

Figure 9.17 Band diagram along the 1-D slice in Figure 9.11. V_{N+P} is the quasi-Fermi level splitting across the EB junction, V_P is the hole quasi-Fermi level change across the base region, and V_{PN+} is the quasi-Fermi level splitting across the CB junction.

Fortunately we have already solved for $\psi(x)$ and for $\Delta p(x)$. Inserting Equations 9.92 and 9.86 into Equation 9.95 and working through the algebra, we obtain

$$\psi_P(x) = \psi(0) + \frac{kT}{q}\left\{\frac{(\mu_N - \mu_P)}{(\mu_N + \mu_P)} \ln\left[\frac{\sinh\left(\frac{W_D - x}{L_A}\right) + \theta}{\sinh(W_D/L_A) + \theta}\right]\right.$$
$$+ \ln\left[\frac{J_{TOTAL}L_A}{2qD_N n_i}\frac{\sinh\left(\frac{W_D - x}{L_A}\right)}{\cosh(W_D/L_A)} + \frac{N_A^-}{n_i}\right] - \frac{4\mu_N}{(\mu_N + \mu_P)}\frac{\cosh(W_D/L_A)}{\sqrt{\theta^2 + 1}}$$
$$\left.\times\left[\tanh^{-1}\left(\frac{\theta\tanh\left(\frac{W_D - x}{2L_A}\right) - 1}{\sqrt{\theta^2 + 1}}\right) - \tanh^{-1}\left(\frac{\theta\tanh\left(\frac{W_D}{2L_A}\right) - 1}{\sqrt{\theta^2 + 1}}\right)\right]\right\} \quad (9.96)$$

Equation 9.96 describes the variation in hole quasi-Fermi potential across the drift layer, where θ is given by Equation 9.90. The total voltage drop across the drift layer can be written

$$V_P = \psi_P(W_D) - \psi_P(0)$$
$$= \frac{kT}{q}\left\{\frac{(\mu_N - \mu_P)}{(\mu_N + \mu_P)}\ln\left[\frac{\theta}{\sinh(W_D/L_A) + \theta}\right] - \ln\left[\frac{J_{TOTAL}L_A}{2qD_N N_A^-}\tanh\frac{W_D}{L_A} + 1\right] - \frac{4\mu_N}{(\mu_N + \mu_P)}\right.$$
$$\left.\times\frac{\cosh(W_D/L_A)}{\sqrt{\theta^2 + 1}}\left[\tanh^{-1}\left(\frac{-1}{\sqrt{\theta^2 + 1}}\right) - \tanh^{-1}\left(\frac{\theta\tanh(W_D/2L_A) - 1}{\sqrt{\theta^2 + 1}}\right)\right]\right\} \quad (9.97)$$

Before calculating other voltage drops in the structure, we pause for a moment to consider the common-emitter current gain, or β, of the internal npn BJT. Since we now have expressions for the carrier densities and electric field as a function of x in the base, we can evaluate these expressions at the collector edge $x = W_D$ to determine the collector and base currents in the internal BJT. We can then use the ratio J_C/J_B to determine β. The electron and hole currents at $x = W_D$ can be written

$$J_N(W_D) = q\mu_N n(W_D)E(W_D) + qD_N \left.\frac{\partial n}{\partial x}\right|_{x=W_D}$$
$$J_P(W_D) = q\mu_P p(W_D)E(W_D) - qD_P \left.\frac{\partial p}{\partial x}\right|_{x=W_D} \tag{9.98}$$

Inserting $E(W_D)$ from Equation 9.89 and setting $n(W_D) = 0$ and $p(W_D) = N_A^-$, we can write

$$J_N(W_D) = -\frac{\mu_N J_{TOTAL}}{(\mu_N + \mu_P)}\left[1 + \frac{\mu_P}{\mu_N}\frac{1}{\cosh\left(W_D/L_A\right)}\right]$$
$$J_P(W_D) = -\frac{\mu_P J_{TOTAL}}{(\mu_N + \mu_P)}\left[1 - \frac{1}{\cosh\left(W_D/L_A\right)}\right] \tag{9.99}$$

The hole current is negative, since the factor in square brackets is always positive. A negative hole current corresponds to holes moving *into* the p− drift region at $x = W_D$. These holes are not flowing across the CB junction, since this junction is reverse biased. Rather, they are flowing into the drift region through the p-channel MOSFET, and this constitutes the base current of the BJT. The electron current is also negative, and corresponds to electrons flowing across the CB junction into the collector. This is the collector current of the BJT, so the current gain can be written

$$\boxed{\beta = \frac{J_C}{J_B} = \frac{J_P(W_D)}{J_N(W_D)} = \frac{\left(\dfrac{\mu_N}{\mu_P}\right)\cosh\left(\dfrac{W_D}{L_A}\right) + 1}{\cosh(W_D/L_A) - 1}} \tag{9.100}$$

Figure 9.18 shows the current gain given by Equation 9.100 as a function of base width. For the parameters used in Figures 9.14–9.16, the current gain is 8.44. This may seem a low value, but a high β is not necessary for a successful IGBT, since the base current is not supplied from an external circuit but rather is "borrowed" from the collector current through the integral MOSFET, as can be seen from the equivalent circuit in Figure 9.11. We note that Equation 9.100 becomes inaccurate at large values of W_D/L_A, since the carrier densities at large x are no longer high enough to constitute high-level injection.

As an aside, we also note that the base current $J_P(W_D)$ in Equation 9.99 can also be calculated from the total hole charge stored in the drift region. The stored hole charge is found by integrating Equation 9.86 from $x = 0$ to $x = W_D$, resulting in

$$\boxed{Q_{STORED} = \frac{J_{TOTAL}L_A^2}{2D_N}\left[\frac{1}{\cosh\left(W_D/L_A\right)} - 1\right]} \tag{9.101}$$

Setting $J_P(W_D)$ equal to the total recombination current in the base, Q_{STORED}/τ_A, we obtain the same result as in Equation 9.99.

We now return to the task of calculating the total voltage drop across the IGBT as a function of current. The next voltage drop to be determined is the drop across the forward-biased EB junction, V_{N+P}. Employing the law of the junction, the electron density at the edge of the p− drift region at $x = 0$ can be written

$$p(0)n(0) = n(0)^2 = n_i^2 \exp(qV_{N+P}/kT) \tag{9.102}$$

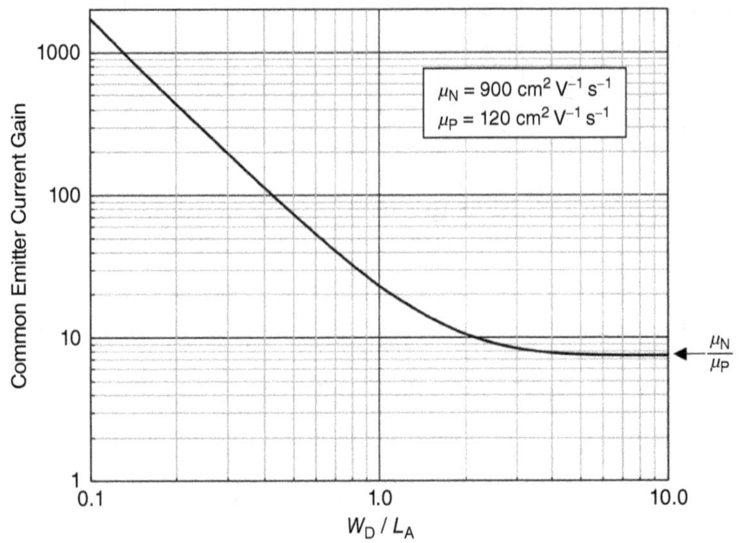

Figure 9.18 Common emitter current gain of the internal BJT in the IGBT as a function of base width using Equation 9.100.

Solving Equation 9.102 for V_{N+P} and inserting $n(0)$ from Equation 9.86 leads to

$$V_{N+P} = 2\frac{kT}{q}\ln\left[\frac{J_{TOTAL}L_A}{2qD_N n_i}\tanh\left(\frac{W_D}{L_A}\right)\right] \tag{9.103}$$

The voltage drop V_{PN+} across the reverse-biased CB junction does not need to be calculated, since this voltage is controlled by the voltage drops across the p-channel MOSFET and the vertical p-channel JFET, as will be discussed next.

Current flowing through the p-channel MOSFET produces two voltage drops: the drop across the MOSFET channel and the drop across the vertical p-channel JFET that is gated by the grounded n+ collector regions. The MOSFET is operating in the linear region and the channel resistance is given by the p-channel analog to Equation 8.52, namely

$$R_{MOS,SP} = \frac{L_{CH} \bullet S}{\mu_P^* C_{OX}(V_G - V_T)} \tag{9.104}$$

where $\mu_P{}^*$ is the inversion layer hole mobility and L_{CH} is the channel length. The voltage drop is obtained by multiplying by the current density passing through the MOSFET, which is the base current given by the second equation in Equation 9.99. The result is

$$V_{MOS} = \frac{\mu_P J_{TOTAL}}{(\mu_N + \mu_P)}\left[\frac{1}{\cosh\left(W_D/L_A\right)} - 1\right]\left[\frac{L_{CH} \bullet S}{\mu_P^* C_{OX}\left(V_G - V_T\right)}\right] \tag{9.105}$$

The first bracketed factor is negative, so the voltage at the drain end of the MOSFET channel will also be negative, as expected for a p-channel MOSFET.

The voltage drop across the vertical JFET region can be obtained by multiplying the base current by the linear region resistance of the JFET. Here we can use Equation 8.11, which gives the current–voltage relationship of a double-gated JFET, with the following modifications: Recalling that the JFET within the IGBT is oriented vertically, we replace L in Equation 8.11 with the channel length L_J and $(2a)$ with the JFET width W_J, where L_J and W_J are illustrated in Figure 9.11. The drain current in Equation 8.11 includes both halves of the channel, but we now wish to calculate the potential drop using only one half of the channel, so we replace I_D in Equation 8.11 with $(2\,I_B)$. Since we now have a p-channel JFET, we replace μ_N by μ_P and N_D by N_A. Finally, we set $V_G = 0$, since the n+ collector regions that serve as the gates of the JFET are grounded. We also need to account for the fact that the source of the JFET is not at ground, as assumed in Equation 8.11, but rather at a negative potential V_{MOS} due to the potential drop across the MOSFET channel. Making the indicated modifications, we can write

$$\frac{I_D}{2} = I_B = q\mu_P N_A^- \frac{W \cdot (W_J/2)}{L_J}\left\{V_{DS} - \frac{4}{3}\frac{\sqrt{2\epsilon_S/qN_A}}{W_J}\left[\left(\psi_{BI} + V_{DS} + V_S\right)^{3/2} - \left(\psi_{BI} - V_S\right)^{3/2}\right]\right\} \quad (9.106)$$

where we have made use of the fact that $V_{DS} = V_D - V_S$. The specific resistance of the JFET is given by

$$R_{JFET,SP} = \frac{W \cdot (W_J/2)}{\left(\partial I_B/\partial V_{DS}\right)\Big|_{V_{DS}=0}} \quad (9.107)$$

Taking the derivative of I_B with respect to V_{DS} using Equation 9.106 and evaluating at $V_{DS} = 0$, then inserting into Equation 9.107 yields

$$R_{JFET,SP} = \frac{L_J}{q\mu_P N_A^-}\left(1 - \sqrt{\frac{2\epsilon_S\left(\psi_{BI} + V_S\right)}{qN_A(W_J/2)^2}}\right)^{-1} \quad (9.108)$$

Equation 9.108 may be compared with Equation 8.18, keeping in mind that Equation 9.108 refers to a vertical JFET with $V_G = 0$ and $V_S \neq 0$, while Equation 8.18 refers to a lateral JFET with $V_S = 0$. Finally, the voltage drop across the JFET can be found by multiplying the on-resistance by the base current density from the second Equation 9.99,

$$J_{JFET} = \frac{\mu_P J_{TOTAL}}{(\mu_N + \mu_P)}\left[\frac{1}{\cosh\left(W_D/L_A\right)} - 1\right]\left[\frac{L_J}{q\mu_P N_A^-}\left(1 - \sqrt{\frac{2\epsilon_S\left(\psi_{BI} + V_S\right)}{qN_A(W_J/2)^2}}\right)^{-1}\right] \quad (9.109)$$

where the source voltage V_S is set equal to the voltage drop across the MOSFET channel given by Equation 9.105. To minimize the voltage drop across the JFET, the doping N_A in the JFET region is often made higher than in the IGBT drift region, and it is important to keep in mind that N_A in the drift region and N_A in the JFET region may be different.

The total voltage drop across the IGBT, that is, the emitter-to-collector voltage, can be found as a function of current by adding the individual voltage drops already computed:

$$V_{EC} = V_{MOS} + V_{JFET} - V_P + V_{N+P} + V_{SUB} \quad (9.110)$$

where Equations 9.105, 9.109, 9.97, and 9.103 are used in Equation 9.110, and V_{SUB} accounts for the voltage drop across the substrate. The negative sign for V_P arises from the polarity assumed in the derivation of Equation 9.97.

It is important to point out the limitations of the above analysis. First, we have assumed linear-region operation of the MOSFET and JFET channels. This is acceptable since we are primarily interested in the on state, which implies a low V_{EC} and hence low V_{DS} across the channels. However, our equations will not describe the full $J_C - V_{EC}$ characteristics. To do this we must replace Equations 9.105 and 9.109 with equations that are valid for large V_{DS}. Secondly, we have implicitly assumed one-dimensional current flow in each section of the device and have not considered the effects of current spreading as carriers flow from one region to another. This is a major shortcoming, and the current–voltage relation given by Equation 9.110 will not provide a quantitatively accurate description of conditions inside a real device; a quantitative description requires full two-dimensional computer simulations. Finally, it is important to consider the effects of operating temperature, particularly with respect to incomplete ionization of dopants, dependence of mobility on temperature, and the strong dependence of lifetime on temperature. Power switching devices operate from ambient temperature to the maximum temperature limit of the device and package, and their performance parameters vary significantly over this range. It is essential to evaluate operation at both the highest and lowest junction temperatures anticipated for the particular application.

9.2.2 *Blocking Voltage*

If the IGBT is in the conducting state and the gate of the IGBT is taken below threshold, current flow within the device ceases and the IGBT enters the forward blocking state. In the blocking state, the device is required to block a large (negative) V_{EC} with minimal leakage. This large V_{EC} is supported by the reverse-biased CB junction whose depletion region extends primarily into the lightly-doped base, as shown in Figure 9.19a. However, the depletion region in the base cannot be allowed to penetrate to the emitter, because this would constitute punch-through, and a large current would flow. The diagram on the left shows the electric field just at the onset of punch-through. The emitter-collector voltage V_{EC} is the area under the electric field profile, and this represents the maximum blocking voltage of the device. In the asymmetrical structure in Figure 9.19b, a thin p+ buffer layer is inserted between the p− base and the n+ emitter, making the electric field profile trapezoidal. Assuming the peak field is equal to the critical field E_C in both cases, the asymmetrical IGBT has a larger area under the electric field profile than the symmetrical IGBT. This results in a higher forward blocking voltage for the same base thickness (of course, to achieve the trapezoidal profile, the base doping in the asymmetrical case will be lower than in the symmetrical structure for the same base thickness).

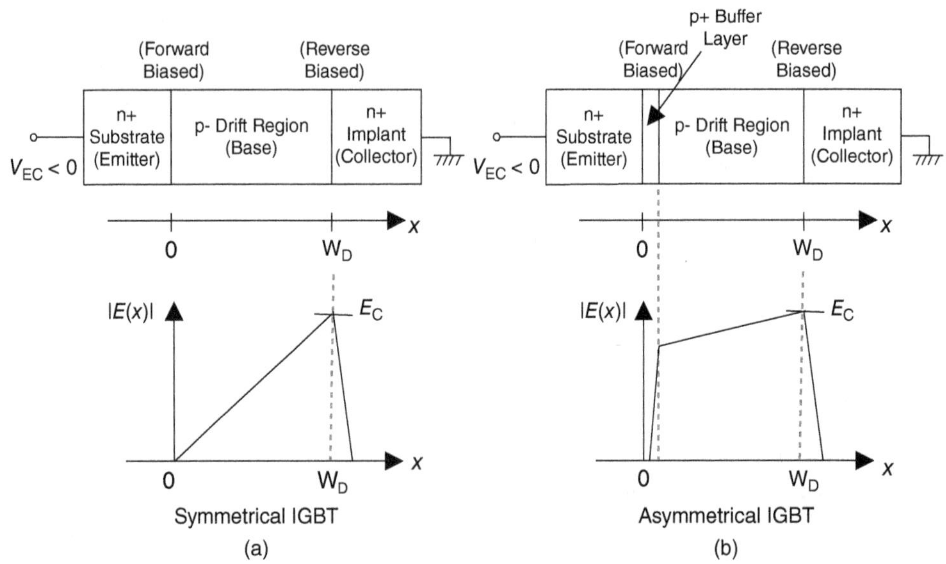

Figure 9.19 Electric field profiles in the blocking state for (a) symmetrical and (b) asymmetrical IGBTs.

As is the case with the power MOSFET, the blocking voltage can also be limited by the field in the gate oxide. This can be seen by reference to the cross-section in Figure 9.11. If the gate is at ground potential and the emitter is at a large negative voltage, a depletion region will be present under the gate oxide over the JFET region (center of the drawing in Figure 9.11). As V_{EC} becomes more negative, the depletion region expands, and the electric field at the oxide/semiconductor interface approaches E_C, the critical field for avalanche breakdown. By Gauss' law, the electric field in the oxide is related to the semiconductor surface field by the ratio of the dielectric constants, Equation 8.80. For SiO_2 on SiC, this ratio is approximately 2.6, meaning the oxide field is 2.6 times higher than the semiconductor field. In SiC, E_C is in the range 1.5–2.5 MV cm^{-1} for dopings of 10^{14}–10^{16} cm^{-3}, so the oxide field will be approximately 4–6.5 MV cm^{-1}. As discussed in Section 8.2.11, the field in SiO_2 must typically be kept below about 4 MV cm^{-1} to prevent gradual degradation of the oxide and premature device failure. Therefore, the blocking voltage of the IGBT is the largest value of V_{CE} that does not cause avalanche breakdown in the CB junction *and* maintains the oxide field below 4 MV cm^{-1}.

As with all power devices, the maximum blocking voltage of the IGBT is usually limited by field crowding at the edges of pn junctions. This can be mitigated by the use of edge terminations, as will be discussed in Section 10.1. The designer can also take advantage of two-dimensional effects to increase the blocking voltage, such as when pn junctions are used to shield the oxide from high electric fields. In the IGBT, this is accomplished by reducing the spacing between n+ collector regions, that is, by reducing the JFET width W_J in Figure 9.11. This allows field lines to terminate on the grounded n+ collectors rather than penetrate the oxide to the grounded gate, thereby reducing the oxide field. However, this must be balanced against the increase in on-resistance of the narrower JFET channel. In all such cases, accurate analysis involves two-dimensional effects that require computer solutions.

9.2.3 Switching Characteristics

A major consideration with IGBTs is the transient power dissipated during the switching event or the *switching loss*. The problem is particularly acute during the turn-off transient, due to the large minority carrier charge that must be removed when the device is turned off.

The switching power is proportional to the frequency of switching events and the energy dissipated per event, as shown by Equation 7.15. Equation 7.16 expresses the switching energy as the sum of the energy dissipated during the turn-off transient E_{OFF} and during the turn-on transient E_{ON}. In the IGBT the turn-on energy loss is small, and we will therefore concentrate on the turn-off loss.

During on-state operation, the thick base region of the BJT is flooded with holes and electrons, and operates in high-level injection. The total charge stored in the base is given by Equation 9.101. When the gate is taken below threshold, the MOSFET no longer supplies holes to the base of the BJT, but the holes already stored in the base continue to undergo recombination until they are all removed. Since the base current is equal to the total recombination current in the base, we can regard these stored holes as a source of base current. The turn-off transient is therefore intimately related to the time required for all the stored holes to recombine. Since $\Delta n(x) \approx \Delta p(x)$ for charge neutrality, and since $V_{EB} = kT/q \ln[\Delta n(0)/n_{P0}]$ by the law of the junction, the presence of holes will keep the EB junction forward biased, and current will continue to flow through the BJT portion of the IGBT until all the holes have been removed. We will now consider the transient removal of holes in more detail.

To illustrate the different phases of the turn-off transient, we consider an IGBT whose base width W_D is large compared to the ambipolar diffusion length [3, 4]. Figure 9.20 shows the cross-section of the device to be examined, which is driving a clamped inductive load, represented here by a current source. The IGBT base is 175 μm thick and doped 2×10^{14} cm^{-3}, resulting in a theoretical plane-junction breakdown of 25 kV. The ambipolar diffusion length in the base is 22.5 μm, so $W_D/L_A \approx 7.8$. The IGBT incorporates a p-type current spreading layer (CSL), similar to the advanced double-diffused metal-oxide-semiconductor field effect transistor (DMOSFET) of Section 8.2.7. The clamped inductive

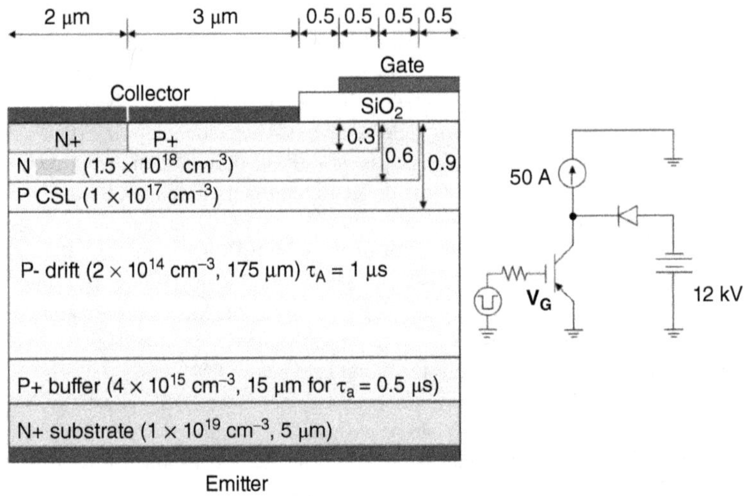

Figure 9.20 Cross-section of a p-channel IGBT driving a clamped inductive load. The IGBT incorporates a p-type current-spreading layer (CSL) and a non-punch-through p+ buffer layer ([3] reproduced with permission from IEEE).

load circuit is modeled by a 50 A current source with a clamping voltage of 12 kV, and the device active area is assumed to be 1 cm^2.

Figure 9.21 shows simulated current and voltage waveforms at 175 °C during the turn-off transient. At $t = 4$ μs the gate voltage is switched from −20 to 0 V. The turn-off transient begins at 5 μs, and consists

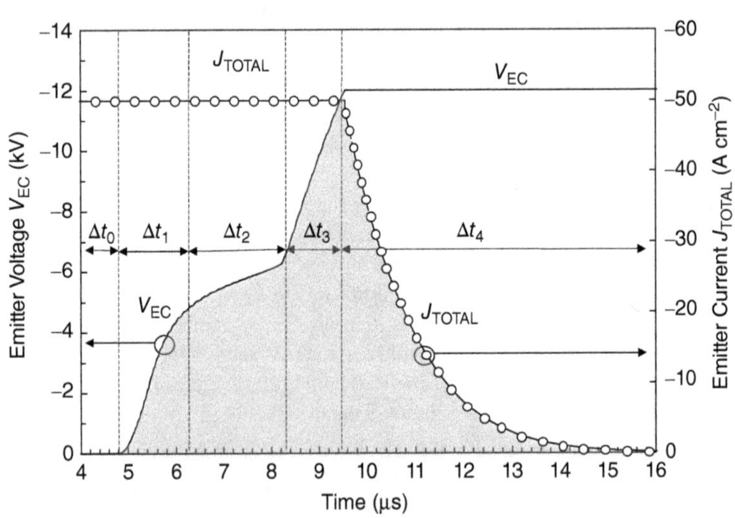

Figure 9.21 Instantaneous voltage and current waveforms during the turn-off transient. The instantaneous power is the product of the current and voltage waveforms, and the switching energy is the integral of the power, suggested by the shading in the figure ([3] reproduced with permission from IEEE).

of four phases, labeled $\Delta t_1 - \Delta t_4$. We will discuss each phase in more detail below. We notice that the current remains constant at 50 A cm^{-2} as the emitter voltage rises to -12 kV. The instantaneous power dissipated in the device is the product of the current and voltage waveforms. The total switching energy E_{OFF} is the integral of the instantaneous power, represented by the shaded region in the figure. The peak power dissipation is about 600 kW cm^{-2} occurring at 9.5 μs, and the total switching energy is 2.4 J cm^{-2}.

We will now discuss each phase of the transient. Figure 9.22 shows the electron density $n(x)$ in the p– base and p+ buffer layer as a function of time during the transient. In time period Δt_1 the emitter voltage rises rapidly. During this period the electron density in the portion of the base closest to the collector falls rapidly, and by 6.0 μs the base is depleted to a distance of 145 μm from the collector junction. The reverse voltage across the collector junction is the integral of the electric field, which is shown in Figure 9.23. The base-collector voltage at 6.0 μs is approximately –3.8 kV.

In time period Δt_2 the emitter voltage rises more slowly. This can be understood as follows. The emitter current is held constant at 50 A cm^{-2} by the current source, and the base current is given by $J_B = J_C/\beta \approx J_E/\beta$. During time period Δt_1 the neutral base becomes progressively narrower, and by the end of Δt_1 the neutral base width W is approximately equal to the diffusion length $L_A = 22.5$ μm. The dependence of β on base width is given by Equation 9.100 and plotted in Figure 9.18. We see that when W is large compared to the diffusion length, β is constant, but as W drops below the diffusion length, β rises rapidly. If β increases but J_C remains constant, J_B must decrease. Recall that J_B represents the total recombination in the base. So during time period Δt_2, β is increasing and the recombination current is decreasing. As a result, the charge density in the base decreases more slowly and the depleted portion of the base expands more slowly. This results in a slower increase in V_{BC} (and V_{EC}), as seen in Figure 9.21. By the end of Δt_2 at 8.3 μs, the base is completely depleted (Figure 9.22b) and the electric field profile is shown in Figure 9.23. The BC voltage at 8.3 μs is approximately –5.5 kV.

During time period Δt_3 the emitter voltage rises rapidly as the depletion region penetrates the more heavily doped buffer layer. The electric field profile is now trapezoidal, as shown in Figure 9.23. Because the doping in the buffer layer is high, the electric field in the base increases rapidly as the depletion region spreads into the buffer layer, and V_{BC} and V_{EC} rise rapidly.

By 9.66 μs, the integral of the electric field has reached -12 kV and the clamping diode becomes forward biased. This is the beginning of time period Δt_4. During this period the forward-biased diode clamps the emitter voltage at -12 kV, and the emitter current falls as the remaining electrons and holes recombine. The emitter current is proportional to the slope of the electron density in the buffer layer, and Figure 9.22d shows this slope gradually decreasing as the last electrons and holes recombine. The 50 A current is increasingly carried by the clamping diode, and the IGBT eventually turns off.

We now consider how the stored charge, instantaneous power, and switching energy are affected by the lifetimes in the base and the p+ buffer layer. Carrier lifetimes are a strong function of temperature, as will be discussed below. The lifetime values we will quote are room-temperature lifetimes, even though all simulations are conducted at 175 °C. The upper portion of Figure 9.24 shows that as the drift region (base) lifetime is increased from 1 to 10 μs, the stored charge increases slightly and the turn-off time also increases slightly. However, the strongest effect comes from the lifetime in the buffer layer, as shown in the lower portion of Figure 9.24. As the buffer lifetime is decreased from 500 to 20 ns, the stored charge drops dramatically and the turn-off transient becomes much shorter. When the lifetime falls below about 100 ns, the ambipolar diffusion length in the buffer becomes shorter than the buffer layer thickness, and an increasing fraction of the injected electrons recombine in the buffer layer before reaching the drift layer. Consequently, the stored charge decreases, shortening the turn-off time. Of course, less stored charge means less conductivity modulation of the drift region, and the on-state power dissipation is higher.

How can we evaluate the trade-off between switching loss and on-state loss as we vary the lifetimes? This can be done using the procedure described in Section 7.1.3. The total power dissipation, including on-state loss and switching loss, depends on the on-state current density and switching frequency, and is given by Equation 7.17. The total power dissipation is set equal to the package power limit, say 300 W cm^{-2}, and Equation 7.17 is solved by iteration at each frequency to determine the on-state current that produces a power dissipation equal to the package limit. This gives us a plot of maximum current

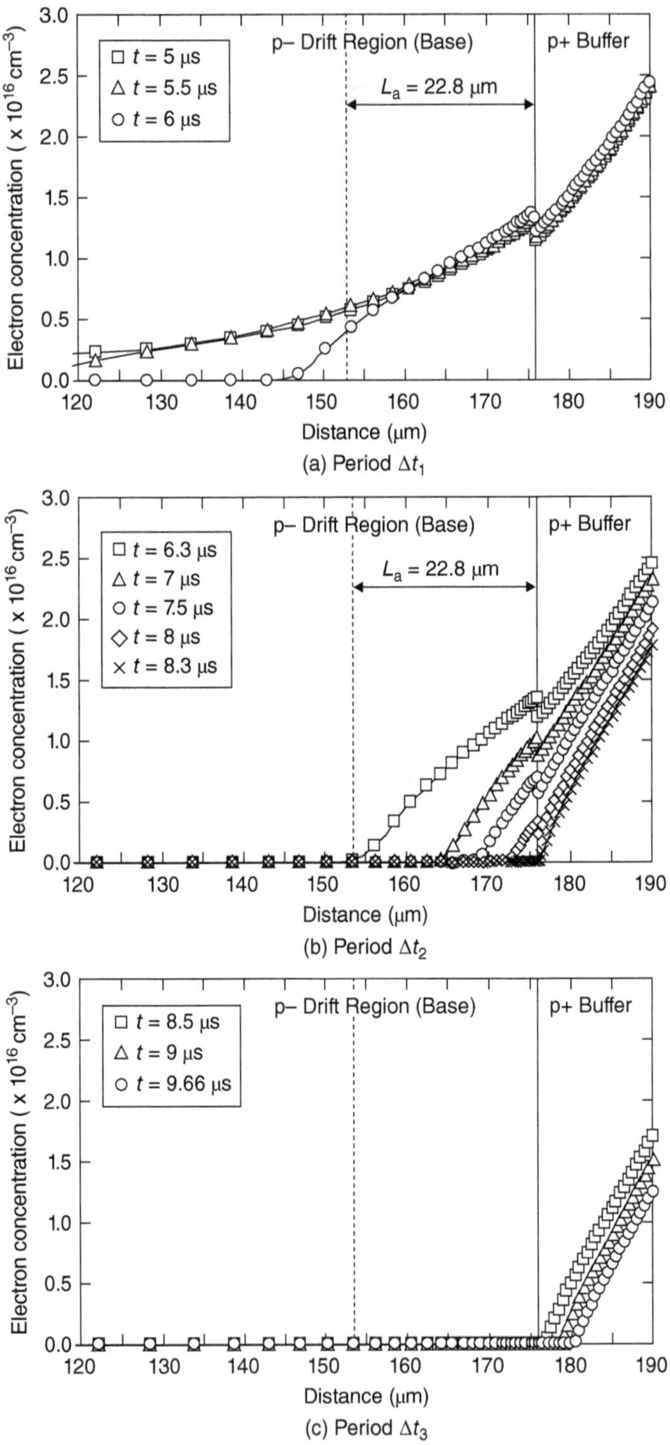

Figure 9.22 Electron density in the base and p+ buffer layer as a function of time during the turn-off transient. (a) Period Δt_1, (b) period Δt_2, (c) period Δt_3, and (d) period Δt_4 ([3] reproduced with permission from IEEE).

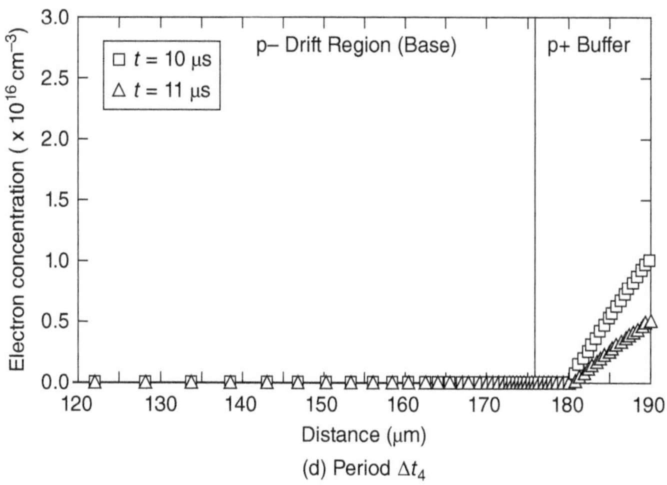

Figure 9.22 *(Continued)*

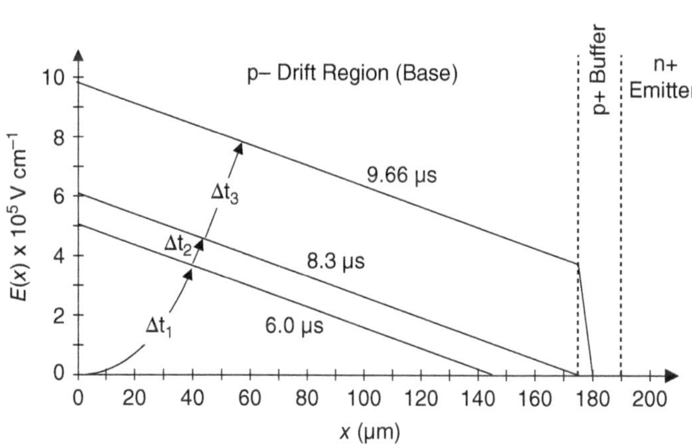

Figure 9.23 Electric field profile in the base during the turn-off transient. The base-collector voltage is the area under the electric field.

density versus frequency. We then repeat this procedure for different assumed values of buffer layer and drift layer lifetimes, and the results for the example device at a junction temperature of 175 °C are shown in Figure 9.25 [4]. Here we see that increasing the base lifetime leads to a higher maximum current at low switching frequencies, where switching loss is less important, but the maximum current falls rapidly with frequency, since switching loss is proportional to frequency. The maximum current of an optimized DMOSFET with the same theoretical blocking voltage is shown for comparison. The MOSFET maximum current is almost independent of frequency, since the MOSFET switching energy is low. Figure 9.25 also shows that decreasing the buffer lifetime from 500 to 20 ns increases the maximum current. This is because of the drastically reduced switching loss associated with the reduced charge storage shown in Figure 9.24. Comparing the p-channel IGBT and the n-channel MOSFET at the specified blocking voltage and temperature, we conclude that the IGBT is superior at switching frequencies below 0.5–1 kHz, while the MOSFET is superior at frequencies above 0.5–1 kHz.

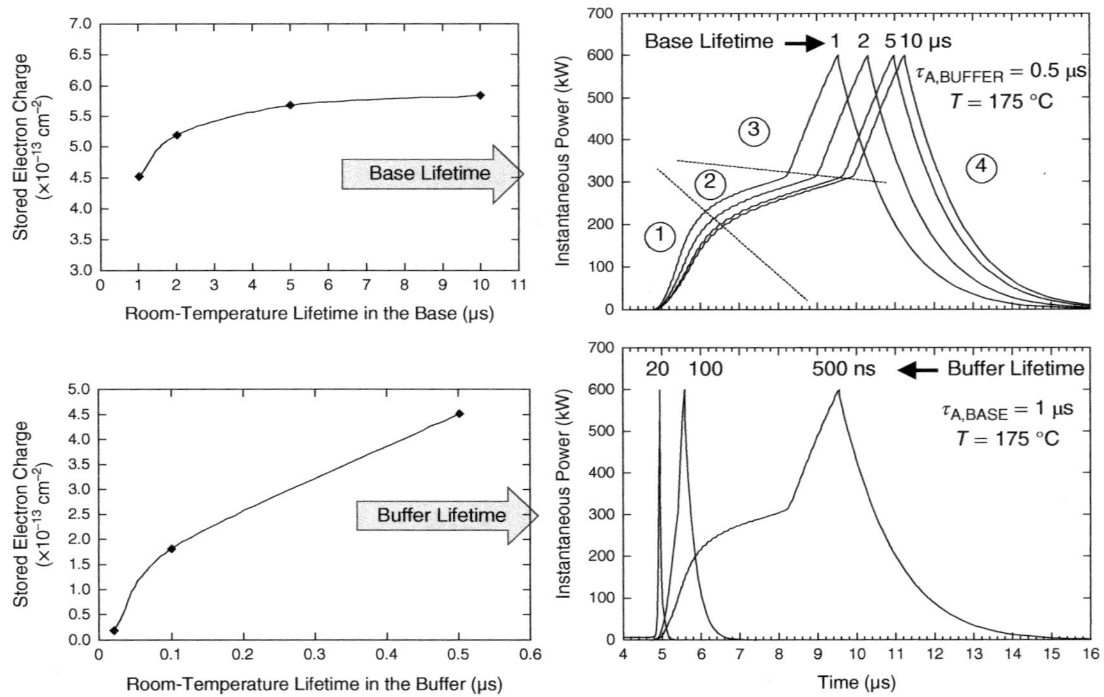

Figure 9.24 The effect of room-temperature ambipolar lifetime in the base and the buffer layer on the stored charge and the instantaneous power waveforms. The buffer layer lifetime has a strong effect on both stored charge and the switching speed ([3] reproduced with permission from IEEE).

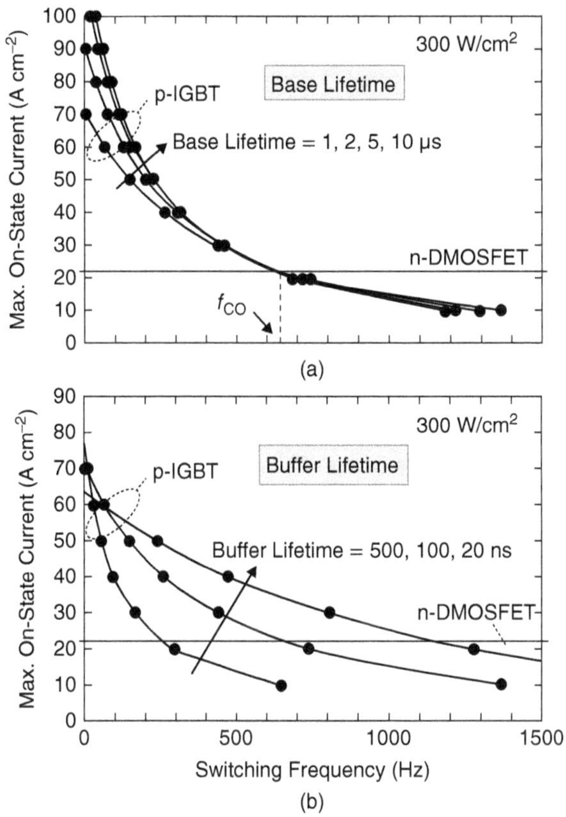

Figure 9.25 Maximum on-state current as a function of frequency for an IGBT for different base lifetimes (a) and different buffer lifetimes (b). A 300 W cm^{-2} package power dissipation limit is assumed ([3] reproduced with permission from IEEE).

9.2.4 Temperature Dependence of Parameters

Almost all our critical device parameters are functions of temperature, and their variation has a strong influence on device performance. Table 9.1 gives equations for several important 4H-SiC parameters as functions of absolute temperature and doping density (where applicable).

Figure 9.26 shows the ambipolar diffusion coefficient, ambipolar lifetime, and ambipolar diffusion length in lightly-doped 4H-SiC as a function of temperature. For comparison, the electron mobility is also shown. The ambipolar diffusion coefficient decreases with temperature due to increased phonon scattering, but this is more than compensated by the increase in lifetime with temperature. As a result, the diffusion length increases slightly with temperature. This makes the on-state performance of the IGBT almost independent of temperature. By comparison, the power MOSFET on-resistance degrades significantly with temperature, since the decreasing electron mobility increases the resistance of the unmodulated drift region. This is the reason why the performance comparison in the last section was conducted at 175 °C.

Table 9.1 Temperature and doping dependence of semiconductor parameters in 4H-SiC.

Parameter	Equation	Units
Electron mobility μ_N	$\mu_N = \dfrac{1141(T/300)^{-2.8}}{1+[(N_A+N_D)/1.94\times10^{17}]^{0.61}}$	$cm^2\ V^{-1}\ s^{-1}$
Hole mobility μ_P	$\mu_P = \dfrac{124(T/300)^{-2.8}}{1+[(N_A+N_D)/1.76\times10^{19}]^{0.34}}$	$cm^2\ V^{-1}\ s^{-1}$
Ambipolar lifetime τ_A	$\tau_A = 57.9\ \tau_A\vert_{300\ K}\exp(-0.105/kT)$	μs
Ionized donor density N_D^+	$N_D^+ = \dfrac{N_D}{1+2\exp[(F_N-E_C+0.066\ eV-1.9\times10^{-8}N_D^{1/3})/kT]}$	cm^{-3}
Ionized acceptor density N_A^-	$N_A^- = \dfrac{N_D}{1+4\exp[(E_V-F_P+0.191\ eV-3.0\times10^{-8}N_A^{1/3})/kT]}$	cm^{-3}
Bandgap energy E_G	$E_G = 3.23+7.036\times10^{-4}[49.751-T^2/(T+1509)]$	eV

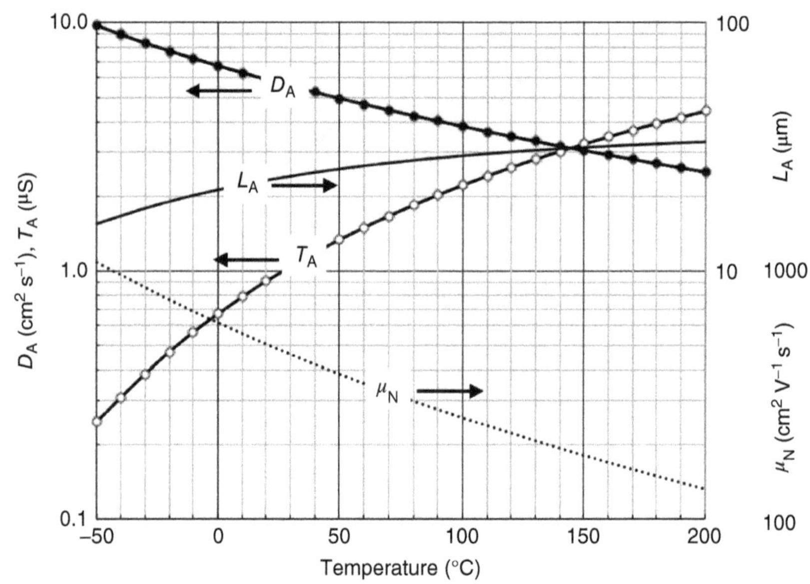

Figure 9.26 Ambipolar diffusion coefficient, lifetime, and diffusion length in lightly-doped 4H-SiC as a function of temperature. For comparison, electron mobility is also shown. A room temperature lifetime of 1.0 μs is assumed in these calculations.

9.3 Thyristors

The final power device to be considered is the pnpn thyristor. The thyristor is a four-layer, three-junction power switch that can be thought of as two merged bipolar transistors, as illustrated in Figure 9.27. The pnp BJT has a narrow base, and its emitter serves as the *anode*. The npn BJT has a wide, lightly-doped base, and its emitter acts as the *cathode*. The base of the pnp BJT is provided with an external contact

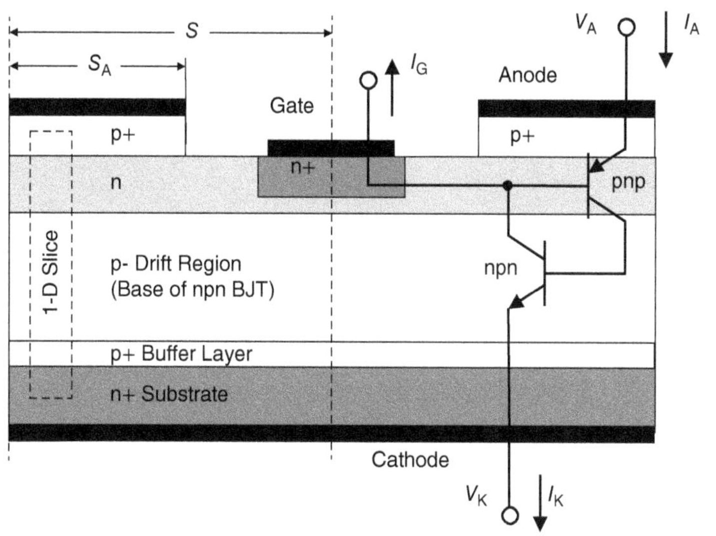

Figure 9.27 Cross-section of a pnpn thyristor with a p+ buffer layer to prevent punch-through in the blocking state. The dashed box identifies a one-dimensional slice that will be used for analysis.

called the *gate*, whose purpose will be described below. SiC thyristors are typically fabricated on n+ substrates because of the high resistance of p+ substrates.

As seen in Figure 9.27, the collector of the pnp BJT is the base of the npn BJT, and the collector of the npn is the base of the pnp. Consequently, when both BJTs are conducting there is no need to supply external base current, since the collector of each BJT supplies base current to the other BJT, and conduction through the four-layer structure is self-sustaining. If both BJTs are off and there is no external supply of base current, the structure is non-conducting. This gives rise to two distinct regimes for $V_{AK} > 0$, as illustrated in Figure 9.28. These are the *forward blocking* mode and the *forward conducting* mode. In the blocking mode the thyristor holds off a large V_{AK} with very little current, while in the conducting mode the thyristor carries a large current with very low voltage drop. Forward operation is therefore bi-stable, since the device can be in either of two regimes at the same V_{AK}. The thin dotted lines depict a negative resistance regime where the anode–cathode voltage snaps back to a low value once a triggering threshold is reached. The triggering threshold can be varied by supplying external base current to the pnp BJT through the gate terminal. Once the BJTs begin to conduct and conduction becomes self-sustaining, the gate current can be removed and the device will remain in conducting state until the current falls below the *holding current* I_H, or V_{AK} is reduced below the *holding voltage* V_H. We will now consider the two forward modes in more detail.

9.3.1 Forward Conducting Regime

In the forward conducting regime both BJTs are operating in saturation, and all three junctions in the thyristor are forward biased. As stated above, forward conduction is self-sustaining, but this requires that the increase in current around the loop formed by the interconnected collector and base regions is unity. The increase in current in each BJT is the ratio I_C/I_B, which is the common-emitter current gain β, so we require that $\beta_{NPN} \cdot \beta_{PNP} = 1$. Since $\alpha = \beta/(\beta + 1)$, this is equivalent to requiring that the sum of the alphas is unity, or $\alpha_{NPN} + \alpha_{PNP} = 1$.

To analyze the conduction in more detail, we refer to the one-dimensional slice shown in Figure 9.29. Here we illustrate the hole and electron densities as a function of position throughout the device. As

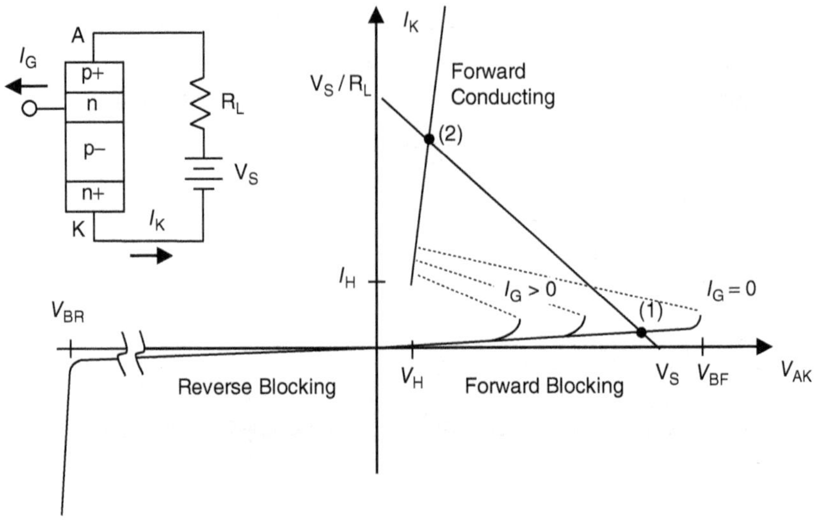

Figure 9.28 Current–voltage characteristics of a pnpn thyristor.

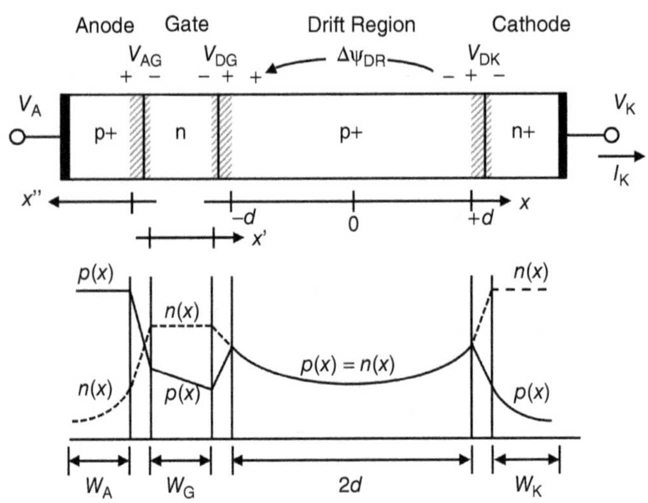

Figure 9.29 Carrier densities in the thyristor in the forward conducting state, showing voltage polarities, and coordinate axes assumed in the analysis. All three junctions are forward biased, and both BJTs operate in saturation.

noted, all three junctions are forward biased and injecting carriers. The p− drift region is in high-level injection, but we assume all the other regions are in low-level injection. This is reasonable, since their doping levels are orders of magnitude higher than the drift region. We also assume that the gate region is short compared to the minority-carrier diffusion length, that is, $W_G \ll L_{PG}$.

The total anode-cathode voltage V_{AK} is the difference in the quasi-Fermi levels from anode to cathode, as illustrated in the band diagram of Figure 9.30. With reference to the band diagram, we can write V_{AK} as

$$V_{AK} = (V_{AG} - V_{DG}) + V_P(-d) + \Delta\psi_{DR} + V_N(+d) \tag{9.111}$$

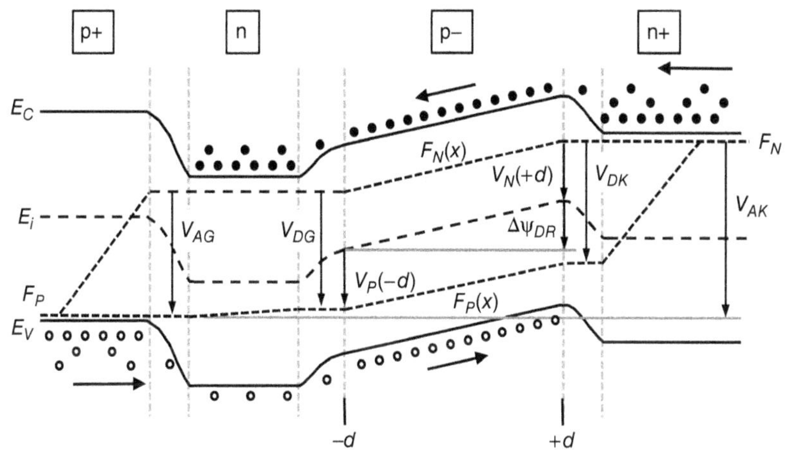

Figure 9.30 Band diagram through the thyristor in the forward conducting regime, showing the different potentials and voltages used in the analysis.

We first consider the potential drop across the drift region, $\Delta\psi_{DR}$. Since this region is in high-level injection, we can set $n(x) = p(x)$ and obtain an expression for $n(x) = p(x)$ by solving the ambipolar diffusion equation, subject to the appropriate boundary conditions. We assume unity injection efficiency at both ends of the drift region, which allows us to set $J_P(+d) = 0$ and $J_N(-d) = 0$. Setting $J_N(-d) = 0$ is equivalent to assuming the hole current diffusing across the n-type gate region from the anode is much larger than any electron current injected into the drift region from the gate. Under these assumptions, the development is the same as for the pin diode in Section 7.3, and the carrier densities are given by Equation 7.46, which can be written

$$\boxed{\Delta n(x) = \Delta p(x) = \frac{\tau_{AD} J_K}{2qL_{AD}}\left[\frac{\cosh(x/L_{AD})}{\sinh(d/L_{AD})} - \frac{(\mu_{ND} - \mu_{PD})}{(\mu_{ND} + \mu_{PD})}\frac{\sinh(x/L_{AD})}{\cosh(d/L_{AD})}\right]} \tag{9.112}$$

Here τ_{AD}, L_{AD}, μ_{ND}, and μ_{PD} refer to parameters in the lightly-doped drift region. To obtain the potential drop across the drift region, we again proceed as with the pin diode: We write an expression for the total current as a function of the electric field, solve for the electric field as a function of current, then integrate the electric field to obtain the potential as a function of position using Equation 9.112 for the carrier densities. We then evaluate this expression at the two ends of the drift region $x = \pm d$ to calculate the potential drop. The result is the same as Equation 7.51, and can be written

$$\boxed{\begin{aligned}\Delta\psi_{DR} = \frac{kT}{q}\Bigg\{ & \frac{8b}{(b+1)^2}\frac{\sinh(d/L_{AD})}{\sqrt{1 - B^2\tanh^2(d/L_{AD})}}\tan^{-1}\left[\sqrt{1 - B^2\tanh^2(d/L_{AD})}\sinh(d/L_{AD})\right] \\ & + B\ln\left[\frac{1 + B\tanh^2(d/L_{AD})}{1 - B\tanh^2(d/L_{AD})}\right]\Bigg\}\end{aligned}} \tag{9.113}$$

where $b = \mu_{ND}/\mu_{PD}$ and $B = (\mu_{ND} - \mu_{PD})/(\mu_{ND} + \mu_{PD})$. $\Delta\psi_{DR}$ is the change in electrostatic potential, or the change in the intrinsic Fermi level E_i, across the drift region.

We will next obtain an expression for the sum $V_P(-d) + V_N(+d)$. Referring to the band diagram in Figure 9.30, the electron density at $x = +d$ can be written

$$n(+d) = n_i \exp[qV_N(+d)/kT] \tag{9.114}$$

where $V_N(+d)$ is the difference between the electron quasi-Fermi level and the intrinsic level at $x = +d$, a quantity known as the *chemical potential for electrons*. Likewise, at $x = -d$ the hole density can be written

$$p(-d) = n_i \exp[qV_P(-d)/kT] \tag{9.115}$$

where $V_P(-d)$ is the difference between the intrinsic level and the hole quasi-Fermi level at $x = -d$, or the *chemical potential for holes*. Solving Equations 9.114 and 9.115 for $V_N(-d)$ and $V_P(+d)$ and adding yields

$$V_p(-d) + V_N(+d) = \frac{kT}{q} \ln\left[\frac{n(-d)\,n(+d)}{n_i^2}\right] \tag{9.116}$$

When evaluated in equilibrium, Equation 9.116 gives the *built-in potential* of the junction. We can express Equation 9.116 in terms of current using Equation 9.112 to evaluate $n(-d)$ and $n(+d)$, resulting in

$$\boxed{V_P(-d) + V_N(+d) = \frac{2kT}{q} \ln\left(\frac{J_{\mathrm{TOTAL}}}{2qn_i\theta}\frac{\tau_{\mathrm{AD}}}{L_{\mathrm{AD}}}\right)} \tag{9.117}$$

where θ is

$$\theta = \left\{\left[\coth\left(\frac{d}{L_{\mathrm{AD}}}\right) + B\tanh\left(\frac{d}{L_{\mathrm{AD}}}\right)\right]\left[\coth\left(\frac{d}{L_{\mathrm{AD}}}\right) - B\tanh\left(\frac{d}{L_{\mathrm{AD}}}\right)\right]\right\}^{-1/2} \tag{9.118}$$

We note that this treatment parallels that of the pin diode, and Equation 9.118 is simply a re-statement of Equation 7.59.

The quasi-Fermi level splitting at the GD junction V_{DG} can be obtained from the law of the junction evaluated at $x = -d$,

$$p(-d)n(-d) = n(-d)^2 = n_i^2 \exp(qV_{\mathrm{DG}}/kT) \tag{9.119}$$

Solving for V_{DG} and employing Equation 9.112 for $n(-d)$ yields

$$\boxed{V_{\mathrm{DG}} = 2\frac{kT}{q} \ln\left\{\left(\frac{J_{\mathrm{TOTAL}}}{2qn_i}\frac{\tau_{\mathrm{AD}}}{L_{\mathrm{AD}}}\right)\left[\coth\left(\frac{d}{L_{\mathrm{AD}}}\right) + B\tanh\left(\frac{d}{L_{\mathrm{AD}}}\right)\right]\right\}} \tag{9.120}$$

Finally, we need to calculate the quasi-Fermi level splitting at the AG junction, V_{AG}. The anode and gate are assumed to be in low-level injection and the electric fields in the quasi-neutral regions are negligible, so current flow in these regions is due to diffusion. The total current crossing the AG junction is the sum of electron and hole diffusion currents, and can be written

$$J_{\mathrm{TOTAL}} = -qD_{\mathrm{PG}} \left.\frac{\partial p}{\partial x'}\right|_{x'=0} - qD_{\mathrm{NA}} \left.\frac{\partial n}{\partial x''}\right|_{x''=0} \tag{9.121}$$

where D_{PG} is the hole diffusion coefficient in the gate region and D_{NA} is the electron diffusion coefficient in the anode. The electron and hole densities in the anode and gate regions are obtained by solving the minority carrier diffusion equations in each region. If the anode is long compared to the electron diffusion length, the electron density in the anode can be written

$$n(x'') \approx \frac{n_{i\mathrm{A}}^2}{N_{\mathrm{AA}}^-} \exp\left(\frac{qV_{\mathrm{AG}}}{kT}\right) \exp\left(-\frac{x''}{L_{\mathrm{NA}}}\right) \tag{9.122}$$

From this it follows that

$$\left.\frac{\partial n}{\partial x''}\right|_{x''=0} = -\frac{n_{iA}^2}{L_{NA}N_{AA}^-}\exp\left(\frac{qV_{AG}}{kT}\right) \tag{9.123}$$

Since the gate region is short compared to the hole diffusion length, the hole density in the gate is a linear function of position, and we can write

$$\left.\frac{\partial p}{\partial x'}\right|_{x'=0} = \frac{p(w_G) - p(0)}{W_G} \tag{9.124}$$

Using the law of the junction, the pn product throughout the GD depletion region can be written

$$p(W_G)n(W_G) = p(W_G)N_{DG}^+ = n_i^2\exp(qV_{DG}/kT) = n(-d)^2 \tag{9.125}$$

Thus

$$p(W_G) = n(-d)^2/N_{DG}^+ \tag{9.126}$$

and we can write Equation 9.124 as

$$\left.\frac{\partial p}{\partial x'}\right|_{x'=0} = \frac{n(-d)^2 - n_i^2\exp(qV_{AG}/kT)}{W_G N_{DG}^+} \tag{9.127}$$

Inserting Equations 9.123 and 9.127 into Equation 9.121 and using Equation 9.112 for $n(-d)$ yields

$$\begin{aligned} J_{TOTAL} = &\frac{qD_{PG}}{W_G N_{DG}^+}\left\{n_i^2\exp\left(\frac{qV_{AG}}{kT}\right) - \left(\frac{J_{TOTAL}\tau_{AD}}{2qL_{AD}}\right)^2\left[\coth\left(\frac{d}{L_{AD}}\right) + B\tanh\left(\frac{d}{L_{AD}}\right)\right]^2\right\} \\ &+ q\frac{n_{iA}^2}{N_{AA}^-}\frac{D_{NA}}{L_{NA}}\exp\left(\frac{qV_{AG}}{kT}\right) \end{aligned} \tag{9.128}$$

Equation 9.128 is a quadratic equation in J_{TOTAL}, and can be written more compactly as

$$C_1 J_{TOTAL}^2 + J_{TOTAL} = C_2\exp(qV_{AG}/kT) \tag{9.129}$$

where

$$C_1 = \frac{qD_{PG}}{W_G N_{DG}^+}\left(\frac{\tau_{AD}}{2qL_{AD}}\right)^2\left[\coth\left(\frac{d}{L_{AD}}\right) + B\tanh\left(\frac{d}{L_{AD}}\right)\right]^2 \tag{9.130}$$

$$C_2 = q\left(\frac{D_{PG}n_i^2}{W_G N_{DG}^+} + \frac{D_{NA}n_{iA}^2}{L_{NA}N_{AA}^-}\right) \tag{9.131}$$

We note that C_1 and C_2 are constants that depend only on device parameters. Solving Equation 9.129 for J_{TOTAL} gives

$$J_{TOTAL} = \sqrt{\left(\frac{1}{2C_1}\right)^2 + \frac{C_2}{C_1}\exp\left(\frac{qV_{AG}}{kT}\right)} - \left(\frac{1}{2C_1}\right) \tag{9.132}$$

We now express V_{AG} in Equation 9.132 in terms of V_{AK} using Equation 9.111,

$$V_{AG} = V_{AK} + V_{DG} - \Delta\psi_{DR} - [V_P(-d) + V_N(-d)] \tag{9.133}$$

Inserting Equation 9.120 for V_{DG} and Equation 9.117 for $[V_P(-d) + V_N(-d)]$ into Equation 9.133 allows us to write

$$\exp(qV_{AG}/kT) = C_3\exp(qV_{AK}/kT) \tag{9.134}$$

where

$$C_3 = \exp\left(\frac{-q\Delta\psi_{DR}}{kT}\right)\theta^2\left[\coth\left(\frac{d}{L_A}\right) + B\tanh\left(\frac{d}{L_A}\right)\right]^2 \tag{9.135}$$

Once again we note that the new parameter C_3 is a computable constant that depends only on device parameters. Inserting Equation 9.134 into Equation 9.132 gives the desired expression for the current–voltage characteristics in the forward conducting regime:

$$J_K = J_{TOTAL} = \sqrt{\left(\frac{1}{2C_1}\right)^2 + \frac{C_2C_3}{C_1}\exp\left(\frac{qV_{AK}}{kT}\right)} - \left(\frac{1}{2C_1}\right) \tag{9.136}$$

At the high current densities at which thyristors typically operate, the exponential term in the square root dominates all other terms, and we can write

$$\boxed{J_K \approx J_0 \exp(qV_{AK}/2kT)} \tag{9.137}$$

where

$$\boxed{J_0 = \sqrt{1 + \frac{D_{NA}}{D_{PG}}\frac{W_G}{L_{NA}}\frac{N_{DG}^+}{N_{AA}^-}\frac{n_{iA}^2}{n_i^2}}\left(2qn_i\frac{L_{AD}}{\tau_{AD}}\theta\right)\exp\left(\frac{-q\Delta\psi_{DR}}{2kT}\right)} \tag{9.138}$$

Here θ and $\Delta\psi_{DR}$ are constants that depend only on device parameters, and are given by Equations 9.118 and 9.113 respectively.

The result of this long derivation is perhaps somewhat anticlimactic. Equation 9.137 tells us that the thyristor current in the forward conduction regime increases exponentially with V_{AK}, with an exponential slope of 1/2. This is the same dependence as the pin diode, Equation 7.57. Moreover, the prefactor J_0 in Equation 9.138 is identical to the prefactor of the pin diode in Equation 7.58 except for the square root factor, which has a value close to unity. Thus we conclude that *the forward I–V characteristics of the thyristor are almost identical to those of a comparable pin diode.*

The above analysis assumed that the anode is thick compared to an electron diffusion length, $W_A \gg L_{NA}$. If the anode is thinner than a diffusion length, we simply replace L_{NA} with W_A in all the foregoing equations.

9.3.2 *Forward Blocking Regime and Triggering*

We turn now to the forward blocking regime and consider how the thyristor is switched from the forward blocking to the forward conducting regimes. Figure 9.28 contains a simple circuit with the thyristor connected to a resistive load. Kirchoff's current and voltage laws tell us there are two stable operating points, formed by the intersections of the resistive load line (dashed) with the thyristor I–V characteristics. If the supply voltage $V_S = 0$, the thyristor is in the forward blocking mode at the origin. As V_S is increased with $I_G = 0$, the operating point (1) moves laterally along the forward blocking characteristics until V_S becomes greater than the forward blocking voltage V_{BF}, when the circuit abruptly switches to operating point (2). As we shall see, the switching threshold can be reduced in a controllable manner by drawing current out of the gate.

The switching threshold is best understood by considering the internal currents in two cross-coupled BJTs, as illustrated in Figure 9.31, where the npn BJT is designated BJT-1 and the pnp BJT is designated BJT-2. The cathode current can be written

$$I_K = I_{C1} + I_{C2} = \alpha_1 I_{E1} + \alpha_2 I_{E2} = \alpha_1 I_K + \alpha_2 I_A = \alpha_1 I_K + \alpha_2(I_K + I_G) \tag{9.139}$$

Solving for I_K gives

$$I_K = \frac{\alpha_2 I_G}{1 - (\alpha_1 + \alpha_2)} \tag{9.140}$$

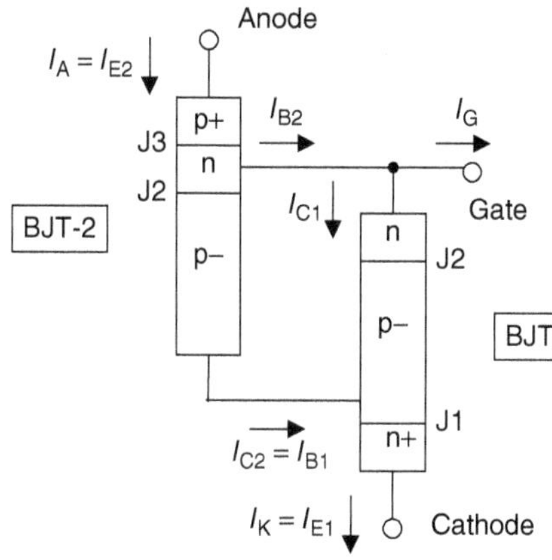

Figure 9.31 Visualization of the thyristor as two cross-coupled BJTs. The collector of each BJT supplies base current to the other BJT, and the gate terminal allows base current to be supplied to BJT-2 externally.

Equation 9.140 shows that the cathode current becomes infinite when the sum of the alphas equals one. The alphas are evaluated in the forward blocking mode where very little current flows and all regions are in low-level injection. Equation 9.10 defines the common-base current gain α as the product of the base transport factor α_T and the emitter injection efficiency γ. For wide-base npn BJT-1 the base transport factor α_T is given by Equation 9.11,

$$\alpha_T = \frac{1}{\cosh(W/L_{NB})} \tag{9.11}$$

The emitter injection efficiency γ is the fraction of emitter current due to electrons injected into the base, so $\gamma = I_{EN}/I_E$. Equation 9.10 specifies the emitter current I_E as $(I_{EN} + I_{EP})$, but when the currents are small, as in the forward blocking regime, I_E must also take into account the current due to recombination in the EB depletion region. With this in mind, γ can be written

$$\gamma = \frac{I_{EN}}{I_E} = \frac{I_{EN}}{I_{EN} + I_{EP} + I_R} \approx \frac{I_{EN}}{I_{EN} + I_R} \tag{9.141}$$

Here I_R is the recombination current in the depletion region of the EB junction, and $I_{EP} \ll I_{EN}$ due to the large doping asymmetry between the n+ emitter and the p− base. For an n+/p− one-sided step junction, the recombination current can be written

$$I_R \approx 2\sqrt{2}A\left(\frac{n_i}{\tau_G}\right)\frac{kT}{\sqrt{qN_{AB}(\psi_{BI} - V_{BE})/\epsilon_S}}\exp\left(\frac{qV_{BE}}{2kT}\right) \tag{9.142}$$

where A is the junction area and the generation lifetime $\tau_G = \tau_N + \tau_P$ is the same as the ambipolar lifetime τ_A. Note that N_{AB} is the total dopant concentration in the base, rather than the ionized concentration. Combining Equation 9.142 with Equation 9.7 and assuming the CB junction is strongly reverse biased, we can write Equation 9.141 in the form

$$\gamma \approx \frac{1}{1 + \dfrac{2\sqrt{2}\,kT/q}{\sqrt{qN_{AB}(\psi_{BI} - V_{BE})/\epsilon_S}}\left(\dfrac{L_{NB}}{D_{NB}\tau_G}\right)\left(\dfrac{N_{AB}^{-}}{n_i}\right)\tanh\left(\dfrac{W}{L_{NB}}\right)\left[\dfrac{\exp\left(qV_{BE}/2kT\right)}{\exp(qV_{BE}/kT) - 1}\right]} \tag{9.143}$$

Equation 9.143 can now be combined with Equations 9.10 and 9.11 to calculate α.

Examining Equation 9.143, we see that γ in the low-current regime is no longer a constant, but instead depends on V_{BE} and hence on the current. When V_{BE} is small, γ approaches zero. As V_{BE} increases, γ increases, and at sufficiently high $V_{BE}, \gamma \to 1$. Figure 9.32a is a plot of γ versus J_{EN} for several values of W/L_{NB}, using Equation 9.143 for γ and Equation 9.7 for J_{EN}. As seen, γ (and therefore α) increases monotonically with current. The same general arguments apply to the narrow-base pnp BJT-2.

The dependence of α on current can also be illustrated using a *Gummel plot,* shown in Figure 9.32b. This presentation depicts J_C and J_B on a log scale as a function of V_{BE}, and we have also included α and β to aid the discussion. For $V_{BE} < 2.4$ V, J_B is dominated by recombination in the depletion region, Equation 9.142, whereas for $V_{BE} > 2.4$ V, J_B is dominated by hole injection into the emitter, Equation 9.2. Recombination in the neutral base is small, since the base is short compared to a diffusion length. At high currents the ratio $\beta = J_C/J_B$ is constant, independent of current, but at low currents β and α increase with increasing current.

The thyristor's triggering threshold is reached when the sum of the alphas equals one, as shown by Equation 9.140. Considering the thyristor in the forward blocking regime, we can now ask how the alphas of the internal BJTs depend on V_{AK}. As V_{AK} is increased, the alphas increase through two mechanisms. The first mechanism is base width modulation. Figure 9.33 illustrates the depletion regions in the device in the forward blocking mode. The middle junction is reverse biased, and its depletion region spreads mainly into the lightly-doped drift region, the base of BJT-1. As V_{AK} increases, x_P expands, W_1 shrinks, and α_1 increases. The second mechanism is more subtle. As V_{AK} increases and x_P widens, the thermal generation current within the depletion region of the middle junction increases. The generation current is identified in Figure 9.33, and can be written

$$I_{GEN} = qA\left(\frac{n_i}{\tau_G}\right)x_P \approx qA\left(\frac{n_i}{\tau_G}\right)\sqrt{\frac{2\epsilon_S}{qN_{AB}}V_{AK}} \tag{9.144}$$

where we have assumed all the applied voltage V_{AK} appears across the reverse-biased middle junction. As seen, the generation current increases as the square root of V_{AK}. Since $I_G = 0$, Figure 9.33 shows that this generation current must flow through the emitters of both BJTs. When the currents are small, as they are in the forward-blocking mode, α is an increasing function of current, as shown in Figure 9.32a. Thus, increasing V_{AK} increases I_{GEN}, which increases the current, which increases α.

Although the above analysis of the triggering threshold appears reasonable, it turns out that triggering does not actually require I_K to become infinite, as suggested by Equation 9.140. Instead, triggering only requires that $\partial I_K/\partial I_G$ becomes infinite. When this condition is met, any small fluctuation in gate current will produce an unlimited increase in cathode current, and the thyristor switches on. By reference to Figure 9.31 we can write

$$\partial I_K = \partial I_{C1} + \partial I_{C2} = \frac{\partial I_{C1}}{\partial I_{E1}}\partial I_K + \frac{\partial I_{C2}}{\partial I_{E2}}\partial I_A = \widetilde{\alpha}_1\partial I_K + \widetilde{\alpha}_2\partial I_A \tag{9.145}$$

where $\widetilde{\alpha}_1$ and $\widetilde{\alpha}_2$ are *small-signal* alphas defined by

$$\widetilde{\alpha} = \frac{\partial I_C}{\partial I_E} = \frac{\partial}{\partial I_E}(\alpha I_E) = \alpha + I_E\frac{\partial\alpha}{\partial I_E} \tag{9.146}$$

If $\partial\alpha/\partial I_E > 0$, the small-signal $\widetilde{\alpha}$ will be greater than the DC α, and this is usually the case. Returning to Figure 9.31, we can also write

$$\partial I_A = \partial I_K + \partial I_G \tag{9.147}$$

Substituting Equation 9.147 into Equation 9.145 yields

$$\frac{\partial I_K}{\partial I_G} = \frac{\widetilde{\alpha}_2}{1-(\widetilde{\alpha}_1+\widetilde{\alpha}_2)} \tag{9.148}$$

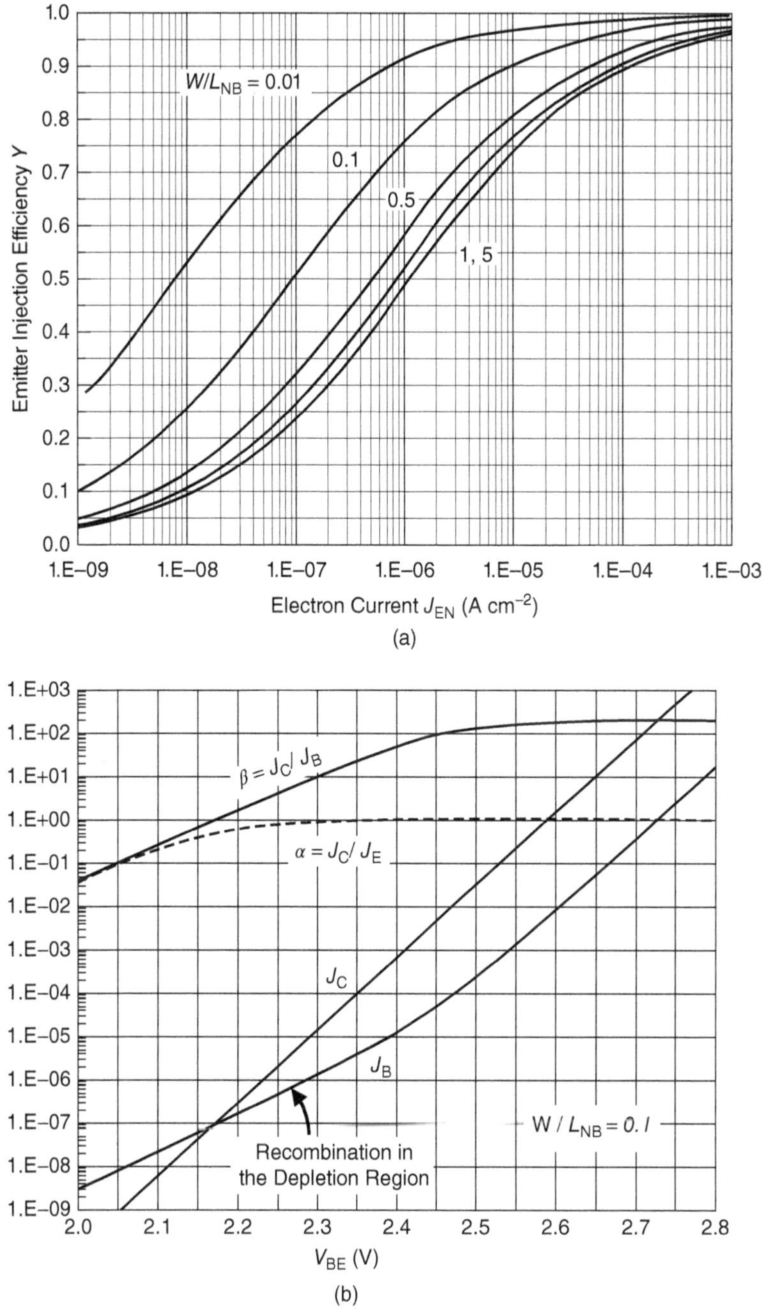

Figure 9.32 (a) Emitter injection efficiency as a function of current, calculated using Equation 9.143 with J_{EN} given by Equation 9.7. In this example, $N_{AB} = 1 \times 10^{14}$ cm^{-3}, $N_{DE} = 1 \times 10^{19}$ cm^{-3}, and $\tau_G = 1$ μs. (b) Gummel plot for the BJT of (a) with $W = 0.1\ L_{NB}$. At low currents, J_B is dominated by recombination in the depletion region, and α and β depend on current.

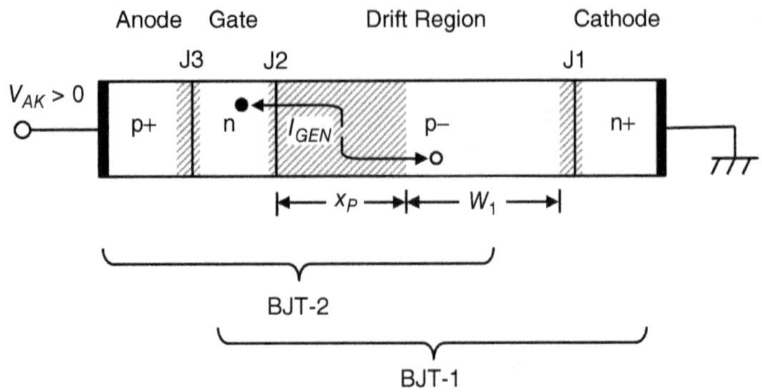

Figure 9.33 One-dimensional slice through the thyristor in the forward blocking state, showing the generation current in the depletion region of the middle junction. The depletion regions are shaded.

Equation 9.148 indicates that the thyristor actually triggers when the sum of the *small-signal* alphas equals one. The small-signal alphas are defined by Equation 9.146, but how do we calculate $\partial\alpha/\partial I_E$? We know from Equation 9.143 that the emitter injection efficiency increases with current, so alpha also increases with current. Thus $\partial\alpha/\partial I_E > 0$, as postulated, and the small-signal alphas are indeed slightly higher than the DC alphas. The expressions for the small-signal alphas can be calculated by writing Equation 9.143 in terms of current and then differentiating. The resulting expressions have been derived by Yang and Voulgaris [5], and the reader is referred to this reference for the details.

Once the triggering threshold is reached, the thyristor switches quickly from the blocking to the conducting state, and the internal BJTs go from their forward-active regimes to their saturation regimes. This process can be understood by reference to Figure 9.34. Here we assume the thyristor is driving a resistive load, and the gate current is zero. As the supply voltage V_S is increased from zero, the thyristor is initially in the blocking mode, and the operating point moves along the lower part of the characteristic, as illustrated by points (1) and (2). The internal BJTs likewise move along their $I_B = 0$ lines from point (1) to point (2), as shown. When the supply voltage reaches the switching threshold, which happens just beyond point (2), the thyristor switches to the conducting state and comes to rest at point (5). Once at point (5), if V_S is increased or decreased, the operating point moves along the conducting characteristic, so long as the current remains above the holding current. For reference, the carrier densities and band diagrams in the thyristor at points (1) and (5) are illustrated in Figure 9.35.

We will now examine the trajectory in more detail. While the thyristor is in the blocking state, the product $\beta_1 \cdot \beta_2 < 1$ (remember the npn BJT has a very wide base, and the emitter injection efficiencies are very low when the currents are low). As V_S approaches the switching threshold, three effects occur. First, the current gain of the npn BJT increases due to base width modulation as W_1 shrinks. This increases the product $\beta_1 \cdot \beta_2$. Second, the widening depletion region of the middle junction increases the leakage current I_G, which increases the injection efficiencies of both BJTs according to Equation 9.143. This also increases the product $\beta_1 \cdot \beta_2$. Third, avalanche multiplication in the reverse-biased middle junction adds more current, further increasing the injection efficiencies and increasing $\beta_1 \cdot \beta_2$. At some point $\beta_1 \cdot \beta_2$ will exceed 1. When this happens, the current around the loop formed by the interconnected collectors and bases begins to increase. This corresponds to the BJT operating point moving from (2) to (3), (4) and (5) as the base currents increase, as shown in Figure 9.34b. The situation stabilizes when the internal BJTs enter saturation at point (5). Why is this point stable? In saturation the *effective* β, the ratio of *actual* collector current to base current in a BJT, is lower than in the forward active region. This reduces the β product to 1 and operation stabilizes at point (5). To see why the circuit is stable at point (5), imagine that the cathode current tries to increase. With a higher cathode current, the voltage drop across the load increases, the voltage drop across the thyristor V_{AK} therefore decreases,

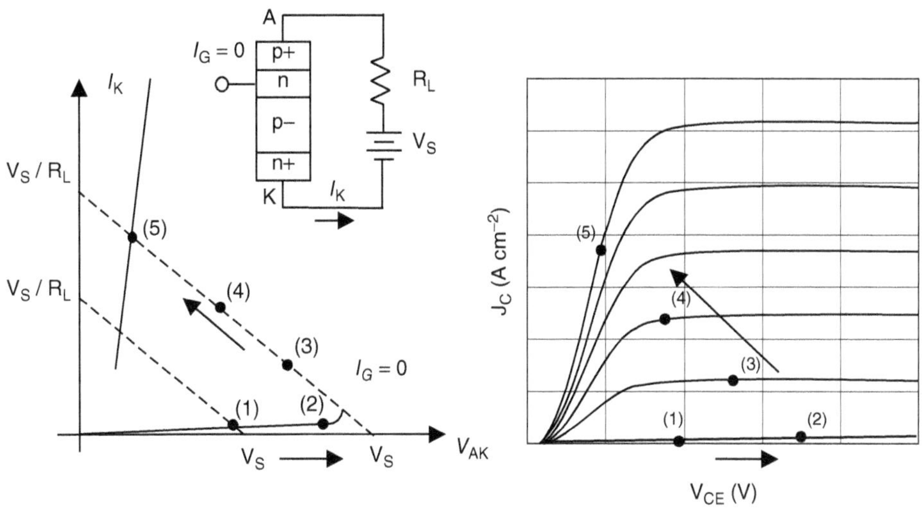

Figure 9.34 Illustration of the trajectory of the thyristor and its internal BJTs during the turn-on transient.

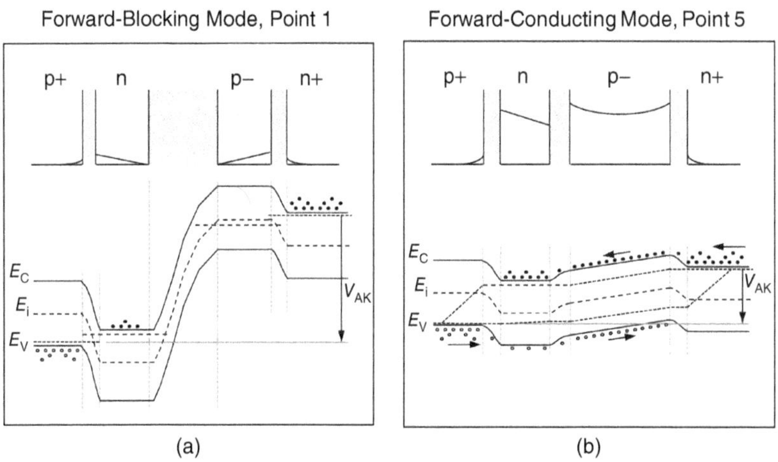

Figure 9.35 Band diagrams and minority carrier densities in the thyristor in the forward-blocking mode (a) and forward-conducting mode (b).

the depletion width x_P decreases, the base width W_1 increases, and β_1 decreases, reducing the current. If the cathode current tries to decrease, the opposite happens. These forces maintain the circuit at point (5).

The switching threshold can be reduced by supplying an external gate current $I_G > 0$. This acts as base current to BJT-2, and the collector of BJT-2 then supplies base current to BJT-1. These higher currents increase the betas, allowing $\beta_1 \cdot \beta_2$ to reach 1 at a lower value of V_{AK}. This makes it possible to trigger the thyristor at a specific $V_{AK} < V_{BF}$ under the control of an external circuit.

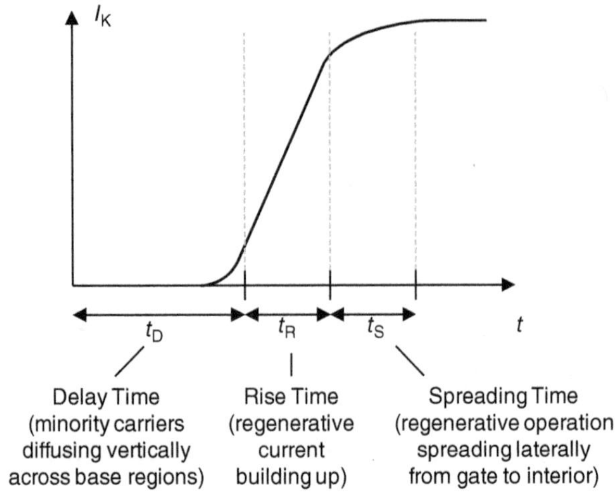

Figure 9.36 Three phases of the turn-on transient.

9.3.3 *The Turn-On Process*

The turn-on process occurs in three phases that can be considered sequentially, as illustrated in Figure 9.36. The three critical times are (i) the delay time, (ii) the rise time, and (iii) the spreading time. We will discuss each of these in turn.

Let us assume the thyristor is initially in the forward-blocking mode, and is triggered to switch on by a gate current pulse I_G that begins at $t = 0$. Conditions inside the device prior to $t = 0$ are as indicated in Figure 9.35a; both BJTs are in their forward-active modes but there is essentially no minority carrier injection into either base region. Referring to Figure 9.31, the gate current pulse that begins at $t = 0$ supplies base current to pnp BJT-2. This causes injection of holes from the p+ emitter into the narrow n-type base. The holes diffuse across the base, and the first holes arrive at the collector of BJT-2 in a base transit time τ_{B2} given by

$$\tau_{B2} = \frac{W_{B2}^2}{2D_{P,B2}} \tag{9.149}$$

where W_{B2} is the width of the neutral base and $D_{P,B2}$ is the diffusion coefficient for holes in the base. Once the holes reach the collector of BJT-2, they act as base current to npn BJT-1. This causes the injection of electrons from the n+ emitter of BJT-1 into the wide p− base. These electrons diffuse toward the collector of BJT-1, and the first electrons reach the collector in a transit time τ_{B1} given by

$$\tau_{B1} = \frac{W_{B1}^2}{2D_{N,B1}} \tag{9.150}$$

where W_{B1} is the width of the neutral base and $D_{N,B1}$ is the diffusion coefficient for electrons in the wide p− base. At this point, the regenerative build-up of current within the thyristor begins. The delay time is then approximately given by the sum of the base transit times, or

$$\boxed{t_D \approx \tau_{B1} + \tau_{B2}} \tag{9.151}$$

The second phase is characterized by regenerative current feedback between the two BJTs, and the current rapidly builds up until a steady state is reached. The current rise time t_R can be calculated by considering the build-up of stored charge in the base regions of the BJTs. We assume the build-up of stored charge is due only to carriers that flow in as base current, and we neglect recombination since the rise time t_R is short compared to the minority carrier lifetimes in the base. Again referring to Figure 9.31, the build-up of charge in the base of BJT-1 can be written

$$\frac{\partial Q_{B1}}{\partial t} \approx I_{B1} = I_{C2} = \alpha_2 I_{E2} = \alpha_2 I_A \tag{9.152}$$

Likewise, the build-up of charge in the base of BJT-2 can be written

$$\frac{\partial Q_{B2}}{\partial t} \approx I_{B2} = I_{C1} + I_G = \alpha_1 I_{E1} + I_G = \alpha_1 I_K + I_G \tag{9.153}$$

Focusing on Equation 9.152, we can write $\alpha_2 I_A$ as

$$\alpha_2 I_A = I_{C2} = Q_{B2}/\tau_{B2} \tag{9.154}$$

where the collector current is expressed as the base minority carrier charge divided by the diffusion transit time across the base (note that in the forward-active mode, $Q_{B2} = qAp(0)W_{B2}/2$, so Equation 9.154 is equivalent to setting $J_C = qD_{P,B2}\,p(0)/W_{B2} = qD_{P,B2}\,\partial p/\partial x$, which is the usual expression for diffusion current in the base). Inserting Equation 9.154 into Equation 9.152 and differentiating yields

$$\frac{\partial^2 Q_{B1}}{\partial t^2} = \frac{1}{\tau_{B2}}\frac{\partial Q_{B2}}{\partial t} = \frac{1}{\tau_{B2}}(\alpha_1 I_K + I_G) \tag{9.155}$$

where we have used Equation 9.153 for $\partial Q_{B2}/\partial t$. However, we can also write

$$\alpha_1 I_K = I_{C1} = Q_{B1}/\tau_{B1} \tag{9.156}$$

so Equation 9.155 becomes

$$\frac{\partial^2 Q_{B1}}{\partial t^2} = \frac{Q_{B1}}{\tau_{B1}\tau_{B2}} + \frac{I_G}{\tau_{B2}} \tag{9.157}$$

Equation 9.157 is a second-order differential equation describing the build-up of minority carrier charge in the base of BJT-1, and the solution can be written

$$Q_{B1}(t) = (\alpha_1 I_{K,SS} - I_G)\tau_{B1}\exp[(t - t_R)/\sqrt{\tau_{B1}\tau_{B2}}] - \tau_{B1} I_G \tag{9.158}$$

where $I_{K,SS}$ is the steady-state value of the cathode current. Equation 9.158 tells us that the charge in the thyristor base builds up exponentially with a time constant given by the square root of the product of the base transit times: $(\tau_{B1}\tau_{B2})^{1/2}$. The rise time t_R can be found by setting the base charge Q_{B1} to zero at $t = 0$, resulting in

$$\boxed{t_R = \sqrt{\tau_{B1}\tau_{B2}}\ln(\alpha_1 I_{K,SS}/I_G + 1)} \tag{9.159}$$

In a real thyristor, two-dimensional effects play a major role, and the third phase of the turn-on process represents the lateral spread of regenerative action from the region near the gate contact into the interior of the device. When the gate pulse is applied, charge injection occurs initially in the region near the gate contact, and regenerative action begins there first. As the charge builds up, this conducting region supplies base current to adjacent regions, and regenerative action spreads across the device until nearly homogeneous current flow is established. The time required for this stabilization is the spreading time t_S. It has been found that the spreading is characterized by a *spreading velocity* v_S, and the spreading time can be written

$$\boxed{t_S \approx S_A/v_S} \tag{9.160}$$

where S_A is the half-width of the anode, as shown in Figure 9.27. In silicon, the spreading velocity has been found to be proportional to $(J_A\tau_A)^{1/2}/W_{B1}$, with a typical value in the range 5×10^3 to 1×10^4 cm s^{-1}. The spreading velocity therefore increases with anode current density and ambipolar lifetime, and is inversely proportional to the thickness of the wide p-base region. The base-width dependence means that the spreading velocity will be lower for devices designed for high blocking voltages.

SiC thyristors differ from silicon thyristors in several important respects. First, due to incomplete ionization of acceptors in the heavily-doped p+ anode, the emitter injection efficiency of the pnp BJT, γ_2, is less than 1. This is different from silicon, where the injection efficiencies are routinely assumed to be 1. As temperature is increased, the acceptor ionization increases, as shown in Figure A.1, and γ_2 increases. One result is that the rise time t_R decreases with increasing temperature due to the increased injection efficiency. Secondly, as noted by Levinshtein *et al.* [6], the turn-on process in SiC thyristors is more homogeneous than in silicon devices. In silicon, the three phases of the turn-on transient have the relationship $t_R < t_D \ll t_S$, and the turn-on transient is dominated by the spreading time t_S, which is 50 to 500 μs for a 4 kV device. In SiC, the three phases of turn-on transient are nearly equal, $t_R \approx t_D \approx t_S \approx 50-100$ ns for the same 4 kV device. Thus the SiC thyristor switches to the conducting state about 3 orders-of-magnitude faster than a comparable silicon thyristor.

9.3.4 dV/dt *Triggering*

When the thyristor is in the forward-blocking mode, a rapid rise in anode-cathode voltage V_{AK} can cause the device to switch on prematurely. This is known as dV/dt triggering, and can be understood as follows. The middle junction is reverse biased, and its depletion region extends primarily into the lightly-doped p– drift region, as shown in Figure 9.33. As V_{AK} is increased, the depletion region widens and the neutral base of the npn BJT-1 shrinks, increasing α_1. In addition, the majority carriers moving out of the widening depletion region represent an internal current of magnitude $C_M dV_{AK}/dt$, where C_M is the capacitance of the middle junction. This can be viewed as base current to both BJT-1 and BJT-2, which in turn induces additional emitter current, causing the alphas to rise, as shown in Figure 9.32a. The increasing alphas and increasing base currents can enable the coupled BJTs to reach a self-sustaining condition, whereupon the thyristor switches on. The effect increases with both the magnitude and the first derivative of V_{AK}.

The dV/dt effect can be reduced by (i) reverse biasing the gate with respect to the anode to prevent injection from the emitter of the pnp BJT, (ii) reducing the lifetime in the base regions of the BJTs to reduce the alphas, although this degrades on-state performance, or (iii) providing anode shorts. Anode shorts have the desirable effect of reducing the effective current gain of the pnp BJT when the current is low, while allowing the gain to increase rapidly at higher currents. The basic concept is equivalent to placing a resistor across the EB junction of the BJT-2, as shown in Figure 9.37. The effective current gain of this composite structure can be written

$$\alpha_{EFF} = \frac{I_{C2}}{I_A} = \frac{I_{C2}}{I_{E2}}\frac{I_{E2}}{I_A} = \alpha_2\frac{I_{E2}}{I_{E2}+I_{SHUNT}} = \frac{\alpha_2}{1+I_{SHUNT}/I_{E2}} \tag{9.161}$$

As shown by the $I-V$ characteristics in Figure 9.37b, at low currents $I_{SHUNT} > I_{E2}$ and $\alpha_{EFF} < \alpha_2$. However as V_{AG} increases, I_{E2} increases more rapidly than I_{SHUNT}, and $\alpha_{EFF} \to \alpha_2$. The use of anode shorts also has the desirable effect of increasing the forward blocking voltage, as will be discussed below.

Anode shorts can be implemented by simply extending the top metal layer across the entire cell area in Figure 9.27, thereby connecting the anode ohmic contacts to the gate ohmic contacts. The lateral spreading resistance of the n-type gate layer provides the desired resistance, which can be adjusted by the spacing of the anode and gate contacts. The anode shorts do not interfere with turn-on initiated by a gate current pulse, provided the shunt current I_R is small compared to the gate current I_G.

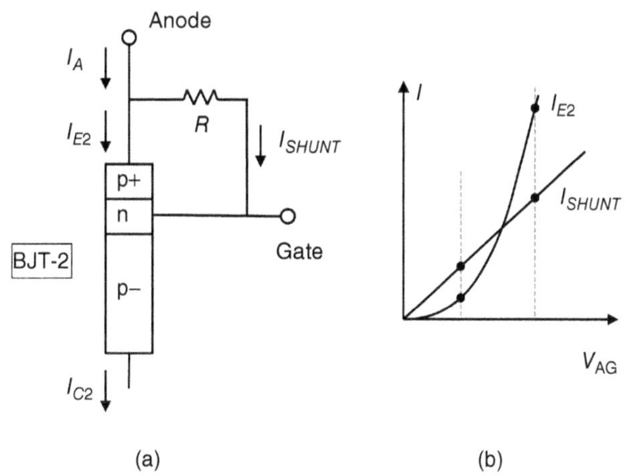

Figure 9.37 Illustration of the effect of anode shorts on thyristor operation. (a) The basic concept and (b) the $I-V$ characteristics.

9.3.5 *The* dI/dt *Limitation*

When the turn-on process is initiated by a gate current pulse, regenerative action occurs first in the region adjacent to the gate contacts, then spreads laterally into areas further removed from the gate contacts. The time required for the thyristor to stabilize is the *spreading time* t_S. During the initial part of the spreading transient, conduction occurs mainly within a very small area near the gate contacts, and the high local current density can lead to extreme local heating that can destroy the device. While this is an important limitation in silicon thyristors, it is much less severe in SiC devices for several reasons: (i) the thermal conductivity of SiC is twice as high as in silicon, (ii) the maximum allowable temperature is higher than in silicon, due to the lower intrinsic carrier concentration and more robust material properties, and (iii) the plasma spreading time in SiC thyristors is orders-of-magnitude shorter than in silicon devices, leading to more homogeneous turn-on.

9.3.6 *The Turn-Off Process*

There are two principal ways in which the thyristor may be turned off. If the supply voltage V_S is reversed, the thyristor naturally transitions into the reverse-blocking mode. The cathode current immediately reverses direction, and a large reverse current flows until all stored charge is extracted from the interior of the device. This form of turn-off is used in AC applications where V_S reverses every half cycle, and is the only method of turn-off for two-terminal thyristors (*semiconductor-controlled rectifiers*, or SCRs). In DC applications the supply voltage remains positive, and to quench the current the thyristor must transition to the forward-blocking mode. This can be accomplished by a suitable negative pulse applied to the gate. This method of turn-off requires careful design so that the reverse gate current can break the self-sustaining current feedback of the internal cross-coupled BJTs. Such a device is known as a *gate turn-off* thyristor or GTO. We will first describe the turn-off process by V_S reversal, and then consider gate-controlled turn-off in the GTO.

9.3.6.1 Turn-Off due to Voltage Reversal

Figure 9.38 shows the current and voltage waveforms for turn-off by supply voltage reversal. The turn-off process can be described qualitatively as follows. The thyristor is initially in the forward-conducting

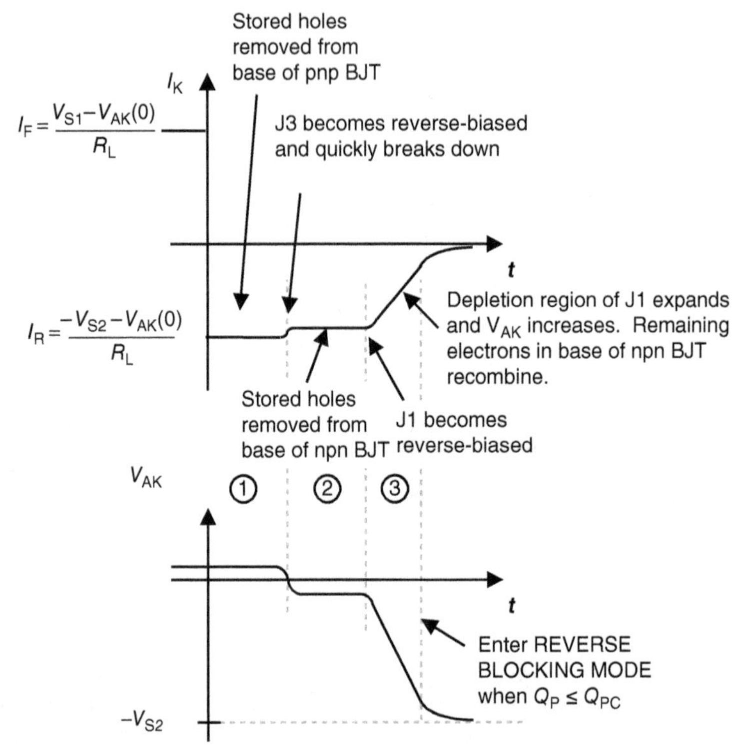

Figure 9.38 Three phases of a turn-off transient initiated by supply voltage reversal.

mode and driving a resistive load, as illustrated in Figure 9.39. At time $t = 0$ the supply voltage is suddenly switched from $+V_{S1}$ to $-V_{S2}$. The cathode current will immediately change sign, as stored carriers are drawn out of the device. The reversal of current only requires a change in the *slope* of the carrier distributions at the edges of J1 and J3, as illustrated in Figure 9.40a, and this can happen very quickly. However both junctions remain forward biased until the carrier densities at the junction edges drop below their equilibrium values. This can be understood by recalling the law of the junction, Equation 7.25,

$$pn = n_i^2 \exp(qV_J/kT) \tag{7.25}$$

which shows that V_J remains positive until the pn product at the junction edges falls below n_i^2. As a result, the voltage across the thyristor V_{AK} remains positive, although the current switches immediately from $(+V_{S1} - V_{AK})/R_L$ to $(-V_{S2} - V_{AK})/R_L$, as shown in Figures 9.38 and 9.39. The current and voltage then remain almost constant during the first phase of the turn-off, shown as ① in the figure, as carriers are removed from the base regions of the BJTs. Since the pnp BJT has a narrow base, it discharges more rapidly than the wide base npn BJT, and at the end of phase 1 the carrier density at the edge of J3 drops below its equilibrium value, and J3 becomes reverse biased, Figure 9.40a. This marks the beginning of phase 2.

Because the emitter and base are heavily doped, J3 enters avalanche breakdown at a low reverse voltage, and V_{AK} drops to approximately $-V_{BR,J3}$, a negative value that is small compared to $-V_{S2}$, as shown in Figure 9.38. The cathode current drops slightly as the voltage drop across R_L decreases. The current and voltage remain constant during phase 2, shown as ② in the figure, and the next major event comes at

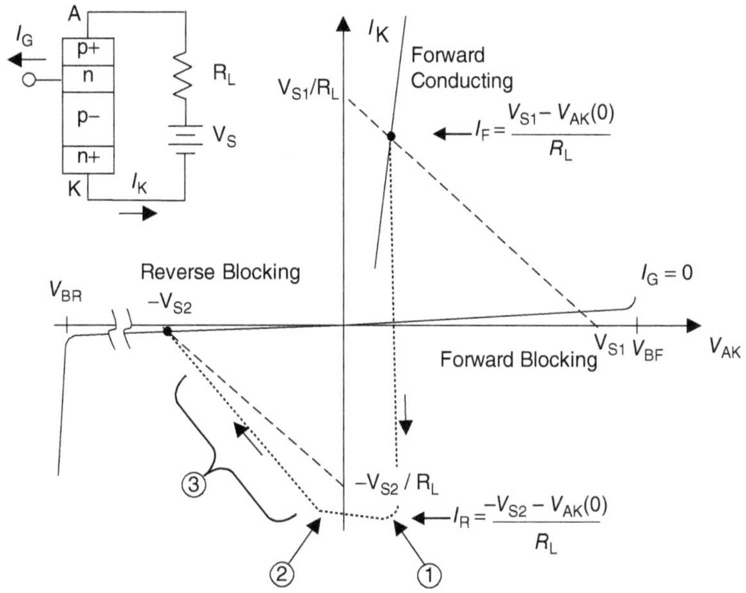

Figure 9.39 Operating points and load lines for a thyristor in the forward-conducting mode and the reverse-blocking mode. If V_S changes instantly from V_{S1} to $-V_{S2}$, the thyristor operating trajectory is depicted by the dotted line.

the end of phase 2 when the carrier density at J1 falls below its equilibrium value and J1 becomes reverse biased, Figure 9.40b. This marks the beginning of phase 3.

Since the p– drift region is wide and lightly doped, J1 can support a large reverse voltage, and V_{AK} becomes increasingly negative as the depletion region expands during phase three, Figure 9.40c,d. As V_{AK} becomes more negative, the thyristor supports more of the negative supply voltage, reducing the voltage drop across R_L, and the current falls, as shown in Figures 9.38 and 9.39. During the entire turn-off transient, J2 remains forward biased and the npn BJT operates in the inverse-active mode, with J2 forward biased and J1 reverse biased. The n-type collector of the npn BJT (the gate layer of the thyristor) injects electrons into the p– base, while the reverse-biased J1 sweeps electrons out of the base. Because of the low injection efficiency for holes from the base into the collector, very few holes are injected into the collector, and the hole charge in the base decays mainly by recombination. Recombination is a relatively slow process, and this phase tends to be the longest portion of the turn-off transient.

An estimate of the recombination time can be obtained from the charge-control equation for base charge,

$$\frac{\partial Q_P}{\partial t} = -\frac{Q_P}{\tau_{AD}} \tag{9.162}$$

where Q_P is the total hole charge in the p– base and τ_{AD} is the ambipolar lifetime in the drift region. This assumes that the majority of the base charge is removed by recombination, and not by diffusion. Equation 9.162 has the solution

$$Q_P(t) = Q_P(0)\exp(-t/\tau_{AD}) \tag{9.163}$$

where $Q_P(0)$ is the total charge stored in the conducting state, and can be written

$$Q_P(0) = \tau_{AD}\alpha_1 I_F \tag{9.164}$$

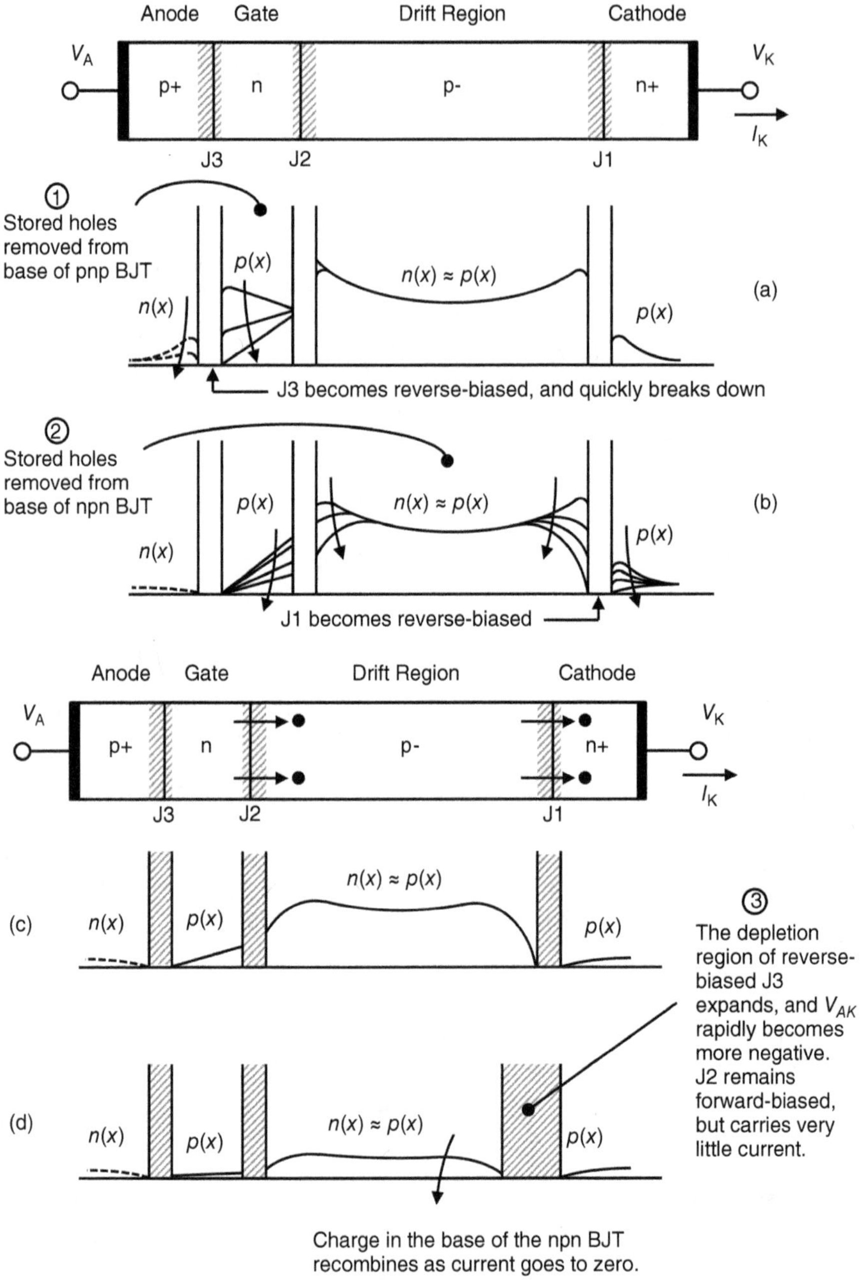

Figure 9.40 (a–d) Illustration of conditions in the thyristor during the turn-off transient.

We now define the *critical charge* as the base charge corresponding to the holding current in the forward-conducting mode,

$$Q_{PC} \equiv \tau_{AD}\alpha_1 I_H \tag{9.165}$$

The turn-off can be considered complete when the base charge falls below the critical charge, since at that point the regenerative feedback is insufficient to sustain forward operation, and the device is incapable of spontaneously turning on. Combining Equations 9.163–9.165 and setting t_{REC} equal to the time when $Q_P(t) = Q_{PC}$, we can write

$$\boxed{t_{REC} \approx \tau_{AD} \ln(I_F/I_H)} \tag{9.166}$$

t_{REC} is the approximate duration of the third period of the thyristor turn-off, which typically dominates the transient.

In the above discussion, the supply voltage is suddenly switched from a positive to a negative value at $t = 0$. However, in most practical applications the thyristor is operated with an AC power source, and the voltage reversal occurs at the zero crossing of the sinusoidal $V_S(t)$ waveform. This tends to obscure the first two periods of the turn-off transient, and a triangular reverse current waveform is observed, followed by a recombination tail, as illustrated in Figure 9.41 for two values of t_{REC}. To shorten the current tail, we could reduce the lifetime τ_{AD} in the p− base region as shown by Equation 9.166, but this would increase the forward voltage drop in the conducting state and also increase the leakage current that triggers avalanche breakdown in the forward-blocking mode.

9.3.6.2 Turn-Off due to a Negative Gate Pulse

In DC applications the supply voltage remains positive, and the thyristor turns off by transitioning to the forward-blocking mode, moving from point (2) to point (1) in Figure 9.28. This is accomplished in the *gate-turn-off (GTO)* thyristor by a negative pulse to the gate that diverts base current from the pnp BJT and breaks the regenerative feedback that sustains forward conduction.

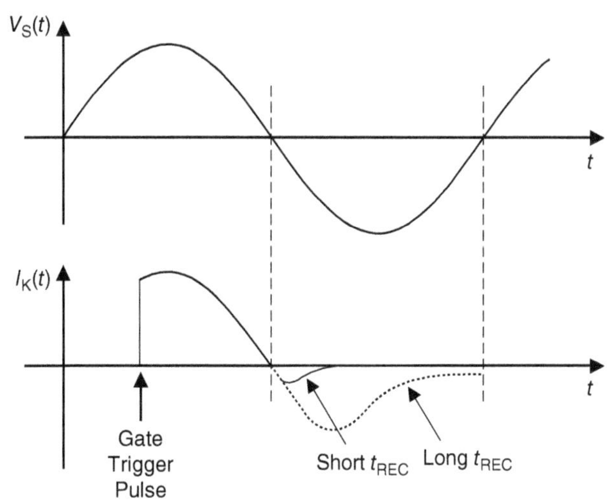

Figure 9.41 Thyristor waveforms when operating with an AC power source. Current tails are illustrated for two values of recombination time t_{REC}.

We will first consider the GTO turn-off process using a simplified one-dimensional model, then we will discuss two-dimensional effects. The first question is: "What magnitude of gate current is needed to turn the device off?" To answer, we note with reference to Figure 9.31 that when the thyristor is stabilized in forward conduction, the base current of the pnp BJT is given by

$$I_{B2} = I_A - I_{C2} = I_A - \alpha_2 I_A = (1 - \alpha_2)I_A \tag{9.167}$$

Since this is the base current required to sustain stable conduction, we require a negative gate pulse I_G^- applied to the base of BJT-2 that brings I_{B2} below this value, namely,

$$I_{B2} = I_{C1} - I_G^- = \alpha_1 I_K - I_G^- < (1 - \alpha_2)I_A = (1 - \alpha_2)(I_K - I_G^-) \tag{9.168}$$

Note that I_G^- has the opposite polarity to I_G in Figure 9.31. Solving for I_G^- yields

$$I_G^- > \left(\frac{\alpha_1 + \alpha_2 - 1}{\alpha_2}\right) I_K \tag{9.169}$$

The negative gate current required to turn off the thyristor is proportional to the cathode current I_K, and this ratio can be expressed in the form of a *turn-off gain,*

$$\beta_{OFF} = I_K / I_G^- = \frac{\alpha_2}{\alpha_1 + \alpha_2 - 1} \tag{9.170}$$

A large turn-off gain is desirable to simplify operation, and this requires that α_2 be close to unity and α_1 be small. A high α_2 occurs naturally in the GTO, since the base of the pnp BJT is narrow and the emitter is heavily doped. However, to take advantage of the high α_2 we must avoid anode shorts. α_1 is naturally low due to the wide base of the npn BJT, and it can be reduced further by inserting a thin heavily-doped p+ buffer between the p– drift region and the n+ substrate, as shown in Figure 9.27. This layer reduces the injection efficiency of the npn emitter, thereby reducing α_1.

Since the GTO is designed for DC operations under a positive supply voltage, a high reverse blocking capability is not needed. This allows us to optimize the forward blocking voltage without regard to the reverse blocking voltage. The p+ buffer layer has a beneficial effect in this regard, since it permits a punch-through design, as illustrated in Figure 9.42b. Here the p+ buffer layer prevents the depletion

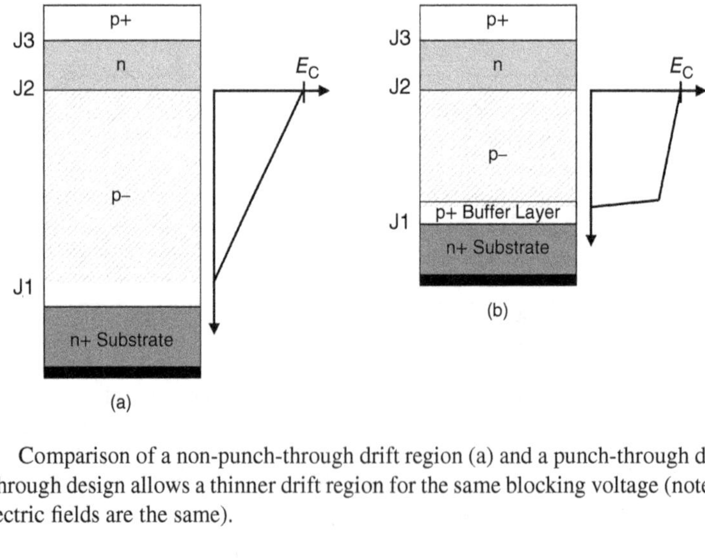

Figure 9.42 Comparison of a non-punch-through drift region (a) and a punch-through drift region (b). The punch-through design allows a thinner drift region for the same blocking voltage (note that the areas under the electric fields are the same).

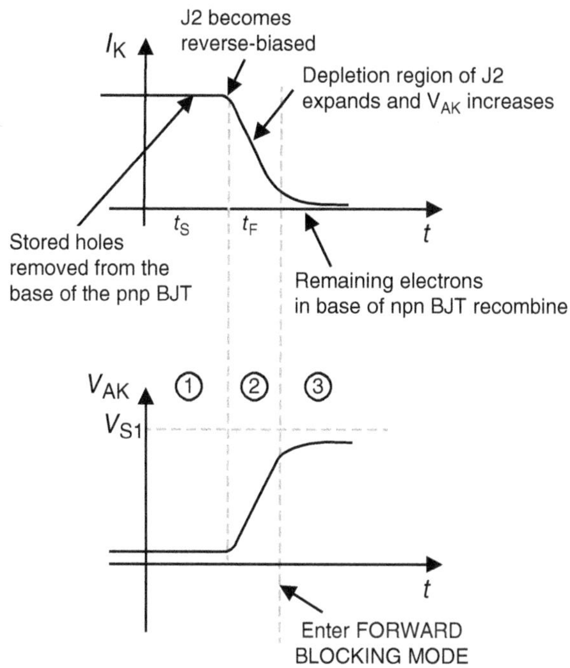

Figure 9.43 Three phases of a GTO turn-off transient, initiated by a negative gate pulse.

region of the J2 junction from reaching the substrate in the forward-blocking mode, allowing the desired V_{BF} to be achieved with a thinner p− base. This reduces the forward voltage drop of the thyristor in the conducting state, as shown by Equation 9.113 and Figure 7.13.

The turn-off process in the GTO can be described in three phases, as shown in Figure 9.43. Conditions inside the thyristor during the transient are illustrated in Figure 9.44. In the first phase, stored holes are removed from the base of the pnp BJT by the negative gate current. The process is as follows: A negative gate current corresponds to the flow of majority electrons out of the base. This reduces the injection of electrons from the base into the emitter, which reduces V_{BE}. The reduced V_{BE} reduces the injection of holes from the emitter, lowering the hole density in the base. At the end of the storage phase, junction J2 becomes reverse biased and both BJTs enter their forward-active regions. During the second phase, the depletion region of J2 spreads into the p− base layer, and the reverse voltage supported by J2 increases. As a result, the voltage drop across the thyristor V_{AK} rises, and the current, limited by the load resistance, falls. The final phase corresponds to the recombination of the remaining electrons in the base of the npn BJT.

We now consider each of the three phases in turn. During the first (storage) phase, the injection of holes from the pnp emitter is reduced until J2 becomes reverse biased and the BJTs enter their forward-active modes. This process is inherently two-dimensional, since injection is first reduced in the regions adjacent to the gate contacts, and the quenching process spreads laterally into areas further removed from the gate contacts. This is the inverse of the turn-on process, where the electron–hole plasma spreads laterally from the gate into the interior with a spreading velocity v_S. Here we talk of a *squeezing velocity*, as the injection is squeezed into a smaller and smaller region under the center of the anode before being quenched entirely. The situation can be visualized by reference to Figure 9.27. Since gate current flows laterally through the n-type base, a lateral voltage drop develops due to the sheet resistivity of the base, given by Equation 9.78, and the voltage drop across the EB junction of the pnp BJT is reduced as we

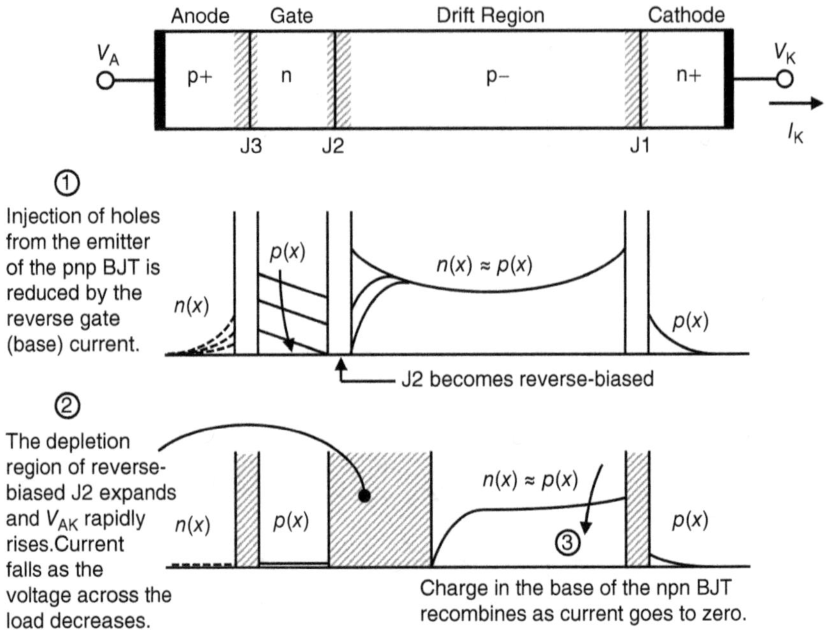

Figure 9.44 Illustration of conditions in the GTO during the turn-off transient, assuming a one-dimensional structure.

proceed toward the edges of the anode. This reduces the injection near the anode edges, squeezing the current into a small region under the center of the anode. The effect of base spreading resistance in BJTs was discussed in Section 9.1.9.

An approximate analysis of the storage time given by Wolley [7] shows that t_S increases with turn-off gain β_{OFF}, and can be approximated by

$$\tau_S \approx \frac{W_G^2}{2D_{PG}}(\beta_{OFF} - 1)\ln\left(\frac{S_A L_{PG}/W_G^2 + 2L_{PG}^2/W_G^2 - \beta_{OFF} + 1}{4L_{PG}^2/W_G^2 - \beta_{OFF} + 1}\right) \tag{9.171}$$

Here D_{PG} is the hole diffusion coefficient in the gate layer, L_{PG} is the hole diffusion length in the gate layer, and W_G is the thickness of the gate layer, that is, the base of BJT-2. Equation 9.171 shows that there is a conflict between the desire for a high β_{OFF} to reduce drive requirements, and a low β_{OFF} to speed the turn-off process.

At the end of the storage phase, the entire area of junction J2 is reverse biased. During the second phase, the depletion region of J2 expands into the base of the npn BJT. As this occurs, the reverse voltage supported by J2 increases. With more voltage dropped across the thyristor, the current (limited by the load resistance) falls. An approximate expression for the time for the current to fall to 10% of its initial value can be derived by considering the rate of expansion of the depletion region [8]. The result is

$$\tau_F \approx \cosh^{-1}(\sqrt{10})\frac{qAn^*\sqrt{V_S}}{(V_S/R_L)}\sqrt{\frac{2\epsilon_S}{qN_{AD}}} = 1.82\frac{qn^*\sqrt{V_S}}{J_K(0)}\sqrt{\frac{2\epsilon_S}{qN_{AD}}} \tag{9.172}$$

where n^* is the average electron density in the undepleted portion of the drift region. We note that the fall time is proportional to the square root of supply voltage and inversely proportional to the initial cathode current density, $J_K(0)$.

9.3.7 Reverse-Blocking Mode

The reverse-blocking mode is important for thyristors that operate under AC conditions, where the supply voltage oscillates between positive and negative values. We will discuss avalanche breakdown and blocking voltage in Chapter 10.

References

[1] Miyake, H., Kimoto, T. and Suda, J. (2011) 4H-SiC BJTs with record current gains of 257 on (0001) and 335 on (000$\bar{1}$). *IEEE Electron. Device Lett.*, **32** (7), 841–843.
[2] Huang, C.F. and Cooper, J.A. (2003) High current gain 4H-SiC npn bipolar junction transistors. *IEEE Electron. Device Lett.*, **24** (6), 396–398.
[3] Tamaki, T., Walden, G.G., Sui, Y. and Cooper, J.A. (2008) Numerical study of the turn-off behavior of high-voltage 4H-SiC IGBTs. *IEEE Trans. Electron. Devices*, **55** (8), 1928–1933.
[4] Tamaki, T., Walden, G.G., Sui, Y. and Cooper, J.A. (2008) Optimization of on-state and switching performance for 15–20 kV 4H-SiC IGBTs. *IEEE Trans. Electron. Devices*, **55** (8), 1920–1927.
[5] Yang, E.S. and Voulgaris, N.C. (1967) On the variation of small-signal alphas of a p-n-p-n device with current. *Solid-State Electron.*, **10** (7), 641–648.
[6] Levinshtein, M.E., Ivanov, P.A., Agarwal, A.K. and Palmour, J.W. (2005) On the homogeneity of the turn-on process in high-voltage 4H-SiC thyristors. *Solid-State Electron.*, **49** (2), 233–237.
[7] Wolley, D.E. (1966) Gate turn-off in p-n-p-n devices. *IEEE Trans. Electron. Devices*, **13** (7), 590–597.
[8] Kao, Y.C. and Brewster, J.B. (1974) A description of the turn-off performance of gate controlled switches. IEEE Conference Record of the Industry Applications Society (IAS) Annual Meeting, pp. 689–693.

10

Optimization and Comparison of Power Devices

10.1 Blocking Voltage and Edge Terminations for SiC Power Devices

An essential requirement of all semiconductor power devices is the ability to withstand a large terminal voltage with minimal leakage current in the off state. The maximum terminal voltage a device can withstand is called the *blocking voltage*. The blocking voltage is generally determined by material properties as well as device design. The limiting mechanism may be any of the following: (i) punch-through of the base region in a MOSFET (metal-oxide-semiconductor field effect transistor), BJT (bipolar junction transistor), IGBT (insulated-gate bipolar transistor), or thyristor; (ii) avalanche breakdown in a reverse-biased pn or Schottky junction, either as a discrete rectifier or as a part of a switching transistor or thyristor; (iii) excessive leakage current in a reverse-biased pn or Schottky junction; or (iv) excessive electric field in the oxide of an MOS-based power device such as a MOSFET or IGBT.

Punch-through can be avoided by making the doping-thickness product of the base region large enough that the base cannot be completely depleted before the onset of avalanche breakdown. This requires

$$N \cdot W > \varepsilon_S E_C / q \tag{10.1}$$

where N and W represent the doping and width of the base region and E_C is the critical field for avalanche breakdown, to be discussed below. Since this restriction is typically observed for all power switching devices, punch-through will not be considered further.

In an MOS-based device such as a power MOSFET or IGBT, the blocking voltage is often limited by the oxide field. As discussed in Sections 8.2.6 and 8.2.11, the oxide field must be kept below about 4 MV cm^{-1} for long-term device reliability. Care must be exercised in the design of MOSFETs and IGBTs to make sure the oxide field does not exceed this value prior to avalanche breakdown of the blocking junction. This typically requires numerical simulations due to the two-dimensional nature of the fields, particularly in trench-gate or UMOS structures.

In well-designed devices, the blocking voltage is ultimately limited by avalanche breakdown of the reverse-biased blocking junction. The blocking voltage is reduced by two-dimensional field crowding at the edges of the device, which can be mitigated by the use of special edge terminations. In the sections below, we will first consider avalanche breakdown in planar SiC junctions using a one-dimensional analysis. We will then discuss two-dimensional field crowding and show how it can be minimized by different edge termination techniques.

Fundamentals of Silicon Carbide Technology: Growth, Characterization, Devices, and Applications, First Edition.
Tsunenobu Kimoto and James A. Cooper.

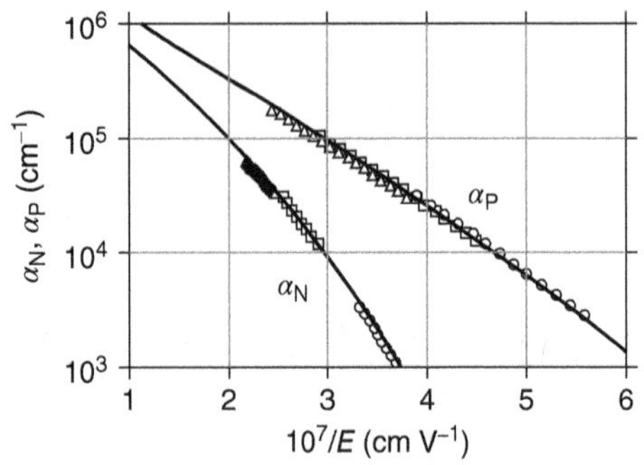

Figure 10.1 Ionization rate coefficients in 4H-SiC measured at room temperature ([1] reproduced with permission from AIP Publishing LLC).

10.1.1 Impact Ionization and Avalanche Breakdown

Avalanche breakdown is caused by the impact ionization of electrons and holes in a high-field region. The impact ionization process can be understood by considering the behavior of electrons and holes as they move through the crystal. As charged entities, electrons and holes are accelerated by the electric field, and their kinetic energy increases until they undergo a collision. The collision is a scattering event that takes the electron (or hole) into a lower energy state, with its energy typically transferred to the crystal lattice as heat. Immediately following the scattering event, the electron or hole is again accelerated, and the process is repeated as holes and electrons transit the high-field region. If the field is sufficiently high, the electron or hole may acquire enough kinetic energy between collisions that the energy liberated is sufficient to break a covalent bond, creating a new hole–electron pair. This process is called *impact ionization*.

The number of impact ionization events initiated by an electron (or hole) per unit length traveled is known as the *ionization rate for electrons* α_N or the *ionization rate for holes* α_P. α_N and α_P are strong functions of the electric field. Figure 10.1 shows the ionization rates for transport along the *c*-axis in 4H-SiC at room temperature, measured by Konstantinov *et al.* [1] This data can be represented by the empirical expressions [2]

$$\alpha_N(E) \approx 1.69 \times 10^6 \text{ (cm}^{-1}\text{)} \exp\left[-\left(\frac{9.69 \times 10^6 \text{ (V cm}^{-1}\text{)}}{E}\right)^{1.6}\right] \tag{10.2}$$

and

$$\alpha_P(E) \approx 3.32 \times 10^6 \text{(cm}^{-1}\text{)} \exp\left[-\left(\frac{1.07 \times 10^7 \text{ (V cm}^{-1}\text{)}}{E}\right)^{1.1}\right] \tag{10.3}$$

Figure 10.2 compares the ionization rates for silicon and 4H-SiC. The ionization rates for 4H-SiC at a given field are orders-of-magnitude lower than for silicon.

Avalanche breakdown in 4H-SiC is typically initiated by holes, since the ionization rates are higher for holes than for electrons. Figure 10.3 shows the ionization rates for holes in 4H-SiC for several temperatures [3]. The lines represent the empirical equation

$$\alpha_P(E) \approx (6.09 \times 10^6 - 9.2310^3 T) \text{ (cm}^{-1}\text{)} \exp\left[-\left(\frac{8.90 \times 10^6 - 4.95 \times 10^3 T}{E}\right)^{1.09}\right] \tag{10.4}$$

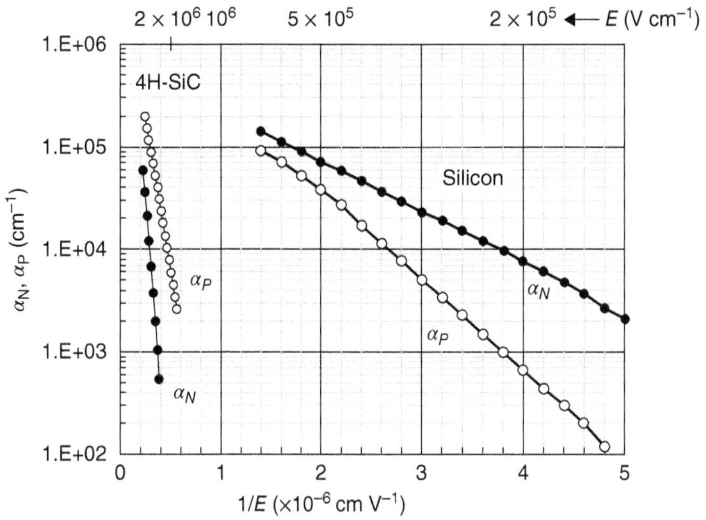

Figure 10.2 Comparison of ionization rates in silicon and 4H-SiC at room temperature. At a given field, the rates for SiC are orders-of-magnitude below those of silicon.

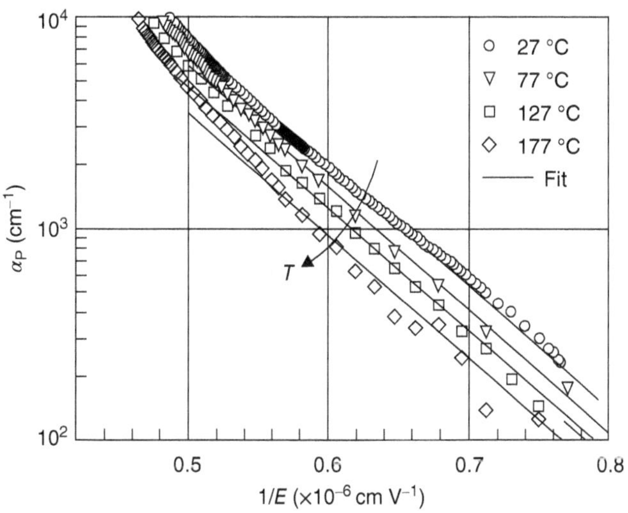

Figure 10.3 Temperature dependence of the hole ionization rate in 4H-SiC. The lines are generated using empirical Equation 10.4 ([3] reproduced with permission from Trans Tech Publications).

where T is the absolute temperature. The decrease in ionization rates with temperature occurs because of increased phonon scattering, making it less likely that a hole can acquire enough kinetic energy between collisions to create a hole–electron pair.

We will now apply the ionization rate expressions to calculate the breakdown voltage of a reverse-biased p + n one-sided step junction, such as shown in Figure 7.1. The depletion region extends

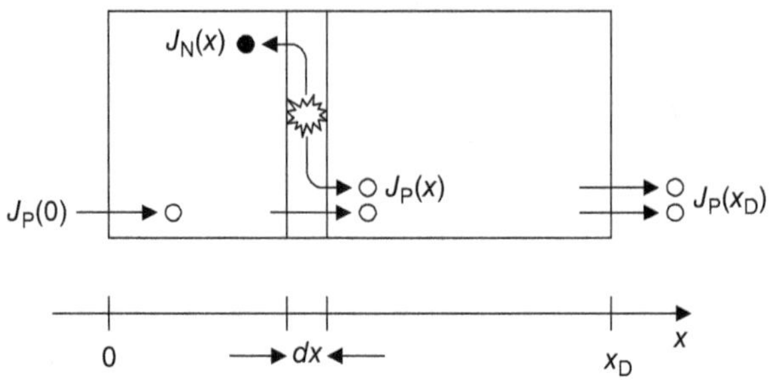

Figure 10.4 One-dimensional cross-section of a high-field region, illustrating generation of new electron–hole pairs by impact ionization.

from $x = 0$ to x_D and the electric field decreases from a peak value at $x = 0$ to zero at $x = x_D$. Any electrons and holes in the depletion region are subject to an electric field and can potentially initiate impact ionization events. Consider the cross-section of the depletion region shown in Figure 10.4. We assume hole current $J_P(0)$ is entering at $x = 0$ and hole current $J_P(x)$ is flowing at point x. Impact ionization in the differential slice at x creates additional holes and electrons, and the increase in hole current across the dx region is

$$dJ_P = \alpha_P J_P(x)dx + \alpha_N J_N(x)dx \tag{10.5}$$

which we can write as

$$\frac{dJ_P}{dx} = \alpha_P J_P(x) + \alpha_N J_N(x) \tag{10.6}$$

Since $J_P(x) + J_N(x) = J_{TOTAL}$ is not a function of x, we can write Equation 10.6 as

$$\frac{dJ_P}{dx} = (\alpha_P - \alpha_N)J_P(x) + \alpha_N J_{TOTAL} \tag{10.7}$$

Since α_P and α_N are functions of electric field and the electric field is a function of position, Equation 10.7 can be written in the form

$$\frac{df}{dx} = P(x)f(x) + Q(x) \tag{10.8}$$

where $f(x)$, $P(x)$, and $Q(x)$ are functions of x. The differential Equation 10.8 has the formal solution

$$f(x) = \frac{\int_0^x Q(x)\exp\left[-\int_0^x P(x')\,dx'\right]dx + f(0)}{\exp\left[-\int_0^x P(x)\,dx\right]} \tag{10.9}$$

Substituting for $f(x)$, $P(x)$, and $Q(x)$ in Equation 10.9 yields

$$J_P(x) = \frac{\int_0^x \alpha_N(x)J_{TOTAL}\exp\left[\int_0^x (\alpha_N(x') - \alpha_P(x'))\,dx'\right]dx + J_P(0)}{\exp\left[\int_0^x (\alpha_N(x) - \alpha_P(x))\,dx\right]} \tag{10.10}$$

We define the *hole multiplication factor* M_P as the fractional increase in hole current across the depletion region due to impact ionization, $M_P = J_P(x_D)/J_P(0)$. Since we assumed no electron current at

$x = x_D, J_N(x_D) = 0$ and we can set $J_P(x_D) = J_{TOTAL}$. Under these assumptions, if we now evaluate Equation 10.10 at $x = x_D$ we can write

$$J_{TOTAL} = J_P(x_D) = J_{TOTAL} \frac{\int_0^{x_D} \alpha_N(x) \exp\left[\int_0^x \left(\alpha_N(x') - \alpha_P(x')\right) dx'\right] dx + 1/M_P}{\exp\left[\int_0^{x_D} \left(\alpha_N(x) - \alpha_P(x)\right) dx\right]} \tag{10.11}$$

The complicated fraction on the right-hand side must be equal to 1, so M_P is given by

$$M_P = \frac{1}{\exp\left[\int_0^{x_D} \left(\alpha_N(x) - \alpha_P(x)\right) dx\right] - \int_0^{x_D} \alpha_N(x) \exp\left[\int_0^x \left(\alpha_N(x') - \alpha_P(x')\right) dx'\right] dx} \tag{10.12}$$

The exponential factor in the second term in the denominator can be rewritten using the identity

$$\exp\left[\int_0^{x_D} g(x)\,dx\right] = \exp\left[\int_0^{x} g(x)\,dx\right] \exp\left[\int_x^{x_D} g(x)\,dx\right] \tag{10.13}$$

resulting in

$$M_P = \frac{\exp\left[\int_0^{x_D} \left(\alpha_P(x) - \alpha_N(x)\right) dx\right]}{1 - \int_0^{x_D} \alpha_N(x) \exp\left[\int_x^{x_D} \left(\alpha_P(x') - \alpha_N(x')\right) dx'\right] dx} \tag{10.14}$$

Avalanche breakdown occurs when the multiplication factor M_P goes to infinity, or when

$$\int_0^{x_D} \alpha_N(x) \exp\left[\int_x^{x_D} \left(\alpha_P(x') - \alpha_N(x')\right) dx'\right] dx \to 1 \tag{10.15}$$

Equation 10.15 is known as the *ionization integral for holes*. The above analysis could also have been conducted assuming only an electron current entering at $x = x_D$. In this case we would have defined an electron multiplication factor M_N, and avalanche breakdown would occur when M_N goes to infinity, or when

$$\int_0^{x_D} \alpha_P(x) \exp\left[\int_0^{x} \left(\alpha_N(x') - \alpha_P(x')\right) dx'\right] dx \to 1 \tag{10.16}$$

Considering that both holes and electrons are present in the device, avalanche breakdown will occur at the lowest voltage where either Equation 10.15 or 10.16 is satisfied.

It is customary to designate the peak field E_M at the onset of breakdown as the *critical field* for avalanche breakdown, E_C. E_C can be computed from the ionization integrals by an iterative procedure, as follows: For a given doping of the n− layer, select a value of reverse voltage and calculate the ionization integrals using Equations 10.15 and 10.16. If both integrals are less than 1, chose a slightly higher voltage and repeat the calculation. The lowest voltage for which either Equation 10.15 or 10.16 is satisfied is the breakdown voltage for that doping, and the peak field at the junction is the critical field for that doping. The critical field for a one-sided step junction in 4H-SiC at room temperature can be approximated using an empirical expression given by Konstantinov *et al.* [1],

$$E_C \approx \frac{2.49 \times 10^6 V\,(\text{cm})}{1 - 0.25\log_{10}(N\,(10^{16}\text{cm}^{-3}))} \tag{10.17}$$

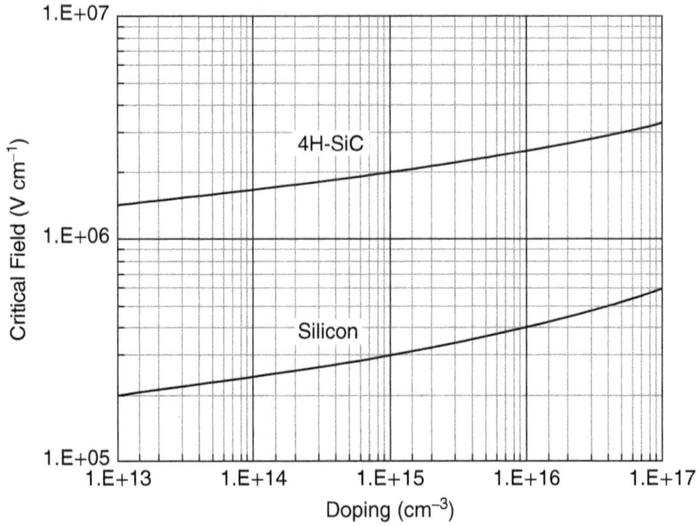

Figure 10.5 Critical field for room temperature avalanche breakdown in a 4H-SiC one-sided step junction, as given by Equation 10.17. Also shown for comparison is the critical field for silicon.

where N is the doping of the lightly-doped side of the junction. Figure 10.5 shows the critical field in silicon and 4H-SiC as a function of doping. The critical field is only a weak function of doping, and at doping levels below 10^{15} cm^{-3} the critical field in 4H-SiC is approximately seven times higher than in silicon. Since the power-device figure of merit scales inversely as the *cube* of critical field (see Section 7.1), this represents about a 350 × advantage for 4H-SiC.

As might be expected from the temperature dependence of the ionization rates, the avalanche breakdown voltage in 4H-SiC increases with temperature, as shown in Figure 10.6 [4]. A positive temperature

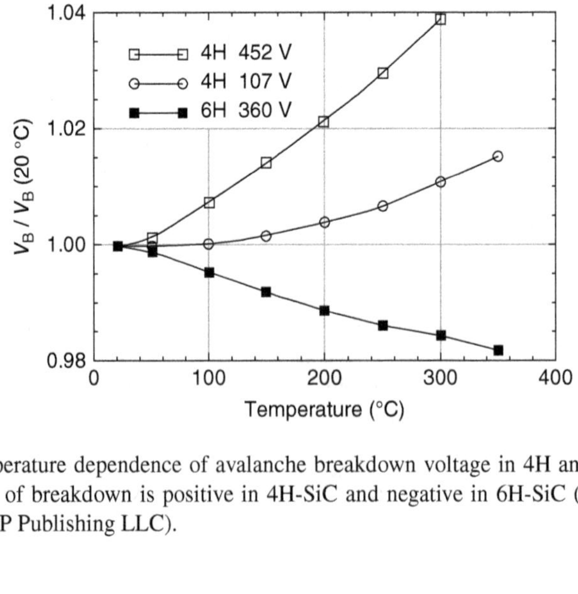

Figure 10.6 Temperature dependence of avalanche breakdown voltage in 4H and 6H-SiC. The temperature coefficient of breakdown is positive in 4H-SiC and negative in 6H-SiC ([4] reproduced with permission from AIP Publishing LLC).

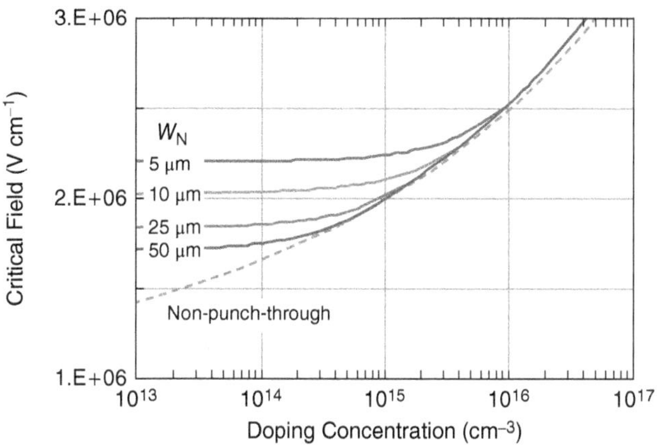

Figure 10.7 Critical field for room temperature avalanche breakdown in a 4H-SiC one-sided step junction with a punch-through structure. The critical field for a non-punch-through structure, given by Equation 10.17, is shown as a dashed line ([2] reproduced with permission from Dallas T. Morisette).

coefficient of breakdown is desirable, since devices with negative temperature coefficient of breakdown are potentially unstable.

It is important to recognize that the critical field in Equation 10.17 was calculated assuming a *non-punch-through* structure such as in Figure 7.1. In the *punch-through* structure of Figure 7.2, evaluation of the ionization integral yields a critical field that depends not only on doping, but also on the thickness of the lightly-doped region. Figure 10.7 shows the critical field in 4H-SiC calculated using the ionization integrals for both punch-through and non-punch-through structures [2]. The dashed line is the non-punch-through critical field given by Equation 10.17. In punch-through designs, the critical field for breakdown is higher than the value given by Equation 10.17.

10.1.2 Two-Dimensional Field Crowding and Junction Curvature

In most of our discussions to this point, we have considered only one-dimensional slices within our devices. This allows us to calculate carrier densities, current flows, and electrostatic potentials using one-dimensional analyses. In some cases this approach gives quantitatively accurate answers, but it often happens that the two-dimensional (or three-dimensional) nature of real devices necessitates a computer simulation. This is certainly true in the calculation of blocking voltage. The results in the preceding subsection were obtained using a one-dimensional analysis, but in real devices the blocking voltage is invariably limited by two-dimensional field crowding at the periphery of the device. Next, we will discuss how to evaluate the field crowding and consider techniques to mitigate it.

Consider the cylindrical p + /n− one-sided step junction illustrated in Figure 10.8. The p+ region has radius r_J and the depletion edge has radius r_D. Poisson's equation in cylindrical coordinates is

$$\frac{1}{r}\frac{\partial}{\partial r}\left(r\frac{\partial\psi}{\partial r}\right) = \frac{\rho}{\varepsilon_S} = -\frac{qN_D}{\varepsilon_S} \tag{10.18}$$

where r is the radial coordinate, $r_J \le r \le r_D$, and ρ is the charge per unit volume in the depletion region. Integration with respect to r yields

$$r\frac{\partial\psi}{\partial r} = -\frac{qN_D}{2\varepsilon_S}r^2 + C \tag{10.19}$$

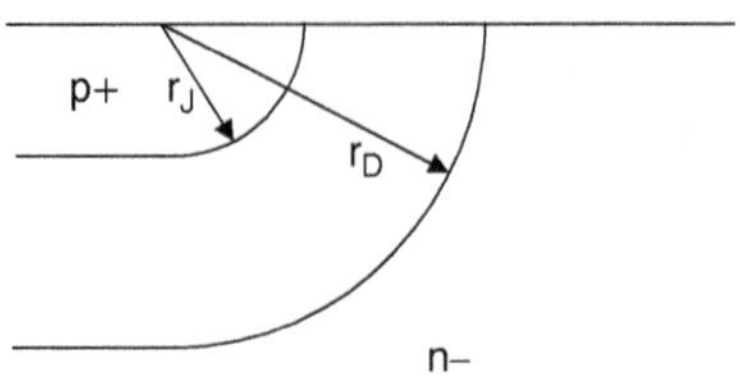

Figure 10.8 Cross-section of a p + /n− one-sided step junction with cylindrical geometry, such as occurs at the edges of real structures. r_J is the radius of the junction and r_D is the radius of the depletion region edge.

Noting that $\partial\psi/\partial r = -E(r)$ and choosing the constant of integration such that $E(r_D) = 0$, we can write

$$E(r) = \left(-\frac{qN_D}{2\varepsilon_S}\right)\left(\frac{r_D^2 - r^2}{r}\right) \tag{10.20}$$

The peak field occurs at the metallurgical junction at $r = r_J$,

$$E_M = E(r_J) = \left(-\frac{qN_D}{2\varepsilon_S}\right)\left(\frac{r_D^2 - r_J^2}{r_J}\right) \tag{10.21}$$

Integrating Equation 10.20 with respect to r to calculate the potential ψ (r), and setting $\psi(r_J) = 0$, we obtain

$$\psi(r) = \left(\frac{qN_D}{2\varepsilon_S}\right)\left[\left(\frac{r_J^2 - r^2}{2}\right) + r_D^2 \ln\left(\frac{r}{r_J}\right)\right] \tag{10.22}$$

The total voltage across the depletion region is the potential difference $\psi(r_D) - \psi(r_J)$,

$$V_R = \psi(r_D) = \left(\frac{qN_D}{2\varepsilon_S}\right)\left[\left(\frac{r_J^2 - r_D^2}{2}\right) + r_D^2 \ln\left(\frac{r_D}{r_J}\right)\right] \tag{10.23}$$

The peak field can be plotted as a function of reverse voltage by choosing successively larger values of r_D and evaluating Equations 10.21 and 10.23. Figure 10.9 shows the effect of junction curvature on the peak field for several values of r_J at an assumed doping of 2×10^{15} cm^{-3}. The peak field of a planar junction of the same doping is also shown for comparison. Reducing the junction radius r_J increases the peak field at a given reverse voltage, and this will significantly reduce the breakdown voltage. Because of this effect, the blocking voltage of practical devices is normally limited by edge breakdown unless special edge termination techniques are employed.

Edge terminations used in SiC devices fall into five general classes: trench isolation, beveled junctions, junction termination extension (JTE), floating field rings (FFRs), and multiple floating zone (MFZ-) or space-modulated (SM-) JTE. We will discuss each of these below.

10.1.3 Trench Edge Terminations

Trench isolation is useful when the reverse voltage is below about 2 kV and the depletion width at breakdown is on the order of 10 μm or less. Under these conditions it is possible to etch an isolation trench extending ~15–20% through the lightly-doped drift region, as illustrated in Figure 10.10. This simple technique is often sufficient because it tends to linearize field lines near the p + /n− junction where the field is highest. Although field crowding occurs at the corner of the trench, the field here is considerably

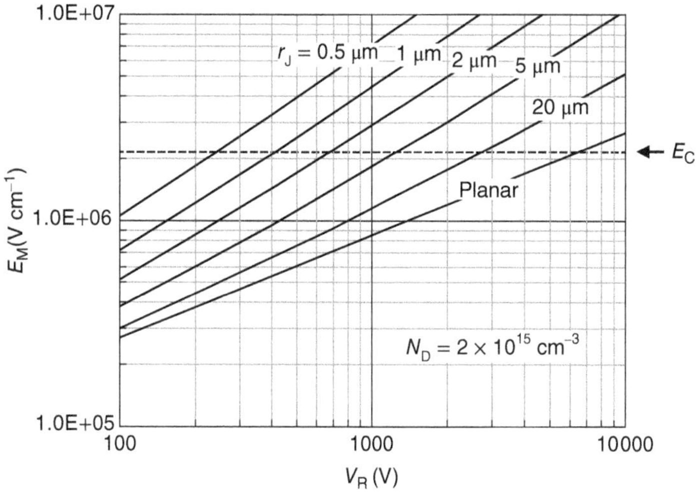

Figure 10.9 Peak field at the cylindrical junction as a function of reverse voltage, with junction radius as a parameter.

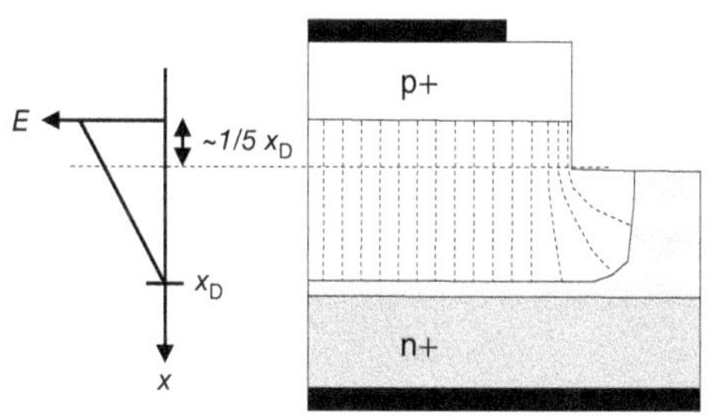

Figure 10.10 Cross-section of a p + /n− one-sided step junction terminated by trench isolation.

lower than the peak field, reducing the negative impact of field crowding. A high-quality passivation layer is required to control surface charges and prevent surface leakage.

10.1.4 Beveled Edge Terminations

At higher blocking voltages, an effective way to provide edge termination is the use of a beveled junction profile. A *positive bevel* is one in which the junction area decreases with distance into the lightly-doped side, as illustrated in Figure 10.11 for a p + /n− junction. The requirement for charge balance on either side of the junction causes an increase in the depletion width on the lightly-doped side near the beveled surface. In this case, the charge balance is restored by replacing the positive charge missing from area Q_1 by a roughly equal additional charge Q_2. As a result, the depletion width along the beveled surface W_S is larger than the depletion width W in the bulk. Since the potential drop across the junction is the same everywhere, the electric field along the surface is reduced relative to that in the bulk. Numerical simulations confirm that the surface field is reduced below the bulk field for all bevel angles $0 < \theta < 90°$ [5]. In practice, most SiC power devices are fabricated with the heavily-doped layer on top, making it difficult to achieve a positive bevel in real devices.

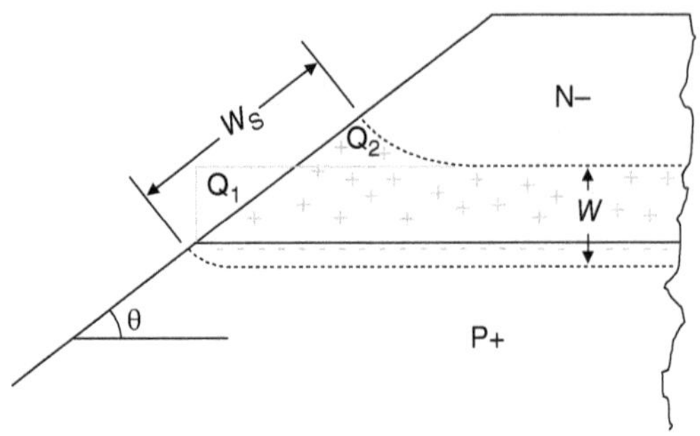

Figure 10.11 Charge distribution in a p + /n− junction with a positive bevel.

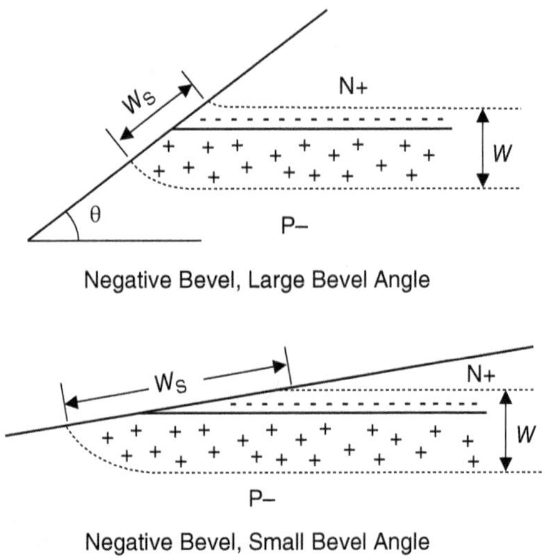

Figure 10.12 Charge distribution in an n + /p− junction with a negative bevel.

A *negative bevel* is one in which the junction area increases with distance into the lightly-doped side, as illustrated in Figure 10.12 for an n + /p− junction. In this case the depletion region on the lightly-doped side shrinks near the surface to balance the missing charge in the heavily-doped side. This tends to make the depletion width along the surface W_S smaller than the depletion width W in the bulk, increasing the surface field above the bulk field. However, at very small bevel angles the shallow slope makes the increase in depletion width on the heavily-doped side exceed the decrease in depletion width on the lightly-doped side, as illustrated in the lower part of Figure 10.12. When this occurs, the peak surface field is once again lower than the peak field in the bulk.

The exact conditions under which a negative bevel reduces the surface field can be predicted by a two-dimensional numerical solution of Poisson's equation. Adler and Temple have studied negative bevels in silicon junctions where the heavily-doped layer is formed by diffusion [6]. After extensive simulations, they found that for negative beveled junctions the normalized breakdown voltage (as a fraction of the ideal breakdown voltage) can be represented by a universal curve that depends on an *effective* bevel angle θ_{EFF} given by

$$\theta_{EFF} = 0.04\theta(x_{DB}^{-}/x_{DB}^{+})^2 \tag{10.24}$$

Here θ is the actual bevel angle, $x_{DB}{}^{-}$ is the depletion width on the lightly-doped side at breakdown, and $x_{DB}{}^{+}$ is the depletion width on the heavily-doped side at breakdown. Figure 10.13 shows normalized

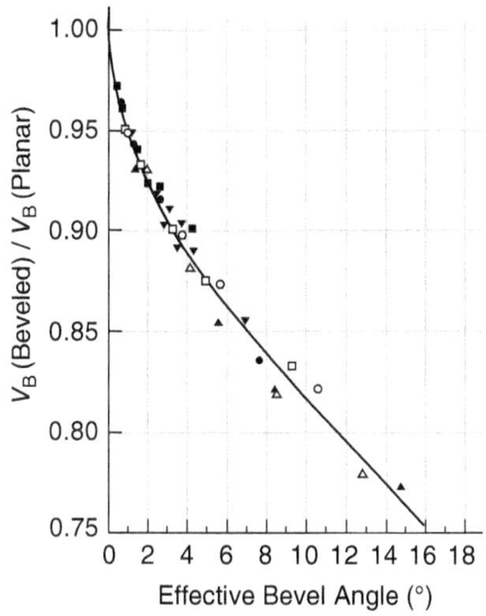

Figure 10.13 Normalized breakdown voltage as a function of effective bevel angle θ_{EFF} for a negatively-beveled silicon junction with a graded doping profile ([6] reproduced with permission from IEEE).

breakdown voltage as a function of effective bevel angle. For sufficiently small angles the breakdown approaches the ideal planar-junction value. However, these simulations were carried out for graded, diffused junctions with a complementary error function doping profile, and the results may not be accurate for the abrupt junctions typically formed in SiC.

Figure 10.14 shows the maximum field at the surface and the maximum field in the bulk material, normalized to the ideal breakdown field, as a function of effective bevel angle [6]. At small bevel angles the maximum surface field is well below the ideal (planar) breakdown field, but the maximum field in the bulk is slightly higher than the ideal breakdown field. This means that avalanche breakdown first occurs in the bulk near the surface, rather than at the surface itself, and the breakdown voltage is therefore lower than the ideal (planar) value. As the effective bevel angle is increased, both the maximum surface field and the maximum bulk field increase and the breakdown voltage is further reduced.

10.1.5 Junction Termination Extensions (JTEs)

The third form of edge termination used in SiC is junction termination extension (JTE) [7]. This termination consists of one or more concentric p-type rings of carefully-controlled dopant concentration surrounding the main junction, as shown in Figure 10.15. The key requirement is that the total dopant concentration per unit area in each ring be low enough that the ring will be completely depleted before avalanche breakdown occurs at the edge of the ring. The exposed acceptor atoms in the depleted rings then terminate field lines that otherwise would crowd into the corner of the main junction. The field lines in this situation are illustrated schematically in Figure 10.16.

Because of the two-dimensional nature of the fields, optimization of JTE ring design is best performed by computer simulations. Figure 10.17 shows breakdown voltage as a function of dose for single-zone JTE terminations of various widths on a 100 μm n− SiC drift region doped 6×10^{14} cm^{-3} [8]. The breakdown voltage peaks at the dose where the ring becomes depleted just before reaching breakdown. For higher doses, the ring does not fully deplete and breakdown occurs at the outer edge of the ring. For lower doses, the ring depletes without breaking down, and breakdown occurs at the edge of the main junction.

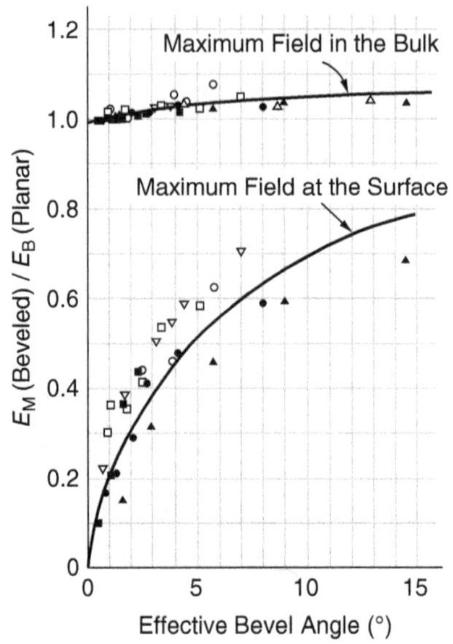

Figure 10.14 Maximum field at the surface and maximum field in the bulk material as a function of effective bevel angle for negatively-beveled silicon junctions with a graded doping profile ([6] reproduced with permission from IEEE).

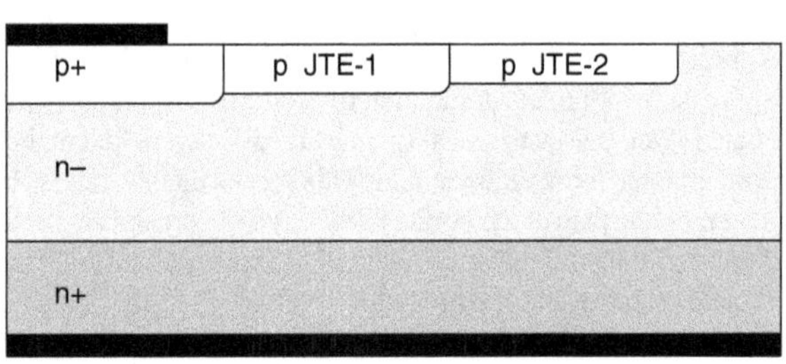

Figure 10.15 Cross-section of a p + /n– one-sided step junction protected by a two-zone junction termination extension (JTE).

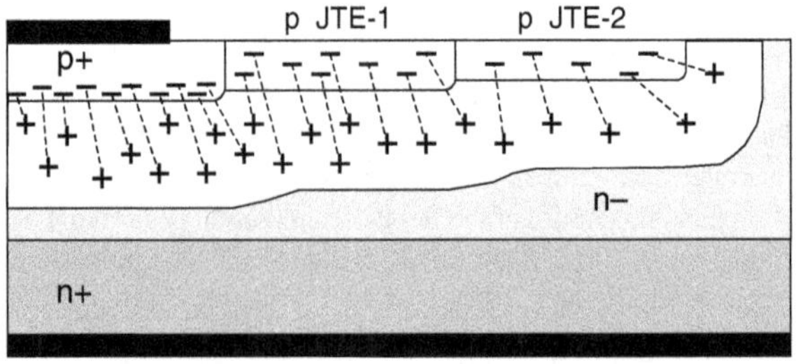

Figure 10.16 Illustration of field lines in a two-zone JTE termination under reverse bias.

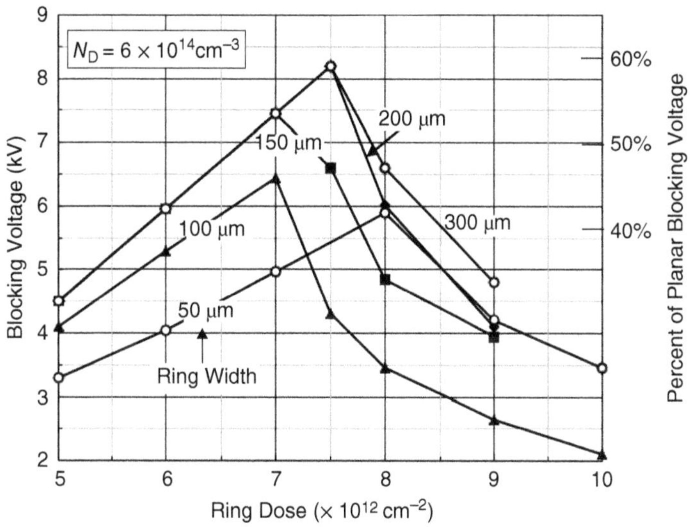

Figure 10.17 Blocking voltage as a function of ring dose (dopant concentration per unit area) for single-zone JTE terminations with ring width as a parameter. The n− drift region has a theoretical planar breakdown voltage of 13.8 kV ([8] reproduced with permission from Imran A. Khan).

For a given ring width, the breakdown voltage falls rapidly for doses higher than the optimum dose. For this reason, it is customary to select a dose that is approximately 75% of the optimum, to allow for variation in implant activation percentage during processing. The maximum breakdown voltage increases with ring width until the width is about twice the thickness of the depletion region at breakdown.

The narrow dose window to achieve a high breakdown voltage can be mitigated by the use of multiple JTE rings. Figure 10.18 shows the breakdown voltage for a three-zone JTE system on a 100 μm n− drift region doped 8×10^{14} cm^{-3} [9]. The dose of the inner ring is fixed at three times that of the outer ring, and the dose of the middle ring is twice that of the outer ring. The breakdown voltage exhibits three peaks as a function of dose. These correspond to transitions in the location of the breakdown point within the structure. At high doses, none of the rings deplete and breakdown occurs at the edge of the outer ring. As the outer ring dose falls below 1.1×10^{13} cm^{-2}, this ring becomes completely depleted before breaking down and the breakdown point shifts to the edge of the middle ring. When the outer ring dose falls below 7×10^{12} cm^{-2}, corresponding to a middle ring dose of 1.4×10^{13} cm^{-2}, the middle ring also depletes before breaking down and breakdown shifts to the edge of the inner ring. When the outer ring dose falls below 4.5×10^{12} cm^{-2}, corresponding to an inner ring dose of 1.3×10^{13} cm^{-2}, the inner ring depletes and breakdown moves to the edge of the main junction. By including multiple rings, the range of dose over which a high breakdown voltage is achieved is widened considerably.

Simulations indicate that the depth and dopant profile of the JTE implant have little effect on performance [9], but surface charge can play an important role and should be controlled by the use of a high-quality oxide passivation layer.

10.1.6 Floating Field-Ring (FFR) Terminations

Floating field-ring (FFR) terminations consist of a series of isolated concentric p+ rings surrounding the main junction, and are usually formed in the same processing step as the main junction so that no additional processing is required. Since the rings are heavily doped, the performance of a floating field-ring system does not require precise control of activated implant dose.

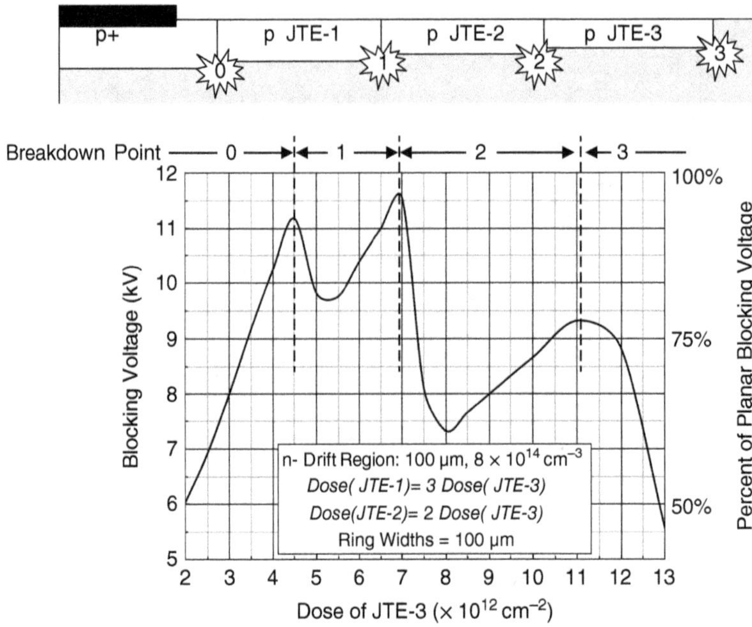

Figure 10.18 Blocking voltage as a function of ring dose for a three-zone JTE termination. The n− drift region is 100 μm thick, doped 8×10^{14} cm^{-3}, with a theoretical planar breakdown voltage of 12 kV ([9] reproduced with permission from Trans Tech Publications).

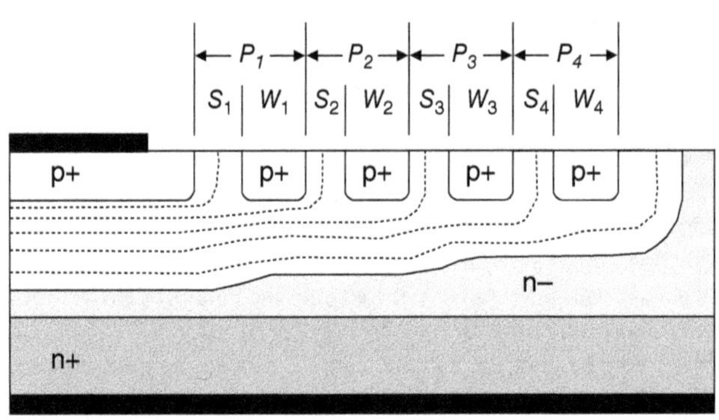

Figure 10.19 Cross-section of a p + /n− one-sided step junction protected by four floating field rings (FFRs). Several equipotential lines under reverse bias are illustrated.

Figure 10.19 illustrates a four-ring FFR termination along with equipotential lines under reverse bias. As the reverse voltage is increased from zero, the depletion region of the main junction gradually expands until it reaches the first field ring. This ring then acts as an equipotential region and the depletion region continues to expand outward from this ring. The equipotential lines in the figure may be taken as representative of the successive positions of the depletion edge as the reverse voltage is gradually increased. The effect of the FFRs is to spread out the potential distribution laterally along

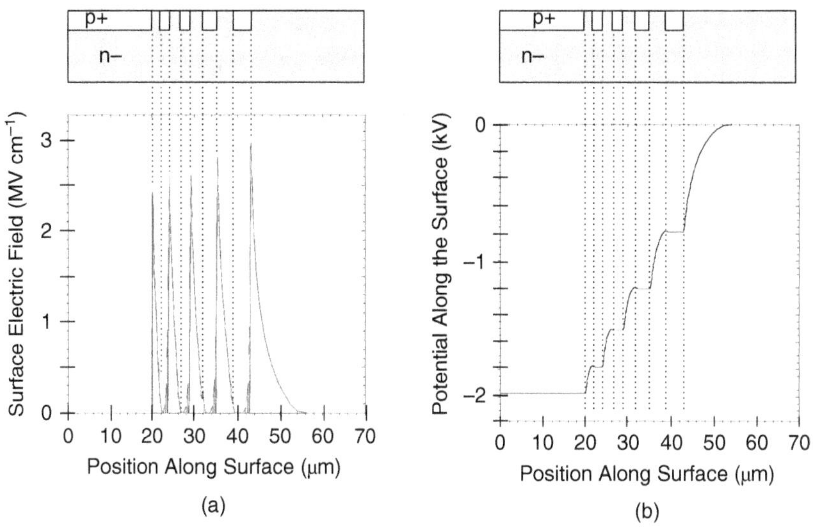

Figure 10.20 Electric field (a) and electrostatic potential (b) at the surface of a four-ring FFR termination on a 25 μm n– drift region doped 3.4 × 10^{15} cm^{-3}. The reverse voltage is 1.975 kV and the ring parameters are $S_1 = 2$ μm, $W/S = 1$, and $X_{FFR} = 1.25$ ([10], reproduced with permission from James A. Cooper).

the surface, reducing the lateral electric field that would otherwise initiate avalanche breakdown at the main junction.

Figure 10.20 shows electric field and potential along the surface under 1975 V reverse bias for a four-ring FFR termination on a 25 μm drift region doped 3.4 × 10^{15} cm^{-3} [10]. The highest field values occur at the outer edges of the field rings. The ring system spreads the potential laterally along the surface, as seen from the potential plot.

Since the field rings are all heavily doped, doping is not a design variable, but the number of rings, the width of each ring, and the spacing of each ring from the next inner ring are all design parameters. To be effective, the ring system should extend laterally at least twice the depth of the depletion region at breakdown. For high-voltage devices it is not unusual to incorporate several tens of concentric rings, and this introduces an inordinate number of free design parameters. Therefore, it is helpful to adopt a systematic method of specifying ring widths and spacings. For example, one approach is to require that the width and spacing of every ring have the same ratio W/S, and that the width (and spacing) increase with successive rings by a fixed *expansion ratio* defined as

$$X_{FFR} = W_{i+1}/W_i = S_{i+1}/S_i \tag{10.25}$$

In this implementation, the period $P_i = W_i + S_i$ also increases with successive rings by the expansion ratio X_{FFR}. Under these constraints, one can evaluate FFR system performance as a function of the initial spacing of the first ring S_1, the W/S ratio, the expansion ratio X_{FFR}, and the total number of rings in the system. Figure 10.21 shows breakdown voltage obtained by numerical simulation for a number of different ring systems on a 25 μm SiC n-type drift layer doped 3.5 × 10^{15} cm^{-3} [10]. This layer has a theoretical planar breakdown voltage of 3.5 kV. In the plot, each point denotes a specific number of rings and total width of the ring system for a particular design, and the expressions in parentheses give the initial spacing S_1, expansion ratio X_{FFR}, and width-to-spacing ratio W/S.

Although no specific algorithm for optimum design emerges, we can make some general observations. First, increasing the total width of the ring system increases the breakdown voltage, but the improvement

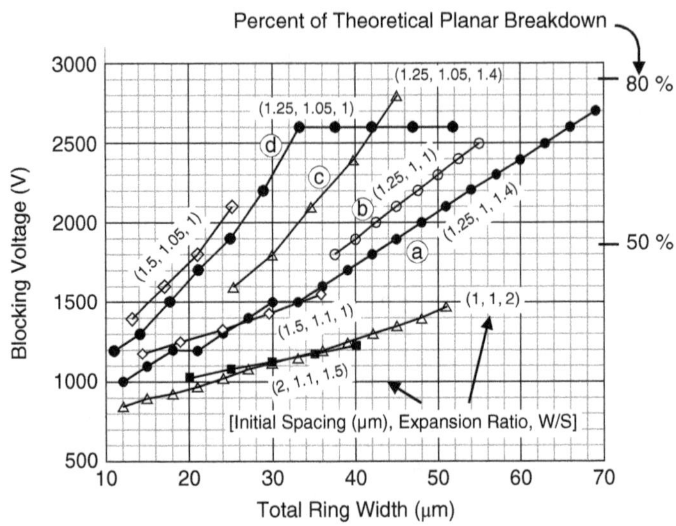

Figure 10.21 Blocking voltage versus total ring width for a variety of FFR terminations on a 25 μm n– drift region doped 3.5×10^{15} cm^{-3} [9]. Each point represents a possible ring system, with the total width of the ring system given on the horizontal axis ([10], reproduced with permission from James A. Cooper).

saturates when the ring system extends beyond the lateral depletion width at breakdown. Second, comparing curve (a) to (c) and curve (b) to (d) we conclude that an expansion ratio of 5% (1.05) gives a higher blocking voltage than a uniform ring spacing at a given total FFR width. Third, comparing (a) to (b) and (c) to (d) we see that in the regime where breakdown voltage is increasing with ring width, a W/S ratio of 1 gives a higher blocking voltage at a given total FFR width than a W/S ratio of 1.4. Finally, comparing (c) to (d) we conclude that a W/S ratio of 1.4 may allow a higher saturation value of blocking voltage than a W/S ratio of 1.0.

To summarize, floating field rings can provide blocking voltages of 75–80% of the ideal planar breakdown. They are not sensitive to the activated dopant concentration, and can be fabricated without any additional processing steps. High-voltage SiC devices have been fabricated with as many as 50 concentric rings in their edge termination structure.

10.1.7 Multiple-Floating-Zone (MFZ) JTE and Space-Modulated (SM) JTE

As discussed above, single-zone JTE is sensitive to the activated dopant concentration, and the tight processing tolerance is a challenge in manufacturing. Multi-zone JTE reduces the sensitivity to activated dopant concentration, but brings added processing complexity and cost. The tight processing tolerance of JTE can be avoided with floating field rings, since they are insensitive to the activated dopant concentration. However, it is difficult to identify an algorithm for optimizing FFR terminations.

A fifth termination method combines concepts from both JTE and FFR terminations. Multiple floating zone (MFZ) JTE [11] and space-modulated (SM) JTE [12] are closely-related techniques that achieve the broad dose window of multi-zone JTE with a single implant step. MFZ-JTE employs a series of concentric floating rings having a dose that allows each ring to fully deplete before reaching breakdown. The structure, shown in Figure 10.22, is similar to the FFR structure in Figure 10.19, but the rings are

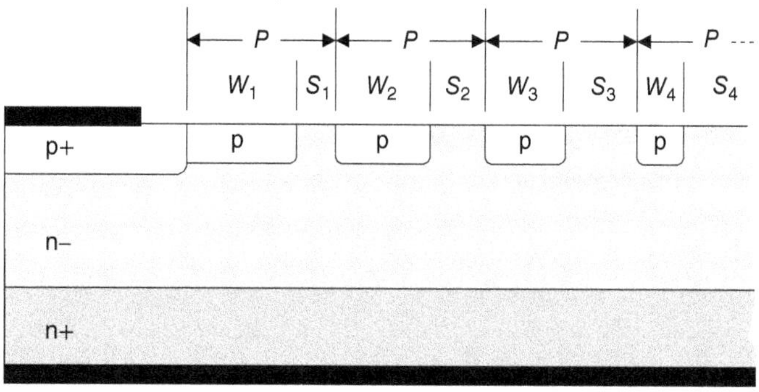

Figure 10.22 Cross-section of a p + /n– one-sided step junction protected by multiple-floating-zone JTE termination. The period of each zone is the same, but the W/S ratio decreases from the inner to the outer ring.

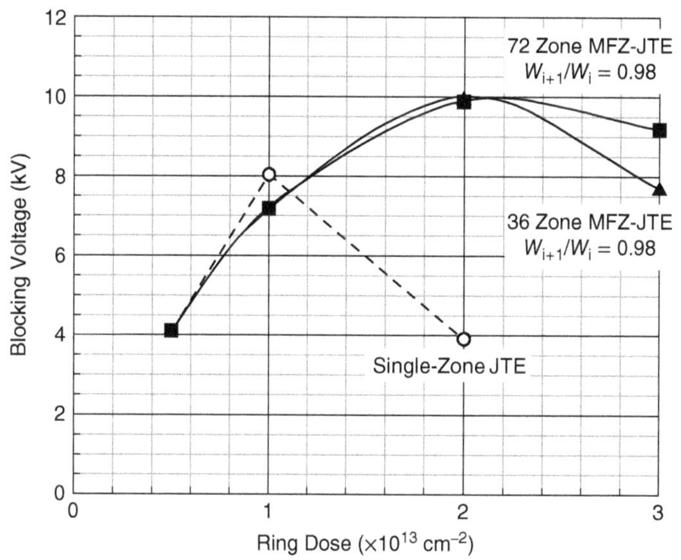

Figure 10.23 Blocking voltage as a function of ring dose for single-zone JTE and two MFZ-JTE terminations on a 120 μm n– drift region doped 8.9×10^{14} cm^{-3} ([11] reproduced with permission from IEEE).

more lightly doped. A careful examination will also reveal that each zone has the same period $P = W_i + S_i$, but the W/S ratio of successive zones decreases as we move away from the main junction. In this way the effective termination charge in each zone can be tapered smoothly from the full dose at the main junction to zero at the edge of the ring system. Figure 10.23 shows the breakdown voltage on a 120 μm, 8.9×10^{14} cm^{-3} n-type drift layer for single-zone JTE, a 36-zone MFZ-JTE and a 72-zone MFZ-JTE, as a function of ring dose [11]. All three terminations have a total width of 450 μm. The MFZ-JTE system provides a much broader range of acceptable doses than the single-zone JTE.

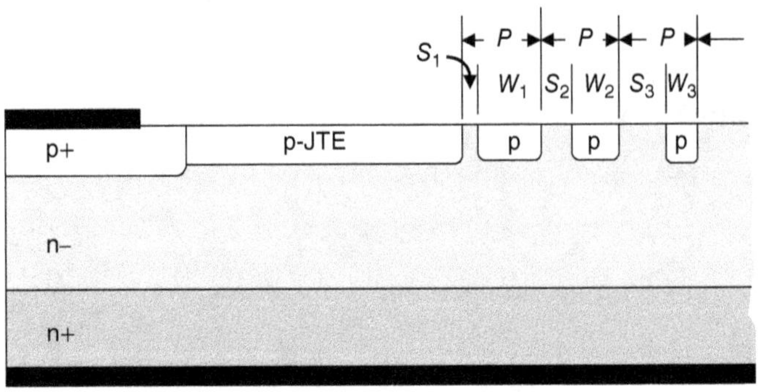

Figure 10.24 Cross-section of a p + /n– one-sided step junction protected by a space-modulated JTE termination, consisting of a single JTE ring whose outer edge is split into a number of concentric floating rings.

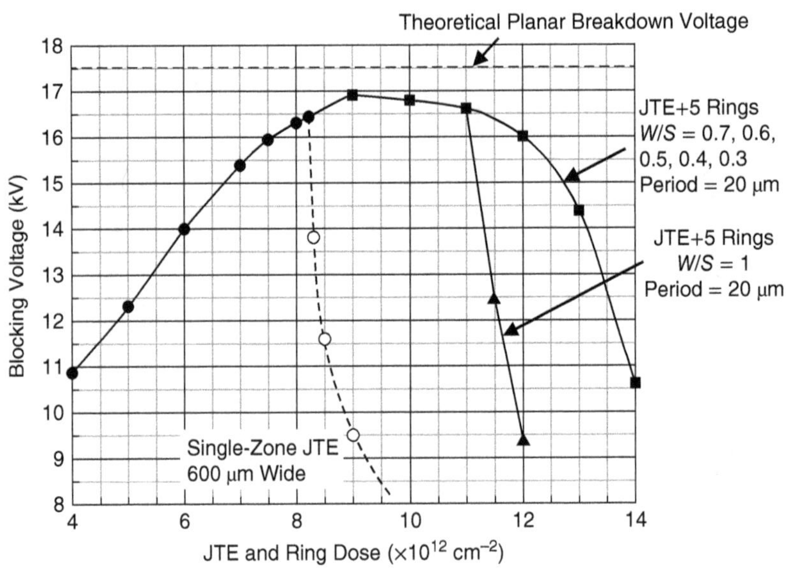

Figure 10.25 Blocking voltage as a function of ring dose for a single-zone JTE and two SM-JTE terminations on a 120 μm n– drift region doped 1×10^{14} cm^{-3} ([11] reproduced with permission from IEEE). The theoretical planar breakdown voltage for this drift layer is 17.5 kV.

Space-modulated (SM) JTE consists of a single broad JTE ring whose outer edge is broken into a number of isolated concentric rings of the same dose, as shown in Figure 10.24. Figure 10.25 shows the breakdown voltage on a 120 μm, 1×10^{14} cm^{-3} n-type drift layer for a single-zone JTE, a five-ring SM-JTE with constant W/S ratio, and a five-ring SM-JTE with a decreasing W/S ratio [12]. Both SM-JTE terminations have a constant ring period $P = W_i + S_i = 20$ μm, and all three terminations have a total width of 600 μm. The five-ring SM-JTE with decreasing W/S ratio provides the broadest range of acceptable doses.

The termination techniques discussed above can be applied with only minor modifications to all the SiC power devices discussed in this book.

10.2 Optimum Design of Unipolar Drift Regions

Unipolar devices such as the Schottky diode, JFET and MOSFET have negligible stored charge, and their switching loss is small compared to their on-state power dissipation. The on-state power dissipation of a unipolar device was given in Equation 7.5 as

$$P_{ON} = R_{ON,SP} J_{ON}^2 \tag{7.5}$$

where $R_{ON,SP}$ is the specific on-resistance and J_{ON} is the on-state current density. The unipolar device figure of merit was defined in Equation 7.9 as

$$\text{FOM} = A\sqrt{P_{MAX} V_B^2 / R_{ON,SP}} \tag{7.9}$$

where A is the area of the device, P_{MAX} the maximum allowable power dissipation, and V_B the blocking voltage. The area A is limited by material, yield, and cost considerations, and the maximum power dissipation P_{MAX} is limited by the thermal capability of the device and package. The remaining quantity $V_B{}^2/R_{ON,SP}$ is the figure of merit for the device, and it is the goal of the designer to maximize this quantity.

10.2.1 Vertical Drift Regions

All power devices support the terminal voltage in the blocking state by a reverse-biased junction, and for most vertical SiC power devices this is a p + /n− one-sided step junction, such as shown in Figure 7.1. If we neglect field crowding and assume uniform doping in the n− region, the blocking voltage can be written

$$V_B = \begin{cases} (\varepsilon_S E_C^2)/(2qN_D), & x_{DB} \le W_N \\ \left(E_C^* - \dfrac{qN_D W_N}{2\varepsilon_S}\right) W_N, & x_{DB} > W_N \end{cases} \tag{10.26}$$

Here x_{DB} is the depletion width at breakdown for a p + /n− one-sided step junction with an infinitely wide n− region, such as shown in Figure 7.1, and $E_C{}^*$ will be defined below. Assuming a triangular field profile, x_{DB} is given by

$$x_{DB} = \varepsilon_S E_C/(qN_D) \tag{10.27}$$

The critical field E_C is given as a function of doping in Equation 10.17 and is plotted in Figure 10.5. We define $E_C{}^*$ as the effective critical field in a truncated p + /n − /n+ junction with a trapezoidal field profile, such as shown in Figure 7.2, and its dependence on doping and width of the drift region is shown in Figure 10.7. The resulting dependence of V_B on doping and width of the drift region is given by Equation 10.26 and is plotted in Figure 7.3.

The specific on-resistance of a unipolar device is the sum of all the resistance elements between the terminals, but if the blocking voltage is high the on-resistance is dominated by the resistance of the drift region. The specific on-resistance of the drift region can be written

$$R_{ON,SP} = W_N/(q\mu_N N_D^+) \tag{10.28}$$

where $N_D{}^+$ is the *ionized* dopant density in the drift region. The mobility parallel to the *c*-axis in 4H-SiC can be described by the empirical expression [2]

$$\boxed{\mu_N = \frac{1141(T/300)^{-2.8}}{1 + (N_D/1.94 \times 10^{17})^{0.61}}} \tag{10.29}$$

which is plotted in Figure 10.26.

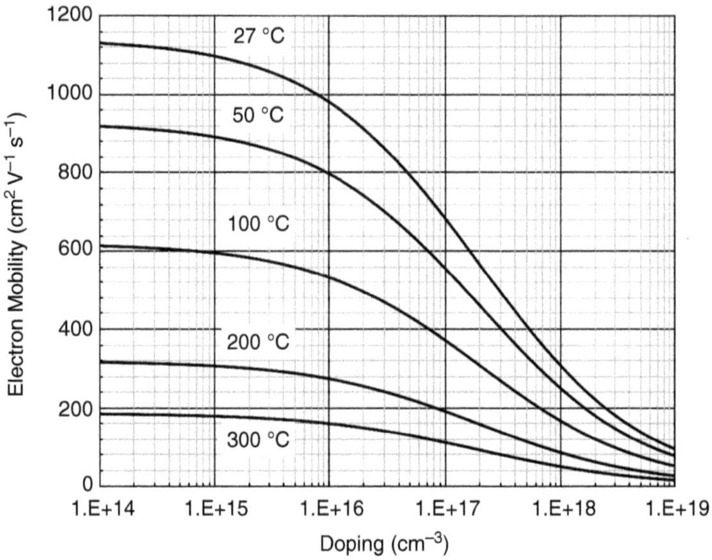

Figure 10.26 Electron mobility parallel to the *c*-axis in 4H-SiC as a function of doping and temperature, as given by Equation 10.29.

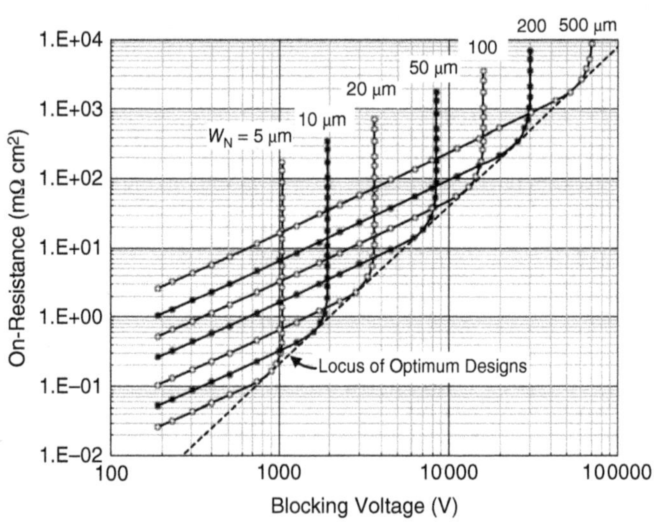

Figure 10.27 Specific on-resistance of an n− drift region in 4H-SiC at room temperature as a function of blocking voltage with drift region width as a parameter.

For the unipolar drift region, Equations 10.26–10.29 establish a relationship between on-resistance and blocking voltage. We can examine this relationship by stepping the doping through a series of values and calculating $R_{ON,SP}$ and V_B at each doping, with the mobility and critical field re-evaluated at each doping. The on-resistance can then be plotted as a function of blocking voltage, as shown in Figure 10.27. The points on each curve correspond to different values of doping, decreasing from 2×10^{17} cm^{-3} on the left to 1.5 × 10^{13} cm^{-3} on the right (in the sequence 2.0, 1.5, 1.0, 0.7, 0.5, 0.3, 0.2). As the doping is decreased, the depletion region width at breakdown x_{DB} increases, following Equation 10.27, and the blocking voltage increases according to Equation 10.26. Eventually the depletion region extends through the entire drift region and the blocking voltage increases more slowly, finally saturating at the value $E_C^*\ W_N$, as shown by Equation 10.26. The electric field profiles are illustrated in Figure 7.2. Note that even after the blocking voltage saturates, the on-resistance continues to increase as doping is reduced, as shown by Equation 10.28, and the $R_{ON,SP}-V_B$ characteristic becomes almost vertical.

The optimum design point for the unipolar drift region is the doping – thickness combination that produces the lowest on-resistance at a specified blocking voltage. The dashed line tangent to the curves in Figure 10.27 is the locus of optimum design points and can be described by the empirical equation [2]

$$R_{ON,SP}(\mathrm{opt}) \approx 2.8 \times 10^{-8}(T/300)^{2.8} V_B^{2.29}\ (\mathrm{m\Omega\ cm^2}) \tag{10.30}$$

To illustrate the use of these curves, suppose the specified blocking voltage is 3 kV. Then Figure 10.27 tells us the optimum design has a drift region width of 20 μm and a doping of about 5 × 10^{15} cm^{-3} (the doping can be found by counting points on the 20 μm curve from 2 × 10^{17} cm^{-3} on the left to 5 × 10^{15} cm^{-3} at the optimum point, in the sequence 2.0, 1.5, 1.0, etc.). To facilitate identification of the optimum doping, Figure 10.28 plots the figure of merit as a function of doping, with drift region width as a parameter. For a 20 μm drift region, the optimum doping is very close to 5 × 10^{15} cm^{-3}. The optimum doping and drift region width for a desired blocking voltage are independent of temperature and are given by the empirical equations [2]

$$N_D(\mathrm{opt}) \approx 1.1 \times 10^{20} V_B^{-1.27}\ (\mathrm{cm^{-3}}) \tag{10.31}$$

$$W_N(\mathrm{opt}) \approx 2.62 \times 10^{-3} V_B^{1.12}\ (\mu\mathrm{m}) \tag{10.32}$$

For easy reference, Figure 10.29 plots the optimum doping and drift region width from Equations 10.31 and 10.32 as a function of blocking voltage.

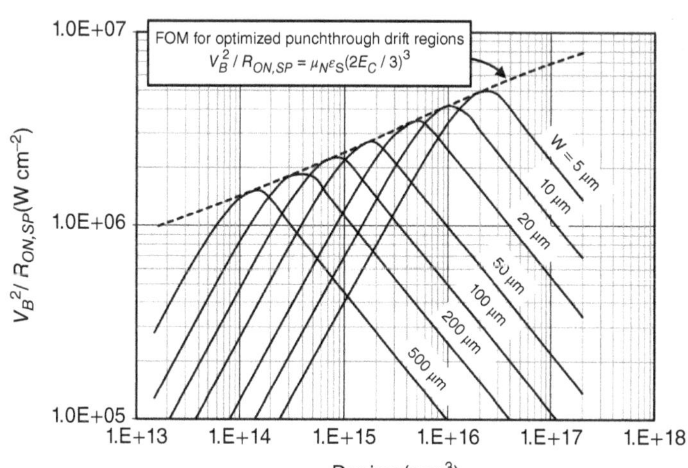

Figure 10.28 Unipolar device figure of merit $V_B^2/R_{ON,SP}$, evaluated at room temperature as a function of doping with drift region width as a parameter. For a given width, the optimum doping is the doping for which the FOM is a maximum.

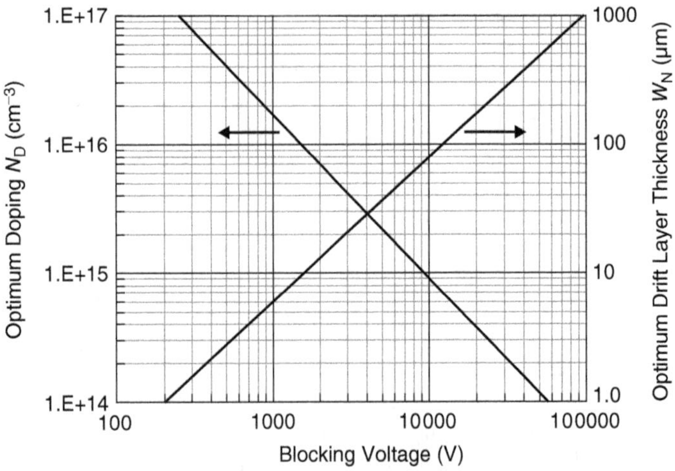

Figure 10.29 Optimum drift region doping and width as a function of blocking voltage for n-type drift regions in 4H-SiC.

10.2.2 Lateral Drift Regions

When the blocking voltage is not too high, it is feasible to implement power devices using a lateral structure rather than a vertical structure. This places all electrical terminals, the source, gate, and drain, on the top surface. The lateral structure is particularly suited to *power integrated circuits,* where the power transistor is monolithically integrated with control electronics on the same chip.

Figure 10.30 shows the cross-section of a lateral MOSFET that employs the reduced-surface-field (RESURF) concept [13]. The lightly-doped drain (LDD) extends laterally a distance L along the surface and has thickness T. In the blocking state, a reverse bias exists between the n-type drain and the grounded p-type body layer underneath, and the LDD is designed so that it completely depletes before the field at the junction reaches the critical field for avalanche breakdown. This requires

$$qN_D T < \varepsilon_S E_C \tag{10.33}$$

where N_D is the LDD doping and E_C is the critical field. The significant feature of this structure is that when the LDD is fully depleted, all the field lines from donor charges in the LDD extend vertically, terminating on acceptor charges in the base layer below, as illustrated in Figure 10.31. Since all the

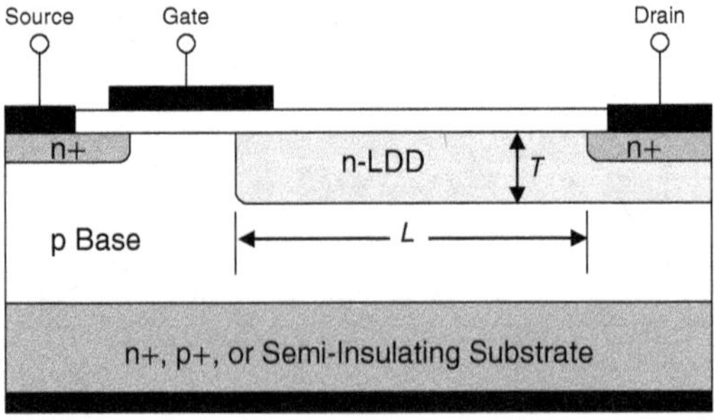

Figure 10.30 Cross-section of a lateral power MOSFET using the reduced-surface-field (RESURF) concept. The p-type base layer is at ground potential.

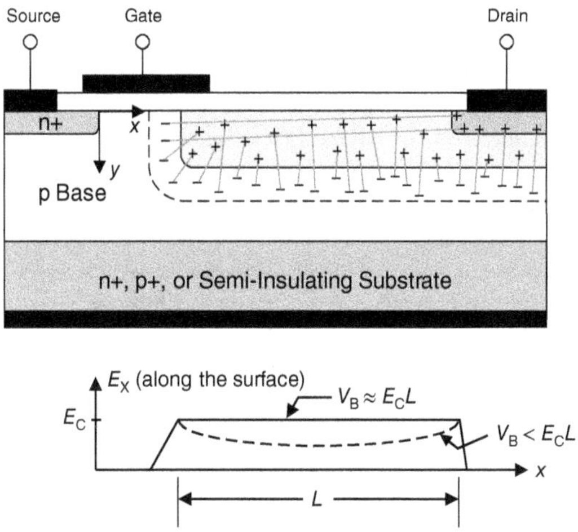

Figure 10.31 Illustration of electric field lines in a lateral RESURF MOSFET in the blocking state. The surface electric field is plotted below the figure, and the dashed line indicates the approximate field when two-dimensional effects are included.

donor charges in the LDD are terminated on acceptors in the base layer, x-directed field lines do not terminate on charges in the LDD layer, and there is no field taper in the x-direction, as indicated in the plot. This means that we can increase the drain voltage until E_X reaches the critical field E_C, at which point the blocking voltage is given by

$$V_B \approx E_C L \tag{10.34}$$

The specific on-resistance of the LDD is

$$R_{ON,SP} = R \cdot A = \left(\frac{\rho L}{WT}\right)(WL) = \frac{L^2}{q\mu_N N_D T} \tag{10.35}$$

where ρ is the resistivity of the LDD layer and W is the width of the device. Note that the on-resistance depends only on the doping-thickness product $N_D T$ and the length of the LDD region.

We can calculate the figure of merit for the RESURF structure, assuming the LDD resistance dominates the device resistance. Using Equations 10.33–10.35 we can write

$$\frac{V_B^2}{R_{ON,SP}} = V_B^2\left(\frac{q\mu_N N_D T}{L^2}\right) = V_B^2\left(\frac{\mu_N \varepsilon_S E_C}{V_B^2/E_C^2}\right) = \mu_N \varepsilon_S E_C^3 \tag{10.36}$$

Comparing Equation 10.36 to Equation 7.13, we find that the theoretical limit of a RESURF device is actually four times larger than that of a comparable vertical unipolar power device. However, the situation is not as simple as our analysis would indicate, since we have neglected two-dimensional effects. In practice, field crowding at the ends of the LDD produces field spikes, and the true field is like the dashed line in Figure 10.31. Nevertheless, lateral RESURF devices are a viable alternative to vertical devices in cases where the blocking voltage is not too high.

Although Figures 10.30 and 10.31 illustrate a lateral MOSFET, the same discussion applies to any unipolar device with a lateral drift region, such as a lateral JFET. A modified version of the RESURF principle is also utilized in silicon super-junction MOSFETs, where the drift regions are vertical. To date, no vertical super-junction devices have been reported in SiC.

10.3 Comparison of Device Performance

We conclude this chapter by considering how the different types of power devices may be compared in terms of performance. A fair comparison requires that we evaluate all devices as a function of both blocking voltage and switching frequency, since switching loss is a major limitation on device performance. As discussed in Section 7.1, our figure of merit is the on-state current density J_{ON} that each device can carry at a specified switching frequency and blocking voltage, subject to the requirement that the total power dissipation in the device be less than the package limit, which we will arbitrarily take as 300 W cm^{-2}. We designate this current density as J_{300}, and the device with the highest J_{300} at a given blocking voltage and switching frequency is the preferred device for that application.

The above relationships can be visualized using the construction of Figure 10.32, which compares a power MOSFET and an IGBT in a three-dimensional $J_{ON}-V_B-f$ parameter space. The two surfaces represent the loci of constant power dissipation equal to 300 W cm^{-2}. Let's first examine how these surfaces depend on blocking voltage and frequency.

To achieve a higher blocking voltage, it is necessary to use a thicker and more lightly-doped drift region. This increases the on-resistance of the device, and hence the on-state power dissipation. To keep the total power below the assumed package limit of 300 W cm^{-2}, we must reduce the current density as the blocking voltage is increased. This explains why the surfaces slope downward as V_B is increased.

Frequency enters the picture through the switching loss. The switching power dissipation is directly proportional to switching frequency. If we are required to operate at a higher frequency where the switching loss is higher, we must again decrease the current to keep the total power dissipation below 300 W cm^{-2}. This explains the slope of the surfaces downward as frequency is increased. The effect is stronger for bipolar devices such as the IGBT, since they dissipate more energy per switching event.

The constant-power surfaces of the MOSFET and the IGBT intersect along a line that can be projected down to the V_B-f plane. This line represents the design points where the two devices have equal performance, in that they permit the same maximum on-state current for operating points along this line. In this example, the IGBT carries higher current at low switching frequencies, while the MOSFET carries higher current at high frequencies. Constructions such as this enable us to visualize the relative performance of different devices in the full current–voltage–frequency parameter space.

The above visualization can be made quantitative as follows. For unipolar devices we can usually assume (i) the I–V characteristics near the origin are linear, and hence can be described by an

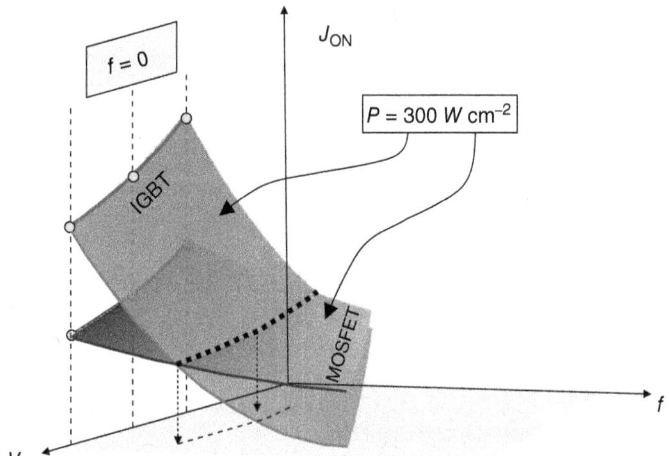

Figure 10.32 Surfaces of constant power dissipation of a MOSFET and an IGBT in three-dimensional $J_{ON}-V_B-f$ parameter space ([15] reproduced with permission from Trans Tech Publications).

on-resistance, and (ii) the switching power is small compared to the on-state power dissipation. These assumptions led to Equations 7.8 and 7.9, and the figure of merit for unipolar devices was taken to be $V_B^2/R_{ON,SP}$. However, in bipolar devices the current passes through one or more forward-biased pn junctions, and these inject minority carriers that must be removed during the turn-off transient. Hence we must consider the energy dissipated during each switching cycle E_{SW}. The switching power dissipation is proportional to the switching energy and the frequency, as given in Equation 7.15. Moreover, since the current path in bipolar devices goes through forward-biased diodes whose I–V characteristics are nonlinear, the on-state I–V characteristics often cannot be described by a simple resistance. (This statement depends on the *number* of forward-biased junctions in the current path. The pin diode, IGBT, and thyristor have an *odd* number of forward-biased junctions in the current path, and their I–V characteristics are nonlinear near the origin. The BJT has an *even* number of forward-biased junctions, and its I–V characteristics near the origin are linear.)

To account for both the switching loss and the nonlinear I–V characteristics, we use the procedure described in Section 7.1, which is summarized as follows:

1. Determine the dependence of switching energy E_{SW} on on-state current J_{ON} using two-dimensional transient computer simulations. These simulations should include the energy dissipated in the external circuit during switching, and hence a specific load circuit must be assumed.
2. Determine the dependence of on-state power P_{ON} on J_{ON} using two-dimensional steady-state computer simulations and obtain a value for P_{OFF} at the assumed blocking voltage.
3. For a given switching frequency f, adjust J_{ON} using Equation 7.17 until the total power dissipation P_{TOTAL} equals 300 W cm^{-2}. The resulting current is the desired figure of merit J_{300} for that particular blocking voltage and switching frequency.

This procedure has been used to compare the performance of power MOSFETs and IGBTs at blocking voltages of 15 and 20 kV [14, 15] and will serve as an example of the technique. First, an n-channel DMOSFET and a p-channel DMOS IGBT were individually optimized for the desired blocking voltage, 15 or 20 kV, using computer simulations. The on-state I–V characteristics of the 20 kV devices are shown in Figure 10.33 at two temperatures, 27 and 175 °C. The I–V characteristics of the IGBT are

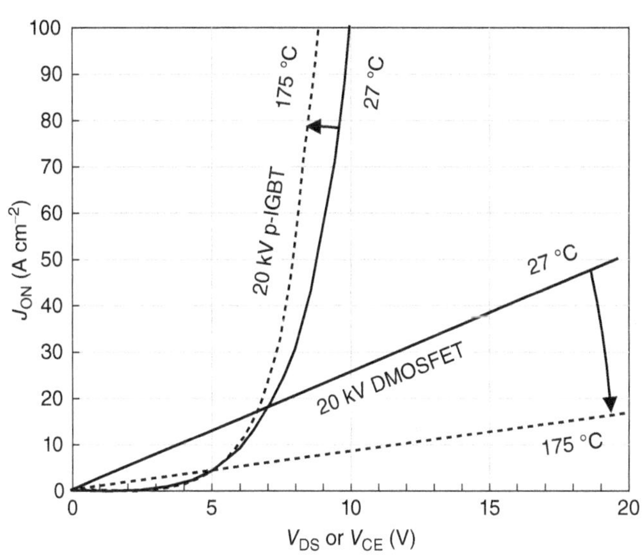

Figure 10.33 Current–voltage characteristics at 27 °C and 175 °C of an optimized DMOSFET and an optimized p-channel IGBT designed to block 20 kV ([15] reproduced with permission from Trans Tech Publications).

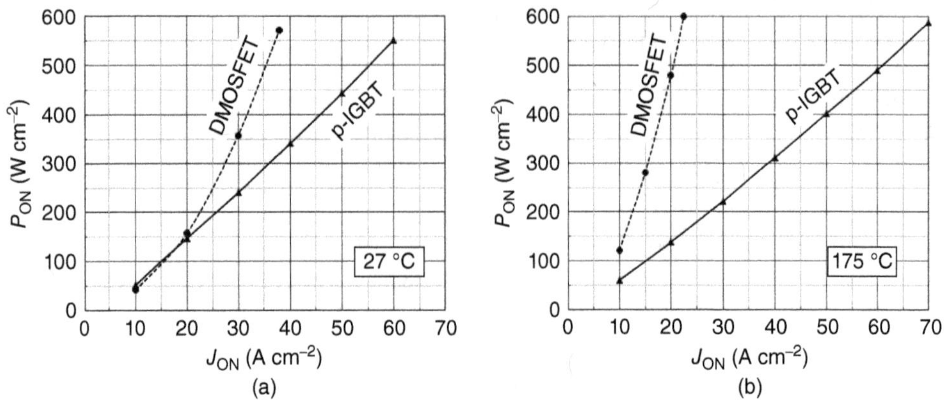

Figure 10.34 On-state power dissipation of the DMOSFET and IGBT of Figure 10.33 as a function of on-state current density, (a) at 27 °C and (b) at 175 °C ([14] reproduced with permission from Tomohiro Tamaki).

nonlinear, and cannot be represented by an on-resistance. The MOSFET I-V characteristics are linear, but the on-resistance degrades with temperature due to the decrease in mobility with temperature shown in Figure 10.26. In contrast, the IGBT is relatively insensitive to temperature. This is because the ambipolar diffusion length depends on both the diffusion coefficient and the lifetime. As shown in Figure 9.26, the lifetime increases faster with temperature than the diffusion coefficient decreases, and the diffusion length increases slightly with temperature, improving device performance.

The on-state power $P_{ON} = J_{ON}V_{DS}$ or $P_{ON} = J_{ON}V_{CE}$ can be calculated at any J_{ON} directly from Figure 10.33, and is plotted as a function of current in Figure 10.34. The switching energy is obtained as a function of on-current from transient simulations using a clamped inductive load (modeled by a current source of magnitude J_{ON}), and the resulting E_{SW} is plotted as a function of current in Figure 10.35. These plots are used in conjunction with Equation 7.17 to determine the J_{ON} that corresponds to a total power dissipation of 300 W cm^{-2}. This is the desired figure of merit J_{300}. Figure 10.36 shows J_{300} for each device as a function of switching frequency. We observe that at low switching frequencies, the IGBT provides higher current than the MOSFET. This is due to conductivity modulation of the drift region,

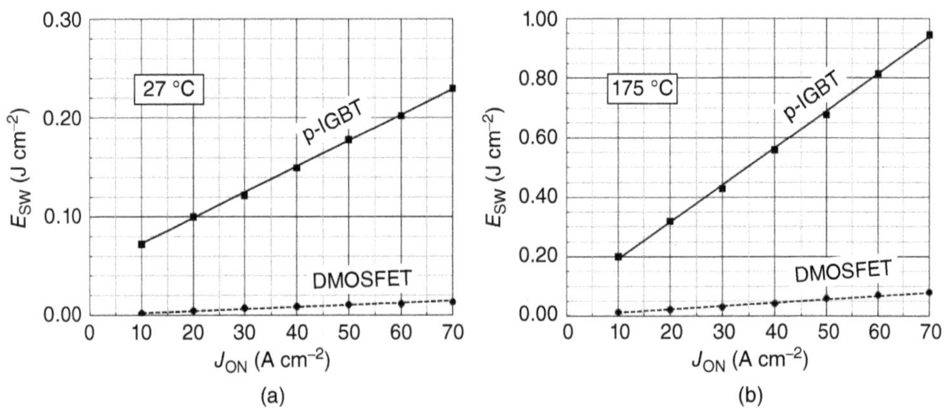

Figure 10.35 Switching energy of the DMOSFET and IGBT of Figure 10.33 as a function of on-state current density, as obtained from two-dimensional transient simulations using a clamped inductive load, (a) at 27 °C and (b) at 175 °C ([14] reproduced with permission from Tomohiro Tamaki).

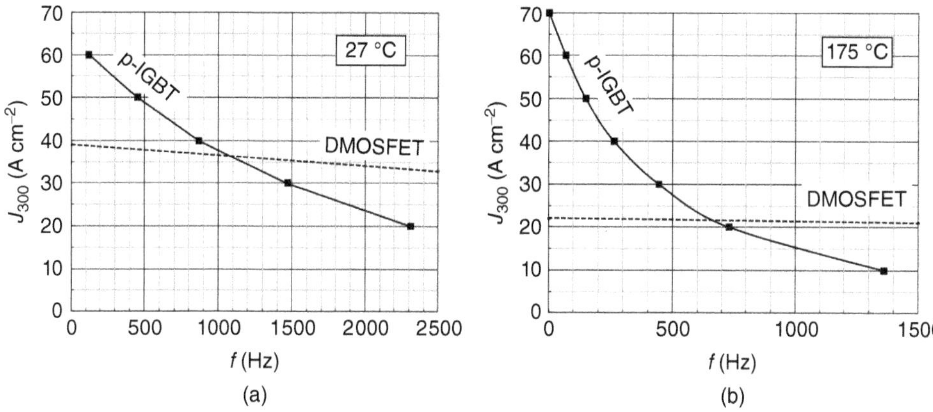

Figure 10.36 Current density of the DMOSFET and IGBT of Figure 10.33 assuming a total power dissipation of 300 W cm^{-2}, plotted as a function of switching frequency, (a) at 27 °C and (b) at 175 °C ([13] reproduced with permission from Tomohiro Tamaki).

which reduces the forward voltage drop at high currents, as seen in Figure 10.33. However, the minority carriers that provide the conductivity modulation must be removed during each turn-off event, and the reverse current transient gives rise to a power dissipation that is proportional to frequency. As frequency is raised, the on-current J_{ON} must be reduced to keep the total power dissipation below 300 W cm^{-2}, and at higher switching frequencies the on-current of the IGBT falls below that of the MOSFET. Thus the MOSFET is the preferred device for high-frequency applications, while the IGBT is superior at low frequencies.

The methodology above can be used to compare bipolar or unipolar devices for any desired blocking voltage and switching frequency. In general, bipolar devices perform better at high blocking voltages and low frequencies, while unipolar devices are superior at low blocking voltages and high frequencies. However, at sufficiently low blocking voltages, MOSFETs and JFETs are superior to IGBTs and thyristors, regardless of the frequency. Why is this? There are two reasons. First, at low blocking voltages the drift region resistance is relatively small, and does not need to be reduced by conductivity modulation. Second, IGBTs and thyristors have an odd number of pn junctions in their current paths. When all junctions are forward biased, a net voltage drop of one forward-biased diode appears in series with the remainder of the device. This adds an additional component to the static power dissipation of the IGBT and thyristor. (In contrast, the BJT has an even number of junctions in its current path, and the voltage drops of two oppositely-directed junctions cancel out.)

At sufficiently high frequencies, unipolar devices are superior at all blocking voltages, as seen in the comparison of pin and Schottky diodes in Figure 7.9. This is because the switching loss of bipolar devices is proportional to frequency and becomes the dominant loss at high frequencies, forcing a continuous reduction in J_{ON} to keep P_{TOTAL} below 300 W cm^{-2}.

In all comparisons involving power devices, it is important to consider both high and low temperature extremes, since device performance is sensitive to temperature. A device dissipating 300 W cm^{-2} will have a junction temperature well above ambient. For this reason, analyses performed only at room temperature provide an incomplete picture of device performance.

References

[1] Konstantinov, A.O., Wahab, Q., Nordell, N. and Lindefelt, U. (1997) Ionization rates and critical fields in 4H silicon carbide. *Appl. Phys. Lett.*, **71** (1), 90–92.

[2] Morisette, D.T. (2001) Development of robust power Schottky barrier diodes in silicon carbide. PhD thesis. Purdue University.
[3] Loh, W.S., David, J.P.R., Ng, B.K. *et al.* (2009) Temperature dependence of hole impact ionization coefficient in 4H-SiC photodiodes. *Mater. Sci. Forum*, **615–617**, 311–314.
[4] Konstantinov, A.O., Nordell, N., Wahab, Q. and Lindefelt, U. (1998) Temperature dependence of avalanche breakdown for epitaxial diodes in 4H-silicon carbide. *Appl. Phys. Lett.*, **73** (13), 1850–1852.
[5] Davies, R.L. and Gentry, F.E. (1964) Control of electric field at the surface of p-n junctions. *IEEE Trans. Electron. Devices*, **11** (7), 313–323.
[6] Adler, M.S. and Temple, V.K. (1978) Maximum surface and bulk electric fields at breakdown for planar and beveled devices. *IEEE Trans. Electron. Devices*, **25** (10), 1266–1270.
[7] Temple, V.A.K. (1977) Junction termination extension (JTE): a new technique for increasing avalanche breakdown voltage and controlling surface electric fields in p-n junctions. International Electron Devices Meeting Technical Digest, pp. 423–426.
[8] Khan, I.A. (2002) High voltage SiC MOSFETs. PhD thesis. Purdue University.
[9] Wang, X. and Cooper, J.A. (2004) Optimization of JTE edge termination for 10 kV power devices in 4H-SiC. *Mater. Sci. Forum*, **457–460**, 1257–1262.
[10] Singh, S., Tjandra, E. and Cooper, J. A. (2003) High-voltage termination of SiC devices using floating field rings. unpublished, reproduced with permission from James A. Cooper.
[11] Sung, W., Van Brunt, E., Baliga, B.J. and Huang, A.Q. (2011) A new edge termination technique for high-voltage devices in 4H-SiC-multiple-floating-zone junction termination extension. *IEEE Electron. Device Lett.*, **32** (7), 880–882.
[12] Feng, G., Suda, J. and Kimoto, T. (2012) Space-modulated junction termination extension for ultra-high voltage p-i-n diodes in 4H-SiC. *IEEE Trans. Electron. Devices*, **59** (2), 414–418.
[13] Appels, J.A. and Vaes, H.M.J. (1979) High voltage thin layer devices (RESURF) devices. IEEE International Electron Devices Meeting Technical Digest, pp. 238–241.
[14] Tamaki, T., Walden, G. G., Sui, Y., and Cooper, J. A. (2007) On-state and switching performance of high-voltage 4H-SiC DMOSFETs and IGBTs. International Conference on Silicon Carbide and Related Materials, Otsu, Japan.
[15] Tamaki, T., Walden, G.G., Sui, Y. and Cooper, J.A. (2009) On-state and switching performance of high-voltage 4H-SiC DMOSFETs and IGBTs. *Mater. Sci. Forum*, **600–603**, 1143–1146.

11

Applications of Silicon Carbide Devices in Power Systems

11.1 Introduction to Power Electronic Systems

The topic of power electronic systems is both broad and deep, and we will only present an overview in this chapter. Our objective is to consider those systems in which the substitution of silicon carbide devices may produce significant advantages in performance, efficiency, reliability, and/or overall system cost. The discussion will be limited to the basic circuit topology and device requirements, and will not explore second-order effects that are also important for a practical design. For these, the reader is referred to one of the specialty texts on power systems [1], and thence to the literature.

A block diagram of a generic power processing system is shown in Figure 11.1. This system provides an interface between two ports, typically a source of electric power and a load to which electric power is delivered. In the general case, the power processor may consist of three elements: an electronic converter connected to port 1, an electronic converter connected to port 2, and an energy storage element between the two converters. The converters may include one or more power semiconductor devices, along with passive components such as resistors, inductors, and capacitors. The energy storage element between the converters is typically either an inductor or a capacitor. In most cases the power processor is designed to be unidirectional, with power flowing from the source port to the load port, but in some cases the power flow can be bidirectional, as in motor drives for electric vehicles where regenerative braking is used to return kinetic energy to a storage device.

Electronic converters may be classified based on whether their inputs and outputs are DC or AC. The four possible input–output combinations are listed in Table 11.1. Converters may also be classified according to the switching mode upon which their operation is based. The four possible switching modes are

1. Uncommutated (e.g., diode rectifiers)
2. Line-frequency commutated (e.g., thyristor rectifiers and inverters)
3. Switch-mode (e.g., pulse-width modulated waveform generators)
4. Resonant (switching occurs at a zero crossing of the voltage or current waveform).

Converters may employ any of the semiconductor devices discussed in Chapters 7–10, with the type of device depending on the application and the circuit topology employed. Often, the designer has the choice of several possible devices. For example, when a switching device is required, the designer might choose a JFET (junction field-effect transistor), a MOSFET (metal-oxide-semiconductor field effect transistor),

Fundamentals of Silicon Carbide Technology: Growth, Characterization, Devices, and Applications, First Edition.
Tsunenobu Kimoto and James A. Cooper.

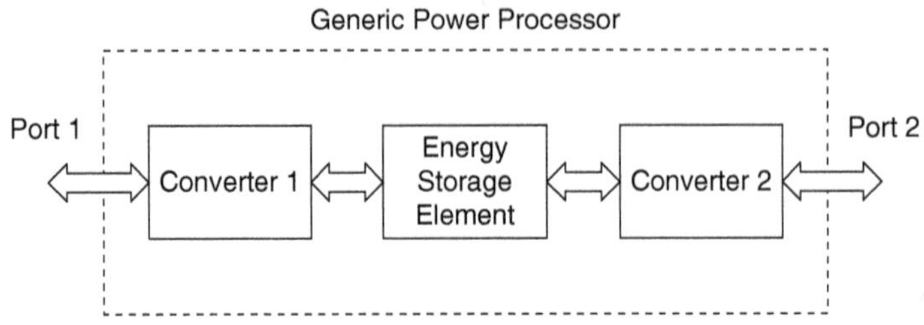

Figure 11.1 Schematic of a generic power processor.

Table 11.1 Classification of electronic power converters.

Input	Output	Designation	Possible switching modes			
			Uncommutated (e.g., diodes)	Line-frequency commutated	Switch mode	Resonant
AC	DC	Rectifier	X	X	–	X
DC	AC	Inverter	–	X	X	X
AC	AC	AC converter	–	X		X
DC	DC	DC converter	–	–	X	–

a BJT (bipolar junction transistor), or an IGBT (insulated-gate bipolar transistor), depending on the requirements of the application.

This chapter is organized as follows. In Section 11.2 we introduce three basic converter circuits: (i) line-frequency-commutated rectifiers and inverters, (ii) switch-mode DC converters and power supplies, and (iii) switch-mode inverters. In Section 11.3 we discuss motor drives for DC motors, induction motors, synchronous motors, and hybrid and electric vehicles. Section 11.4 covers the applications of SiC power devices in renewable energy, and Section 11.5 deals with switch-mode power supplies. Finally, in Section 11.6 we summarize the present state-of-the-art of SiC power devices, as compared to the silicon devices with which they compete.

11.2 Basic Power Converter Circuits

11.2.1 Line-Frequency Phase-Controlled Rectifiers and Inverters

Line-frequency phase-controlled converters are used to transfer power between a line-frequency AC environment and a controlled DC environment. Thyristor-based line-frequency converters are used primarily in high-power three-phase applications, especially in cases where bidirectional power flow is desired. Examples include high-voltage DC transmission systems and high-power AC and DC motor drives where regenerative braking is employed. In line-frequency converters, turn-off of the thyristors occurs at zero crossings of the thyristor current, which are naturally synchronized with the terminal voltage of the AC port.

A basic thyristor converter driving a resistive load is illustrated in Figure 11.2, along with its operational waveforms. The thyristor is triggered at an arbitrary phase angle $0 \leq \alpha \leq \pi$ by a short gate pulse. Once triggered, the thyristor remains in its forward-conducting mode until the cathode voltage changes

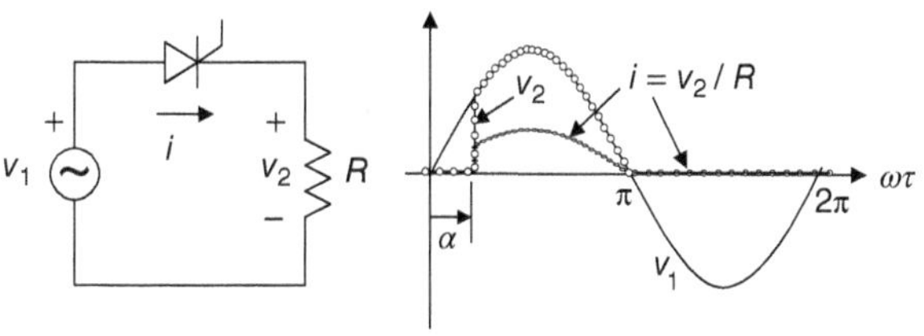

Figure 11.2 Schematic of a simple thyristor-based line-frequency phase-controlled converter driving a resistive load.

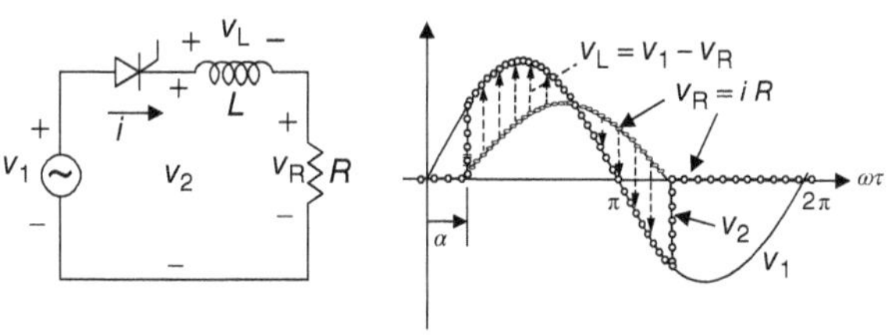

Figure 11.3 A line-frequency phase-controlled converter driving an inductive load.

sign at $\omega\tau = \pi$, whereupon it enters its reverse-blocking mode. When the cathode voltage again becomes positive at $\omega\tau = 2\pi$, the thyristor enters its forward-blocking mode until the next gate trigger pulse at $\omega\tau = 2\pi + \alpha$. In this analysis the forward voltage drop of the thyristor is neglected, and the load voltage υ_2 is equal to the source voltage during the period when the thyristor is conducting. The current waveform is a truncated half-sinusoid, and the average power delivered to the load can be varied from zero to $(\upsilon_{1,\text{RMS}})^2/(2R)$ by adjusting the phase angle α of the triggering pulse.

Figure 11.3 shows a thyristor driving an inductive load from a sinusoidal source. Prior to triggering of the thyristor, the current is zero. Once the thyristor is triggered, a current begins to flow and the inductor voltage depends on the current according to

$$\upsilon_L = L\frac{\partial i}{\partial t} = \upsilon_i - iR \tag{11.1}$$

In Figure 11.3, the inductor voltage is shown graphically as the difference between the source voltage υ_1 and the resistor voltage $\upsilon_R = i\,R$ when the thyristor is on. As long as υ_L is positive, $\partial i/\partial t$ is positive and the current increases. When the resistor voltage υ_R equals the source voltage υ_1, the inductor voltage υ_L changes sign and the current begins to decrease. When the current reaches zero, the thyristor enters its reverse-blocking mode and the current remains zero until the source voltage becomes positive again and the next trigger pulse arrives. When the source voltage is negative and the current is positive, reactive power stored in the inductor is being returned to the source. Stored power is also delivered to the resistive load during this period, since the resistor voltage and current are both positive.

Figure 11.4 shows a thyristor driving a load consisting of an inductor in series with a voltage source. This type of circuit is representative of a DC motor, where the voltage source E_2 represents the back-emf induced in the stator windings by the rotating magnetic field of the rotor. The current is initially zero and the thyristor voltage is $\upsilon_1 - E_2$. As υ_1 increases during the first half-cycle, $\upsilon_1 - E_2$ eventually becomes positive and the thyristor enters its forward-blocking mode. The current remains zero until the thyristor is triggered, at which point the load is effectively connected to the source ($\upsilon_2 = \upsilon_1$) and the current begins to increase. The inductor voltage is given by

$$v_L = L\frac{\partial i}{\partial t} = v_2 - E_2 \tag{11.2}$$

and is shown in Figure 11.4. When $v_L = v_2 - E_2$ becomes negative, $\partial i/\partial t$ is negative and the current decreases. When the current reaches zero, the thyristor enters its reverse-blocking mode. With zero current, all the source voltage develops across the thyristor, and $v_2 \rightarrow E_2$. The current remains zero until the thyristor is triggered after the start of the next AC cycle.

The converter of Figure 11.4 only delivers power to the load during the first half-cycle. A converter capable of delivering power during both half-cycles is shown in Figure 11.5. We assume that the load inductance is large, and may be represented by an equivalent DC current source I_2. The operation can be understood as follows. During the half-cycle preceding $\omega\tau = 0$, thyristors 3 and 4 are conducting and the load voltage v_2 is equal to $-v_1$, since T3 and T4 cross-connect the load to the source. When the source voltage goes positive at $\omega\tau = 0$, thyristors 1 and 2 enter their forward-blocking modes and the load voltage v_2 goes negative through the conduction of T3 and T4 (the ideal current source develops

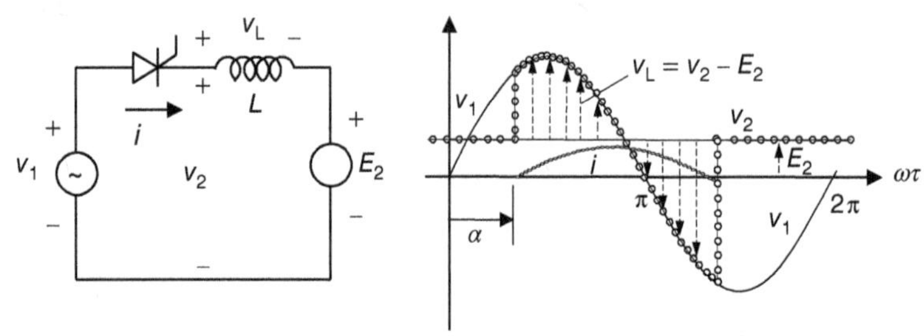

Figure 11.4 A line-frequency phase-controlled converter driving an inductive load and an opposing voltage source.

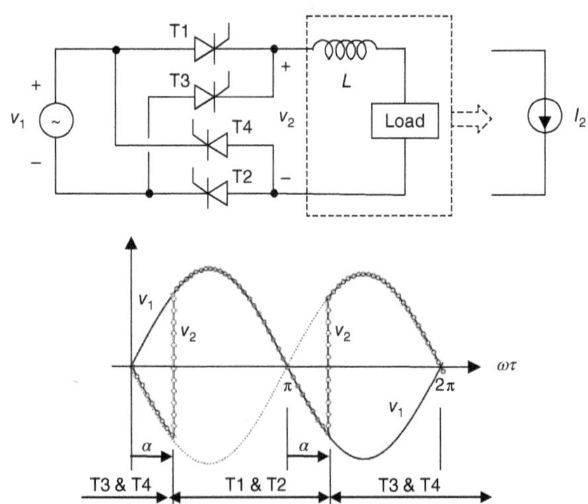

Figure 11.5 A line-frequency phase-controlled converter that delivers power during both half cycles to an inductive load, represented by a DC current source.

whatever voltage is necessary to maintain a constant current I_2). T1 and T2 are triggered at $\omega\tau = \alpha$. With T1 and T2 conducting, the load voltage v_2 switches rapidly to $+v_1$, and T3 and T4 enter their reverse-blocking modes.

The line-frequency phase-controlled converters of Figures 11.2–11.5 can operate in two quadrants of the V_2–I_2 plane as either rectifiers or inverters, as illustrated in Figure 11.6. During the portion of the ac cycle where v_2 and i_2 are positive, power is delivered from port 1 to port 2, and the converter is operating as a rectifier. For the portion of the cycle where v_2 is negative and i_2 is positive, power is delivered from port 2 to port 1, and the converter is operating as an inverter. The portions of the AC cycle over which rectification and inversion occur are determined by the triggering angle α.

Consider the converter of Figure 11.5. For triggering angles $0 \le \alpha \le \pi/2$ the converter operates in the rectifier mode, while for triggering angles $\pi/2 \le \alpha \le \pi$ it operates in the inverter mode. In the rectifier mode, the converter can be used as a battery charger or a DC motor drive. In this case, the generalized load, Figure 11.7a, is replaced by a voltage source E_2 representing the battery or the back-emf in the stator windings of the motor, as shown in Figure 11.7b. The same converter can be used in the inverter mode to transfer power from an energy source such as a solar array to the AC power grid. In this case the

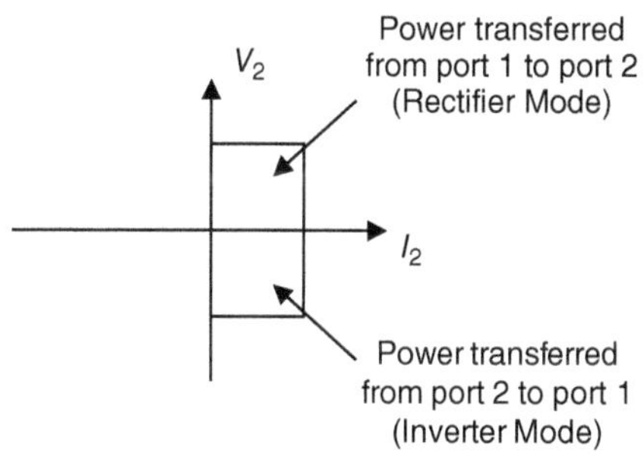

Figure 11.6 The V_2–I_2 plane. Operation in the first quadrant means that port 2 is absorbing power from port 1, and represents rectification. Operation in the fourth quadrant means that port 2 is delivering power to port 1, and corresponds to the inverter mode of operation. Bidirectional converters may operate in either quadrant, depending on the triggering angle.

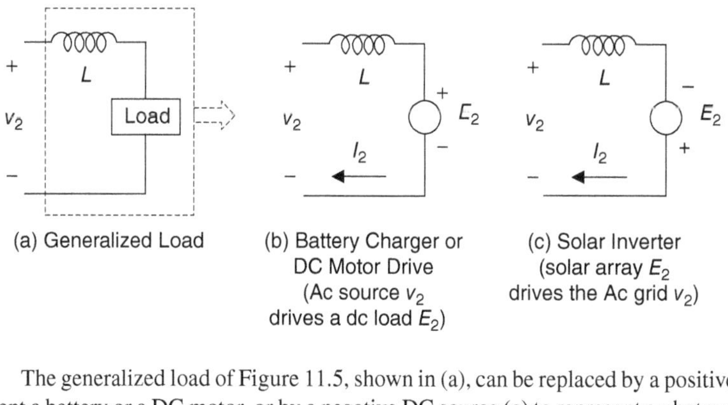

Figure 11.7 The generalized load of Figure 11.5, shown in (a), can be replaced by a positive DC source (b) to represent a battery or a DC motor, or by a negative DC source (c) to represent a photovoltaic power source.

load is replaced by a DC source of the opposite polarity as shown in Figure 11.7c. This configuration can also be used as a motor drive for high-power AC synchronous motors.

11.2.2 Switch-Mode DC–DC Converters

DC–DC converters are used to transfer power between two DC environments. Typical applications include switch-mode DC power supplies and DC motor drives. Consider the generic power processor of Figure 11.1 where the source at port 1 is the AC line and the load at port 2 requires regulated DC power. In this case, a switch-mode DC converter would be used as converter 2 and a rectifier used as converter 1. In this section we will focus on converter 2, which converts unregulated DC to regulated DC. The circuit topologies to be considered are (i) step-down (buck) converters, (ii) step-up (boost) converters, (iii) step-down/step-up (buck/boost) converters, and (iv) full-bridge DC converters.

Figure 11.8 shows a schematic of a step-down or buck converter. In this figure we use a generic circuit symbol for a transistor switch, keeping in mind that the actual switching device might be a JFET, MOSFET, BJT, or IGBT, depending on the application. As the name implies, the buck converter delivers regulated DC power to the load at a *lower voltage* than the unregulated DC power at the source. The regulation is accomplished by adjusting the duty factor of a periodic rectangular waveform applied to the control electrode of the transistor. When the transistor is on, current flows from the source to the load through a low-pass filter formed by the inductor and capacitor. When the transistor is off, the current through the inductor cannot change instantaneously and the current path is through the inductor, the load, and the diode. If the period of the switching waveform is short compared to the time constant of the filter, the load voltage V_2 and load current I_2 may be regarded as DC quantities. In this case, the load voltage is simply the source voltage multiplied by the duty factor: $V_2 = \delta\, V_1$, where $0 \le \delta \le 1$.

Figure 11.9 shows a step-up or boost converter. With the transistor on, current flows through the inductor, storing reactive energy in the inductor. When the transistor turns off, the inductor current cannot

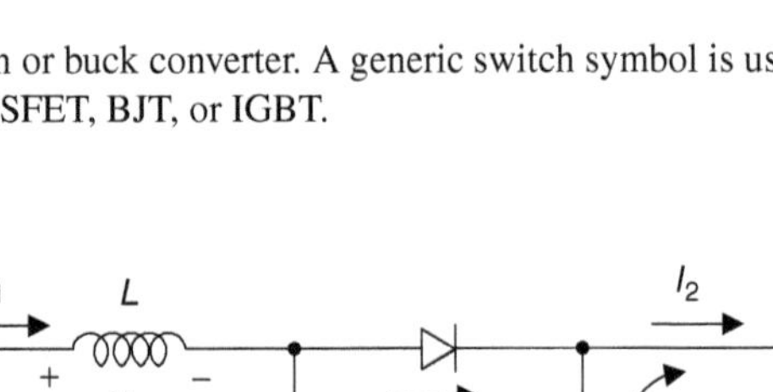

Figure 11.8 A step-down or buck converter. A generic switch symbol is used for the transistor, which may be either a JFET, MOSFET, BJT, or IGBT.

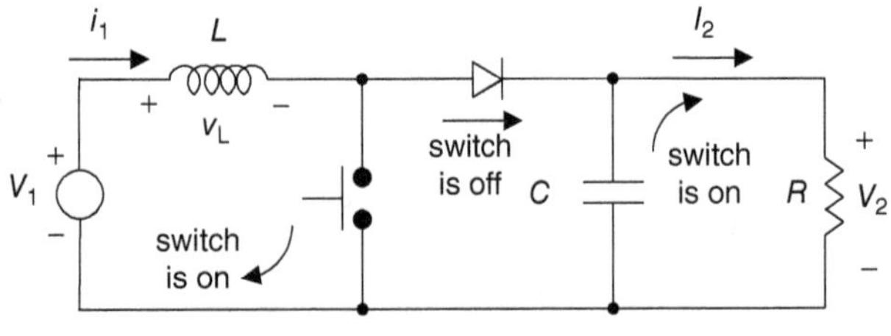

Figure 11.9 A step-up or boost converter.

change instantaneously, and flows through the diode to the capacitor and load resistor. With the transistor on, the diode prevents the capacitor from discharging through the transistor, and the capacitor instead supplies current to the load. In this way the inductor and capacitor function as a low-pass filter, keeping the current through the load constant, provided the period of the switching waveform is short compared to the RC and LC time constants of the circuit.

In steady state, the integral of the inductor voltage over one period must be zero. To see this, recall that

$$v_L = L\frac{\partial i_1}{\partial t} \tag{11.3}$$

Cross-multiplying and integrating over one period yields

$$\int_0^T v_L \mathrm{d}t = L\int_{i_1(0)}^{i_1(T)} \mathrm{d}i_1 = 0 \tag{11.4}$$

since in steady state $i_1(T) = i_1(0)$. When the transistor is on, $v_L = V_1$, and when the transistor is off, $v_L = V_1 - V_2$ (neglecting voltage drops across the transistor and diode). Thus Equation 11.4 can be written

$$(V_1)(\delta)T + (V_1 - V_2)(1 - \delta)T = 0 \tag{11.5}$$

from which we obtain

$$V_2 = \frac{V_1}{(1 - \delta)} \tag{11.6}$$

Since $\delta \geq 0$, we are guaranteed that $V_2 \geq V_1$, hence the name "boost converter." Note that V_2 can become arbitrarily large as δ approaches unity.

It is sometimes necessary to provide an output voltage that may be either larger or smaller than the input voltage. This can be done using the step-down/step-up or buck/boost converter shown in Figure 11.10. This converter is obtained from the circuit of Figure 11.8 by interchanging the inductor and diode. When the transistor is on, current flows through the inductor, storing reactive energy. When the transistor turns off, the inductor current cannot change instantaneously, and flows through the capacitor and load resistor and back through the diode. With the transistor on, the diode prevents source current from flowing to the load, and the load current is supplied by the capacitor. Note that the polarity of the output voltage is opposite to the two previous converters. Since the integral of the inductor voltage over one period must be zero, we can write

$$(V_1)(\delta)T - (V_2)(1 - \delta)T = 0 \tag{11.7}$$

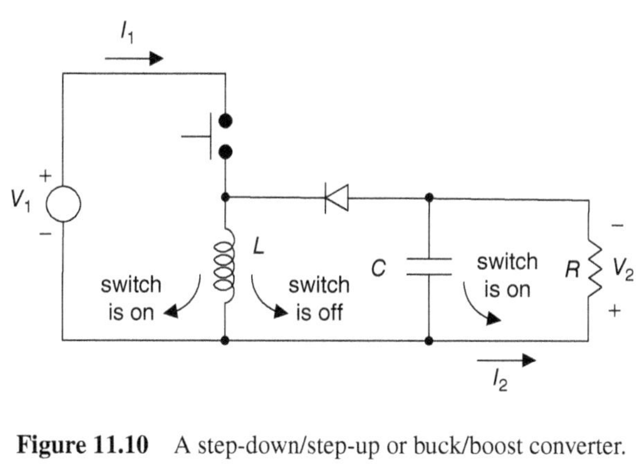

Figure 11.10 A step-down/step-up or buck/boost converter.

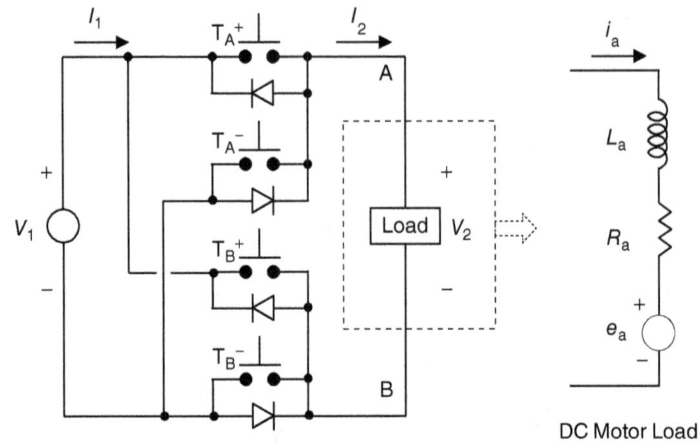

Figure 11.11 A full-bridge DC–DC converter.

where we have again neglected voltage drops across the transistor and diode. Solving for V_2 gives

$$V_2 = \left(\frac{\delta}{1-\delta}\right) V_1 \tag{11.8}$$

If $0 \le \delta < 0.5$, $V_2 \le V_1$ and the circuit functions as a buck converter, while if $0.5 < \delta \le 1$, $V_2 \ge V_1$ and the circuit functions as a boost converter.

The final DC converter to be considered is the full-bridge converter of Figure 11.11. This same basic circuit configuration appears often in power electronics, and is used in switch-mode inverters to be discussed in the next section. Here we consider only the conversion of unregulated DC to regulated DC. In this implementation, one of the transistors in the T_A^+/T_A^- pair is on and the other off at all times, and similarly for the transistors in the T_B^+/T_B^- pair. Switching of transistors in the T_A^+/T_A^- pair occurs simultaneously with switching of transistors in the T_B^+/T_B^- pair. In this way, the load is continually connected to the source, either directly connected through T_A^+ and T_B^-, or cross-connected through T_A^- and T_B^+. The diodes are used to clamp excursions in the load voltage that exceed the source voltage, either positive or negative.

In DC converter applications, the load contains one or more energy storage elements, such as the inductance of a DC motor winding illustrated in Figure 11.11. The transistors in the converter are switched at a frequency whose period is short compared to the RL time constant of the load. If the load is a DC motor, this ensures that the load current and the emf e_A induced in the motor winding are DC quantities.

The magnitude and polarity of V_2 are determined by the duty factor δ of the switching waveforms. The time-average load voltage can be written

$$V_2 = \frac{(V_1)(\delta)T - (V_1)(1-\delta)T}{T} = (2\delta - 1)V_1 \tag{11.9}$$

Equation 11.9 shows that the average load voltage varies linearly with the duty factor, increasing from $-V_1$ when $\delta = 0$, to $+V_1$ when $\delta = 1$. Thus, both polarities of output voltage can be obtained simply by varying the duty factor, independent of the direction of the current. Consider the DC motor load shown in Figure 11.11. If the induced emf e_A exceeds the source voltage, as happens during regenerative braking, the current is in the opposite direction to that indicated in the figure, and power flows from the load back to the source. Thus the full-bridge converter is capable of operation in all four quadrants of the I_2–V_2 plane, as illustrated in Figure 11.12.

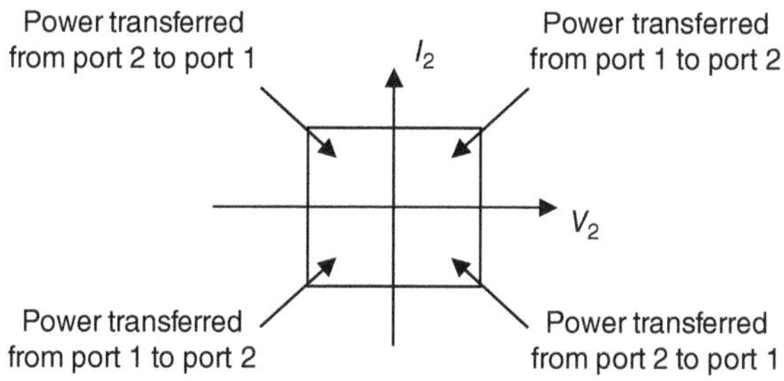

Figure 11.12 Operation of the full-bridge DC–DC converter in the I_2-V_2 plane. In the first or third quadrants power is transferred from port 1 to port 2, while in the second or fourth quadrants power is transferred from port 2 to port 1.

As an aside, we note that four-quadrant operation is available in this converter even if the transistors have a preferential direction for current flow. For example, BJTs and IGBTs have much higher gain in the forward direction than in the reverse direction. In the full-bridge converter, BJTs or IGBTs are connected so that their preferential current direction is opposite to that of their parallel diode. This way the diode carries most of the current when the current flow is opposite to the preferential direction of the transistor.

11.2.3 Switch-Mode Inverters

Switch-mode inverters convert unregulated DC into sinusoidal AC of variable amplitude and frequency. Typical applications are in AC motor drives and uninterruptible AC power supplies. If the application calls for conversion from line-frequency AC, the generic power processor of Figure 11.1 will consist of a rectifier as converter 1 and a switch-mode inverter as converter 2. The switch-mode inverter uses pulse-width-modulated (PWM) switching to synthesize a sine wave output.

Figure 11.13 shows a single-phase half-bridge switch-mode inverter. The DC source V_1 is bridged by two equal capacitors, with each capacitor charging to a voltage of $\frac{1}{2} V_1$. We assume the capacitors are large enough that their voltages remain essentially constant during one cycle. Transistors T_A^+ and T_A^- are switched with opposite polarity signals so that at any given time one transistor in the pair is on and

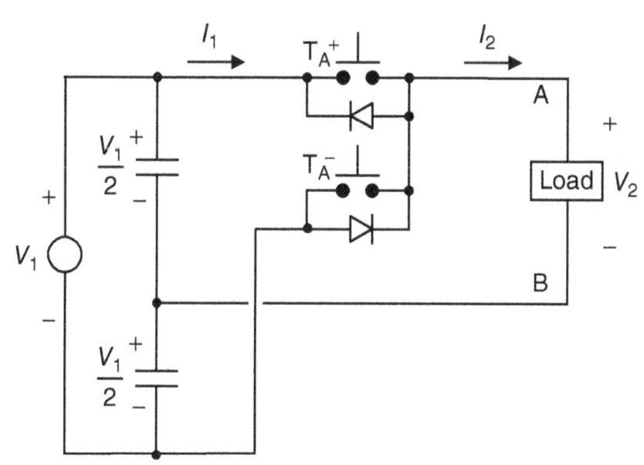

Figure 11.13 A single-phase half-bridge switch-mode inverter. The switching waveform is modulated to synthesize a sine wave across the load, as shown in Figure 11.14.

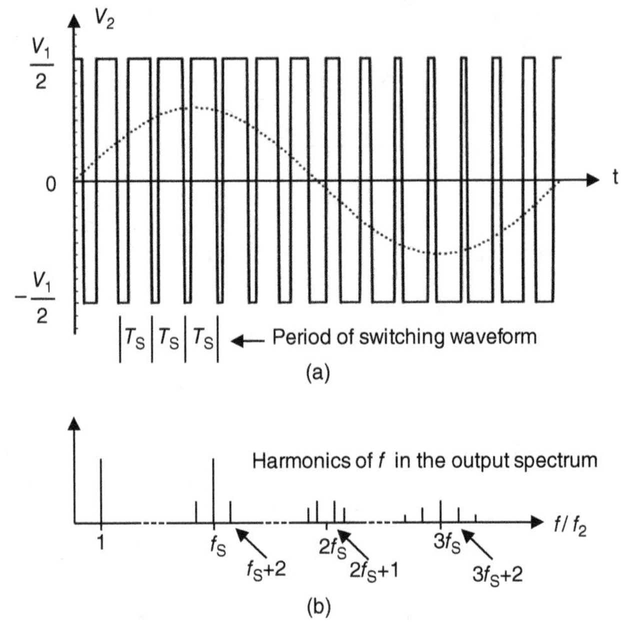

Figure 11.14 Waveforms of the half-bridge switch-mode inverter (a) and its frequency spectrum (b). The fundamental Fourier component of the output waveform is the dashed sinusoid at frequency f_2.

the other off. When T_A^+ is on, the A terminal of the load is connected to $+V_1$ and the B terminal to the capacitor midpoint at $\frac{1}{2}V_1$. When T_A^- is on, the A terminal of the load is connected to $-V_1$ and the B terminal to the capacitor midpoint at $\frac{1}{2}V_1$. Thus the load voltage V_2 switches between $+\frac{1}{2}V_1$ and $-\frac{1}{2}V_1$, as shown in Figure 11.14. The V_2 output waveform is pulse-width modulated at frequency f_S to synthesize a sine wave of frequency f_2 at the output.

The harmonic content of the output waveform can be obtained by Fourier analysis and contains components at the fundamental frequency f_2, and at integral multiples of f_2 with the (much higher) switching frequency f_S, along with sidebands as shown in the figure. It is desirable that $f_S \gg f_2$ so that these harmonics are well above the response capability of the load being driven. When this is the case, the load responds as if driven by the fundamental Fourier component.

The frequency f_S of the switching waveform should satisfy the following criteria:

1. $f_S \gg f_2$, where f_2 is the frequency of the synthesized sine wave. Higher f_S values push the harmonics to higher frequencies and make the load respond as if driven by a pure sinusoid at the fundamental frequency f_2.
2. For $(f_S/f_2) \leq 21$, f_S should be an *odd integer multiple* of f_2. This eliminates the even harmonics of f_S from the Fourier analysis (as shown in Figure 11.14) so that only odd harmonics are present, that is, $f_S, 3f_S, 5f_S$ … Although the even harmonics are removed, their sidebands remain, but they have lower amplitude and are less disruptive.
3. For $(f_S/f_2) > 21$ the harmonics are small, and f_S does not have to be an integer multiple of f_2, that is, the switching waveform and the output waveform may be asynchronous. This makes it possible to vary the frequency of the output waveform without changing the switching waveform. (An exception occurs when driving an AC motor, since even small sub-harmonics can produce undesirably large stator currents.)

4. f_s should lie outside the audible frequency range. In most cases f_S is chosen to be either below 6 kHz or above 20 kHz. Higher f_S produces a higher quality sine wave, but leads to proportionally higher switching losses in the transistors of the converter.

For low-frequency applications ($f_2 \leq 200$ Hz), f_S/f_2 may be in the range 9–15, whereas for high-frequency applications ($f_2 \geq 20$ kHz), f_S/f_2 may be larger than 100. The switching frequency is an important parameter in selecting the optimum device for the application, especially when high blocking voltages are required, since switching loss in bipolar devices such as BJTs, IGBTs, and thyristors is often the dominant loss, and is proportional to f_S.

Figure 11.15 shows a single-phase full-bridge switch-mode inverter. This is the same circuit as the full-bridge DC converter of Figure 11.11, with the only operational difference being the PWM waveforms applied to the transistors. Like the converter of Figure 11.11, the full-bridge inverter of Figure 11.15 is capable of operation in all four quadrants, permitting bidirectional power flow. The full-bridge inverter can also be obtained from the half-bridge inverter of Figure 11.13 by connecting the B terminal of the load to the positive and negative terminals of the source through a second pair of switching transistors T_B^+ and T_B^-.

There are two possible PWM schemes for the full-bridge inverter. In *bipolar* modulation, the waveforms applied to transistors T_A^+ and T_A^- in the full-bridge inverter are the same as for the half-bridge inverter, while the waveforms applied to transistors T_B^+ and T_B^- are the inverse. The control waveform thus consist of two sub-periods. During the first sub-period, transistors T_A^+ and T_B^- are on while transistors T_A^- and T_B^+ are off. This connects the load directly across the source so that $V_2 = +V_1$. During the second sub-period, transistors T_A^+ and T_B^- are off while transistors T_A^- and T_B^+ are on. This cross-connects the load to the source so that $V_2 = -V_1$. As a result, the output voltage V_2 alternates between $+V_1$ and $-V_1$ (hence the name "bipolar"), and the fraction of time allotted to each state is modulated to synthesize a sine wave output similar to that of Figure 11.14, with one important difference: Because of the symmetrical transistor configuration, the output voltage swing of the full-bridge inverter, shown in Figure 11.16, is twice as large ($\pm V_1$) as the half-bridge inverter ($\pm \frac{1}{2} V_1$). Thus the full-bridge inverter can deliver the same output power using half the current. This is an important advantage, since it reduces the need for paralleling devices in high-power applications.

The second PWM scheme for the full-bridge inverter is called *unipolar* modulation. This scheme has the same effect as doubling the switching frequency, in terms of the harmonics present in the output waveform, without actually changing the frequency f_S at which the transistors are switched. Stated

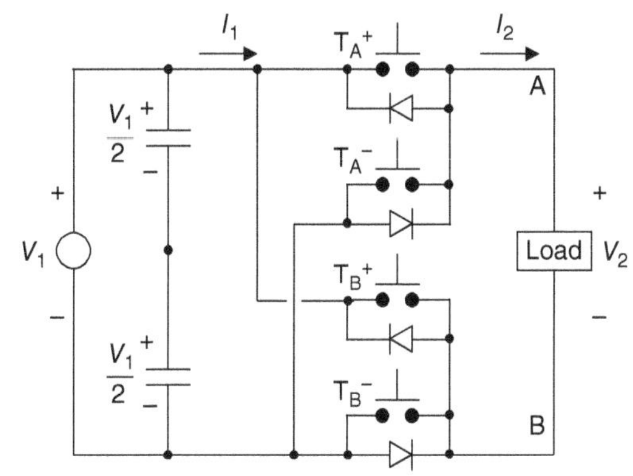

Figure 11.15 A single-phase full-bridge switch-mode inverter. The control waveforms are designed to synthesize a sinusoidal output, shown dashed in Figure 11.16.

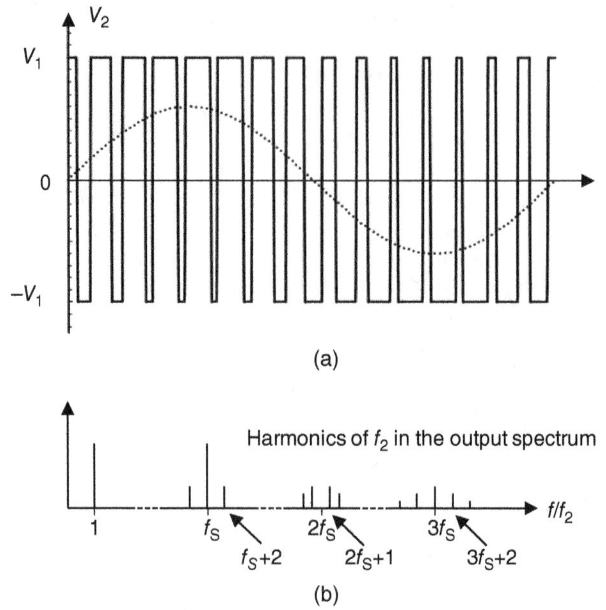

Figure 11.16 Waveforms (a) and output spectrum (b) of the full-bridge switch-mode inverter when operated using bipolar PWM.

another way, with unipolar modulation one can achieve the same harmonic content at *half* the switching frequency of bipolar modulation. This is a significant advantage, since transistor switching loss is proportional to switching frequency. Unipolar modulation requires separately-timed control signals to transistors T_A^+ and T_A^- and transistors T_B^+ and T_B^-, rather than inverse-polarity signals with the same timing, as in bipolar modulation. The resulting output waveform, shown in Figure 11.17, steps between $+V_1$ and zero during the first half-cycle and between $-V_1$ and zero during the second half-cycle, hence the name "unipolar". In unipolar modulation, the ratio (f_S/f_2) should be an even integer, as compared to an odd integer in bipolar modulation. With this choice, all odd harmonics are eliminated from the output spectrum along with their sidebands, as seen in Figure 11.17. What remains are the *sidebands* of the even harmonics at $2(f_S f_2)$, $4(f_S f_2)$, and so on. Note that the principal harmonics at $2(f_S f_2)$, $4(f_S f_2)$, and so on are also suppressed, and only their sidebands remain.

Many applications, such as uninterruptible AC power supplies and AC motor drives, require three-phase AC outputs. Figure 11.18 shows a three-phase switch-mode inverter driving a three-phase AC motor. The three-phase inverter can be envisioned as three single-phase half-bridge inverter sections of the type shown in Figure 11.13, driven by waveforms of the type shown in Figure 11.14, where the control waveforms for the a, b, and c phases are displaced 120° with respect to each other. Unlike the single-phase half-bridge inverter of Figure 11.13, the line-to-neutral voltages of the three-phase inverter V_{AN}, V_{BN}, and V_{CN} switch between $+V_1$ and zero rather than from $+\frac{1}{2} V_1$ to $-\frac{1}{2} V_1$. Because of the 120° relationship between phases, the line-to-line voltages V_{AB}, V_{BC}, and V_{CA} swing between $+V_1$ and $-V_1$. Since one of the two switches in each leg is always on at any instant in time, the output voltage is independent of the magnitude and direction of the output current. Thus, this inverter is capable of operating in all four quadrants, permitting bidirectional power flow.

In three-phase inverters, the only harmonics of concern are those of the line-to-line voltages. The frequency spectra of the line-to-neutral voltages are the same as the half-bridge inverter shown in

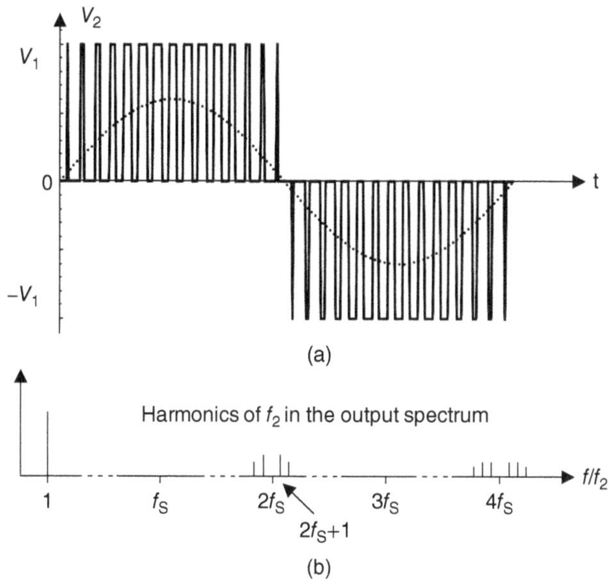

Figure 11.17 Waveforms (a) and output spectrum (b) of the full-bridge switch-mode inverter when operated using unipolar PWM.

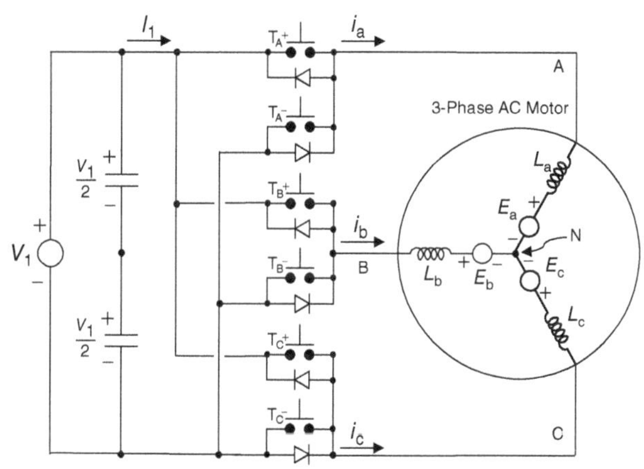

Figure 11.18 A three-phase switch-mode inverter driving a three-phase AC motor.

Figure 11.14. However, when the line-to-neutral signals are combined algebraically to obtain the line-to-line voltages, their 120° phase shift results in cancellations that eliminate some of the harmonics. This is particularly true if the ratio f_S/f_2 is an odd integer multiple of three (i.e., 3, 9, 15 ...), since this removes all principal harmonics of f_2 from the spectrum, leaving only their sidebands. An example of a line-to-line waveform and the associated frequency spectrum are shown in Figure 11.19.

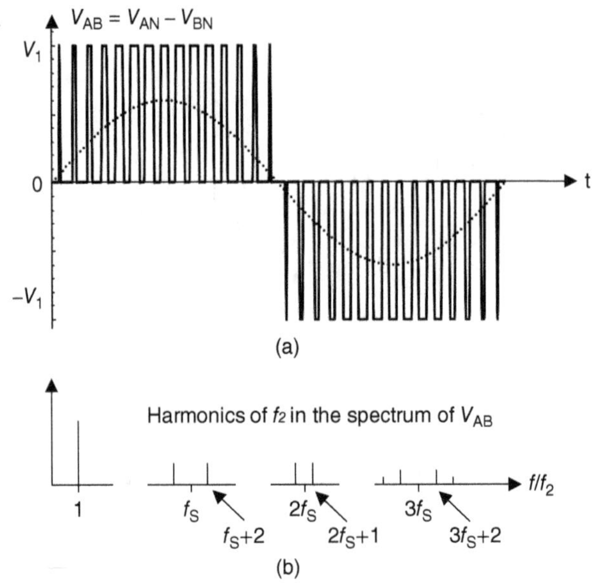

Figure 11.19 Line-to-line waveforms (a) and output spectrum (b) of the three-phase switch-mode inverter of Figure 11.18.

The switching waveform of the three-phase switch-mode inverter should satisfy the following criteria:

1. For $(f_S/f_2) \leq 21$, (f_S/f_2) should be an odd integer multiple of 3 (i.e., 3, 9, 15, or 21). This is required to cancel all direct harmonics of f_2 from the output spectrum.
2. For $(f_S/f_2) > 21$ the harmonics are small, and f_S does not have to be an integer multiple of f_2, that is, the switching waveform and the output waveform may be asynchronous. This makes it possible to vary the frequency of the output waveform without changing the switching waveform. (An exception occurs when driving an AC motor, since even small sub-harmonics can produce undesirably large stator currents.)

11.3 Power Electronics for Motor Drives

11.3.1 Introduction to Electric Motors and Motor Drives

Electric motors can be classified into three primary types: DC motors, induction (or asynchronous) motors, and synchronous motors. The drive requirements for the three types are different, and will be considered below. Electric motor applications range from low power (a few watts) to very high power (megawatts), and from high precision, such as servo drives for robotics, to less critical applications, such as adjustable speed drives for pumps and fans. The applications may call for single-quadrant operation (motoring), two-quadrant operation (motoring plus regenerative braking), or four-quadrant operation (reversible motoring and braking). All of these factors play a role in the design and performance specifications of the electronic motor drive.

In general, the current rating of the motor drive is dictated by the torque required of the motor in the particular application, since electromechanical torque is proportional to current. The voltage rating of the motor drive is determined by the rotational speed and controllability requirements, based on the following

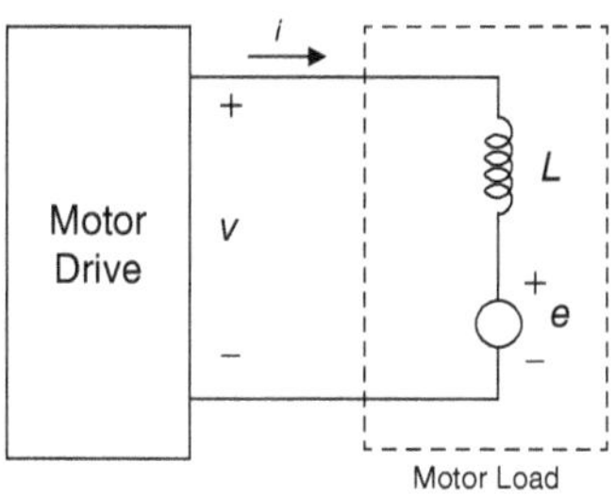

Figure 11.20 Generic equivalent circuit of a motor and its associated drive circuit.

considerations. In both DC and AC motors, rotation produces a back-emf in the motor windings, and the equivalent circuit presented by the motor to the drive circuit can be represented as a voltage source (the back-emf) in series with the winding inductance, as illustrated in Figure 11.20. The rate of change in current (and therefore torque) is given by

$$\frac{\partial i}{\partial t} = \frac{v - e}{L} \tag{11.10}$$

where v is the output voltage of the drive, e is the back-emf of the motor, and L is the inductance of the motor windings. The back-emf, in turn, is proportional to the rotational speed of the rotor. To achieve a short response time to speed and position commands, the output voltage of the drive v must exceed the back-emf e by a sufficient margin. Thus the voltage rating of the motor drive is determined by the speed of the motor (through the back-emf) and the rate at which motor torque needs to be changed. We now consider drive circuits for the three primary types of motors.

11.3.2 DC Motor Drives

DC motors are typically used for speed and position control in applications where low initial cost and good performance characteristics are desired. In DC motors, the stator establishes a stationary magnetic field using either permanent magnets or stator field windings. When the field is provided by windings, the stator current controls the field flux ϕ_F. If magnetic saturation is neglected, the field flux is proportional to the field current,

$$\phi_F = k_F I_F \tag{11.11}$$

where k_F is the field constant of the motor. The rotor carries the armature windings that supply variable power to the motor and the load. The armature windings are connected to a segmented copper commutator that rotates with the shaft and is contacted by stationary carbon brushes mounted on the stator.

In a DC motor, the electromechanical torque is produced by an interaction between the stator's field flux and the rotor's armature flux. The rotor flux is proportional to the armature current, and the electromechanical torque can be written

$$\boxed{T_{EM} = k_T \phi_F i_A} \tag{11.12}$$

where k_T is the torque constant of the motor. In addition, a back-emf is induced in the armature windings by their rotation through the stator field. This back-emf is proportional to the field flux and the angular velocity,

$$\boxed{e_A = k_E \phi_F \omega_M} \tag{11.13}$$

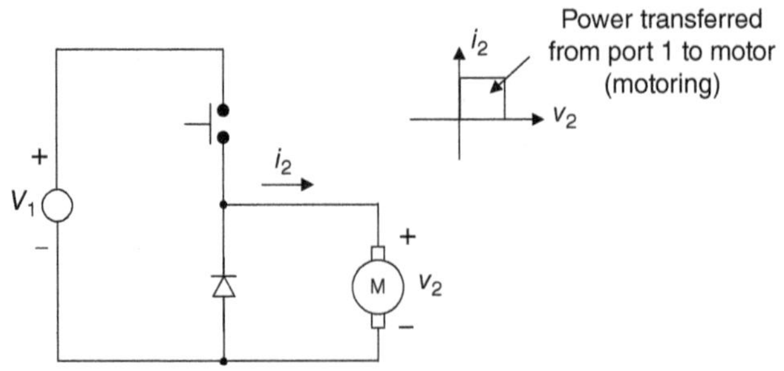

Figure 11.21 A DC motor with single-quadrant operation driven by a simple buck converter similar to that of Figure 11.8.

where k_E is the voltage constant of the motor and ω_M is the angular velocity of the rotor. Setting the electrical power delivered to the motor ($e_A\, i_A$) equal to the mechanical power delivered by the motor to the load ($\omega_M\, T_{EM}$), we find that $k_T = k_E$, with k_T in units of [N m / A Wb] and k_E in units of [V s / Wb].

The right side of Figure 11.11 shows the equivalent circuit of the armature windings in a DC motor, where R_A is the winding resistance, L_A the winding self-inductance, and e_A the back-emf given by Equation 11.13. In the normal mode of operation, i_A and e_A are positive, and the motor produces positive torque, Equation 11.12, and a positive rotational velocity, Equation 11.13, delivering mechanical power $\omega_M\, T_{EM}$ to the load. However, it is often desirable to use the motor for regenerative breaking. To do this, the terminal voltage V_2 is reduced below the induced emf e_A so that the direction of current is reversed, that is, i_A becomes negative. This has the effect of reversing the torque, Equation 11.12, thereby slowing the rotation of the motor and the load. In addition, the mechanical power delivered to the load $\omega_M\, T_{EM}$ and the electrical power drawn from the source $e_A\, i_A$ both become negative, which represents net power taken from the kinetic energy of the load and returned to the source (note that e_A remains positive, since the rotational velocity ω_M has not changed sign). Eventually the back-emf is reduced to zero when the motor comes to a stop. If the terminal voltage V_2 is made negative, the torque is negative, and the motor rotates in the opposite direction, producing a negative back-emf. Thus it is possible to reverse the direction of a DC motor simply by reversing the voltage and current polarities applied to the armature, and the DC motor is capable of operating in all four quadrants of Figure 11.12.

The selection of a power converter to drive a DC motor depends on whether single-quadrant, two-quadrant, or four-quadrant operation is desired. If the rotation is unidirectional and braking is not required, single-quadrant operation can be provided by the simple buck converter of Figure 11.21. If rotation is unidirectional but braking is required, two-quadrant operation can be obtained using the converter of Figure 11.22. In this circuit, T_A and T_B are switched so that only one is on at any time. When T_A is on and T_B is off, i_2 and v_2 are positive (motoring). When T_B is on and T_A is off, the back-emf of the motor causes i_2 to reverse direction (braking).

Applications that require reversible-speed operation at moderate power, along with regenerative braking, call for a four-quadrant converter such as the full-bridge DC converter of Figure 11.11. An approach for high-power, fully reversible applications is to connect two line-frequency phase-controlled inverters of the type shown in Figure 11.5 in anti-parallel to achieve four-quadrant operation, as shown in Figure 11.23. For forward rotation, converter 1 operates in the rectifier mode for motoring, while converter 2 operates in the inverter mode for braking. For reverse rotation, converter 2 operates in the rectifier mode for motoring and converter 1 operates in the inverter mode for braking.

11.3.3 Induction Motor Drives

Induction motors or "asynchronous motors" are AC motors that supply power to the rotor by electromagnetic induction rather than by brushes or slip rings. Induction motors are preferred in applications

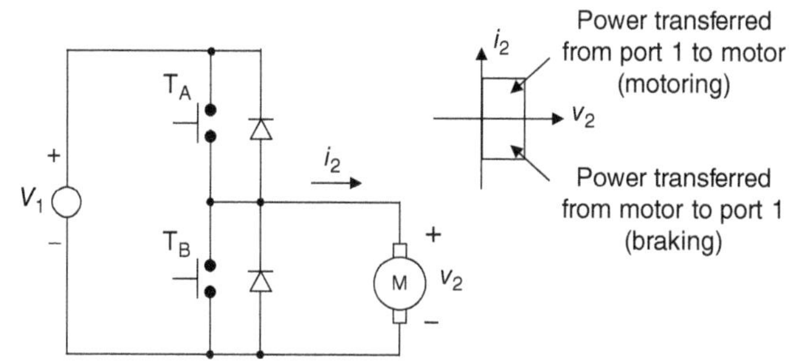

Figure 11.22 A DC motor with unidirectional rotation and regenerative braking, driven by a simple two-quadrant converter.

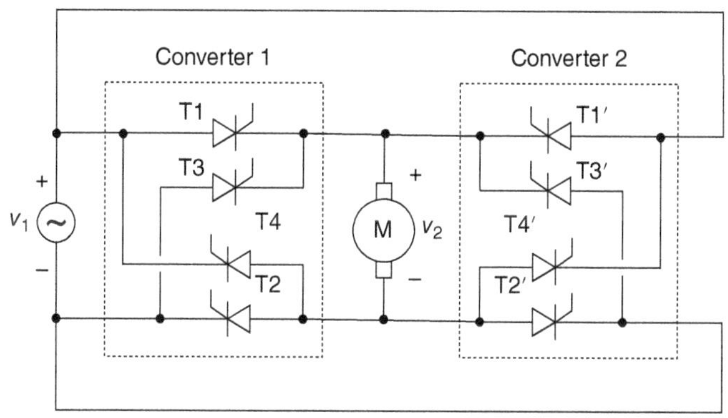

Figure 11.23 A DC motor with reversible rotation and regenerative braking, driven by two anti-parallel line-frequency phase-controlled converters of the type shown in Figure 11.5.

where low cost and rugged construction are desired. They operate at nearly constant rotational speed determined by the angular frequency of the AC drive signal.

Most induction motors are driven by a three-phase AC power source. The stator of an induction motor usually contains multiple poles of three-phase windings, as illustrated in Figure 11.24. The wiring diagram for the stator windings is shown at the top. Phase 1 produces four poles, two "N" and two "S". In the motor diagram, stator current is flowing away from the reader in the "1+" segments and toward the reader in the "1−" segments, producing magnetic field lines that penetrate the rotor. The rotor itself has no external electrical connections, and can be one of three types. Squirrel-cage rotors have a series of conducting bars around the periphery, oriented parallel to the rotor axis and shorted at each end by conducting rings, thereby forming a cage-like structure. Slip-ring rotors have windings connected to slip rings that replace the bars of a squirrel-cage design. Solid-core rotors are made from magnetically soft steel.

In operation, the stator windings create a rotating magnetic field that rotates at the synchronous speed ω_S given by

$$\omega_S = 2\omega/p \tag{11.14}$$

where ω is the angular frequency of the driving voltage and p is the number of poles. The field lines from the stator penetrate the rotor, inducing currents in the bars or windings of the rotor. These currents, in turn, create a rotor field that rotates at the synchronous speed with respect to the stator. However, the rotor itself does not rotate at the synchronous speed, because if it did there would be no relative motion

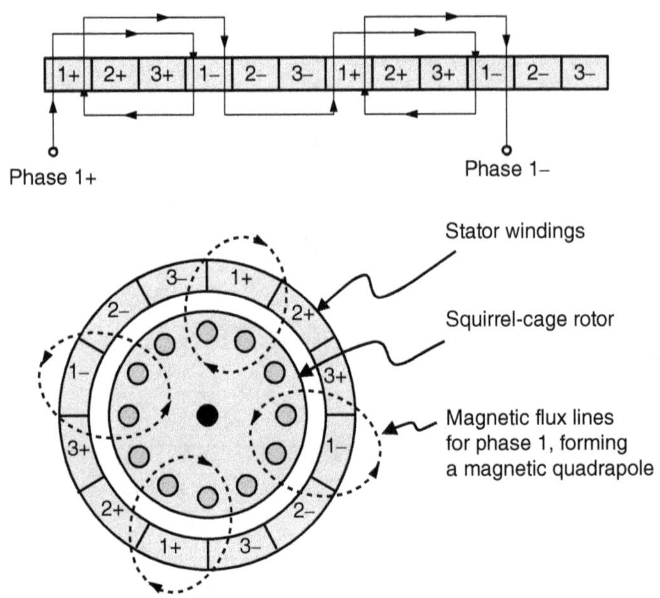

Figure 11.24 Schematic diagram of a four-pole three-phase squirrel-cage induction motor.

between the rotor and the rotating stator field, and hence no induced currents in the rotor. Instead, the rotor rotates in the same direction as the stator field, but at a speed ω_R that is slightly less than ω_S. This means the rotor is "slipping" with respect to the stator field at a relative speed, called the "slip speed" ω_{SL}, given by

$$\omega_{SL} = \omega_S - \omega_R \tag{11.15}$$

It is customary to refer to the "slip" of the motor, where the slip s is the normalized slip speed defined by

$$s \equiv (\omega_S - \omega_R)/\omega_S \tag{11.16}$$

The rotor field is synchronous with the stator field, but rotates at speed ω_{SL} with respect to the rotor, since the rotor is slipping by that amount with respect to the stator field.

The electrical response of the motor can be represented by the per-phase equivalent circuit of Figure 11.25, where V_S is the rms line-to-line voltage of the three-phase motor drive and I_S is the rms phase current. Here E_{AG} is the back-emf induced in the stator windings by the rotor field. The stator current I_S consists of two components: I_M is the portion of the stator current that establishes the air-gap flux, and I_R represents the interaction with the rotor field that produces the electromechanical torque. R_S and L_S are the resistance and self-inductance of the stator windings, and L_M is the magnetizing inductance of the stator windings. R_R and L_R are the resistance and inductance of the rotor. R_R/s is the effective resistance of the rotor as reflected back into the stator circuit, where s is the slip between the rotor and the stator fields. For small values of slip corresponding to normal operation, the rotor current I_R is small compared to the magnetizing current I_M, and R_S and L_S can be neglected, making $V_S \approx E_{AG}$.

The electromagnetic torque is produced by the interaction between the rotating stator field flux ϕ_{AG} and the flux of the rotor field. Since the rotor field is proportional to the rotor current, we can write

$$T_{EM} \approx k_T \phi_{AG} I_R \tag{11.17}$$

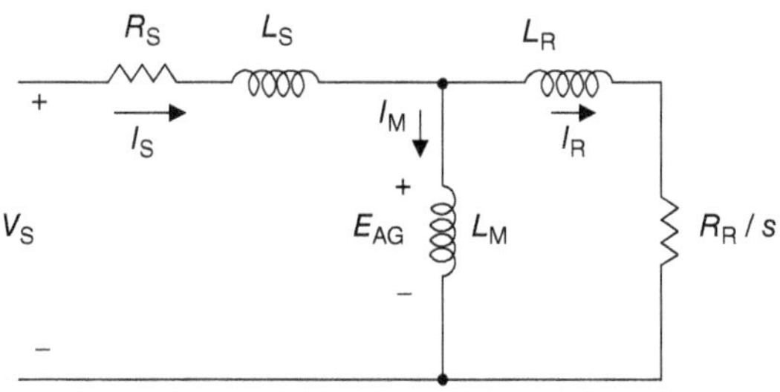

Figure 11.25 Per-phase equivalent circuit of an induction motor.

where k_T is the torque constant of the motor. Equation 11.17 for the induction motor is analogous to Equation 11.12 for the DC motor. In the induction motor, the induced rotor current I_R is proportional to the air-gap flux ϕ_{AG} and the slip frequency ω_{SL}, so Equation 11.17 can also be written

$$\boxed{T_{EM} \approx k'_T \phi_{AG}^2 \omega_{SL} = k'_T \phi_{AG}^2 s \omega_S} \tag{11.18}$$

For zero slip (synchronous rotation), the effective resistance R_R/s in the rotor leg of the equivalent circuit of Figure 11.25 is infinite and no rotor current flows, meaning no torque is delivered to the load, as indicated by Equations 11.17 and 11.18. Torque is only produced when the rotor speed ω_R is less than the synchronous speed ω_S, and Equations 11.18 and 11.16 tells us the torque is proportional to this difference.

Under normal operating conditions, the voltage drops across the stator winding resistance R_S and self-inductance L_S are small compared to the back-emf E_{AG}. The back-emf, in turn, is proportional to the stator field and the angular frequency. Thus, we can write

$$\boxed{V_S \approx E_{AG} = k_E \phi_{AG} \omega_S} \tag{11.19}$$

where k_E is the voltage constant of the motor. Equation 11.19 for the induction motor is analogous to Equation 11.13 for the DC motor.

The torque-speed relation given by Equation 11.18 is only valid for small values of slip, which is the normal operating regime of the motor. The full torque-speed curve is illustrated in Figure 11.26. The *rated torque* and *rated speed* are normally specified on the nameplate of the motor, and correspond to the upper limit of torque (and lower limit of speed) for which the linear relationship in Equation 11.18 applies. The steady-state speed is determined by the intersection of the torque-speed curves of the load and the motor, as shown in Figure 11.27 for different values of synchronous frequency ω_S. The synchronous frequency is controlled by the power electronic converter driving the motor. For the torque provided by the motor to equal the rated torque at any frequency, Equation 11.18 requires that the air-gap flux remain constant as frequency is varied. Equation 11.19 then shows that the source voltage V_S provided by the drive electronics must scale directly with frequency ω_S.

Figure 11.27 also allows us to visualize how an induction motor accelerates from a dead stop. At zero rotation ($\omega_R = 0$) the motor torque exceeds the load torque by a wide margin, resulting in angular acceleration of the motor and load. As the rotor speed increases, the motor and load torques eventually balance, resulting in a steady-state rotation at a speed determined by the synchronous frequency of the drive electronics.

Induction motors are also capable of electromagnetic braking, which occurs when the rotor speed ω_R exceeds the rotational speed of the stator field ω_S. In this mode the induction motor acts as a generator, transferring power from the rotating shaft back to the source. This can be understood by extending the

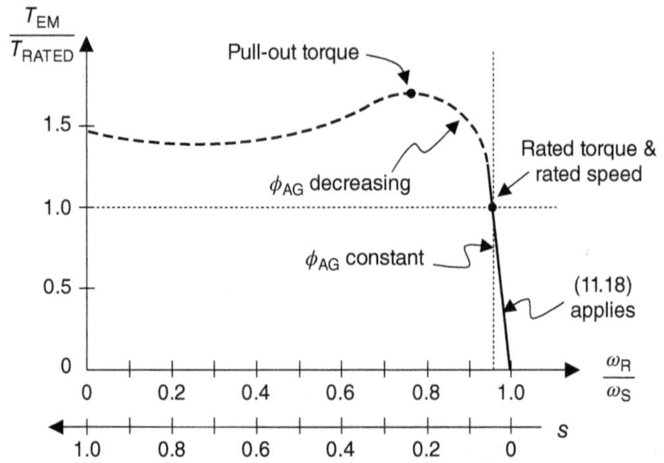

Figure 11.26 Torque-speed curve of an induction motor. The normal operating region is the linear regime at the far right, where the slip is small and the air-gap flux is constant.

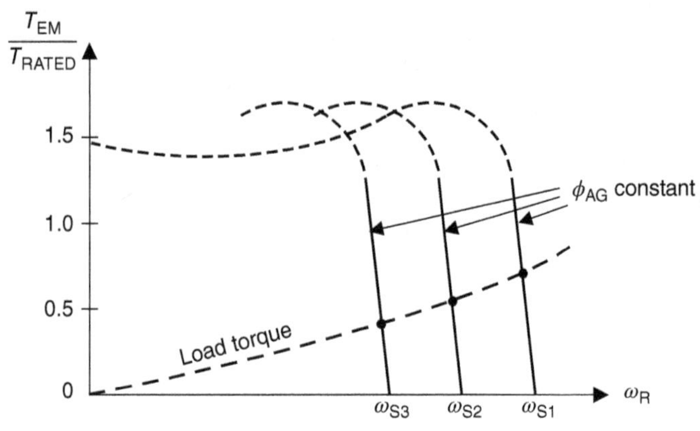

Figure 11.27 Torque-speed diagram of an induction motor and its load. The steady-state operating point is the intersection of the load torque and motor torque curves, and can be modulated by varying the frequency ω_S of the driving waveform.

torque-speed curve of Figure 11.26 to rotational speeds greater than ω_S, as shown in Figure 11.28. For $\omega_R > \omega_S$ the torque is negative, which decelerates the rotor and transfers inertial energy back into the power source. Consider a specific example. Suppose the motor is initially operating with positive torque at a steady-state speed $\omega_R = 0.95\,\omega_S$. If the drive frequency is reduced to $\omega_S' = 0.9048\,\omega_S$, the rotor speed (assuming it has not yet changed) is given by $\omega_R = 0.95\,\omega_S = 0.95\,(\omega_S'/0.9048) = 1.05\,\omega_S'$. The rotor speed now exceeds the synchronous speed of the stator field, placing the motor in the negative-torque regime, and the motor decelerates until the torque again becomes positive. The motor can be decelerated smoothly to a full stop if the stator frequency is reduced gradually.

Based on the above considerations, power electronic converters for induction motor drives must produce three-phase AC with variable frequency and magnitude. The frequency is adjusted to control the

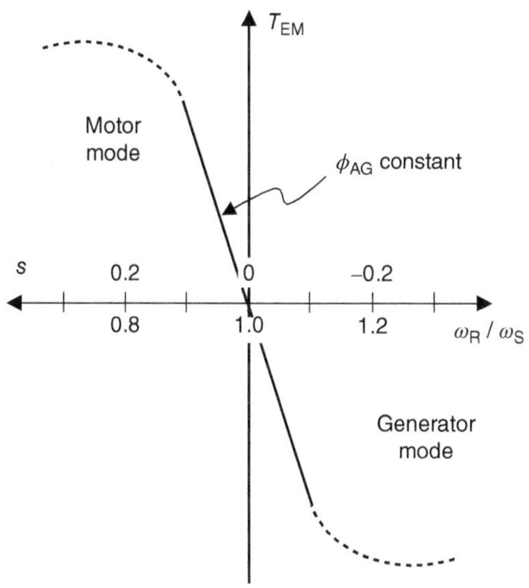

Figure 11.28 Torque-speed diagram of an induction motor for rotor speeds above and below the synchronous speed. For $\omega_R > \omega_S$ the rotor currents, torque, and slip are negative, and the motor is in the generator mode, transferring inertial energy back to the source.

speed of the motor, and the magnitude is adjusted to insure constant air-gap flux as the speed is varied, in accordance with Equation 11.19. If electromagnetic braking is desired, the converter must be capable of at least two-quadrant operation.

Referring to the generic power processor of Figure 11.1, the induction motor drive may utilize either a diode rectifier or a four-quadrant switch-mode converter as converter 1, depending on whether regenerative braking is required. The three-phase switch-mode inverter of Figure 11.18 can be used as converter 2 to provide variable-frequency, variable-magnitude AC power to the motor. This inverter is capable of four-quadrant operation, allowing for electromagnetic braking. During braking, inertial energy is absorbed from the load and transferred back to the power electronics through converter 2. This energy must either be dissipated within the power processor or returned to the AC source. If dissipative braking is used, a resistor is switched in parallel with the filter capacitor in the power processor during braking, as illustrated in Figure 11.29a. If regenerative braking is employed, a four-quadrant switch-mode converter must be used as converter 1, allowing power to be returned to the AC line, as shown in Figure 11.29b.

11.3.4 Synchronous Motor Drives

Synchronous motors are AC motors that employ a permanent-magnet rotor or a DC-wound rotor along with a three-phase AC stator, where the rotor rotates at the synchronous speed of the stator field. Synchronous motors are used as servo drives in robotic applications and as adjustable-speed drives in high-power applications. Low-power applications usually employ permanent-magnet synchronous motors, also called "brushless DC" motors, while high-power applications use wound-rotor synchronous motors, also called "salient-pole" motors. Figure 11.30 illustrates these two motor types.

In both brushless and wound-rotor motors, the rotor field flux ϕ_F is stationary with respect to the rotor, and rotates at the synchronous speed ω_S with respect to the stator and motor housing. The angular frequency of the stator currents is $\omega = (p/2)\,\omega_S$, where p is the number of poles on the rotor (two are

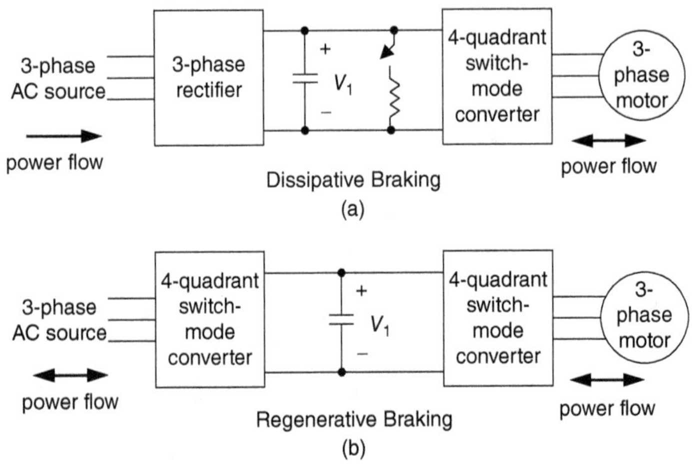

Figure 11.29 Induction motor drives capable of dissipative braking (a) and regenerative braking (b).

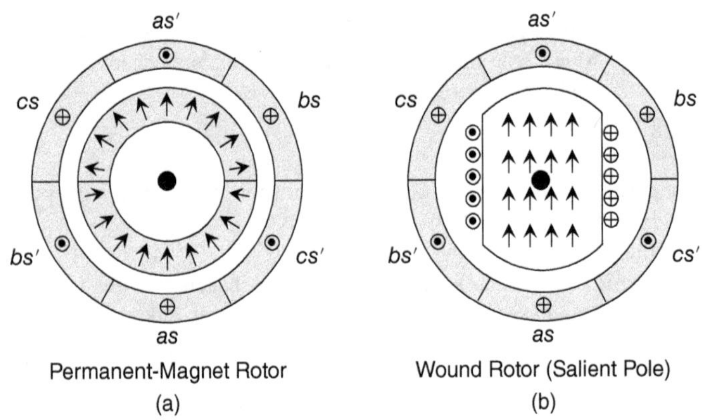

Figure 11.30 Schematic diagram of synchronous motors using a permanent magnet rotor (a) and a wound rotor (b). The rotor establishes a DC magnetic field that rotates at the synchronous speed of the stator field.

shown in Figure 11.30). The rotor field induces a voltage in the stator windings that is proportional to the rotor flux and the synchronous speed. For stator phase "a", the rms value of the induced voltage is given by

$$\boldsymbol{E}_{\mathrm{FA}} = \frac{\omega N_{\mathrm{S}}}{\sqrt{2}} \boldsymbol{\phi}_{\mathrm{F}} \tag{11.20}$$

where N_{S} is the equivalent number of turns in the stator phase windings.

The three-phase stator windings produce a stator field ϕ_{S} that rotates at the synchronous speed ω_{S} with an amplitude proportional to the fundamental frequency component of the stator current. Like the rotor field, the rotating stator field induces a voltage in the stator windings. For phase "a", the induced voltage due to all three stator windings can be written

$$\boldsymbol{E}_{\mathrm{SA}} = j\omega L_{\mathrm{A}} \boldsymbol{I}_{\mathrm{A}} = \omega L_{\mathrm{A}} I_{\mathrm{A}} \exp(j\delta) \tag{11.21}$$

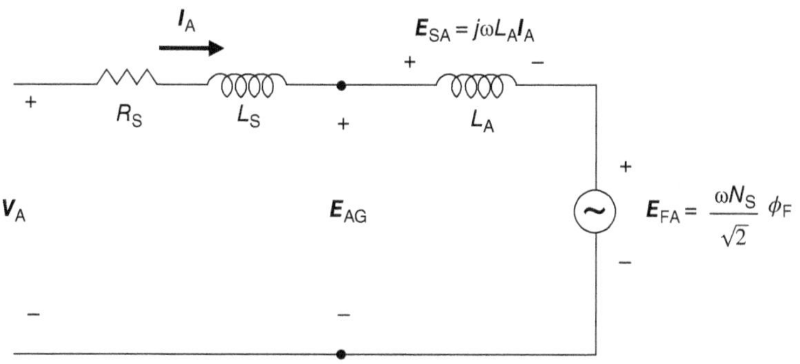

Figure 11.31 Per-phase equivalent circuit of a synchronous motor.

where L_A is the armature inductance (3/2 the self-inductance of phase "a"), $\boldsymbol{I}_A$ is the stator current in phase "a," and $\boldsymbol{E}_{SA}$ is a phasor having a *torque angle* δ with respect to $\boldsymbol{E}_{FA}$. The fluxes of the stator and rotor add in phasor fashion to produce the air gap flux $\boldsymbol{\phi}_{AG}$. Similarly, the induced voltages $\boldsymbol{E}_{FA}$ and $\boldsymbol{E}_{SA}$ add in phasor fashion to produce the net air-gap voltage $\boldsymbol{E}_{AG}$ in the stator windings. These relationships can be understood using the per-phase stator equivalent circuit shown in Figure 11.31, where R_S and L_S are the resistance and self-inductance of the stator phase windings and $\boldsymbol{E}_{FA}$ and $\boldsymbol{E}_{SA}$ are induced emfs due to the rotor and stator fields, respectively. The per-phase stator terminal voltage is

$$\boldsymbol{V}_A = \boldsymbol{E}_{AG} + (R_S + j\omega L_S)\boldsymbol{I}_A \tag{11.22}$$

The electromagnetic torque is proportional to the phasor product of the rotor and stator fluxes, and can be written

$$T_{EM} = k_T \phi_F I_A \sin(\delta) \tag{11.23}$$

where k_T is the torque constant of the motor and δ is the torque angle between the rotor and stator flux phasors.

Synchronous motors are often operated with sinusoidal induced-voltage waveforms, particularly in servo applications. In this case the torque angle δ is maintained at 90°, and for a fixed field flux ϕ_F, the electromagnetic torque is proportional to the stator current, as given by Equation 11.23. The stator current waveforms are sinusoidal with a 120° phase relationship, and are synchronized to the position of the rotor, which is monitored by a rotational position sensor.

Synchronous motors may also be designed with magnetic structures that produce trapezoidal induced-voltage waveforms. In this case the drive waveforms are rectangular, with constant amplitude $+I_S$ for 120° of rotation, zero amplitude for the next 60° of rotation, constant amplitude $-I_S$ for the next 120° of rotation, then zero amplitude for another 60° of rotation. As before, the three drive waveforms have a 120° phase relationship with each other, resulting in a constant torque with minimum ripple.

For either sinusoidal or trapezoidal operation, the three-phase current-regulated voltage-source inverter of Figure 11.18 may be used to drive the stator windings, and the waveforms $i_a(t)$, $i_b(t)$, and $i_c(t)$ are either sinusoidal or rectangular, depending on the type of motor. The complete drive circuit includes position feedback from a sensor monitoring the rotor position, along with electronics to generate the control signals to the inverter. Since the timing of the stator currents is synchronized to the instantaneous rotor position, the torque angle is always maintained at its optimal value of 90°, and the torque is determined by the magnitude of the stator drive currents according to Equation 11.23.

In very high power applications (>1000 hp), thyristor-based load-commutated inverters are used. Figure 11.32 shows a thyristor-based current-source inverter (CSI) driven by a line-voltage-commutated thyristor-based rectifier. The phase waveforms $i_a(t)$, $i_b(t)$, and $i_c(t)$ are rectangular, as described above

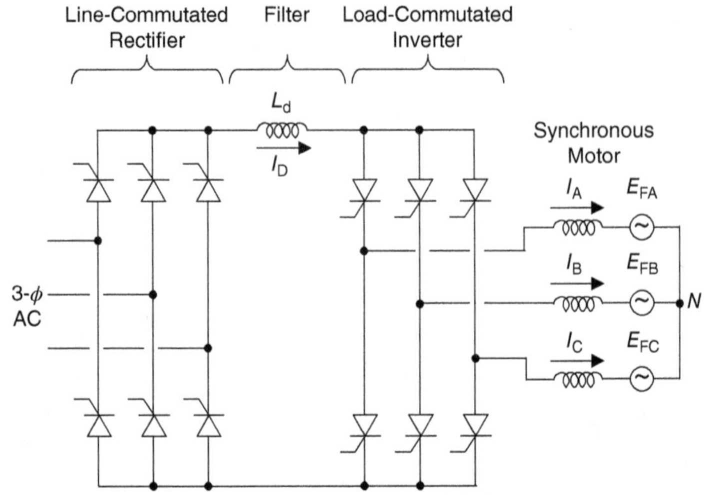

Figure 11.32 Thyristor-based motor drive used for high-horsepower synchronous motors. The drive employs a line-commutated rectifier as converter 1 and a load-commutated current-source inverter as converter 2. The per-phase equivalent circuit of both the rectifier and inverter is given in Figure 11.5.

for the trapezoidal motor operation, and although the induced-voltage waveforms in each phase are trapezoidal, the line-to-line voltages are sinusoidal. The line-voltage-commutated rectifier has the per-phase equivalent circuit of Figure 11.5, and the load-commutated inverter has the same per-phase circuit, but is operated in the inverter mode, as in Figure 11.7c.

11.3.5 Motor Drives for Hybrid and Electric Vehicles

With the growing emphasis on energy independence and environmental sustainability, electric vehicles and hybrid electric vehicles are assuming increased importance. This represents a large potential market for SiC power devices.

In a pure electric vehicle (EV), the wheels are driven by an AC electric motor powered by a rechargeable storage battery. In a hybrid electric vehicle (HEV), an internal combustion engine is combined with an electric drive system. Figure 11.33 presents block diagrams of a pure internal combustion drive, a pure electric drive, and two forms of hybrid drives: the *parallel* hybrid and the *series* hybrid. The parallel hybrid drive obtains motive power from either the internal combustion engine or from an electric motor powered by a battery. The series hybrid drive obtains motive power from an electric motor energized either by the battery or by a generator driven by the internal combustion engine. Both series and parallel drives are capable of regenerative braking, where kinetic energy of motion is returned to the battery.

Figure 11.34 shows a high-capacity series drive (a) typical of diesel-electric locomotives or military tanks, where separate traction motors are placed on each wheel. Figure 11.34 also shows a *combined* hybrid drive (b) used in automotive applications. The combined hybrid allows energy to flow along four different pathways: (i) from the internal combustion engine through the transmission to the wheels, (ii) from the internal combustion engine through a generator and power converter to the battery, (iii) from the battery through a power converter to the electric motor and thence to the wheels, and (iv) from the wheels to the battery by regenerative braking, with the electric motor operating in the generator mode and the bi-directional converter operating in the rectifier mode.

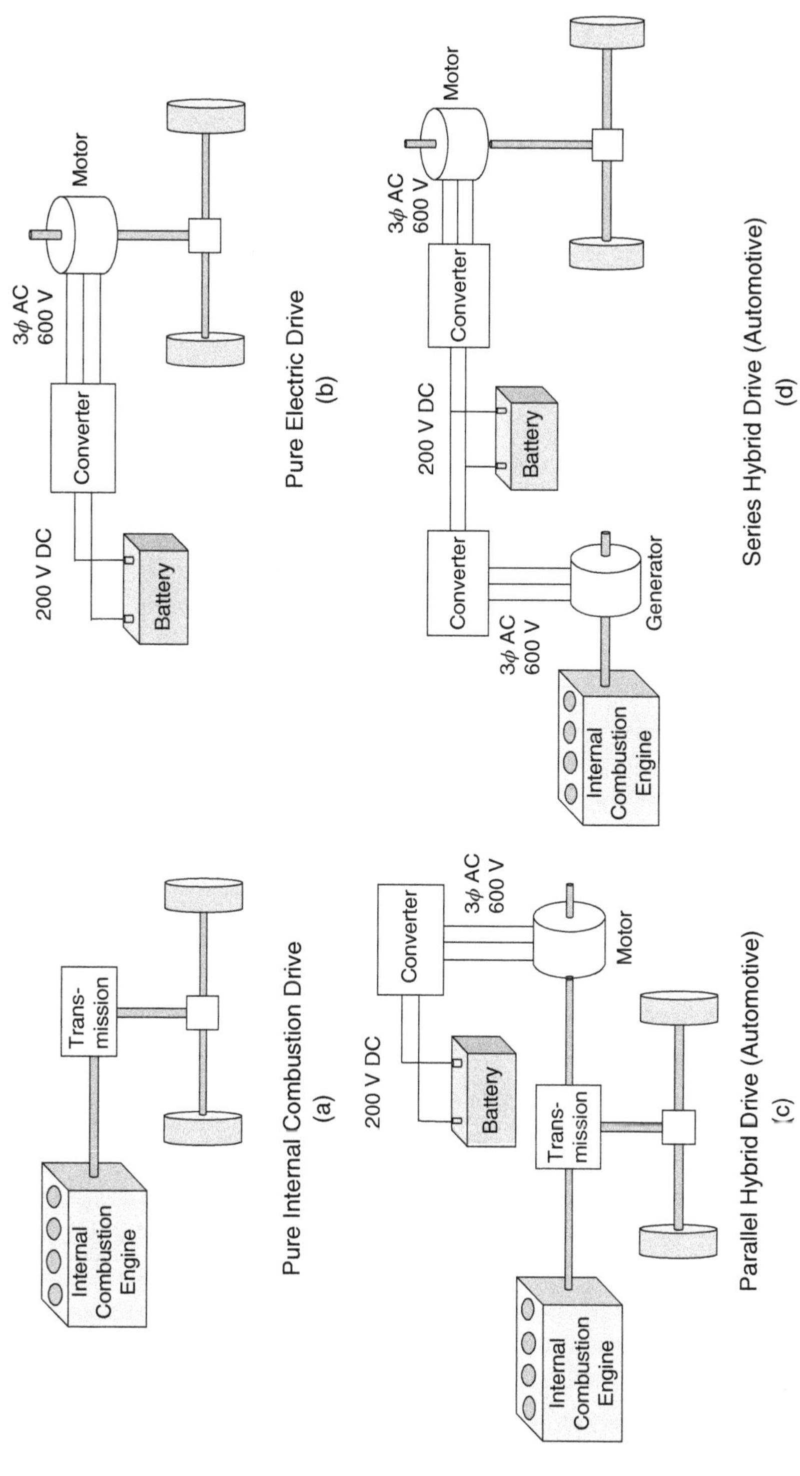

Figure 11.33 Block diagrams of a conventional internal combustion drive (a), a pure electric drive (b), a parallel hybrid drive (c), and a series hybrid drive (d).

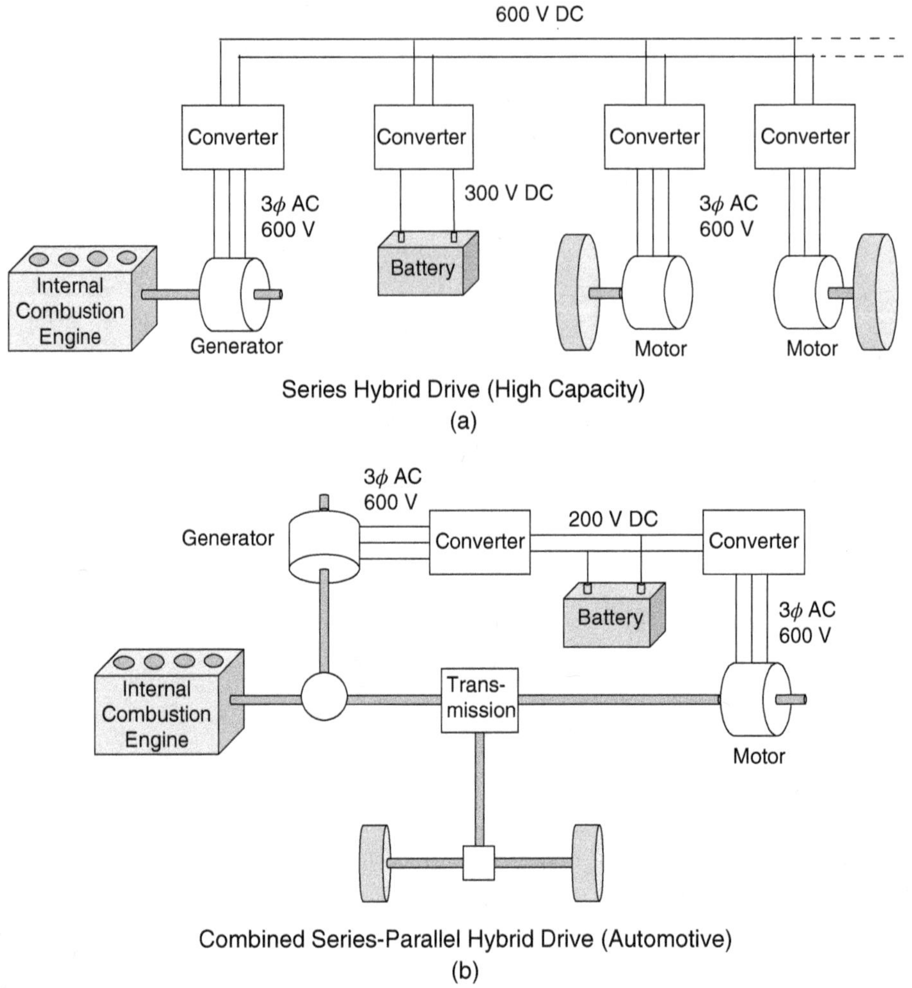

Figure 11.34 Block diagrams of a high-capacity series hybrid drive (a) and a combined series-parallel hybrid drive (b).

The drive motor in electric vehicles is typically either an AC induction motor (or "asynchronous motor") or an AC synchronous motor with a permanent-magnet rotor (also called a "brushless DC" motor). Neither motor requires electrical connection to the rotor. In the induction motor, the rotor field is established by currents induced in the rotor by the rotating field of the stator. In the permanent-magnet synchronous motor, the rotor field is established by a permanent magnet. Both motors require three-phase AC drive to the stator windings to establish a rotating magnetic field in the air gap. This drive is supplied by the three-phase converter located between the battery and the motor in Figures 11.33 and 11.34. The frequency of the AC drive is adjusted to control the speed of the motor, and the magnitude is adjusted to control the torque produced by the motor, as discussed below.

The control circuits for the converter will be different depending on whether the inverter is driving an induction motor or a permanent-magnet synchronous motor. In an induction motor, the rotor slips with respect to the stator field, and the torque is given by Equation 11.18. For constant slip s, the torque is proportional to the square of the air-gap flux, which is established by the stator drive current. In a permanent-magnet synchronous motor, the rotor turns at synchronous speed established by the stator and the torque is given by Equation 11.23. If the synchronous motor is operated with sinusoidal waveforms, the torque angle δ is maintained at 90° by position sensors, and the torque is proportional to the stator drive current.

Electric motors in EVs and HEVs operate at AC drive voltages in the 600 V range, while the battery voltage is typically 200–300 V. The power converter between the battery and electric motor in Figures 11.33 and 11.34 employs the bidirectional three-phase switch-mode inverter shown in Figure 11.18. A 600 V output voltage requires switching transistors with blocking voltages around 1200 V. In the past, these converters have been implemented with silicon IGBTs. However, as will be discussed in Section 11.6, SiC unipolar power switching devices (MOSFETs and JFETs) have performance parameters superior to silicon MOSFETs and IGBTs at 600 V and above, and are expected to replace silicon components in many EV and HEV applications.

11.4 Power Electronics for Renewable Energy

Solar and wind-energy farms provide renewable energy with minimal pollution, and are rapidly growing components of worldwide energy production. Both these energy sources require power electronics to convert DC power (solar) or asynchronous AC power (wind) to AC power synchronized to the electric utility grid.

11.4.1 Inverters for Photovoltaic Power Sources

Solar cells are large-area pn diodes specially designed to admit light into the depletion region, where photogenerated carriers are separated by the built-in electric field. Electrons and holes flow to their majority-carrier sides of the junction, giving rise to a reverse photocurrent I_{PH}. The I–V characteristics are given by the Shockley diode equation, Equation 7.26, with an additive term that represents the reverse photocurrent,

$$I = I_0[\exp(qV_J/kT) - 1] - I_{PH} \approx I_0[\exp(qV_J/kT) - 1] - k_{PH}\Phi_{PH} \tag{11.24}$$

The photocurrent I_{PH} is proportional to the photon flux Φ_{PH}, and the proportionality constant k_{PH} is only a weak function of junction voltage. The I–V characteristics are illustrated schematically in Figure 11.35. The photocurrent subtracts from the ideal diode current, causing the I–V characteristics to enter the fourth quadrant, where power is delivered by the cell to the external circuit. Silicon solar cells operate with an open-circuit voltage in the range 0.5–0.65 V, depending on photon flux and temperature. The power delivered by the cell is given by $P = I \cdot V$, and the maximum power point is obtained when the I–V product is a maximum, illustrated by the points in Figure 11.35. Commercial solar arrays are maintained at the optimum power point by a perturb-and-adjust technique, where slight adjustments are made to the operating point every few seconds to automatically seek the maximum power point.

Since the solar array produces low-voltage DC, power electronics is needed to convert this to high-voltage sinusoidal AC that is synchronized to the utility grid at near-unity power factor. For power levels below a few kilowatts, such as produced by single-home solar arrays, the connection is typically made to a single-phase AC line. The conversion occurs in four stages: (i) low-voltage DC from the solar array is converted to high-frequency low-voltage AC by a switch-mode inverter, (ii) this is passed through a step-up transformer to produce high-voltage high-frequency AC, (iii) which is rectified and filtered, yielding high-voltage DC, (iv) which is converted to high-voltage line-frequency AC, synchronized to the AC utility grid. Figure 11.36 shows a solar converter consisting of a switch-mode

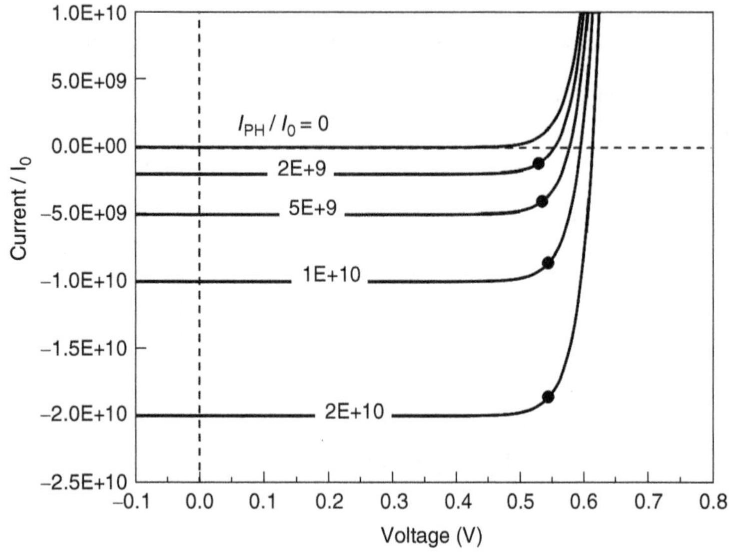

Figure 11.35 Current–voltage characteristics of a solar cell under different illumination intensities, expressed as the ratio of photocurrent I_{PH} to diode saturation current I_0.

inverter, a high-frequency step-up transformer, a diode rectifier, and a thyristor-based line-commutated inverter. The thyristor-based inverter is the same converter shown in Figure 11.5, now operating in the inverter mode. For single-phase utility connections, the converter output voltage is typically in the range 208–240 V.

At power levels above a few kilowatts, such as produced by commercial solar arrays, the interface to the utility grid is a three-phase connection. Figure 11.37 shows a solar converter for this application. This circuit is similar to the single-phase converter of Figure 11.36 except the output inverter has been implemented as a three-phase switch-mode inverter similar to that shown in Figure 11.18. For three-phase connections, the converter output is typically 480 V, and is stepped to 12 kV or higher by a line-frequency transformer connected to the converter output (not shown).

11.4.2 Converters for Wind Turbine Power Sources

A growing fraction of renewable energy worldwide is supplied by wind farms. Each wind turbine drives a three-phase AC generator with a permanent-magnet rotor. The power produced is proportional to the cube of wind velocity, and the optimum shaft speed varies with wind conditions, making it impractical to generate AC at a constant frequency. Therefore, power electronics is used to interconnect this variable-frequency power to the utility grid.

Figure 11.38 shows a three-phase AC converter suitable for moderate-power wind turbines or small hydroelectric power sources. In this converter, variable-frequency three-phase AC is converted to DC by a diode rectifier, filtered, and converted to three-phase line-frequency AC by a switch-mode inverter similar to that of Figure 11.18. The inverter is controlled to ensure that its output is synchronized to the utility grid at near-unity power factor.

The permanent-magnet AC generators in large commercial wind turbines have voltage outputs of 3–5 kV and power ratings up to 10 MVA. The converter interface to the utility grid is also around 3–5 kV. At these power levels, thyristor-based converters are typically used. Figure 11.39 shows a

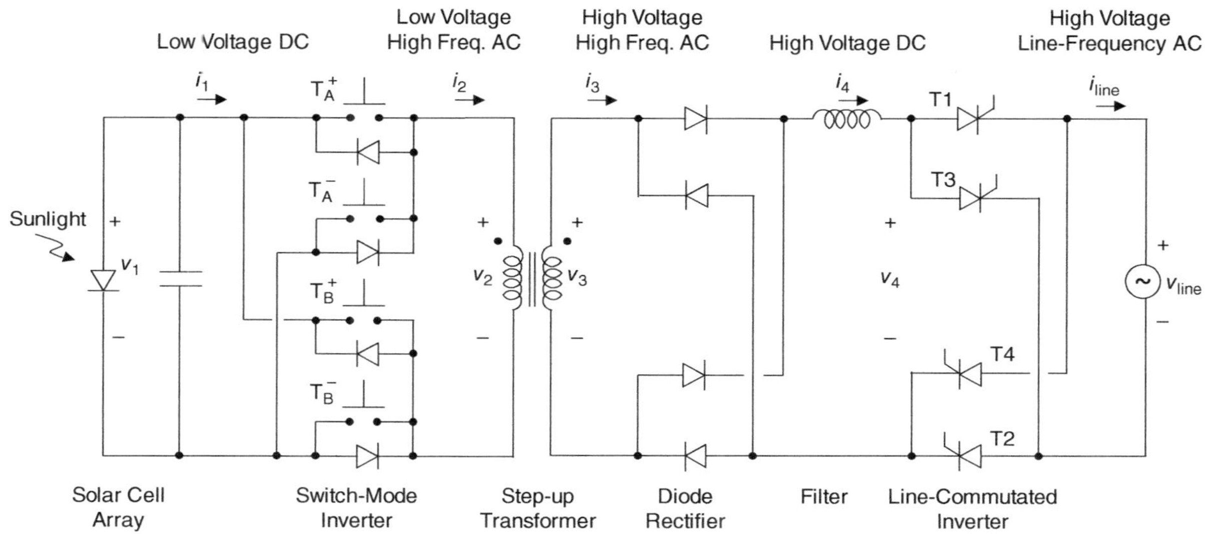

Figure 11.36 A single-phase solar converter consisting of a switch-mode inverter, a high-frequency step-up transformer, a diode rectifier and filter, and a thyristor-based line-commutated inverter.

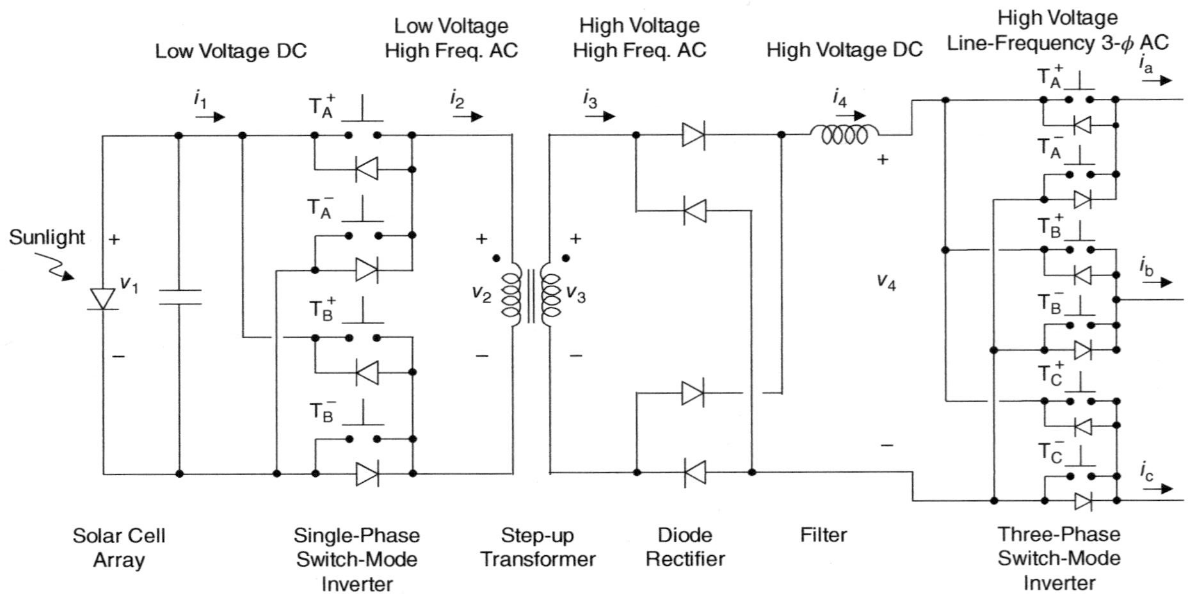

Figure 11.37 A three-phase solar converter. This converter is similar to the single-phase converter of Figure 11.36, but employs a three-phase switch-mode inverter at the output.

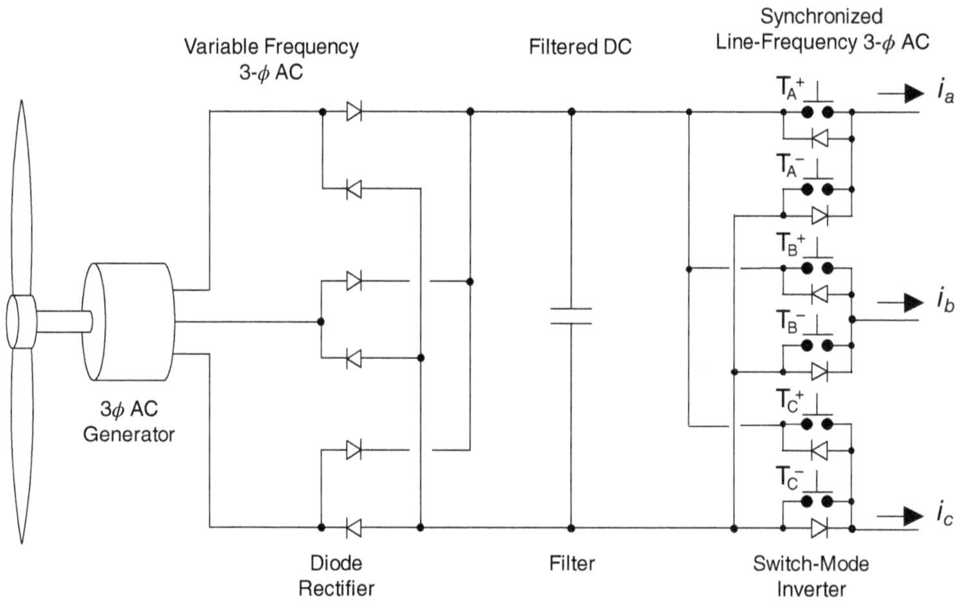

Figure 11.38 A three-phase wind turbine converter for use at moderate power levels.

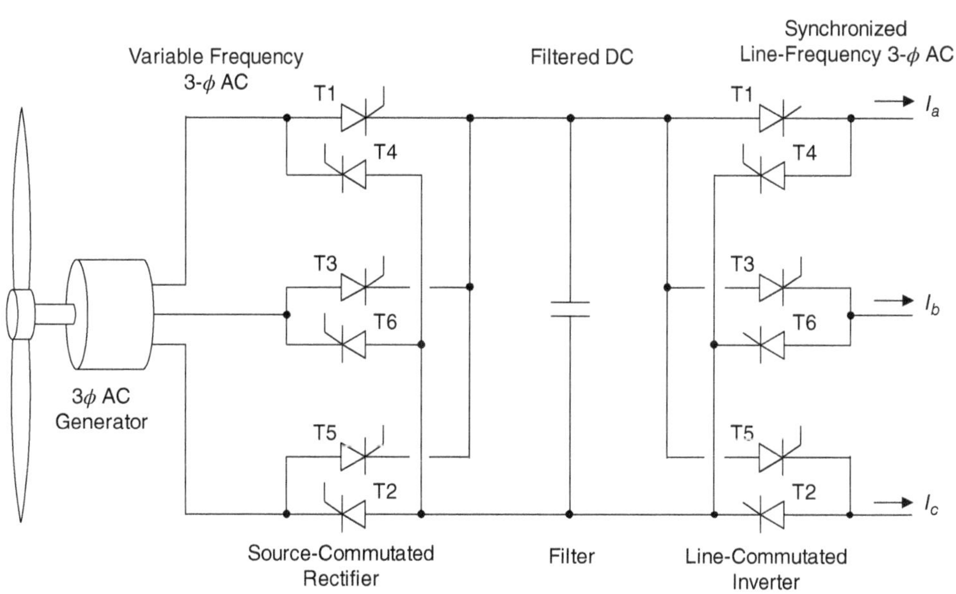

Figure 11.39 A three-phase wind-energy converter suitable for high-power turbine systems.

high-power converter that incorporates a three-phase source-commutated thyristor rectifier followed by a three-phase line-commutated thyristor inverter. Both the rectifier and inverter are three-phase versions of the phase-commutated converter shown in Figure 11.5. Commercial units currently use silicon gate-turn-off thyristors (GTOs) or integrated gate-commutated thyristors (IGCTs).

11.5 Power Electronics for Switch-Mode Power Supplies

Regulated power supplies deliver stable DC power to a load. Over a range of output loading, the power supply acts as an ideal voltage source, providing a controlled voltage that is independent of the load current. In most cases the output must be electrically isolated from the input, that is, the output must be floating with respect to the ground reference of the input supply. In the past, most regulated power supplies used analog circuits, but the availability of advanced semiconductor devices has made it possible to implement *switch-mode* power supplies that are smaller, lighter, and more efficient than analog supplies.

Figure 11.40 shows a generic block diagram of a regulated switch-mode power supply. Line-frequency AC is rectified and filtered to produce unregulated DC, which is inverted to high-frequency AC, passed through a high-frequency isolation transformer, rectified, and filtered to produce regulated DC at the output.

In this section we will focus on the DC–DC conversion portion of the power supply, and consider five topologies: (i) flyback converters, (ii) forward converters, (iii) push-pull converters, (iv) half-bridge converters, and (v) full-bridge converters. The first two converters take the isolation transformer only into the first quadrant of its B–H loop, because the magnetizing current in the primary winding is always positive. Depending on the topology of the converter, this can lead to residual magnetization of the core that must be eliminated by a demagnetizing winding. The last three converters produce bidirectional magnetizing currents in the primary winding, taking the transformer alternately into the first and third quadrants of its B–H loop, and residual magnetization is not an issue.

We begin with the flyback converter, shown in Figure 11.41. This topology is derived from the buck-boost converter of Section 11.2.2, shown in Figure 11.10. In the flyback converter, the simple inductor has been replaced by a two-winding inductor that functions as an isolation transformer, shown in Figure 11.41 along with its magnetizing inductance L_M. The operation of the circuit can be understood as follows. When the transistor is on, current flows through the primary winding of the transformer, and magnetic energy is stored in the core of the transformer. During this time, the diode is reverse

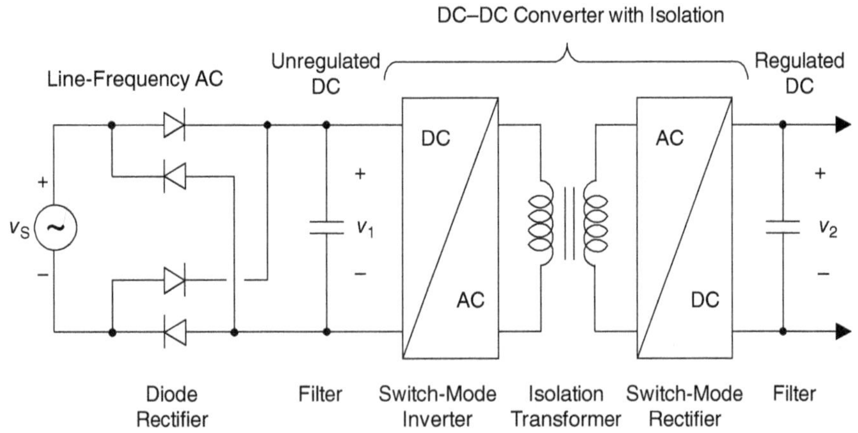

Figure 11.40 Block diagram of a switch-mode power supply.

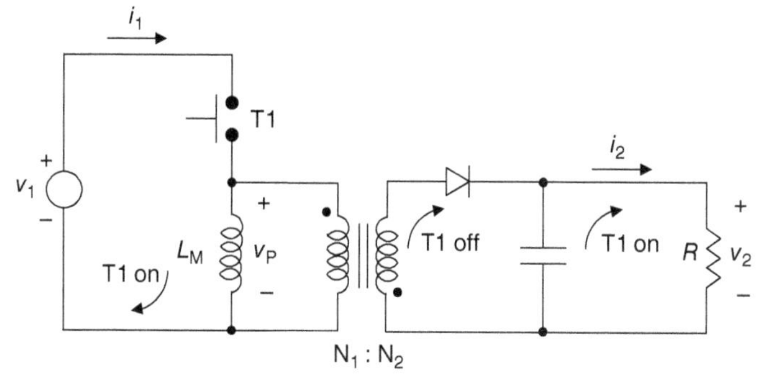

Figure 11.41 A flyback converter derived from the buck-boost converter of Figure 11.10.

biased and the capacitor supplies current to the load R. The flux ϕ in the transformer can be found using Ampere's law,

$$N_1\phi = L_M i_1 \tag{11.25}$$

where N_1 is the number of turns in the primary winding. From Faraday's law we obtain the voltage across the primary winding,

$$v_P = N_1\frac{d\phi}{dt} = L_M\frac{di_1}{dt} \tag{11.26}$$

Using Equations 11.25 and 11.26 we can obtain an expression for the build-up of flux as a function of time,

$$\phi(t) = \phi(0) + \frac{L_M}{N_1}i_1 = \phi(0) + \frac{1}{N_1}\int v_P dt \tag{11.27}$$

where $\phi(0)$ is the flux at the start of the switching cycle. During the time when the transistor is on, $v_P = v_1$ and Equation 11.27 can be written

$$\phi(t) = \phi(0) + \frac{v_1}{N_1}t \tag{11.28}$$

Equation 11.28 shows that the flux increases linearly with time while the transistor is conducting. The maximum flux is attained at the point when the transistor switches off,

$$\phi_{MAX} = \phi(0) + \frac{v_1}{N_1}t_{ON} \tag{11.29}$$

where t_{ON} is the length of time the transistor is on. When the transistor is off, the voltage across the primary is determined by the turns ratio of the transformer N_1/N_2 times the secondary voltage v_2. In analogy with Equation 11.28, since the primary voltage is now negative, the flux decreases as

$$\phi(t) = \phi_{MAX} - \frac{v_2}{N_2}(t - t_{ON}) \tag{11.30}$$

In steady state, the flux at the end of the switching period $\phi(T)$ is equal to the flux at the start of the switching period $\phi(0)$. Inserting Equation 11.29 into Equation 11.30, we can write

$$\phi(T) = \phi_{MAX} - \frac{v_2}{N_2}(T - t_{ON}) = \phi(0) + \frac{v_1}{N_1}t_{ON} - \frac{v_2}{N_2}(T - t_{ON}) = \phi(0) \tag{11.31}$$

Using the relation in the last equality to solve for v_2 yields

$$v_2 = \frac{N_2}{N_1}\left(\frac{\delta}{1-\delta}\right)v_1 \tag{11.32}$$

where δ is the duty factor, $\delta = t_{ON}/T$. Equation 11.32 is analogous to Equation 11.8 for the non-isolated buck-boost converter. In the present circuit, the output voltage is related to the input voltage by the turns ratio of the transformer and the duty cycle of the switching waveform.

Using Equation 11.32, the maximum voltage across the transistor in the off state can be written

$$v_{T,MAX} = v_1 - v_P = v_1 + \frac{N_1}{N_2}v_2 = \left(\frac{1}{1-\delta}\right)v_1 \quad (11.33)$$

Equation 11.33 specifies the voltage rating of the switching transistor in the flyback converter.

The transformer in the flyback converter serves a dual purpose. It functions as an energy storage device, like the inductor in the buck-boost converter, and it also provides the desired input–output isolation. When the transistor is on, current flows in the primary while the diode blocks reverse current in the secondary. Energy is stored in the magnetic field while the transistor is conducting, and that energy is delivered to the load when the transistor is off. The longer the transistor is on, the greater the energy that is transferred to the load. At a 50% duty factor, the output voltage is equal to v_1, and the output voltage can be controlled to be above or below v_1 by adjusting the duty factor.

The second converter topology is the forward converter, shown in Figure 11.42. This circuit is derived from the buck converter of Figure 11.8 by inserting a transformer in series with the transistor. When the transistor is on, current flows through the secondary of the transformer, through the series diode and the inductor to the load. The inductor voltage during this period is

$$v_L = \frac{N_2}{N_1}v_1 - v_2, \quad 0 < t < t_{ON} \quad (11.34)$$

When the transistor is off, the inductor current flows through the shunt diode, and the voltage across the inductor is

$$v_L = -v_2, \quad t_{ON} < t < T \quad (11.35)$$

Since

$$v_L = L\frac{di_L}{dt} \quad (11.36)$$

we can write

$$\int_0^T v_L dt = L\int_{i(0)}^{i(T)} di_L = 0 \quad (11.37)$$

Using Equations 11.34 and 11.35 in Equation 11.37 yields

$$\left(\frac{N_2}{N_1}v_1 - v_2\right)t_{ON} + (-v_2)(T - t_{ON}) = 0 \quad (11.38)$$

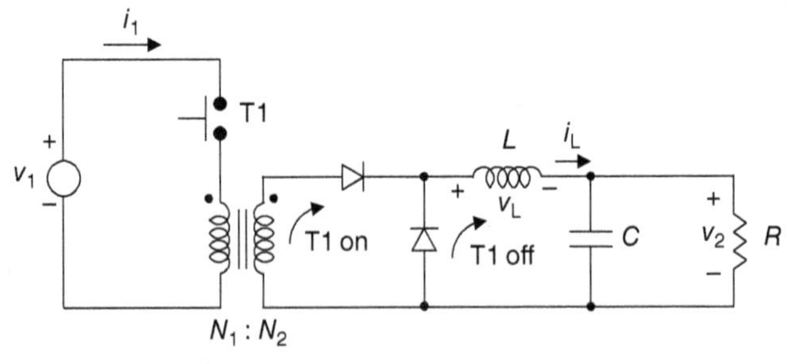

Figure 11.42 A forward converter derived from the buck converter of Figure 11.8.

from which we obtain

$$v_2 = \frac{N_2}{N_1}\delta v_1 \tag{11.39}$$

Because of the current flow directions in the transformer, the forward converter tends to build up residual magnetization in the transformer core. This magnetization can be removed by including a third demagnetizing winding on the transformer core, wound in the opposite direction to the primary and connected across v_1 through a diode (not shown). Alternatively, the core can be demagnetized by connecting a zener diode across the transistor.

The remaining three converters establish bi-directional current flow in the primary winding of the transformer, taking it into both the first and third quadrant of its B–H loop. A two-transistor push-pull converter is shown in Figure 11.43. Transistors T1 and T2 are turned on alternately, each for a time t_{ON}, with a short blanking interval Δt during which both are off. Hence $t_{ON} + \Delta t = T/2$. We define the duty factor $\delta = t_{ON}/T$ and note that with this definition, $0 < \delta < 0.5$. During the first half-cycle, T1 is on and current flows through the inductor through the upper diode, as shown in the figure. The inductor voltage during this period is given by Equation 11.34. During the blanking interval Δt when both transistors are off, the inductor current splits through the two secondary windings and the inductor voltage is given by

$$v_L = -v_2, \quad t_{ON} < t < t_{ON} + \Delta t \tag{11.40}$$

Setting the integral of the inductor voltage over one half-cycle to zero and using Equations 11.34 and 11.40 allows us to solve for v_2, yielding

$$v_2 = 2\frac{N_2}{N_1}\delta v_1 \tag{11.41}$$

The half-bridge converter is shown in Figure 11.44. The switching waveforms are the same as the push-pull converter just described. Transistors T1 and T2 are turned on alternately, each for a time t_{ON}, with a blanking interval Δt during which both are off. Again, $t_{ON} + \Delta t = T/2$ and the duty factor $\delta < 0.5$. During the first half-cycle, T1 is on and current flows through the inductor through the upper diode, as shown in the figure. The inductor voltage during the period when T1 is on is given by

$$v_L = \frac{N_2}{N_1}\frac{v_1}{2} - v_2, \quad 0 < t < t_{ON} \tag{11.42}$$

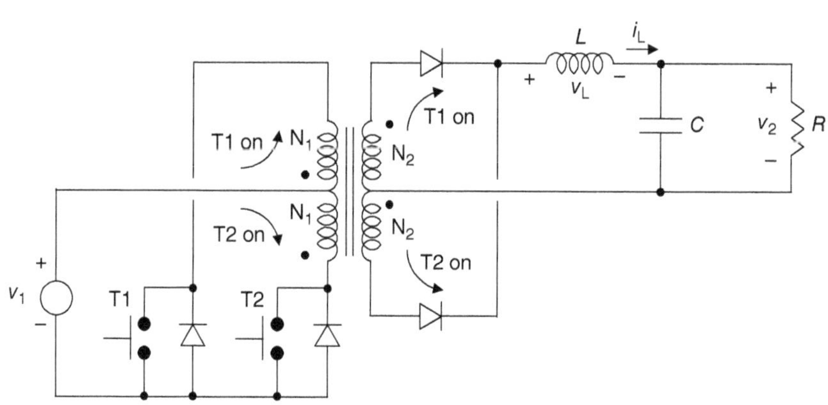

Figure 11.43 A push-pull converter, derived from the buck converter discussed earlier. This converter requires two transistors and a center-tapped transformer.

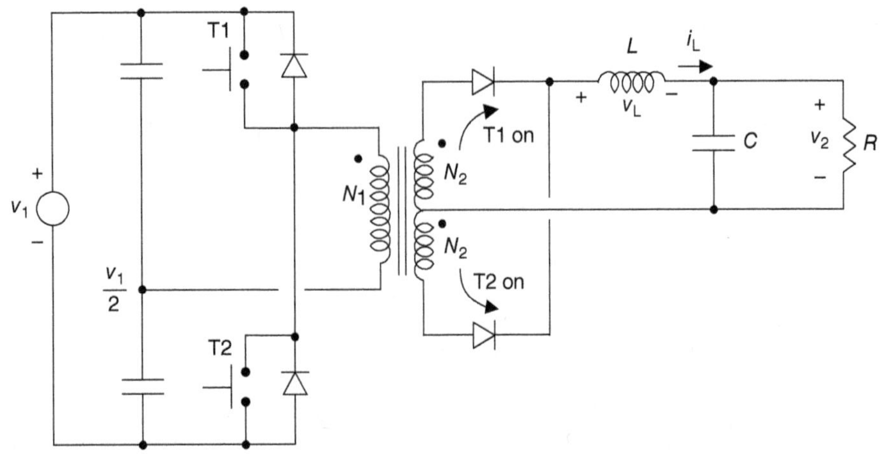

Figure 11.44 A half-bridge converter. This converter has the same output circuit as the push-pull converter of Figure 11.43.

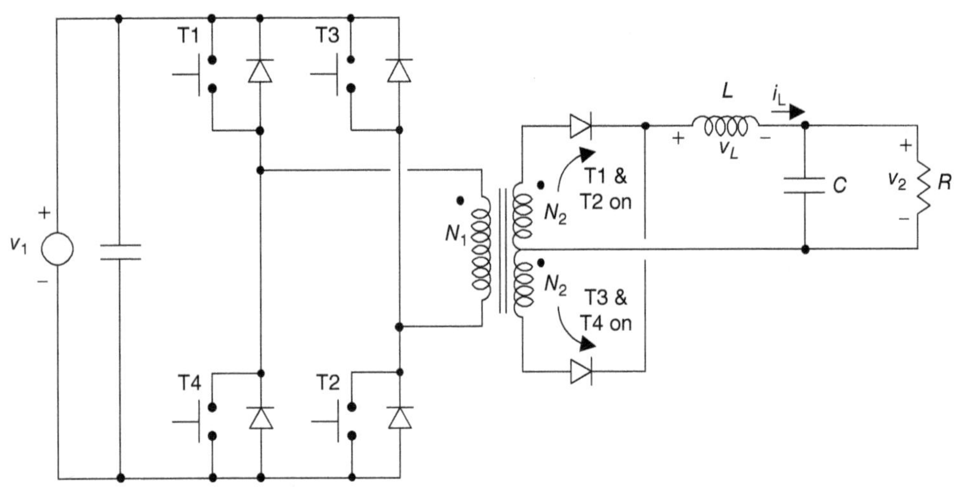

Figure 11.45 A full-bridge converter. This converter is similar to the half-bridge converter of Figure 11.44, and employs four transistors, each with half the current rating of the half-bridge converter.

During the blanking interval both transistors are off, and the inductor voltage is given by Equation 11.40. Setting the integral of the inductor voltage over one half-cycle to zero and using Equations 11.40 and 11.42 allows us to solve for v_2,

$$v_2 = \frac{N_2}{N_1} \delta v_1 \tag{11.43}$$

The full-bridge converter is shown in Figure 11.45. Transistors T1 and T2 are switched together, and transistors T3 and T4 are switched together, using the same switching waveforms as the two previous converters. The analysis of the full-bridge converter proceeds in the same way as the previous converters, and the output voltage is again given by Equation 11.41 with the duty factor $\delta < 0.5$. The full-bridge converter has the advantage that the current flowing through the transistors is half that of the half-bridge converter, allowing the use of smaller transistors. However, the transistors in both converters must support the full supply voltage v_1.

11.6 Performance Comparison of SiC and Silicon Power Devices

In the previous sections, we reviewed the fundamentals of power electronic systems, discussed basic converter circuit topologies, and considered several power systems that can benefit from SiC devices. Having established the general device requirements for various systems, it is natural to ask how the performance of today's SiC power devices compares to the existing technology. In this section we will present actual performance data for several types of SiC power devices and compare their performance to existing silicon devices and emerging GaN devices (where available). We caution that the discussion in this section represents only a snapshot in time. Silicon technology is mature and approaching its fundamental limits dictated by material parameters. SiC is in its adolescence and experiencing rapid growth. GaN is in its infancy, with great promise for the future. Because of the continuing evolution of all three technologies, the comparisons presented in this section are fleeting, and the reader should consult the literature for the latest information.

The most straightforward device comparison deals only with on-state performance, and neglects switching losses. This approach is acceptable when comparing unipolar devices such as Schottky diodes, JFETs, and MOSFETs, but a more sophisticated approach is needed to evaluate bipolar devices such as pin diodes, BJTs, IGBTs, and thyristors.

As discussed in Section 7.1, the unipolar device figure of merit (FOM) is $V_B{}^2/R_{ON,SP}$. This FOM has a theoretical maximum dictated by the mobility, dielectric constant, and breakdown field of the material, and is given by Equation 7.13, repeated here for a punch-through design:

$$V_B^2/R_{ON,SP} = \frac{\mu_N \varepsilon_S E_C^3}{(3/2)^3} \tag{7.13}$$

On a logarithmic plot of $R_{ON,SP}$ versus V_B, the theoretical maximum unipolar FOMs are diagonal lines with a slope close to 2 (a precise calculation deviates slightly from 2 due to the dependence of mobility and critical field on doping). Figure 11.46 shows the unipolar limits for silicon, SiC, and GaN devices (note that these limits must be modified in the case of superjunction devices, which are vertical RESURF structures whose FOM has a slope of 1 instead of 2). Highest performance corresponds to the lower right corner of the figure, and all real devices have on-resistances above their respective limit lines. The right-hand axis shows maximum on-state current density at a power dissipation of 300 W cm^{-2}, calculated using Equation 7.5. Owing to their higher critical field, the theoretical performance limits for SiC and GaN are significantly higher than for silicon, which accounts for the great interest in these technologies.

The first SiC devices to enter commercial production were Schottky barrier diodes (SBDs) and their cousins, JBS (junction-barrier Schottky) diodes. These are used as clamping diodes across transistor switches in all the half-bridge and full-bridge converter circuits discussed in this chapter. Inserting SiC SBDs and JBS diodes instead of silicon pin diodes reduces the switching loss in these circuits due to the faster turn-off of unipolar diodes, as discussed in Chapter 7. This can lead to significant increase in efficiency in switch-mode power supplies used, for example, in file servers in large data centers. (It is not practical to use silicon SBDs for these applications because of their large reverse leakage currents, arising from the lower barrier heights of Schottky metals on silicon.)

Figure 11.47 shows the FOMs for a number of SiC SBDs reported to date [2–7]. Since Schottky diodes have an offset voltage in their I–V characteristics, the maximum current cannot be calculated directly from Equation 7.5, which assumes no offset voltage. The power dissipation must include both the offset

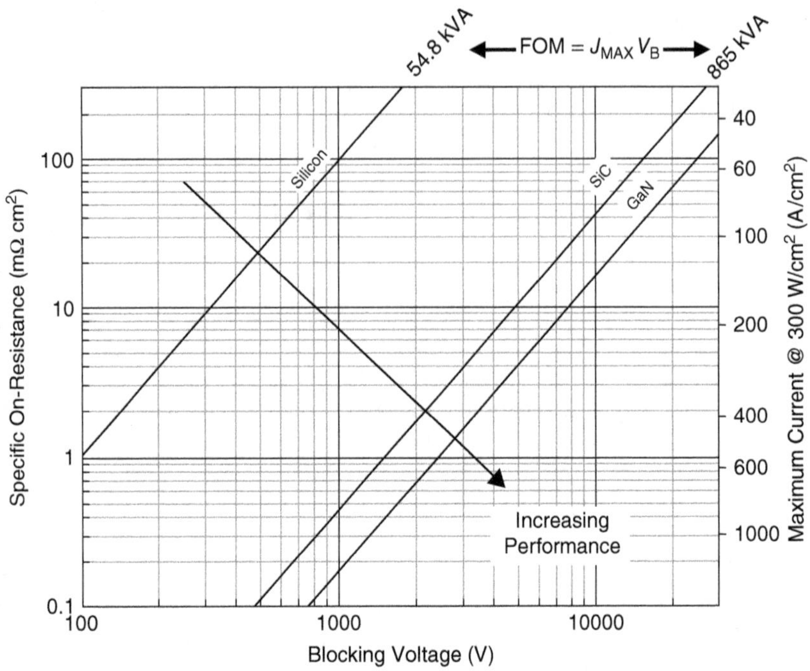

Figure 11.46 A standard device performance plot relating specific on-resistance to blocking voltage. The theoretical maximum unipolar FOMs are shown for silicon, SiC, and GaN.

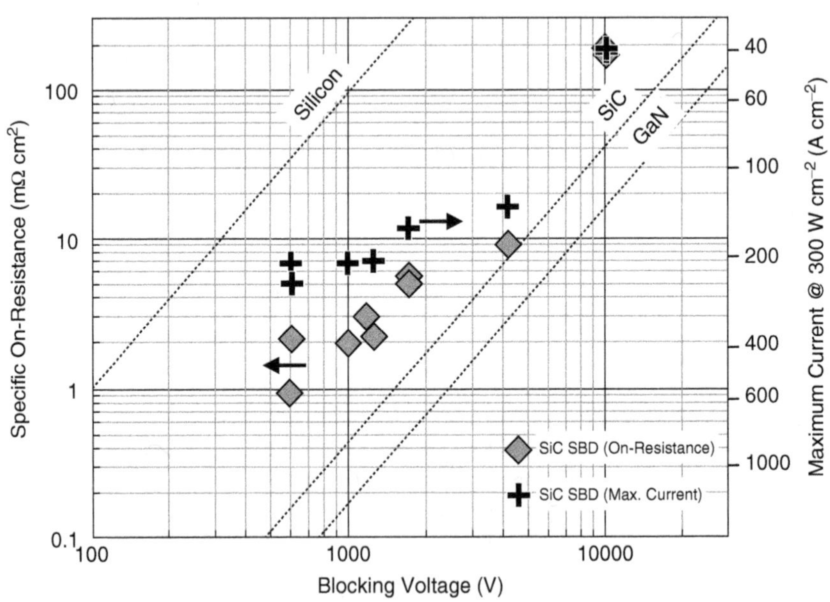

Figure 11.47 On-resistance (diamonds) and maximum current (crosses) for several SiC Schottky barrier and JBS diodes.

voltage and the voltage across the differential on-resistance and can be written

$$P_{ON} = J_{ON}V_F = J_{ON}(V_{OFFSET} + J_{ON}R_{ON,SP}) = J_{ON}V_{OFFSET} + J_{ON}^2 R_{ON,SP} \tag{11.44}$$

To calculate the on-state current for a given power dissipation, Equation 11.44 is solved for J_{ON}, yielding

$$J_{ON} = \sqrt{\left(\frac{V_{OFFSET}}{2R_{ON,SP}}\right)^2 + \frac{P_{ON}}{R_{ON,SP}}} - \left(\frac{V_{OFFSET}}{2R_{ON,SP}}\right) \tag{11.45}$$

The maximum on-state current is obtained from Equation 11.45 by setting the on-state power to some limiting value, say 300 W cm^{-2}, and the offset voltage to the appropriate value for the device, which for SiC SBDs is around 1 V. In the plot of Figure 11.47, the gray diamonds are the on-resistances of the SBDs reading to the left-hand axis, and the black crosses are the maximum on-state currents of the same SBDs reading to the right-hand axis.

Figure 11.48 shows representative performance metrics for silicon LDMOSFETs, superjunction MOSFETs, and IGBTs [8]. IGBTs are bipolar devices and are not subject to the unipolar device theoretical limits, since their drift regions are conductivity modulated. Moreover, IGBTs have an offset voltage of about one forward-biased diode drop in their $I-V$ characteristics, so Equation 11.45 is used to calculate their maximum current, assuming an offset voltage of 0.8 V (for silicon). In the plot of Figure 11.48, the light gray crosses are the on-resistances of silicon IGBTs reading to the left-hand axis, and the black crosses are the maximum on-state currents of the same IGBTs reading to the right-hand axis. The MOSFET points may be read to both the left- and right-hand axes, since they have no offset voltage.

Figure 11.49 presents the reported on-state performance for a number of SiC power DMOSFETs and UMOSFETs [8]. For blocking voltages in the range from 600 V to 2 kV, both types of SiC MOSFETs

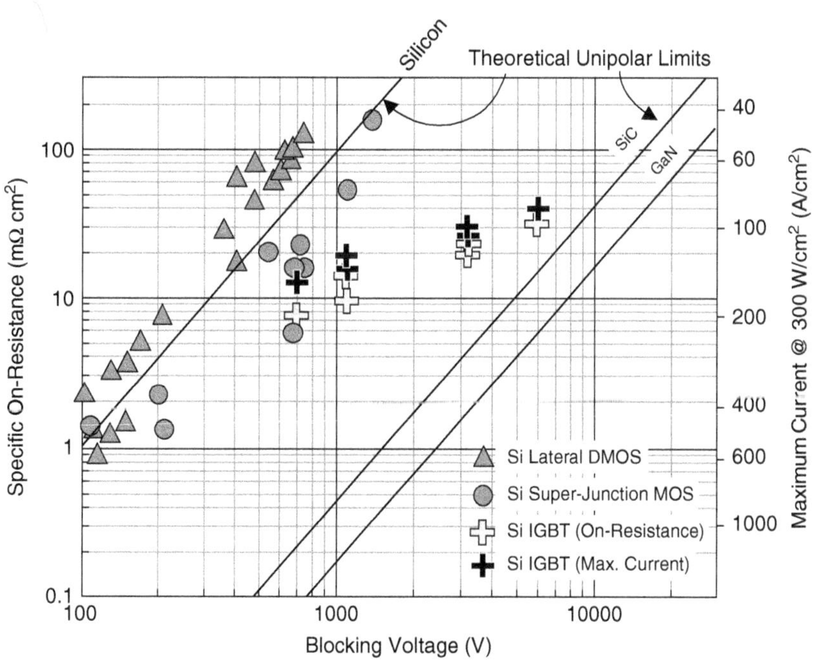

Figure 11.48 State-of-the-art performance of silicon lateral DMOSFETs, superjunction MOSFETs, and IGBTs ([8] reproduced with permission from IEEE).

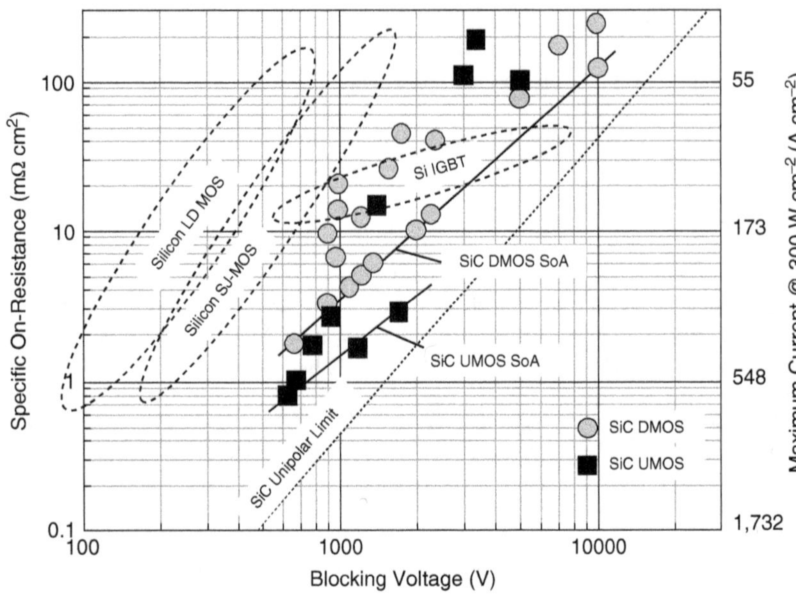

Figure 11.49 State-of-the-art performance of SiC DMOSFETs and UMOSFETs, with silicon MOSFET and IGBT regimes indicated ([8] reproduced with permission from IEEE).

have maximum currents much higher than silicon devices. For example, at a blocking voltage of 1 kV, SiC UMOSFETs provide current densities as much as 3 × higher than silicon IGBTs at the same power dissipation. We also note that the best SiC UMOSFETs carry about 50% higher on-current than SiC DMOSFETs, as indicated by the SoA lines in the figure. Above 2 kV the on-state performance of SiC DMOSFETs approaches that of silicon IGBTs, but extrapolation of the UMOS SoA line to higher blocking voltages suggests that SiC UMOSFETs may retain their advantage over silicon IGBTs to 5 kV or higher.

It is important to emphasize that any comparison of MOSFETs and IGBTs must take into account the *switching losses* of both devices. The switching loss in IGBTs is substantial, due to the large stored minority charge that must be removed before the device can fully turn off. This loss mechanism is absent in unipolar devices such as MOSFETs and JFETs. Switching loss is proportional to switching frequency, as shown by Equation 7.15, and switching loss can become the dominant power dissipation mechanism at high frequencies. This can force a reduction in maximum current to keep the total power dissipation below 300 W cm^{-2}. The reader is referred to Section 10.3 for a more complete discussion.

Figure 11.50 shows performance metrics for SiC JFETs, SiC BJTs, and GaN HEMTs (high-electron-mobility transistors) [8–13]. SiC JFETs have achieved blocking voltages up to 10 kV with on-state currents similar to SiC MOSFETs [10, 11]. Because they do not rely on a gate oxide, JFETs avoid issues related to oxide reliability, and are particularly suited for high-temperature operation. SiC BJTs have been demonstrated with blocking voltages as high as 21 kV and performance very close to the SiC unipolar limit [9]. Like JFETs, BJTs do not depend on a gate oxide and are therefore capable of operation at temperatures well above 200 °C. A drawback of BJTs is their base drive requirement, which adds a power dissipation term that has been neglected in this figure. However, BJTs with common-emitter current gains as high as 257 have recently been reported [14], easing the demands on base drive circuits.

GaN HEMTs have demonstrated impressive performance in the lower blocking voltage regime. GaN technology is in an earlier stage of development than SiC, but has the potential for significant improvement in the future. The reader is referred to the current literature for details.

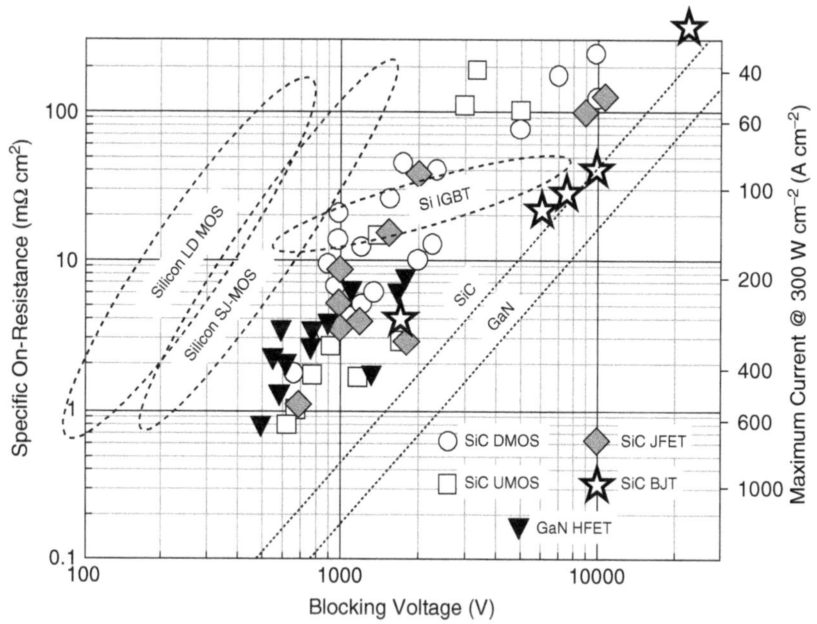

Figure 11.50 State-of-the-art performance of SiC JFETs, BJTs, and GaN heterojunction field-effect transitors (HFETs), with SiC MOSFETs in the background ([8] reproduced with permission from IEEE).

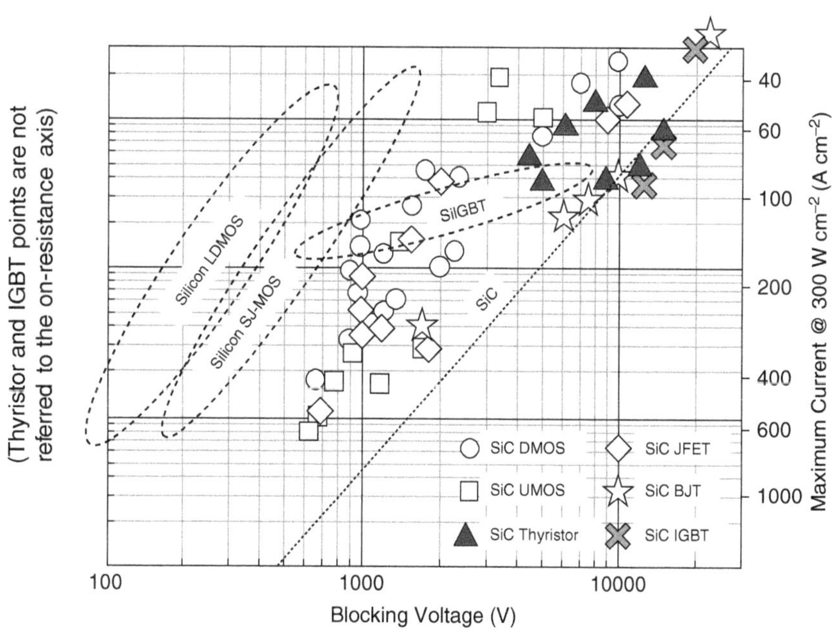

Figure 11.51 On-state performance of SiC IGBTs and thyristors, with SiC MOSFETs, JFETs, and BJTs in the background ([8] reproduced with permission from IEEE).

Figure 11.51 shows on-state performance for selected SiC IGBTs and thyristors reported to date [15–19]. The on-resistance axis has been removed from this figure, since the plotted points indicate only the maximum current density. Like silicon IGBTs, SiC IGBTs and thyristors have offset voltages, and their maximum on-state current is calculated using Equation 11.45 with an offset voltage of 2.8 V. From the figure we see that SiC IGBTs and thyristors have performance metrics similar to the best SiC BJTs. All three are bipolar devices, capable of operation beyond the SiC unipolar limit. It is expected that as material quality improves and more applications arise at blocking voltages above 10 kV, development of high-voltage SiC BJTs, IGBTs, and thyristors will accelerate.

References

[1] Mohan, N., Undeland, T.M. and Robbins, W.P. (2003) *Power Electronics*, 3rd edn, John Wiley & Sons, Inc., Hoboken, NJ.

[2] Raghunathan, R., Alok, D. and Baliga, B.J. (1995) High-voltage 4H-SiC Schottky barrier diodes. *IEEE Electron. Device Lett.*, **16** (6), 226–227.

[3] Rupp, R., Treu, M., Mauder, A. *et al.* (1999) Performance and reliability issues of SiC-Schottky diodes. *Mater. Sci. Forum*, **338–342**, 1167–1170.

[4] Singh, R., Cooper, J.A., Melloch, M.R. *et al.* (2002) SiC power Schottky and PiN diodes. *IEEE Trans. Electron. Devices*, **49** (4), 665–672.

[5] Zhao, J.H., Alexandrov, P. and Li, X. (2003) Demonstration of the first 10-kV 4H-SiC Schottky barrier diodes. *IEEE Electron. Device Lett.*, **24** (6), 402–404.

[6] Nakamura, T., Miyanagi, T., Kamata, I. *et al.* (2005) A 4.15 kV 9.07 mΩ-cm^2 4H-SiC Schottky-barrier diode using Mo contact annealed at high temperature. *IEEE Electron. Device Lett.*, **26** (2), 99–101.

[7] Hull, B.A., Sumakeris, J.J., O'Loughlin, M.J. *et al.* (2008) Performance and stability of large-area 4H-SiC 10-kV junction barrier Schottky rectifiers. *IEEE Trans. Electron. Devices*, **55** (8), 1864–1870.

[8] Ohashi, H. (2012) Power devices then and now, strategy of Japan, in *24th International Symposium on Power Semiconductor Devices and ICs (ISPSD 2012)*, IEEE, pp. 9–12.

[9] Miyake, H., Okuda, T., Niwa, H. *et al.* (2012) 21-kV SiC BJTs with space-modulated junction termination extension. *IEEE Electron. Device Lett.*, **33** (11), 1598–1600.

[10] Veliadis, V., Stewart, E.J., Hearne, H. *et al.* (2010) A 9-kV normally-on vertical-channel SiC JFET for unipolar operation. *IEEE Electron. Device Lett.*, **31** (5), 470–472.

[11] Zhao, J.H., Alexandrov, P., Zhang, J. and Li, X. (2004) Fabrication and characterization of 11-kV normally off 4H-SiC trenched-and-implanted vertical junction FET. *IEEE Electron. Device Lett.*, **25** (7), 474–476.

[12] Zhao, J.H., Kiyoski, T., Alexandrov, P. *et al.* (2003) 1710-V 2.77 mΩ-cm^2 4H-SiC trenched and implanted vertical junction field-effect transistors. *IEEE Electron. Device Lett.*, **24** (2), 81–83.

[13] Zhang, Y., Sheng, K., Su, M. *et al.* (2007) 1000-V 9.1 mΩ-cm^2 normally off 4H-SiC lateral RESURF JFET for power integrated circuit applications. *IEEE Electron. Device Lett.*, **28** (5), 404–407.

[14] Miyake, H., Kimoto, T. and Suda, J. (2011) 4H-SiC BJTs with record current gains of 257 on (0001) and 335 on (000-1). *IEEE Electron. Device Lett.*, **32** (7), 841–843.

[15] Ryu, S.-H., Cheng, L., Dhar, S. *et al.* (2012) Development of 15 kV 4H-SiC IGBTs. *Mater. Sci. Forum*, **717–720**, 1135–1138.

[16] Sui, Y., Wang, X. and Cooper, J.A. (2007) High voltage self-aligned p-channel DMOS IGBTs in 4H-SiC. *IEEE Electron. Device Lett.*, **28** (8), 728–730.

[17] Sugawara, Y., Takayama, D., Asano, K. *et al.* (2004) 12.7 kV ultra high voltage SiC commutated gate turn-off thyristor: SICGT. International Symposium on Power Semiconductor Devices and ICs, Kitakyushu, Japan, pp. 365–368.

[18] Zhang, Q.J., Agarwal, A., Capell, C. *et al.* (2012) 12 kV, 1 cm^2 SiC GTO thyristors with negative bevel termination. *Mater. Sci. Forum*, **717-720**, 1151–1154.

[19] Cheng, L., Agarwal, A.K., Capell, C. *et al.* (2013) 15 kV, large-area (1 cm^2), 4H-SiC p-type gate turn-off thyristors. *Mater. Sci. Forum*, **740-742**, 978–981.

12

Specialized Silicon Carbide Devices and Applications

Because of its wide bandgap, high thermal stability, and resistance to corrosive environments, SiC is an enabling technology for many applications that cannot be served by conventional semiconductors. Among these applications are high-power microwave devices for commercial and military systems; high-temperature electronics for automotive, aerospace, and well-logging; rugged MEMSs (micro-electro-mechanical sensor) devices for hostile environments; gas and chemical sensors for internal combustion engines, furnaces, and boilers; and solar-blind UV photodetectors. We will discuss each of these applications in this chapter.

12.1 Microwave Devices

SiC has a higher breakdown field, higher saturation drift velocity, and higher thermal conductivity than silicon or GaAs, making it an ideal material for microwave power generation at frequencies in the L and S bands (1–2 and 2–4 GHz). SiC devices are extremely robust, and both microwave MESFETs (metal-semiconductor field-effect transistors) and static induction transistors (SITs) have entered commercial production.

12.1.1 Metal-Semiconductor Field-Effect Transistors (MESFETs)

SiC microwave MESFETs were developed between 1995 and 2002 as replacements for GaAs microwave field-effect transistors (FETs) [1]. Figure 12.1 shows the cross-section of a SiC MESFET, along with typical dopings and dimensions. The gate metal forms a Schottky contact with the n-type channel layer, and negative gate voltages create a depletion region under the gate that extends into the channel, reducing the current. At sufficiently negative gate voltages the depletion region reaches completely through the channel, and the drain current is pinched off. The p-type buffer layer prevents channel electrons from entering the semi-insulating substrate where they could be trapped in deep levels. Figure 12.2 shows the $I_D - V_{DS}$ characteristics of a typical SiC microwave MESFET [2]. MESFET $I - V$ characteristics are derived in many textbooks [3, 4] and the reader is referred to these for the relevant equations.

The performance characteristics of microwave power transistors are typically summarized in plots of microwave output power, power gain, and power-added efficiency (PAE) as a function of input power at a

Fundamentals of Silicon Carbide Technology: Growth, Characterization, Devices, and Applications, First Edition. Tsunenobu Kimoto and James A. Cooper.

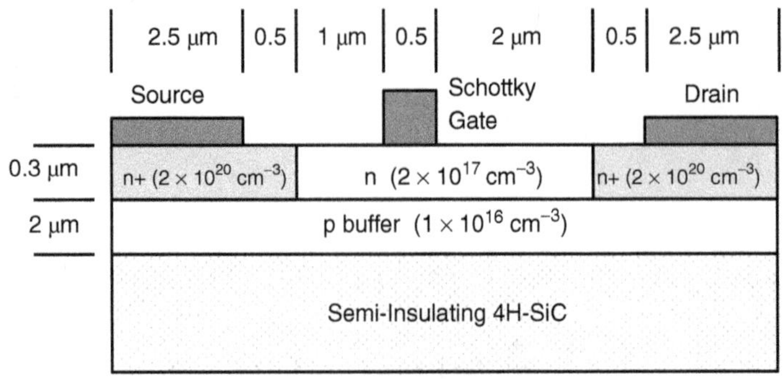

Figure 12.1 Cross-section of a SiC microwave MESFET, along with typical dimensions.

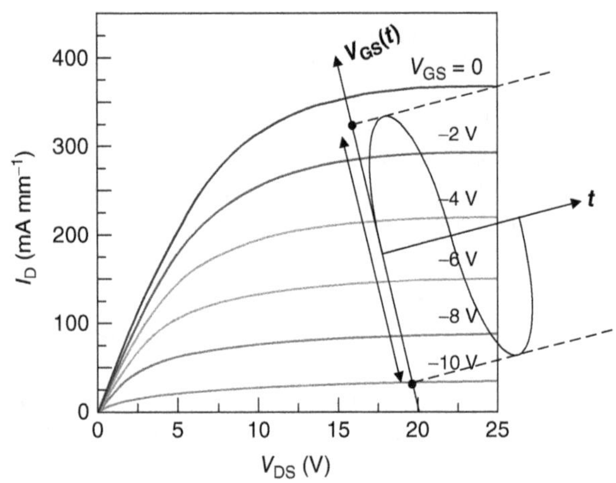

Figure 12.2 Conceptual $I_D - V_{DS}$ characteristics of a microwave MESFET. The sketch illustrates the input waveform as it moves the operating point along a hypothetical load line ([2] reproduced with permission from The Electrochemical Society (ECS)).

fixed frequency. To interpret such a plot, it will be helpful to review a few basic concepts from microwave theory.

Microwave power gain and output power are typically expressed in units of dB and dBm, respectively. The power gain in dB is given by

$$G\ (\text{dB}) = 10\log_{10}(P_{\text{OUT,RF}}/P_{\text{IN,RF}}) \tag{12.1}$$

which can be inverted to yield the power ratio,

$$G = P_{\text{OUT,RF}}/P_{\text{IN,RF}} = 10^{G(\text{dB})/10} \tag{12.2}$$

The output power in dBm is given by

$$P_{\text{OUT,RF}}\ (\text{dBm}) = 10\log_{10}\left(\frac{P_{\text{OUT,RF}}\ (\text{mW})}{1\ (\text{mW})}\right) \tag{12.3}$$

Table 12.1 Power gain and output power expressed in decibels and in absolute units.

G(dB)	G(ratio)	$P_{OUT,RF}$(dBm)	$P_{OUT,RF}$(mW)
50	10^5	50	10^5
40	10^4	40	10^4
30	10^3	30	10^3
20	10^2	20	10^2
10	10^1	10	10^1
0	10^0	0	10^0
−10	10^{-1}	−10	10^{-1}

which can be inverted to yield,

$$P_{OUT,RF}\ (\text{mW}) = (1\ \text{mW})10^{\frac{P_{OUT,RF}\ (\text{dBm})}{10}} \tag{12.4}$$

Table 12.1 gives power gain G and output power $P_{OUT,RF}$ in both numerical values and in dB units.

Power-added efficiency (PAE) is the increase in RF power produced by the amplifier, divided by the total DC input power,

$$\text{PAE} \equiv 100\left(\frac{P_{OUT,RF} - P_{IN,RF}}{P_{IN,DC}}\right) \tag{12.5}$$

The *drain efficiency* is the rf (radio frequency) output power divided by the DC input power

$$\text{Drain Efficiency} \equiv \frac{P_{OUT,RF}}{P_{IN,DC}} = \frac{P_{OUT,RF}}{V_{DC}I_{DC}} \tag{12.6}$$

The *total efficiency* or *overall efficiency* is the rf output power divided by the total input power

$$\text{Total Efficiency} = \frac{P_{OUT,RF}}{P_{IN,DC} + P_{IN,RF}} = \frac{P_{OUT,RF}}{V_{DC}I_{DC} + P_{IN,RF}} \tag{12.7}$$

Referring to Figure 12.2 we note that as the amplitude of the input signal increases, the transistor is eventually driven into the nonlinear portion of the $I_D - V_{DS}$ characteristics, and the output power begins to saturate. This is accompanied by a reduction in gain and the introduction of harmonics into the output waveform due to the nonlinearities of the amplification. Therefore, as the input power is increased we expect the output power to eventually saturate and the PAE and gain to decrease.

Figure 12.3 shows output power, power gain, and PAE measured on a 4H-SiC microwave MESFET with 0.25 mm gate width, operated at 10 GHz [1]. The maximum output power is 30.4 dBm (1.1 W or 4.3 W mm^{-1}) at an input power of 21.4 dBm (138 mW) and PAE of 20%. The power gain is flat at 10.6 dB (gain = 11.5) up to an input power of 19.6 dBm. Figure 12.4 shows output power and PAE for a much larger device with a total gate width of 48 mm, operated at 3.1 GHz [1]. The total output power reaches 49.1 dBm (80 W) at a PAE of 38%.

12.1.2 Static Induction Transistors (SITs)

The static induction transistor (SIT) was introduced by J. I. Nishizawa in 1950 [5–7], and the first SiC SIT was reported in 1995 [8, 9]. Figure 12.5 shows the cross-section of a vertical microwave SIT with a recessed Schottky gate. Electrons flow from the n+ source regions on the top surface to the n+ drain

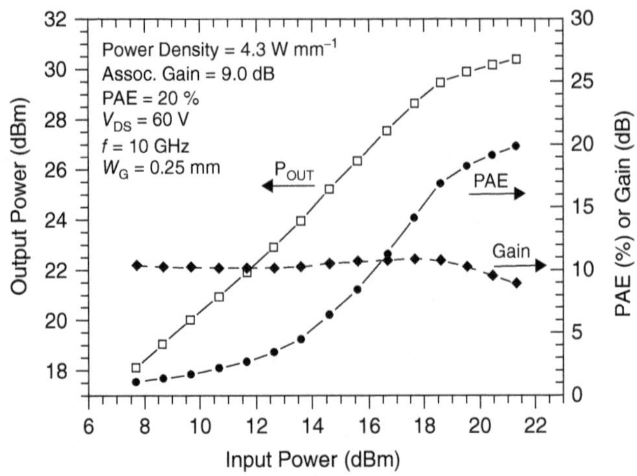

Figure 12.3 Output power, power gain, and power-added efficiency for a 4H-SiC microwave MESFET with 0.25 mm gate width, operated at 10 GHz ([1] reproduced with permission from Cambridge University Press).

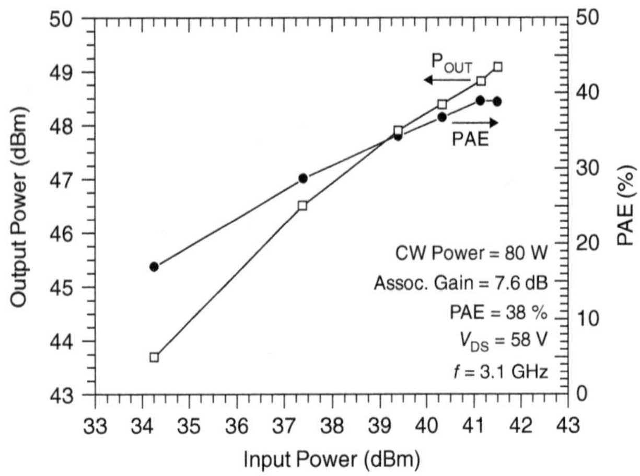

Figure 12.4 Output power and power-added efficiency for a 4H-SiC microwave MESFET with 48 mm gate width, operated at 3.1 GHz ([1] reproduced with permission from Cambridge University Press).

region on the bottom, with the current modulated by depletion regions extending from the embedded Schottky gates. In this sense, the SIT functions like a vertical JFET or MESFET, but with a critical difference: the gate length L_G, defined by the thickness of the gate electrodes, is small compared to the width of the channel between gates (i.e., $L_G \ll 2a$). When this is the case, the current does not saturate, as in a MESFET or JFET, but increases with increasing drain voltage for a fixed gate voltage, as illustrated in Figure 12.6.

We can identify three conduction regimes in the characteristics of Figure 12.6. At gate voltages near zero, regime (1), the depletion region surrounding the gate is small and the vertical channel is not pinched

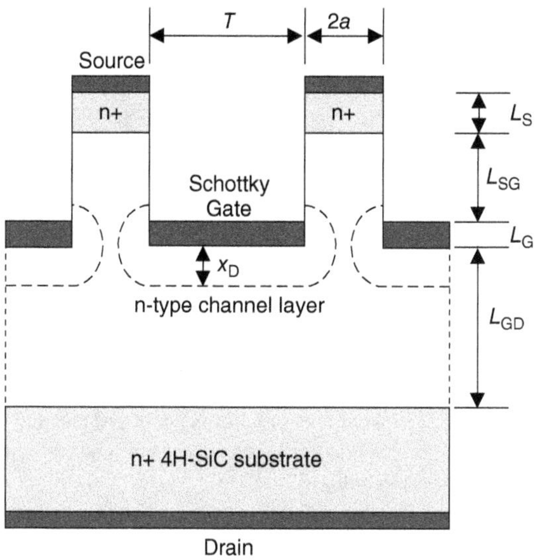

Figure 12.5 Cross-section of a SiC microwave static induction transistor (SIT) ([10] reproduced with permission from Andreas Przadka).

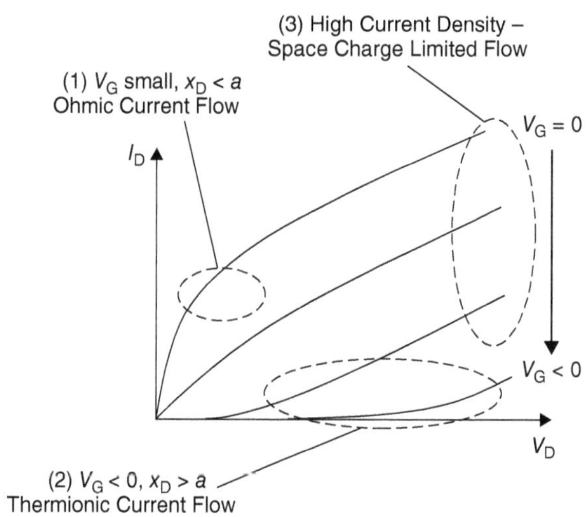

Figure 12.6 Illustration of the I_D-V_{DS} characteristics of a microwave SIT ([10] reproduced with permission from Andreas Przadka).

off ($x_D < a$). Under these conditions the current flow is ohmic, controlled primarily by the gate voltage. At large negative gate voltages, regime (2), the channel is fully depleted and current flow is by thermionic emission over a potential barrier at the saddle point of the channel, as illustrated in Figure 12.7. Here the current increases exponentially as the potential barrier is lowered. In the SIT geometry, the potential barrier is modulated by both the gate voltage and the drain voltage. This is different from a MESFET

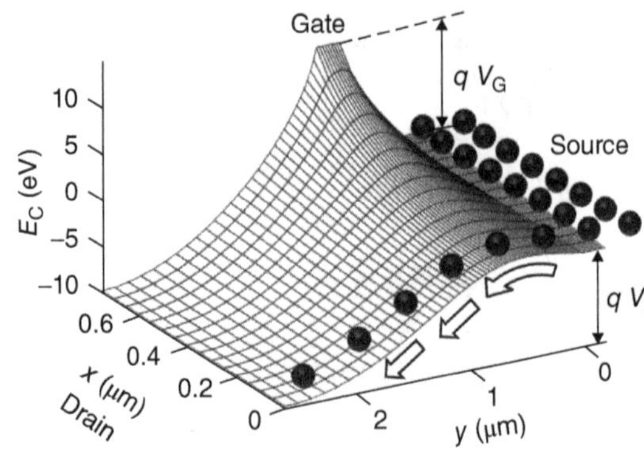

Figure 12.7 Illustration of the electrostatic potential in the depleted channel between gates (only half the channel is shown). The source is grounded, the gate is negative, and the drain is positive ([10] reproduced with permission from Andreas Przadka).

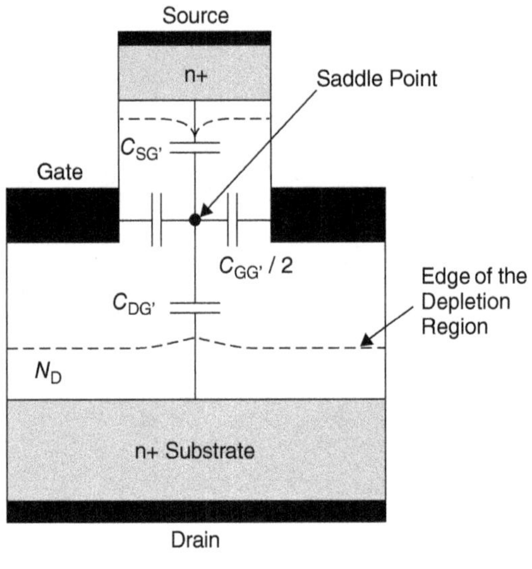

Figure 12.8 Capacitive model showing how the electrostatic potential at the saddle point is determined jointly by the source, gates, and drain ([10] reproduced with permission from Andreas Przadka).

or JFET, where $L_G \gg 2a$ and the gate has sole control over the channel. The behavior of the SIT in this regime can be understood by the simple equivalent circuit of Figure 12.8. Here the potential at the saddle point G' is determined by capacitive coupling from the gate, source, and drain. In the SIT, the very small L_G reduces the gate coupling, allowing the drain to modulate the potential barrier, and an increase in drain voltage is accompanied by an exponential increase in current. Finally, at high currents, regime (3) in Figure 12.6, the electron density in the channel exceeds the doping density, and the electrons drift at their saturated drift velocity. Consequently, the current here becomes space-charge limited.

The static $I_D - V_{DS}$ relationship of the SIT is complicated by two-dimensional effects and the competing conduction mechanisms, and a straightforward analytical solution is not available. For this reason, the

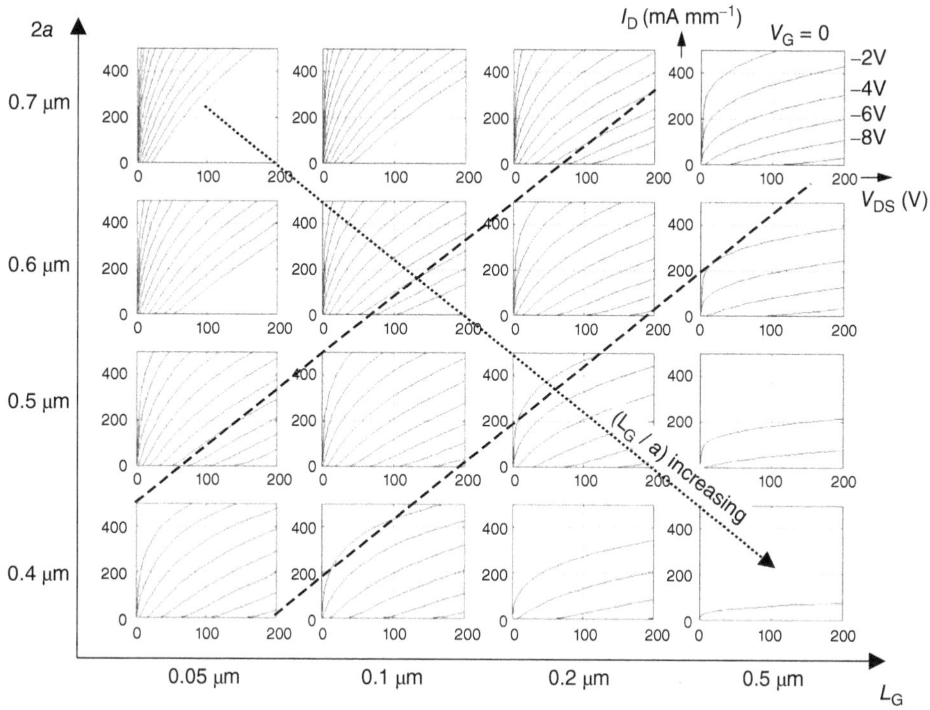

Figure 12.9 Static $I_D - V_{DS}$ characteristics of a microwave SIT as a function of gate length L_G and channel width $2a$, obtained by two-dimensional numerical simulation. For all plots, $N_D = 5 \times 10^{16}$ cm^{-3}, $L_{SG} = 0.5$ μm, $L_{GD} = 2$ μm, $L_S = 0.5$ μm, and $T = 1.5$ μm ([10] reproduced with permission from Andreas Przadka).

behavior is best studied by two-dimensional numerical simulations. Figure 12.9 shows how the $I_D - V_{DS}$ characteristics depend on L_G and a [10]. As the gate length is reduced or the channel thickness is increased (moving toward the upper left in the figure), the gate loses control over the potential barrier, and the current increases rapidly with drain voltage. On the other hand, as the (L_G/a) ratio is increased (moving toward the lower right in the figure), the gate exerts more control and the characteristics approach those of a conventional MESFET or JFET. SIT-like behavior corresponds roughly to the region between the dashed lines.

Practical SITs have hundreds of parallel channels or mesas. To reduce the gate-source capacitance and enhance microwave performance, the source contacts on top of the mesas are typically interconnected with airbridges. Figure 12.10 shows a photograph of a SIT employing airbridge source connections [11]. The mesa width is 0.5 μm and the trench width is 1 μm. The airbridge is formed by electroplating gold to a thickness of 1.2 μm.

SiC SITs deliver the highest pulsed power density of any microwave transistor from UHF through S-band. Table 12.2 lists performance figures for several devices reported in the literature. Power densities range from 1 to 2 W mm^{-1}, with PAEs of 30–50% and power gains of 7–10 dB.

The maximum power available from a SIT is limited by thermal considerations. To gain insight into the thermal limitation, it is instructive to consider a specific example. From Table 12.2 we see that a typical rf output power $P_{OUT,RF}$ is about 1.5 W mm^{-1}, a typical power gain G is about 9.5 dB (gain = ~9 : 1), and a typical PAE is about 45%. Inserting these values into Equation 12.5 allows us to solve for $P_{IN,DC}$, giving

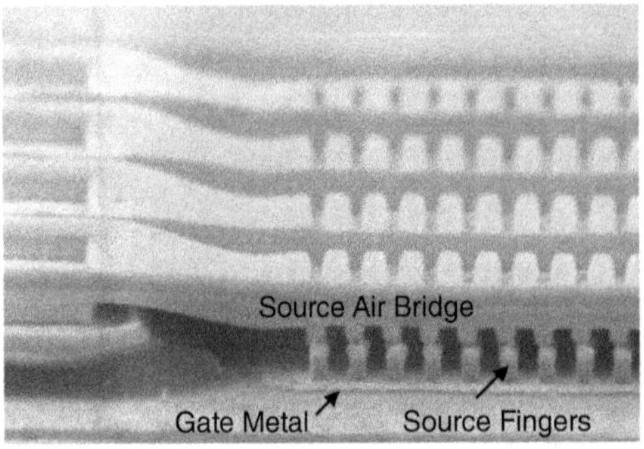

Figure 12.10 Photograph of a C-band SiC SIT with source airbridge interconnects ([11] reproduced with permission from IEEE).

Table 12.2 Performance figures for SiC microwave static induction transistors.

Band	Frequency (GHz)	$P_{OUT,RF}$(W mm^{-1})	PAE or (drain efficiency) (%)	Gain (dB)	References
UHF	0.6	1.35	47	8.7	[8]
UHF	0.95	0.6	27	14	[11]
L	1.3	1.67	55	7.7	[12]
L	1.3	1.55	(52)	7.0	[13]
S	1.3	1.6	(40)	–	[14]
S	2.9	1.55	(40)	–	[11]
S	3.0	1.2	42	9.5	[15]
S – C	3–4	1.27	40	9.5	[16]
S – C	4.0	0.91	(30)	7.0	[11]
S – C	4.0	–	–	10	[13]

about 3 W mm^{-1} of DC power dissipated within the device. The finger pitch ($2a + T$) of the device in Figure 12.10 is 1.5 μm, so the power per unit area would be (3 W mm^{-1})/(1.5 μm) ≈ 2×10^5 W cm^{-2}. This power is dissipated in heating the device, and this heat must be removed by the thermal management system.

To continue the calculation, suppose the SIT is operated in the pulsed mode, and we wish to find the maximum pulse duration to keep the temperature rise across the SiC wafer below 200 °C (to simplify the problem, we assume the time between pulses is long enough for cooling to take place). Figure 12.11 shows numerical calculations of transient temperature rise at the surface of a uniform semiconductor slab of thickness L whose surface is subjected to a heat flux F_0 within a circle of radius R_0 [17]. The left-hand axis represents normalized temperature rise, with K the thermal conductivity of the slab. The bottom axis is normalized time, with κ the thermal diffusivity of the slab. The right-hand axis and top axis are absolute temperature rise and absolute time for a specific device, as will be described below.

The different curves in the figure correspond to different heated radii R_0. Early in the pulse the surface temperature increases as t^2, but eventually the heat distribution in the slab reaches steady state and the surface temperature stabilizes. For large-area devices ($R_0/L \gg 1$), the heat flow is essentially one-dimensional and the final temperature saturates at $\Delta T = (L/K)F_0$ (upper curve), just the heat flux times the thermal resistance of the slab. For small-area devices ($R_0/L \ll 1$), the final temperature saturates at $\Delta T \approx (L/K)\,(R_0/L)F_0$ (one of the lower curves).

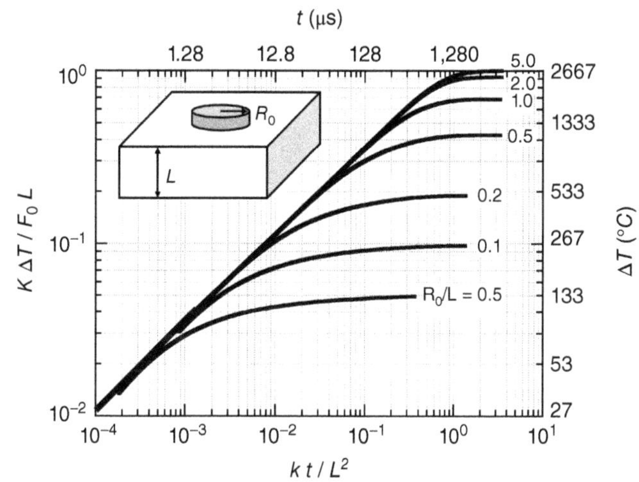

Figure 12.11 Transient temperature rise at the top of a cylindrical slab as a function of heated radius R_0 and thickness L ([17] reproduced with permission from James A. Cooper).

If we insert numerical values for the thermal conductivity and thermal diffusivity of 4H-SiC ($K = 3.0$ W cm^{-1} K^{-1}, $\kappa = 1.25$ cm^2 s^{-1}) and if we assume a substrate thickness L and heat flux F_0, we can relabel the axes in Figure 12.11 in terms of absolute temperature and time. Taking $L = 400$ μm and $F_0 = 2 \times 10^5$ W cm^{-2} (from the calculations above) gives us the labels on the right and top axes of Figure 12.11. If we wish to keep the temperature rise across the SiC wafer less than 200 °C, and assuming the lateral dimensions of the SIT are large compared to the wafer thickness ($R_0 \gg L$), the top axis tells us the duration of the rf pulse must be less than 5.8 μs. In reality, we must use an even shorter pulse if we take into consideration incomplete cooling of the slab between pulses. Alternatively, if we wish to operate the SIT in continuous-wave (cw) mode, again assuming the lateral dimensions are large compared to the wafer thickness, then in steady state we can write $\Delta T = (L/K)F_0$. To keep $\Delta T < 200$ °C we must restrict F_0 to less than 1.5×10^4 W cm^{-2}. The corresponding rf output power, using the gain and PAE values assumed earlier, would be 0.112 W mm^{-1}, far less than the full 1.5 W mm^{-1} output power of which the device is capable. Note that these calculations only consider the thermal resistance of the SiC chip, and do not include the thermal resistance of the package.

Another limitation on maximum available power comes from the requirement for impedance matching at the output. For most microwave systems the characteristic impedance is 50 Ω. Figure 12.12 illustrates the $I_D - V_{DS}$ of a SIT operating along a 50 Ω load line (here we neglect reactive excursions). Assuming that under rf conditions the SIT can follow the DC $I_D - V_{DS}$ characteristics, the rf output power is proportional to the gray shaded area under the load line, and is given by

$$P_{OUT,RF} = (V_{DS,RMS})(I_{D,RMS}) = \left(\frac{1}{\sqrt{2}}\frac{1}{2}\Delta V\right)\left(\frac{1}{\sqrt{2}}\frac{1}{2}\Delta I\right) = \frac{1}{8}\Delta V \Delta I \tag{12.8}$$

The load line in Figure 12.12 represents the equation

$$R_L = \Delta V / \Delta I \tag{12.9}$$

Combining Equations 12.8 and 12.9, we can write

$$\boxed{P_{OUT,RF} = \frac{1}{8}\frac{\Delta V^2}{R_L}} \tag{12.10}$$

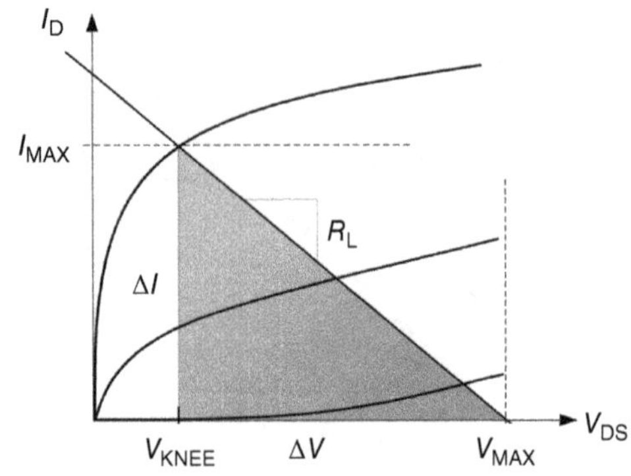

Figure 12.12 Power triangle illustrating rf operation of a SIT with a load line of resistance R_L ([10] reproduced with permission from Andreas Przadka).

Equations 12.8–12.10 can be interpreted as follows: ΔV is determined by the maximum voltage V_{MAX} supported by the device, which is independent of device area. However, ΔI is proportional to the device area. Since R_L is typically 50 Ω and $\Delta V < V_{MAX}$, Equation 12.9 imposes an upper limit on ΔI, and hence on the size of the device to achieve impedance matching at the output. Similarly, Equation 12.10 represents an upper limit on the output power available from the device, subject to impedance matching. SiC can support higher voltages ΔV than silicon or GaAs due to its higher critical field, and the ΔV^2 factor in Equation 12.10 explains why SiC SITs are capable of much higher power densities.

12.1.3 Impact Ionization Avalanche Transit-Time (IMPATT) Diodes

Impact-ionization avalanche transit-time (IMPATT) diodes are two-terminal microwave oscillators that generate rf power by means of a transit-time effect. Structurally, an IMPATT diode is a reverse-biased pin diode where the lightly-doped drift region includes a thin "avalanche region" with higher doping adjacent to the p+ region. The avalanche region is totally depleted, and the higher doping results in an electric field that is sharply peaked at the metallurgical junction. IMPATT diodes produce the highest cw microwave power of any semiconductor device at millimeter-wave frequencies (30–300 GHz). The theory and operation of IMPATT diodes are discussed in several texts [4], and the reader is referred to these for operational details.

The advantage of SiC for IMPATT diodes can be summarized in two important figures of merit. The electronic limit for IMPATT diodes is given by [4]

$$\boxed{P_{OUT,RF} f^2 = \frac{(E_C v_{sat})^2}{4\pi X_C}} \qquad (12.11)$$

where E_C is the critical field for avalanche breakdown and v_{SAT} is the saturation drift velocity. X_C is the capacitive reactance of the diode, given by

$$X_C = 1/(2\pi f C_D) \qquad (12.12)$$

where C_D is the capacitance of the depletion region. Owing to the higher critical field and higher saturation velocity, the electronic limit for 4H-SiC IMPATT diodes is approximately 5 × higher than for silicon and 19 × higher than for GaAs.

The thermal limit for IMPATT diodes is given by [4]

$$P_{\mathrm{OUT,RF}} f = \frac{K \Delta T}{2\pi \varepsilon_S X_C} \tag{12.13}$$

where K is the thermal conductivity and ΔT is the maximum temperature rise across the device. Because of its higher thermal conductivity, higher permissible ΔT, and higher critical field (leading to lower X_C), the thermal limit for 4H-SiC IMPATT diodes is approximately $400 \times$ higher than for silicon and $350 \times$ higher than for GaAs.

Although SiC appears to be a superior material for IMPATT diodes, only limited development has taken place to date [18].

12.2 High-Temperature Integrated Circuits

With the increasing emphasis on energy efficiency, especially in the manufacturing and transportation sectors, distributed sensing and control electronics are being seen as key enablers for progress. In many cases the optimum location for these electronic elements is in a high-temperature environment. For example, in automobiles powered by internal combustion engines, the placement of sensors and control electronics directly on the engine can lead to increased efficiencies and lower emissions. Incorporating control electronics on individual brakes enables automatic stability algorithms that enhance safety and prevent accidents. Both of these applications require electronics that operate at temperatures above the capability of silicon. As a robust wide-bandgap semiconductor, SiC is well suited for these applications. The status and prospects for high-temperature SiC integrated circuits have been reviewed by Neudeck *et al.* [19].

The economic benefit of high-temperature electronics to an overall system can be orders of magnitude greater than the added cost of the specialized components, especially when amortized over the lifetime of the system. In well drilling operations, for example, the use of down-hole electronics to relay directional and compositional information to the surface can pay dividends in more precise and effective drilling. In commercial passenger aircraft, sensors and associated electronics on the hot sections of turbine engines can be used to optimize the combustion process, enabling increases in fuel efficiency that translate to significant savings over the multi-decade lifetime of the aircraft.

From a systems perspective, there is considerable benefit in co-locating amplifiers and control electronics along with the sensors and actuators in the high-temperature environment. Consider again the case of commercial passenger aircraft. Although it is possible to place only the sensor or actuator in the hot zone, keeping the control electronics in a cooler environment, this requires more wiring and longer wiring runs. Several widely-reported aviation accidents involving numerous fatalities have been traced to electrical fires or explosions triggered by damaged wiring insulation. The solution to both the weight and reliability concerns is to partition the system with the interface electronics located in the hot zone. This makes it possible to replace noisy analog interconnects with digital interconnects, or to implement optical data links that further reduce weight and eliminate sparking hazards. All these considerations provide impetus to the development of analog and digital integrated circuits that can be located in high-temperature environments.

Some systems simply would not be feasible without high-temperature electronics. Examples include space missions to the inner planets, where surface temperatures on Venus and Mercury can approach 450 °C, or *in situ* sensors and controls for nuclear reactors, where temperatures can also reach 450 °C. These environments can only be accessed by electronics based on a high-temperature semiconductor such as SiC.

Figure 12.13 charts several end-use applications for high-temperature electronics, along with the capabilities of bulk silicon and silicon-on-insulator (SOI) technology, SiC MOSFET technology, and SiC JFET and BJT technology. Silicon integrated electronics has several limitations. The intrinsic carrier

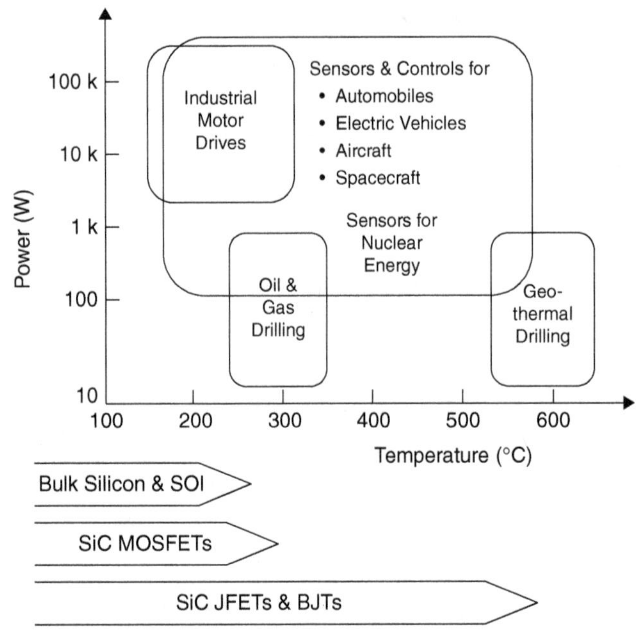

Figure 12.13 Illustration of the operating regimes of several high-temperature systems, along with the temperature ranges for bulk silicon and silicon-on-insulator (SOI), and SiC MOSFET, JFET, and BJT technologies.

concentration n_i is a strong function of temperature, and because of its lower bandgap, the intrinsic carrier concentration in silicon approaches normal dopant densities at temperatures as low as 250 °C. In addition, thermal generation in depleted regions is proportional to n_i, so the leakage in reverse-biased silicon pn junctions also becomes a limitation, which is the main reason for utilizing SOI. Based on these considerations, bulk silicon and SOI integrated circuits are generally limited to temperatures below 250–300 °C.

As discussed in Section 8.2.11, MOS devices in SiC (and in silicon) are limited in temperature by oxide reliability issues. These include not only catastrophic failure of the oxide, but also more subtle mechanisms such as threshold drift and interface degradation, leading to out-of-spec performance. MESFETs do not have a critical gate oxide, but suffer from excessive gate leakage at elevated temperatures. This effectively limits SiC MESFETs to temperatures below about 250 °C.

Much higher temperature operation can be achieved using SiC JFETs or BJTs, since these devices do not rely on a critical gate oxide and do not suffer from the high gate leakage of a Schottky barrier. Here, the most serious limitation is the thermal stability of metallurgy used for interconnects, ohmic contacts, and packaging. The conventional ohmic contact to p-type SiC involves aluminum, but aluminum is unsuited for high-temperatures because of its high reactivity. At present there is no known metallurgy that provides good ohmic contacts to p-type SiC with long-term reliability at temperatures above 300 °C. This is a serious limitation for SiC BJTs because they require low-resistance contacts to both n- and p-type layers. N-channel JFETs are more forgiving in this regard, since a p-type contact is made only to the p+ gate, which carries little current. Ohmic contacts to n-type SiC are robust at elevated temperatures. Multilayer combinations of Ti/$TaSi_2$/Pt have demonstrated stable ohmic behavior to n+ SiC for over 1000 hours at 600 °C [20, 21]. These considerations tend to favor SiC JFETs over SiC BJTs for extreme temperature applications.

Several bipolar and JFET integrated circuits have been developed in SiC. Bipolar integrated circuits based on the transistor–transistor logic (TTL) family were reported by Lee *et al.* [22]. Singh and Cooper [23] fabricated TTL integrated circuits on semi-insulating 4H-SiC, achieving stage delays of 9.8 ns at room temperature and 11.7 ns at 355 °C. Lanni *et al.* [24] reported SiC bipolar integrated circuits based on emitter-coupled logic (ECL) that operate at 300 °C with a stage delay of 62 ns. In spite of these demonstrations, long-term stability of bipolar circuits at elevated temperature has not yet been demonstrated.

The most impressive results to date have been achieved using SiC JFET technology. NASA Glenn Research Center has demonstrated stable operation of analog and digital JFET integrated circuits for thousands of hours at 500 °C [25]. Although circuit failures began to occur after a few thousand hours, all could be traced to interconnect failures, and no device failures were observed in tests exceeding 6000 hours at 500 °C. This demonstrates the inherent stability of SiC JFET devices, but also shows the need for continued improvement in metallization systems for elevated temperatures.

12.3 Sensors

In the most general sense, a sensor is any device that provides information on some environmental variable. Examples include devices to measure temperature or humidity; mechanical sensors to measure pressure, force, acceleration, or displacement; chemical sensors to measure the composition of gases or fluids; biological sensors to detect the presence or concentration of biological agents; optical sensors to detect light or perform optical or infrared imaging; and radiation detectors.

Silicon carbide has a number of unique material properties that make it suitable for sensor applications beyond the capabilities of silicon. These include a bandgap (for the 4H-polytype) 2.9 × higher than silicon, critical electric field 7 × higher than silicon, thermal conductivity 2.5 × higher than silicon, and Young's modulus 3.5 × higher than silicon. The sublimation temperature of SiC is 2830 °C, compared to a melting point of 1412 °C for silicon, and SiC does not undergo thermally-induced plastic deformation observed in silicon at temperatures above 500 °C. SiC is one of the hardest materials known to man. It is impervious to most room-temperature chemical etchants, including the strongest acids. Because of its chemical stability, SiC is suitable for applications in corrosive environments, but the lack of chemical etchants makes it difficult to form complex three-dimensional structures in the material.

To date, SiC sensor development has focused mainly on applications that cannot be easily served by silicon devices. These typically fall into one of three categories: mechanical sensors, gas sensors, and optical sensors. SiC sensor technology has been reviewed in several publications [26, 27].

12.3.1 Micro-Electro-Mechanical Sensors (MEMS)

Micro-electro-mechanical systems (MEMS) take advantage of the excellent mechanical properties and inherent high-temperature capability of SiC, as cited above. Many of these applications do not require single-crystal material, and can be fabricated from cheaper polycrystalline SiC. MEMS devices are generally fabricated using one of two techniques: bulk micromachining or surface micromachining. In bulk micromachining, suspended mechanical structures are formed by the selective removal of the underlying substrate. In surface micromachining, suspended mechanical structures are formed by the selective removal of an underlying thin film. An excellent review of SiC MEMS technology has been given by Zorman and Parro [28].

In one form of bulk micromachining, a thin film of 3C-SiC is epitaxially grown on a silicon wafer, and portions of the silicon wafer are selectively removed to produce free-standing SiC structures. This approach has several advantages. First, silicon can be easily removed by wet chemical etching, leaving the thin SiC regions untouched. This makes it easy to form complex free-standing SiC structures that would be difficult if formed from bulk SiC. In addition, silicon substrates are larger and cheaper than SiC wafers, leading to further cost savings.

The second form of bulk micromachining makes use of single-crystal or polycrystalline SiC wafers that are patterned and etched in a series of steps to form the required mechanical structures. Vertical features such as pits, mesas, and trenches can be fabricated by conventional dry etching using inductively-coupled plasma (ICP) or deep reactive-ion etching (DRIE) techniques, but the greater challenge is to form complex free-standing structures that require lateral undercutting of SiC layers. This can be done using doping-selective electrochemical or photoelectrochemical etching, as discussed in Section 6.2.3. In these techniques, selectively-doped regions are formed by patterned ion implantation, broad-area epitaxy, or repeated sequences of implantation and epitaxy, depending on the three-dimensional geometry to be formed. p-type regions can be selectively removed using electrochemical etching. In this process the sample is immersed in a dilute HF solution, and the presence of holes stimulates anodic oxidation at the interface between p-type SiC and the solution, where the oxide is removed by the dilute HF. The chemical process proceeds by the following reactions (although the first reaction appears to be dominant) [29]:

$$\mathrm{SiC} + 2\mathrm{H_2O} + 4\mathrm{h}^+ \rightarrow \mathrm{SiO} + \mathrm{CO} + 4\mathrm{H}^+$$

$$\mathrm{SiC} + 4\mathrm{H_2O} + 8\mathrm{h}^+ \rightarrow \mathrm{SiO_2} + \mathrm{CO_2} + 8\mathrm{H}^+ \tag{12.14}$$

where h^+ denotes a hole in the valence band of the SiC. The etching process is selective due to the absence of holes in n-type material.

A modified version of electrochemical etching, called photoelectrochemical etching, can be used to selectively remove n-type layers [21]. In this process the sample is immersed in a dilute HF solution and anodically etched under UV illumination. Photogenerated holes collect at the surface of n-type regions due to the surface band bending, leading to oxidation and etching. However, the etching is not uniform, but instead forms porous SiC. The pores have diameters from 10 to 30 nm, spaced 10 to 50 nm apart, resulting in a porosity of about 53% [21]. Since holes are depleted from the surface of p-type regions, these regions experience negligible etching. The porous n-type layers can be subsequently removed by thermal oxidation followed by an HF dip.

A wide variety of MEMS sensors have been built in the 3C-SiC-on-silicon system because of its ease of fabrication and lower cost. Pressure sensors that make use of the piezoresistance of SiC were developed for the automotive industry in 1999 [30]. These structures utilize 3C-SiC grown on SOI wafers, and take advantage of the high-temperature electrical stability of SiC. Pressure sensors can also be based on changes in capacitance rather than piezoresistance [31]. In these devices, pressure causes a deformation of a thin SiC membrane that serves as one plate of a capacitor, and the deformation is detected by a change in capacitance. 3C-SiC is attractive in these applications because of its excellent high-temperature mechanical properties.

Although micromachining of bulk SiC wafers is considerably more difficult than SiC-on-silicon, substantial progress has been made. Photoelectrochemical etching has been used to fabricate pressure sensors in bulk 6H-SiC that operate from room temperature to 500 °C [32]. High-g piezoresistive accelerometers in 6H-SiC have been fabricated using deep reactive-ion etching (DRIE) [33]. These devices function at temperatures well beyond the capabilities of silicon parts.

Because of its chemical inertness, SiC is also of interest in the emerging field of biological MEMS or bio-MEMS. SiC is biocompatible and sterilizable [34], and porous SiC is being explored for biofiltration applications because of its resistance to biofouling [35]. These same properties make SiC MEMS suitable for implantable medical devices for both therapeutics and diagnostics [36].

12.3.2 Gas Sensors

Gas sensors are a subset of chemical sensors that detect the presence of specific molecules in a gas stream. The most important application of gas sensors is in the field of combustion control. For example, gas sensors in the exhaust stream of gasoline and diesel engines can enable closed-loop control systems that optimize the combustion process, increasing fuel efficiency and minimizing pollution. Similar applications exist in the operation of boilers and furnaces used for residential and commercial heating. The high-temperature capability of SiC sensors make them ideal for these applications.

In automotive applications, SiC gas sensors have response times in the millisecond range, short enough to distinguish the exhaust products of individual cylinders. These sensors can monitor the fuel/air mixture and detect molecules such as carbon monoxide, nitrous oxides, and sulfur oxides, all of which contribute to atmospheric pollution. In diesel engines, the major pollutants are nitrous oxides. These are removed from the exhaust stream by the introduction of ammonia ahead of a catalytic converter. Monitoring the ammonia content at the input and output of the catalytic converter can enable more efficient control of the conversion process.

SiC gas sensors have been implemented using Schottky barrier diodes, MOS capacitors, and MOSFETs. The sensing action of all three devices is based on the chemical reactions that occur at high temperatures between a catalytic metal such as Pt or Pd, and hydrogen- or oxygen-containing species in the gas stream. These species are adsorbed on the surface of the catalytic metal where they dissociate, releasing hydrogen or oxygen atoms that diffuse through the metal to the interface with SiC or SiO_2. At high temperatures, atomic hydrogen and oxygen diffuse rapidly, often reaching the interface in milliseconds to microseconds, where they chemically alter the metal work function. This can be detected through the $I-V$ characteristics of a Schottky diode, the $C-V$ characteristics of an MOS capacitor, or the I_D-V_{DS} characteristics of a MOSFET.

In Schottky diode sensors, the catalytic metal forms a Schottky barrier with the SiC. The metal work function is directly related to the Schottky barrier height, as illustrated in Figure 7.4. Thus we can write

$$\Phi_B = \Phi_M - \chi \tag{12.15}$$

where Φ_M is the metal work function and χ is the electron affinity of the semiconductor. The current in a Schottky diode depends exponentially on Φ_B, as can be seen by reference to Equation 7.18, and the current is therefore a sensitive measure of the work function. Figure 12.14 shows the effect of alternately exposing a SiC Schottky diode gas sensor to atmospheres containing oxygen or propane at 550 °C [37]. The decomposition of propane releases atomic hydrogen that diffuses through the catalytic metal to the interface, reducing the barrier height. Exposure to oxygen removes hydrogen through the production of

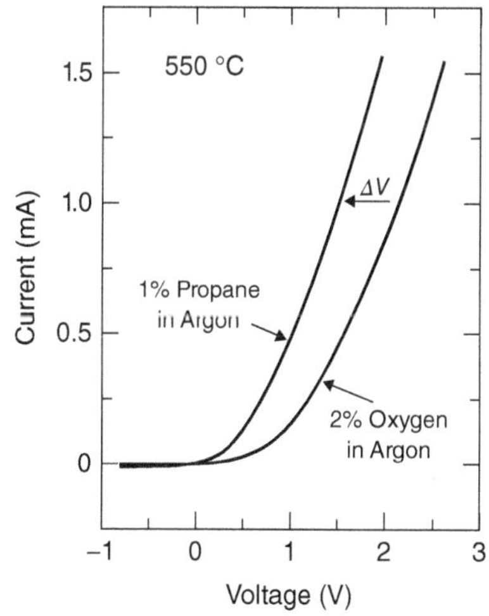

Figure 12.14 Current–voltage characteristics of a SiC Schottky diode gas sensor with a catalytic Schottky metal after exposure to oxygen and propane (hydrogen) atmospheres at 550 °C ([37] reproduced with permission from Wiley-VCH GmbH).

water vapor, resulting in an increase in barrier height. The $I-V$ characteristics can be repeatedly cycled between the two cases by alternating exposure to oxygen and propane.

The sensing process for all types of SiC gas sensors can be summarized as follows: a catalytic reaction that releases atomic hydrogen decreases the effective work function of the metal gate or anode. This occurs with any hydrocarbon molecule that can be catalytically decomposed at high temperatures. Any reaction that releases atomic oxygen increases the effective work function of the metal. The change in work function can be detected through its influence on the terminal characteristics of the device.

Turning now to the MOS capacitor sensor, the flat-band voltage is given by Equation 8.56, repeated below

$$V_{FB} = \frac{\Phi_M}{q} - \frac{Q_F}{C_{OX}} - \frac{Q_{IT}(\psi_S = 0)}{C_{OX}} \tag{12.16}$$

Here Q_F is the area density of fixed charges at the oxide/semiconductor interface and $Q_{IT}(\psi_S)$ represents the charge trapped in interface states, which varies with surface potential ψ_S. The main point is that the flat-band voltage scales directly with the metal work function Φ_M. Since the effective work function of a catalytic metal gate can be modified by exposure to hydrogen- or oxygen-containing gases, the capacitance–voltage curve of the MOS capacitor will be shifted along the voltage axis as a result of such exposure. This makes it possible to detect the presence of hydrocarbons or oxygen-containing species in a gas stream by monitoring the capacitance at a constant gate voltage, or by varying the gate voltage to maintain a constant capacitance. Figure 12.15 shows $C-V$ curves of a $Pt/SiO_2/SiC$ MOS capacitor on n-type 6H-SiC after exposure to oxidizing and reducing ambients at 800 K (527 °C) [38]. Hydrogen exposure reduces the metal work function and hence the flat-band voltage, while exposure to oxygen increases the work function and flat-band voltage. Oxygen exposure also creates new interface states that broaden the $C-V$ curve along the voltage axis. Removing the oxygen by exposure to hydrogen passivates the interface states, reducing the width of the $C-V$ curve. The $C-V$ curve can be repeatedly cycled between the two cases by alternating exposure to hydrogen and oxygen.

MOS capacitor sensors require external circuitry to measure capacitance, either to monitor its change or to maintain a constant capacitance by adjusting the gate voltage. These circuits are unnecessary if we measure the drain current in a MOSFET with a catalytic metal gate [29]. In MOSFET sensors, the channel doping is adjusted so that the device is normally on (depletion mode), and the gate is connected to the source, resulting in a two-terminal device. The drain current is monitored at a constant drain voltage that places the MOSFET in saturation, that is, $V_D > (V_G - V_T) = -V_T$ (note that $V_G = 0$ and $V_T < 0$).

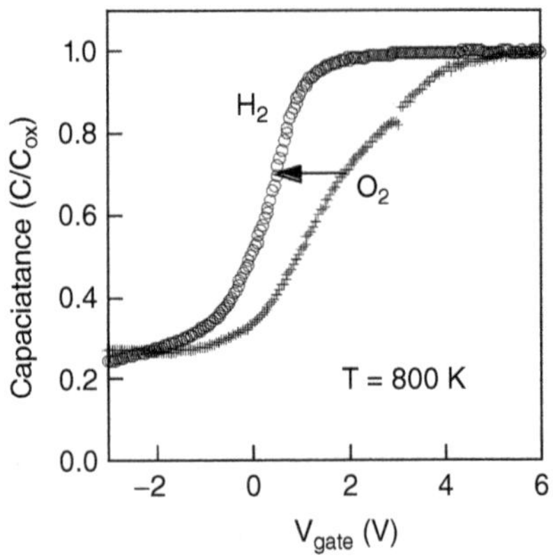

Figure 12.15 Capacitance–voltage curves of a $Pt/SiO_2/6H-SiC$ MOS capacitor gas sensor after exposure to hydrogen- and oxygen-containing atmospheres at 800 K ([38] reproduced with permission from IEEE).

The saturation drain current can be found from Equations 8.47 and 8.49. For the case where the gate is connected to the source ($V_G = V_S = 0$), and assuming the threshold voltage $V_T < 0$, the saturation current can be written

$$I_{D,SAT} \approx \mu_N^* C_{OX} \frac{W}{2L} V_T^2 \tag{12.17}$$

where V_T is given by Equation 8.59, rewritten below with $V_S = 0$,

$$V_T = \left[\frac{\Phi_M}{q} - \frac{Q_F}{C_{OX}} - \frac{Q_{IT}\left(\psi_S = 2\psi_F\right)}{C_{OX}} \right] + 2\psi_F + \sqrt{2V_0(2\psi_F)} \tag{12.18}$$

Since the MOSFET is depletion-mode, V_T is negative, implying that the negative terms in Equation 12.18 are larger than the positive terms. As we introduce hydrogen-bearing compounds to the catalytic gate, atomic hydrogen is released, reducing Φ_M, which is positive. As a result, the threshold voltage becomes more negative and the saturation current increases. Conversely, oxygen-bearing molecules increase Φ_M, making V_T less negative, and the saturation current decreases. This can be seen in Figure 12.16, where a depletion-mode SiC MOSFET is repeatedly exposed to nitric oxide at 525 °C [39]. The NO decomposes, releasing oxygen that increases Φ_M, reducing the drain current. A gradual reduction in the baseline signal is observed over time, suggesting that more work is needed to obtain stable performance under repeated cycling.

Each of the sensor types described above, Schottky diodes, MOS capacitors, and MOSFETs, have their limitations. Schottky diodes suffer from gradual degradation of characteristics under long-term operation due to chemical changes at the metal/semiconductor interface. MOS capacitors and MOSFETs are limited by oxide reliability issues at elevated temperatures, as discussed in Section 8.2.11. In all cases, the repeatability and stability of sensor response needs to be improved to allow quantitative measurements to be made with a high level of confidence.

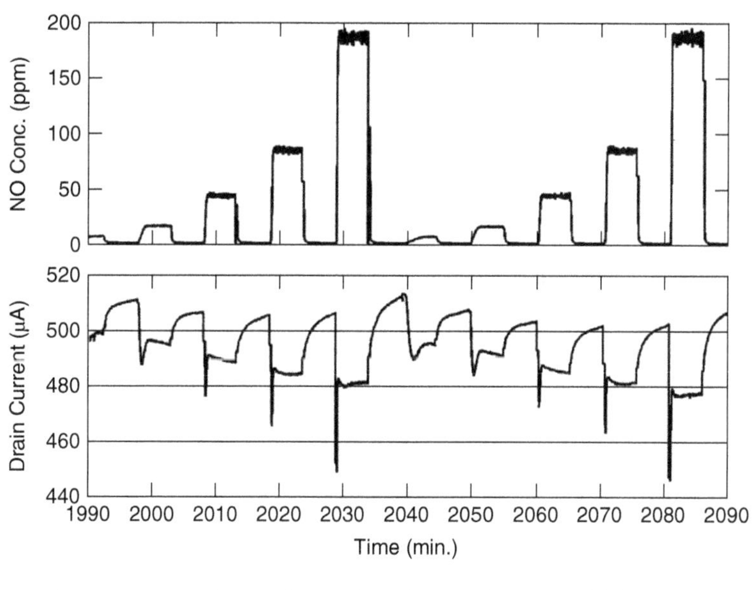

Figure 12.16 Drain current variation of an n-channel depletion-mode 4H-SiC MOSFET gas sensor with a 40 nm Pt gate under exposure to 10–200 ppm of nitric oxide gas in a background of synthetic air at 525 °C. The gate is shorted to the source, and the drain is held at 8 V ([39] reproduced with permission from Cambridge University Press).

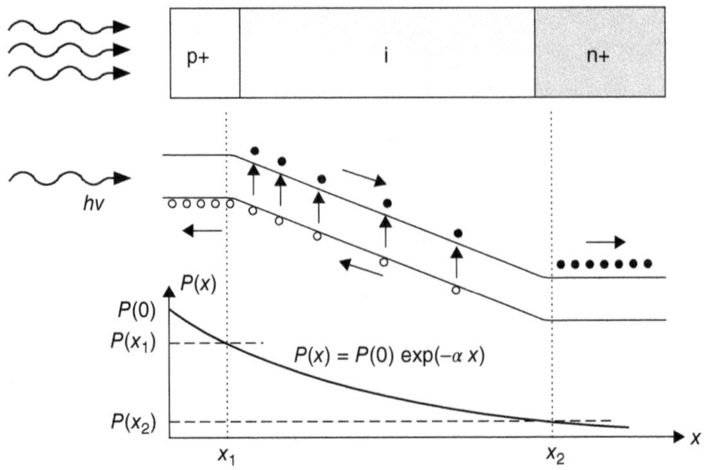

Figure 12.17 Schematic diagram illustrating the operation of a pin photodiode.

12.3.3 *Optical Detectors*

The wide bandgap of SiC makes it transparent to visible light, since photons in the visible spectrum have energies less than the bandgap and are not absorbed. This is an advantage in applications where the desired optical signal may be overwhelmed by ambient light in the visible spectrum. As a result, SiC is primarily used for ultraviolet or "solar-blind" photodetectors. The most common photodetectors in SiC are pin photodiodes and avalanche photodiodes (APDs). We will begin our discussion with pin photodiodes, then consider the modifications required to convert them into APDs.

The basic structure of a pin photodiode, shown in Figure 12.17, consists of an n+ substrate with a thick lightly-doped region (the "i" region) and a thin heavily-doped p+ region at the surface. Ohmic contacts to the p+ region are spaced apart on the surface to permit the entry of light, and the surface is coated with an anti-reflective coating. In normal operation, the photodiode is reverse biased so that the i-region is fully depleted, and the electric field sweeps electrons toward the n+ region and holes toward the p+ region. Incident photons having energies greater than the bandgap can generate electron–hole pairs, and carriers generated in the i-region or within a diffusion length of the i-region are swept out of the device as photocurrent.

The optical power per unit area penetrating the semiconductor decreases exponentially with distance from the surface, and can be written

$$P(x) = P(0)\exp(-\alpha x) \tag{12.19}$$

where α is the *absorption coefficient*, measured in units of cm^{-1}. The absorption coefficient is a strong function of photon energy (or wavelength). Photons with energy less than the bandgap cannot generate electron–hole pairs, and hence are not absorbed. Photons with energy equal to the bandgap have enough energy to promote electrons from the top of the valence band to the bottom of the conduction band, but all such transitions must satisfy momentum conservation (k conservation). Since SiC is an indirect bandgap semiconductor, states at the bottom of the conduction band do not lie at the same k value as states at the top of the valence band, and direct transitions between the two are not possible. At photon energies somewhat greater than the bandgap, pairs of valence and conduction band states can be found at the same k values with ΔE equal to the photon energy. The lowest energy at which this occurs represents the *absorption edge* of SiC, and the absorption coefficient increases rapidly at higher energies as more states that satisfy k-conservation become available. Figure 12.18 plots the absorption coefficient for several semiconductors

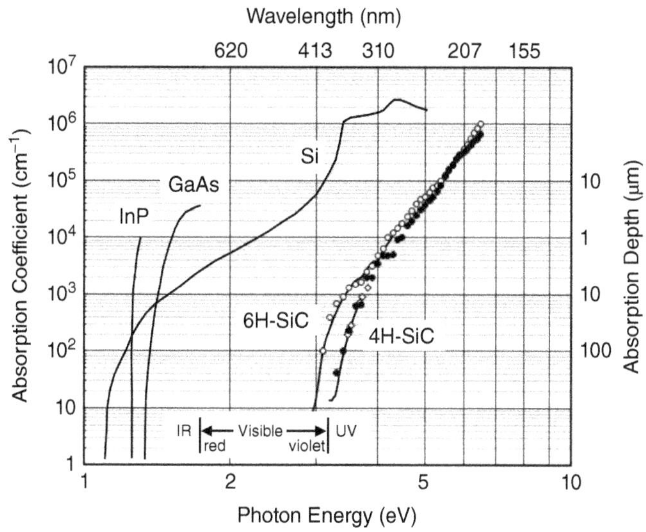

Figure 12.18 Room temperature absorption coefficients for several semiconductors. Measurements below 350 nm are difficult, and the data should be regarded as approximate.

as a function of photon energy [40–44]. The *absorption depth* $1/\alpha$ is shown on the right-hand axis. This is the depth at which the optical power has fallen to $1/e$ of its value at the surface. The visible portion of the spectrum lies between about 1.8 and 3.2 eV. Compared to the more conventional semiconductors, 4H- and 6H-SiC are relatively non-absorptive in the visible spectrum, which makes them ideal for use as UV photodetectors where response to visible light is to be minimized.

Returning to Figure 12.17, we now wish to define some relevant figures of merit for the photodiode. The *responsivity* R is defined as the photocurrent flowing out of the terminals, divided by the incident optical power reaching the surface, $AP(0)$. We assume all electron–hole pairs generated in the depletion region and in the neutral regions within a diffusion length of the depletion edge are separated by the electric field and flow out of the terminals as photocurrent. This defines the *collection region,* indicated by x_1 and x_2 in Figure 12.17. Applying Equation 12.19, the optical power per unit area absorbed in the collection region is

$$P_{\text{ABS}} = P(x_1) - P(x_2) = P(0)[\exp(-\alpha x_1) - \exp(-\alpha x_2)] \tag{12.20}$$

This may be written as

$$P_{\text{ABS}} = P(0)\exp(-\alpha x_1)[1 - \exp(-\alpha \Delta x)] \tag{12.21}$$

where $\Delta x = x_2 - x_1$ is the width of the collection region. The factors preceding the square brackets represent the optical power reaching the collection region at $x = x_1$, and the bracketed factor represents the fraction of that power absorbed within the collection region. Since all the electrons and holes generated within the collection region flow out of the terminals, the photocurrent may be written

$$I_{\text{PH}} = qA\frac{P_{\text{ABS}}}{h\nu} \tag{12.22}$$

where h is Planck's constant, υ is the photon frequency, and $h\upsilon$ is the photon energy E_{PH}. Inserting Equation 12.21 into Equation 12.22 yields

$$I_{\text{PH}} = qA\frac{P(0)}{h\nu}\exp(-\alpha x_1)[1 - \exp(-\alpha \Delta x)] = qA\frac{P(0)}{h\nu}\eta \tag{12.23}$$

where η is the *quantum efficiency*,

$$\boxed{\eta = \exp(-\alpha x_1)[1 - \exp(-\alpha \Delta x)]} \tag{12.24}$$

The quantum efficiency can also be interpreted as the photocurrent, measured in electrons per second, divided by the optical flux, measured in photons per second. Using Equation 12.23, the responsivity can be written as

$$\boxed{R = \frac{I_{PH}}{AP(0)} = \frac{q\eta}{h\nu} = \frac{q\lambda}{hc}\eta} \quad [\text{A/W}] \tag{12.25}$$

where c is the speed of light in vacuum. In using Equations 12.24 and 12.25, we must keep in mind that the quantum efficiency and the responsivity depend on the wavelength, through the energy dependence of the absorption coefficient shown in Figure 12.18. It is also useful to recall that energy and wavelength are related by

$$E_{PH} = \frac{hc}{\lambda} = \frac{1.24\ (\text{eV})}{\lambda\ (\mu\text{m})} \tag{12.26}$$

The wavelength dependence of responsivity can be understood as follows. At short wavelengths (high photon energies) the absorption coefficient α is large, and most of the incident photons are absorbed near the surface. Consequently, few photons reach the collection region. Although photogeneration occurs in the near-surface region, there is no electric field to separate the carriers, and they recombine before they can contribute to current. At long wavelengths the photons have insufficient energy to generate electron–hole pairs, and the absorption coefficient is very small. Consequently, most of the photons pass completely through the collection region without being absorbed. At intermediate wavelengths such that the absorption depth $1/\alpha$ lies within the collection region, significant photocurrent can be obtained, and the responsivity is a maximum.

Figure 12.19 shows the measured responsivity of a 4H-SiC photodiode [45]. This particular device is an APD, which will be discussed below, but the figure is used here to illustrate the relationship between responsivity R and quantum efficiency η. For a constant value of quantum efficiency, responsivity increases linearly with wavelength, as shown by Equation 12.25, so lines of constant quantum efficiency correspond to the diagonal lines in the figure. Two experimental curves are shown, one for a bare photodiode and one for a photodiode with a HfO_2/SiO_2 dielectric coating that acts as a narrow-band optical filter. The filtered diode achieves a quantum efficiency of 40% over a narrow band from 260 to 280 nm, and has a rejection ratio of 10^6 to photons in the visible spectrum.

Avalanche photodiodes (APDs) resemble the pin photodiode of Figure 12.17, but have a region where photogenerated carriers can impact ionize, increasing the current. Figure 12.20 shows a lo–hi–lo APD structure where a thin, heavily-doped n layer is inserted partway through the n-collection region. Under reverse bias, the electric field is relatively constant in the lightly-doped regions, but rises rapidly through the heavily-doped layer, resulting in the field profile shown. Electron–hole pairs generated in the collection region are separated by the electric field, with the electrons drifting toward the n+ contact and holes drifting toward the high-field region. If the field in the avalanche region is close to the critical field, holes can impact ionize as they drift through this layer, creating new electron–hole pairs that add to the current.

It is important that holes are drawn toward the avalanche region, and not electrons. Holes have a higher ionization coefficient than electrons, as shown in Figure 10.1, and they undergo impact ionization at lower fields than electrons. With this APD design, the field in the avalanche region can be adjusted so that holes ionize but electrons do not. This ensures that electrons generated in the avalanche region do not initiate new ionization events, which could lead to runaway breakdown.

It is possible to construct very sensitive photodetectors using the APD structure, since many electron–hole pairs may be produced for each incident photon. Unfortunately, the avalanche process

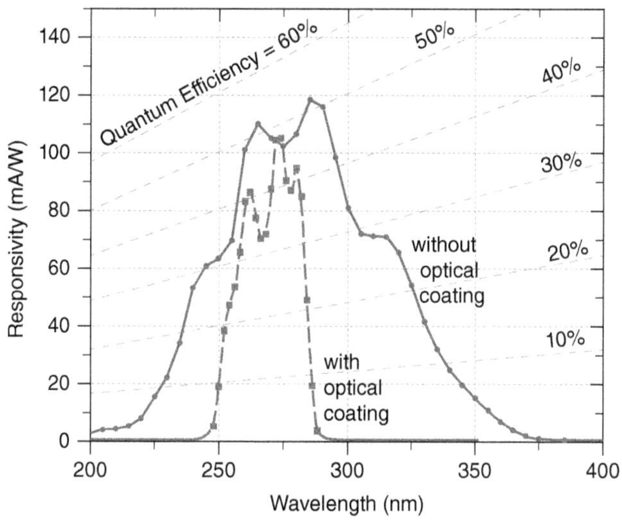

Figure 12.19 Responsivity of a 4H-SiC avalanche photodiode, with and without an optical coating. The diagonal lines are loci of constant quantum efficiency η ([45] reproduced with permission from Trans Tech Publications).

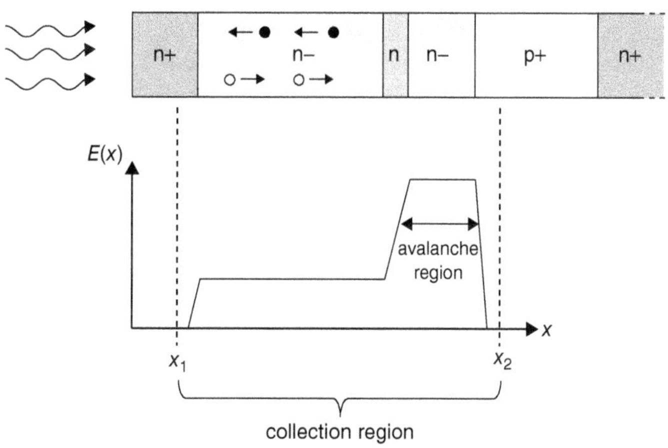

Figure 12.20 Structure of a lo–hi–lo avalanche photodiode.

is inherently noisy, due to the random nature of impact ionization. However, it often happens that the photodiode noise is overshadowed by noise in the circuits used to amplify the optical signal, so that introducing electrical gain into the photodiode using the avalanche process may not degrade the overall signal-to-noise ratio of the system. By careful design it is possible, using SiC APDs, to achieve performance that exceeds the capability of photomultiplier tubes, as we shall see in a moment.

To discuss detector sensitivity, we need to introduce two new parameters. The *noise equivalent power* (NEP) is defined as the optical power that produces a detector current equal to the rms noise current of the detector per unit bandwidth. Stated a different way, the NEP is the optical power in the signal required to produce a unity signal-to-noise ratio when measured over a unit bandwidth. Thus, the signal-to-noise ratio may be expressed as the ratio of the actual incident optical power (the signal) to the noise-equivalent optical power,

$$\text{SNR} = P(0)/\text{NEP} \tag{12.27}$$

Clearly, a smaller value of NEP (i.e., lower noise) results in a larger signal-to-noise ratio for a given optical signal.

The noise current in the photodiode arises primarily from shot noise due to both the signal current and the thermally-generated leakage current in the diode (dark current). The thermal generation rate in a semiconductor is given in Equation 7.35. In the depletion region of a reverse-biased diode, the electron and hole concentrations are negligible and, assuming midgap traps, Equation 7.35 reduces to

$$G \approx n_i/(\tau_N + \tau_P) \tag{12.28}$$

This tells us that the dark current is proportional to the intrinsic carrier concentration. Since n_i in SiC is many orders-of-magnitude lower than in silicon, the dark current and associated noise in a SiC photodiode are also orders-of-magnitude lower than in silicon. This leads to a lower NEP, and is a significant advantage for the detection of single-photon events.

The *specific detectivity* D^* is defined as the inverse of the NEP for a diode of unit area measured over unit bandwidth, or

$$D^* = \frac{\sqrt{AB}}{NEP}\ (\text{cm Hz}^{1/2}\ \text{W}^{-1}) \tag{12.29}$$

where A is the area of the diode and B is the measurement bandwidth. The low NEP of SiC photodiodes is expected to lead to higher values of detectivity. Figure 12.21 shows the specific detectivity for several types of photodetectors, including SiC pin and Schottky diodes. Also shown for comparison is a conventional photomultiplier tube [46]. The figure shows that the best SiC photodiodes have performance comparable to photomultiplier tubes. By operating a SiC APD very near breakdown, multiplication gains of 10^6 have been demonstrated, making it possible to achieve single-photon detection [38].

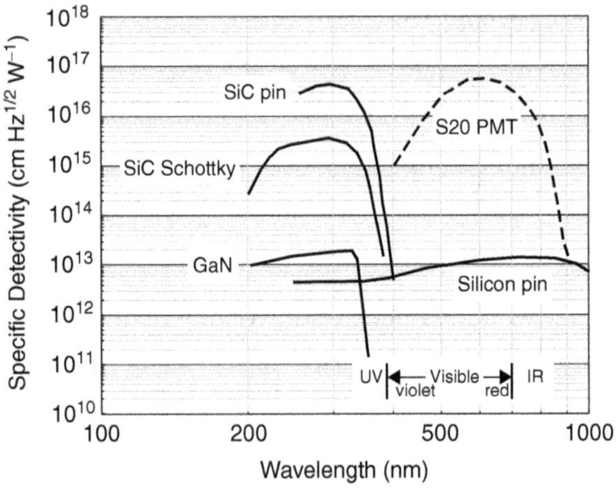

Figure 12.21 Specific detectivity for several types of photodetectors, including SiC pin and Schottky diodes and conventional photomultiplier tubes ([46] reproduced with permission from Trans Tech Publications).

References

[1] Allen, S.T., Sheppard, S.T., Pribble, W.L. *et al.* (1999) Recent progress in SiC microwave MESFETs, in *Materials Research Society Symposium Proceedings*, vol. **572** (eds S.C. Binari, A.A. Burk, M.R. Melloch and C. Nguyen), Materials Research Society, Warrendale, PA, pp. 15–22.

[2] Cai, S., Li, L., Wang, L. *et al.* (2012) S-band 300 W output SiC MESFET. *ECS Trans.*, **50** (3), 333–339.

[3] Pierret, R.F. (1996) *Semiconductor Device Fundamentals*, Addison-Wesley, Reading, PA.

[4] Sze, S.M. and Ng, K.K. (2007) *Physics of Semiconductor Devices*, 3rd edn, John Wiley & Sons, Inc., Hoboken, NJ.

[5] Watanabe Y. and Nishizawa J. (1950) Japan Patent 205060, Published no. 28-6077, Dec. 1950.

[6] Nishizawa, J. (1972) A low impedance field effect transistor. International Electron Devices Meeting Technical Digest, IEEE, New York, pp. 144–147.

[7] Nishizawa, J.I., Terasaki, T. and Shibata, J. (1975) Field-effect transistor versus analog transistor (static induction transistor). *IEEE Trans. Electron. Devices*, **22** (4), 185–197.

[8] Siergiej, R. R., Clarke, R. C., Agarwal, A. K. *et al.* (1995) High power 4H-SiC static induction transistors. International Electron Devices Meeting Technical Digest, IEEE, New York, pp. 353–356.

[9] Sriram, S., Siergiej, R.R., Clarke, R.C. and Agarwal, A.K. (1997) SiC for microwave power transistors. *Phys. Status Solidi A*, **162**, 441–457.

[10] Przadka, A. (1999) High frequency, high power static induction transistors in silicon carbide: simulation and fabrication. PhD thesis. Purdue University, West Lafayette.

[11] Henning, J.P., Przadka, A., Melloch, M.R. and Cooper, J.A. (2000) A novel self-aligned fabrication process for microwave static induction transistors in silicon carbide. *IEEE Electron. Device Lett.*, **21** (12), 578–580.

[12] Bojko, R.J., Siergiej, R.R., Eldridge, G.W. *et al.* (1998) Recent progress in 4H-SiC static induction transistors for high frequency power generation. IEEE Device Research Conference Digest, Charlottesville, VA, pp. 96–97.

[13] Agarwal, A.K., Chen, L.-S., Eldridge, G.W. *et al.* (1998) Ion-implanted static induction transistors in 4H-SiC. IEEE Device Research Conference Digest, Charlottesville, VA, pp. 94–95.

[14] Clarke, R.C., Morse, A.W., Esker, P., and Curtice, W.R. (2000) A 16 W, 40% efficient, continuous wave, 4H-SiC, L-band SIT. Proceedings of 2000 IEEE/Cornell Conference on High Performance Devices, pp. 141–143.

[15] Morse, A.W., Esker, P.M., Clarke, R.C. *et al.* (1996) Application of high power silicon carbide transistors at radar frequencies. IEEE MTT-S Digest, pp. 677–680.

[16] Siergiej, R.R., Morse, A.W., Esker, P.M. *et al.* (1997) High gain 4H-SiC static induction transistors using novel sub-micron airbridging. IEEE Device Research Conference Digest, Boulder, CO, pp. 136–137.

[17] Cooper, J.A. (1992) Design Curves for Transient Heat Flow in Semiconductor Devices. ECE Technical Reports 284, Purdue University, West Lafayette.

[18] Yuan, L., Cooper, J. A., Melloch, M. R. and Webb, K. J. (2001) Experimental demonstration of a silicon carbide IMPATT oscillator. *IEEE Electron. Device Lett.* **22** (6) 266 – 268.

[19] Neudeck, P.G., Okojie, R.S. and Chen, L.-Y. (2002) High-temperature electronics – a role for wide bandgap semiconductors? *Proc. IEEE*, **90** (6), 1065–1076.

[20] Okojie, R.S., Lukco, D., Chen, Y.L. *et al.* (2001) Reaction kinetics of thermally stable contact metallization on 6H-SiC. *Mater. Res. Soc. Symp. Proc.*, **640**, H7.5.1–H7.5.6.

[21] Okojie, R.S., Lukco, D., and Spry, D. (2001) Reliability of Ti/TaSi /Pt ohmic contacts on 4H-and 6H-SiC after 1000 hours in air at 600 C. Proceedings of the Electronic Materials Conference, Notre Dame, IN, p. 6.

[22] Lee, J.-Y., Singh, S. and Cooper, J.A. (2008) Demonstration and characterization of bipolar monolithic integrated circuits in 4H-SiC. *IEEE Trans. Electron. Devices*, **55** (8), 1946–1953.

[23] Singh, S. and Cooper, J.A. (2011) Bipolar integrated circuits in 4H-SiC. *IEEE Trans. Electron. Devices*, **58** (4), 1084–1090.

[24] Lanni, L., Ghandi, R., Malm, B.G. *et al.* (2012) Design and characterization of high-temperature ECL-based bipolar integrated circuits in 4H-SiC. *IEEE Trans. Electron. Devices*, **59** (4), 1076–1083.

[25] Neudeck, P.G., Garverick, S.L., Spry, D.J. *et al.* (2009) Extreme temperature 6H-SiC JFET integrated circuit technology. *Phys. Status Solidi A*, **206** (10), 2329–2345.

[26] Wright, N.G. and Horsfall, A.B. (2007) SiC sensors: a review. *J. Phys. D*, **40** (20), 6345–6354.

[27] Wright, N.G., Horsfall, A.B. and Vassilevski, K. (2008) Prospects for SiC electronics and sensors. *Mater. Today*, **11** (1–2), 16–21.

[28] Zorman, C.A. and Parro, R.J. (2008) Micro- and nanomechanical structures for silicon carbide MEMS and NEMS. *Phys. Status Solidi B*, **245** (7), 1401–1424.

[29] Shor, J.S. and Kurtz, A.D. (1994) Photoelectrochemical etching of 6H-SiC. *J. Electrochem. Soc.*, **141** (3), 778–781.

[30] Ziermann, R., von Berg, J., Obermeier, E. *et al.* (1999) High temperature piezoresistive β-SiC-on-SOI pressure sensor with on chip SiC thermistor. *Mater. Sci. Eng., B*, **61–62** (7), 576–578.

[31] Young, D.J., Du, J., Zorman, C.A. *et al.* (2004) High-temperature single-crystal 3C-SiC capacitive pressure sensor. *IEEE Sens. J.*, **4** (4), 464–470.

[32] Okojie, R.S., Ned, A.A. and Kurtz, A.D. (1998) Operation of α(6H)-SiC pressure sensor at 500 °C. *Sens. Actuators, A*, **66** (1–3), 200–204.

[33] Atwell, A.R., Okojie, R.S., Kornegay, K.T. *et al.* (2003) Simulation, fabrication, and testing of bulk micromachined 6H-SiC high-g piesoresistive accelerometers. *Sens. Actuators, A*, **104** (1), 11–18.

[34] Kotzar, G., Freas, M., Abel, P. *et al.* (2002) Evaluation of MEMS materials of construction for implantable medical devices. *Biomaterials*, **23** (13), 2737–2750.

[35] Rosenbloom, A.J., Sipe, D.M., Shishkin, Y. *et al.* (2004) Nanoporous SiC: a candidate semi-permeable material for biomedical applications. *Biomed. Microdevices*, **6** (4), 261–267.

[36] Gabriel, G., Erill, I., Caro, J. *et al.* (2007) Manufacturing and full characterization of silicon carbide-based multi-sensor micro-probes for biomedical applications. *Microelectron. J.*, **38** (3), 406–415.

[37] Lloyd Spetz, A., Uneus, L., Svenningstorp, H. *et al.* (2001) SiC based field effect gas sensors for industrial applications. *Phys. Status Solidi A*, **185** (1), 15–25.

[38] Ghosh, R.N., Tobias, P., Ejakov, S.G. and Golding, B. (2002) Interface states in high temperature SiC gas sensing. *Proc. IEEE*, **2**, 1120–1125.

[39] Matocha, K., Tilak, V., Sandvik, P. and Tucker, J. (2005) High-temperature SiC MOSFET gas sensors. Material Research Society Symposium Proceedings, vol. **828**, pp. A7.9.1–A7.9.6.

[40] Philipp, H.R. and Taft, E.A. (1960) in *Silicon Carbide – A High Temperature Semiconductor* (eds J.K. O'Connor and J. Smieltens), Pergamon Press, Oxford, London, New York, Paris, p. 366.

[41] Panferov, A. and Kurinec, S.K. (2011) Modeling quantum efficiency of ultraviolet 6H-SiC photodiodes. *IEEE Trans. Electron. Devices*, **58** (11), 3976–3983.

[42] Sridhara, S.G., Devaty, R.P. and Choyke, W.J. (1998) Absorption coefficient of 4H silicon carbide from 3900 to 3250 Å. *J. Appl. Phys.*, **84** (5), 2963–2964.

[43] Sridhara, S.G., Eperjesi, T.J., Devaty, R.P. and Choyke, W.J. (1999) Penetration depths in the ultraviolet for 4H-, 6H-, and 3C-silicon carbide at seven common laser pumping wavelengths. *Mater. Sci. Eng., B*, **61–62** (7), 229–233.

[44] Zollner, S., Chen, J.G., Duda, E. *et al.* (1999) Dielectric functions of bulk 4H and 6H SiC and spectroscopic ellipsometry studies of thin SiC films on Si. *J. Appl. Phys.*, **85** (12), 8353–8361.

[45] Vert, A., Soloviev, S. and Sandvik, P. (2010) Performance of silicon carbide photodiode arrays and photomultipliers. *Mater. Sci. Forum*, **645–648**, 1069–1072.

[46] Yan, F., Xin, X., Alexandrov, P. *et al.* (2006) Development of ultra high sensitivity UV silicon carbide detectors. *Mater. Sci. Forum*, **527–529**, 1461–1464.

Appendix A

Incomplete Dopant Ionization in 4H-SiC

Dopants atoms in silicon carbide are incorporated substitutionally in place of either a silicon or carbon atom in the hexagonal crystal lattice. Due to the stacking sequence of the polytype, not all silicon or carbon sites are equivalent in terms of their surroundings, and each donor or acceptor can exhibit multiple site-dependent energy levels. Dopants on cubic sites typically have higher ionization energies than dopants on hexagonal sites.

Aluminum is the principal p-type dopant in 4H-SiC, and occupies either a hexagonal or a cubic silicon site, having ionization energies of 197.9 and 201.3 meV respectively [1]. The primary n-type dopants are nitrogen and phosphorus. Nitrogen substitutes for carbon, and on a hexagonal C-site it has an ionization energy of 61.4 meV. Phosphorus substitutes for silicon, and on a cubic Si-site it has an ionization energy of 60.7 meV [1].

Because of its high ionization energy, aluminum acceptors in neutral regions of 4H-SiC are not fully ionized at room temperature. The density of ionized acceptors (and donors) is denoted by N_A^- (and N_D^+), respectively. Incomplete ionization has a profound effect on device performance, and must be taken into consideration in our equations. Nitrogen and phosphorus donors have lower ionization energies and tend to be fully ionized at room temperature, so in most cases $N_D^+ = N_D$. However, it is important to include the correct expression for the ionized acceptor density in all equations.

In depletion regions, the band bending moves the donor and acceptor energy levels far enough from the Fermi energy that even deep aluminum acceptors are fully ionized (occupied by electrons) at room temperature. In determining whether to use the ionized concentration N_A^- or the total concentration N_A in an equation, we must decide whether the expression refers to the *charge density in a depletion region* (use N_A), or to the *carrier density in a neutral region* (use N_A^-). In all equations developed in this book, care has been taken to distinguish these two situations and include the correct representations.

We now consider numerical values typical of 4H-SiC. The equilibrium hole and electron concentrations in extrinsic material ($N_A \gg N_D$ or $N_D \gg N_A$) at moderate temperatures ($n_i \ll N_{A,D}$) are given by $p = N_A^-$ (p-type material) or $n = N_D^+$ (n-type material). The ionized dopant concentrations N_A^- or N_D^+ in a neutral region can be calculated from the charge neutrality condition, and the equilibrium density of holes in p-type material can be written [2]

$$p = N_A^- = \frac{\eta}{2}\left(\sqrt{1 + \frac{4N_A}{\eta}} - 1\right) \tag{A1}$$

Fundamentals of Silicon Carbide Technology: Growth, Characterization, Devices, and Applications, First Edition.
Tsunenobu Kimoto and James A. Cooper.

where η is given by

$$\eta = \frac{N_V}{g_A} \exp\left(-\frac{E_A - E_V}{kT}\right) \tag{A2}$$

Here E_A is the energy level of the acceptor impurity, g_A is the degeneracy factor for acceptors (typically taken as 4), and N_V is the effective density of states in the valence band, given by

$$N_V = 2\left(\frac{2\pi m_{dh}^* kT}{h^2}\right)^{3/2} \tag{A3}$$

where m_{dh}^* is the density-of-states effective mass for holes and h is Planck's constant. Similarly, the equilibrium density of electrons in n-type material is given by

$$n = N_D^+ = \frac{\gamma}{2}\left(\sqrt{1 + \frac{4N_D}{\gamma}} - 1\right) \tag{A4}$$

where

$$\gamma = \frac{N_C}{g_D} \exp\left(-\frac{E_C - E_D}{kT}\right) \tag{A5}$$

Here E_D is the donor energy level, g_D is the degeneracy factor for donors (typically taken as 2), and N_C is the effective density of states in the conduction band, given by

$$N_C = 2\left(\frac{2\pi m_{de}^* kT}{h^2}\right)^{3/2} \tag{A6}$$

where m_{de}^* is the density-of-states effective mass for electrons.

Figure A.1 shows the ionization fraction for aluminum acceptors in neutral regions of 4H-SiC, computed using an ionization energy of 200 meV. Room temperature is indicated by the dashed line. At a

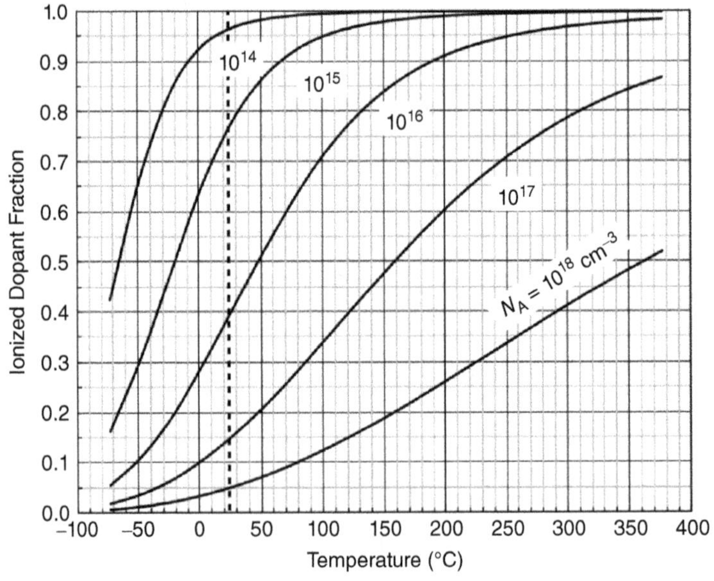

Figure A.1 Ionization fraction for aluminum acceptors in 4H-SiC, computed using an ionization energy of 200 meV. Room temperature is indicated by the dashed line.

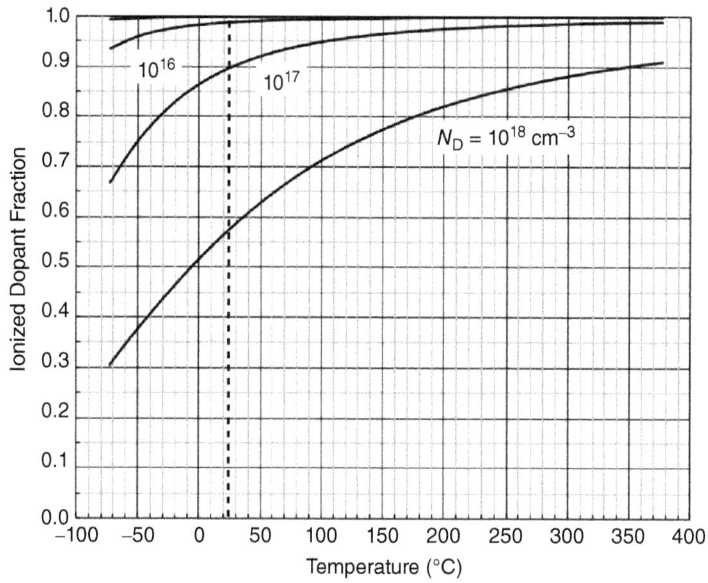

Figure A.2 Ionization fraction for nitrogen or phosphorus donors in 4H-SiC, computed using an ionization energy of 61 meV. Room temperature is indicated by the dashed line.

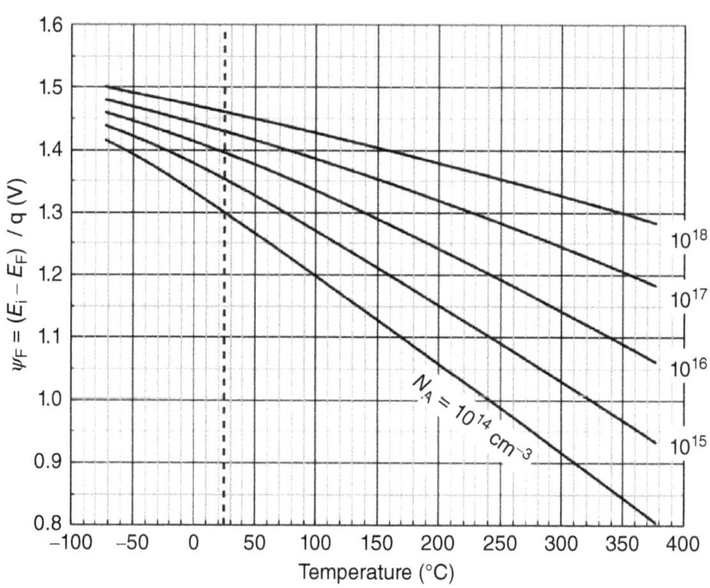

Figure A.3 Fermi potential in aluminum-doped 4H-SiC as a function of doping and temperature.

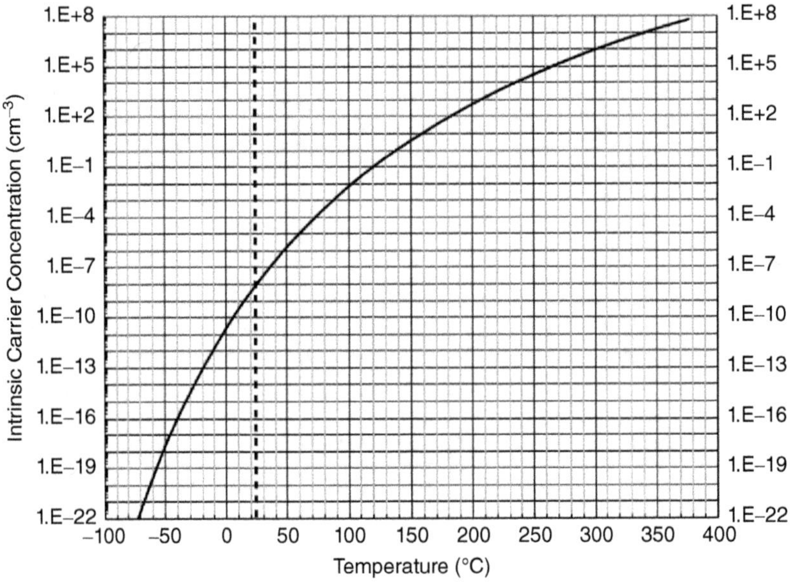

Figure A.4 Intrinsic carrier concentration in 4H-SiC as a function of temperature. Room temperature is indicated by the dashed line.

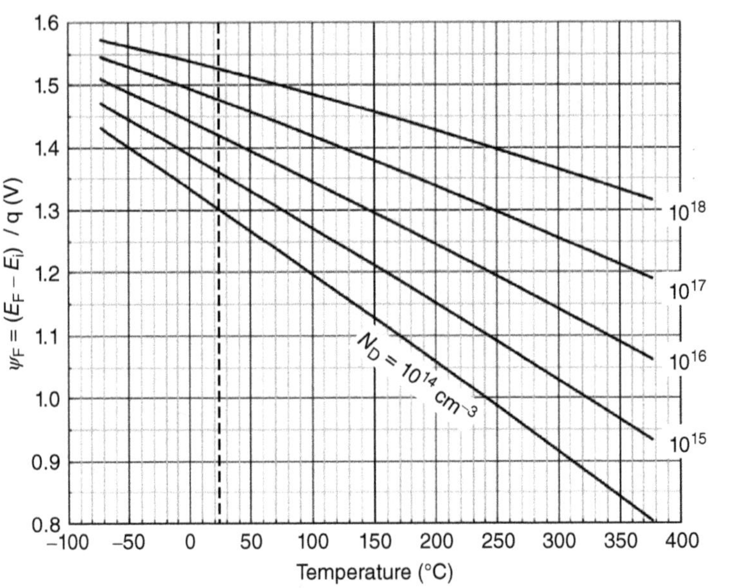

Figure A.5 Fermi potential in nitrogen or phosphorus-doped 4H-SiC as a function of doping and temperature.

doping of 1×10^{17} cm^{-3}, only about 15% of the acceptors are ionized at room temperature, and the equilibrium hole concentration is only about 1.5×10^{16} cm^{-3}. The ionization fraction increases with temperature, reaching about 75% at 300 °C.

Figure A.2 shows the ionization fraction for nitrogen or phosphorus donors in 4H-SiC, computed using an ionization energy of 61 meV. At a doping of 1×10^{17} cm^{-3}, approximately 90% of the donor atoms are ionized at room temperature.

Figure A.3 shows the Fermi potential ψ_F for aluminum-doped 4H-SiC calculated using Equations 8.21 and A1–A3, including the dependence of E_G on temperature. As temperature increases, the Fermi level moves closer to the midgap, and the change is more dramatic for lighter dopings. The reduction in ψ_F is caused by the rapid increase of n_i with temperature, as shown in Figure A.4. Figure A.5 shows similar calculations for the Fermi potential in nitrogen or phosphorus-doped 4H-SiC.

The above equations assume that the doping is not so high as to create an impurity band. However, at doping levels above 10^{19} cm^{-3} the mean spacing between dopant atoms is less than 5 nm, and electron wavefunctions from adjacent atoms overlap, giving rise to an impurity band that reduces the effective bandgap. This reduces the dopant ionization energies $E_{A,D}$, leading to more complete ionization than predicted by Equations A1 and A4.

References

[1] Ivanov, I.G., Henry, A. and Janzen, E. (2005) Ionization energies of phosphorus and nitrogen donors and aluminum acceptors in 4H silicon carbide from the donor-acceptor pair emission. *Phys. Rev. B*, **71** (24), 241201-1–241201-4.

[2] Pierret, R.F. (2003) *Advanced Semiconductor Fundamentals*, Pearson Education, Inc., Upper Saddle River, NJ.

Appendix B

Properties of the Hyperbolic Functions

Hyperbolic functions occur frequently when describing conduction in bipolar devices, and it is helpful to list some of their important properties here. The hyperbolic functions are defined as follows:

$$\begin{aligned}
\sinh(\theta) &= \frac{\exp(\theta) - \exp(-\theta)}{2} \\
\cosh(\theta) &= \frac{\exp(\theta) + \exp(-\theta)}{2} \\
\tanh(\theta) &= \frac{\sinh(\theta)}{\cosh(\theta)} = \frac{\exp(\theta) - \exp(-\theta)}{\exp(\theta) + \exp(-\theta)} \\
\coth(\theta) &= \frac{1}{\tanh(\theta)} = \frac{\cosh(\theta)}{\sinh(\theta)} = \frac{\exp(\theta) + \exp(-\theta)}{\exp(\theta) - \exp(-\theta)} \\
\mathrm{sech}(\theta) &= \frac{1}{\cosh(\theta)} = \frac{2}{\exp(\theta) + \exp(-\theta)} \\
\mathrm{csch}(\theta) &= \frac{1}{\sinh(\theta)} = \frac{2}{\exp(\theta) - \exp(-\theta)}
\end{aligned} \tag{B1}$$

These functions can be easily visualized, as shown in Figures B.1 and B.2. From these figures we see that *sinh, tanh, coth,* and *csch* are odd functions, positive for positive arguments and negative for negative arguments. *cosh* and *sech* are even functions, positive for both positive and negative arguments. Thus we can write

$$\begin{aligned}
\sinh(-\theta) &= -\sinh(\theta) \\
\cosh(-\theta) &= \cosh(\theta) \\
\tanh(-\theta) &= -\tanh(\theta) \\
\coth(-\theta) &= -\coth(\theta) \\
\mathrm{sech}(-\theta) &= \mathrm{sech}(\theta) \\
\mathrm{csch}(-\theta) &= -\mathrm{csch}(\theta)
\end{aligned} \tag{B2}$$

Fundamentals of Silicon Carbide Technology: Growth, Characterization, Devices, and Applications, First Edition.
Tsunenobu Kimoto and James A. Cooper.

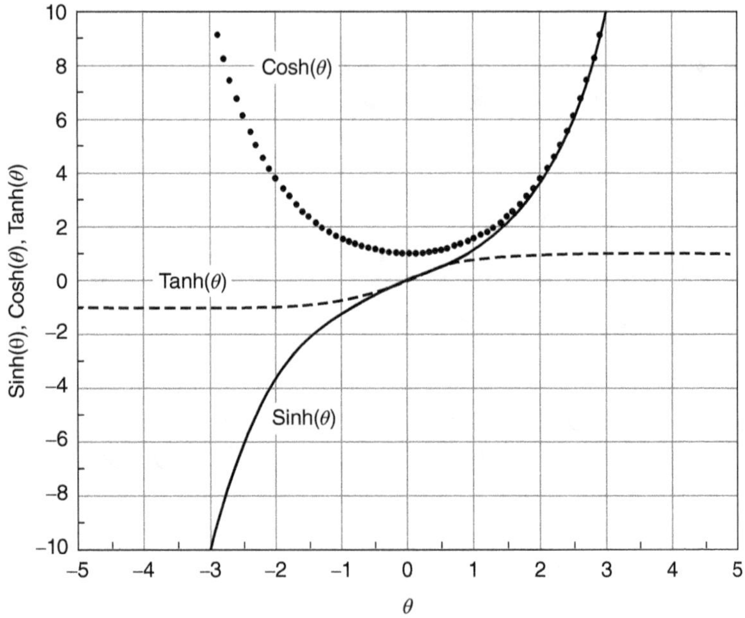

Figure B.1 Hyperbolic sine, cosine, and tangent functions.

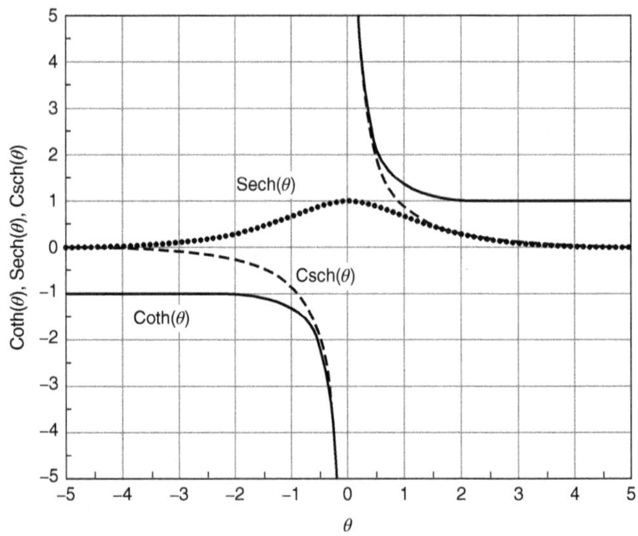

Figure B.2 Hyperbolic cotangent, secant, and cosecant functions.

It is also useful to note that

$$\begin{aligned}
\cosh^2(\theta) - \sinh^2(\theta) &= 1 \\
\coth^2(\theta) - \mathrm{csch}^2(\theta) &= 1 \\
\tanh^2(\theta) + \mathrm{sech}^2(\theta) &= 1
\end{aligned} \tag{B3}$$

Hyperbolic functions of the sum of two arguments can be written

$$\begin{aligned}
\sinh(\theta + \phi) &= \cosh(\theta)\sinh(\phi) + \sinh(\theta)\cosh(\phi) \\
\cosh(\theta + \phi) &= \sinh(\theta)\sinh(\phi) + \cosh(\theta)\cosh(\phi)
\end{aligned} \tag{B4}$$

The inverse hyperbolic functions can be written in terms of logarithms as

$$\begin{aligned}
\sinh^{-1}(\theta) &= \ln\left(\theta + \sqrt{\theta^2 + 1}\right) \\
\cosh^{-1}(\theta) &= \ln\left(\theta + \sqrt{\theta^2 - 1}\right), \ \theta \geq 1 \\
\tanh^{-1}(\theta) &= \frac{1}{2}\ln\left(\frac{1+\theta}{1-\theta}\right), \ |\theta| < 1 \\
\coth^{-1}(\theta) &= \frac{1}{2}\ln\left(\frac{1+\theta}{1-\theta}\right), \ |\theta| > 1 \\
\mathrm{sech}^{-1}(\theta) &= \ln\left(\frac{1+\sqrt{1-\theta^2}}{\theta}\right), \ 0 < \theta \leq 1 \\
\mathrm{csch}^{-1}(\theta) &= \ln\left(\frac{1}{\theta} + \frac{\sqrt{1+\theta^2}}{|\theta|}\right), \ \theta \neq 0
\end{aligned} \tag{B5}$$

The derivatives of the hyperbolic functions are

$$\begin{aligned}
\frac{d}{d\theta}\sinh(\theta) &= \cosh(\theta) \\
\frac{d}{d\theta}\cosh(\theta) &= \sinh(\theta) \\
\frac{d}{d\theta}\tanh(\theta) &= \mathrm{sech}^2(\theta) = \frac{1}{\cosh^2(\theta)} = 1 - \tanh^2(\theta) \\
\frac{d}{d\theta}\coth(\theta) &= -\mathrm{csch}^2(\theta) = -\frac{1}{\sinh^2(\theta)} = 1 - \coth^2(\theta) \\
\frac{d}{d\theta}\mathrm{csch}(\theta) &= -\coth(\theta)\mathrm{csch}(\theta) \\
\frac{d}{d\theta}\mathrm{sech}(\theta) &= -\tanh(\theta)\mathrm{sech}(\theta)
\end{aligned} \tag{B6}$$

The equations dealing with conduction in bipolar devices often involve integrals of hyperbolic functions. Some useful integrals are listed below:

$$\int \frac{d\theta}{[\sinh(\theta) + a]} = \frac{2\tanh^{-1}\left[\dfrac{a\tanh(\theta/2) - 1}{\sqrt{a^2+1}}\right]}{\sqrt{a^2+1}}, \ a > 0, \ \theta > 0 \tag{B7}$$

$$\int \frac{d\theta}{\cosh(\theta) + a\sinh(\theta)} = \frac{2\tan^{-1}\left[\dfrac{a + \tanh(\theta/2)}{\sqrt{1-a^2}}\right]}{\sqrt{1-a^2}} \tag{B8}$$

$$\int \frac{\sinh(\theta) + a\cosh(\theta)}{\cosh(\theta) + a\sinh(\theta)} d\theta = \ln[\cosh(\theta) + a\sinh(\theta)] \tag{B9}$$

$$\int \frac{[\sinh(\theta) + b]\cosh(\theta)}{[\sinh(\theta) + a][\sinh(\theta) + c]} d\theta = \frac{(a-b)\ln[\sinh(\theta) + a] + (b-c)\ln[\sinh(\theta) + c]}{a-c} \tag{B10}$$

Appendix C

Major Physical Properties of Common SiC Polytypes

C.1 Properties

All measurements at room temperature if not specified otherwise. Values taken from [1–19].

Properties/polytype	3C-SiC	4H-SiC	6H-SiC
Stacking sequence	ABC	ABAC	ABCACB
Bandgap (eV)	2.36	3.26	3.02
Exciton gap (eV), 2 K	2.390	3.265	3.023
Lattice constant			
a (Å)	4.3596	3.0798	3.0805
c (Å)	–	10.0820	15.1151
Density (g cm^{-3})	3.21	3.21	3.21
Electron effective mass			
$m_{//}(m_0)$	0.67	0.33	2.0
$m_{\perp}(m_0)$	0.25	0.42	0.48
Hole effective mass			
$m_{//}(m_0)$	~1.5	1.75	1.85
$m_{\perp}(m_0)$	~0.6	0.66	0.66
Number of conduction band minima	3	3	6
Effective density of states in the conduction band (cm^{-3})	1.5×10^{19}	1.8×10^{19}	8.8×10^{19}
Effective density of states in the valence band (cm^{-3})	1.9×10^{19}	2.1×10^{19}	2.2×10^{19}

(*continued overleaf*)

Fundamentals of Silicon Carbide Technology: Growth, Characterization, Devices, and Applications, First Edition.
Tsunenobu Kimoto and James A. Cooper.

Properties/polytype	3C-SiC	4H-SiC	6H-SiC
Intrinsic carrier density (cm^{-3})	0.1	5×10^{-9}	1×10^{-6}
Electron mobility ($cm^2\ V^{-1}\ s^{-1}$) (at low doping)			
μ perpendicular to *c*-axis	~1000	1020	450
μ parallel to *c*-axis	**~1000**	**1200**	**100**
Hole mobility ($cm^2\ V^{-1}\ s^{-1}$) (at low doping)	100	120	100
Electron saturated drift velocity ($cm\ s^{-1}$)	$\sim 2 \times 10^{7a}$	2.2×10^7	1.9×10^7
Hole saturated drift velocity ($cm\ s^{-1}$)	$\sim 1.3 \times 10^{7a}$	$\sim 1.3 \times 10^{7a}$	$\sim 1.3 \times 10^{7a}$
Breakdown electric field ($MV\ cm^{-1}$) (at $N_D = 3 \times 10^{16}\ cm^{-3}$)			
E_B perpendicular to *c*-axis	1.4	2.2	1.7
E_B parallel to *c*-axis	**1.4**	**2.8**	**3.0**
Ionization energy of dopants (meV)			
Nitrogen (hexagonal/cubic)	55	61/126	85/140
Aluminum (hexagonal/cubic)	250	198/201	240
Relative dielectric constant			
ε_s perpendicular to *c*-axis	9.72	9.76	9.66
ε_s parallel to *c*-axis	9.72	10.32	10.03
Thermal conductivity ($W\ cm^{-1}\ K^{-1}$)	3.3–4.9[a]	3.3–4.9	3.3–4.9
Young modulus (GPa)	310–550	390–690	390–690
Poisson's ratio	0.24	0.21	0.21

[a]Estimated values.

C.2 Temperature and/or Doping Dependence of Major Physical Properties

1. *Temperature dependence of bandgap* [4]

$$E_g(T) = E_{g0} - \frac{\alpha T^2}{T + \beta} \tag{C1}$$

E_{g0}: bandgap at 0 K, T: absolute temperature,

$$\alpha = 8.2 \times 10^{-4} \text{eV K}^{-1}, \beta = 1.8 \times 10^3 \text{K}.$$

2. *Doping dependence of carrier mobility (perpendicular to the c-axis)* [12–17]

$$\mu_e(\text{4H–SiC}) = \frac{1020(T/300)^{-n}}{1 + \left(\dfrac{N_D + N_A}{1.8 \times 10^{17}}\right)^{0.6}}\ \text{cm}^2\ \text{V}^{-1}\ \text{s}^{-1} \tag{C2}$$

$$\mu_e(\text{6H–SiC}) = \frac{450(T/300)^{-n}}{1+\left(\dfrac{N_D+N_A}{2.5\times10^{17}}\right)^{0.6}}\ \text{cm}^2\ \text{V}^{-1}\ \text{s}^{-1} \tag{C3}$$

$$\mu_h(\text{4H–SiC}) = \frac{118(T/300)^{-m}}{1+\left(\dfrac{N_D+N_A}{2.2\times10^{18}}\right)^{0.7}}\ \text{cm}^2\ \text{V}^{-1}\ \text{s}^{-1} \tag{C4}$$

$$\mu_h(\text{6H–SiC}) = \frac{98(T/300)^{-m}}{1+\left(\dfrac{N_D+N_A}{2.4\times10^{18}}\right)^{0.7}}\ \text{cm}^2\ \text{V}^{-1}\ \text{s}^{-1} \tag{C5}$$

N_D: donor density, N_A: acceptor density given in units of cm^{-3}.
n = 2.4~2.8 for lightly-doped material (10^{14}~10^{15} cm^{-3}), 240 K < T < 600 K
= 1.8~2.4 for moderately-doped material (10^{16}~10^{17} cm^{-3}), 280 K < T < 600 K
m = 2.2~2.5 for lightly-doped material (10^{14}~10^{15} cm^{-3}), 240 K < T < 600 K
= 1.8~2.2 for moderately-doped material (10^{16}~10^{17} cm^{-3}), 280 K < T < 600 K

3. *Doping dependence of critical electric field strength (4H-SiC, parallel to the c-axis)* [19]

$$E_C(\text{4H-SiC}) = \frac{2.49\times10^6}{1-0.25\log_{10}\left(\dfrac{N}{1\times10^{16}}\right)}\ \text{V cm}^{-1} \tag{C6}$$

N: doping density of the lightly-doped side of the junction (cm^{-1}).

4. *Impact ionization coefficients* [19, 20]

$$\alpha_N(E) = 1.69\times10^6\ \text{cm}^{-1}\exp\left[-\left(\frac{9.69\times10^6\ \text{V cm}^{-1}}{E}\right)^{1.6}\right]\ \text{(electron)} \tag{C7}$$

$$\alpha_P(E) = 3.32\times10^6\ \text{cm}^{-1}\exp\left[-\left(\frac{1.07\times10^7\ \text{V cm}^{-1}}{E}\right)^{1.1}\right]\ \text{(hole)} \tag{C8}$$

E: electric field strength.

References

[1] Levinshtein, M.E., Rumyantsev, S.L. and Shur, M.S. (2001) *Properties of Advanced Semiconductor Materials: GaN, AlN, InN, BN, SiC, SiGe*, John Wiley & Sons, Inc., New York.
[2] Harris, G.L. (1995) *Properties of Silicon Carbide*, INSPEC.
[3] Adachi, S. (2005) *Properties of Group-IV, III-V, and II-VI Semiconductors*, John Wiley & Sons, Ltd, Chichester.
[4] Choyke, W.J. (1969) Optical properties of polytypes of SiC: interband absorption, and luminescence of nitrogen-exciton complexes. *Mater. Res. Bull.*, **4**, S141–S152.
[5] Devaty, R.P. and Choyke, W.J. (1997) Optical properties of silicon carbide polytypes. *Phys. Status Solidi A*, **162** (1), 5–38.
[6] Choyke, W.J. and Devaty, R.P. (2004) Optical properties of SiC: 1997–2002, in *Silicon Carbide – Recent Major Advances* (eds W.J. Choyke, H. Matsunami and G. Pensl), Springer, pp. 413–435.
[7] Egilsson, T., Ivanov, I.G., Son, N.T. *et al.* (2004) Exciton and defect photoluminescence from SiC, in *Silicon Carbide, Materials, Processing, and Devices* (eds Z.C. Feng and J.H. Zhao), Taylor & Francis, pp. 81–120.
[8] Janzen, E., Gali, A., Henry, A. *et al.* (2008) Defects in SiC, in *Defects in Microelectronic Materials and Devices*, Taylor & Francis, pp. 615–669.

[9] Chen, W.M., Son, N.T., Janzen, E. *et al.* (1997) Effective masses in SiC determined by cyclotron resonance experiments. *Phys. Status Solidi A*, **162** (1), 79–94.
[10] Volm, D., Meyer, B.K., Hofmann, D.M. *et al.* (1996) Determination of the electron effective-mass tensor in 4H SiC. *Phys. Rev. B*, **53** (23), 15409–15412.
[11] Son, N.T., Persson, C., Lindefelt, U. *et al.* (2004) Cyclotron resonance studies of effective masses and band structure, in *SiC, Silicon Carbide – Recent Major Advances* (eds W.J. Choyke, H. Matsunami and G. Pensl), Springer, pp. 437–492.
[12] Schaffer, W.J., Negley, G.H., Irvine, K.G. and Palmour, J.W. (1994) Conductivity anisotropy in epitaxial 6H and 4H SiC. *Mater. Res. Soc. Symp. Proc.*, **339**, 595–600.
[13] Iwata, H., Itoh, K.M. and Pensl, G. (2000) Theory of the anisotropy of the electron Hall mobility in n-type 4H– and 6H–SiC. *J. Appl. Phys.*, **88** (4), 1956–1961.
[14] Pernot, J., Zawadzki, W., Contreras, S. *et al.* (2001) Electrical transport in n-type 4H silicon carbide. *J. Appl. Phys.*, **90** (4), 1869–1878.
[15] Kagamihara, S., Matsuura, H., Hatakeyama *et al.* (2004) Parameters required to simulate electric characteristics of SiC devices for n-type 4H–SiC. *J. Appl. Phys.*, **96** (5), 5601–5606.
[16] Matsuura, H., Komeda, M., Kagamihara, S. *et al.* (2004) Dependence of acceptor levels and hole mobility on acceptor density and temperature in Al-doped -type 4H-SiC epilayers. *J. Appl. Phys.*, **96** (5), 2708–2715.
[17] Koizumi, A., Suda, J. and Kimoto, T. (2009) Temperature and doping dependencies of electrical properties in Al-doped 4H-SiC epitaxial layers. *J. Appl. Phys.*, **106** (1), 013716.
[18] Khan, I.A. and Cooper, J.A. Jr., (2000) Measurement of high-field electron transport in silicon carbide. *IEEE Trans. Electron Devices*, **47** (2), 269–273.
[19] Konstantinov, A.O., Wahab, Q., Nordell, N. and Lindefelt, U. (1997) Ionization rates and critical fields in 4H silicon carbide. *Appl. Phys. Lett.*, **71** (1), 90–92.
[20] Morisette, D.T. (2001) Development of robust power Schottky barrier diodes in silicon carbide. PhD Thesis. Purdue University.

Index

Fundamentals of Silicon Carbide Technology: Growth, Characterization, Devices, and Applications, First Edition.
Tsunenobu Kimoto and James A. Cooper.

www.ingramcontent.com/pod-product-compliance
Lightning Source LLC
LaVergne TN
LVHW082328270326
834612LV00003B/200